Ecology: The Economy of Nature

Ecology: The Economy of Nature

Eighth Edition

Rick Relyea
Rensselear Polytechnic Institute

Robert Ricklefs
University of Missouri–St. Louis

w.h.freeman
Macmillan Learning
New York

Vice President, STEM: *Ben Roberts*
Director, Editorial Program Mangement: *Andrew Dunaway*
Program Manager: *Jennifer Edwards*
Director of Development: *Lisa Samols*
Executive Marketing Manager: *Will Moore*
Marketing Assistant: *Savannah DiMarco*
Development Editor: *Randi Rossignol*
Manuscript Prep Editor: *Crissy Dudonis*
Media Editor: *Jennifer Compton*
Lead Content Developer: *Mary Tibbets*
Editorial Assistant: *Alexandra Hudson*
Media Permissions Manager: *Christine Buese*
Photo Researcher: *Richard Fox*
Director of Design, Content Management: *Diana Blume*
Senior Design Manager: *Blake Logan*
Cover and Text Design: *Gary Hespenheide*
Director, Content Management Enhancement: *Tracey Kuehn*
Managing Editor: *Lisa Kinne*
Content Project Manager: *Pamela Lawson*
Media Project Manager: *Jodi Isman*
Art Manager: *Matthew McAdams*
Illustration Coordinator: *Janice Donnola*
Illustrations: *MGMT Design and Nicolle R. Fuller, Sayo-Art LLC*
Senior Workflow Project Supervisor: *Susan Wein*
Project Management: *Lumina Datamatics, Inc.*
Composition: *Lumina Datamatics, Inc.*
Printing and Binding: *King Printing Co., Inc.*
Cover Photo: *Rolf Nussbaumer/Nature Picture Library*

Library of Congress Control Number: 2016962077

ISBN-13: 978-1-319-28268-4
ISBN-10: 1-319-28268-7

© 2018, 2014, 2008, 2001 by W. H. Freeman and Company

Printed in the United States of America

Macmillan Learning
One New York Plaza
Suite 4600
New York, NY 10004-1562
www.macmillanlearning.com

Brief Contents

Contents

Part II: Organisms

Part IV: Species Interactions

Part VI: Global Ecology

22 Landscape Ecology and Global Biodiversity 506

About the Authors

Christine Relyea

RICK RELYEA is the Director of the Darrin Fresh Water Institute at Rensselaer Polytechnic Institute. He also serves as Director of the Jefferson Project at Lake George, a groundbreaking partnership between Rensselaer, IBM, and the FUND for Lake George. For the project, Relyea leads a team of scientists, engineers, computer scientists, and artists who are using the latest in science and technology to understand, predict, and enable resilient ecosystems.

From 1999 to 2014, Relyea was at the University of Pittsburgh. In 2005, he was named the Chancellor's Distinguished Researcher and in 2014 he received the Tina and David Bellet Award for Teaching Excellence. From 2007 to 2014, Relyea served as the director of the university's field station, the Pymatuning Laboratory of Ecology, where he oversaw a diverse set of ecological field courses and facilitated researchers from around the world.

Rick has taught thousands of undergraduate students in introductory ecology, behavioral ecology, and evolution. His research is recognized throughout the world and has been published in *Ecology, Ecology Letters, American Naturalist, Environmental Pollution, PNAS*, and other leading ecological journals. The research spans a wide range of ecological and evolutionary topics, including animal behavior, sexual selection, ecotoxicology, disease ecology, community ecology, ecosystem ecology, and landscape ecology. Currently, Relyea's research focuses on aquatic habitats and the diverse species that live in these ecosystems.

Maria W. Pil

ROBERT RICKLEFS is Curators' Professor of Biology at the University of Missouri–St. Louis, where he has been a member of the faculty since 1995. His teaching at Missouri, and previously at the University of Pennsylvania, has included courses in introductory and advanced ecology, biogeography, evolution, and biological statistics. Bob's research has addressed a broad range of topics in ecology and evolutionary biology, from the adaptive significance of life-history traits of birds, to island biogeography and the community relationships of birds, herbivorous insects, and forest trees. In particular, he has championed the importance of recognizing the impact of large-scale processes on local ecological assemblages of species. Bob has published in numerous journals, including *Science, Nature, PNAS, Evolution, Ecology, Ecology Letters*, and the *American Naturalist*. His contributions have been recognized by honorary doctorates from the Université Catholique de Louvain (Belgium), Aarhus University (Denmark), and the University of Burgundy (France). He is a member of the American Academy of Arts and Sciences and the National Academy of Sciences of the United States. Bob published the first edition of *The Economy of Nature* in 1976.

Preface

From the Authors…

Since *The Economy of Nature* debuted in 1976, it has enjoyed a long and loyal following. Instructors and students have appreciated the book's vision: present the material that instructors want to teach in a way that will excite students and encourage an appreciation of how science is done. One of the things we've done to achieve this goal, is to structure the book so that chapters focus on four to six key concepts. They key concepts are presented as learning goals at the beginning of each section, repeated in section headings, and then revisited in the chapter summary. This structure allows instructors and students to focus on the essential messages of each chapter. Another key difference in our presentation is that the text offers students more experience in seeing and interpreting scientific data from a wide range of systems. We have also increased our coverage of species interactions, community ecology, and ecosystem ecology to provide a broader scope for an introductory course.

We understand that ecological applications are valued for their relevance, their ability to make the material more interesting to students, and their inherent interest. For these reasons, multiple studies with real-world applications are included in every chapter. Our goal here is to underscore the value of understanding ecology and help students understand why ecology is relevant to their lives.

We strive to include global change throughout our ecology course. To make this possible, we discuss global change issues in every chapter. These changes help instructors and students make the connections between basic ecology and real ecological issues that affect their lives.

We know that students appreciate clear writing and strong visuals. To make the text an appealing study tool that students will want to read, we worked with superb editors who helped us put the science into plain language without diminishing the complexity of the concepts. The key concepts are accompanied by spectacular photos, beautiful illustrations, easy-to-interpret graphs.

Ecology: The Economy of Nature, Eighth Edition

OPTIONS FOR STUDENTS

The eighth edition is now available as both a textbook and as a fully integrated e-Book in **Sapling Plus**. SaplingPlus combines the powerful multimedia resources for *Ecology: The Economy of Nature* with an integrated e-Book and the robust assessment library, creating an extraordinary new learning resource for students.

HALLMARK FEATURES OF *ECOLOGY: THE ECONOMY OF NATURE*

Fostering ecological literacy with enhanced pedagogical features.

Learning Objectives begin each chapter, making the learning goals in each chapter clear for students. Each Learning Objective corresponds to a section of the chapter and each section now concludes with **Concept Checks** to allow students to check their understanding before moving to the next section. Additionally, the end of each chapter **Critical Thinking Questions** provide students with the opportunity to apply their understanding of the Learning Objectives.

CONCEPT CHECK

1. What are the different ways in which parasites can be transmitted between hosts?
2. What is the process that allows a parasite to jump to a new species?
3. How does a reservoir species help a parasite population persist over time?

LEARNING OBJECTIVES

After reading this chapter, you should be able to:

15.1 Identify the many different types of parasites affecting the abundance of host species.

15.2 Describe how parasite and host dynamics are determined by the parasite's ability to infect the host.

15.3 Illustrate how parasite and host populations c fluctuate in regular cycles.

15.4 Explain the process of parasites evolving offer strategies, while hosts evolve defensive strate

---

CRITICAL THINKING QUESTIONS

1. Compare and contrast the advantages and disadvantages of life as an ectoparasite versus an endoparasite.

2. Why might parasites that are not very harmful to hosts in their native range be useful in controlling non-native hosts in an introduced range?

3. If a parasite has a reservoir host species, how effectively will the parasite population be controlled by immunizing a susceptible host species?

4. Given that there is currently no cure for mad cow disease, what is likely to be the most effective action to reduce its transmission?

5. Why might we continue to discover new emerging infectious diseases?

6. Compare and contrast horizontal versus vertical transmission of a parasite.

7. Using the basic S-I-R model of parasite and host dynamics, explain why the proportion of infected individuals in the population declines over time.

8. In the S-I-R model of parasite and host dynamics, how does the outcome change if we allow new susceptible individuals to be born into the population?

9. When using a *t*-test, what factors make it more likely that you will find a significan difference?

10. Explain why Dutch elm disease might become less lethal to its host over time.

New Tutorial Homework Questions, new to the eighth edition, are available in SaplingPlus. Every Learning Objective has tutorial-style questions, with targeted hints and feedback. These assessments are designed to guide students as they study ecology and provide instructors with immediate feedback on student performance.

New Tutorial Videos, called EcoTV, are also new to the eighth edition. We've identified the concepts in ecology that are the most difficult for students to learn. From El **Niño** to meta-population dynamics, we've developed over 25 tutorial videos called EcoTV, available here: http://www.macmillanlearning.com/ricklefsvideos. These are also available with assessment questions in SaplingPlus.

A learning-by-doing approach to using data and basic quantitative tools.

We know that many students need help applying basic quantitative tools. There are a number of features in *Ecology: The Economy of Nature* that will help your students develop and apply their quantitative skills.

Analyzing Ecology Videos In addition to the 25+ EcoTV videos on challenging topics in ecology, we've enhanced the popular Analyzing Ecology feature with its own Eco TV videos. These videos focus on helping students learn basic statistical and mathematical techniques that real ecologists use every day. We show students how to do the math and then challenge them to apply it in Your Turn. We make a point of integrating these tools with research studies that are discussed in the chapters.

Graphing the Data Exercises, found at the end of each chapter, give students additional practice with quantitative skills, particularly with creating and interpreting graphs. Graphs are used liberally throughout the text to present and describe actual research data.

To provide immediate feedback to students, answers to both Analyzing Ecology and Graphing the Data exercises are provided at the back of the book, and we've included additional exercises in SaplingPlus as homework.

ANALYZING ECOLOGY

Understanding Statistical Significance

Eco TV — macmillanlearning.com/ricklefsvideo

In examining adaptations of prey, counter-adaptations of predators, or any other ecological measurements, we often consider experiments in which researchers find differences in the outcomes of experimental manipulations. Until now, we have not explored how ecologists assess when such differences are meaningful versus when the differences are due to chance.

For any group of measurements, such as the concentration of toxins in ladybugs fed high and low amounts of food, there will be variation among the individuals of each group. If we were to sample ladybugs at random from the high- and low-food treatment groups, we would find that the mean toxin concentration is higher in the high-food treatment group. However, the measurements taken on some of the individuals in the high-food treatment group might overlap with the measurements taken on some of the individuals in the low-food treatment group. When the means are similar and the distribution of the data from two groups is almost entirely overlapping, we would have to conclude that the two groups are nearly identical in whatever we are measuring, as shown in the figure below.

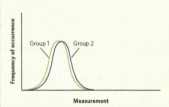

In contrast, when the means are very far apart and the distribution of the data from two groups shows no overlap, as in the case of the next graph, we would feel confident that the two groups are completely different in whatever we are measuring.

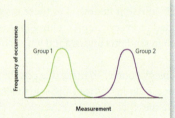

Although it is rare for two groups to have completely overlapping or completely nonoverlapping distributions, we need to know if the degree of overlap between the two data sets is acceptable in order to conclude that the groups are different from each other, with regard to the variable being measured.

Scientists agree that two distributions can be considered "significantly different" if we can sample the two distributions many times and find that the means of those distributions overlap less than 5 percent of the time. This somewhat arbitrary, but widely accepted, cutoff value is known as alpha (α). Thus, we say that our cutoff for statistical significance is $\alpha < 0.05$. Determining that something has statistical significance is not the same as stating that a difference between two means is large, substantial, or important. In other words, the everyday use of "significant" is not synonymous with the scientific use of "significantly different."

YOUR TURN In Chapter 2, "Analyzing Ecology: Standard Deviation and Standard Error," we mentioned that when data have a normal distribution, about 68 percent of the data fall within 1 standard deviation of the mean, 95 percent of the data fall within 2 standard deviations of the mean, and 99.7 perc[...] dard deviations of the m[...] if you had two groups of[...] distributions of data, app[...] deviations apart would t[...] two groups to be conside[...]

GRAPHING THE DATA The Functional Response of Wolves

In Gates of the Arctic National Park in Alaska, researchers monitored the densities of wolves and their major prey, including caribou (*Rangifer tarandus*). To understand whether the wolves could potentially regulate the growth of the caribou population, they wanted to know the shape of the wolves' functional response. They could do this by determining the number of caribou killed by wolves in different areas and at different times of the year.

Using the data from this study, plot the relationship between caribou density and the number of caribou killed per wolf. Then plot the relationship between caribou density and the proportion of caribou killed per wolf. Based on your graphs, what type of functional response do the wolves have?

WOLF AND CARIBOU DATA		
CARIBOU DENSITY (NUMBER/km²)	NUMBER OF CARIBOU KILLED PER WOLF (PER DAY)	PROPORTION OF CARIBOU KILLED PER WOLF (PER DAY)
0.1	0.50	1.80
0.2	0.70	0.90
0.3	0.90	0.50
0.4	0.95	0.30
0.5	0.98	0.22
1.0	1.00	0.15
1.5	1.01	0.10
2.0	1.02	0.07
2.5	1.03	0.05
3.0	1.03	0.04

Scientific discovery presented through a global lens.

Global Climate Change is a topic critical to a modern approach to ecology. In the eighth edition, global climate change, and its many effects, are considered in every chapter. This brings to the forefront how real ecological issues impact students' daily lives.

Encouraging ecological literacy through environmental applications.

As ecology instructors, we want to make the material interesting and relevant to our students—most of whom will not go on to become professional ecologists. The text's *hundreds of applied studies demonstrate the relevance of ecology in students' lives.*

Chapter-Opening Case Studies highlight important, relevant, and current research to pique student interest in the topics to be covered by the chapter. The chapter openers foreshadow the concepts of each chapter while also conveying the dynamic nature of ecology as a modern science in which discoveries continue to be made today.

Ecology Today: Applying the Concepts end every chapter with examples of applied ecology that bring together the major concepts of the chapter and demonstrate their practical importance in a variety of arenas, including human health, conservation, and managing our environment.

Every chapter of *Ecology: The Economy of Nature,* Eighth Edition, has been carefully edited to include new examples, applications and updates, and most importantly, to create an accessible, successful learning experience for a wide range of students. Integrated applications to the environment, medicine, and public health show the relevance of ecology to contemporary issues that are of importance to all people.

Plant Flowering

Plants are also susceptible to climate change, which has the potential to alter the initiation of flower production. One of the longest studies began in the nineteenth century with the writer Henry David Thoreau, who is best known for having spent a year in a small cabin at Walden Pond in Concord, Massachusetts, and for his numerous essays about the natural world. Thoreau kept data on more than 500 species as their cue for flowering and day length has not changed. Other species, such as highbush blueberry (*Vaccinium corymbosum*) and yellow wood sorrel (*Oxalis europaea*), flower 3 to 4 weeks earlier now than they did in 1852. These unique data collected over a century and a half indicate that a seemingly small change in average annual temperature has been associated with dramatic changes in initial flowering time.

Consequences of Altered Breeding Events

The changing breeding seasons of plants and animals in response to global warming do not by themselves cause any problems to the species that are responding. Problems can arise, however, when a species depends on the environment to provide the necessary resources with an altered breeding season. The pied flycatcher (*Ficedula hypoleuca*), for example, is a bird that breeds in Europe each spring. In 1980, researchers in the Netherlands found that the date of egg hatching of the flycatcher began just a few days before the peak abundance of caterpillars, which are a major prey item for the flycatcher chicks. As spring temperatures warmed over the next 2 decades, however, tree leaves appeared 2 weeks earlier and the caterpillars reached their peak abundance 2 weeks earlier. The pied flycatcher, however, retained its normal date of egg hatching, which was 2 weeks later than the new time for peak caterpillar abundance. As a result, the chicks of the flycatcher no longer have a major source of food and the pied flycatcher population has declined by 90 percent.

Figure 8.15 Hatching survival in leatherback sea turtles. At one of four sites where sea turtles breed, researchers monitored precipitation and the proportion of turtle eggs that hatched from 1982 to 2010. Data from P. S. Tomillo et al, Global analysis of the effect of local climate on the hatching output of leatherback turtles, *Nature Scientific Reports* 5 (2015). doi:10.1038/srep16789

Figure 23.12 Changes in forest cover. While some regions of the world experienced a decline in forest cover from 2001 to 2015, other regions experienced an increase. Data from Global Forest Watch, http://www.globalforestwatch.org/map. Source: Hansen/UMD/Google/USGS/NASA, accessed through Global Forest Watch

Tree Cover
Tree Loss
Tree Gain

Resources to Enhance the Teaching Experience

We have thoroughly revised, refreshed, and expanded the extensive set of online learning tools for *Ecology: The Economy of Nature*. All the media resources for this edition will be available in our new SaplingPlus. SaplingPlus is a comprehensive and robust online teaching and learning platform that also incorporates all instructor students and resources along with Gradebook functionality.

STUDENT RESOURCES AVAILABLE IN SAPLINGPLUS

- **LearningCurve** Put "testing to learn" into action. Based on educational research, LearningCurve really works: Game-like quizzing motivates students and adapts to their needs based on their performance. It is the perfect tool to get students engaged before class and for review after class. Additional reporting tools and metrics help instructors get a handle on how students are progressing.

- **Online Homework** Created and supported by educators, SaplingPlus's instructional online homework drives student success and saves educators time. Every homework problem contains hints, answer-specific feedback, and solutions to ensure that students find the help they need. With SaplingPlus, every problem counts.

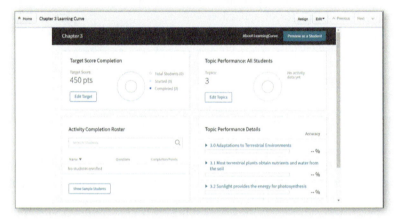

- **Pre-Lecture Reading Quizzes** Quizzes available in SaplingPlus for each concept heading help ensure that students read the material before attending class.

- **e-Book** A complete online version of the textbook that allows the reader to highlight, bookmark, and add notes. For the first time, the e-Book is also available through an app that allows students to read offline and have the book read aloud to them, in addition to the highlighting, note-taking, and keyword search.

■ **EcoTV (Ecology Tutorial Videos)** Over 50 tutorial videos provide an office-hour-like experience for students on the most difficult concepts in ecology. They are provided online (http://www.macmillanlearning.com/ricklefsvideos) and can be used in class or for students to access any time they need help. EcoTV tutorial videos can also be assigned in SaplingPlus with assessment so students may check their understanding. All student responses are recorded in the instructor gradebook, giving instructors insight on student comprehension.

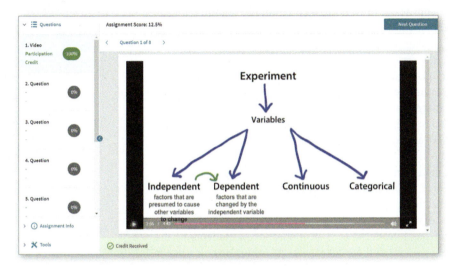

■ **Graphing the Data Activities** As a supplement to the Graphing the Data exercises at the end of each chapter in the book, SaplingPlus offers students additional practice with creating and interpreting graphs.

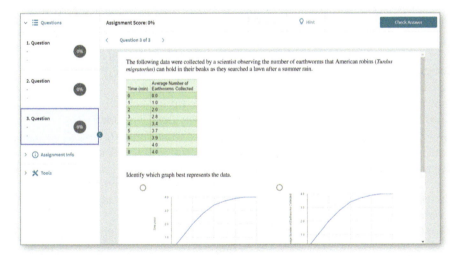

Instructor Resources Available in SaplingPlus

■ **Enhanced Lectures for PowerPoint Presentations** We provide lecture slides containing all the photos and figures to help instructors create their lectures more rapidly. Photos and illustrations from the text are available in both JPEG and PowerPoint formats.

■ **Test Bank** More than 1,400 questions test student understanding and integration of the concepts at six different levels, from knowledge and comprehension checks to application, analysis, synthesis, and evaluation.

■ **Activities** Classroom activities use proven active learning techniques to engage students with the material and to encourage critical thinking.

Acknowledgments

We were incredibly fortunate to work with a tremendous group of people who made this new edition possible. As with every textbook, each chapter began with a text manuscript and an art manuscript that moved through many revisions to make the final version clear and interesting to undergraduate students. We had the privilege of working with Editorial Program Director Andrew Dunaway, Acquisitions Editor Jennifer Edwards, and Developmental Editor Randi Rossignol. These individuals played a major role in revising this new edition. Alexandra Hudson and Crissy Dudonis handled all the reviewing and surveys. Blake Logan, Senior Design Manager, and Diana Blume, Director of Design, can be credited with the text's beautiful design. Other individuals who played key roles in the book's success include Illustration Coordinator Janice Donnola, Content Project Manager Pamela Lawson, and Senior Workflow Supervisor Susan Wein. Richard Fox and Christine Buese were tireless in hunting down just the right photo to meet each of our requests. We are grateful to the Senior Project Managers at Lumina Datamatics, Andrea Stefanowicz and Misbah Ansari, and to Patti Brecht, copyeditor, for keeping the project on track and for their hard work. Thanks to Executive Marketing Manager Will Moore and to the Macmillan Sales force for all their hard work, ensuring that instructors get the most from the new edition and the media that's been developed.

For eight editions, countless colleagues and instructors have helped shape *Ecology: The Economy of Nature* into a book that has introduced tens of thousands of students to the wonders of ecology. We are extremely grateful to these colleagues. As we undertook a major revision with this eighth edition, we once again received extensive help at every stage of development from many of our colleagues and fellow instructors. We extend our sincere thanks to the following people who graciously offered their time:

Eddie Alford, *Arizona State University*
Loreen Allphin, *Brigham Young University*
Marty Anderies, *Arizona State University*
Tom Arsuffi, *Texas Tech University*
Betsy Bancroft, *Southern Utah University*
Paul Bartell, *Pennsylvania State University*
David Baumgardner, *Texas A&M University*
Christopher Beatty, *Santa Clara University*
Marc Bechard, *Boise State University*
Mark Belk, *Brigham Young University*
Michael F. Benard, *Case Western Reserve University*
Ritin Bhaduri, *Spelman College*
Andrew Blaustein, *Oregon State University*
Steve Blumenshine, *California State University–Fresno*
Michelle D. Boone, *Miami University of Ohio*
Jennifer Borgo, *Coker College*
Victoria Borowicz, *Illinois State University*
Alison Boyer, *University of Tennessee*
Judith Bramble, *DePaul University*
Shannon Bros-Seemann, *San José State University*
Ken Brown, *Louisiana State University*
Romi Burks, *Southwestern University*
Willodean Burton, *Austin Peay State University*
David Byres, *Florida State College–Jacksonville*
Daniel Capuano, *Hudson Valley Community College*
Walter P. Carson, *University of Pittsburgh*

J. Chadwick Johnson, *Arizona State University*
Michael F. Chislock, *Auburn University*
George Cline, *Jacksonville State University*
Clay Corbin, *Bloomsburg University of Pennsylvania*
Douglas Crawford-Brown, *University of North Carolina at Chapel Hill*
William Currie, *University of Michigan*
Richard Deslippe, *Texas Tech University*
Jacqueline M. Doyle, *Towson University*
Hudson DeYoe, *University of Texas–Pan American*
Joe D'Silva, *Norfolk State University*
James Dunn, *Grand Valley State University*
Kenneth Ede, *Oklahoma State University–Tulsa*
James Elser, *Arizona State University*
Rebecca Ferrell, *Metropolitan State University of Denver*
Kerri Finlay, *University of Regina*
Ben Fitzpatrick, *University of Tennessee*
Lloyd Fitzpatrick, *University of North Texas*
Matt Forister, *University of Nevada*
Norma Fowler, *University of Texas at Austin*
Steven J. Franks, *Fordham University*
Rachel E. Gallery, *University of Arizona Tucson*
Danielle Garneau, *State University of New York at Plattsburgh*
Pamela Geddes, *Northeastern Illinois University*
Linda Green, *Georgia Institute of Technology*
Danny Gustafson, *The Citadel*

Monika Havelka, *University of Toronto–Mississauga*
Floyd Hayes, *Pacific Union College*
Stephen Hecnar, *Lakehead University*
Colleen Hitchcock, *Boston College*
Gerlinde Hoebel, *University of Wisconsin–Milwaukee*
Claus Holzapfel, *Rutgers University–Newark*
Robert Howard, *Middle Tennessee State University*
Jon Hubbard, *Gavilan College*
Rebecca Penny Humphrey, *Aquinas College*
Anthony Ippolito, *DePaul University*
John Jaenike, *University of Rochester*
Steven Juliano, *Illinois State University*
Thomas Jurik, *Iowa State University*
Kristen M. Kaczynski, *California State University, Chico*
Doug Keran, *Central Lakes College*
Tigga Kingston, *Texas Tech University*
Christopher Kitting, *California State University–East Bay*
Catherine Kleier, *Regis University*
Jamie Kneitel, *California State University–Sacramento*
Ned J. Knight, *Linfield College*
Andrew M. Kramer, *University of Georgia*
William Kroll, *Loyola University of Chicago*
Hugh Lefcort, *Gonzaga University*
Mary Lehman, *Longwood University*
Dale Lockwood, *Colorado State University*
Eric Long, *Seattle Pacific University*
Genaro Lopez, *University of Texas–Brownsville*
C. J. Lortie, *York University*
Lisa L. Manne, *College of Staten Island*
Terri J. Matiella, *The University of Texas San Antonio*
Marty D. Matlock, *University of Arkansas*
Robert McGregor, *Douglas College*
L. Maynard Moe, *California State University–Bakersfield*
Don Moll, *Missouri State University*
Peter Morin, *Rutgers University*
Patrick Osborne, *University of Missouri–St. Louis*
Peggy Ostrom, *Michigan State University*
Michael Palmer, *Oklahoma State University*
Mitchell Pavao-Zuckerman, *University of Arizona*
William Pearson, *University of Louisville*
Bill Perry, *Illinois State University*
Kenneth Petren, *University of Cincinnati*
Raymond Pierotti, *University of Kansas*
David Pindel, *Corning Community College*
Craig Plante, *College of Charleston*
Thomas Pliske, *Florida International University*

Diane Post, *University of Texas–Permian Basin*
Mark Pyron, *Ball State University*
Laurel Roberts, *University of Pittsburgh*
Robert Rosenfield, *University of Wisconsin–Stevens Point*
Tatiana Roth, *Coppin State University*
Arthur N. Samel, *Bowling Green State University*
Nate Sanders, *University of Tennessee*
Mark Sandheinrich, *University of Wisconsin–La Crosse*
Thomas Sasek, *University of Louisiana at Monroe*
Kenneth Schmidt, *Texas Tech University*
Robert Schoch, *Boston University*
Erik Scully, *Towson University*
Kathleen Sealey, *University of Miami*
Kari A. Segraves, *Syracuse University*
David Serrano, *Broward College*
Chrissy Spencer, *Georgia Institute of Technology*
Janette Steets, *Oklahoma State University*
Juliet Stromberg, *Arizona State University*
Stephen Sumithran, *Eastern Kentucky University*
Keith Summerville, *Drake University*
Carol Thornber, *University of Rhode Island*
David Tonkyn, *Clemson University*
William Tonn, *University of Alberta*
James Traniello, *Boston University*
Stephen Vail, *William Paterson University*
Michael Vanni, *Miami University*
Eric Vetter, *Hawaii Pacific University*
Joe von Fischer, *Colorado State University*
Mitch Wagener, *Western Connecticut State University*
Diane Wagner, *University of Alaska–Fairbanks*
Sean Walker, *California State University–Fullerton*
Xianzhong Wang, *Indiana University–Purdue University Indianapolis*
John Weishampel, *University of Central Florida*
Carrie Wells, *University of North Carolina at Charlotte*
Marcia Wendeln, *Wright State University*
Tom Wentworth, *North Carolina State University*
Yolanda Wiersma, *Memorial University*
Frank Williams, *Langara College*
Susan Willson, *St. Lawrence University*
Alan E. Wilson, *Auburn University*
Ben Wodika, *Truman State University*
Kelly Wolfe-Bellin, *College of the Holy Cross*
Lan Xu, *South Dakota State University*
Todd Yetter, *University of the Cumberlands*

Ecology: The Economy of Nature

Introduction:
1 Ecology, Evolution, and the Scientific Method

Searching for Life at the Bottom of the Ocean

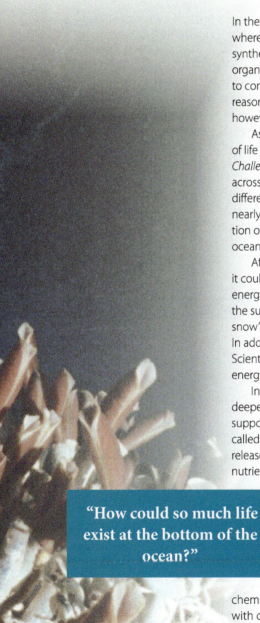

In the early 1800s, scientists hypothesized that deep ocean waters—depths greater than 275 m where sunlight cannot penetrate—were devoid of life. Without sunlight there can be no photosynthesis, and without photosynthesis there can be no plants or algae to serve as food for other organisms. The cold temperatures and extreme pressures of deep ocean waters were also thought to contribute to the absence of deep-sea life. Given that ocean depths can exceed 10,000 m, it was reasonable to hypothesize that the deepest areas of the ocean could not support life. At the time, however, scientists were not able to explore the deepest regions of the ocean.

As exploration continued throughout the nineteenth century, scientists' ideas about the limits of life began to change. In an 1873 expedition, scientists aboard the British research ship HMS *Challenger* dragged a large, open-sided heavy box suspended from long ropes behind the ship across the floor of the Atlantic Ocean. This box—known as a dredge—sampled the sea floor in different parts of the ocean at depths of up to 4,572 m. The scientists were astonished to discover nearly 5,000 new species. When it became clear that life flourished at depths beyond the penetration of light, scientists were forced to reject their earlier hypothesis that no life existed in the deep ocean waters.

After discovering this rich abundance of deep-sea life, scientists needed to understand how it could exist. The lack of light suggested that deep-sea organisms were somehow sustained by energy that did not come from photosynthesis on the ocean floor. Scientists had observed that the surface waters of the ocean produced a steady descent of tiny particles—known as "marine snow"—that were produced by the death and decomposition of organisms living near the surface. In addition to marine snow, when large organisms such as whales died, they fell to the ocean floor. Scientists hypothesized that marine snow and the decaying bodies of large organisms must be the energy source needed to sustain organisms in deep ocean waters.

In the 1970s, scientists were finally able to send small submarines to take a firsthand look at the deepest ocean areas. Their discoveries were shocking. They confirmed that much of the ocean floor supported living organisms, and furthermore that areas near openings in the floor of the ocean, called *hydrothermal vents*, contained a great diversity of deep-sea species. Hydrothermal vents release plumes of hot water with high concentrations of sulfur compounds and other mineral nutrients. A tremendous number of species surrounded these hydrothermal vents, including clams, crabs, and fish. Indeed, the total amount of life at these depths rivaled that seen in some of the most diverse places on Earth. It became clear that the amount of energy contained in the descending marine snow was not sufficient to support such a diverse and abundant set of life forms. That hypothesis now had to be rejected.

> "How could so much life exist at the bottom of the ocean?"

The observation that life existed near the hydrothermal vents suggested that the vents were somehow supplying the energy for these species. Scientists had known for a long time that some species of bacteria could obtain their energy from chemicals rather than from the Sun. The bacteria use the energy in chemical bonds, combined with carbon dioxide (CO_2), to produce organic compounds—a process known as *chemosynthesis*—similar to the way that plants and algae use the energy of the Sun and CO_2 to produce

A deep-sea vent. In some regions of the ocean floor, hot water containing sulfur compounds is released from the ground. The sulfur compounds provide energy for chemosynthetic bacteria, which then serve as food for many other species that live near the vents, including these rust-colored tube worms (*Tevnia jerichonana*) that have been stained orange by iron compounds emitted from the vents. Photo by EMORY KRISTOF/National Geographic Creative.

organic compounds through photosynthesis. Based on this knowledge, scientists hypothesized that the hot vents, which release water with dissolved hydrogen sulfide gas and other chemicals, provided a source of energy for bacteria and that these bacteria could be consumed by the other organisms living around the vents.

After several years of investigations, researchers found that the immediate area around the hot vents contained a group of organisms known as tubeworms, which can grow to more than 2 m long. These animals possess specialized organs that house vast numbers of chemosynthetic bacteria that live in a close relationship with the tubeworms. The tubeworms capture the sulfide gases and CO_2 from the surrounding water and pass these compounds to the bacteria. The bacteria then use the sulfide gases and CO_2 to produce organic compounds. Some of these organic compounds are passed to the tubeworms, which use the organic compounds as food. These bacteria also represent a food source for many of the other animals that live near the vents. In turn, these bacteria-consuming animals can be consumed by larger animals, such as fish.

Research on deep-sea vents continues today, and our hypotheses continue to be revised. Searching for deep-sea vents has historically been difficult and researchers estimated that the vents were relatively rare and widely spaced at 12 to 200 km apart. In 2016, researchers using advanced sensors that could detect the chemicals produced by the vents discovered that the deep-sea vents are actually three to six times more abundant than previously thought and are spaced only 3 to 20 km apart.

The story of the deep-sea vents demonstrates how scientists work: They make observations, devise hypotheses, test the hypotheses to confirm or reject them, and, if a hypothesis is rejected, devise a new hypothesis. As you will see throughout this chapter and subsequent chapters, science is an ongoing process that often leads to fascinating discoveries about how nature works.

SOURCES:

Baker, E. T., et al. 2016. How many vent fields? New estimates of vent field populations on ocean ridges from precise mapping of hydrothermal discharge locations, *Earth and Planetary Science Letters* 449:186–96.

Dubilier, N., et al. 2008. Symbiotic diversity in marine animals: The art of harnessing chemosynthesis, *Nature Reviews Microbiology* 6:725–740.

Dunn, R. R. 2002. *Every Living Thing* (HarperCollins).

LEARNING OBJECTIVES

After reading this chapter, you should be able to:

1.1 Illustrate how ecological systems exist in a hierarchical organization.

1.2 Understand how ecological systems are governed by physical and biological principles.

1.3 Describe how different organisms play diverse roles in ecological systems.

1.4 Distinguish the many approaches scientists use to studying ecology.

1.5 Explain how humans influence ecological systems.

The story of deep-sea vents offers an excellent introduction to the science of *ecology*. **Ecology** is the scientific study of the interactions among organisms and the environment.

Although Charles Darwin never used the word *ecology* in his writings, he appreciated the importance of beneficial and harmful interactions among species. In his 1859 book *On the Origin of Species*, Darwin compared the large number of interactions among species in nature to the large number of interactions among consumers and businesses in human economic systems. With this metaphor in mind, he described species interactions as "the economy of nature," which a century later inspired the title of this ecology textbook.

The word *ecology* came into general use in the late 1800s. Since that time, the science of ecology has grown and diversified. Professional ecologists now number in the tens of thousands and have produced an immense body of

knowledge about the world around us. Ecology is an active, modern science that continues to yield fascinating new insights about the environment and our impact on it. As we saw in the chapter opening story about life in the ocean depths, science is an ongoing process through which our understanding of nature constantly changes. Scientific investigation uses a variety of tools to understand how nature works. This understanding is never complete or absolute. It changes constantly as scientists make new discoveries and as growing human populations and technological advances cause major environmental changes, often with dramatic consequences. With the knowledge that ecologists provide through their study of the natural world, we are in a better

Ecology The scientific study of the abundance and distribution of organisms in relation to other organisms and environmental conditions.

position to develop effective policies to manage environmental concerns related to land use, water, natural catastrophes, and public health.

This chapter will start you on the road to thinking like an ecologist. Throughout this book, we will consider the full range of **ecological systems**—biological entities that have their own internal processes and interact with their external surroundings. Ecological systems exist at many different levels, ranging from an individual organism to the entire globe. Despite tremendous variations in size, all ecological systems obey the same principles with regard to their physical and chemical attributes and the regulation of their structure and function.

We begin this journey by examining the many different levels of organization for ecological systems, the physical and biological principles that govern ecological systems, and the different roles species play. Once we understand these basics, we will consider the many approaches to studying ecology and then consider the importance of understanding ecology when faced with the wide variety of ways that humans affect ecological systems.

1.1 Ecological systems exist in a hierarchy of organization

An ecological system may be an *individual,* a *population,* a *community,* an *ecosystem,* or the entire *biosphere.* As you can see in **Figure 1.1**, each ecological system is a subset of the next larger one, forming a hierarchy. In this section, we will examine the individual components of ecological systems and how we study ecology at different levels in the ecological hierarchy.

INDIVIDUALS

An **individual** is a living being—the most fundamental unit of ecology. Although smaller units in biology exist—for example, an organ, a cell, or a macromolecule—none of them has a separate life in the environment. Every individual has a membrane, or other covering, across which it exchanges energy and materials with its environment. This boundary separates the internal processes and structures of the ecological system from the external resources and conditions of the environment. In the course of its life, an individual transforms energy and processes materials. To accomplish this, it must acquire energy and nutrients from its surroundings and rid itself of unwanted waste products. This process alters the conditions of the environment and affects the resources available for other organisms. It also contributes to the movement of energy and chemical elements.

POPULATIONS AND SPECIES

Scientists assign organisms to particular *species.* Historically, the term **species** was defined as a group of organisms that naturally interbreed with each other and produce fertile offspring. Over time, scientists have realized that this definition does not fit all species. In fact, no single definition can apply to all organisms. For example, some species of salamanders are all female and only produce daughters, which are clones of their mothers. In this case, individuals do not interbreed, but we consider these individuals to be of the same species because they are genetically very similar to each other. In addition, some organisms that we consider distinct species can interbreed. In such cases, we cannot use a lack of interbreeding to draw the line between species.

Defining a species becomes even more complicated when we consider organisms such as bacteria. Scientists increasingly appreciate that bacteria can transfer bits of their DNA to other bacteria that are not closely related, a process known as *horizontal gene transfer.* This can happen in a number of ways: when a bacterium engulfs genetic material from the environment, when two bacteria come into contact and exchange genetic material, or when a virus transfers genetic material between two bacteria. Such cases make it difficult to group bacteria into distinct species. Despite these difficulties, the term *species* has still proven useful to ecologists.

A **population** consists of individuals of the same species living in a particular area. For example, we might talk about a population of catfish living in a pond, a population of wolves living in Canada, or a population of tubeworms living near a hydrothermal vent on the ocean floor. The boundaries that determine a population can be natural, for example, where a continent meets the ocean. Alternatively, a population might be defined by other criteria such as a political boundary. For example, a scientist might want to study the population of bald eagles (*Haliaeetus leucocephalus*) that resides in Pennsylvania, whereas biologists of the U.S. Fish and Wildlife Service might wish to study the bald eagle population of the entire United States.

Populations have five distinct properties that are not exhibited by individuals: geographic range, abundance, density, change in size, and composition. The *geographic range* of a population—also known as its distribution—is the extent of land or water within which a population lives. For example, the geographic range of the North American grizzly bear (*Ursus arctos*) includes western Canada, Alaska, Montana, and Wyoming. The *abundance* of a population refers to the total number of individuals. The *density* of a population refers to the number of individuals per unit of area. For instance, we might count the grizzly bears in an area and determine that there is 1 bear/100 km^2. The *change in size* of a population refers to increases and decreases in the number of individuals in an area over time. Finally, *composition* of a

---

Ecological systems Biological entities that have their own internal processes and interact with their external surroundings.

Individual A living being; the most fundamental unit of ecology.

Species Historically defined as a group of organisms that naturally interbreed with each other and produce fertile offspring. Current research demonstrates that no single definition can be applied to all organisms.

Population The individuals of the same species living in a particular area.

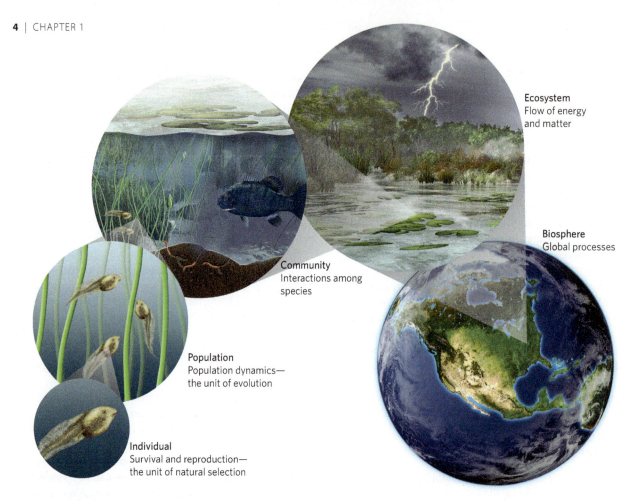

Figure 1.1 The hierarchy of organization in ecological systems. At each level of complexity, ecologists study different processes.

population describes the makeup of the population in terms of gender, age, or genetics. For example, we can ask what proportion of the grizzly population is male versus female, or juvenile versus adult.

COMMUNITIES

At the next level of the ecological hierarchy, we identify an ecological **community,** which is composed of all populations of species living together in a particular area. The populations in a community interact with each other in various ways. Some species eat other species, while others, for example, bees and the plants they pollinate, have cooperative relationships that benefit both parties. These types of interactions influence the number of individuals in each population. A community may cover large areas, such as a forest, or may be enclosed within a very small area, such as the community of tiny organisms that live in the digestive systems of animals, or in the tiny amount of water found in tree holes. In practical terms, ecologists who study communities do not study every organism in the community. Instead, they generally study a subset of the species in the community, such as the trees, the insects, or the birds, as well as the interactions among those species.

The boundaries that define a community are not always rigid. For example, if you were to examine the community of plants and animals that live at the base of a mountain versus the top of a mountain in Colorado, you would find that the two locations contain many different species. However, if you were to walk up the mountain, you would notice that some species of trees, such as Douglas firs (*Pseudotsuga menziesii*), are abundant at the beginning of your hike and then gradually dwindle as you move higher. Other species, such as subalpine firs (*Abies lasiocarpa*), begin to take their place as the number of Douglas firs declines. In other words, the boundaries of the upper and lower forest communities are not distinct. Because of this, scientists must often decide on the boundaries of a community they want to study. For instance, an ecologist might decide to study the community of plants and animals on a large desert ranch in New Mexico, or the community of aquatic organisms that lives along a designated stretch of coastline in California. In these cases, no distinct boundary separates the studied community from the area that surrounds it.

ECOSYSTEMS

From communities we move on to *ecosystems.* An **ecosystem** is composed of one or more communities of living organisms interacting with their nonliving physical and chemical

Community All populations of species living together in a particular area.

Ecosystem One or more communities of living organisms interacting with their nonliving physical and chemical environments.

environments, which include water, air, temperature, sunlight, and nutrients. Ecosystems are complex ecological systems that can include thousands of different species living under a great variety of conditions. For example, we may speak of the Great Lakes ecosystem or the Great Plains ecosystem.

At the ecosystem level, we typically focus on the movement of energy and matter between physical and biological components of the ecosystem. Most energy that flows through ecosystems originates with sunlight and eventually escapes Earth as radiated heat. In contrast, matter cycles within and between ecosystems. With the exception of places such as deep-sea vents where energy is acquired through chemosynthesis, the energy for most ecosystems comes from the Sun and is converted to organic compounds by photosynthetic plants and algae. These organisms can then be eaten by *herbivores*—animals that eat plants—which, in turn, are eaten by *carnivores*—animals that eat other animals. In addition, dead organisms and their waste products can be consumed by *detritivores,* which themselves can be consumed by other animals. In all these cases, each step results in some of the energy originally assimilated from sunlight being converted into growth or reproduction of consumers; the remainder of the energy is lost to the surroundings as heat and is eventually radiated back into space.

When considering matter in an ecosystem, we often look at the most common elements that organisms use, such as carbon, oxygen, hydrogen, nitrogen, and phosphorus. These elements comprise a major portion of the most important compounds for living organisms, including water,

carbohydrates, proteins, and DNA. These elements can be held in many different places, or *pools,* on Earth, including in living organisms, and in the atmosphere, water, and rocks. The movement of these elements among these pools is known as the *flow* of matter. For instance, many nutrients that are in the soil are taken up by plants, and these plants are consumed by animals. The nutrients exist in an animal's tissues, and many leave as excreted waste. When the animal dies, the nutrients in its tissues are returned to the soil, thereby completing the cycling of nutrients.

The boundaries of ecosystems, like those of populations and communities, are often not distinct. Scientists generally distinguish ecosystems by their relative isolation with respect to flows of energy and materials, but, in reality, few ecosystems are completely isolated. Even aquatic and terrestrial ecosystems exchange materials and energy by runoff from the land and the harvesting of aquatic organisms by terrestrial consumers, as when bears capture salmon on their upstream spawning runs.

THE BIOSPHERE

At the highest level of the ecological hierarchy is the **biosphere,** which includes all the ecosystems on Earth. As **Figure 1.2** illustrates, distant ecosystems are linked together by exchanges of energy and nutrients carried by currents of wind and water and by the movements of organisms, such as migrating whales, birds, and fish. Such movement connects

---

Biosphere All the ecosystems on Earth.

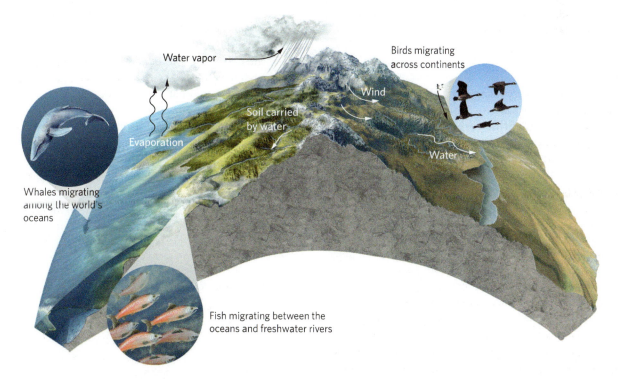

Figure 1.2 The biosphere. The biosphere consists of all ecosystems on Earth, which are linked together by movements of air, water, and organisms.

terrestrial, freshwater, and marine ecosystems by carrying soil, nutrients, and organisms.

The biosphere is the ultimate ecological system. All transformations of the biosphere are internal, with two exceptions: the energy that enters from the Sun, and the energy that is lost to space. The biosphere holds practically all materials that it has ever had, and retains whatever waste materials we generate.

STUDYING ECOLOGY AT DIFFERENT LEVELS OF ORGANIZATION

Each level in the hierarchy of ecological systems is distinguished by unique structures and processes. As a result, ecologists have developed different approaches for exploring these levels and for answering the questions that arise. The five approaches to studying ecology match the different levels of hierarchy: the *individual approach*, the *population approach*, the *community approach*, the *ecosystem approach*, and the *biosphere approach*.

The **individual approach** to ecology emphasizes the way in which an individual's morphology (the size and shape of its body), physiology, and behavior enable it to survive in its environment. This approach also seeks to understand why an organism lives in some environments but not in others. For example, an ecologist studying plants at the organism level might ask why trees are dominant in warm, moist environments, whereas shrubs with small, tough leaves are dominant in environments with cool, wet winters and hot, dry summers.

Ecologists who use the individual approach are often interested in **adaptations**—the characteristics of an organism that make it well-suited to its environment. For example, desert animals have enhanced kidney function to help them conserve water. The cryptic coloration of many animals helps them avoid detection by predators. Flowers are shaped and scented to attract certain kinds of pollinators. Adaptations are the result of evolutionary change through the process of *natural selection*, which we will consider later in this chapter.

The **population approach** to ecology examines variation over time and space in the number of individuals, the density of individuals, and the composition of individuals, which includes the sex ratio, the distribution of individuals among different age classes, and the genetic makeup of a population. Changes in the number or density of individuals can reflect the balance of births and deaths within a population, as well as immigration and emigration of individuals from a local population. This can be influenced by a number of factors, including interactions with other species and the physical conditions of the environment, such as temperature or the availability of water. In the process of evolution, genetic mutations may alter birth and death rates, genetically distinct types of individuals may become common within a population, and the overall genetic makeup of the population may change. Because other species might serve as food, pathogens, or predators, interactions among species can also influence the births and deaths of individuals within a population.

Figure 1.3 The community approach to ecology. Ecologists using the community approach study interactions among plants and animals that live together. For example, on an African plain, ecologists might ask how cheetahs affect the abundance of gazelles and how the gazelles affect the abundance of the plants that they consume. Photo by Michel & Christine Denis-Huot/Science Source.

The **community approach** to ecology is concerned with understanding the diversity and relative abundances of different kinds of organisms living together in the same place. The community approach focuses on interactions between populations, which can either promote or limit the coexistence of species (**Figure 1.3**). For example, in studying the Serengeti Plains of Africa, an ecologist taking the community approach might ask how the presence of zebras, which consume grasses, might affect the abundance of other species, such as gazelles, which also consume the grasses.

The **ecosystem approach** to ecology describes the storage and transfer of energy and matter, including the various chemical elements essential to life, such as oxygen, carbon, nitrogen, phosphorus, and sulfur. These movements of energy and matter occur through the activities of organisms and through the physical and chemical transformations that occur in the soil, atmosphere, and water.

The **biosphere approach** to ecology is concerned with the largest scale in the hierarchy of ecological systems. This

---

Individual approach An approach to ecology that emphasizes the way in which an individual's morphology, physiology, and behavior enable it to survive in its environment.

Adaptation A characteristic of an organism that makes it well suited to its environment.

Population approach An approach to ecology that emphasizes variation over time and space in the number of individuals, the density of individuals, and the composition of individuals.

Community approach An approach to ecology that emphasizes the diversity and relative abundances of different kinds of organisms living together in the same place.

Ecosystem approach An approach to ecology that emphasizes the storage and transfer of energy and matter, including the various chemical elements essential to life.

Biosphere approach An approach to ecology concerned with the largest scale in the hierarchy of ecological systems, including movements of air and water—and the energy and chemical elements they contain—over Earth's surface.

approach tackles the movements of air and water—and the energy and chemical elements they contain—over Earth's surface. Ocean currents and winds carry the heat and moisture that define the climates at each location on Earth, which, in turn, govern the distributions of organisms, the dynamics of populations, the composition of communities, and the productivity of ecosystems.

We have described these five approaches as distinct. However, most ecologists use multiple approaches to study the natural world. A scientist who wants to understand how an ecosystem will respond to a drought, for example, will likely want to know how individual plants and animals respond to a lack of water, how these individual responses affect the populations of plants and animals, how a change in the populations might affect interactions among species, and how a change in species interactions might affect the flow of energy and matter.

CONCEPT CHECK

1. What is ecology?
2. Why do ecologists consider both individuals and ecosystems to be ecological systems?
3. What are the unique processes that are examined when taking the individual, population, community, and ecosystem approaches to studying ecology?

1.2 Ecological systems are governed by physical and biological principles

Although ecological systems are complex, they are governed by a few basic principles. Life builds on the physical properties and chemical reactions of matter. The diffusion of oxygen across body surfaces, the rates of chemical reactions, the resistance of vessels to the flow of fluids, and the transmission of nerve impulses all obey the laws of thermodynamics. Within these constraints, life can pursue many options and has done so with astounding innovation. In this section, we briefly review the three major biological principles that you may recall from your introductory biology course: *conservation of matter and energy, dynamic steady states,* and *evolution.*

CONSERVATION OF MATTER AND ENERGY

The **law of conservation of matter** states that matter cannot be created or destroyed, but can only change form. For example, as you drive a car, gasoline is burned in the engine; the amount of fuel in the tank declines, but you have not destroyed matter. The molecules that comprise gasoline are converted into new forms, including carbon monoxide (CO), carbon dioxide (CO_2), and water (H_2O).

Another important biological principle, the **law of conservation of energy**—also known as the **first law of thermodynamics**—states that energy cannot be created or destroyed. Like matter, energy can only be converted into

different forms. Living organisms must constantly obtain energy to grow, maintain their bodies, and replace energy lost as heat.

These two laws imply that ecologists can track the movement of matter and energy as it is converted into new forms through organisms, populations, communities, ecosystems, and the biosphere. At every level of organization, we should be able to determine how much energy and matter enter the system and account for its movement. For example, consider a field full of cattle (*Bos taurus*) eating grass. At the organismal level, we can determine how much energy an individual animal consumes and then calculate the proportion of this energy that is converted into growth of its body, the maintenance of its physiology, and its waste. At the population level, we can calculate how much energy the entire herd of cattle consumes by eating grass. At the community level, we can evaluate how much energy each species of grass creates via photosynthesis and how much of this energy is passed on to cattle and other plant-eating species, such as rabbits, that might coexist with the cattle. At the ecosystem level, we can estimate how elements such as carbon flow from the grasses to herbivores (cattle and rabbits) and then on to predators. We can then track how dead grass, the waste products of herbivores and predators, and the dead bodies of herbivores and predators all decompose and return to the soil. At the biosphere level, we can examine how the energy flows among the many ecosystems and moves around the world.

DYNAMIC STEADY STATES

Although matter and energy cannot be created or destroyed, ecological systems continuously exchange matter and energy with their surroundings. When gains and losses are in balance, ecological systems are unchanged and the system is said to be in a **dynamic steady state.** For example, birds and mammals continuously lose heat in a cold environment. However, this loss is balanced by heat gained from the metabolism of foods, so the animal's body temperature remains constant. Similarly, the proteins of our bodies are continuously broken down and replaced by newly synthesized proteins, so our appearance remains relatively unchanged.

The principle of the dynamic steady state applies to all levels of ecological organization, as illustrated in **Figure 1.4**. For individual organisms, assimilated food and energy must balance energy expenditure and metabolic breakdown of tissues. A population increases with births and immigration, and it decreases with death and emigration. At the community level, the number of species living in a community decreases when a species becomes extinct, and increases when a new species colonizes the area. Ecosystems and the biosphere receive energy from the Sun, and this gain of

---

Law of conservation of matter Matter cannot be created or destroyed; it can only change form.

Law of conservation of energy Energy cannot be created or destroyed; it can only change form. *Also known as* **The First law of thermodynamics**.

Dynamic steady state When the gains and losses of ecological systems are in balance.

Inputs	Level	Outputs

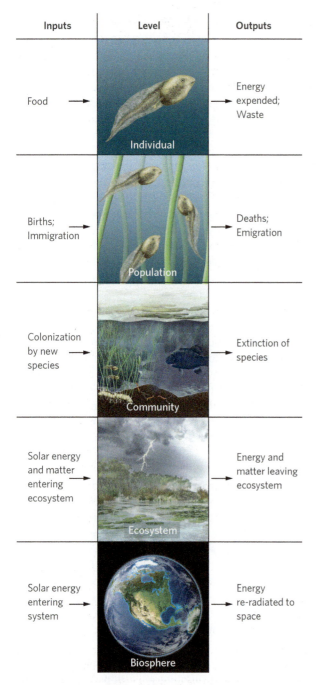

Food →	Individual	→ Energy expended; Waste
Births; Immigration →	Population	→ Deaths; Emigration
Colonization by new species →	Community	→ Extinction of species
Solar energy and matter entering ecosystem →	Ecosystem	→ Energy and matter leaving ecosystem
Solar energy entering system →	Biosphere	→ Energy re-radiated to space

Figure 1.4 Dynamic steady states. At all levels of organization, the inputs to the systems must equal the outputs.

energy is balanced by heat energy radiated by Earth back out into space. One of the most important questions ecologists ask is how the steady states of ecological systems are maintained and regulated. We will return to this question frequently throughout this book.

An understanding of dynamic steady states helps provide insights regarding the inputs and outputs of ecological systems. Of course, ecological systems also change. Organisms grow, populations vary in abundance, and abandoned fields revert to forest. Yet all ecological systems have mechanisms that tend to maintain a dynamic steady state.

EVOLUTION

Although matter and energy cannot be created or destroyed, what living systems do with matter and energy is as variable as all the forms of organisms that have ever existed on Earth. To understand the variation among organisms—the diversity of life—we turn to the concept of *evolution.*

An attribute of an organism, such as its behavior, morphology, or physiology, is the organism's **phenotype.** A phenotype is determined by the interaction of the organism's **genotype,** or the set of genes it carries, with the environment in which it lives. For example, your height is a phenotype that is determined by your genes and the nutrition you received in the environment where you were raised.

Over the history of life on Earth, the phenotypes of organisms have changed and diversified dramatically. This is the process of **evolution,** which is a change in the genetic composition of a population over time. Evolution can happen through a number of different processes that we will discuss in detail in later chapters. Perhaps the best known process is evolution by **natural selection,** which is a change in the frequency of genes in a population through differential survival and reproduction of individuals that possess certain phenotypes. As outlined by Charles Darwin in his book *On the Origin of Species,* evolution by natural selection depends on three conditions:

1. Individual organisms vary in their traits.
2. Parental traits are inherited by their offspring.
3. The variation in traits causes some individuals to experience higher **fitness,** which we define as the survival and reproduction of an individual.

When these three conditions exist, an individual with higher survival and reproductive success will pass more copies of his or her genes to the next generation. Over time, the genetic composition of a population changes as the most successful phenotypes come to predominate. As a result, the population becomes better suited to the surrounding environmental conditions. Phenotypes that are well suited to their environment and, in turn, confer higher fitness are known as adaptations. Consider the example in **Figure 1.5** in which some individuals in a population of caterpillars are colored in such a way that they blend in with their surroundings and escape the notice of predators while other individuals are not. If color is inherited, over time the population will consist of a progressively larger proportion of caterpillars that blend in with their environment.

Phenotype An attribute of an organism, such as its behavior, morphology, or physiology.

Genotype The set of genes an organism carries.

Evolution Change in the genetic composition of a population over time.

Natural selection Change in the frequency of genes in a population through differential survival and reproduction of individuals that possess certain phenotypes.

Fitness The survival and reproduction of an individual.

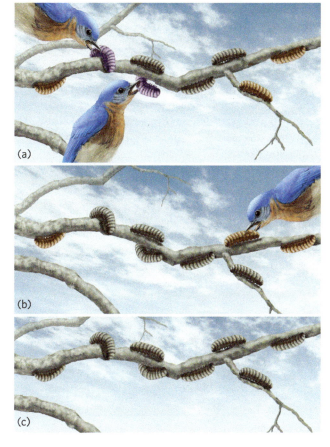

Figure 1.5 Evolution by natural selection. In this example, the caterpillar population is initially quite variable in color **(a)**. Individuals that better match the twig are less obvious to the bird hunting for food and therefore more likely to survive. If color is genetically inherited, the next generation **(b)** of the caterpillar population will be better matched to resemble twigs. As this natural selection continues over many generations, the color of the caterpillar population will closely match the twigs **(c)**. At this stage, the color of the caterpillar represents an adaptation against predation.

Species do not evolve in isolation. Rather, evolution in one species opens up new possibilities for other species with which the evolving species interacts. For example, milkweed plants have evolved the ability to produce a sap that is toxic to most herbivores. However, caterpillars of the monarch butterfly (*Danaus plexippus*) have evolved the ability to eat the leaves of milkweed plants and tolerate the toxic chemicals. Monarch caterpillars not only tolerate these toxic compounds, but also sequester these compounds in their bodies and use the toxins to defend themselves, as larvae as well as adults, against bird predators. In addition, both larvae and adults have evolved conspicuous "warning" coloration to advertise their toxicity. After the caterpillar species evolved these defensive abilities, predatory birds evolved a new ability to discriminate between caterpillars and butterflies that were toxic and those that were edible. In this case, the evolution of toxic chemicals in milkweed plants led to the evolution of chemical tolerance by the monarch butterflies, which further led to birds

evolving the ability to discriminate between the unpalatable monarch butterflies and other species of palatable butterflies.

As you can see from these examples, the complexity of ecological communities and ecosystems builds on, and is fostered by, existing complexity. Ecologists seek to understand how these complex ecological systems came into being and how they function in their environmental settings.

CONCEPT CHECK

1. Describe how ecological systems are governed by physical and biological principles.
2. What does it mean when we say that ecological systems are in a dynamic steady state?
3. What are the three conditions required for evolution by natural selection to occur?

1.3 Different organisms play diverse roles in ecological systems

Transformations of matter and energy in ecological systems are performed by both small and large forms of life, and these different life forms can play unique roles in ecological systems. In this section, we will examine how organisms interact with each other and the environment. We will see how ecologists categorize species based on how they obtain their energy and how they interact with other species. We will also describe the major groups of organisms that have evolved, discuss the diverse ecological roles found within each group, and examine the concepts of *habitat* and *niche*.

BROAD EVOLUTIONARY PATTERNS

Early in the history of Earth, ecosystems were dominated by bacteria. The ancient evolution of these groups is still debated by scientists, but a leading hypothesis, illustrated in **Figure 1.6**, is that bacteria are the oldest group of organisms. Over time, bacteria gave rise to the archaea. Bacteria and archaea are both *prokaryotic* organisms, which are single-cell organisms that do not possess distinct cellular organelles, such as a nucleus. These evolutionary events likely happened in the ocean and may have occurred near the deep-sea vents that we discussed in the chapter opener. If so, it may be that the first bacteria used chemosynthesis and this later gave rise to the evolution of photosynthesis.

Over time, *eukaryotic* organisms evolved, which possess distinct cellular organelles. The key event in the evolution of eukaryotes occurred when one bacterium engulfed another. The engulfed bacterium became what we now know as the *mitochondrion,* an important organelle for cellular respiration in eukaryotic organisms. This ancestor subsequently gave rise to all modern organisms that contain mitochondria, including red algae, green algae, plants, fungi, and animals. As the evolution of the eukaryotes progressed, there was a second key event. A eukaryotic cell engulfed another bacterium that was capable of photosynthesis, which evolved

Figure 1.6 The evolution of life on Earth. Bacteria are the oldest forms of life on Earth. The engulfing of one bacterial species by another led to the rise of eukaryotes containing cell organelles such as mitochondria and chloroplasts.

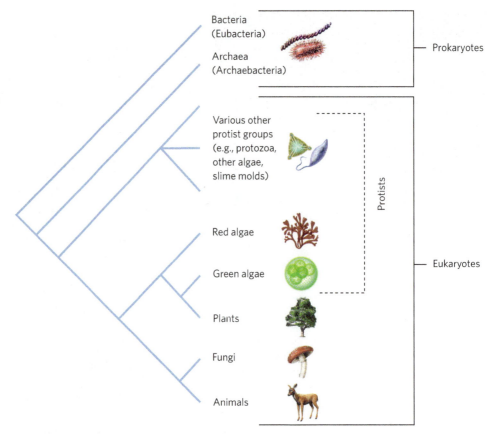

into what we now know as the chloroplast. The group of eukaryotic organisms that contained chloroplasts subsequently gave rise to modern-day red algae, green algae, and plants. Those species that didn't contain chloroplasts gave rise to modern-day fungi and animals.

Bacteria also modified the biosphere, making it possible for other forms of life to exist. For example, photosynthetic bacteria that were present more than 3 billion years ago produced oxygen as a by-product of photosynthesis. The higher concentrations of oxygen in the atmosphere and oceans favored the evolution of additional life forms that required oxygen, such as plants and animals. Despite all these changes, bacteria have persisted to the present day. As we shall see, their unique biochemical capabilities allow them to consume resources that their more complex descendants cannot use, and to tolerate ecological conditions that are beyond the capacities of other organisms.

Ecosystems depend on the activities of many forms of life. Each major group fills a unique and necessary role in the biosphere. We will now briefly review the major groups of organisms, including bacteria, protists, plants, fungi, and animals.

Bacteria

Although bacteria are microscopic, their enormous range of metabolic capabilities enables them to accomplish many unique biochemical transformations and to occupy parts of the biosphere where plants, animals, fungi, and most protists cannot survive. Some bacteria can assimilate molecular nitrogen (N_2) from the atmosphere, which they use to synthesize proteins and nucleic acids. Other species of bacteria, such as those living in deep-sea hydrothermal vents, can use inorganic compounds such as hydrogen sulfide (H_2S) as sources of energy in chemosynthesis. Furthermore, many bacteria can live under anaerobic conditions, such as in mucky soils and sediments, in which free oxygen is lacking, where their metabolic activities release nutrients that can be taken up by plants. Finally, some bacteria, including cyanobacteria (colloquially known as blue-green algae), can conduct photosynthesis. Cyanobacteria account for a large fraction of the photosynthesis that occurs in aquatic ecosystems. When bodies of water contain high amounts of nutrients, cyanobacteria can form dense populations that turn the water green, an event known as an **algal bloom** (**Figure 1.7**). Later in the book, we will have much more to say about the special place of bacteria in ecosystems.

Protists

Protists are a highly diverse group of mostly single-celled eukaryotic organisms that includes the algae, slime molds, and protozoans. This bewildering variety of protists fills almost every ecological role. For instance, algae are the primary photosynthetic organisms in most aquatic ecosystems. Some algae can form large plantlike structures, such

Algal bloom A rapid increase in the growth of algae in aquatic habitats, typically due to an influx of nutrients.

Figure 1.7 Cyanobacteria. Also known as blue-green algae, cyanobacteria are bacteria capable of conducting photosynthesis. These organisms can grow rapidly under highly fertile conditions, producing large floating mats in the water that can be toxic to animals, as illustrated in this photo of Lake Mendota, Wisconsin. Photos by Lee Wilcox (top) and SINCLAIR STAMMERS/ Getty Images (inset).

as the seaweed known as kelps, which can grow up to 100 m in length. Because of this large size, regions of the ocean containing large amounts of kelps, such as that shown in **Figure 1.8**, are called kelp forests. Although kelps may resemble large plants, the actual organization of kelp tissues is less structurally complex than that of trees and other plants.

Other protists are not photosynthetic. Foraminifera and radiolarians are protists that feed on tiny particles of organic matter or absorb small dissolved organic molecules. Some of the ciliate protozoans are effective predators of other microorganisms. Many protists live in the guts or other tissues of a host organism where they might be helpful or cause harm. For example, termites are a type of insect that consume large amounts of cellulose. Cellulose is a very difficult substance for animals to digest, but the termite has a community of protists—as well as bacteria—in its gut that are very effective at breaking down the cellulose. Some of the best known harmful protists include *Plasmodium,* which causes malaria in humans, and *Trypanosoma brucei,* which causes sleeping sickness.

Figure 1.8 Kelp forests. Although most protists are very small organisms, some protists such as seaweeds can grow very large and look like large plants. This kelp forest is off the coast of Southern California. Photo by Mark Conlin/Image Quest Marine.

Plants

Plants are well known for their role in using the energy of sunlight to synthesize organic molecules from CO_2 and H_2O. On land, most plants have structures with large exposed surfaces—their leaves—to capture the energy from sunlight (**Figure 1.9a**). Leaves are thin because surface area is more important than leaf thickness for capturing light energy. To obtain carbon, terrestrial plants take up gaseous CO_2 from the atmosphere. At the same time, they lose large amounts of water by evaporation from their leaf tissues to the atmosphere. Thus, plants need a steady supply of water to replace the water lost during photosynthesis. Most plants are firmly rooted in the ground and in constant contact with water in the soil. Others, including orchids and a variety of tropical "air plants" (*epiphytes*), typically grow by attaching themselves to other plants—often the trunks of trees—and can photosynthesize only in humid environments that are bathed in clouds (Figure 1.9b).

Although we typically envision plants as organisms that obtain their energy from sunlight, plants can also obtain energy in other ways. For example, several groups of plants have evolved to be simultaneously photosynthetic and carnivorous. These plants include the Venus fly trap (*Dionaea muscipula*), several species of sundews, and several species of pitcher plants (Figure 1.9c). They often live in locations that are low in nutrients, so the invertebrates they trap and consume provide an additional source of nutrients and energy. In addition, more than 400 species of plants—including more than 200 species of orchids—lack chlorophyll and therefore cannot photosynthesize to obtain energy. Scientists once thought that these plants were acting as decomposers and obtained their organic carbon from dead organic matter, but we now know that many of these plants are, in fact, acting as parasites on fungi, which are the real decomposers in the ecosystem. These parasitic plants obtain the vast majority of their organic carbon from fungi. Other plants, such as dodder, have little chlorophyll and no roots (Figure 1.9d). Instead, the stringy plant—also known as strangle weed or devil's guts—winds around other plants,

(a)

(b)

(c)

(d)

Figure 1.9 Plants. Plants can play many roles in an ecosystem. **(a)** Most plants, such as this garlic mustard (*Alliaria petiolata*), are rooted in the soil and make organic compounds by photosynthesis. **(b)** Epiphytes, such as this *Haraella odorata*, also conduct photosynthesis but grow above the ground and are attached to other plants. **(c)** Carnivorous plants, such as this Venus flytrap (*Dionaea muscipula*), can both photosynthesize and obtain nutrients by trapping and digesting invertebrates. **(d)** Some plants, such as dodder, act as parasites by taking nutrients from other plants. Photos by (a) Zoonar/Lothar Hinz/AGE Fotostock; (b) Kriz Petr/AGE Fotostock; (c) Zigmund Leszczynski/Animals Animals-Earth Scenes; and (d) A Jagel/AGE Fotostock.

penetrates their tissues, and sucks up water, nutrients, and products of photosynthesis. The dodder is a serious pest for many farm crops.

Fungi

Fungi assume unique roles in the biosphere because of their distinctive growth form. Whereas some fungi such as yeasts are unicellular, most other fungi are multicellular. Most fungal organisms consist of threadlike structures called *hyphae* that are only a single cell in diameter. These hyphae may either form a loose network, which can invade plant or animal tissues or dead leaves and wood on the soil surface, or grow together into reproductive structures such as mushrooms (**Figure 1.10**). Because fungal hyphae are able to penetrate deeply into tissue, they readily decompose dead plant material, eventually making nutrients available to other organisms. Fungi digest their foods externally and secrete acids and enzymes into their immediate surroundings. Such digestion allows fungi to decompose dead organisms and dissolve nutrients from soil minerals.

Although most fungi function as decomposers, they can interact with other species in both positive and negative roles. Many fungi live in mutualistic relationships with plants, living either around or within the roots of plants.

Using their extensive network of hyphae, they obtain scarce nutrients from the surrounding soil and provide them to the plant. In exchange, the plant provides the products of photosynthesis. Other fungi, though, can act as pathogens. Several closely related species of fungus cause Dutch elm disease, which has caused widespread death in several species of elm trees throughout North America and Europe during the past 100 years. Fungal pathogens are also a major problem for many crops, including potatoes, wheat, and rice.

Animals

Animals play a wide range of roles as consumers in ecological systems. Some animals, for example, elephants, gazelles, and voles, eat plants. Other animals, such as mountain lions, rattlesnakes, and frogs, eat other animals. Ticks, lice, and tapeworms are animals that live on or in other organisms. Finally, animals such as flies, bees, butterflies, moths, and bats can serve as important plant pollinators and seed dispersers.

CATEGORIZING SPECIES BASED ON SOURCES OF ENERGY

Ecologists often categorize organisms according to how they obtain energy, as illustrated in **Figure 1.11**. Organisms that use photosynthesis to convert solar energy into

Figure 1.10 Fungi. The mushrooms produced by this sulphur tuft fungus (*Hypholoma fasciculare*) in Belgium are fruiting bodies of much larger, unseen masses of threadlike hyphae that penetrate decaying wood and leaf litter. Fungi are effective decomposers. Photos by Philippe Clement/naturepl.com (top) and Steve Gschmeissner/Science Source (inset).

organic compounds or use chemosynthesis to convert chemical energy into organic compounds are known as **producers** or **autotrophs.** Organisms that obtain their energy from other organisms are known as **consumers** or **heterotrophs.** There are many different kinds of heterotrophs. Some consume plants, some consume animals, and some consume dead organic matter. In the next section, we will discuss these various interactions in more detail.

Not all species fit neatly into categories of autotrophs or heterotrophs. Some species can obtain their sources of carbon through a variety of ways. Because these species take a mixed approach to obtaining their energy, they are called **mixotrophs.** Mixotrophs are quite common in nature. For example, some bacteria can switch back and forth between photosynthesis and chemosynthesis. In addition, many species of algae can photosynthesize and obtain organic carbon by engulfing bacteria, protists, and bits of organic carbon that exist in the water. Other mixotrophs include the carnivorous plants that we discussed earlier in this chapter. These plants obtain their energy both from photosynthesis and from consuming invertebrates.

TYPES OF SPECIES INTERACTIONS

In considering the diversity of species that exist on Earth, we are often interested in the roles that they play. Ecologists categorize species by the types of interactions they have with other species, as you can see from the examples in **Figure 1.12.** Below is a brief introduction to these interactions, beginning with the various types of consumers, each of which will be covered in much greater detail in later chapters.

Predation

Predators are organisms that kill and partially or entirely consume another individual. The mountain lion, for example, is a predator that kills white-tailed deer (*Odocoileus virginianus*) and many other species of prey.

Producer An organism that uses photosynthesis to convert solar energy into organic compounds or uses chemosynthesis to convert chemical energy into organic compounds. *Also known as* **Autotroph**.

Consumer An organism that obtains its energy from other organisms. *Also known as* **Heterotroph**.

Mixotroph An organism that obtains its energy from more than one source.

Predator An organism that kills and partially or entirely consumes another individual.

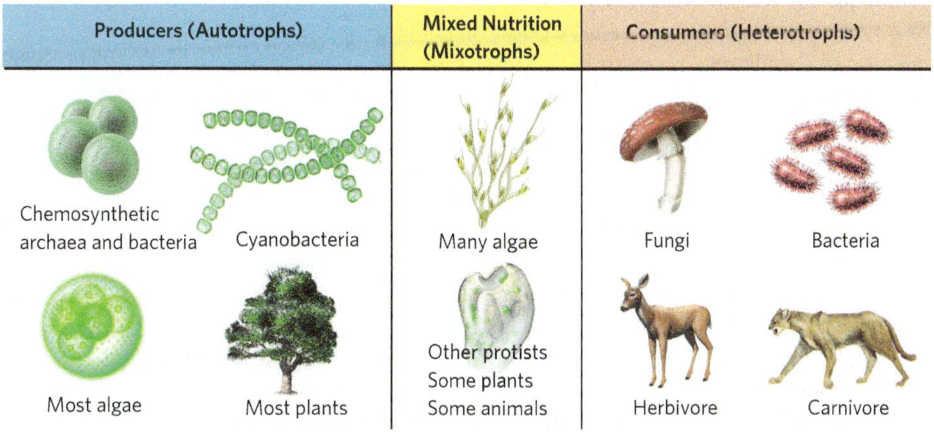

Producers (Autotrophs)		Mixed Nutrition (Mixotrophs)	Consumers (Heterotrophs)	
Chemosynthetic archaea and bacteria	Cyanobacteria	Many algae	Fungi	Bacteria
Most algae	Most plants	Other protists Some plants Some animals	Herbivore	Carnivore

Figure 1.11 Categories of species based on their energy sources. Species that obtain their energy from photosynthesis or chemosynthesis are known as producers or autotrophs. Species that obtain their energy from consuming other species are heterotrophs. Species that can take a mixed strategy of being producers and heterotrophs are mixotrophs.

Figure 1.12 The four types of consumers. Predators can be broken down into predators such as mountain lions, parasitoids such as braconid wasps shown here on a tomato hornworm, parasites such as winter ticks, and herbivores such as bison.

Parasitoids represent a special kind of predator. Parasitoids lay their eggs on or inside other animals, particularly insects, and the eggs hatch into larvae that consume the host individual from the inside, eventually killing it. Most parasitoid species are wasps and flies.

Parasitism

Parasites are organisms that live in or on another organism, called the *host.* An individual parasite rarely kills its host, although some hosts die when they are infected by a large number of parasites. Common parasites include tapeworms and ticks. When a parasite causes a disease, it is called a **pathogen.** Pathogens include several species of bacteria, viruses, protists, fungi, and a group of worms called *helminths.*

Herbivory

Herbivores are organisms that consume producers, such as plants and algae. When an herbivore consumes a plant, it typically consumes only a small portion of the plant and does not kill the plant. For example, caterpillars consume a few leaves or parts of leaves, which the plant can regenerate. Cattle consume the tops of grass leaves but do not destroy the growing region, located at the base of the plant.

Competition

Competition can be defined as an interaction with negative effects between two species that depend on the same limiting resource to survive, grow, and reproduce. For example, two species of grass might compete for nitrogen in the soil. As a result, the survival, growth, and reproduction of each are reduced when living with the other species of grass in the same area compared to when they are living in the area alone. Similarly, coyotes and wolves might compete for the same prey animals in the forest, such that they survive, grow, and reproduce better when they are living alone than when the other species is present. Competition for limited resources is a very common interaction in nature.

Mutualism

When two species interact in a way such that each species receives benefits from the other, their interaction is a **mutualism.** The lichens in **Figure 1.13**, for example, are composed of a fungus living together with green algal cells or cyanobacteria as a single organism. The fungus provides nutrients to the algae and the algae provide carbohydrates from photosynthesis to the fungus. Other examples of mutualisms include the bacteria that help digest plant material in the guts of cattle, fungi that help plants extract mineral nutrients from the soil in return for carbohydrate energy from the plant, and honeybees that pollinate flowers as they obtain nectar.

Parasitoid An organism that lives within and consumes the tissues of a living host, eventually killing it.

Parasite An organism that lives in or on another organism, but rarely kills it.

Pathogen A parasite that causes disease in its host.

Herbivore An organism that consumes producers such as plants and algae.

Competition An interaction resulting in negative effects between two species that depend on the same limiting resource to survive, grow, and reproduce.

Mutualism An interaction between two species in which each species receives benefits from the other.

Algal cells

Fungal hyphae

Figure 1.13 **Mutualism.** A lichen is a symbiotic association of a fungus and algal cells. Photo by VAUGHN A FLEMING/SCIENCE PHOTO LIBRARY/Getty Images.

Commensalism

Commensalisms are interactions in which two species live in close association and one species receives a benefit, while the other experiences neither a benefit nor a cost. For example, plants such as burdock (*Arctium lappa*) produce fruits containing tiny barbs that stick to the hair of mammals that brush up against it. The burdock receives a benefit of having its seeds dispersed, while the mammal is neither helped nor harmed by carrying these fruits on its body.

Because organisms are specialized for particular ways of life, many different types of organisms are able to live together in close association. A close physical relationship between two different types of organisms is referred to as a **symbiotic relationship.** Many parasites, parasitoids, mutualists, and commensal organisms live in symbiotic relationships.

When considering the different types of interactions among species, it can be helpful to categorize interactions between the two participants as positive (+), negative (−), or neutral (0). **Table 1.1** provides a summary of species interactions using this approach.

TABLE 1.1
The Outcome of Interactions Between Two Species

Type of interaction	Species 1	Species 2
Predation/parasitoidism	+	−
Parasitism	+	−
Herbivory	+	−
Competition	−	−
Mutualism	+	+
Commensalism	+	0

Interactions that provide a benefit to a species are indicated by a "+" symbol, interactions that cause harm to a species are indicated by a "−" symbol, and interactions that have no effect on a species are indicated by a "0" symbol.

Consumers of Dead Organic Matter

Consumers of dead organic matter—including *scavengers, detritivores,* and *decomposers*—also play important roles in nature. **Scavengers,** such as vultures, consume dead animals. **Detritivores,** such as dung beetles and many species of millipedes, break down dead organic matter and waste products—known as *detritus*—into smaller particles. **Decomposers,** such as many species of mushrooms, break down dead organic material into simpler elements and compounds that can be recycled through the ecosystem.

HABITAT VERSUS NICHE

In addition to knowing how a species makes its living by interacting with other species, we also need to consider the physical setting in which it lives. For example, if you were to walk through a meadow in the eastern or central United States, you would likely come across the eastern cottontail rabbit (*Sylvilagus floridanus*). This species thrives in abandoned farm fields that are full of grasses and other tall wildflowers interspersed with shrubs. These plants provide food for the rabbit and protection from its many predators, including coyotes (*Canis latrans*) and several species of hawk. In considering species in nature, ecologists find it useful to distinguish between where an organism lives and what it does.

The **habitat** of an organism is the place, or physical setting, in which it lives. In the case of the rabbit, the habitat consists of old fields that contain grasses, wildflowers, and shrubs. Habitats are distinguished by physical features, often including the predominant form of plant or animal life. Thus, we speak of forest habitats, desert habitats, stream habitats, and lake habitats (**Figure 1.14**).

During the early years of ecological research, scientists developed a complex system of classifying habitats. For example, they began by distinguishing between terrestrial

---

Commensalism An interaction in which two species live in close association and one species receives a benefit, while the other experiences neither a benefit nor a cost.

Symbiotic relationship When two different types of organisms live in a close physical relationship.

Scavenger An organism that consumes dead animals.

Detritivore An organism that feeds on dead organic matter and waste products that are collectively known as detritus.

Decomposer Organisms that break down dead organic material into simpler elements and compounds that can be recycled through the ecosystem.

Habitat The place, or physical setting, in which an organism lives.

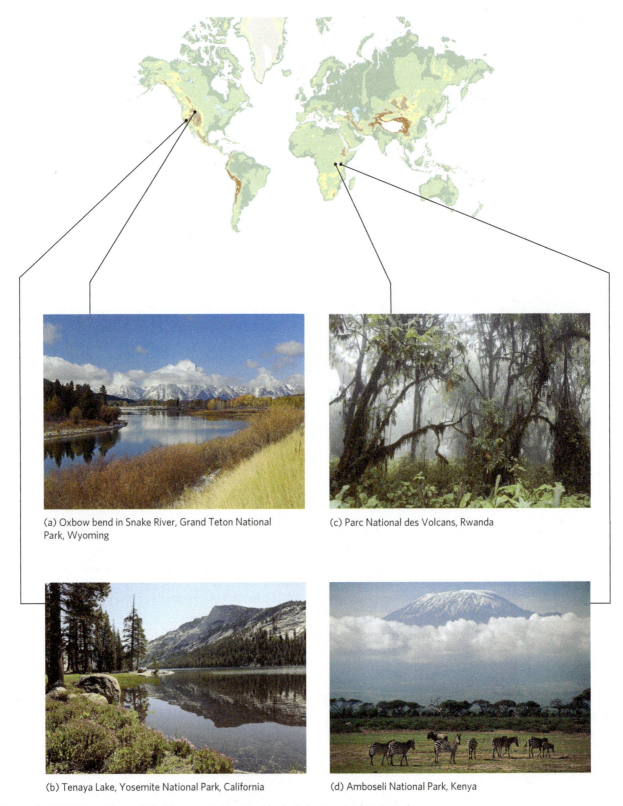

(a) Oxbow bend in Snake River, Grand Teton National Park, Wyoming

(c) Parc National des Volcans, Rwanda

(b) Tenaya Lake, Yosemite National Park, California

(d) Amboseli National Park, Kenya

Figure 1.14 Habitats. Terrestrial habitats are distinguished by their dominant vegetation, whereas aquatic habitats are distinguished by their depth and the presence or absence of flowing water. **(a)** Freshwater streams contain flowing water. **(b)** Lakes are typically large bodies of water that have very little flow. **(c)** In the tropical rainforest, warm temperatures and abundant rainfall support the highest productivity and biodiversity on Earth. **(d)** Tropical grasslands, which develop where rainfall is sparse, support vast herds of grazing herbivores during the productive rainy season. Photos by (a) George Sanker/naturepl.com; (b) McPHOTO/AGE Fotostock; (c) Michel Gunther/Science Source; and (d) Staffan Widstrand/naturepl.com.

and aquatic habitats. Among aquatic habitats, they identified freshwater and marine habitats. Among the marine habitats, they classified coastal habitats, the open ocean, and the ocean floor. As such classifications became more finely divided, the distinctions began to break down since scientists found that habitat types overlap and that absolute distinctions rarely exist. However, the idea of habitat is nonetheless useful because it emphasizes the variety of conditions to which organisms are exposed. For example, inhabitants of extreme ocean depths and tropical rainforest canopies experience altogether different conditions of light, pressure, temperature, oxygen concentration, moisture, and salt concentrations, as well as differences in food resources and predators.

An organism's **niche** includes the range of abiotic and biotic conditions it can tolerate. In the case of the cottontail rabbit, its niche includes the ranges of temperature and humidity that it can tolerate, the plants that it eats, and the predatory hawks and coyotes with which it coexists. An important ecological principle states that each species has a distinct niche. No two species have exactly the same niche because each has distinctive attributes of form and function that determine the conditions it can tolerate, how it feeds, and how it escapes enemies. Consider the hundreds of insect species that might live in a garden; each has a unique niche in terms of the food that it eats (**Figure 1.15**). For example,

Niche The range of abiotic and biotic conditions that an organism can tolerate.

Figure 1.15 Niche. Even within a group of similar organisms, such as insects, each species has a distinct niche. In the case of insects, the food they consume is only one aspect of their niche. **(a)** The European corn borer is specialized to feed on corn plants. **(b)** The Colorado potato beetle is specialized to feed on the leaves of potato plants. **(c)** The caterpillar of the cabbage white butterfly is specialized to feed on the leaves of cabbage, broccoli, and cauliflower. Photos by (a) Scott Sinklier/AGE Fotostock; (b) blickwinkel/Alamy; and (c) Nigel Cattlin/Alamy.

the caterpillar of the cabbage white butterfly (*Pieris rapae*) feeds on the group of plants that have been cultivated from the wild mustard plant (*Brassica oleracea*), including cabbage, broccoli, and cauliflower. However, the Colorado potato beetle (*Leptinotarsa decemlineata*) feeds almost exclusively on the leaves of potato plants (*Solanum tuberosum*). Similarly, the European corn borer (*Ostrinia nubilalis*) feeds primarily on corn plants (*Zea mays*). The variety of habitats and niches holds the key to much of the diversity of living organisms.

CONCEPT CHECK

1. How do the sources of energy acquired by plants, animals, and fungi differ?
2. What are the major types of species interactions?
3. Compare and contrast an organism's habitat and an organism's niche.

1.4 Scientists use several approaches to studying ecology

Scientists have investigated the diverse roles that organisms play in the environment for more than a century. Ecologists investigate their subject matter using a systematic procedure, often referred to as the scientific method. The three steps of this process, shown in **Figure 1.16**, are (1) observations regarding a pattern in nature, (2) development of a hypothesis and its associated predictions, and (3) testing the hypothesis.

OBSERVATIONS, HYPOTHESES, AND PREDICTIONS

Most research begins with a set of observations about nature that invite explanation. Usually, these observations identify and describe a consistent pattern. As we learned in the history of research on the deep-sea vents, once it was discovered that the diversity and abundance of species living around the vents could not be sustained by the relatively small amount of organic matter drifting down from the sunlit surface, new hypotheses regarding chemosynthetic bacteria had to be developed and tested. In such cases, some hypotheses will be supported, while others will be rejected and require new hypotheses. This process is the scientific method.

To help you better understand the scientific method, imagine that you are walking around a pond on a warm,

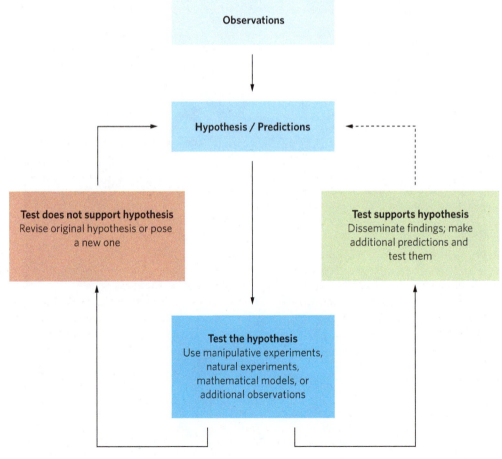

Figure 1.16 The scientific method. The scientific method begins with observing patterns in nature and developing a hypothesis that explains how or why the pattern exists. Predictions from a hypothesis are tested with manipulative experiments, natural experiments, mathematical models, or additional observations.

spring night following a rainstorm. You would likely hear male frogs making frog calls. If you returned to the same pond on cooler nights following a dry period, you would be less likely to hear frogs calling. If you traveled to many different ponds, you would observe this pattern over and over again. That is, you would observe and describe a consistent pattern in nature. Repeated natural patterns lead scientists to hypothesize about the causes of these patterns.

Hypotheses are ideas that potentially explain a repeated observation. In the case of the frogs, we consistently observed that they called only on warm nights after a rainstorm. Having established the existence of this pattern, we want to understand it better. We might want to explain *how* frogs sense changes in temperature and rainfall, and how sensing these environmental changes stimulates frogs to call. We also might want to explain *why* frogs call on warm nights after it rains. How do the frogs benefit from calling—perhaps by attracting mates—and what, if any, are the costs of calling?

Hypotheses about how and why organisms respond to the environment represent different types of explanations. The "how" explanation addresses the details of the animal's sensory perception and changes in its hormone concentrations, nervous system, and muscular system. In the case of the frogs, we might hypothesize that the frog's nervous system detects warm temperatures and rain. This initiates changes in a male frog's hormones and physiology that cause him to contract muscles that make him call.

Hypotheses that address the immediate changes in an organism's hormones, physiology, nervous system, or muscular system are known as **proximate hypotheses.** If these hypotheses are supported by our observations, then we can make *predictions*. **Predictions** are statements that arise logically from hypotheses. For example, if our hypothesis about how a rainy night causes male frogs to call is correct, then we can predict that any frogs exposed to warm rain will respond by changing the concentration of specific hormones that stimulate the brain to send a signal to the muscles of the vocal apparatus to contract.

Ultimate hypotheses address why an organism has evolved to respond in a certain way to its environment— that is, the fitness costs and benefits of a particular response. For example, we might hypothesize that male frogs call to attract females. Furthermore, if we suspect that male frogs call to attract females, then perhaps males sing on warm nights after a rainstorm because such nights produce the best conditions for laying eggs, which is when females are most interested in mating. The males benefit from calling on a warm, wet night because they will be more likely to attract females and therefore father more offspring. If the males call at other times, they will attract fewer female mates and receive a much smaller benefit. But what about costs? We might hypothesize that when male frogs call to attract females, they risk attracting the attention of predators. The increased risk of death represents a high fitness cost to male frogs.

We have now generated a number of predictions that logically follow from our ultimate hypotheses about male frog calling: (1) Males that call will attract females; (2) females actively search for males only on warm, wet nights because that produces the best conditions for laying eggs; and (3) if singing imposes a cost, then males should save their singing for times when it will provide the greatest benefit.

TESTING HYPOTHESES WITH MANIPULATIVE EXPERIMENTS

A particular hypothesis can be supported by our observations, but it can rarely be confirmed beyond a doubt. However, our confidence increases as we continue testing a hypothesis and repeatedly find that our observations support it. Although the methods of acquiring scientific knowledge appear to be straightforward, many pitfalls exist. For example, an observed relationship between two factors does not necessarily mean that one factor causes the other to change. The cause must be determined independently. To accomplish this, we can design **manipulative experiments** in which a hypothesis is tested by altering the factor that is hypothesized to be an underlying cause of the phenomenon.

To understand the process of a manipulative experiment, consider the observation that herbivorous insects often consume less than 10 percent of a plant's tissues. Ecologists have proposed several hypotheses to explain this. One hypothesis is that predators consume insect herbivores at such a high rate that insect populations remain low. This low insect population cannot eat very much of the plant tissue. This seems like a reasonable hypothesis, but how do we test it with a manipulative experiment?

Researchers working on this question decided to explore whether the predation hypothesis applies to insects feeding on oak trees in Missouri. They observed that birds consume many insects on oak leaves and then hypothesized that birds reduce the populations of insect herbivores. If this hypothesis is correct, when birds are absent, insect populations should increase and consume more leaf biomass. Confirmation of this prediction would support the hypothesis; a lack of confirmation would lead them to reject the hypothesis and propose a new one.

To test the hypothesis that predation by birds lowers the abundance of insects on oak trees, the researchers decided to conduct a manipulative experiment in which they used

---

Hypothesis An idea that potentially explains a repeated observation.

Proximate hypothesis A hypothesis that addresses the immediate changes in an organism's hormones, physiology, nervous system, or muscular system.

Prediction A logical consequence of a hypothesis.

Ultimate hypothesis A hypothesis that addresses why an organism has evolved to respond in a certain way to its environment in terms of the fitness costs and benefits of the response.

Manipulative experiment A process by which a hypothesis is tested by altering a factor that is hypothesized to be an underlying cause of the phenomenon.

cages that excluded birds from the trees (**Figure 1.17a**). The **manipulation,** also known as the **treatment,** is the factor that we want to vary in an experiment. Often one of the manipulations used is a *control.* A **control** is a manipulation that includes all aspects of an experiment except the factor of interest. In the oak tree experiment, the caged trees served as the treatment, whereas the uncaged trees served as the control.

Once we decide on which manipulations we wish to do, we have to assign each manipulation to a specific **experimental unit.** An experimental unit is the object to which we apply the manipulation. In the case of the oak tree experiment, the researchers decided that they would use groups of three white oak tree saplings as their experimental units. After making this decision, each experimental unit was either caged—by surrounding the group of three saplings with bird-proof netting—or left uncaged to allow bird access.

A manipulation to a single experimental unit might provide exciting results, but the results may not be reliable unless the experiment is repeated and demonstrates a similar outcome. Being able to produce a similar outcome multiple times is known as **replication,** which is an integral feature of most experimental studies. In the oak tree study, the investigators decided to add cages to 10 groups of trees and leave 10 groups of trees uncaged. In doing so, they replicated their experiment 10 times.

When we assign different manipulations to our experimental units, the assignments must be made using **randomization,** which means that every experimental unit has an equal chance of being assigned a particular manipulation. In the oak tree experiment, the researchers randomly assigned groups of trees to be caged or left uncaged as controls. In this way, they could be assured that the caged trees were initially no different from the controls.

Once the researchers set up the experiment, they collected data on the number of herbivorous insects and the percentage of leaf tissue that had been consumed. They found that caged trees had about twice as many insect herbivores as the control trees. Moreover, the percentage of leaf tissue consumed at the end of the growing season was nearly twice as high on caged trees as on the control trees (Figure 1.17b). These findings led the researchers to conclude that the experiment supported their hypothesis.

Although many experiments are conducted in natural settings such as oak forests or lakes (**Figure 1.18a**), other experiments use smaller experimental venues. Many experiments make use of **microcosms,** which are simplified ecological systems that attempt to replicate the essential features of an ecological system in a laboratory or field setting. In the case of experiments that study aquatic systems, for example, microcosms might consist of large outdoor tanks of water. These tanks would include many of the features of natural water bodies, including soil, vegetation, and a diversity of organisms (Figure 1.18b). The use of microcosms assumes that a response to manipulations in a microcosm is representative of responses that would occur in a natural habitat. For example, we might wish to understand how species of fish compete for food. Observing competition among the fish

(a)

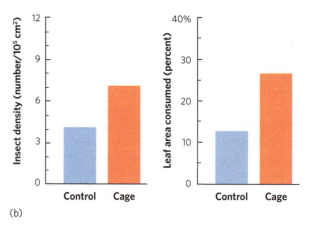

(b)

Figure 1.17 A manipulative experiment. Manipulative experiments provide the strongest tests of hypotheses. **(a)** In a study that tested if birds are an important factor in determining the number of insects on oak trees in Missouri, ecologists placed cages around some white oak saplings to exclude birds and left other oak saplings uncaged to serve as a control. **(b)** From this experiment, the researchers measured the number of insect herbivores per leaf and the amount of leaf tissue that was consumed in each of the two treatments. After R. J. Marquis and C. J. Whelan, Insectivorous birds increase growth of white oak through consumption of leaf-chewing insects, *Ecology* 75 (1994): 2007–2014. Photo by Chris Whelan, University of Illinois.

Manipulation The factor that we want to vary in an experiment. *Also known as* **Treatment**.

Control A manipulation that includes all aspects of an experiment except the factor of interest.

Experimental unit The object to which we apply an experimental manipulation.

Replication Being able to produce a similar outcome multiple times.

Randomization An aspect of experiment design in which every experimental unit has an equal chance of being assigned to a particular manipulation.

Microcosm A simplified ecological system that attempts to replicate the essential features of an ecological system in a laboratory or field setting.

Figure 1.18 Experimental venues. The choice of experimental venue often represents a trade-off between the complexity of natural conditions and the more highly controlled conditions of a laboratory experiment. **(a)** Manipulative experiments of entire lakes, such as this lake in Wisconsin, include natural conditions, but such experiments are difficult to replicate. **(b)** Microcosm experiments, such as this experiment at the University of Pittsburgh's Pymatuning Lab of Ecology, include many features of a lake by containing communities of aquatic organisms. Using microcosms allows the manipulations to be replicated many times. **(c)** Laboratory experiments, such as this pesticide experiment conducted in Petri dishes, allow researchers to conduct highly controlled experiments, but they are conducted under very unnatural conditions. Photos (a) courtesy of Carl Watras; (b) and (c) courtesy of Rick Relyea.

species in a murky lake might not be feasible, but a large tank of water that included many features of the lake might work well, providing that the fish behave similarly under both conditions. If so, the results of the microcosm experiment may yield results that can be generalized to the larger, natural system. Experiments can also be conducted at very small scales, such as Petri dishes in the laboratory (Figure 1.18c). Choosing the proper venue for a given experiment represents a trade-off to researchers from the very natural outdoor experiments in which many factors are difficult to manipulate, to the very artificial but highly controlled lab experiments where a wide variety of factors can be manipulated.

ALTERNATIVE APPROACHES TO MANIPULATIVE EXPERIMENTS

Many hypotheses cannot be tested by experiments, either because the amount of area or length of time needed to test the hypotheses is simply too large, or because it is not possible to isolate particular variables and devise a suitable control. These limitations are common when we are trying to understand patterns that have occurred over long periods, or in systems such as entire populations or ecosystems that are too large to be manipulated.

Several different hypotheses might explain a particular observation equally well, so investigators must make predictions that distinguish among the alternatives. For example, many ecologists have observed a decrease in the number of species as one moves north or south, away from the equator. This repeated pattern has many potential explanations. As one travels north from the equator, average temperature and precipitation decrease, sunlight decreases, and seasonality increases. Each of these factors alone or together could affect the number of species that can coexist in a specific locality. Indeed, dozens of hypotheses have been proposed to explain the observed decrease in the number of species as one moves away from the equator. Isolating the effect of each factor has proved difficult because all the other factors change together.

Ecologists have several alternative approaches that address these difficulties. One option, the **natural experiment,** relies on natural variation in the environment to test a hypothesis. For example, consider the hypothesis that the number of species on an island is influenced by the size of the island because larger islands have more available

Natural experiment An approach to hypothesis testing that relies on natural variation in the environment.

ANALYZING ECOLOGY

Why Do We Calculate Means and Variances?

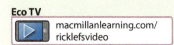
Eco TV
macmillanlearning.com/ricklefsvideo

As we saw in the oak tree experiment, when testing hypotheses, ecologists make **observations,** including measurements that are collected from organisms or the environment. These observations, also known as **data,** are then used to test hypotheses. In the case of the oak trees, the researchers collected data on the density of herbivorous insects and the amount of leaf tissue they consumed. In asking questions in ecology, we often want to know the average value, or *mean,* of the data collected from different treatments or obtained under different conditions. In the case of the oak tree experiment, the researchers wanted to compare the mean density of insects on caged trees versus uncaged trees to determine whether birds depressed the numbers of insects consuming the tree leaves.

While a comparison of the different means tells us about the central tendencies of the data, ecologists often want to know if the data used to generate the mean have high or low variability. For example, if the mean density of insects on leaves was 10 insects per square meter of leaf surface in both of the following sets of data, which group is more variable?

Group A: 10, 9, 11, 10, 8, 12, 9, 11, 8, 12

Group B: 10, 5, 15, 10, 6, 14, 5, 15, 7, 13

Although both groups of data have the same mean, observations in Group A range from 8 to 12, whereas in Group B they range from 5 to 15. Therefore, the data in Group B are more variable.

Why do we care about the variability of the data that we collect? Given that each mean is calculated from a set of data that has either a narrow or wide range, the variability gives us an idea of how much the distributions of data overlap with each other. If two groups of data have different means but the distributions of data overlap a great deal, then we cannot be confident that the two groups are actually different from each other. In contrast, if two groups of data have different means but the distributions of data do not overlap, then we might be very confident that the two groups are different.

One way to measure how widely the data points are spread around the mean is to calculate the *variance of the mean.* The **variance of the mean** is a measurement that indicates the spread of data around the mean of a population when every member of the population has been measured. Data points that are more widely spread around the mean will have a higher variance. The easiest way to calculate the variance in a set of data (denoted as σ^2) is to do it in two steps:

1. Square each observed value (denoted as χ) and calculate the mean of these squared values (where E indicates that we are taking the mean of several values):

$$E[\chi^2]$$

Observations Information, including measurements, that is collected from organisms or the environment. *Also known as* **Data**.

Variance of the mean A measurement that indicates the spread of data around the mean of a population when every member of the population has been measured.

---

niches, support larger populations that resist extinction, and are easier for organisms to find and colonize. A manipulative experiment to test this hypothesis would be impossible since it would require both a massive manipulation of many islands as well as the ability to observe a difference in the number of colonizing species over hundreds or even thousands of years. Instead, we can test the hypothesis by comparing the number of species living on islands of different sizes that have been formed over shorter periods by changes in sea or lake levels. Although a manipulative experiment is not possible in such cases, a natural experiment like this still allows researchers to determine if patterns in nature are consistent or inconsistent with hypotheses about the underlying causes.

Ecologists also use *mathematical models* to explore the behavior of ecological systems. In a **mathematical model,** an investigator designs a representation of a system with a set of equations that corresponds to the hypothesized relationships among each of the system's components. For example, we might use a mathematical model to represent how births and immigration add to the growth rate of a population, and how deaths and emigration subtract from that growth rate. In this sense, a mathematical model is a hypothesis; it provides an explanation of the observed structure and functioning of the system.

We can test the accuracy of a mathematical model by comparing the predictions it yields with observations in nature. For example, epidemiologists have developed models to describe the spread of communicable diseases. These models include such factors as the proportions of a population that are susceptible, exposed, infected, and recovered from infections. The models also include the rates of transmission and the probability that the organism will cause a disease in an

---

Mathematical model A representation of a system with a set of equations that correspond to hypothesized relationships among the system's components.

2. From this mean, subtract the square of the mean observed value:

$$\sigma^2 = E[\chi^2] - [E(\chi)]^2$$

In words, $E[\chi^2]$ is the mean of the squared observed values and $[E(\chi)]^2$ is the square of the mean observed value.

When we calculate the variance of the mean, the calculation is based on the assumption that we have measured every member of a population. In reality, we often cannot measure every member, but instead measure a sample of the population. In the oak tree study, for example, the researchers did not measure the insects on all the oak trees, but instead used a sample of 10 groups of oak trees. When we measure a sample of the population, the variation in the data is called the **sample variance.** The sample variance is very similar to the variance of the mean, except that we now account for how many samples of the population we measured (denoted as n). The sample variance (denoted as s^2) is calculated as

$$s^2 = \frac{n}{n-1} \times \sigma^2$$

or

$$s^2 = \frac{n}{n-1} \times (E[\chi^2] - [E(\chi)^2])$$

As you might notice from this equation, as the number of samples becomes very large, the value of the sample variance approaches the value of the variance of the mean for the entire population.

To help you understand how to calculate the sample variance, consider the following set of observations on the abundance of insects per leaf on caged and uncaged trees:

Caged trees	Uncaged trees
8	4
6	3
7	2
9	4
5	2

For the caged trees, we can calculate the mean of the values as

$$(8 + 6 + 7 + 9 + 5) \div 5 = 7$$

and the mean of the squared values as

$$(8^2 + 6^2 + 7^2 + 9^2 + 5^2) \div 5 = 51$$

We can then calculate the sample variance for the caged-tree data as

$$s^2 = \frac{n}{n-1} \times (E[\chi^2] - [E(\chi)^2])$$

$$s^2 = \frac{5}{5-1} \times (51 - (7)^2)$$

$$s^2 = (1.25) \times (51 - 49) = 2.5$$

YOUR TURN Using the data from the five replicates of uncaged trees, calculate the mean and sample variance of insect abundance.

Sample variance A measurement that indicates the spread of data around the mean of a population when only a sample of the population has been measured.

---

infected organism. By including all these factors, such models can make predictions about the frequency and severity of disease outbreaks. These predictions can then be tested by comparing them with real-world observations of disease outbreaks. This approach is being used for a number of important wildlife diseases, including the transmission of rabies in such animals as bats, raccoons, skunks, and foxes, and the transmission of Lyme disease in wildlife and human populations.

Mathematical models can be used on any scale. For example, at a larger scale, ecologists have created mathematical models to investigate how burning fossil fuels affects the CO_2 content of the atmosphere. To manage human impacts on our environment, it is critically important to understand this relationship. Models of global carbon content include, among other factors, equations that describe the uptake of CO_2 by plants and the dissolution of CO_2 in the oceans. The earlier versions of these models failed to match observations and overestimated the annual increase in atmospheric CO_2 concentrations. The Earth evidently contains CO_2 "sinks" such as regenerating forests that remove CO_2 from the atmosphere. By including the effects of these CO_2 sinks, the refined carbon models more accurately describe observed atmospheric data and are more likely to predict future changes accurately.

For any model, we can support or reject the hypothesis by comparing the model's predictions against our observations. Rejected models can be further refined to incorporate additional complexities and better fit our observations.

CONCEPT CHECK

1. Explain the scientific method.
2. Compare and contrast proximate hypotheses versus ultimate hypotheses.
3. In what ways do manipulative experiments differ from natural experiments?

Figure 1.19 Human impacts on ecological systems. The growth of human population, particularly over the past two centuries, has altered much of the planet. Humans have destroyed habitats, converted land to agriculture, created air and water pollution, burned large amounts of fossil fuel, and overharvested plants and animals.

1.5 Humans influence ecological systems

For more than a century, ecologists have labored passionately to understand how nature works, from the level of the organism to the level of the biosphere. The wonders of the natural world summon our curiosity about life and our environment. For many ecologists, a curiosity about how nature works is reason enough to study ecology. Increasingly, however, ecologists find themselves struggling to understand how the rapidly growing human population, now more than 7 billion people, is affecting the planet. Our need to understand nature is becoming more and more urgent as the growing human population stresses the functioning of ecological systems. Environments dominated by humans or created by them—including urban and suburban living places, agricultural fields, tree farms, and recreational areas—are also ecological systems. The welfare of humanity depends on maintaining the proper functioning of these systems.

The human population currently consumes massive amounts of energy and resources and produces large amounts of waste. As a result, virtually the entire planet is strongly influenced by human activities (**Figure 1.19**). These influences include the degradation of the natural environment and the disruption of many important functions that natural environments provide to humans. Growing human consumption of natural resources has caused a number of ecological problems. For example, removing plants from their natural environment to use as house plants and exploiting animals for human consumption and the pet trade have caused the decline of many species in their native habitats. The species affected are diverse, including the cacti of the American Southwest that are collected for sale as house plants, several species of reptiles and amphibians that are sold in the pet trade, and many species of fish and whales that are overexploited by commercial fishing.

As commerce has become more global, species have been introduced unintentionally to new locations at an increasing rate. Some of these species, such as rats, snakes, and pathogens, can have devastating effects on local species. To feed 7 billion people, we have converted a large amount of land for agricultural use. This conversion has brought with it a number of challenges, including loss of natural habitats, pollution from fertilizers and pesticides, and questions about growing genetically modified crops. Some crops, such as corn, are now increasingly being used as sources of fuel, also known as *biofuels,* causing even more land to be converted to agricultural use. Humans also need land for housing, business, and industry. This has further reduced the amount of natural habitat available for other species and has been a major contributor to the decline and extinction of many species. We will deal with these issues in greater detail throughout the book.

Another suite of ecological challenges is caused by wastes produced by human activity. For example, untreated sewage and industrial processes can damage the air, water, and soil. In addition, the use of nuclear power plants to generate electricity produces substantial amounts of nuclear waste. Of all human wastes, perhaps none has a higher public awareness than the *greenhouse gases* responsible for global warming. **Greenhouse gases** are compounds in the atmosphere that absorb the infrared heat energy emitted by Earth and then emit some of the energy back toward Earth. In doing so, the gases prevent much of this energy radiated from the surface of Earth from escaping into space. Greenhouse gases include many different compounds, but an important player is CO_2, which is produced by burning fossil fuels in the cars we drive and in the fossil fuel–powered, electricity-generating plants that provide electricity to so many of our homes and businesses. As the human population continues to grow and demands for energy increase, we burn more fossil fuels and produce more greenhouse gases. The more greenhouse gases we put into the atmosphere, the warmer our Earth becomes.

Because ecological systems are inherently complex, it is difficult to predict and manage the effects of a growing human population on ecological systems at every level. At the level of the organism, for example, we might want to know how a pesticide sprayed in the environment could affect each of the many tissues and organ systems of an animal's body, leading to changes in behavior, growth, and reproduction. At the community level, we might ask how a decrease in abundance of one species caused by commercial harvesting could affect the populations of many other species in that community. At the biosphere level, we would like to quantify the large number of sources that emit CO_2 into the atmosphere and understand the processes that take CO_2 out of the atmosphere. Each of these cases presents a set of complex questions that are not easy to answer. Yet we need a solid understanding of how the ecological system operates before we can predict the outcome of human impacts on the system and recommend ways to minimize damages. In the chapters that follow, you will develop that understanding.

THE ROLE OF ECOLOGISTS

The plight of individual species headed toward extinction arouses us emotionally. However, ecologists increasingly realize that the only effective means of preserving the species of the world is through the conservation of ecosystems and the management of large-scale ecological processes. Individual species, including those that humans rely on for food and other products, are themselves dependent on the maintenance of environmental support systems. Local effects of human activities on ecological systems can often be managed once we understand the underlying mechanisms responsible for change. Increasingly, however, our activities have led to multiple, widespread effects that are more difficult for scientists

to characterize and for legislative and regulatory bodies to control. For this reason, a sound scientific comprehension of environmental problems is a necessary prerequisite to action.

The media is filled with reports of environmental problems: disappearing tropical forests, depleted fish stocks, emerging diseases, global warming, and wars that cause environmental tragedies and human suffering. But it is important to know that there are success stories as well. Many countries have made great strides in cleaning up their rivers, lakes, and air. Fish are once again migrating up major rivers in North America and Europe to spawn. Acid rain has decreased, thanks to changes in the combustion of fossil fuels. The release of chlorofluorocarbons, which damage the ozone layer that shields the surface of Earth from ultraviolet radiation, has decreased dramatically. The inevitability of global warming caused by increasing atmospheric CO_2 concentrations has provoked global concern and set off an international research effort. Conservation programs, including breeding endangered species in captivity, have saved some animals and plants from certain extinction. They have also heightened public awareness of environmental issues, and sometimes sparked public controversy.

These successes would not have been possible without a general consensus founded on evidence produced by scientific study of the natural world. Understanding ecology will not by itself solve our environmental problems, because these problems also have political, economic, and social dimensions. However, as we contemplate the need for global management of natural systems, our effectiveness in this enterprise critically depends on our understanding of their structure and functioning—an understanding that depends on knowing the principles of ecology.

This book introduces you to the study of ecology by building an understanding of all aspects of the discipline. We begin by looking at the individual level, including how species have adapted to the challenges of the aquatic and terrestrial environments. We will then explore the topic of evolution, including how species have evolved various strategies for mating, reproducing, and living in social groups. Next, we move to the population level with a discussion of population distributions, population growth, and population dynamics over space and time. With a firm understanding of populations, we move on to examine species interactions, communities, and ecosystems. Finally, we consider ecology at the global level and investigate global patterns of biodiversity and global conservation.

CONCEPT CHECK

1. In what ways have humans altered ecological systems?
2. How can our knowledge of ecological systems help humans manage these systems?

Greenhouse gases Compounds in the atmosphere that absorb the infrared heat energy emitted by Earth and then emit some of the energy back toward Earth.

ECOLOGY TODAY   CONNECTING THE CONCEPTS

The California Sea Otter

The California sea otter. This once abundant marine mammal has experienced large fluctuations in numbers as a result of human activities during the past three centuries. Photo by Neil A. Fisher/Vancouver Aquarium.

In this first chapter, we have examined a wide range of topics, including the hierarchy of perspectives in ecology, the biological and physical principles that govern natural systems, the variety of roles that different species play, the multiple approaches to studying ecology, and the influence of humans on ecological systems. To help you see how these topics interconnect, let's examine a case study of the sea otter (*Enhydra lutris*) off the Pacific coast. Humans have affected sea otter populations for hundreds of years. Several scientific approaches have been taken to understand these impacts and to help reverse them.

The sea otter was once abundant, with a geographic range that extended around the northern Pacific Rim, from Japan up to Alaska and down to Baja California. However, in the 1700s and 1800s, intense hunting for otter pelts reduced the population to near extinction and the fur industry subsequently collapsed. When a small population was discovered off the coast of central California in the 1930s, the otters were placed under protection. As a result, the population increased to several thousand individuals by the 1990s, though in more recent years, the otter has again experienced population declines. These changes in the size of otter populations presented an opportunity for scientists to examine a natural experiment in action.

Ecologists quickly realized that to understand the causes and consequences of the sea otter's fluctuations in abundance, they needed to use a range of ecological approaches, from the individual to the ecosystem. Taking an individual approach, ecologists established that the sea otter was a predator on a wide range of prey species, including abalone, spiny lobsters, small fish, crabs, sea urchins, and small snails. Among these prey items, observations of otter feeding behavior revealed that otters prefer certain prey such as abalone, a large species of sea snail. They will only eat other small species of snails when their preferred prey becomes rare.

Once scientists understood the sea otter's niche, they were better able to protect it. However, not everyone was happy about the resurgence of sea otters through the 1990s. California anglers became upset; they argued that the growing otter population would cause a dramatic change in the marine community, including a drastic reduction in the populations of commercially valuable fish, abalone, and spiny lobsters—all harvested for human consumption. However, scientists who took a community approach to ecology found that an increasing otter population was also having other dramatic effects on the marine community. For example, the otter's consumption of sea urchins—marine invertebrates that eat kelps—was causing an increase in kelps (Figure 1.8). Kelps can be harvested for making fertilizer, food, and pharmaceuticals. Thus, the growing otter population caused sea urchins to decrease, kelps to increase, and the commercial harvesting of kelps to increase. It turns out that the increase in kelps also provided young fish with a refuge from predators and a place to feed. Thus, the sea otter plays a key role in determining the community composition of coastal marine ecosystems.

In the 1990s, the sea otter population mysteriously began to decline. To understand these declines, scientists used individual, community, and ecosystem approaches. In 1998, researchers showed that populations of otters in the vicinity of the Aleutian Islands, Alaska, had declined precipitously during the 1990s. The reason was that killer whales, or orcas (*Orcinus orca*), which previously had not preyed on otters, had begun to come close to shore where they consumed large numbers of otters. Why did killer whales adopt this new behavior? The researchers pointed out that populations of the principal prey of killer whales—seals and sea lions—collapsed during the same period, perhaps causing the whales to hunt the otters as an alternative food source. Why did the seals and sea lion populations decline? One can only speculate at this point, but intense human fisheries have reduced fish stocks exploited by the seals to levels low enough to seriously threaten seal populations.

There also were declines in otter populations along the California coast. Initially, declines in sea otters were attributed to the use of gill nets along the coast to exploit a new fishery that inadvertently killed otters in substantial numbers. Subsequent legislation moved the fishery farther offshore to help protect the otters. In this same region, the otters were also dying from infections by two protist parasites, *Toxoplasma gondii* and *Sarcocystis neurona*. These parasites cause a lethal inflammation of the brain. In 2010, for example, 40 dead and dying sea otters were found near Morro Bay, California, and 94 percent were infected with *S. neurona*. This was a surprising observation because the only known hosts of these parasites are opossums (*Didelphis virginiana*) and several species of cats. Given that these mammals live on land, how did sea otters become infected?

Scientists hypothesized links between the terrestrial and marine ecosystems that allowed the parasites to infect sea otters. To date, two potential links have been suggested. First, house cats that spend time outside defecate on land and their feces contain the parasites; when it rains, the parasites get washed into local streams and rivers, and eventually make their way to the ocean. Second, when humans flush cat feces and kitty litter down the toilet and into the sewer system, the waste water eventually enters the ocean. Although manipulative experiments found that the protists do not infect marine invertebrates and cause illness, the invertebrates can take the parasites into their bodies inadvertently while feeding. When invertebrates infested with parasites are consumed by otters, the otters get infected. New research indicates that abalone do not carry the parasites, whereas small marine snails

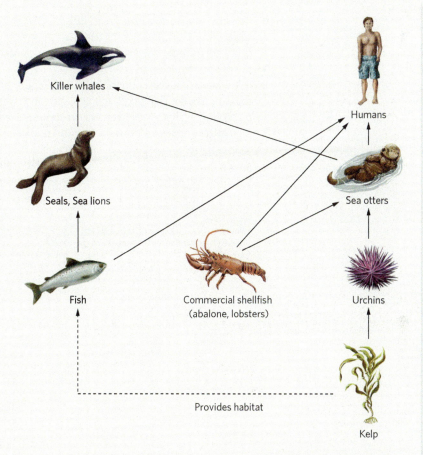

Sea otters and the species with which they interact. Once scientists determined the major species in the ocean that affected the abundance of otter populations, they could better protect the otter from extinction. Solid arrows indicate consumption of one species by another.

do. Thus, when otters have an abundance of their preferred food, such as abalone, they have a low risk of being infected by the deadly parasite. When abalone is scarce, however, the otters are forced to feed on small snails that carry the parasite, which dramatically increases the risk of infection and death.

The story of the sea otter highlights the importance of understanding ecology from multiple approaches using both manipulative and natural experiments. It also underscores the multiple roles that species can play in communities and ecosystems and how humans can dramatically influence the outcome. This understanding can then be used to take action to reverse harmful impacts on the environment. In the case of the sea otter, education campaigns now encourage the public to keep their cats inside more and to put used cat litter into the trash rather than flushing it down the toilet.

SOURCES:

Johnson, C. K., et al. 2009. Prey choice and habitat use drive sea otter pathogen exposure in a resource-limited coastal system, *Proceedings of the National Academy of Sciences* 106:2242–2247.

Miller, M. A. 2010. A protozoal-associated epizootic impacting marine wildlife: Mass mortality of southern sea otters (*Enhydra lutris nereis*) due to *Sarcocystis neurona* infection. *Veterinary Parasitology* 172:183–194.

Estes, J. A. 2016. *Serendipity: An Ecologist's Quest to Understand Nature* (University of California Press).

SUMMARY OF LEARNING OBJECTIVES

1.1 Ecological systems exist in a hierarchical organization.
The hierarchy begins with individual organisms and moves up through higher levels of complexity, including populations, communities, ecosystems, and the biosphere. At each of these levels, ecologists study different types of processes.

Key Terms: Ecology, Ecological systems, Individual, Species, Population, Community, Ecosystem, Biosphere, Individual approach, Adaptation, Population approach, Community approach, Ecosystem approach, Biosphere approach

1.2 Ecological systems are governed by physical and biological principles.
These principles include the conservation of matter and energy, dynamic steady states, a requirement to expend energy, and the evolution of new phenotypes and new species.

Key Terms: Law of conservation of matter, Law of conservation of energy (First law of thermodynamics), Dynamic steady state, Phenotype, Genotype, Evolution, Natural selection, Fitness

1.3 Different organisms play diverse roles in ecological systems.
The major groups of organisms are plants, animals, fungi, protists, and bacteria. These organisms are involved in numerous species interactions, including competition, predation, mutualism,

and commensalism. Each organism lives in specific habitats and has a particular niche.

Key Terms: Algal bloom, Producer (Autotroph), Consumer (Heterotroph), Mixotroph, Predator, Parasitoid, Parasite, Pathogen, Herbivore, Competition, Mutualism, Commensalism, Symbiotic relationship, Scavenger, Detritivore, Decomposer, Habitat, Niche

1.4 Scientists use several approaches to studying ecology.
Like all scientists, ecologists use the scientific method of developing and testing hypotheses. The testing of proximate and ultimate hypotheses can be accomplished using manipulative experiments, natural experiments, or mathematical models.

Key Terms: Hypothesis, Proximate hypothesis, Prediction, Ultimate hypothesis, Manipulative experiment, Manipulation, Control, Experimental unit, Replication, Randomization, Microcosm, Observations (Data), Variance of the mean, Sample variance, Natural experiment, Mathematical model

1.5 Humans influence ecological systems.
The rapid growth of the human population during the past two centuries has increased human influence on ecological systems, particularly as the result of resources they consume and waste products they release.

Key Term: Greenhouse gases

CRITICAL THINKING QUESTIONS

1. How might understanding one level of ecological organization help us understand processes occurring at a higher level of ecological organization?

2. At the population level, what would happen to a population of animals that was not in a dynamic steady state over long time periods?

3. Why are phenotypes the product of both their genes and their environments?

4. When we consider the major forms of life on Earth in Figure 1.6, what are the characteristics that connect the various types of organisms in a given group and suggest that they share a common ancestor?

5. If natural selection favors adaptive phenotypes, in what ways might prey populations evolve if they experience predators over many generations?

6. In the experiment on herbivore insects consuming the leaves of oak trees (p. 22), describe how the researchers could have also

conducted a natural experiment in addition to their manipulative experiment.

7. In the Northern Hemisphere, many species of birds fly south during the autumn months. Propose a proximate and an ultimate cause for this behavior.

8. When experimental manipulations are conducted to test a hypothesis, what is the purpose of including a control?

9. Given the difficulty in conducting a manipulative experiment to identify the effects of elevated CO_2 across the globe, how might we be able to validate the mathematical models that have been created?

10. Using the data for the caged trees from the "Analyzing Ecology" exercise (p. 24), calculate the sample variance if the variance of the mean stayed the same but the sample size (n) were to increase from 10 to 100 to 1,000. How does sample size affects the estimate of the sample variance relative to the variance of the mean?

2 Adaptations to Aquatic Environments

The Evolution of Whales

Life on Earth began in the water. Of the many species that live in water, some of the most fascinating are the whales—a group that is particularly well suited to aquatic life. Surprisingly, the ancestor of modern whales can be traced back to a terrestrial mammal related to cattle, pigs, and hippos. Scientists first proposed an evolutionary relationship between whales and this group of land mammals in 1883, based on observations of similarities in their skeletons. In the 1990s, DNA technology confirmed that the groups are also related genetically. In 2007, scientists discovered a critical link between hippos and whales: the fossilized bones of a previously unknown, large terrestrial mammal, in the genus *Indohyus*, that may have spent at least some of its time in the water. Scientists speculate that over the subsequent 50 to 60 million years, selection imposed by the challenges of living in an aquatic environment led to the evolution of whales as we know them today.

Modern whales have evolved a wide range of adaptations for aquatic life. One of the most obvious challenges is the ability to swim efficiently. For example, the killer whale is capable of swimming up to 48 km per hour. Such speeds are possible only with a highly streamlined body. Over evolutionary time, natural selection would have favored any individuals that had a more streamlined body, including bodies with reduced hindlimbs. In modern whales, tiny remnants of the hindlimb bones can be found entirely within the whale's body. During the course of the whale's evolutionary history, whale ears became internalized. Whether this was due to selection for a streamlined body or for some other reason is not known, but the outcome made for a more streamlined body.

Obtaining oxygen is another challenge for whales because they need to dive for long periods in search of food. Over time, there was an evolutionary change in the location of the nostrils from the front of the head to the top. While scientists cannot be certain of the selective forces that caused this change, one hypothesis is that changes in nostril position may have been favored over time to the point that the modern whale can more easily grab a breath of air when surfacing. In addition, whales can hold their breath for long periods. Sperm whales, for example, dive to depths of 500 m and may stay below the surface for more than an hour as they search for fish, squid, and other food. During a dive, the sperm whale relies on oxygen stored in its body. It might surprise you to learn that very little of this oxygen resides in the lungs; most is bound to hemoglobin in the blood or to a similar oxygen storage molecule, myoglobin, in the muscles. Under water, whales slow their metabolism by reducing blood flow to nonvital organs such as the skin, intestines, lungs, and kidneys, while blood flow to the brain and heart continues. Consequently, during a dive the temperature of all but a few key organs drops, the heart rate slows, and demand for oxygen is reduced.

> "Surprisingly, the ancestor of modern whales can be traced back to a terrestrial mammal related to cattle, pigs, and hippos."

Whale ancestor. The ancestor of modern whales, in the genus *Indohyus*, was a terrestrial animal that spent part of its time in the water. Over time, the descendants of this animal developed numerous adaptations for living in water that are found in modern-day whales.

Sperm whale. Modern whales, such as these sperm whales (*Physeter macrocephalus*) swimming off the coast of Portugal, have a number of adaptations that enable them to live in an aquatic environment. Photo by Willyam Bradberry/Shutterstock.

Regulating body temperature is yet another challenge. Because heat loss occurs much faster in water than in air, a thick layer of fat under the skin insulates most oceanic mammals living in cold waters. Like a warm coat, this insulation slows the loss of heat generated by its internal organs. Whales also maintain a higher metabolic rate than land mammals of similar size, and this helps generate extra heat. The whale's vascular structure also helps maintain heat; the veins and arteries of a whale's flippers and tail are next to each other. This allows the warm arterial blood that travels away from the heart to transfer heat to cooler blood in adjacent veins as it returns from a whale's extremities to the heart.

A final challenge is how to find food in the water, and two different strategies evolved long ago: Baleen whales have long plates in their mouths to filter tiny prey out of the water, whereas toothed whales have teeth in their mouth to grab prey. Modern toothed whales also use echolocation, which involves emitting a sound into the water and then listening for the sound to bounce back as an "echo" when it intercepts objects such as food. Since all toothed whales use echolocation, but baleen whales do not, scientists suspected that echolocation must have evolved long ago in an ancestor of toothed whales. In 2016, researchers reported that they examined a whale skull that was found in South Carolina and estimated to be 27 million years old. Their striking discovery was that the skeleton contained a specially shaped inner ear, which indicated the whale was capable of echolocation, confirming that this adaptation occurred very early in the evolution of toothed whales.

The evolution of whales took place over a period of 50 million years. While the selective forces occurring over this time cannot be known with certainty, the fossils found to date suggest that an ancient terrestrial mammal ultimately evolved a number of adaptations that gave rise to modern-day whales. In this chapter, we examine how the challenges posed by living in an aquatic environment have caused a wide variety of organisms to evolve adaptations.

SOURCES:

New fossil evidence suggests echolocation evolved early in whales, *Science News,* August 5, 2016, https://www.sciencenews.org/article/new-fossil-suggests-echolocation-evolved-early-whales.

Valley of the whales, *National Geographic Magazine,* August 2010, http://ngm.nationalgeographic.com/2010/08/whale-evolution/mueller-text.

Whales descended from tiny deer-like ancestors, *Science Daily,* December 21, 2007, http://www.sciencedaily.com/releases/2007/12/071220220241.htm.

LEARNING OBJECTIVES

After reading this chapter, you should be able to:

2.1 Describe some of the many properties of water that make it favorable to life.

2.2 Explain the challenges faced by aquatic animals and plants of maintaining an internal balance of water and salt concentrations.

2.3 Describe how the uptake of gases from water is limited by diffusion.

2.4 Explain how temperature limits the occurrence of aquatic life.

Scientists generally agree that life began in the ocean and that the first forms of life were simple bacteria. Over millions of years, these bacteria gave rise to an incredible diversity of organisms, many of which still live in the water. Other species, as we saw in the case of the whale, evolved terrestrial forms that later evolved to return to life in the ocean. In this chapter, we begin an exploration of ecology at the individual level by examining the ways in which the properties of water support and constrain aquatic organisms and direct the evolution of adaptations.

2.1 Water has many properties favorable to life

Water is abundant over most of Earth's surface. Because water has an immense capacity to dissolve inorganic compounds, it is an excellent medium for the chemical processes of living systems. In fact, it is hard to imagine a form of life that could exist without water. In this section, we will look at how water makes life possible, including water's thermal properties, its density and viscosity, and its function as a solvent for inorganic nutrients.

THERMAL PROPERTIES OF WATER

On Earth, water can be found as a solid (ice), as a liquid, and as a gas (water vapor). Pure water—water not containing any dissolved minerals or other compounds—becomes a solid below 0°C and becomes a gas above 100°C at sea level. At higher elevations, the freezing point of water changes very little, but the boiling point can be several degrees lower. Within the temperature range organisms usually encounter, it is present in liquid form.

When water contains dissolved compounds, such as salts, its freezing temperature drops below 0°C. This is why road salts are applied to ice- and snow-covered roads; the salts allow the ice and snow to melt at a lower temperature than they otherwise would. Dissolved compounds also raise the boiling point of water above 100°C.

The temperature of water remains relatively steady even when heat is removed or added rapidly, as can happen at the air–water interface or at an organism's surface, such as on the surface of a whale in the ocean. This is because water has a high specific heat, which is the amount of heat required to increase its temperature by 1°C. In addition, water transfers heat rapidly, causing heat to spread evenly throughout a body of water, which also slows localized changes in temperature.

Water also resists changing from one state to another. For example, raising the temperature of 1 kg of liquid water by 1°C requires the addition of 1 Calorie of heat. However, converting 1 kg of liquid water into water vapor requires the addition of 540 Calories of heat. Similarly, lowering the temperature of 1 kg of liquid water by 1°C requires the removal of 1 Calorie of heat, but converting that amount of liquid water to ice requires the removal of 80 Calories of heat. In short, liquid water is very resistant to changing states. This resistance helps prevent large bodies of water from freezing solid during winter.

Another curious, yet fortunate, thermal property of water is the way it changes density with changes in temperature. You may recall from previous courses in biology or chemistry that most compounds become denser under cooler temperatures. However, water achieves its highest density—that is, its molecules are the most packed together—at 4°C. Above and below 4°C, the water molecules are less tightly packed, so the water becomes less dense. Below 0°C, pure water is converted into ice, which is less dense than liquid water, as you can see in **Figure 2.1**. As a result of its lower density, ice floats on the surface of liquid water. This means that lakes experiencing cold winters will generally have a layer of 4°C water at the bottom. Above this layer will be water that is cooler than 4°C and on top of that a layer of ice.

Water's unusual thermal properties are especially important to aquatic plants and animals. In large bodies of fresh water, such as lakes, the bottom of the lake does not freeze, in part because of the insulation that ice provides from very cold air temperatures. The salt in ocean water lowers the freezing point of the water to well below 0°C, which helps prevent most oceans from freezing. In both cases, the available liquid water provides a refuge for organisms during periods of cold temperatures.

DENSITY AND VISCOSITY OF WATER

The adaptations of aquatic organisms often exploit the density and viscosity of water. For example, animals and plants are made up of bone, proteins, and other materials that are somewhat denser than salt water, and much denser than fresh water. However, organisms can also contain fats and oils that are less dense than water. In some cases, they also

Figure 2.1 The density of water. As water cools, the molecules contract and become more dense. Below 4°C, they begin to expand and become less dense. Below 0°C, pure water is converted into ice that is even less dense. As a result of its lower density, ice floats on the surface of liquid water. Photo by Zoonar/Christa Kurtz/ AGE Fotostock.

possess pockets of air, such as the lungs of the whales that were described at the beginning of this chapter. The combination of the materials that compose an animal's body and the presence of air pockets determine whether an organism will float or sink in water.

For those organisms that are denser than their surrounding water, a variety of adaptations can either reduce an organism's density or retard its rate of sinking. For example, many fish species have a gas-filled swim bladder that can adjust size to make the density of the fish's body equal to that of the surrounding water. Human divers use this same concept when they wear air-filled inflatable vests to help match the density of the water. When the vest is full of air, divers float on the surface of the water. As they release some of the air, they begin to sink. By adjusting the amount of air in the vest, divers can make their density equal to the water at a particular depth, making it easier to swim because they are not spending energy resisting their body's attempt to float or sink. Some large kelps, like those we saw in Figure 1.8, have gas-filled bulbs that cause their leaflike blades to float up into the sunlit surface waters. The whales discussed earlier become buoyant when they take a breath of air, but a slow release of air bubbles will help them sink to a particular depth. At the other end of the size spectrum, many of the microscopic unicellular algae that float in great numbers in the surface waters of lakes and oceans use droplets of oil as flotation devices (**Figure 2.2**). Because oils are less dense than water, the algae can use the oil droplets to help offset their natural tendency to sink.

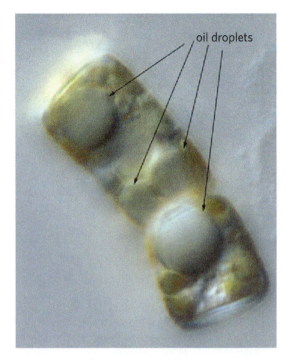

oil droplets

Figure 2.2 Adaptation to water density. These algae (*Cyclotella cryptica*) are able to float near the water's surface by using oil droplets that have a lower density than water. Photo by Bigelow Laboratory for Ocean Sciences on behalf of the Provasoli-Guillard National Center for Marine Algae and Microbiota.

Aquatic organisms also possess adaptations to deal with the high *viscosity* of water. We can think of **viscosity** as the thickness of a fluid that causes objects to encounter resistance as they move through it. In response to living in water, fast-moving aquatic animals such as fish, penguins, and whales have evolved highly streamlined shapes that reduce the drag caused by the high viscosity of water (**Figure 2.3**).

Figure 2.3 Streamlined shape. Large, fast-moving aquatic organisms like the barracuda (*Sphyraena barracuda*) have evolved highly streamlined shapes to help them move through the highly viscous water. Photo by GEORGE GRALL/National Geographic Creative.

Figure 2.4 Adaptation to water viscosity. Some small aquatic organisms exploit the high viscosity of water by evolving large appendages, such as the antenna and feathery projections of this tiny marine crustacean. These appendages help slow down movement through the viscous water and thereby retard sinking. Photo by Solvin Zankl/naturepl.com.

The viscosity of water is higher in cold water than in warm water, which can make swimming in cold water more difficult. Movement in water is even more difficult for smaller animals. However, the same high viscosity that impedes the progress of tiny organisms as they swim in the water also impedes them from sinking. Because these organisms are slightly denser than water, they are prone to sinking due to the force of gravity. To take advantage of water's viscosity, many tiny marine animals have evolved long, filamentous appendages that cause greater drag in the water. The appendages function like a parachute slowing the fall of a body through air (**Figure 2.4**).

DISSOLVED INORGANIC NUTRIENTS

Both aquatic and terrestrial organisms require a variety of nutrients to build necessary biological structures and maintain life processes. Large amounts of hydrogen, carbon, and oxygen are necessary for building most compounds found in organisms. Nitrogen, phosphorus, and sulfur are used in building proteins, nucleic acids, phospholipids, and bones. Other major nutrients—including potassium, calcium, magnesium, and iron—play important roles as solutes and as structural components of bones, woody plant cells, enzymes, and chlorophyll. Certain organisms need additional minor nutrients. For example, diatoms are a group of algae that need silica to construct their glassy shells (**Figure 2.5**).

Viscosity The thickness of a fluid that causes objects to encounter resistance as they move through it.

Figure 2.5 Use of inorganic nutrients. Diatoms, such as this species of *Arachnoidiscus,* are a type of algae that require minor nutrients such as silica to build a hard, glassy shell. Image is magnified 175 times. Photo by Steve Gschmeissner/Science Source.

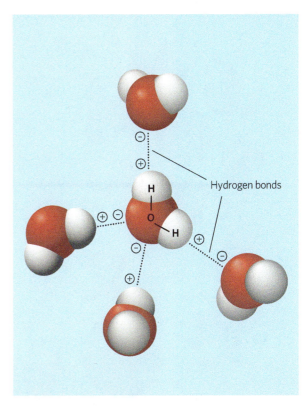

Figure 2.6 Water molecules. Because of the configuration of water molecules, they are negatively charged on the oxygen end and positively charged on the hydrogen end. The attractive forces of these opposite charges, known as hydrogen bonds, allow water molecules to be attracted to each other and to the charged ions of other compounds such as salts and sugars.

Similarly, some species of bacteria require the element molybdenum, which makes up part of the enzyme they use to convert nitrogen from the atmosphere (N_2) into ammonia (NH_3).

The Solubility of Minerals

Water is a powerful solvent with an impressive capacity to dissolve substances, which makes them accessible to living systems. Because of this property, water also provides a medium in which substances can react chemically to form new compounds.

Water acts as a solvent because of its molecular structure. As you can see in **Figure 2.6**, water molecules consist of an oxygen atom in the middle and two hydrogen atoms connected in a V-shaped arrangement. This arrangement results in an unequal sharing of electrons: The oxygen end of the water molecule has a slightly negative charge and the hydrogen end has a slightly positive charge. When the two ends of the molecule possess opposite charges, we say that the molecule is *polar.* Water is a polar molecule: The negative oxygen end of one water molecule is strongly attracted to the positive hydrogen end of another. These forces of attraction are known as *hydrogen bonds.*

The polar nature of water molecules also allows them to be attracted to other polar compounds. Some solid compounds consist of electrically charged atoms, or groups of atoms, called **ions.** For example, common table salt—sodium chloride (NaCl)—contains positively charged sodium ions (Na^+) and negatively charged chloride ions (Cl^-). In their solid form, these ions are arranged in a crystal lattice. In water, however, the charged sodium and chloride ions are attracted by the charges of the water molecules. As shown

in **Figure 2.7**, the attraction of these ions to water molecules is stronger than the attraction that holds the crystal together. As a result, when sodium chloride is added to water, its crystal lattice breaks apart, and water molecules surround the salt ions. In other words, when you put salt in water, it dissolves. This solvent ability of water is not restricted to ionic compounds such as salts; it occurs with any polar compound, including the various types of sugars that organisms commonly use. In contrast, water is not a good solvent for oils and fats because they are nonpolar compounds.

The solvent properties of water explain the presence of minerals in streams, rivers, lakes, and oceans. When water vapor in the atmosphere condenses to form clouds, the condensed water in the atmosphere is nearly pure. However, as it falls back to Earth as rain or snow, water acquires some minerals from dust particles in the atmosphere. Precipitation that hits land comes into contact with rocks and soils, and it dissolves some of their minerals. These dissolved minerals are carried toward the ocean with the rainwater.

Ions Atoms or groups of atoms that carry an electric charge.

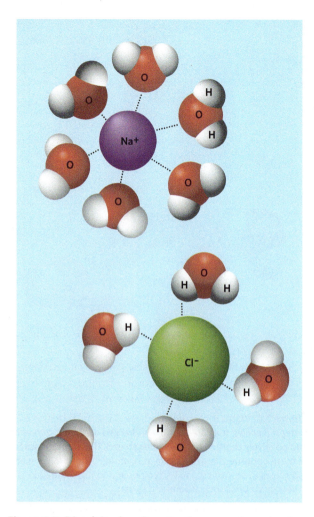

Figure 2.7 Dissolving ions in water. Because water molecules have negative and positive ends, they attract the negatively and positively charged ions, such as the sodium and chlorine ions found in sodium chloride. The forces of attraction to water molecules are stronger than the forces of attraction within the crystal, so the ions separate and become surrounded by water molecules.

Water in most lakes and rivers contains a dissolved mineral concentration of 0.01 to 0.02 percent, whereas ocean water contains a dissolved mineral concentration of 3.4 percent. Oceans have much higher concentrations of dissolved minerals because mineral-laden water continuously flows in from streams and rivers. The constant evaporation from the ocean's surface removes pure water and leaves the minerals behind. Over billions of years, this process has caused an increase in the concentrations of minerals in the oceans.

Every mineral has an upper limit of solubility in water, known as **saturation.** This limit generally increases with higher temperatures. After a mineral achieves saturation, water cannot hold any more of the mineral and it begins to precipitate out of solution. For some minerals, such as sodium, ocean concentrations are far below saturation. Most of the sodium that is washed into ocean basins remains dissolved and its concentration in seawater continues to increase over time. In contrast, the concentrations of other

minerals in the oceans commonly exceed their saturation concentrations. For example, calcium ions (Ca^{2+}) in the water readily combine with dissolved CO_2 to form calcium carbonate ($CaCO_3$), which has a low solubility in ocean water. Over millions of years, the excess calcium carbonate that has washed into the oceans from streams and rivers has subsequently precipitated out of the water. This precipitated calcium carbonate, combined with the calcium carbonate from the bodies of countless tiny marine organisms, has resulted in massive limestone sediments (**Figure 2.8**). Today, these sediments are important sources of limestone for construction applications such as stone blocks and concrete, for agricultural uses such as fertilizer, and for numerous industrial processes.

Hydrogen Ions

Among dissolved substances in water, hydrogen ions (H^+) deserve special mention because they are extremely reactive with other compounds. In pure water, a small fraction of the water molecules (H_2O) break apart into their hydrogen (H^+) and hydroxide (OH^-) ions. The concentration of hydrogen ions in a solution is referred to as its **acidity.** Acidity is commonly measured as **pH,** which is defined as the negative logarithm of hydrogen ion concentration (as measured in moles/L, where 1 mole = 6.02×10^{23} molecules):

$$pH = -\log (H^+ \text{ concentration})$$

As you can see in the pH scale shown in **Figure 2.9**, water containing a high concentration of hydrogen ions has a low pH value, whereas water containing a low concentration of hydrogen ions has a high pH value. Therefore, we categorize water with low pH values as *acidic,* water with a mid-range value of 7 as *neutral,* and water with a high pH value as *basic* or *alkaline.* As we will see, natural rain or snow can vary a great deal in pH, depending on the presence of different chemical compounds in the atmosphere that affect the pH value.

Hydrogen ions, because of their high reactivity, dissolve minerals from rocks and soils, enhancing the natural solvent properties of water. For example, in the presence of hydrogen ions, the calcium carbonate found in limestone dissolves readily, according to the following chemical equation:

$$H^+ + CaCO_3 \rightarrow Ca^{2+} + HCO_3^-$$

Calcium ions are important to life processes, and their presence is vital to organisms such as snails, mussels, and clams that form shells made of calcium carbonate. As a result, these animals are less abundant in streams and lakes that are low in calcium. Therefore, hydrogen ions

Saturation The upper limit of solubility in water.

Acidity The concentration of hydrogen ions in a solution.

pH A measure of acidity or alkalinity; defined as pH = $-\log$ [H^+].

Figure 2.8 The formation of limestone sediments. The continuous addition of calcium minerals into oceans from streams and rivers causes calcium to combine with CO_2 to become calcium carbonate. Because calcium carbonate is not very soluble in water, it precipitates out of the water to form massive sediments over millions of years. This site of limestone sediments located in Victoria, Australia, was once under water but is now above the water due to changes in ocean depth. Photo by Phillip Hayson/ Science Source.

are essential for making certain nutrients available for life processes. At high concentrations, however, hydrogen ions negatively affect the activities of most enzymes. In addition, high concentrations of hydrogen ions cause many heavy metals to begin dissolving in water. These heavy metals,

including arsenic, cadmium, and mercury, are highly toxic to most aquatic organisms.

The normal range of pH of lakes, streams, and wetlands is between 5 and 9. However, some bodies of water can have even lower pH values. Sometimes lower pH conditions have a natural cause. Bogs, for example, are aquatic habitats with vegetation such as sphagnum mosses that release H^+ ions into the water and thereby make the water more acidic and unsuitable for many other species of plants.

Other bodies of water have low pH as a result of human influences. For example, the release of sulfur dioxide (SO_2) and nitrogen dioxide (NO_2) from coal-powered industrial plants became a major environmental issue in the 1960s. At that time, ecologists in Russia, China, northern Europe, the United States, and Canada began to notice that many bodies of water were becoming more acidic and less hospitable to numerous species of fish and other aquatic organisms. They also noticed that trees were dying, particularly in the spruce-fir forests that existed at high elevations in these regions of the world.

It turned out that the areas with more acidic bodies of water and dying trees were all far downwind of industrial areas with coal-powered factories that had tall smokestacks. Years of data collection revealed that the sulfur dioxide and nitrogen dioxide emitted into the atmosphere by these smokestacks were converted to sulfuric acid and nitric acid in the atmosphere. These acids were transported downwind and then came back down to Earth as **acid deposition,** also known as **acid rain.** The acid deposition occurred in two forms: as gases and particles that stuck to plants and soil, a form called *dry acid deposition,* and as rain and snow, a form called *wet acid deposition.* Acid deposition lowered the pH of precipitation and, as a result, water of unusually low pH was entering streams, lakes, and forests. Most aquatic species

pH value	H⁺ ion concentration (moles per liter)	
0	1	
1	10^{-1}	Stomach acid
2	10^{-2}	
3	10^{-3}	Carbonated beverages
4	10^{-4}	Acid rain
5	10^{-5}	
6	10^{-6}	
7	10^{-7}	Most streams and rivers
8	10^{-8}	Human blood / Oceans
9	10^{-9}	
10	10^{-10}	
11	10^{-11}	
12	10^{-12}	Alkaline lake
13	10^{-13}	
14	10^{-14}	

Acidic
Neutral
Basic

Figure 2.9 The relationship between pH and hydrogen ion concentration in water. The pH scale of hydrogen ion concentration extends from 0 (highly acidic) to 14 (highly alkaline). The pH of rainfall can vary a great deal around the world.

Acid deposition Acids deposited as rain and snow or as gases and particles that attach to the surfaces of plants, soil, and water. *Also known as* **Acid rain.**

cannot tolerate water with a pH lower than 4, so these bodies of water became toxic to many aquatic organisms, which included insects and fish.

In forests, acid deposition has several effects. First, it leaches the calcium out of the needles of conifer trees such as spruce. It also causes increased leaching of soil nutrients that trees require, including calcium, magnesium, and potassium. Finally, acid deposition causes aluminum to dissolve in the water. Although aluminum naturally occurs in the soil, it is typically not in a form that is available to plants. Dissolved aluminum can negatively affect a plant's ability to take up nutrients. Collectively, these effects of acid deposition make trees more susceptible to the harmful effects of natural stressors that include drought, diseases, and extreme temperatures. In short, while the trees do not die directly from acid deposition, they become more susceptible to other causes of death. Because acid deposition interacts with so many other natural stressors, scientists recognize that acid deposition has contributed to the death of trees in North America, Europe, and Asia. However, the complexity of the interactive effects has made it difficult to accurately estimate how much of the observed tree death is directly attributable to acid deposition.

Once scientists understood the causes and consequences of acid deposition, they began to explore solutions. In the United States, legislation required the installation of smokestack scrubbers that force smokestack gases through a slurry of limestone and water, which remove the gases. The use of these scrubbers has resulted in a major reduction in the amount of acidic compounds going into the atmosphere. The U.S. Environmental Protection Agency (EPA) reports that from 1980 to 2015, emissions of sulfur dioxide declined by 84 percent. At the end of this chapter, in "Connecting the Concepts: The Decline of Coral Reefs," we discuss another example of how understanding environmental problems related to pH can help us develop effective solutions.

CONCEPT CHECK

1. What is unique about water with regard to how temperature affects its density?
2. How can the viscosity of water both hinder and facilitate movement in aquatic animals?
3. Describe the changes in mineral content of water as it moves from rainwater to lake water and, eventually, to ocean water.

2.2 Aquatic animals and plants face the challenge of water and salt balance

THE CHALLENGE OF SALT AND WATER BALANCE

It might surprise you to learn that organisms in an aquatic environment need specialized mechanisms to maintain the proper amount of water in their bodies. To understand why, we need to recognize that the water surrounding an aquatic organism and the water inside its body contain dissolved substances, known as **solutes.** These solutes affect the movement of water in and out of an organism. The movement of water occurs at the cellular level, where water passes across cell membranes from regions of low solute concentration to regions of high solute concentration. At the same time, solutes attempt to move across membranes to equalize the concentration of solutes. Cell membranes typically do not allow the free movement of large solute molecules such as carbohydrates and most proteins. Membranes that allow only particular molecules to pass through—for example, water and small solute ions and molecules—are known as **semipermeable membranes.**

Solutes can move across semipermeable membranes using either *passive* or *active transport*. **Passive transport** occurs when ions and small molecules move through a membrane along a concentration gradient, from a location with many solutes to a location with few solutes. In contrast, **active transport** occurs when cells transport ions and small molecules through a membrane against a concentration gradient to maintain their concentrations. Active transport expends energy because it requires the cell to work against the concentration gradient of solutes.

When the water inside a cell has a higher concentration of solutes than the water outside, water tries to move into the cell from the surrounding environment. In contrast, when the water inside a cell has a lower concentration of solutes than the water outside, water tries to move out of the cell. The movement of water across a semipermeable membrane is called **osmosis.** The force with which an aqueous solution attracts water by osmosis is known as its **osmotic potential,** which is expressed in megapascals (MPa), a unit of pressure. The osmotic potential generated by an aqueous solution depends on its solute concentration, which is measured as the number of solute molecules in a given volume of water.

The challenge for most aquatic organisms is that they live in water with a solute concentration that differs from the solute concentration of their bodies. This difference causes water and solutes to attempt to move in or out of the organism's body, which makes it difficult for organisms to maintain the proper amount of water and solutes in their bodies. Maintaining a particular solute concentration in the body is important because solute concentrations affect

---

Solute A dissolved substance.

Semipermeable membrane A membrane that allows only particular molecules to pass through.

Passive transport The movement of ions and small molecules through a membrane along a concentration gradient, from a location with many solutes to a location with few solutes.

Active transport The movement of molecules or ions through a membrane against a concentration gradient.

Osmosis The movement of water across a semipermeable membrane.

Osmotic potential The force with which an aqueous solution attracts water by osmosis.

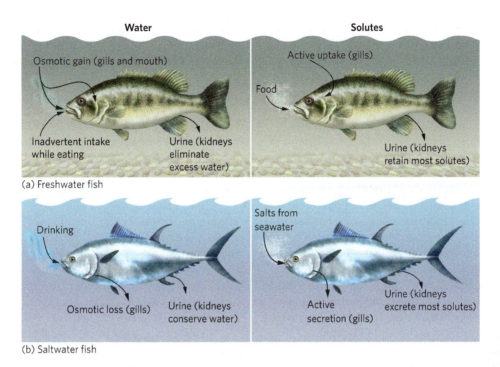

Water

Osmotic gain (gills and mouth)

Inadvertent intake while eating

Urine (kidneys eliminate excess water)

(a) Freshwater fish

Solutes

Active uptake (gills)

Food

Urine (kidneys retain most solutes)

Drinking

Osmotic loss (gills)

Urine (kidneys conserve water)

(b) Saltwater fish

Salts from seawater

Active secretion (gills)

Urine (kidneys excrete most solutes)

Figure 2.10 Osmoregulation in fish. Exchanges of water and solutes differ between freshwater and marine fish. **(a)** Freshwater fish are hyperosmotic: They have a higher salt concentration in their body than exists in the surrounding water. To maintain salt balance, they must excrete large amounts of water and use their gills and kidneys to actively retain solutes. **(b)** Saltwater fish are hyposmotic: They have a lower salt concentration in their body than exists in the surrounding water. To maintain salt balance, they must excrete large amounts of solutes, so their gills and kidneys actively exclude solutes.

the way proteins interact with other molecules. In short, altered solute concentrations can disrupt cell functions. If they take in too many solutes, they must rid themselves of the excess. If they take in too few solutes, they must make up the deficit. Because solutes determine the osmotic potential of body fluids, the mechanisms that organisms use to maintain a proper solute balance are referred to as **osmoregulation.**

ADAPTATIONS FOR OSMOREGULATION IN FRESHWATER ANIMALS

We have seen that the water balance of aquatic animals is closely tied to the concentrations of salts and other solutes in their body tissues and in the environment. Freshwater animals, such as bass and trout, that have higher solute concentrations in their tissues than the surrounding water are said to be **hyperosmotic** compared to their freshwater environment.

Because water and solutes move in the direction that will equalize the concentrations of solutes on both sides of a membrane, a hyperosmotic organism faces a constant challenge: Water attempts to enter its body while solutes attempt to leave. **Figure 2.10a** shows how hyperosmotic fish regulate the balance of solutes in their tissues. Freshwater fish continuously gain water both when they consume food and when osmosis occurs across the mouth and gills, which are the most permeable tissues exposed to the water. Fish respond to

this influx of water by eliminating the excess water through their urine. They add solutes to their bloodstream by using their gill cells to actively transport solutes into the body from the water. In addition, their kidneys remove ions from their urine.

ADAPTATIONS FOR OSMOREGULATION IN SALTWATER ANIMALS

Saltwater animals—such as whales, sardines, and plankton—have lower solute concentrations in their tissues than exist in the surrounding water. Such organisms are said to be **hyposmotic** compared to their saltwater environment. Hyposmotic organisms face a constant challenge to maintain the balance of water and solutes in their tissues. As shown in Figure 2.10b, water tries to leave their bodies and solutes try to enter. To replace the loss of water, saltwater animals drink large amounts of saltwater and release only small amounts of urine. To counteract the accompanying influx of solutes, the excess solutes are actively excreted out of the body using the kidneys and, in the case of fish, the gills.

Osmoregulation The mechanisms that organisms use to maintain proper solute balance.

Hyperosmotic When an organism has a higher solute concentration in its tissues than the surrounding water.

Hyposmotic When an organism has a lower solute concentration in its tissues than the surrounding water.

Some sharks and rays have evolved a unique adaptation to the challenge of water balance in a saltwater environment. Like all vertebrate animals, when sharks and rays digest proteins, they produce ammonia as a by-product, as shown in **Figure 2.11**. Aquatic vertebrates excrete ammonia in their urine, whereas terrestrial vertebrates generally convert this ammonia to urea, which they then excrete at high concentrations in their urine. Interestingly, sharks and rays convert ammonia into urea, too, but they do not excrete all of it. Unlike terrestrial vertebrates, which maintain concentrations of urea below 0.03 percent in their bloodstream, sharks and rays actively retain urea and allow its concentration to increase to 2.5 percent in their bloodstream. Retaining more than 80 times more urea raises the osmotic potential of their blood to that of seawater without any increase in the concentrations of sodium and chloride ions. Consequently, the movement of water across the animal's body surface becomes balanced in relation to the surrounding saltwater, with neither gain nor loss. This adaptation frees sharks and rays from having to drink salt-laden water to replace water lost by osmosis.

Most vertebrates do not retain much urea in their bloodstream because urea impairs protein function. However, sharks, rays, and many other marine organisms that use urea to maintain their water balance have an additional adaptation: They accumulate high concentrations of a compound called trimethylamine oxide to protect proteins from the harmful effects of urea. Freshwater species of rays do not accumulate urea in their blood, though they are similar to saltwater rays in other respects. This confirms the importance of urea for osmoregulation for species of sharks and rays that live in the ocean.

Certain environments pose unusual osmotic challenges. For instance, the salt concentrations in some landlocked bodies of water greatly exceed those of the ocean. This is particularly common in arid regions where, because evaporation outpaces precipitation, very high concentrations of solutes accumulate in the water. The Great Salt Lake in Utah, for example, contains 5 to 27 percent salt, depending on the water level. This is up to eight times more salt than seawater. The osmotic potential of the water in the Great Salt Lake would cause most organisms to shrivel. However, a few aquatic creatures, such as brine shrimp, have adaptations that allow them to thrive in these conditions by excreting salt at a very high rate. This level of excretion comes at a high energetic cost, which they meet by feeding on the abundant photosynthetic bacteria that live in the lake.

The ability of an organism to cope with the osmotic potential of its environment reflects the outcome of evolutionary processes. As we will discuss in Chapter 4, some environments naturally experience large fluctuations in osmotic potential and the organisms living in these environments for long periods of time have evolved ways to adjust to these fluctuations. However, when changes in osmotic concentrations are not within the natural range that organisms have experienced over evolutionary time, individuals typically lack appropriate adaptations and can be harmed. For example, in the northern United States, various mixtures of salts are spread on roads to melt the ice and snow to provide safer driving conditions during winter weather. With the arrival of warmer spring weather, however, all of this salt has to go somewhere. In 2008, researchers reported on the concentrations of road salt in ponds inhabited by amphibians at different distances from salt-treated roads. Because salt ions allow electricity to be conducted through water, the concentration of salt can be measured in units of microsiemens (μS). As you can see in **Figure 2.12a**, ponds close to roads had salt concentrations up to 3,000 μS, which corresponds to 0.12 percent salt. Ponds at least 200 m from a road had essentially 0 μS. The researchers then conducted experiments in which they exposed larval wood frogs (*Rana sylvatica*) and spotted salamanders (*Ambystoma maculatum*) to a range of relevant salt concentrations, from 0 μS to 3,000 μS. The data, shown in Figure 2.12b, revealed that increases in salt concentration caused the larvae of both species to die in high numbers. Not having been exposed to high salt concentrations during their evolutionary history, these freshwater organisms are not adapted to these stressful conditions and are unable to survive in them.

The continued contamination of freshwater habitats by road salt has raised the possibility that some species might be able to evolve increased tolerance to salt over multiple generations. In 2017, researchers reported a case of evolving higher salt tolerance in zooplankton, which are tiny crustaceans

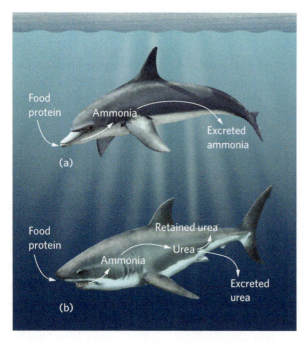

Figure 2.11 Unique adaptations to saltwater. When aquatic vertebrates digest proteins, they produce ammonia as a by-product. **(a)** Most aquatic organisms, such as dolphins, excrete this ammonia in their urine. **(b)** Sharks and rays convert this ammonia into urea and then retain some of the urea in their bloodstream. The result is a higher solute concentration that helps offset the challenge of being hyposmotic in saltwater.

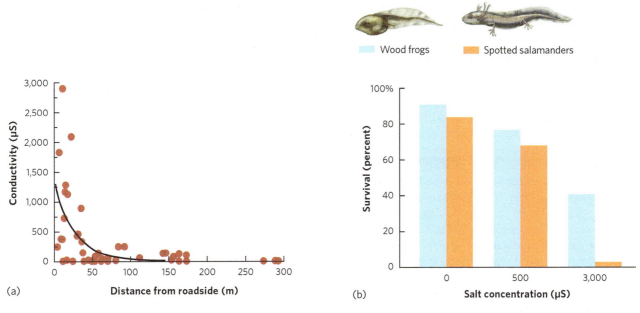

(a)

(b)

Figure 2.12 Effect of salt concentration on amphibians. (a) Based on a sample of ponds in the Adirondack Mountains of New York State, ponds closer to roads had higher conductivity, a measure of salt concentration. **(b)** At this range of salt concentrations, spotted salamanders and wood frogs experienced a decline in survival. Data from: N. E. Karraker et al., Impacts of road deicing salt on the demography of vernal pool-breeding amphibians, *Ecological Applications* 18 (2008): 724–734.

that live in freshwater lakes. When a particular species of zooplankton, known as water fleas (*Daphnia pulex*), was exposed to a range of increased salt concentrations for several generations, those raised under high salt concentrations were subsequently more tolerant to salt. While such studies offer hope that some species can rapidly evolve salt tolerance, we do not know how common this evolutionary ability is among the wide diversity of animals that live in freshwater habitats.

ADAPTATIONS FOR OSMOREGULATION IN AQUATIC PLANTS

Some aquatic plants also face major challenges of salt balance. Mangrove trees, for example, grow on coastal mudflats that are inundated by saltwater during high tides (**Figure 2.13a**).

Not only does this habitat impose a high salt load on the trees, but the high osmotic potential of the saltwater environment also makes it difficult for the roots to take up water. To counter these osmotic problems, many mangroves maintain high concentrations of organic solutes—various amino acids and small sugar molecules—in their roots and leaves to increase their osmotic potential so that water will diffuse into the plant's tissues. In addition, mangroves possess salt glands in their leaves that can secrete salt by active transport to the exterior surface of the leaf (Figure 2.13b). Many mangrove species also exclude salts from their roots by active transport. Because relatively few species of terrestrial plants have evolved this adaptation, mangrove forests do not contain numerous species of plants.

(a)

(b)

Figure 2.13 Salt balance in mangrove trees. (a) The roots of mangrove trees are frequently submerged in saltwater at high tide. These trees are from Palau, an island in the South Pacific. **(b)** Specialized salt glands in the leaves of the button mangrove (*Conocarpus recta*) excrete a salty solution. As the solution evaporates, it leaves behind salt crystals on the outer surface of the leaves. Photos by (a) Reinhard Dirscherl/Alamy, and (b) Ulf Mehlig.

Standard Deviation and Standard Error

Eco TV

macmillanlearning.com/
ricklefsvideo

When the researchers tested the effects of road salt on larval amphibians, they exposed groups of larvae to three salt concentrations and replicated the experiment five times. In the previous chapter, we discussed how ecologists use data from such manipulative experiments to determine how different factors affect the means and variance of the variables that are measured. While the variance is a useful measure of how consistent measurements are among replicates, ecologists also use several other related measures of variation, including the *sample standard deviation* and the *standard error of the mean*. Each of these can be calculated from a measure of sample variance (s^2), as we discussed in Chapter 1.

When data are collected from limited samples of a much larger distribution of data, we can gain additional information about the data by calculating a *sample standard deviation*. The **sample standard deviation** gives us a standardized way of measuring how widely our data are spread from the mean. Large sample standard deviations indicate that many of our data are spread far from the mean. Small sample standard deviations indicate that most of our data points are close to the mean value.

If data are normally distributed—that is, if they follow a bell-shaped curve like the one in the figure at the right—then about 68 percent of our data will fall within 1 standard deviation of the mean. Moreover, about 95 percent of our data will fall within 2 standard deviations of the mean and 99.7 percent of our data will fall within 3 standard deviations of the mean. For data that have a wide distribution of frequencies, as in the upper figure (a), the standard deviation value will be large. For data that have a narrow distribution of frequencies, as in the lower figure (b), the standard deviation value will be small.

The sample standard deviation, denoted as *s*, is defined as the square root of the sample variance:

$$s = \sqrt{s^2}$$

The **standard error of the mean** is a useful measurement of variation in our data because it takes into account the number of replicates that were used to measure the standard deviation. The higher the number of replicates, the more precise the estimate we should have of the mean. As a result, an increase in the number of replicates in a given experiment produces a decrease in the standard error of the mean. As we will see in later

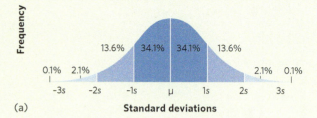

(a)

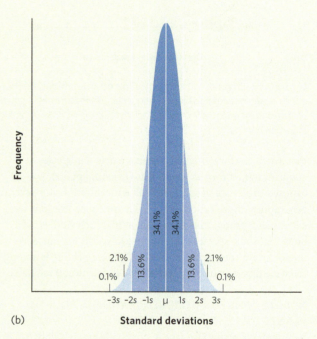

(b)

The normal distribution. In a normal distribution, the most frequent observations fall near the mean and the least frequent observations occur farther away from the mean. The dark blue areas represent 1 standard deviation from the mean and account for 68.3 percent of all data. The medium and dark blue areas combined represent 2 standard deviations from the mean and account for 95.5 percent of all data. The light, medium, and dark blue areas combined represent 3 standard deviations from the mean and account for 99.7 percent of all data. The remaining small amount of data falls outside of 3 standard deviations. **(a)** Data that have a wide distribution have a large standard deviation. **(b)** Data that have a narrow distribution have a small standard deviation.

Sample standard deviation A statistic that provides a standardized way of measuring how widely data are spread from the mean.

Standard error of the mean A measurement of variation in data that takes into account the number of replicates that were used to measure the standard deviation.

chapters, the standard error of the mean is commonly used to determine whether two means are significantly different from each other.

The standard error of the mean (denoted as SE) is defined as the sample standard deviation divided by the square root of the number of observations or replicates (denoted as n):

$$SE = s \div \sqrt{n}$$

Consider the following set of observations on the percentage of surviving wood frog tadpoles that were exposed to salt concentrations of 0 µS or 3,000 µS:

Replicate	0 µS	3,000 µS
1	88	32
2	90	37
3	91	41
4	92	45
5	94	50
Mean	91	41
Variance	5.0	48.5

Using the data from the five replicates that exposed wood frog tadpoles to 0 µS, we see that the mean survival is 91 percent and the sample variance is 5 percent. Using this sample variance, we can calculate the sample standard deviation as

$$s = \sqrt{s^2} = \sqrt{5} = 2.2$$

and we can calculate the standard error of the mean as

$$SE = s \div \sqrt{n} = 2.2 \div \sqrt{5} = 1.0$$

YOUR TURN Use the data collected from the five replicates that exposed wood frog tadpoles to 3,000 µS and calculate the standard deviation of the mean and the standard error of the mean. Explain why the values of s and SE differ.

CONCEPT CHECK

1. What is the difference between passive and active transport of solutes?
2. Compare and contrast the terms hyperosmotic and hyposmotic.
3. Describe one adaptation for osmoregulation in freshwater animals, saltwater animals, and saltwater plants.

2.3 The uptake of gases from water is limited by diffusion

Almost 21 percent of Earth's atmosphere is oxygen, but because we live on land, we rarely think about the task of obtaining this required element. Aquatic organisms also require oxygen to support their metabolism, but obtaining a sufficient supply can be a problem for them because of the limited solubility of oxygen in water. The same is true of the CO_2 required by aquatic plants for photosynthesis. Organisms obtain sufficient quantities of these necessary gases in the aquatic environment through several adaptations that we examine in this section.

CARBON DIOXIDE

Getting enough CO_2 for photosynthesis is a particular challenge for aquatic plants and algae. The solubility of CO_2 in fresh water is about 0.0003 liters of gas per liter of water, which is 0.03 percent by volume, which is about the same as its concentration in the atmosphere. The problem for aquatic plants is that CO_2 diffuses very slowly through water, and plants can use the CO_2 close to the surface of their leaves faster than it arrives by diffusion.

When CO_2 dissolves in water, most of the molecules combine with water and are quickly converted to a compound called carbonic acid (H_2CO_3):

$$CO_2 + H_2O \rightarrow H_2CO_3$$

As shown in **Figure 2.14**, carbonic acid can increase to high concentrations and provide a reservoir of carbon required for photosynthesis. Depending on the acidity of the water, carbonic acid molecules can release hydrogen ions (H^+) to form either **bicarbonate ions (HCO_3^-)** or **carbonate ions (CO_3^{2-})**:

$$H_2CO_3 \rightarrow H^+ + HCO_3^- \rightarrow 2H^+ + CO_3^{2-}$$

Although some of the carbonate ions can combine with calcium ions to form calcium carbonate, as we saw in Figure 2.8, bicarbonate ions dissolve readily in water, so

Bicarbonate ion (HCO_3^-) An anion formed by the dissociation of carbonic acid.

Carbonate ion (CO_3^{2-}) An anion formed by the dissociation of carbonic acid.

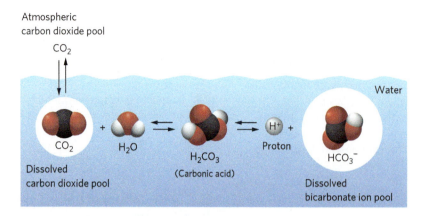

Atmospheric
carbon dioxide pool

Figure 2.14 An equilibrium reaction for carbon in water. The reaction converting CO_2 into bicarbonate ions (HCO_3^-) is an equilibrium reaction. When photosynthetic organisms preferentially use CO_2, because it is used more efficiently, the amount of dissolved CO_2 in the water declines. As the amount of CO_2 declines, some of the bicarbonate ions are converted to CO_2 to replenish the supply. The size of each circle represents the relative size of the carbon pool.

bicarbonate ion is the most common form of inorganic carbon in aquatic habitats. The result, as shown in Figure 2.14, is a concentration of bicarbonate ions equivalent to 0.03 to 0.06 L of CO_2 gas per liter of water (3 to 6 percent)—more than 100 times the concentration of CO_2 in air. In short, this means that most of the CO_2 that is dissolved into water is rapidly converted to bicarbonate ions in aquatic systems.

Dissolved CO_2 and bicarbonate ions are in a chemical equilibrium, which represents the balance achieved between H^+ and HCO_3^- on one hand, and CO_2 and H_2O on the other. When CO_2 is removed from the water by plants and algae that are performing photosynthesis, some of the abundant bicarbonate ions combine with hydrogen ions to produce more CO_2 and H_2O. In essence, the reaction now moves in the reverse direction:

$$H^+ + HCO_3^- \rightarrow CO_2 + H_2O$$

Although the bicarbonate ion is the most common form of inorganic carbon under moderate pH conditions, CO_2 is the most common form under more acidic conditions, such as in bogs. This may be why some species of plants and algae that live in aquatic habitats with low pH can only use CO_2 for photosynthesis. In contrast, because CO_2 and bicarbonate ions are both abundant under moderate pH conditions (pH = 5 to 9), many species of aquatic plants and algae can use both CO_2 and bicarbonate ions for photosynthesis. They can either directly uptake the bicarbonate ion or they can use adaptations to convert the bicarbonate ion to CO_2. One way to do this is by secreting an enzyme into the water that is highly effective at converting bicarbonate ions into CO_2, which can then be taken up by the organism. Plants and algae can also obtain CO_2 by secreting hydrogen ions into the surrounding water. This helps drive the chemical equilibrium in a direction that converts more

of the bicarbonate ions into CO_2 that can then be taken up by the organism.

Even when CO_2 and bicarbonate ions are abundant in water, the slow rate at which these carbon sources diffuse through the water prevents organisms from getting access to them. Indeed, carbon dioxide diffuses through unstirred water about 10,000 times more slowly than through air, and diffusion of HCO_3^- is even slower because larger molecules diffuse at a slower rate. Compounding this slow rate of diffusion is the fact that every surface of an aquatic plant, alga, or microbe is surrounded by a *boundary layer*. A **boundary layer** is a region of unstirred air or water that surrounds the surface of an object. In the water, the boundary layer ranges from as little as 10 micrometers (10 μm, or 0.01 mm) for single-celled algae in turbulent waters to 500 μm (0.5 mm) for a large aquatic plant in stagnant water. As **Figure 2.15** illustrates, because this boundary layer is composed of unstirred water, CO_2 and HCO_3^- can be depleted within the boundary layer by uptake—especially in the region closest to the photosynthesizing organism—but the removed gases are slow to be replaced from the surrounding water. Without a boundary layer, the moving water in the surrounding environment would continually provide the plant with a supply of CO_2 and HCO_3^-. So, despite the generally high concentration of bicarbonate ions in water, photosynthesis may still be limited by carbon availability within the boundary layer.

OXYGEN

Oxygen in the atmosphere has a concentration of 0.21 L per liter of air (21 percent by volume). In water, however, the maximum solubility of oxygen is 0.01 L per liter of water (1 percent), under the conditions of fresh water at 0°C.

Boundary layer A region of unstirred air or water that surrounds the surface of an object.

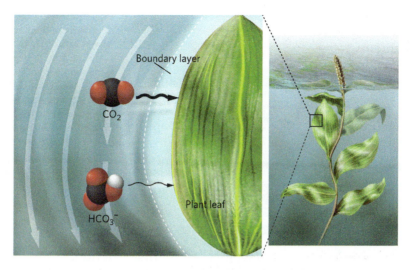

Figure 2.15 Boundary layer. A thin boundary of unstirred water along the surface of photosynthesizing organisms slows the rate of gas diffusion through the water. Smaller molecules, such as CO_2, diffuse faster than larger molecules, such as bicarbonate ions.

Oxygen's low solubility in water can limit the metabolism of organisms in aquatic habitats. For marine mammals such as whales, this is not a problem because they obtain O_2 from the air and store copious amounts in their hemoglobin and myoglobin. For organisms that obtain O_2 from the water, however, the problem of low O_2 concentration is compounded by its slow diffusion in water, similar to that of CO_2. Oxygen is in even shorter supply in waters that cannot support photosynthesis—and therefore do not receive the O_2 produced by photosynthesis—including in deep water that does not receive sunlight and in water-logged sediments and soils. These habitats can become severely depleted of dissolved O_2, making them challenging environments for animals and microbes that use aerobic respiration.

One important adaptation that allows aquatic animals to deal with a limited amount of oxygen involves the direction of blood flow in the gills. Many aquatic animals have gills to extract oxygen from the water. When water passes over the gills, oxygen diffuses across the membranes of the gill cells and enters the capillaries, which are part of the bloodstream. The key to extracting the most oxygen from the water lies in the use of *countercurrent circulation*. In **countercurrent circulation,** two fluids move in opposite directions on either side of a barrier and heat or materials are exchanged. In contrast, **concurrent circulation** involves two fluids moving in the same direction on either side of a barrier and heat or materials are exchanged.

As illustrated in **Figure 2.16,** if blood and water were to flow in the same direction, the concentration of oxygen would quickly come to an intermediate equilibrium. After this region of contact, there is no net movement of oxygen across the membrane. In contrast, when blood and water flow in opposite directions, the concentration of oxygen in the water exceeds the concentration in the blood throughout

most of the region of contact. This happens because even as the capillaries begin to develop high concentrations of oxygen, the adjacent water still has a higher concentration of oxygen. As a result, the oxygen continues to diffuse into the gill capillaries. Thus, countercurrent blood flow in animal gills allows much more oxygen to move from the water to the gills.

Species of animals that live in habitats with low amounts of oxygen have evolved a number of additional adaptations. In the deep oceans, many organisms have very low rates of activity, thereby reducing their need for oxygen. Many species of zooplankton, a group of tiny crustaceans, can increase the amount of hemoglobin in their bodies to the point that their normally transparent bodies turn red. Other animals, such as tadpoles and fish that live in oxygen-depleted swamps, swim to the surface and take gulps of air. Many tadpoles can use this air because they possess primitive lungs in addition to gills. Fish store this air in a swim bladder from which they extract the oxygen into their bloodstream.

One of the most surprising animal adaptations for obtaining oxygen was recently discovered in a species of North American salamander. For more than a century, it was known that eggs of the spotted salamander, which are usually attached to sticks that are submerged in water, have a mutualistic relationship with a species of algae (*Oophila amblystomatis*). The algae obtain a place to live and photosynthesize while the developing embryo obtains

---

Countercurrent circulation Movement of two fluids in opposite directions on either side of a barrier through which heat or dissolved substances are exchanged.

Concurrent circulation Movement of two fluids in the same direction on either side of a barrier through which heat or dissolved substances are exchanged.

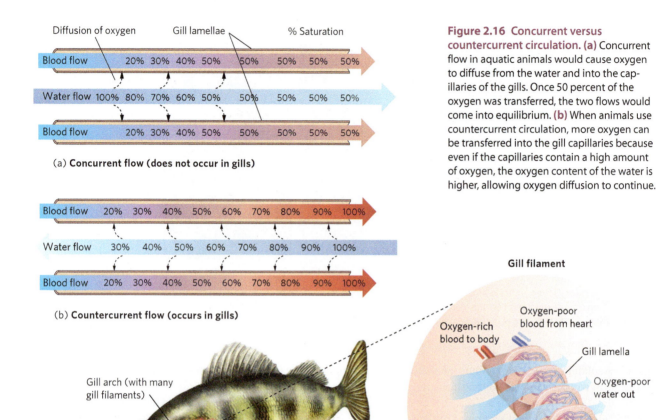

Diffusion of oxygen   Gill lamellae   % Saturation

Blood flow   20% 30% 40% 50% 50% 50% 50% 50%

Water flow 100% 80% 70% 60% 50% 50% 50% 50% 50%

Blood flow   20% 30% 40% 50% 50% 50% 50% 50%

(a) **Concurrent flow (does not occur in gills)**

Blood flow 20% 30% 40% 50% 60% 70% 80% 90% 100%

Water flow 30% 40% 50% 60% 70% 80% 90% 100%

Blood flow 20% 30% 40% 50% 60% 70% 80% 90% 100%

(b) **Countercurrent flow (occurs in gills)**

Gill arch (with many gill filaments)

Water in

Water out

Gill filament

Oxygen-rich blood to body

Oxygen-poor blood from heart

Gill lamella

Oxygen-poor water out

Oxygen-rich water in

Blood flow

Figure 2.16 Concurrent versus countercurrent circulation. (a) Concurrent flow in aquatic animals would cause oxygen to diffuse from the water and into the capillaries of the gills. Once 50 percent of the oxygen was transferred, the two flows would come into equilibrium. **(b)** When animals use countercurrent circulation, more oxygen can be transferred into the gill capillaries because even if the capillaries contain a high amount of oxygen, the oxygen content of the water is higher, allowing oxygen diffusion to continue.

oxygen from the photosynthesizing algae (**Figure 2.17**). This oxygen benefit is important because it allows the salamander embryos to have a higher survival rate and to hatch earlier and larger. In 2011, however, scientists reported that this relationship was much closer than they had recognized. The algae not only live in the fluid of the egg surrounding the embryo, but also move into the embryo, positioning themselves in between the developing embryo's cells. This was the first discovery of an algae living within the tissues of a vertebrate animal. Two years later, in 2013, scientists discovered that the algae not only provide oxygen to the developing embryos but also provide sugars produced by photosynthesis. Together, these two beneficial effects of the algae allow faster growth of the salamander embryos.

When an environment becomes completely devoid of oxygen, it is referred to as **anaerobic** or **anoxic**. Anaerobic conditions pose problems for terrestrial plants rooted in waterlogged soils, such as the many species of mangrove trees that live along coastal mudflats. The roots of these trees need oxygen for respiration, so the plants have evolved special air-filled tissues extending from the roots that rise above

Figure 2.17 Salamander embryos and algae. The embryos of the spotted salamander contain algae that live in the egg sac and within the cells of the embryo. The algae provide extra oxygen to the embryos, which improves embryo survival and growth. Photo courtesy of Roger Hangarter/University of Indiana.

Anaerobic Without oxygen. *Also known as* **Anoxic**.

the waterlogged soils and exchange gases directly with the atmosphere (see Figure 2.13).

Many microbes are able to live in environments without oxygen because they use anaerobic respiration. A common product of anaerobic respiration by bacteria living in anoxic soils is hydrogen sulfide gas (H_2S). This gas is the cause of the rotten-egg smell that occurs when soils saturated with water become anaerobic.

CONCEPT CHECK

1. Why does the boundary layer surrounding a photosynthetic organism make it more difficult to exchange CO_2 and O_2?
2. What is the equilibrium reaction that illustrates the conversion of CO_2 to bicarbonate?
3. Why are deep ocean waters typically low in oxygen?

2.4 Temperature limits the occurrence of aquatic life

At the beginning of this chapter, we saw that whales possess a number of adaptations that allow them to deal with the cold ocean temperatures, including thick layers of fat, high metabolism, and countercurrent circulation that exchanges heat between the warm arteries and cold veins. For aquatic organisms, most physiological processes occur only within the range of temperatures at which water is liquid. Relatively few organisms can survive temperatures above 45°C, which is the upper limit of the physiological range for most eukaryotic organisms. In this section, we will look at how organisms have adapted to hot and cold temperatures.

HEAT AND BIOLOGICAL MOLECULES

Temperature influences physiological processes because of the way in which heat affects organic molecules. Heat imparts kinetic energy to living systems, causing biological molecules to change shape. Heat also accelerates chemical reactions by increasing the rate of molecule movement. In fact, the rate of most biological processes increases two to four times for each 10°C rise in temperature. We can see this when we examine the data for a fish known as the miiuy croaker (*Miichthys miiuy*), shown in **Figure 2.18**. The fish's rate of oxygen consumption approximately doubles when the environmental temperature increases from 15°C to 25°C. To understand the relationship between temperature and physiological process, we calculate the ratio of the physiological rate at one temperature to the physiological rate when the temperature is 10°C cooler, a ratio that is referred to as the $\mathbf{Q_{10}}$ of the physiological process. By knowing the Q_{10} values of different physiological processes, we can better understand which processes are more sensitive to a change in temperature.

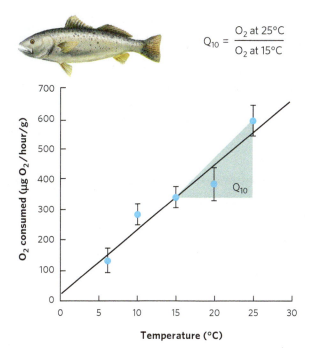

$$Q_{10} = \frac{O_2 \text{ at } 25°C}{O_2 \text{ at } 15°C}$$

Figure 2.18 Oxygen consumption as a function of temperature. For the miiuy croaker, the amount of oxygen it consumes increases as the temperature increases. By dividing the rate of oxygen consumption at 25°C by the rate of oxygen consumption at 15°C, one can arrive at the Q_{10} value for this physiological function. Error bars are standard deviations. Data from: Z. Zheng et al., Effects of temperature and salinity on oxygen consumption and ammonia excretion of juvenile miiuy croaker, Miichthys miiuy (*Basilewsky*), *Aquaculture International* 16 (2008): 581–589.

Higher temperatures allow organisms to do many things more rapidly. They can swim, run, and fly faster. They can also digest and assimilate more food and, as a result, grow and develop faster. Beyond a certain point, however, high temperatures can depress life processes. In particular, proteins and other biological molecules become less stable at higher temperatures, and may not function properly or may lose their structure. The molecular motion caused by heat tends to open up, or *denature,* the structure of these molecules.

Given that proteins denature at high temperatures, scientists have been intrigued by the fact that some organisms, such as **thermophilic** (heat-loving) bacteria, can live at very high temperatures. For example, some photosynthetic bacteria can tolerate temperatures as high as 75°C and some archaebacteria can live in hot springs at temperatures up to 110°C (**Figure 2.19**). The chemosynthetic bacteria that live near deep-sea vents are a group of thermophiles that we discussed in Chapter 1. Researchers have discovered that the

---

$\mathbf{Q_{10}}$ The ratio of the rate of a physiological process at one temperature to its rate at a temperature 10°C cooler.

Thermophilic Heat-loving.

Figure 2.19 Thermophiles. Some species of bacteria and archaebacteria can live under very hot conditions, such as these hot springs in Fly Geyser, Nevada. Thermophiles often come in a variety of bright colors. Photo by Jack Dykinga.

proteins of thermophilic bacteria have higher proportions of particular amino acids that form stronger bonds than the proteins of other, heat-intolerant species. These strong forces of attraction within and between molecules prevent them from being shaken apart under high temperatures, so the proteins do not denature.

Temperature also affects other biological compounds. For instance, the physical properties of fats and oils, which are major components of cell membranes and constitute the energy reserves of animals, depend on temperature. When fats are cold, they become stiff; when they are warm, they become fluid.

Organisms are negatively affected when exposed to temperatures that range above or below those to which they are adapted. Water used to keep nuclear power plants cool provides an example of this problem. Many nuclear power plants extract water from nearby lakes, rivers, or oceans for cooling the power plants and then return the water—which has become much warmer in the process—to its source. But the organisms in these bodies of water have no evolutionary history of being exposed to such high water temperatures and many of them die. In Ohio, for example, a coal-burning power plant raised the temperature of an adjacent stream up to 42°C, which is much warmer than would naturally occur, and warmer than the maximum of 32°C set by the U.S. EPA. Surveys found numerous dead fish in the stream and low numbers of fish downstream where the stream emptied into the Ohio River. Discharging water that is too hot to sustain aquatic species is known as **thermal pollution.** Understanding the thermal optima for aquatic species has led to regulations that restrict

how much a discharging power plant can raise the temperature of a lake or river.

COLD TEMPERATURES AND FREEZING

Temperatures on Earth's surface rarely exceed 50°C, except in hot springs and at the soil surface in hot deserts. However, temperatures below the freezing point of water are common, particularly on land and in small ponds, which may freeze solid during winter. When living cells freeze, the crystal structure of ice disrupts most life processes and may damage delicate cell structures, eventually causing death. Many organisms successfully cope with freezing temperatures either by maintaining their body temperatures above the freezing point of water, or by activating chemical pathways that enable them to resist freezing or to tolerate its effects.

It might surprise you to learn that marine vertebrates are susceptible to freezing in cold seawater. How can blood and body tissues freeze solid while immersed in liquid water? The answer is that dissolved substances lower the temperature at which water freezes. While pure water freezes at 0°C, seawater, which contains about 3.5 percent dissolved salts, freezes at −1.9°C. Because the blood and body tissues of most vertebrates contain about half the salt content of seawater, the animals can freeze solid before the surrounding seawater does.

Marine animals have evolved a number of adaptations to combat the problem of freezing in water. We know that high salt concentrations interfere with many biochemical processes, so raising the concentration of solutes in blood and tissues is not a viable option. Instead, some Antarctic fish prevent freezing by raising their blood and tissue concentrations of nonsalt compounds such as *glycerol*. **Glycerol** is a chemical that prevents the hydrogen bonds of water from coming together to form ice unless the temperatures are well below freezing. A 10 percent glycerol solution in the body lowers the freezing point of water to about −2.3°C without severely disrupting biochemical processes. This is enough to lower the freezing point of the body below the freezing point of seawater. **Glycoproteins** are another group of compounds that can be used to lower the freezing temperature of water. Glycerol and glycoproteins act as antifreeze compounds, similar to the antifreeze used in automobiles, and allow fish such as the Arctic cod (*Boreogadus saida*) to remain active in seawater that is cold enough to cause most fish to freeze solid (**Figure 2.20**). Some terrestrial invertebrates also use the antifreeze approach; their body fluids may contain up to 30 percent glycerol as winter approaches.

Thermal pollution Discharging water that is too hot to sustain aquatic species.

Glycerol A chemical that prevents the hydrogen bonds of water from coming together to form ice unless the temperatures are well below freezing.

Glycoproteins A group of compounds that can be used to lower the freezing temperature of water.

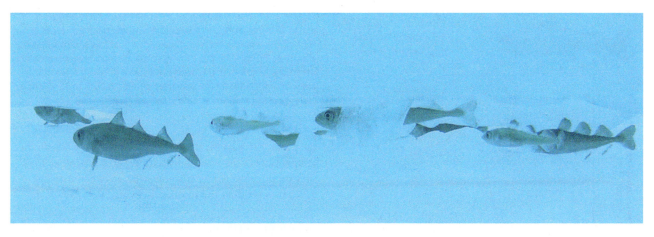

Figure 2.20 Adaptation to different water. The Arctic cod, which is commonly harvested in Russia for human consumption, can live in water that is less than 4°C. Antifreeze compounds in its tissues prevent it from freezing solid. Photo by Elizabeth Calvert Siddon/UAF/NOAA.

Supercooling provides a second physical solution to the problem of freezing. Under certain circumstances, liquids can cool below the freezing point without developing ice crystals. Ice generally forms around an object, called a *seed*, which can be a small ice crystal or other particle. In **supercooling,** however, glycoproteins in the blood impede ice formation by coating any ice crystals that begin to form. In the absence of ice seeds, pure water may cool to more than −20°C without freezing. Such supercooling has been recorded down to −8°C in reptiles and −18°C in invertebrates.

THERMAL OPTIMA

Every organism is best suited to a narrow range of environmental conditions; this range defines the **optimum** environmental conditions. In terms of temperature, most organisms have a **thermal optimum,** meaning the range of temperatures in which they perform best. The thermal optimum is determined by the properties of enzymes and lipids, the structures of cells and tissues, body form, and other characteristics that influence the ability of an organism to function well under the particular conditions of its environment. Returning to the example of fish in the ocean waters of the Antarctic, many species swim actively and consume oxygen at a rate comparable to fish living in much warmer regions near the equator. However, if you put a tropical fish in cold water, it becomes sluggish and soon dies; conversely, Antarctic fish cannot tolerate temperatures warmer than 5°C to 10°C.

Certain adaptations allow fish in cold oceans to swim as actively as fish in warm oceans. Swimming involves a series of biochemical reactions, most of which depend on enzymes. These reactions generally proceed more rapidly at high temperatures, so cold-adapted organisms must either have more of the substrate for a biochemical reaction, more of the enzyme that catalyzes the reaction, or a different version of the enzyme that operates better under colder temperatures. Different forms of an enzyme that catalyze a given reaction are called **isozymes.**

Consider the case of the rainbow trout (*Oncorhynchus mykiss*), a fish that lives in cold streams throughout much of North America. Seasonal changes in temperature are predictable for the trout: During winter, water temperatures may drop close to the freezing point, while summer temperatures can become very warm. In response to these seasonal temperature changes, the trout has evolved the ability to produce different isozymes in winter and in summer. One of these enzymes is acetylcholinesterase, which plays an important role in ensuring proper functioning of the nervous system by binding with the neurotransmitter acetylcholine. To understand how well different isozymes function at different temperatures, we examine the rate of the chemical reaction between acetylcholine and acetylcholinesterase, a measure known as *enzyme-substrate affinity*. The winter isozyme, shown as the blue line in **Figure 2.21**, best catalyzes the reaction at 0°C to 10°C. This affinity drops rapidly at higher temperatures. In contrast, the summer isozyme, shown as the orange line, has a weak affinity for acetylcholine at 10°C and catalyzes reactions the best at 10°C to 20°C, though the affinity drops slowly at higher temperatures. As you might predict,

---

Supercooling A process in which glycoproteins in the blood impede ice formation by coating any ice crystals that begin to form.

Optimum The narrow range of environmental conditions to which an organism is best suited.

Thermal optimum The range of temperatures within which organisms perform best.

Isozymes Different forms of an enzyme that catalyze a given reaction.

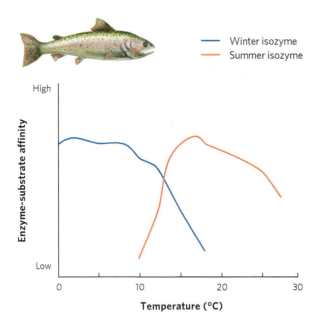

Winter isozyme
Summer isozyme

High

Enzyme-substrate affinity

Low

| 0 | 10 | 20 | 30 |

Temperature (°C)

Figure 2.21 Using isozymes as adaptations to changing water temperatures. In winter, the rainbow trout produces a form of the enzyme acetylcholinesterase that has a high substrate affinity at 0°C to 10°C, but a lower substrate affinity at warmer temperatures. In summer, the trout produces a different form of the enzyme, which has a high substrate affinity at 10°C to 20°C, but a lower substrate affinity at colder temperatures.

the particular isozyme that a trout produces depends on the temperature of the water it lives in. When trout live at 2°C, they produce the winter isozyme, whereas when they live at 17°C, they produce the summer isozyme.

CONCEPT CHECK

1. Explain the adaptation that allows thermophiles to survive in very high temperatures.
2. Describe the adaptations that enable fish to survive in very cold seawater.
3. How do isozymes help organisms function in habitats with a wide range of temperatures?

ECOLOGY TODAY CONNECTING THE CONCEPTS

The Decline of Coral Reefs

Healthy - Dec 2014

Dying - Feb 2015

Dead - Aug 2015

Coral reef diversity. Coral reefs, such as this one off Beqa Island, Fiji, in the South Pacific, are among the most species-diverse places on Earth. Photo by The Ocean Agency/XL Catlin Seaview Survey.

Coral reefs are some of the most beautiful places on Earth and are home to an incredible diversity of species. For instance, the Great Barrier Reef off the eastern coast of Australia contains more than 400 species of coral, 200 species of birds, and 1,500 species of fish. For these reasons, many people become concerned when the biodiversity of coral reefs is threatened by human activity. We have long understood that overfishing and pollution have affected the species that inhabit coral reefs. In the past 20 years, however, scientists have discovered that changes in the abiotic aquatic environment—including changes in temperature, pH, and salinity—are damaging coral reef ecosystems.

Corals are a group of tiny animals that secrete hard exoskeletons made of limestone (calcium carbonate). They can be found worldwide in relatively shallow ocean waters that are low in nutrients and food. Although each individual coral is small—only a few millimeters in size—the limestone bodies of dead corals accumulate over hundreds or thousands of years to form massive coral reefs that can exceed 300,000 km². The corals survive in these low-nutrient waters by living in symbiotic relationships with several species of photosynthetic algae known as *zooxanthallae*. Corals have tubular bodies with tentacles that stick out and catch bits of food and detritus that go by. Their digestion produces CO_2, which can be used by symbiotic algae during photosynthesis. As we learned earlier in this chapter, CO_2 can often be difficult for aquatic producers to obtain. In exchange, the algae produce O_2 and sugars, some of which can be passed to the coral. In short, the algae get a safe place to live and a steady supply of CO_2 for photosynthesis, while the corals get a source of energy in the form of sugars and a steady supply of O_2 for respiration.

During the past two decades, scientists have learned that the symbiotic relationship between the corals and algae is very sensitive to environmental changes. When corals are experiencing stress in their environment, they expel the symbiotic algae from their bodies. Because corals obtain their bright colors from the symbiotic algae, corals that expel their algae often look white and are said to experience **coral bleaching.**

Coral bleaching is associated with unusually high ocean temperatures. As we discussed in this chapter, while increases in water temperature can increase the rate of chemical reactions, temperatures that exceed the thermal optima can be detrimental. Bleaching can begin if summer ocean temperatures are even just 1°C higher than the average maximum. If the temperature rise is brief—a few days or weeks—the algae can recolonize the corals. However, the corals will experience slower growth and reduced reproduction. However, if the temperatures are 2°C to 3°C higher than the average maximum for longer periods of time, corals can die. The rise in ocean temperatures also appears to stress corals in a way that makes them more susceptible to pathogens that can kill them. During the past two decades, scientists have witnessed major bleaching events worldwide: in 1998, 2003, 2005, 2010, and 2016. During the 2016 event, for example, areas in the Great Barrier Reef off the coast of Australia experienced up to 67 percent declines in the survival of shallow-water corals. Fortunately, other regions around Australia experienced much lower death rates. With continuing increases in global temperatures, a topic we will cover in detail in later chapters, temperature-induced coral bleaching is expected to continue.

Changes in salt concentration are also an issue for corals. High ocean temperatures increase the evaporation of water from the ocean, which increases the ocean's salt concentration. In the case of corals, the stress of increased salt concentrations combined with the stress of high temperatures makes them increasingly vulnerable to coral bleaching and coral death.

Another source of decline in corals is a decrease in the pH of seawater. Because atmospheric CO_2 is in equilibrium with dissolved CO_2 in the ocean, the recent increase in atmospheric CO_2 has caused an increase in dissolved CO_2. This, in turn, has caused an increase in

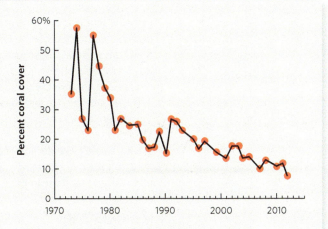

Coral reef decline. Numerous changes in the ocean environment are causing declines in the numbers of live corals. In the Caribbean, for example, a sharp drop in the number of live corals occurred from 1963 to 2012. After J. Jackson et al., Tropical Americas Coral Reef Resilience Workshop, http://cmsdata.iucn.org/downloads/caribbean_coral_report_jbcj_030912.pdf.

Coral bleaching Loss of color in corals as a result of the corals expelling their symbiotic algae.

carbonic acid, and a decline in the pH of the oceans. As we have seen, carbonic acid (H_2CO_3) disassociates to bicarbonate ions (HCO_3^-) and hydrogen ions (H^+). The hydrogen ions can then combine with other carbonate ions to produce bicarbonate, making carbonate ions less available for the corals to create their exoskeleton of calcium carbonate. This decline in pH can also dissolve the calcium carbonate skeletons of the coral.

The decline in coral reefs has become a serious problem. In the Caribbean, for example, the percent of live corals has declined from more than 50 percent in 1977 to less than 10 percent in 2012. Only 20 years ago, scientists debated whether abiotic changes were playing a role in the decline of corals. As more data have been collected, there is now general agreement that changes in temperature, pH, salinity, and pollutants are all contributing. Given that these impacts are expected to rise even more in future decades, scientists currently predict that many species of coral will continue to decline. However, some species of coral may have sufficient genetic variation to allow evolutionary changes that permit them to adapt to the changing environmental conditions and persist.

SOURCES:

McGuirk, R. Great Barrier Reef sees record coral deaths this year, Associated Press, November 29, 2016, http://www.nytimes.com/aponline/2016/11/28/world/asia/ap-as-australia-dying-coral.html.

Renema, W., et al. 2016. Are coral reefs victims of their own past success?, *Science Advances* 2: e1500850, doi:10.1126/sciadv.1500850.

Carpenter, K. E., et al. 2008. One-third of reef-building corals face elevated extinction risk from climate change and local impacts, *Science* 321:560–563.

Hoegh-Guldberg, O., et al. 2007.Coral reefs under rapid climate change and ocean acidification, *Science* 318:1737–1742.

SUMMARY OF LEARNING OBJECTIVES

2.1 Water has many properties favorable to life. These include thermal resistance to changes in temperature, a density and viscosity that select for adaptations for movement, and the ability to dissolve many elements and compounds that are essential to life.

Key Terms: Viscosity, Ions, Saturation, Acidity, pH, Acid deposition

2.2 Aquatic animals and plants face the challenge of water and salt balance. This challenge occurs because animal tissues are generally hyperosmotic or hyposmotic compared to the solute concentration of the surrounding aquatic environment.

Key Terms: Solute, Semipermeable membrane, Passive transport, Active transport, Osmosis, Osmotic potential, Osmoregulation, Hyperosmotic, Hyposmotic, Sample standard deviation, Standard error of the mean

2.3 The uptake of gases from water is limited by diffusion. This limitation can make it difficult for organisms to exchange gases such as CO_2 and O_2. Diffusion is slowed because organisms are surrounded by a thin boundary layer of unstirred water. Although CO_2 can be abundant in water—either as a dissolved gas or in the form of bicarbonate ions—oxygen's low solubility in water makes it less abundant. As a result, many animals have respiratory structures with large surface areas and countercurrent circulation to obtain oxygen from the water.

Key Terms: Bicarbonate ion (HCO_3^-), Carbonate ion (CO_3^{2-}), Boundary layer, Countercurrent circulation, Concurrent circulation, Anaerobic

2.4 Temperature limits the occurrence of aquatic life. Although warmer temperatures increase the rate of chemical reactions, excessively high temperatures cause proteins and other important molecules to become unstable and denature. Cold temperatures also pose a challenge, and many organisms living under near-freezing conditions have evolved adaptations that include the use of glycoproteins and supercooling to prevent the harmful effects of ice crystals growing inside their cells. Organisms that live under a wide range of temperatures often use isozymes to permit proper physiological function at each temperature.

Key Terms: Q_{10}, Thermophilic, Thermal pollution, Glycerol, Glycoproteins, Supercooling, Optimum, Thermal optimum, Isozymes, Coral bleaching

CRITICAL THINKING QUESTIONS

1. As water cools below 4°C, it expands and becomes less dense. Why is this beneficial to organisms living in a lake during a cold winter?

2. Given that heat readily transfers between an organism's body and water, what would you predict about the temperature of the boundary layer around an animal's body compared to the temperature of the surrounding water in cold weather?

3. Based on the properties of water, why do non-polar compounds such as fats and oils not dissolve well in water?

4. In lakes around the world, some have thick deposits of limestone, which is made up of calcium carbonate, and are resistant to the effects of acid rain. What parallels can you draw between these lakes and the smokestack scrubbers being used in coal-powered factories?

5. Both saltwater and freshwater fish have adaptations to control the movement of water and salts across their external surfaces. Describe what would happen without these adaptations.

6. How would the process of natural selection favor the evolution of increased salt tolerance during the decades that cold regions of the world have increased the use of road salt, which ultimately finds its way into streams and lakes?

7. Using your knowledge of how CO_2 dissolves and reacts in water, explain why global increases in atmospheric CO_2 should cause a decrease in the pH of ocean water.

8. How would counter-current selection in fish gills be beneficial for an animal trying to get rid of CO_2?

9. How might the use of isozymes permit an aquatic organism to persist with a gradual increase in thermal pollution but not a rapid increase in thermal pollution?

10. If the algae that live in corals need CO_2 as a carbon source for photosynthesis, why do scientists argue that global increases in CO_2 are playing a role in the decline in coral reefs around the world?

GRAPHING THE DATA   Determining Q_{10} Values in Salmon

Salmon researchers measured the oxygen demands of fish across a range of different water temperatures. Using their data in the following table, and with reference to Figure 2.18, create a graph that demonstrates how oxygen demand changes with temperature. Based on this graph, calculate the Q_{10} value for the salmon between 5 and 15°C. Next, calculate the Q_{10} value for the salmon between 10 and 20°C and compare your two answers.

Temperature (°C)	Oxygen demand (mg O_2/kg/minute)
5	2.0
10	2.7
15	4.0
20	5.6

3 Adaptations to Terrestrial Environments

The Evolution of Camels

When you think of camels, you might envision the iconic animals of the African and Asian deserts. The ancestor of all camel species actually originated in North America about 30 million years ago and camels roamed many parts of North America until about 8,000 years ago. Current evidence suggests that some of these ancestors crossed the Bering land bridge from Alaska about 3 million years ago and made their way into Asia and Africa. These individuals evolved into two modern-day species of camels: the endangered Bactrian camel (*Camelus bactrianus*) and the much more common dromedary camel (*C. dromedarius*). Other ancestors moved from North America down to South America and evolved into a second group: guanacos (*Lama guanicoe*), llamas (*L. glama*), vicuñas (*Vicugna vicugna*), and alpacas (*V. pacos*).

All these animals live in dry environments and have evolved a number of adaptations that help them cope with the harsh conditions. For example, the dromedary camel—the most-studied of the group—risks overheating in the very hot deserts where it lives. During the day, the rays of the Sun strike its body and warm its surface. The camel can respond behaviorally by facing into the Sun and presenting a smaller profile for the Sun's rays to warm. The sandy desert soil is also hot and radiates the heat it absorbs, which makes the camel even hotter. Fortunately, the camel has a very large body relative to its surface area, so these heat inputs raise its body temperature slowly. Although camels, like all mammals, try to maintain a constant body temperature, they can tolerate a rise in body temperature of 6°C before experiencing any harmful effects. In contrast, most mammals can only tolerate an increase of about 3°C. At night, as the air and sand cool rapidly, the camel radiates its excess heat into the air or lies down and transfers its excess body heat to the sand.

One of the many amazing adaptations of camels and other mammals in hot environments is their ability to cool their brains. Whereas much of the body can tolerate short-term increases in temperature, the brain cannot. The camel has evolved an arrangement of veins and arteries that helps solve this problem. As the animal breathes, veins positioned beside the long nasal cavity are cooled by means of evaporating water vapor. The veins that carry cool blood then travel toward the back of the camel's head, where they come into close contact with the arteries that supply blood to the brain. Although the blood does not mix between the veins and arteries, heat is exchanged between the two types of vessels. This cools the arterial blood before it travels to the brain and keeps the brain several degrees cooler than the rest of the body.

Lack of water is also a challenge in the desert environment. As an adaptation, the dromedary camel can store large amounts of water in its body, most of which resides in its tissues. As water is lost from the bloodstream, the water from the tissues enters the blood. As much as 30 to 40 percent of the camel's body mass is water that can be used over several days when it is unable to drink water. In other mammals, a mere 15 percent loss of water can be lethal. Camels also conserve the water that they take in by producing relatively dry feces and urine that is high in waste products and

> "One of the many amazing adaptations of camels and other mammals in hot environments is their ability to cool their brains."

Adaptations of the dromedary camel. Dromedary camels, such as this one in the Hajar Mountains of the United Arab Emirates, have a wide range of adaptations that allow them to live in hot, dry environments. Naftali Hilger/Getty Images.

Guanacos of the Patagonia. Guanacos, from the Patagonian region of Chile, share a common ancestor with Asian camels. As a result of their common ancestry, guanacos and Asian camels also share a number of adaptations for coping with a dry environment. Photo by Morty Ortega.

low in water. In addition, while sweating can be an effective way to cool the body through evaporation, camels that are low on water can reduce the amount that they sweat. Collectively, these adaptations allow the camels to survive when water is scarce.

Camels represent just one case in which terrestrial organisms have evolved numerous adaptations to the challenges posed by terrestrial environments. In this chapter, we will explore the challenges of living on land and the adaptations of terrestrial plants and animals that make their lives possible.

SOURCES

Cain, J. W., et al. 2006. Mechanisms of thermoregulation and water balance in desert ungulates, *Wildlife Society Bulletin* 34:570–581.

Ouajd, S., and B. Kamel, 2009. Physiological particularities of dromedary (*Camelus dromedarius*) and experimental implications, *Scandinavian Journal of Laboratory Animal Science* 36:19–29.

LEARNING OBJECTIVES

After reading this chapter, you should be able to:

3.1 Explain how most terrestrial plants obtain nutrients and water from the soil.

3.2 Illustrate how sunlight provides the energy for photosynthesis.

3.3 Describe the ways in which terrestrial environments pose a challenge for animals to balance water, salt, and nitrogen.

3.4 Understand how adaptations to different temperatures allow terrestrial life to exist around the planet.

As we discussed in Chapter 2, life on Earth probably arose in the water. Following these early origins, life forms evolved adaptations that allowed them to live on land. The transition from water to land posed a number of new challenges. Plants evolved to obtain water and nutrients from the soil, and to conduct photosynthesis under hot and dry conditions. Animals evolved to balance water, salts, and wastes, and to adjust to extreme temperatures in terrestrial environments. In this chapter, we will examine the diversity of adaptations that allow plants and animals to live in terrestrial environments.

3.1 Most terrestrial plants obtain nutrients and water from the soil

A few unusual plants, such as the epiphytes discussed in Chapter 1 (Figure 1.9), obtain necessary water and nutrients without being rooted in the soil. However, the vast majority of plants obtain nutrients and water from the soil through their root systems. As a result, plants have a number of adaptations that help them perform this task.

SOIL NUTRIENTS

In addition to the oxygen, carbon, and hydrogen that plants incorporate into carbohydrates to fuel their survival and growth, plants require many other inorganic nutrients including nitrogen, phosphorus, calcium, and potassium to make proteins, nucleic acids, and other essential organic compounds. Whereas oxygen and carbon are available in the air, other nutrients are obtained as ions dissolved in the water held by the soil around plant roots. Nitrogen exists in soil as ammonium (NH_4^+) and nitrate (NO_3^-) ions, phosphorus exists as phosphate ions (PO_4^{3-}), and calcium and potassium exist as the elemental ions Ca^{2+} and K^+, respectively. The availability of these and other inorganic nutrients varies with their chemical form in the soil, and with temperature, pH, and the presence of other ions. A scarcity of inorganic nutrients, such as nitrogen, often limits plant production in terrestrial environments. We shall have much more to say about nutrient uptake by plants in later chapters.

SOIL STRUCTURE AND WATER-HOLDING CAPACITY

To understand how plants obtain water and nutrients, it is first necessary to understand how water behaves in the soil. The movement of water in the soil can be described in terms of its **water potential,** which is a measure of the water's potential energy. Water potential affects the movement of water in the soil from one location to another and depends on several factors that include gravity, pressure, osmotic

Water potential A measure of water's potential energy.

(a) Saturation (0 MPa) (b) Field capacity (−0.01 MPa) (c) Wilting point (−1.5 MPa)

Figure 3.1 Soil water. (a) Immediately after a rain event, soils can become saturated with water and all spaces between soil particles are filled. **(b)** The field capacity of soil represents the amount of water remaining after it has been drained by gravity. **(c)** The wilting point occurs when the opposing attractive forces of the soil particles prevent plants from extracting any more water.

potential (discussed in Chapter 2), and *matric potential,* so named because the collection of soil particles is known as the *soil matrix.* The **matric** (or **matrix**) **potential** is the potential energy generated by the attractive forces between water molecules and soil particles. It exists because water molecules, which have electrical charges, are attracted to the surfaces of soil particles, which also have electrical charges. This attraction explains why soil is able to retain water against the downward pull of gravity.

Because electrical charges are responsible for the attraction between water molecules and soil particles, those water molecules closest to the surfaces of soil particles adhere the most strongly. When water is plentiful, most of the water molecules are not close to the surfaces of soil particles. As a result, these water molecules are not held tightly and plant roots can easily take up the water. As more of the water is used up, however, the water molecules that remain are positioned close to the soil particles and adhere tightly.

As we learned when discussing osmotic pressure in Chapter 2, scientists quantify water potential in units of pressure, called megapascals (MPa). In a soil that is completely saturated with water, as illustrated in **Figure 3.1a**, the matric potential is 0 MPa. When a saturated soil drains under the force of gravity, the resulting matric potential is about −0.01 MPa. At this point, the force of gravity on the water molecules is equally opposed by the attractive force of soil particles on the water molecules. The maximum amount of water held by soil particles against the force of gravity

is called the **field capacity** of the soil. The field capacity, illustrated in Figure 3.1b, represents the maximum amount of water available to plants.

As water becomes less abundant, such as when plants take up some of the water from the soil, the matric potential values become more negative. Water always moves from areas of higher potential (less negative values) to areas of lower potential (more negative values). So, for plants to extract water from the soil, they must produce a water potential that is lower than that of the soil. As soils dry out, they hold the remaining water ever more tightly because a greater proportion of that water lies close to the surfaces of soil particles. Most crop plants can extract water from soils with water potentials down to about −1.5 MPa. At lower soil water potentials, these plants wilt, even though some water still remains in the soil, as shown in Figure 3.1c. Scientists refer to a water potential of −1.5 MPa as the **wilting point** of soil because this is the lowest water potential at which most plants can obtain water from the soil. There are, however, species of drought-adapted plants that can extract water when the water potential is less than −1.5 MPa.

Matric potential The potential energy generated by the attractive forces between water molecules and soil particles. *Also known as* **Matrix potential**.

Field capacity The maximum amount of water held by soil particles against the force of gravity.

Wilting point The water potential at which most plants can no longer retrieve water from the soil, which is about −1.5 MPa.

The amount of water in soil and its availability to plants depend on the physical structure of the soil. It also explains why the amount of water the soil can hold depends on the soil's surface area; the more surface area a volume of soil has, the more water it can hold. The surface area of soil depends on the sizes of the particles that comprise the soil. Soil particles include sand, silt, and clay, in addition to organic material from decomposing organisms. As shown in **Figure 3.2**, sand particles are the largest, with diameters exceeding 0.05 mm. Silt particles have diameters of 0.002 to 0.05 mm and clay particles are the smallest, with a diameter of less than 0.002 mm. Rarely is a soil composed of a single particle size. Instead, as illustrated in **Figure 3.3**, soils are typically composed of mixtures of different ratios of each particle size. For example, a soil composed of 40 percent sand, 40 percent silt, and 20 percent clay is classified as a *loam* soil. In contrast, a soil containing a higher proportion of silt and a lower proportion of sand is classified as a *silt loam* soil.

(a)

(b)

Figure 3.2 Soil particle size. (a) Soil particles are separated into three sizes: clay, silt, and sand. **(b)** Each soil particle attracts a surface film of water around it. The greater surface area of small clay particles holds a greater total amount of water than the much larger sand particles, which have a much smaller surface area relative to their volume.

Smaller particles have a larger surface area, relative to their volume, compared to larger particles. As a result, the total surface area of particles in a given volume of soil increases as particle size decreases. Therefore, soils with a high proportion of clay particles hold more water than soils with a high proportion of silt particles, which hold more water than soils with a high proportion of sand particles. Soils with a high proportion of sand particles tend to dry out because water quickly drains away, leaving many tiny pockets of air between the large sand particles. Clay soils represent the opposite extreme; each tiny particle of clay can attract a thin film of water on its surface, leaving little space for air pockets. Although clay soils retain a lot of water, clay particles can hold water molecules so tightly that it can be difficult for plants to extract the water from the soil.

In **Figure 3.4**, we can see how soil particle size affects the amount of water in the soil, measured in terms of the percent of the soil volume occupied by water. As we move from sand to silt to clay, there is an increase in field capacity. However, there is also an increase in the wilting point. The difference between the field capacity and the wilting point is the amount of water available to plants. Thus, even when precipitation is frequent, sandy soils cannot retain much of the water that enters the soil. At the other extreme, clay soils can retain a lot of water, but if precipitation is not frequent enough for the soil to reach its field capacity, most of the water in the soil will be unavailable. This means that soils high in sand or high in clay are both poor soils for growing many plants, including crops that humans rely on for food. Instead, soils that contain a mixture of clay, silt, and sand particles—such as loam—are some of the best soils for growing plants.

OSMOTIC PRESSURE AND WATER UPTAKE

In Chapter 2, we noted that osmotic forces cause water molecules to move from areas of low solute concentration to areas of high solute concentration. At the same time, ions and other solutes diffuse through water from regions of high solute concentration to regions of low solute concentration. In the case of a plant, if a root cell has a higher solute concentration than the soil water, osmotic forces can draw the water into the root. It is this osmotic potential in plant roots that causes water to enter the roots from the soil against the attractive forces of soil particles and the downward pull of gravity.

Without any other adaptations, we would expect the solute concentrations within the root cell and in the soil water to eventually come into equilibrium. At this point, the osmotic potentials of the root cell and its surroundings would be equal, and there would be no net movement of water into the plant. However, root cells possess two adaptations that prevent this equalization. First, semipermeable cell membranes prevent larger

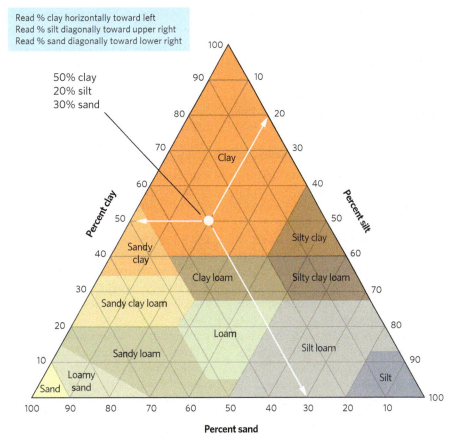

Read % clay horizontally toward left
Read % silt diagonally toward upper right
Read % sand diagonally toward lower right

50% clay
20% silt
30% sand

Figure 3.3 Combinations of soil particle sizes used to categorize soils. Most soils are composed of different percentages of sand, silt, and clay. Each name represents a category with a specific composition of the three particle sizes.

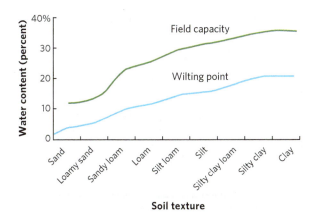

Figure 3.4 Water-holding capacity of different soils. Soils composed of different combinations of sand, silt, and clay differ in their water-holding capacity. Soils containing large amounts of sand have a low field capacity and a low wilting point. In contrast, soils containing large amounts of clay have a high field capacity and a high wilting point.

solute molecules from leaving the plant's root. Second, cell membranes can actively transport ions and small molecules against a concentration gradient into the root cells. These two adaptations maintain high solute concentrations inside the roots and allow strong osmotic forces to continue.

As noted earlier, plants growing in places with very negative water potentials typically have adaptations to help them extract water beyond −1.5 MPa. Plants living in deserts, for example, can lower the water potential of their roots to as much as −6 MPa, thereby overcoming soil water potentials down to −6 MPa. Plants living in very salty environments can also meet the challenge of extracting water from an environment that contains unusually high concentrations of solutes in the form of salt ions. In both situations, plants have evolved adaptations that allow them to increase the concentrations of amino acids, carbohydrates, or organic acids in their root cells. Maintaining these high concentrations of dissolved substances, however, comes at a high metabolic price for the plants because they must divert some of the energy that would normally be used for growth and use it instead to manufacture these additional organic compounds.

Plants lacking the appropriate adaptations grow poorly when exposed to salty conditions. For example, in the deserts of the American Southwest, a large amount of land is irrigated to grow crops, including cotton and orchard trees.

Most well water, however, contains small amounts of salt. As this well water is sprayed over the fields, it moves down into the soil and dissolves any salts. If the irrigation system uses large amounts of water, the water can move deeply into the soil and the salts can be flushed away. If irrigation uses smaller amounts of water—just enough to feed the plant roots—the water stays near the soil's surface. As much of this water is then taken up by plants or evaporates, the salt is left behind on the soil's surface.

With repeated irrigation events, the salt concentration of the soil continually increases. After many years, the soil can have such a high solute concentration that many crop plants cannot create a lower water potential than the soil, and therefore cannot obtain sufficient water. The process of repeated irrigation that causes increased soil salinity is known as **salinization.** High-salt soils occur across 831 million hectares (ha) of land and in 100 countries of the world. It is a particular problem throughout arid areas of the world where irrigation typically is limited to small amounts of water, which concentrates the salts at the soil surface. In 2014, researchers reviewed studies from across the globe and estimated that salinization caused $27 billion in lost crop production around the world. While it is possible to reverse the salinization of soil, it can be costly and take many years.

TRANSPIRATION AND THE COHESION–TENSION THEORY

We have seen how osmotic potential draws water from the soil into the cells of plant roots. How does that water get from the roots to the leaves? Plants conduct water to their leaves through tubular xylem elements, which are the empty remains of xylem cells in the cores of roots and stems, connected to form the equivalent of water pipes. The movement of water through these xylem cells depends on the *cohesion* of water molecules and differences in water potential between the leaves and roots.

The **cohesion** of water is the result of mutual attraction among water molecules. The attraction of hydrogen bonds causes a water molecule moving up the xylem of a plant to pull other water molecules with it. Cohesion of water helps the entire column of water move through the long vessels of a tall tree. The process, shown in **Figure 3.5**, begins when osmotic potential in the roots draws water from the soil into the plant and creates a **root pressure** that forces water into the xylem elements. However, this pressure is counteracted by gravity and the osmotic potential inside living root cells. Because of these two important counteracting forces, root pressure under the best circumstances can raise water to a height of no more than about 20 m, but the tallest trees can achieve heights of more than 100 m.

Fortunately for plants, leaves can also generate water potential as water evaporates from the surfaces of leaf cells inside the leaf and into the tiny air spaces that surround these cells. This evaporated water ultimately moves out of the leaf and into the air in a process known as **transpiration.** The column of water in a xylem element is continuous from the roots to the leaves, since it is held together by hydrogen bonds between the water molecules. Thus, low water potentials in leaves can literally draw a column of bonded water molecules upward through the xylem elements against the osmotic potential of the living root cells and the pull of gravity. The water potential is strong enough under most conditions to pull water up through the roots, xylem, and leaves. As a result, the water potential from transpiration creates a continuous gradient from leaf surfaces in contact with the atmosphere down to the surfaces of root hairs in contact with soil water. The movement of water is due to both water cohesion and water *tension* (which is another name for differences in water potential). This mechanism of water movement from roots to leaves due to water cohesion and water tension is therefore known as the **cohesion–tension theory.**

Based on the cohesion–tension theory of water transport in plants, very tall plants should have a more difficult time moving water up their stems because the movement of a tall column of water in the plant is being opposed by the force of gravity. Recent research estimates that this system limits plants to a maximum height of 130 m. In support of this prediction, the tallest tree that has been reliably measured was a 126-m Douglas fir tree.

Although transpiration generates a powerful force that moves water through a plant, when the soil reaches the wilting point, there is insufficient water moving into the roots to replace the water lost from the leaves. To prevent further water loss from the leaves under dry conditions, plants have various adaptations for reducing transpiration. Most of the cells on the exterior of a leaf are coated with a waxy cuticle that retards water loss. As a result, gas exchange between the atmosphere and the interior of the leaf primarily occurs through small openings on the surface of the leaves, called **stomata** (**Figure 3.6**). The stomata (stoma, singular) are the points of entry for CO_2 and the points of exit for water vapor

---

Salinization The process of repeated irrigation, which causes increased soil salinity.

Cohesion The mutual attraction among water molecules.

Root pressure When osmotic potential in the roots of a plant draws in water from the soil and forces it into the xylem elements.

Transpiration The process by which leaves can generate water potential as water evaporates from the surfaces of leaf cells into the air spaces within the leaves.

Cohesion–tension theory The mechanism of water movement from roots to leaves due to water cohesion and water tension.

Stomata Small openings on the surface of leaves, which serve as the points of entry for CO_2 and exit points for water vapor.

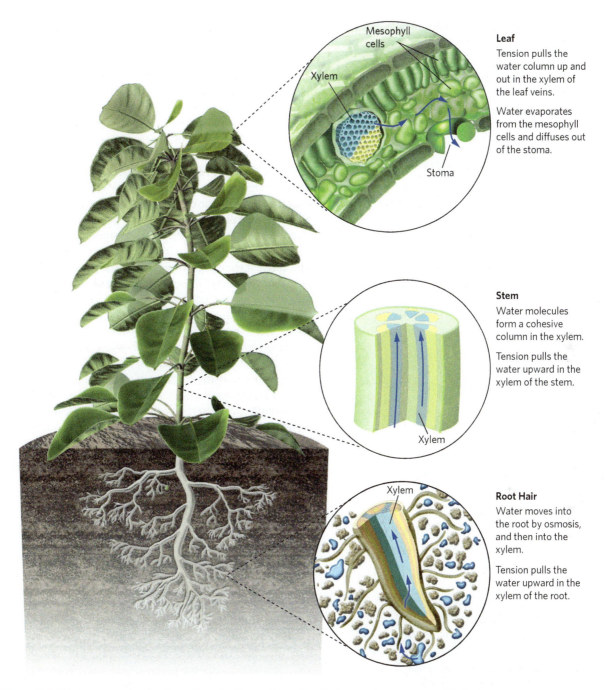

Leaf

Tension pulls the water column up and out in the xylem of the leaf veins.

Water evaporates from the mesophyll cells and diffuses out of the stoma.

Stem

Water molecules form a cohesive column in the xylem.

Tension pulls the water upward in the xylem of the stem.

Root Hair

Water moves into the root by osmosis, and then into the xylem.

Tension pulls the water upward in the xylem of the root.

Figure 3.5 Water movement in plants by cohesion and tension. Differences in water potential, also known as tension, cause water to move from the soil into the roots, from the roots into the stem, and from the stem into the leaves. The cohesion of water causes the water molecules to adhere to each other and to move as a single column of water up the xylem cells.

escaping to the atmosphere by transpiration. When plants experience a scarcity of water, they can reduce water loss to the atmosphere by closing their stomata. As leaf water potential weakens, the *guard cells* that border a stoma collapse slightly, causing the guard cells to press together and close the stoma. Although closing the stomata provides the important benefit of reducing water loss, it comes at the cost of preventing CO_2, needed for photosynthesis, from entering the leaf. As we will soon see, plants living in hot, dry environments have evolved additional adaptations for dealing with this undesirable side effect.

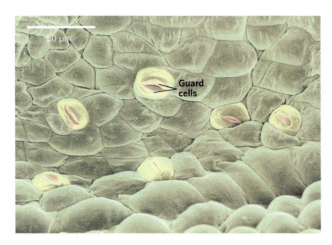

Figure 3.6 Stomata. Stomata are pores in the surfaces of leaves, each bordered by two guard cells. Under conditions of low water availability, the guard cells close the opening and prevent the loss of water from the leaves. Callista Images/Newscom/Cultura/.

CONCEPT CHECK

1. Explain the relationship between soil particle size and the field capacity of soil.
2. Why is the availability of plant water highest in soils with intermediate particle sizes?
3. How can we be sure that root pressure is not sufficient to explain the movement of water in trees?

3.2 Sunlight provides the energy for photosynthesis

Whether in water or on land, energy from the Sun is essential for the existence of most life on Earth. To understand how this energy is captured, we need to examine the energy that is available, the energy that is absorbed, and how this energy is converted into a usable form by photosynthesis. Plants have evolved several adaptive approaches for conducting photosynthesis in terrestrial environments; these adaptations coincide with the environmental conditions of different regions of the world.

AVAILABLE AND ABSORBED SOLAR ENERGY

The energy from the Sun, known as **electromagnetic radiation,** is packaged in small particle-like units called photons. The energy of photons is related positively to their frequency and inversely to their wavelength; the highest energy photons have the highest frequency and the shortest wavelengths. Wavelengths are expressed in units of nanometers (nm), which is one-billionth of a meter. The different wavelengths of light can be separated using a prism. As you can see in **Figure 3.7**, infrared radiation has long wavelengths, which we know contain lower energy. Short wavelengths, such as ultraviolet radiation, contain higher energy. Between the two extremes of infrared and ultraviolet radiation are wavelengths collectively known as **visible light,** which are visible to the human eye. Visible light represents only a small part of the spectrum of electromagnetic radiation.

The visible portion of the spectrum includes the **photosynthetically active region,** which consists of wavelengths of light that are suitable for photosynthesis. This range of wavelengths falls between about 400 nm

---

Electromagnetic radiation Energy from the Sun, packaged in small particle-like units called photons.

Visible light Wavelengths in between infrared and ultraviolet radiation that are visible to the human eye.

Photosynthetically active region Wavelengths of light that are suitable for photosynthesis.

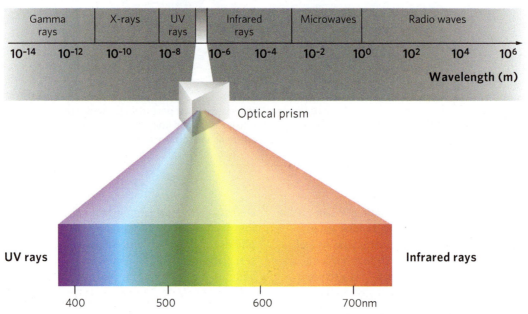

Figure 3.7 The wavelengths of the Sun's energy. The Sun emits electromagnetic radiation that spans a wide range of energy and wavelengths.

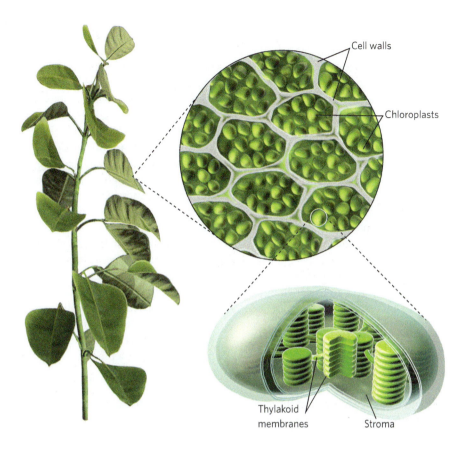

Cell walls

Chloroplasts

Thylakoid membranes

Stroma

Figure 3.8 Chloroplasts. Chloroplasts are the site of photosynthesis. The chloroplasts contain stacks of membranes called thylakoids surrounded by a fluid-filled space known as the stroma.

(violet) and 700 nm (red). Plants, algae, and some bacteria absorb these wavelengths and assimilate their energy by photosynthesis. These also happen to be the wavelengths of greatest intensity at the Earth's surface.

Eukaryotic photosynthetic organisms contain specialized cell organelles known as **chloroplasts.** As you can see in **Figure 3.8**, chloroplasts contain stacks of membranes known as *thylakoids* and a fluid-filled space surrounding the thylakoids called the *stroma*. Embedded within the thylakoid membranes are several kinds of pigments that absorb solar radiation, including *chlorophylls* and *carotenoids*. The patterns of absorption for several of these pigments are shown in **Figure 3.9**.

Chlorophylls, which are primarily responsible for capturing light energy for photosynthesis, absorb red and violet light. Chlorophylls reflect green and blue light, which is why leaves on most plants are predominantly green in color. Over the past 60 years, scientists identified four types of chlorophyll that differ in the wavelengths that they absorb: chlorophyll *a, b, c,* and *d.* Chlorophyll *a* is found in all photosynthesizing organisms and is responsible for the actual steps in photosynthesis. The other types of chlorophyll act as *accessory pigments,* meaning that they capture light energy and then pass it on to chlorophyll *a.* Recently, scientists reported the discovery of a fifth type of chlorophyll that they named *chlorophyll f.* This pigment, discovered in algae that live in shallow rocky pools on the coast of Australia, absorbs light at longer wavelengths than the others.

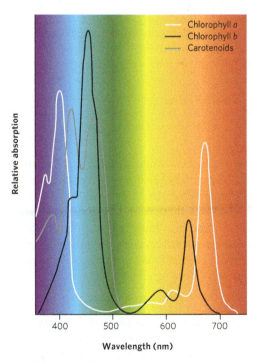

Chlorophyll *a*
Chlorophyll *b*
Carotenoids

Relative absorption

Wavelength (nm)

400 500 600 700

Figure 3.9 Light-absorbing pigments. Photosynthetic organisms contain a variety of photosynthetic pigments including chlorophyll *a* and *b* and carotenoids that serve as accessory pigments that capture the Sun's energy and pass it to the chlorophylls.

Chloroplasts Specialized cell organelles found in photosynthetic organisms.

Carotenoids are also accessory pigments that absorb primarily blue and green light, thereby complementing the absorption spectrum of chlorophyll. Carotenoids, the pigments that give carrots their orange color, reflect yellow and orange light. By containing carotenoids and several types of chlorophyll, producers can absorb a wider range of solar energy and use it to power photosynthesis.

PHOTOSYNTHESIS

In this section, we review the process of photosynthesis and the different pathways that have evolved under different ecological conditions. As you may recall from your introductory biology course, the process of photosynthesis involves photosynthetic pigments absorbing energy from photons of light. This energy is then converted into chemical energy stored in the high-energy bonds of organic compounds. In its simplest form, photosynthesis is the process of combining CO_2, H_2O, and solar energy to produce glucose ($C_6H_{12}O_6$) and oxygen:

$$6\,CO_2 + 6\,H_2O + photons \rightarrow C_6H_{12}O_6 + 6\,O_2$$

This simple equation summarizes a long chain of complex chemical reactions that takes place in two parts: *light reactions* and the *Calvin cycle.*

Light reactions depend on light energy from the Sun and include a series of events from the absorption of light to the production of high-energy compounds and oxygen (O_2). These high-energy compounds are *ATP* (adenosine triphosphate) and *NADPH* (nicotinamide adenine dinucleotide phosphate). The cell uses the energy in these compounds to convert CO_2 into glucose in a process known as the Calvin cycle, which takes place in the stroma of the chloroplast. Over evolutionary time, three distinct biochemical pathways have evolved for the Calvin cycle: *C3, C4,* and *CAM photosynthesis.* As we will soon see, each of these pathways is suited to particular ecological conditions.

C_3 *Photosynthesis*

For most plants, photosynthesis begins with a reaction between CO_2 and a five-carbon sugar known as *RuBP* (ribulose bisphosphate) to produce a six-carbon compound. This reaction is catalyzed by the enzyme **RuBP carboxylase-oxidase** (also known as **Rubisco**). Once the six-carbon compound is created, it immediately splits into two molecules of a three-carbon sugar called *G3P* (glyceraldehyde 3-phosphate). We can represent this process as

$$CO_2 + RuBP \rightarrow 2\,G3P$$

This photosynthetic pathway, in which CO_2 is initially assimilated into the three-carbon compound (G3P), is known as **C_3 photosynthesis.** The vast majority of plants on Earth use C_3 photosynthesis. In most plants, this process occurs in mesophyll cells in leaves.

One of the challenges for plants using C_3 photosynthesis is that Rubisco, the enzyme responsible

for joining CO_2 and RuBP, has a low affinity for CO_2. Consequently, carbon assimilation using Rubisco is quite inefficient at the low concentrations of CO_2 found in the mesophyll cells of plant leaves. To achieve high rates of carbon assimilation, plants must therefore pack their mesophyll cells with large amounts of the enzyme Rubisco. In some plant species, Rubisco can constitute up to 30 percent of the dry weight of leaf tissue.

Rubisco's low affinity for CO_2 is not the only problem plants face. Under certain conditions—such as when temperatures are high, concentrations of O_2 are high, and CO_2 concentrations are low—Rubisco preferentially binds to O_2 rather than CO_2. This occurs when hot, dry conditions cause the leaf stomata to close, preventing new CO_2 from entering to replenish the CO_2 that has been consumed by the Calvin cycle. Closed stomata also prevent O_2 produced by the light reaction from leaving the leaf. Consequently, hot and dry conditions lead to changes in CO_2 and O_2 concentrations that cause Rubisco to preferentially bind to O_2 rather than CO_2. When the Rubisco enzyme binds to O_2, it initiates a series of reactions that reverses the outcome of photosynthesis in a process known as **photorespiration:**

$$2\,G3P \rightarrow RuBP + CO_2$$

The reverse reaction consumes energy and O_2, and produces CO_2. This reverse reaction in plants is called photorespiration because it resembles the process of respiration. In short, what is accomplished by photosynthesis when Rubisco binds to CO_2 is undone by photorespiration when Rubisco binds to O_2.

The problem of photorespiration is caused, in part, by closed stomata. This leads to high O_2 concentrations and low CO_2 concentrations in the leaves. One potential solution is to keep the leaf stomata open. This would permit free gas exchange, allowing CO_2 to enter the leaves and O_2 to exit the leaves. This strategy works as long as plants can replace the water they also lose by transpiration when the stomata are open. However, this solution may be too costly in hot, dry environments where water is scarce. When such costs are too high, natural selection will favor traits that can either reduce the demand for water or reduce the loss of water.

C_4 *Photosynthesis*

We have seen that hot and dry conditions cause stomata to close, which results in a decrease of CO_2, an increase of O_2, and an increase in photorespiration. To address this problem, many herbaceous plants, particularly grasses growing in hot climates, have evolved a modified pathway

---

RuBP carboxylase-oxidase An enzyme involved in photosynthesis that catalyzes the reaction of RuBP and CO_2 to form two molecules of glyceraldehyde 3-phosphate (G3P). *Also known as* **rubisco**.

C_3 photosynthesis The most common photosynthetic pathway, in which CO_2 is initially assimilated into a three-carbon compound, glyceraldehyde 3-phosphate (G3P).

Photorespiration The oxidation of carbohydrates to CO_2 and H_2O by rubisco, which reverses the light reactions of photosynthesis.

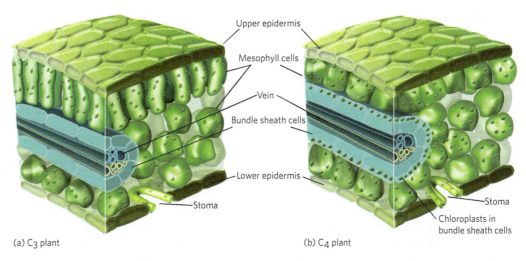

Upper epidermis

Mesophyll cells

Vein

Bundle sheath cells

Lower epidermis

Stoma

Stoma

Chloroplasts in bundle sheath cells

(a) C₃ plant

(b) C₄ plant

Figure 3.10 Leaf cell arrangements for C_3 versus C_4 plants. (a) C_3 plants conduct all steps of photosynthesis in the chloroplasts of the mesophyll cells. **(b)** In C_4 plants, the initial step of carbon assimilation and the light reactions take place in the chloroplasts in mesophyll cells. However, the CO_2 that has been assimilated is then transported to the bundle sheath cells, where the Calvin cycle occurs.

known as *C4 photosynthesis*. Biologists call this modification **C4 photosynthesis** because the first step joins CO_2 with a three-carbon molecule called *PEP* (phosphoenol pyruvate) to produce a four-carbon molecule called *OAA* (oxaloacetic acid):

$$CO_2 + PEP \rightarrow OAA$$

This reaction is the key difference between C_3 and C_4 photosynthesis. It is catalyzed by the enzyme PEP carboxylase, which has a higher affinity for CO_2 than Rubisco. You can see an overview of this process in **Figure 3.10**.

This additional assimilation step occurs in the mesophyll cells of the leaf, which is also the site of the light reaction. In most C_4 plants, however, the Calvin cycle takes place in the bundle sheath cells that surround the leaf veins. This means that the plant must move the CO_2 that has been assimilated in the mesophyll cells over to the bundle sheath cells. To do this, the plant converts OAA into malic acid, which then diffuses into the bundle sheath cells, where another enzyme breaks it down to produce CO_2 and *pyruvate*, a three-carbon compound. In the bundle sheath cells, the chloroplasts use the CO_2 that is brought over from the mesophyll cells for the Calvin cycle. To complete the cycle, pyruvate moves back into the mesophyll cells, where it is converted back to PEP to be used once again. In short, the C_4 pathway is an adaptation that adds a step to the initial assimilation of CO_2 to make it more efficient when CO_2 is present at low concentrations.

This strategy solves the problem of photorespiration by creating concentrations of CO_2 in the bundle sheath cells that are three to eight times higher than is available to C_3 plants. It helps that there are multiple mesophyll cells for each bundle sheath cell, thereby providing a large number of sites for CO_2 assimilation that can provide CO_2 to each

bundle sheath cell. At this higher CO_2 concentration, the Calvin cycle operates more efficiently. In addition, because the enzyme PEP carboxylase has a high affinity for CO_2, it can bind CO_2 at a lower concentration in the cell. This pathway allows the stomata to remain partially or completely closed for longer periods of time, which reduces water loss. However, C_4 photosynthesis has two disadvantages that reduce its efficiency: Less leaf tissue is devoted to photosynthesis, and some of the energy produced by the light reactions is used in the initial C_4 carbon assimilation step.

Whereas C_3 plants are favored in cool, moist climates, C_4 plants are favored in climates that are hot or have less abundant water. When water is abundant, the C_4 pathway does not present a distinct advantage because the costs of the C_3 pathway are relatively low. When water is less abundant, however, the C_4 pathway has an advantage.

C_4 photosynthesis has evolved at least 45 times during the past 30 million years in at least 19 different families of angiosperms. However, only about 4 percent of all plant species on Earth are C_4 plants and they are primarily found in two types of nonwoody plants: *grasses* and *sedges*. C_4 plants dominate tropical and subtropical grasslands and are important components of the plant communities found in arid regions of the world, including the Great Plains of North America. Plants using the C_4 pathway also include many of our most important crop plants such as corn (maize), sorghum, and sugarcane. These plants are highly productive during hot growing seasons. They also account for 20 to 30 percent of all CO_2 fixation and 30 percent of all grain production.

C₄ photosynthesis A photosynthetic pathway in which CO_2 is initially assimilated into a four-carbon compound, oxaloacetic acid (OAA).

As a result, C_4 plants can play substantial roles in the ecosystems in which they live.

CAM Photosynthesis

Certain succulent plants that inhabit water-stressed environments, for example, cacti and pineapple plants, use the same biochemical pathways as C_4 plants. However, rather than separating CO_2 assimilation and the Calvin cycle spatially, using mesophyll and bundle sheath cells, these succulent plants separate the steps in time. Plants that follow this pathway, known as **crassulacean acid metabolism,** or **CAM,** open their stomata for gas exchange during the cool night, when transpiration is minimal, and then conduct photosynthesis during the hot day. The discovery of this arrangement was first made in plants of the family Crassulaceae (the stonecrop family)—which includes the jade plant (*Crassula ovata*).

Like the C_4 plants, CAM plants use an initial step of assimilating CO_2 into OAA, which is then converted to malic acid and stored at high concentrations in vacuoles within the mesophyll cells of the leaf. The enzyme responsible for the assimilation of CO_2 works best at the cool temperatures that occur at night, when the stomata are open. During the day, the stomata close, and the stored organic acids are gradually broken down to release CO_2 to the Calvin cycle. A different enzyme with a higher optimum temperature, geared to promote daytime photosynthesis, regulates the regeneration of PEP from pyruvate following the release of CO_2. Because CAM plants can conduct gas exchange during the night when the air is cooler and more humid, a plant using CAM photosynthesis reduces its water loss. Thus, CAM photosynthesis is an adaptation that results in extremely high water use efficiencies and enables plants that use this pathway to live in very hot and dry regions of the world. While CAM allows photosynthesis to occur in water-limited conditions, it happens at a relatively slow rate. As a result, CAM plants typically grow much more slowly than C_3 or C_4 plants. **Figure 3.11** compares the three alternative photosynthetic pathways.

Plants possessing the C_3 pathway are better adapted to cool, wet conditions, whereas plants possessing the C_4 and CAM pathways are better adapted to warm and arid conditions. However, there is not a clear distinction in where these different types of plants grow. For example, regions that are hot and dry during the summer can be cool and moist during the winter and spring. As a result, such an area can be dominated by short-lived C_4 plants in the summer and short-lived C_3 plants in the winter and spring. Moreover, different photosynthetic pathways represent just one of several adaptations that plants have evolved to handle hot temperatures and a scarcity of water. As we will

Crassulacean acid metabolism (CAM) A photosynthetic pathway in which the initial assimilation of carbon into a four-carbon compound occurs at night.

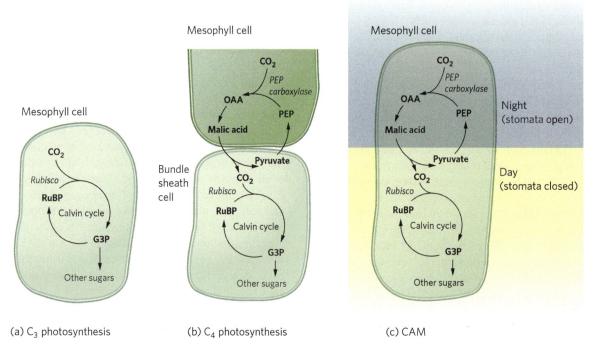

(a) C_3 photosynthesis (b) C_4 photosynthesis (c) CAM

Figure 3.11 Alternative photosynthetic pathways. (a) C_3 plants conduct photosynthesis in mesophyll cells during the day. **(b)** C_4 plants separate the steps of photosynthesis in space. The initial CO_2 assimilation step occurs in the mesophyll cells and the remaining steps occur in the bundle sheath cells. **(c)** CAM plants separate the steps of photosynthesis in time. Assimilation of CO_2 happens at night and the remaining steps occur during the day.

discuss in the next section, many plants also have evolved structural adaptations.

STRUCTURAL ADAPTATIONS TO WATER STRESS

Heat- and drought-adapted plants have anatomical and physiological modifications that improve water uptake and retention, reduce transpiration, and reduce the buildup of heat in their tissues. These adaptations include roots that can take advantage of different water sources, resistance to heat buildup, vein configurations that protect against air blockages, and morphological adaptations in leaves.

Plants living in arid regions often have very shallow or very deep roots, which represent two different adaptive strategies. Plants with very shallow roots, such as many species of cacti, are able to rapidly uptake water from brief rain events in which the rain does not penetrate very far into the soil. Cacti often pair this adaptation with thick, succulent tissues that can hold a great deal of water whenever it becomes available. In contrast to the cacti, some perennial shrubs, such as mesquite, have roots that can extend several meters down into the soil, allowing the shrubs to access water that is very deep below the surface.

Another strategy to combat the effects of heat and drought is to protect plant surfaces from direct sunlight with leaf resins, waxy cuticles, spines, and hairs. You can see many of these adaptations illustrated in **Figure 3.12**. Resins help seal off much of the leaf from water loss, whereas waxy cuticles help make the surfaces of the plant more resistant to losing water. Spines and hairs produce a boundary layer of still air that traps moisture and reduces evaporation. In some cases, to reduce water loss, the stomata are recessed into deep pits that contain hairs. Because thick boundary layers can also retard heat loss, hair-covered surfaces are also prevalent in arid environments that are cool. Long spines can also serve as structures that dissipate excess heat away from the plant.

Some adaptations reduce heat buildup. Plants can reduce their heat loads by producing finely subdivided leaves with a large ratio of edge to surface area. This large amount of leaf edge breaks up the boundary layer surrounding the leaf, which helps dissipate heat from the leaf. Some desert plants have no leaves at all. Many cacti rely entirely on their stems for photosynthesis; their leaves are modified into thorns for protection.

For a long time, scientists observed that plants in habitats with high temperatures generally had smaller leaves than plants in habitats with abundant water. Scientists hypothesized that the smaller leaves represented an adaptation that permitted the dissipation of heat. However, small leaves are not only found in plants from hot, dry places, but also in plants from cool, dry places. Smaller leaves also contain a higher density of large veins that transport and distribute water to the many small veins. In 2011, an international team of scientists discovered that having small leaves with a high density of large veins is actually an adaptation to overcome the problem of air

(a) Silverleaf sunray

(b) Mesquite

(c) Oleander

Thick waxy cuticle

Multiple epidermal layers

Pit with hairs

Stomata

Figure 3.12 Structural adaptations of plants against heat and drought. (a) Spines and hairs on leaf surfaces, such as in the silverleaf sunray plant, shade the plant from direct sunlight and reduce evaporation. **(b)** Finely divided leaves, as found in the mesquite plant, help dissipate any built-up heat. **(c)** Stomata that are recessed into deep pits containing hairs, as in the oleander plant, slow the evaporation rate of water out of the leaf.

bubbles, known as *embolisms,* which can form in large veins. Under severe drought stress, air can move into the stomata and up into the large veins, causing a bubble to form in the vein and blocking water movement. A large density of veins allows the plant to get around this problem by sending water through adjacent veins. This suggests that the small leaf size is actually an adaptation to scarce water in both hot and cool environments, and the fact that small leaves can better dissipate heat in the hot environments may be a valuable secondary benefit. Collectively, these structural adaptations help make it possible for plants to live in regions of the world that have high temperatures or scarce water.

CONCEPT CHECK

1. Explain how light serves as the ultimate source of energy for a meat-eating animal.
2. Why is C_3 photosynthesis inefficient when the concentration of CO_2 in a leaf is low?
3. Explain how plants use structural adaptations to reduce water loss.

3.3 Terrestrial environments pose a challenge for animals to balance water, salt, and nitrogen

While plants have a variety of adaptations for conducting photosynthesis on land, terrestrial life presents many additional challenges, including the need to maintain a balance of water, salt, and nitrogen. This balance is known as *homeostasis.* **Homeostasis** is an organism's ability to maintain constant internal conditions in the face of a varying external environment. All organisms exhibit homeostasis to some degree—for example, the balance of water and salt, or the regulation of body temperature.

Although the occurrence and effectiveness of homeostatic mechanisms vary, all homeostatic systems exhibit **negative feedbacks,** meaning that when the system deviates from its desired state, or set point, internal response mechanisms act to restore that desired state. An example that might be familiar to you is the presence of negative feedbacks in the regulation of body temperature. **Figure 3.13** shows how this works in mammals. The hypothalamus—a gland in the brain—determines whether the body temperature is above or below the desired set point, which differs among species of mammals. If body temperature drops below this set point, the hypothalamus uses neural and hormonal signals to trigger the body to generate more heat. When the body temperature reaches the set point, the hypothalamus triggers the body to stop producing heat. If the body temperature greatly exceeds the set point, the hypothalamus triggers the body to start using cooling mechanisms, including sweating and panting.

In Chapter 2, we saw that to maintain the proper amounts of water and dissolved substances in their bodies, aquatic organisms must balance their gains and losses. This is equally true for organisms on land. Organisms often take in water with a solute concentration that differs from that of

Homeostasis An organism's ability to maintain constant internal conditions in the face of a varying external environment.

Negative feedbacks The action of internal response mechanisms that restores a system to a desired state, or set point, when the system deviates from that state.

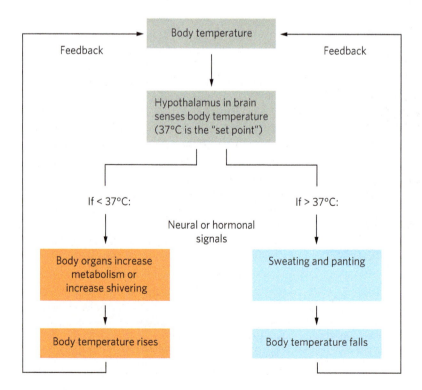

Figure 3.13 Negative feedback for regulating body temperature. In mammals, the hypothalamus acts like a thermostat. When the desired set point differs from the current body temperature, the hypothalamus signals the body to return its temperature to the desired set point.

their bodies, so they must either acquire additional solutes to make up the deficit or rid themselves of excess solutes. If they do not balance the concentration of solutes, many of their physiological functions will not operate correctly. When water evaporates from the surfaces of terrestrial organisms into the atmosphere, solutes are left behind and their concentration in the body tends to increase. Under such circumstances, organisms must excrete excess salts to maintain the proper concentrations in their bodies.

WATER AND SALT BALANCE IN ANIMALS

Water is as important to terrestrial animals as it is to terrestrial plants. Terrestrial animals, with their internalized gas exchange surfaces, are less vulnerable to respiratory water loss than plants. Moreover, because terrestrial animals are not immersed continuously in water, they have little trouble retaining ions. They acquire the mineral ions they need in the water they drink and the food they eat, and they use urine to eliminate excess salts in their bodies. Where fresh water abounds, animals can drink large quantities of water to flush out salts that would otherwise accumulate in the body. Where water is scarce, however, animals make use of adaptations to conserve water.

As one would expect, desert animals have evolved a number of adaptations in response to water scarcity. Kangaroo rats, for example, are a group of small rodents that live in the dry regions of North America (**Figure 3.14**). Behavioral and physiological adaptations allow them to live in these places. The kangaroo rat conserves water by hunting for food during the nights and staying in a cool and humid underground burrow during the hot days—a valuable behavioral adaptation. For both the kangaroo rat and the camel, the kidneys provide an additional physiological adaptation to extreme heat and water scarcity. In all mammals, the kidneys are responsible for removing salts and nitrogenous wastes from the blood. These solutes are dissolved in water, but because water is valuable, a

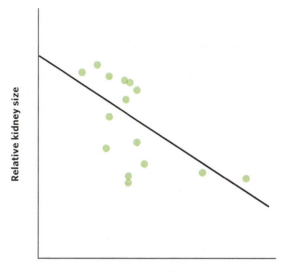

Figure 3.15 Kidney size in South American rodents. Species of South American rodents that live in relatively drier habitats have relatively larger, more efficient kidneys. [After G. B. Diaz et al., Renal morphology, phylogenetic history and desert adaptation of South American hystricognath rodents, *Functional Ecology* 20 (2006): 609–620.]

structure known as the loop of Henle helps recover some of it before the mixture is excreted. Kangaroo rats and camels possess an unusually long loop of Henle, which provides a greater length over which a kidney can recover water from urine prior to excretion. Whereas human kidneys concentrate most solutes in urine to about four times the blood concentration level, the kidneys of kangaroo rats produce urine with solute concentrations as much as 14 times greater.

Although the loop of Henle plays a large role in concentrating the urine of mammals, additional adaptations help conserve water. The efficiency of water use is best determined by the overall size of the kidney relative to the size of a mammal's body. One way to assess the potential importance of relative kidney size as an adaptation for conserving water is to examine how relative kidney size differs among closely related species of rodents that live in habitats with differing amounts of precipitation. Figure 3.15 shows this relationship in a group of South American rodents. Rodents living in habitats with the lowest amount of precipitation have relatively large kidneys, whereas rodents living in habitats with the greatest amount of precipitation have relatively small kidneys.

Because sodium and chloride ions participate in the water conservation mechanism of kidneys, the kidneys do not excrete these ions efficiently. Many animals that lack ready access to fresh water have specialized salt-secreting organs that work on a different principle from the kidneys. In fact, these secretory organs are more like the salt glands of mangrove plants. For example, salt glands of birds and reptiles, which are particularly well developed in marine species, are actually modified tear glands located in the orbit

Figure 3.14 Animal adaptations to conserve water. The Ord's kangaroo rat (*Dipodomys ordii*) lives in dry, desert environments in North America. These environments favor adaptations that allow the kangaroo rat to conserve water, including feeding at night and having large, efficient kidneys. Jim Zipp/Science Source

of the eye that are capable of secreting a concentrated salt solution. These adaptations help animals to balance their salt and water budget on land (**Figure 3.16**).

Figure 3.16 Animal adaptations to expel salt. Many animals that live on land and forage in salt water have evolved specialized glands to expel excess salt. The white crust on top of the head of this marine iguana (*Amblyrhynchus cristatus*) from the Galápagos Islands is salt that was produced in these glands and expelled through the nostrils. Michael Zysman/Shutterstock.

WATER AND NITROGEN BALANCE IN ANIMALS

Most carnivores, whether they eat crustaceans, fish, insects, or mammals, consume excess nitrogen from the proteins and nucleic acids in their diet. When these compounds are metabolized, the excess nitrogen must be eliminated from the body. Most aquatic animals excrete excess nitrogen in the form of ammonia (NH_3) because it is a simple metabolic by-product of nitrogen metabolism. Although ammonia is mildly poisonous to tissues, aquatic animals eliminate it rapidly either in copious dilute urine or directly across the body surface before it reaches a dangerous concentration within the body.

Terrestrial animals, however, rarely have access to large quantities of water to excrete excess nitrogen. Instead, they excrete ammonia in forms that are less toxic than ammonia. This allows them to accumulate higher concentrations of metabolic by-products in their blood and urine without any harmful side effects. Mammals excrete nitrogen in the form of urea ($CO(NH_2)_2$), which is the same substance that we learned sharks produce and retain to achieve osmotic balance in marine environments. Because urea dissolves in water, excreting urea still causes the loss of some water, although the amount depends on the concentrating power of the kidneys. Birds and reptiles excrete

ANALYZING ECOLOGY

Understanding the Different Types of Variables

When we think about testing hypotheses in ecology, we often collect and analyze data to determine whether the hypothesis is supported or refuted. However, before we collect and analyze the data, we have to consider what types of data we are collecting.

The first distinction is between *independent variables* and *dependent variables*. **Independent variables** are factors that are presumed to cause other variables to change. **Dependent variables** are factors that are being changed. For example, in the examination of the relative kidney size of South American rodents (Figure 3.15), the hypothesis was that differences in precipitation have caused the evolution of different kidney sizes. In this case, precipitation is the independent variable and kidney size is the dependent variable.

A second distinction is whether a variable is continuous or categorical. **Continuous variables** can take on any numeric value, including values that are not whole numbers. In the case of the rodents, for example, precipitation represents a continuous variable because it can take on any value (Figure 3.16). Other continuous variables include temperature, salinity, and light. In contrast, **categorical variables**, also known as **nominal variables**, fall into distinct groupings or categories. For example, if we wanted to know how the solute concentration of a kangaroo rat's urine was affected by its diet, we might provide kangaroo rats with one of three different species of seeds and then measure their urine concentrations. In this case, the diet is a categorical variable because the different diets fall into three distinct categories. Other categorical variables include sex (e.g., males and females) and species (e.g., dromedary camels, Bactrian camels, and guanacos).

As we will see in later chapters, these distinctions between dependent versus independent variables and continuous versus categorical variables are important in the statistical analysis of ecological data.

YOUR TURN If you were to conduct an experiment that examined how the water holding capacity differed among soil types, as in Figure 3.4, which would be the independent variable and which would be the dependent variable?

If you were only to compare the water holding capacity of soils that contained 100 percent sand, 100 percent silt, or 100 percent clay, would this soil variable be considered continuous or categorical?

Independent variable A factor that causes other variables to change.

Dependent variable A factor that is being changed.

Continuous variable A variable that can take on any numeric value, including values that are not whole numbers.

Categorical variable A variable that falls into a distinct category or grouping. *Also known as* **Nominal variable.**

nitrogen in the form of uric acid ($C_5H_4N_4O_3$). Uric acid requires even less water and crystallizes out of solution. As a result, it can be excreted as a highly concentrated paste in the urine.

Although excreting urea and uric acid conserves water, it is costly in terms of the energy needed to form these compounds. One method that scientists use to quantify energy costs is to determine the amount of organic carbon consumed to produce the energy necessary for excretion. For example, for each atom of nitrogen excreted in the form of ammonia, no atoms of organic carbon are used. In contrast, to excrete nitrogen in the form of urea requires 0.5 atoms of organic carbon, and uric acid uses 1.25 atoms of organic carbon.

CONCEPT CHECK

1. Why does homeostasis require negative feedbacks?
2. Describe the costs and benefits associated with the different nitrogen products excreted by fish, mammals, and birds.
3. Contrast the concepts of dependent and independent variables.

3.4 Adaptations to different temperatures allow terrestrial life to exist around the planet

On Earth, land temperatures can reach as high as 58°C in northern Africa and as low as −89°C in Antarctica. These extremes can limit the occurrence of life. To understand how organisms are affected by temperature and the adaptations that have evolved to deal with different temperatures, we first need to examine how they gain and lose heat.

SOURCES OF HEAT GAIN AND LOSS

Because body temperature impacts physiological functions, organisms must manage heat gain and heat loss carefully. The ultimate source of heat at the surface of Earth is sunlight, most of which is absorbed by water, soil, plants, and animals and converted to heat. Objects and organisms continuously exchange heat with their surroundings. When the temperature of the environment exceeds the temperature of an organism, the organism gains heat and becomes warmer. When the environment is cooler than the organism, the organism loses heat to the environment and cools. As illustrated in **Figure 3.17**, this exchange of heat can occur through four processes: *radiation, conduction, convection,* and *evaporation.*

Radiation

Radiation is the emission of electromagnetic energy by a surface. The primary source of radiation in the environment is the Sun. As objects in the landscape are warmed by solar radiation, they emit more lower-energy radiation in the form of infrared light. The temperature of the

radiating surface determines how rapidly an object loses energy by radiation to colder parts of the environment. We measure in units of Kelvins (K), also known as absolute temperature, where 0°C = 273°K. The amount of heat radiation increases with the *fourth* power of absolute temperature. For example, we can consider the heat radiation of two small animals such as a mouse and a lizard. If the mammal has a skin temperature of 37°C (310°K) and the lizard has a skin temperature of 17°C (290°K), the difference in heat radiation between the mammal and the lizard is

$$310^4 \div 290^4 = 130\%$$

This means that by having a 20°C higher body temperature, the mammal radiates 30 percent more heat than the lizard.

The relatively high amount of heat radiation produced by animals with a higher body temperature than their external environment has been used by ecologists in a variety of research endeavors, including estimates of population sizes. When biologists need to count the number of moose living in remote regions of Alaska, for example, planes equipped with infrared cameras fly over these regions in the winter and the warm bodies of the moose stand out as a bright signal of infrared radiation against their cold, snowy background. Similar efforts have also been conducted to detect warm birds against a cold background environment, such as the penguins shown in **Figure 3.18**.

Conduction

Conduction is the transfer of the kinetic energy of heat between substances that are in contact with one another. For example, lizards often lie flat on hot rocks to warm their bodies by conduction. Water, because it is so much denser than air, conducts heat more than 20 times faster. As a result, you would lose body heat much faster if you stood in 10°C water than if you stood in 10°C air.

The rate at which heat moves by conduction between an organism and its surroundings depends on three factors: its surface area, its resistance to heat transfer, and the temperature difference between the organism and its surroundings. An organism's surface area helps determine its rate of heat conduction because a greater amount of exposed surface allows a greater surface for the energy transfer to take place. This is why many animals curl up in a ball to lessen the amount of exposed surface when they are trying to stay warm on a cold night.

An organism's resistance to heat transfer is just another way of saying how much insulation the organism has. Thick layers of fat, fur, or feathers have a high resistance to heat transfer and therefore slow the rate of heat

Radiation The emission of electromagnetic energy by a surface.

Conduction The transfer of the kinetic energy of heat between substances that are in contact with one another.

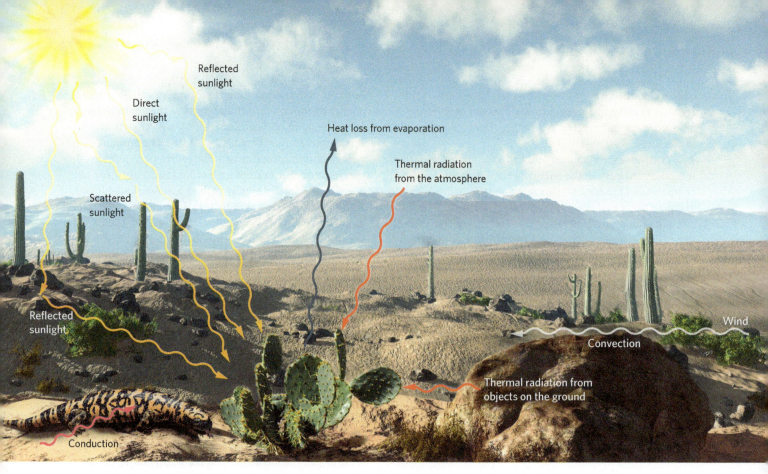

Reflected
sunlight

Direct
sunlight

Scattered
sunlight

Reflected
sunlight

Heat loss from evaporation

Thermal radiation
from the atmosphere

Wind

Convection

Thermal radiation from
objects on the ground

Conduction

Figure 3.17 Sources of heat gain and loss. The Sun is the original source of almost all heat. The heat from the Sun is exchanged among objects across the landscape. In the case of the cactus, heat is gained by direct sunlight, scattered sunlight, and reflected sunlight. Heat can be lost by evaporation of water vapor to the atmosphere. Heat can also be gained or lost by radiation from surrounding objects such as rocks, by conduction where the cactus comes into contact with the soil, and by convection as winds move hot or cold air over the surface of the cactus and disrupt its boundary layer.

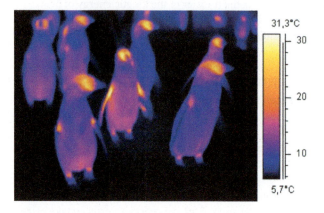

31,3°C

30

20

10

5,7°C

Figure 3.18. Infrared images. Thermal cameras can detect warm animals radiating heat against a cold background, such as these king penguins (*Aptenodytes patagonicus*) in the Blijdrop Zoo, Rotterdam, Netherlands. Photo by Arno Vlooswijk/TService/Science Source.

loss due to conductance. Indeed, this is why you choose to wear insulated boots rather than walking barefoot in snow.

Finally, the rate of heat loss is higher when there are large differences between the temperature of the organism and that of the environment. As we will see in Chapter 4, this feature of conductance is why some hibernating animals lower their body temperatures during the winter. A lower body temperature results in less heat loss to the cold external environment.

Convection

Convection is the transfer of heat by the movement of liquids and gases: Molecules of air or water next to a warm surface gain energy and move away from the surface. In still air, a boundary layer of air forms over a surface of organisms. Having a thicker boundary layer will tend to slow heat transfer between an organism and its environment. When the environment is colder than the organism, the organism tends to warm this boundary layer, which in turn reduces the animal's heat loss. If there is a current of air passing by the animal, the current disrupts the boundary layer and heat can be carried away from the body by convection. This convection of heat away from the body surface is the basis of the wind chill factor we hear about in winter on the evening weather report. Wind on a cold day

Convection The transfer of heat by the movement of liquids and gases.

makes you colder. For example, wind blowing at 32 km per hour in an air temperature of −7°C has the cooling power of still air at −23°C.

In the same way that air movement can remove heat from a warm organism, air movement can add heat to an organism if the boundary layer is cooler than the surrounding air. If you were to stand in a hot desert, and your boundary layer were cooler than the air, for example, a hot wind would disrupt the boundary layer between your skin and the air and make your body even hotter.

Evaporation

Evaporation is the transformation of water from a liquid to a gaseous state with the input of heat energy. Because evaporation removes heat from a surface, it has a cooling effect on an organism. As plants transpire and animals breathe, water evaporates from their exposed gas exchange surfaces, especially at higher temperatures. In dry air, the rate of evaporation nearly doubles with each 10°C increase in temperature.

As summarized in Figure 3.18, all these sources of heat gain and loss can occur simultaneously. Radiation from the Sun can occur as direct sunlight, as well as sunlight that has been scattered as it interacts with gas molecules in the atmosphere or is reflected from clouds and the ground. Plants and animals in contact with rocks, soil, and each other can conduct heat to or from these objects, depending on whether their body temperatures are warmer or colder than the surrounding objects. As winds move the air past the organisms, there can be an additional exchange of heat, depending again on the temperature of the air compared to the temperature of the organism. Finally, organisms that experience evaporation can lose heat because evaporation requires heat energy.

BODY SIZE AND THERMAL INERTIA

Most exchanges of energy and materials between an organism and its environment occur across body surfaces. Therefore, the volume and surface of an organism affect the rate of these exchanges. As an example, let's consider the differences between the body sizes of a mouse and an elephant. The elephant obviously has a much larger volume and it takes much more energy to meet its metabolic needs each day. However, relative to its volume, the elephant has a smaller surface area than the mouse. This relationship becomes more apparent if we make the simplifying assumption that all organisms are shaped like a box with sides of equal length. In this case, the surface area (SA) of an organism increases as the square of its length (L), but the volume (V) of an organism increases as the cube of its length:

$$SA = L^2$$
$$V = L^3$$

In short, as an organism grows larger, its volume grows faster than its surface area.

Of course, organisms are not shaped like boxes, but the same principles apply to the shapes that organisms do have.

Because an organism's metabolic needs are related to its volume and volume increases faster than surface area, an organism's metabolic needs increase faster than the surface area that exchanges energy and materials between the organism and its environment.

The relationship between surface area and volume is particularly relevant when considering heat exchange. Because large organisms have a low surface-to-volume ratio, larger individuals lose and gain heat across their surfaces less rapidly than smaller individuals. In general, larger sizes and lower surface-to-volume ratios make it easier for organisms to maintain constant internal temperatures in the face of varying external temperatures. The resistance to a change in temperature due to a large body volume is known as **thermal inertia.** Although thermal inertia can be an important advantage in cold environments, in hot environments it causes moderately large individuals to have a harder time ridding themselves of excess heat. For this reason, large individuals run a greater risk of overheating. However, very large animals can benefit from thermal inertia under hot environmental conditions because their bodies heat up more slowly. We saw an example of this in the case of the dromedary camels, whose very large bodies slowly added heat during the day but then released this heat during the night.

THERMOREGULATION

The ability of an organism to control the temperature of its body is known as **thermoregulation.** Some organisms, known as **homeotherms,** maintain constant temperature conditions within the cells. Maintaining a constant internal body temperature allows an organism to adjust its biochemical reactions to work most efficiently. In contrast, **poikilotherms** do not have constant body temperatures. These terms tell us whether an organism's temperature is constant or variable, not whether the temperature changes of the body are controlled internally or externally.

ECTOTHERMS

Ectotherms have body temperatures that are largely determined by their external environment. Ectotherms tend to be organisms with low metabolic rates—such as

---

Evaporation The transformation of water from a liquid to a gaseous state with the input of heat energy.

Thermal inertia The resistance to a change in temperature due to a large body volume.

Thermoregulation The ability of an organism to control the temperature of its body.

Homeotherm An organism that maintains constant temperature conditions within its cells.

Poikilotherm An organism that does not have constant body temperatures.

Ectotherm An organism with a body temperature that is largely determined by its external environment.

reptiles, amphibians, and plants—or small body sizes—such as insects—that cannot generate or retain sufficient heat to offset heat losses from their bodies.

Although ectotherms have body temperatures that match the temperature of their environment, they are not powerless to alter their body temperature. Indeed, many species of ectotherms adjust their heat balance behaviorally by moving into or out of shade, by changing their orientation with respect to the Sun, or by adjusting their contact with warm substrates. When horned lizards are hot, for example, they decrease their exposure to the ground surface by standing erect on their legs. When they are cold, they lie flat against the ground and gain heat both by conduction from the ground and from direct solar radiation. This behavior, known as basking, is common among reptiles and insects (**Figure 3.19**). Animals that bask in the radiation of the Sun can effectively regulate their body temperatures. Indeed, their temperatures may rise considerably above that of the surrounding air, well into the range of birds and mammals. Some larger species of ectotherms, such as tuna, can generate a substantial amount of heat by exercising their massive muscles. This flexing of large muscles allows such fish to stay warmer than their external environment, making it possible for them to swim and feed in relatively cold waters.

Some plants can occasionally generate enough heat to make their tissues substantially warmer than the external environment. The skunk cabbage (*Symplocarpus foetidus*), for example, is a foul-smelling plant that lives in wet soils in eastern North America (**Figure 3.20**). The odor attracts insect pollinators such as flies that typically feed on dead, rotting organisms. The skunk cabbage sprouts new leaves in early spring, even when snow still covers the ground. The plant's mitochondria generate enough metabolic heat in its tissues to raise its temperature more than 10°C above the external environment. This incredible achievement requires a great deal of energy, but it provides a number of substantial benefits, including

Figure 3.20 Skunk cabbage. Using mitochondria to generate heat, the skunk cabbage can elevate its temperature by more than 10°C above the environmental temperature. The elevated temperature melts a path through the snow in early spring, making the skunk cabbage one of the first plants to sprout and attract pollinators to its flowers. JAPACK/age fotostock.

earlier flowering in the spring, more rapid development of flowers, and protection from freezing temperatures. In one species of skunk cabbage, scientists have discovered that generating heat also improves the rate of pollen germination and pollen tube growth in the flowers. The heat also benefits the pollinators, which can absorb some of the heat produced by the plant. Collectively, heat generation in plants can be very beneficial to both the plants and their pollinators.

In considering the various adaptations of ectotherms, we should note that the internal temperatures of some species may not vary a great deal. This can occur when the environmental temperature is not highly variable. For example, fish living in the polar oceans experience very cold waters that have little temperature variation. These fish are ectotherms because their body temperatures are determined by the surrounding environment, but their body temperatures are nearly homeothermic.

ENDOTHERMS

Endotherms are organisms that can generate sufficient metabolic heat to raise body temperature higher than the external environment. Most mammals and birds maintain their body temperatures between 36°C and 41°C, even though the temperature of their surroundings may vary from −50°C to +50°C. Maintaining a higher body temperature provides the benefit of accelerated biological activity in colder climates, allowing endotherms an improved ability to find food and escape predators.

Figure 3.19 Basking. Ectotherms such as these painted turtles (*Chrysemys picta*) commonly lie in the sun to increase their body temperature. GEORGE GRALL/National Geographic Creative.

Endotherm An organism that can generate sufficient metabolic heat to raise its body temperature higher than the external environment.

Sustaining internal conditions that differ significantly from conditions in the external environment requires a lot of work and energy. Consider the costs to birds and mammals of maintaining constant high body temperatures in cold environments. As air temperature decreases, the difference between the internal and external environments increases. Recall that heat is lost across body surfaces in direct proportion to this temperature difference. Consider, for example, an animal that maintains its body temperature at 40°C. At an outside temperature of 20°C, it loses heat much faster than it does at an outside temperature of 30°C. The greater the difference between an animal's body temperature and outside temperatures, the greater the heat loss. To maintain a constant body temperature, endothermic organisms must replace heat lost to their environment either by generating metabolic heat or by gaining heat through other means such as solar radiation, conduction, or convection. The rate of metabolism required to maintain a particular body temperature increases in direct proportion to the difference between the temperature of the body and the temperature of the environment.

ADAPTATIONS OF THE CIRCULATORY SYSTEM

You have probably noticed that when walking on a cold day, your hands and feet are the first body parts to become cold. Similarly, because the legs and feet of most birds do not have feathers, these extremities would be major potential sources of heat loss in cold regions if they were held at the same temperature as the rest of the body. Exposed extremities lose heat rapidly, due to their high ratio of surface area to volume. The conduction of heat, particularly from exposed extremities, works against the maintenance of a constant warm body temperature. Ectotherms and endotherms have evolved a number of adaptations to minimize the impact of chilled extremities and thereby to help maintain a warm temperature in the core of the body where many vital organs are located.

One prominent adaptation is *blood shunting*. **Blood shunting** occurs when specific blood vessels can be shut off—at locations called *precapillary sphincters*—so that less of the animals' warm blood flows out to the cold extremities such as the forelimbs and hindlimbs. Instead, much of this blood is redirected into the veins before it ever reaches the extremities. From the veins, it returns to the heart, as shown in **Figure 3.21**. By sending less blood to nonvital areas such as the limbs, the blood experiences less cooling and allows the core of the animal's body to maintain a constant internal temperature while expending less energy.

Blood shunting An adaptation that allows specific blood vessels to shut off so less of an animal's warm blood flows to the cold extremities.

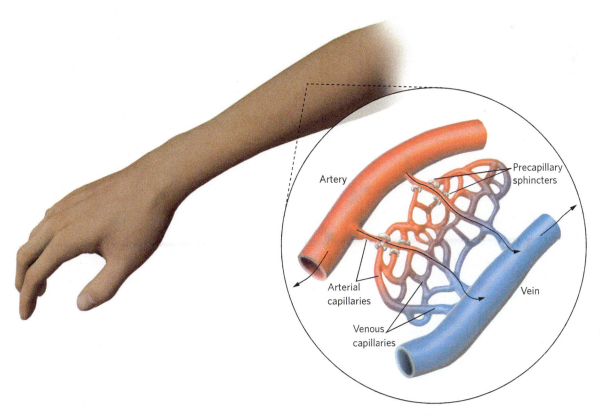

Figure 3.21 Blood shunting. In cold environments, some animals can close certain blood vessels at their precapillary sphincters. This reduces the circular flow of blood from the arteries to the extremities and back to the veins, which limits the amount of chilled blood that returns to the heart.

Another adaptation to cold extremities is countercurrent circulation. In Chapter 2, we saw that fish maximize oxygen uptake by this mechanism; the blood in their gills flows in the opposite direction from water. A similar arrangement occurs with the position of veins and arteries in the extremities of many animals; arteries that carry warm blood away from the heart and toward the extremities are positioned alongside veins that carry chilled blood from the extremities back to the heart. **Figure 3.22** shows an example. When a gull stands on ice or swims with its feet in frigid water, it conserves heat by using countercurrent circulation in its legs. Warm blood in arteries leading to the feet cools as it passes close to veins that return cold blood to the body. Rather than being lost to the environment, heat is transferred from the blood in the arteries to the blood in the veins. The feet themselves are kept only slightly above freezing, which minimizes heat transfer to the environment. The muscles used in swimming and walking are in the upper part of the leg, insulated by feathers that keep the upper legs close to the core body temperature.

The diverse range of adaptations for life on land is a fascinating testament to the ability of natural selection to favor those traits that improve the fitness of organisms. Whether we consider the adaptation of plants to obtain water and nutrients, the variety of scenarios for conducting photosynthesis under different environmental conditions, or the ability of animals to balance water, salt, nitrogen, and heat, these adaptations have evolved to help make the transition from life in water to life on land. Throughout this book, we will return to many of these adaptations as we seek to understand the ecology of communities and ecosystems.

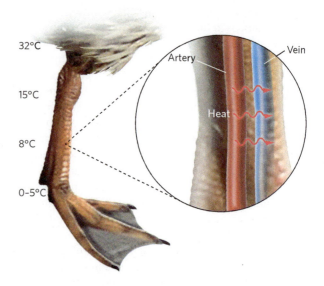

Figure 3.22 Countercurrent blood circulation. The arteries in a gull's leg that carry warm blood from the heart to the feet are positioned next to the veins that carry chilled blood away from the feet and back to the heart. This positioning of the arteries and veins allows the birds to transfer heat from arteries to the veins.

CONCEPT CHECK

1. If a snake is lying on a rock in the desert sun, how is the snake's body temperature affected by radiation, conduction, convection, and evaporation?
2. Why does an animal's surface area increase more slowly than its volume?
3. Under what conditions could an ectotherm be homeothermic?

The Challenge of Growing Cotton

Cotton plants. Cotton has numerous adaptations that help it cope with hot and dry conditions.
Steven Frame/Shutterstock.

Throughout this chapter, we have examined the challenges of the terrestrial environment and the adaptations that organisms have evolved to deal with these challenges. For plants, there are the challenges of extracting water and nutrients from the soil and then photosynthesizing at the cost of water lost due to transpiration. At the same time, plants need to balance their water and salts while also frequently coping with heat gains and losses that can place organisms at the limits of their temperature tolerance. To understand how an organism handles all these challenges, we turn to the common cotton plant.

Cotton is a major crop throughout the world because of its many uses that range from clothing to cottonseed oil. Cotton is typically grown in regions of the world that have warm temperatures with either high or low amounts of precipitation. In the United States, cotton is a $5 billion crop and nearly half of the cotton is grown in the panhandle of Texas. Like many arid regions, the Texas Panhandle experiences droughts every few years. In 2011–2016, Texas experienced one of the worst long-term droughts ever recorded, and this made it a tough place to grow cotton.

Although C_3 plants like cotton are typically associated with moister habitats, cotton can also grow well in relatively dry places. However, extended droughts often cause farmers to irrigate cotton fields because young cotton plants have relatively small and shallow root systems that have a difficult time acquiring water. As we saw with cacti and other succulent plants, shallow root systems can be effective at taking up water from brief rain events, but only if a plant can store excess water in its tissues. However, cotton has a limited ability to store water.

The soils in the Texas Panhandle range from loam to clay. Recall from our earlier discussion and Figure 3.4 that such soils have a moderate to high field capacity. However, they also have a moderate to high wilting point, which makes it hard for cotton to extract water from these soils. In addition, shallow watering in arid regions can lead to soil salinization.

In Texas, cotton is traditionally planted in early May and harvested in July or August. Because July and August can be very dry months, farmers have been experimenting with earlier planting dates. When cotton is planted in April, the cotton plants bloom in late June and produce a larger crop, in part because the plants avoid the drier months of July and August. Plants that are stressed by a lack of water can abort the development of their flowers, and flowers are the source of the cotton fibers. In addition, cotton planted earlier matures in June, when days are longer and the plants have more daylight hours to conduct photosynthesis. In 2010, researchers reported that while an earlier planting date is effective, it only works if the seedling plants are irrigated to supplement scarce natural rainwater. Without irrigation, these seedling plants that have a small root system cannot survive.

To help cotton farmers achieve larger crop yields, scientists have conducted a great deal of research on how to make cotton more drought resistant. Researchers have recently found that when cytokinin, a natural plant hormone, is sprayed onto young cotton plants, it induces the plants to grow larger root systems. By growing more roots that can penetrate deeper into the soil, the cotton is less prone to experience the effects of a water shortage. Cytokinin also stimulates the cotton plants to build a waxy outer coating, which we know makes plants less susceptible to water loss. These two responses provide a 5 to 10 percent increase in crop yields under drought conditions.

Cotton plants also have to deal with hot temperatures. Researchers have found that the ideal temperature for growing cotton is a daytime high of 28°C. However, temperatures where cotton is grown can easily exceed 38°C. At these higher temperatures, the enzyme Rubisco does not work as well. As we discussed, Rubisco is a key enzyme for C_3 photosynthesis; poorly performing Rubisco results in a lower rate of photosynthesis and, in turn, lower cotton yields. Plant breeding efforts have developed varieties of heat-tolerant cotton with variations of the Rubisco enzyme that perform well under higher temperatures.

Other varieties of cotton have been bred to transpire higher amounts of water vapor out of their stomata, which improves the plant's ability to cool itself through the process of evaporation. These varieties of cotton can only transpire more water vapor out of their stomata if they have an abundant supply of water coming in through their roots and must therefore be grown in soils that are either naturally moist or irrigated.

While cotton has been cultivated for more than 6,000 years and bred for desirable qualities for at least 3,000 years, modern science is offering new opportunities to improve the plant's growth and crop yield even more. In 2012 and 2014, researchers in China were able to sequence the genome of two species of cotton (*Gossypium raimondii* and *G. arboretum*). Having the genome should ultimately offer the opportunity to identify the genes responsible for characteristics such as tolerance to drought, heat, and salt.

The story of cotton illustrates how adaptations allow a species to live under a range of challenging environmental terrestrial conditions. It also illustrates how a knowledge of these adaptations can help farmers adjust their practices and aid scientists in cultivating varieties of plants that can better perform in the face of these challenges and produce higher crop yields. However, not all challenges faced by agricultural crops can be solved through plant breeding. When this happens, we are left to plant the crops in regions of the world where their adaptations are better suited to the environmental conditions.

SOURCES:

Fuguang, L., et al. 2014. Genome sequence of the cultivated cotton *Gossypium arboretum*. *Nature Genetics* 46:567–572.

Pettigrew, W. T. 2010. Impact of varying planting dates and irrigation regimes on cotton growth and lint yield production. *Agronomy Journal* 102:1379–1387.

Salvucci, M. E., and S. J. Crafts-Brander. 2004. Inhibition of photosynthesis by heat stress: The activation state of Rubisco as a limiting factor in photosynthesis. *Physiologia Plantarum* 120:179–186.

SUMMARY OF LEARNING OBJECTIVES

3.1 Most terrestrial plants obtain nutrients and water from the soil.
Soil nutrients include nitrogen, phosphorus, calcium, and potassium. The water-holding capacity of soil, known as its field capacity, depends on the particle sizes in the soil. The ability of plants to absorb this water requires that the osmotic potential of the roots be stronger than the matrix potential of the soil. This water moves up the plant stem to its leaves through a combination of osmotic pressure, the cohesion of water molecules, and the force of transpiration.

Key Terms: Water potential, Matric potential, Field capacity, Wilting point, Salinization, Cohesion, Root pressure, Transpiration, Cohesion–tension theory, Stomata

3.2 Sunlight provides the energy for photosynthesis.
Within the full range of electromagnetic radiation produced by the Sun, only a narrow range of wavelengths is used by the photosynthetic pigments of plants. This solar energy is used to power the process of photosynthesis by splitting water molecules and producing molecular oxygen and sugar. There are three pathways of photosynthesis: C_3, C_4, and CAM. Each differs in how it captures CO_2 and each works best in particular environmental conditions. These different pathways are often associated with structural adaptations that help plants from arid regions conserve water.

Key Terms: Electromagnetic radiation, Visible light, Photosynthetically active region, Chloroplasts, RuBP carboxylase-oxidase, C_3 photosynthesis, Photorespiration, C_4 photosynthesis, Crassulacean acid metabolism (CAM)

3.3 Terrestrial environments pose a challenge for animals to balance water, salt, and nitrogen.
Organisms attempt to achieve homeostasis in all these compounds, typically through the use of negative feedbacks. Plants and animals both possess a number of adaptations to balance their concentrations of salt, water, and nitrogenous wastes.

Key Terms: Homeostasis, Negative feedbacks, Independent variable, Dependent variable, Continuous variable, Categorical variable

3.4 Adaptations to different temperatures allow terrestrial life to exist around the planet.
Organisms can gain and lose heat through radiation, conduction, convection, and evaporation. These processes combine to form an individual's heat budget. Temperature can be regulated to different degrees by animals through the process of thermoregulation. Poikilotherms have variable body temperatures, whereas homeotherms have relatively constant body temperatures. The body temperatures of ectotherms are largely determined by their external environment, whereas endotherms can raise their body temperatures to be higher than the external environment. Additional adaptations to assist in thermoregulation include the shunting of blood and countercurrent circulation.

Key Terms: Radiation, Conduction, Convection, Evaporation, Thermal inertia, Thermoregulation, Homeotherm, Poikilotherm, Ectotherm, Endotherm, Blood shunting

CRITICAL THINKING QUESTIONS

1. How do salty soils affect the matric potential and, in turn, affect root pressure and overall water uptake?

2. Based on your knowledge of the relationship between volume and surface area, why do clay soils hold more water than sandy soils?

3. If you were breeding cotton plants to have increased water uptake in dry soils, you could breed plants that contain more amino acids and carbohydrates in the root cells. Why would this be effective and why might crop yield be negatively affected?

4. How do CAM plants solve the problem of obtaining CO_2 for photosynthesis while minimizing water loss?

5. How might the fitness cost of photorespiration in C_3 plants favor the evolution of C_4 plants?

6. How do leaf hairs that surround guard cells reduce water loss through the process of forming a boundary layer?

7. When animals hibernate, they lower their temperature. How would this reduce the rate of heat lost through conduction?

8. In addition to having highly efficient kidneys, what behaviors could you imagine desert animals using to reduce water loss?

9. If you designed an experiment to determine how temperature and the species composition of seeds affect the growth and reproduction of seed-eating birds, what would be the independent and dependent variables?

10. If you were to stand in a river with fast-moving water, which heat exchange processes might occur and why?

GRAPHING THE DATA Relating Mass to Surface Area and Volume

In this chapter, we saw that the surface area of an organism increases approximately as the square of the organism's length, whereas the volume increases approximately as the cube of the organism's length. Using the following data, graph the relationship between length and surface area and the relationship between length and volume. Based on these two graphs, notice how increases in body length affect volume much more rapidly than surface area.

Length (cm)	Surface area (cm²)	Volume (cm³)
10	100	1,000
20	400	8,000
30	900	27,000
40	1,600	64,000
50	2,500	125,000

4 Adaptations to Variable Environments

The Fine-Tuned Phenotypes of Frogs

Every spring the female gray treefrog (*Hyla versicolor*) must choose where she will lay her eggs. The treefrog is a medium-sized frog that lives throughout much of eastern North America and through the central United States to the Gulf Coast of Texas. As adults, they spend most of their time in forests, feeding on insects in trees. However, in the spring the male and female frogs move to the water to breed. Under ideal conditions, females lay their eggs in ponds that remain free of predators throughout the two months it takes for them to hatch into tadpoles and then metamorphose into frogs. Unfortunately, the females have no way to predict whether a pond will contain predators in the weeks ahead. Their offspring, however, have evolved an amazing ability to adjust to a wide range of different predator environments.

After a female frog lays her eggs, the embryos experience rapid growth and development; in just a few days, they are ready to hatch. The timing, however, can change depending on the presence of predators, such as crayfish, that commonly consume frog eggs. Embryos of gray treefrogs, like many species of frogs, can detect the presence of predators by sensing chemical cues that the predators produce. When the embryo detects a nearby predator, development accelerates and it hatches into a tadpole sooner than it would normally, thereby reducing the risk of predation as an embryo. Although it survives the egg predator, it emerges smaller than it would have been if it had remained an embryo for a longer period, so it is more vulnerable to tadpole predators.

The gray treefrog has also evolved an ability to respond to changing environmental conditions after eggs hatch into tadpoles. Like the embryos, gray treefrog tadpoles can sense predators in the water through chemical cues. When they detect the presence of a predator, tadpoles hide at the bottom of the pond, become less active, and start to change shape. Within a few days, the tadpoles develop big red tails. While the reason for the red color remains a mystery, the large tails improve a tadpole's ability to escape from predators because they serve as a large sacrificial target that can be lost to a predator and regrown. However, the energy required to grow (or regrow) a large tail is so great that other body parts cannot grow as fast. Consequently, tadpoles with large tails have smaller mouths and shorter digestive tracts, which limit their ability to eat and grow.

In short, the presence or absence of predators influences the tadpole's phenotype. In an environment without predators, tadpoles become highly active, small-tailed, and fast growing. In the presence of predators, tadpoles become inactive, large-tailed, and slow growing. But the flexibility does not end there. Tadpoles not only detect the presence of predators but also distinguish different species of predators. This allows them to distribute the use of their energy according to the level of risk; they adjust their defenses most strongly to the most dangerous predators, and produce more modest defenses against less dangerous predators. This strategy has the advantage of allowing the tadpole to use its energy where it will make the most difference to its survival.

Additional studies have also shown that tadpoles can even detect what a predator had for lunch. When a predator habitually feeds on tadpoles, the treefrog tadpoles spend more time hiding and undergo changes in shape as we have noted. However, if the predator is feeding on something else, such as snails, the treefrog tadpoles spend less time hiding and only undergo

> **"Tadpoles can even detect what a predator had for lunch."**

Gray treefrog tadpoles. Tadpoles of the gray treefrog that live without predators exhibit high activity and develop relatively small tails that are drab. In contrast, tadpoles raised with predators exhibit low activity and develop large, red tails. John I. Hammond.

small shape changes. In essence, the tadpoles detect that they are in more danger from predators feeding on tadpoles than predators feeding on other prey, and defend themselves accordingly.

The chemical cues emitted by predators feeding on tadpoles could be produced when predators attack and chew the tadpoles or when predators digest the tadpoles. In 2016, researchers isolated each of these possibilities and discovered that chemical cues are emitted during both stages in the predation event. Chemical cues from predators who are only chewing or only digesting tadpoles induce moderate changes in tadpole traits, but predators who are chewing and digesting induce a large change in tadpole traits.

Gray treefrog tadpoles also respond to other environmental conditions, including the presence of intraspecific and interspecific competitors. In ponds without predators, many tadpoles survive and compete for algae, the food on which they depend. In response to the relative scarcity of food, tadpoles develop larger mouths and longer intestines. The larger mouths contain wider rows of toothlike projections, and these improve the ability to scrape algae from rocks and leaves. Longer intestines enable a more efficient extraction of energy from the limited amount of available algae. To grow a large body, however, the tadpole must divert energy away from its tail. As a result, tadpoles that live in an environment with high competition have smaller tails. There is a cost associated with the phenotypic adaptation best suited to survive high competition; if a predator arrives, the tadpoles with smaller tails will be more vulnerable because they do not have the anti-predator phenotype.

The story of the gray treefrog represents a situation in which a species can experience a tremendous amount of environmental variation within and across generations. In response to this variation, the gray treefrog has evolved a wide range of strategies that help to improve its fitness. The responsiveness of the gray treefrog tadpole represents just one example of how organisms have evolved to quickly respond to variation in their environment. In this chapter, we will explore the wide range of environmental variation and look at how species have evolved the ability to alter phenotype in response to changing environments.

SOURCES:

Shaffery, H. M., and R. A. Relyea. 2016. Dissecting the smell of fear: Investigating the processes that induce anti-predator defenses in larval amphibians, *Oecologia* 180:55-65.

Schoeppner, N. M., and R. A. Relyea. 2005. Damage, digestion, and defense: The roles of alarm cues and kairomones for inducing prey defenses, *Ecology Letters* 8:505–512.

LEARNING OBJECTIVES

After reading this chapter, you should be able to:

4.1 Illustrate how variable environments favor the evolution of variable phenotypes.

4.2 Explain the commonly evolved adaptations in response to enemies, competitors, and mates.

4.3 Understand commonly evolved adaptations in response to variable abiotic conditions.

4.4 Describe why migration, storage, and dormancy are strategies to survive extreme environmental variation.

4.5 Explain how variation in food quality and quantity is the basis of optimal foraging theory.

In Chapters 2 and 3, we discussed the range of environmental conditions in aquatic and terrestrial environments and the many adaptations that organisms have evolved to handle these conditions. However, the environmental conditions an organism faces can also vary considerably over time and among different places. In this chapter, we will look at environmental variation and the adaptations that organisms have evolved to respond to these changing environments.

4.1 Environmental variation favors the evolution of variable phenotypes

Most properties of the environment change over time and across space, and they can change at different rates. For example, air temperature can plunge dramatically in a

matter of hours as a cold front passes through a region. On the other hand, ocean water may require weeks or months to cool the same amount. In this section, we will look at temporal and spatial variation in the environment, and then examine how such variation favors the evolution of variable phenotypes.

TEMPORAL ENVIRONMENTAL VARIATION

Temporal environmental variation describes how environmental conditions change over time. Some temporal variation is predictable, including alternation of day and night and seasonal changes in temperature and precipitation.

Temporal environmental variation The description of how environmental conditions change over time.

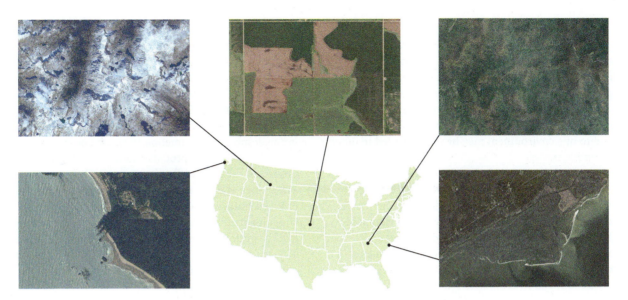

Figure 4.1 Spatial variation in the environment. As one travels from the West Coast to the East Coast of the United States, there is tremendous variation in the natural and human-altered habitats. Photos by (clockwise from top center): DigitalGlobe/ScapeWare3d/Getty Images; DigitalGlobe/ScapeWare3d/Getty Images; DigitalGlobe/ScapeWare3d/Getty Images; DigitalGlobe/ScapeWare3d/Getty Images; DigitalGlobe/ScapeWare3d/Getty Images.

Superimposed on these predictable cycles are irregular and unpredictable variations, including *weather* and *climate*. **Weather** refers to the variation in temperature and precipitation over periods of hours or days. **Climate** refers to the typical atmospheric conditions that occur throughout the year, measured over many years. For example, the climate of Wyoming is typically cold and snowy in the winter but hot and dry in the summer. However, the weather on any particular day cannot be predicted very far in advance; it can vary over intervals of a few hours or days with the movement of cold and warm air masses. Although climate describes the average conditions of a given location throughout the year, climate also can vary over long periods of time. For example, a location might experience a string of years that are much wetter or drier than the average year.

Some types of temporal variation can cause large impacts on ecosystems but occur infrequently in a particular place. For example, droughts, fires, tornadoes, and tsunamis can cause major changes in the landscape, yet their frequency at a particular location is rare. Other sources of temporal variation, such as the current warming of Earth, occur very slowly over decades and centuries. How organisms and populations respond to temporal variation in their environment depends on the severity of the change and how often it occurs. In general, the more extreme the event, the less frequently it occurs.

SPATIAL ENVIRONMENTAL VARIATION

Environmental variation also occurs from place to place due to large-scale variation in climate, topography, and soil type (**Figure 4.1**). If you were to fly across the United States from Oregon to South Carolina, for instance, you would observe a series of major, large-scale environmental changes along the trip: the rocky western coast, northwestern forests, western rangelands, midwestern agricultural fields, eastern forests, and coastal beaches. At smaller scales, environmental variation is generated by the structures of plants, the activities of animals, the composition of soil, and the activities of humans. As with temporal variation, a particular scale of spatial variation may be important to one organism but not to another. The difference between the top and the underside of a leaf, for example, is important to an insect, but not to a moose, which happily eats the whole leaf, insect and all.

Moving through environments that vary in space, an individual experiences environmental variation as a sequence in time. In other words, a moving individual perceives spatial variation as temporal variation. The faster an individual moves and the smaller the scale of spatial variation, the more quickly the individual encounters new environmental conditions and the shorter the temporal scale of the variation. This principle applies to plants as well as to animals. For example, as plant roots grow, they push their way through the soil, and soils commonly contain small-scale variation in moisture and nutrients. If a plant's roots grow quickly in soil with a fine scale of variation, the plant roots will frequently encounter new soil environments. Similarly, wind and animals disperse the seeds of plants. The variety of habitats in which the seeds might land depends on the distance the seeds travel and the scale of spatial variation in the habitat.

Weather The variation in temperature and precipitation over periods of hours or days.

Climate The typical atmospheric conditions that occur throughout the year, measured over many years.

PHENOTYPIC TRADE-OFFS

In Chapter 2, we saw that rainbow trout that express cold-water isozymes in their tissues perform well in cold water but poorly in warm water. In contrast, trout expressing warm-water isozymes perform well in warm water but poorly in cold water (Figure 2.21). Throughout the natural world, we commonly see that phenotypes well suited to one environment may be poorly suited to other environments.

Figure 4.2 illustrates phenotypic fitness in relation to different environments. In Figure 4.2a, an individual possessing phenotype X is well suited to environment X and therefore experiences high fitness. In environment Y, however, the phenotype is no longer well suited to the environment and it therefore experiences reduced fitness. In contrast, an individual that possesses phenotype Y is well suited to environment Y and experiences high fitness in environment Y. However, it is poorly suited to environment X, so it experiences reduced fitness in environment X. When a given phenotype experiences higher fitness in one environment whereas other phenotypes experience higher fitness in other environments, we say that there is a **phenotypic trade-off,** meaning that neither phenotype does well in both environments.

But what if an individual could produce a range of phenotypes and each phenotype could perform well in a specific environment? Individuals with mutations that allow them to produce multiple phenotypes that are uniquely suited to each environment would experience relatively high fitness in both environments and therefore be favored by natural selection. The ability of a single genotype to produce multiple phenotypes is called **phenotypic plasticity.** Phenotypic plasticity is a widespread phenomenon in nature; nearly every organism—bacteria, protists, plants, fungi, and animals—possesses phenotypically plastic traits. Different traits can change at different rates and these environmentally induced traits can be either reversible or irreversible. By changing its traits, an individual often maintains a high level of performance when the environment changes. This means that phenotypically plastic traits often are a mechanism of achieving homeostasis, a concept we discussed in Chapter 3.

Figure 4.2b shows the advantage of being phenotypically plastic. In contrast to the two nonplastic genotypes, labeled as genotype X and genotype Y, the plastic genotype, labeled as genotype Z, has a relatively

Phenotypic trade-off A situation in which a given phenotype experiences higher fitness in one environment, whereas other phenotypes experience higher fitness in other environments.

Phenotypic plasticity The ability of a single genotype to produce multiple phenotypes.

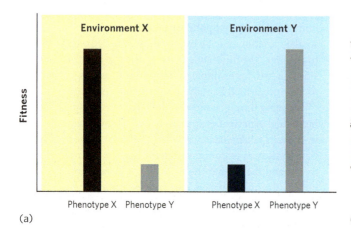
(a)

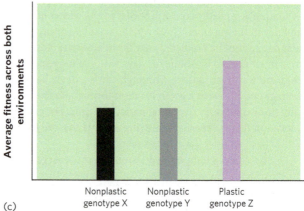
(c)

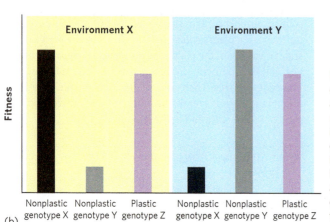
(b)

Figure 4.2 Environments, phenotypes, and fitness. Different environments cause phenotypes to experience different amounts of fitness. **(a)** Phenotypic plasticity evolves because a phenotype has high fitness in one environment and low fitness in another environment. **(b)** Given this trade-off, nonplastic genotypes have high fitness in one environment but low fitness in other environments. In contrast, a plastic genotype can have high fitness in both environments. **(c)** If we consider the average fitness across both environments, we see that the plastic genotype experiences higher average fitness than either of the two nonplastic genotypes.

high fitness in both environments because it can produce a phenotype that is nearly as fit as genotype X in environment X and a phenotype that is nearly as fit as genotype Y in environment Y. If we examine the average fitness of the three different genotypes in Figure 4.2c, we see that it is higher for the plastic genotype. Whenever environmental variation coincides with phenotypic trade-offs across different environments, natural selection will favor the evolution of phenotypic plasticity.

For a long time, scientists applied the concept of phenotypic plasticity to only certain types of traits, such as changes in morphology or physiology. However, we now recognize that many other types of traits—for example, behavior, growth, development, and reproduction—frequently represent phenotypes that can change under different environmental conditions. As a result, the conceptual framework of phenotypic plasticity has expanded in recent years to consider all these types of traits.

We can see the advantage of phenotypic plasticity in the example of the gray treefrog tadpoles discussed at the beginning of this chapter. In environments with predators, the tadpoles produce a phenotype that is well suited to escaping detection and capture. In predator-free environments, the tadpoles produce a phenotype that is well suited for faster growth. If tadpoles only had a single phenotype, they would perform poorly whenever the environment changed. In contrast, a tadpole that can change its behavior and body shape performs relatively well when the environment changes.

The fitness advantage of phenotypic plasticity occurs whenever environmental variation in space or time occurs frequently. If environmental conditions frequently change, then the phenotype favored by natural selection also changes frequently and this gives the plastic genotype a higher average fitness than the nonplastic genotype. If spatial or temporal variation is not common, a single phenotype will be favored; the phenotype with the highest fitness in the stable environment will be favored.

As we have noted in earlier chapters, all phenotypes are the product of genes interacting with environments. As a result, environmentally induced traits have a genetic basis, but reflect the ability of the environment to turn certain genes on or off, which causes different phenotypes to develop. The environments that induce these changes may change so rapidly that they occur within a generation or they may be somewhat slower and vary across generations. The experience of the tree frog is a good example; predators and competitors can differ substantially from pond to pond in a particular year, from year to year in a particular pond, and even from week to week during the time it takes for a generation of tadpoles to metamorphose and leave the pond as frogs.

ENVIRONMENTAL CUES

For an organism to alter its phenotype in an adaptive way, it must first be able to sense its environmental conditions. For example, the gray treefrog tadpole first senses whether the pond contains predators or competitors and then alters its phenotype accordingly to improve its fitness. As we will see throughout the rest of this chapter, environmental cues can take many forms, including smells, sights, sounds, and changes in abiotic conditions. Of the numerous potential cues an organism might use, the best are those that offer the most reliable information about the environment. For example, an animal that requires a reliable cue about competition for food could use the presence of a large number of conspecifics— members of its own species—that will be eating the same thing. But if ample food is available, even a large number of conspecifics would not result in competition for food, so the number of conspecifics might be a poor indicator of the level of competition. If this is the case, then a better environmental cue to indicate high competition for food might be the amount of food that an individual can acquire each day. In such a case where a species has multiple possible environmental cues that it could use, we might expect the species to evolve to use the most reliable cue. When organisms have very reliable cues, they can more accurately produce a phenotype that is well suited to the environment.

RESPONSE SPEED AND REVERSIBILITY

Phenotypically plastic traits respond to changes in the environment at different rates. Some of the trait changes are irreversible. The most rapid responses are typically behavioral traits, which can be altered in seconds. For example, most prey rapidly respond to a predator's pursuit; often it takes less than 1 second for the prey to flee. Physiological plasticity, which is an environmentally induced change in an individual's physiology—sometimes referred to as **acclimation**—can also be relatively rapid. Consider the time it takes humans to acclimate to the low-oxygen conditions that are caused by lower air pressures at high altitudes, or the time required for human skin to tan. Both of these physiological changes can be accomplished in just a few days. In contrast to these behavioral and physiological changes, changes in morphology—including changes in body shape and the size of internal organs—and changes in life history—including time to sexual maturity and number of progeny produced—can take weeks, months, or years.

Differences in response speed have implications for the reversibility of the induced traits. Behavioral traits that are induced by a change in the environment typically can be rapidly reversed if the environment reverts to its original condition. For example, an animal can quickly and easily adjust its food intake as food conditions change over time. Induced changes in morphology and life history are more difficult to reverse. For many organisms such as plants, changes in morphology are difficult or impossible to undo. For example, plants commonly respond to low light

---

Acclimation An environmentally induced change in an individual's physiology.

conditions by growing taller in an attempt to rise above neighboring plants that are casting shade. If the environment suddenly becomes sunny, a plant cannot make itself shorter. Even less reversible are life history decisions such as those related to the timing of reproductive maturity and the number of offspring produced. Once sexual maturity has been achieved, an organism cannot become sexually immature, although it can refrain from reproducing.

The differences in the speed of phenotypic changes, and the ability to reverse phenotypic changes, influence which traits are favored by natural selection. When environments fluctuate rapidly relative to the length of an individual's lifetime, selection should favor plastic behavioral and physiological traits because these traits can often respond rapidly and reverse rapidly. When environments fluctuate more slowly, selection can favor many more types of traits, including morphological and life history traits that are slow to respond and are often much less reversible.

CONCEPT CHECK

1. What is the distinction between weather versus climate?
2. How do phenotypic trade-offs favor the evolution of phenotypic plasticity?
3. Why do phenotypically plastic responses depend on reliable environmental cues?

4.2 Many organisms have evolved adaptations to variation in enemies, competitors, and mates

Many types of environmental variation can induce phenotypic plasticity. Among biotic environments, three of the best-studied types of environmental variation involve the occurrence of enemies, competitors, and mates.

ENEMIES

Because enemies—including predators, herbivores, parasites, and pathogens—pose a major risk to organisms that are consumed, we would expect that many organisms have evolved defenses against their enemies. Like the treefrog tadpoles, many aquatic animals including fish, salamanders, insects, zooplankton, and protists alter their growth and change shape in response to predators in their environment. These changes might improve a prey's ability to escape, make the prey difficult to fit in the predator's mouth, or deter consumption by producing sharp spines.

Ciliates in the genus *Euplotes* are tiny protists that live in lakes and streams. As you can see in **Figure 4.3**, these tiny organisms can sense chemical cues emitted by predators and, within hours, respond by growing "wings" and numerous other projections that make them up to 60 percent larger. The larger size makes it difficult for a predator to fit *Euplotes* into its mouth, so the winged phenotype suffers less

predation. However, because of the considerable amount of energy required to grow these projections, this phenotype takes 20 percent more time to develop. Thus, in a predator-free environment, the superior strategy for *Euplotes* is to have the nonwinged phenotype.

When prey make themselves more difficult to catch or consume, how might predators respond? It turns out that predators can have plastic abilities as well. For example, when the predatory ciliate *Lembadion* is surrounded by the predator-induced, winged *Euplotes*, *Lembadion* grows a larger body with a bigger mouth that can engulf the winged prey. In doing so, *Lembadion* benefits by consuming the winged prey. However, once most of the winged prey have been consumed and only small prey remain, the large *Lembadion* is poorly suited to eat the small prey. As a result, the large *Lembadion* experiences lower fitness than the small *Lembadion*. When this occurs, a large *Lembadion* can go through several cell divisions and revert back to having a smaller phenotype with the smaller mouth.

Prey also use behavioral defenses against predators. They may move away from areas containing predators or become less active to avoid detection. Some animals also congregate in refuges, which are locations that are safe from predation. For example, newly hatched fish take refuge in the dense weeds along a lake shoreline to hide from large predatory fish. Although these behaviors usually reduce the risk of predation, increased safety comes at a price. When prey become less active or congregate in refuges, they spend less time feeding. Also, the supply of food in and around crowded refuges can be depleted quickly. As a result, behavioral defenses commonly come at the cost of slower growth, development, or reproduction. In the absence of predators, prey become more active and leave the refuges to find more food; this change in behavior allows more rapid growth.

Plants also have the ability to respond to the presence of organisms that consume them. The Virginia pepperweed, a member of the mustard family, is eaten by several species of herbivores, including caterpillars and aphids. As **Figure 4.4a** shows, when an herbivore chews on the leaves of pepperweed, the plant rapidly develops extra leaf hairs, called *trichomes,* that make it harder for the leaves to be consumed. It also increases its production of *glucosinolates,* a group of chemicals that gives mustard its strong flavor and functions as a natural insecticide. If previously attacked plants and nonattacked plants are subsequently placed in a garden, the attacked plants, which have more trichomes and more glucosinolates, attract fewer herbivores and exhibit better survival, as shown in Figure 4.4b.

COMPETITION FOR SCARCE RESOURCES

Most organisms face the challenge of scarce resources, which leads to competition. However, the intensity of competition varies across and within habitats. As a result, organisms have evolved a variety of phenotypically

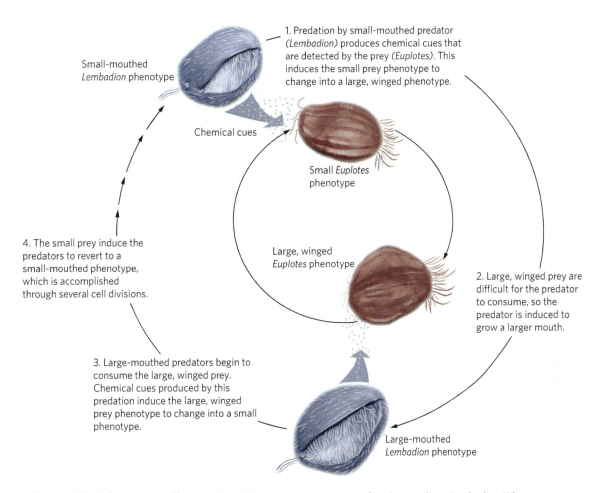

1. Predation by small-mouthed predator (*Lembadion*) produces chemical cues that are detected by the prey (*Euplotes*). This induces the small prey phenotype to change into a large, winged phenotype.

Small-mouthed *Lembadion* phenotype

Chemical cues

Small *Euplotes* phenotype

4. The small prey induce the predators to revert to a small-mouthed phenotype, which is accomplished through several cell divisions.

Large, winged *Euplotes* phenotype

2. Large, winged prey are difficult for the predator to consume, so the predator is induced to grow a larger mouth.

3. Large-mouthed predators begin to consume the large, winged prey. Chemical cues produced by this predation induce the large, winged prey phenotype to change into a small phenotype.

Large-mouthed *Lembadion* phenotype

Figure 4.3 Inducible defenses and offenses. The ciliate *Euplotes* serves as prey for a larger ciliate, *Lembadion*. When *Euplotes* detects the predatory *Lembadion* in the water via chemical cues, it changes shape from an uninduced small phenotype to a predator-induced phenotype that has large "wings" and other projections that make it too large to fit in the predator's mouth. When the predator begins to encounter the large, winged prey, mechanical cues induce it to increase the size of its mouth. This larger mouth can capture the large, winged prey but it is poor at capturing the small prey, which develop in response to chemical cues emitted by large-mouthed predators. As small prey become more abundant, the predatory *Lembadion* rapidly divides several times to once again have a small mouth. After H. W. Kuhlmann and K. Heckmann, *Hydrobiologica* 284 (1994): 219–227; M. Kopp and R. Tollrian, *Ecology* 84 (2003): 641–651; M. Kopp and R. Tollrian, *Ecology Letters* 6 (2003): 742–748.

plastic strategies for high and low competition. As we would expect, responses to high competition often exhibit phenotypic trade-offs that favor the evolution of phenotypically plastic responses.

Jewelweed (*Impatiens capensis*), a flowering plant with beautiful orange flowers, is found in moist habitats throughout much of North America (**Figure 4.5**). In nature, it can grow in clumps that are either very sparse or very dense. The range of densities has an effect on the intensity of competition among the plants for sunlight. When jewelweed is shaded by other plants, it responds by elongating its stems, which allows the plant to become taller and to rise above the competing plants. If competition for light causes the jewelweed to grow taller, we might wonder why the plant doesn't always grow tall. Researchers have discovered that elongated jewelweeds experience greater fitness in high-competition environments, whereas short jewelweeds

experience greater fitness in low-competition environments. Although the researchers were not able to identify the reasons why, it is clear that different phenotypes perform better in different environments, so a plastic phenotype is an effective way to gain high fitness when the intensity of competition varies over time and space.

Animals also respond to competition in a number of fascinating ways. For example, we might expect animals to spend more time searching for food when it is rare than when it is abundant. However, a second option is to find ways to extract more nutrients from the food that is available. One way to do this is to alter the size of the digestive tract, as we saw in the example of gray treefrogs, which respond to competition by increasing the length of their intestine. When the intestine is longer, food spends more time traveling through it and the organism can extract more nutrients from it.

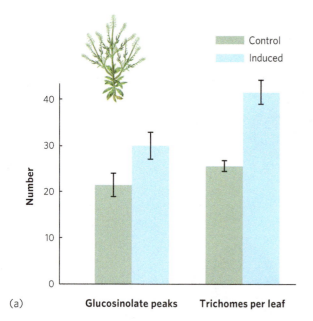

(a)

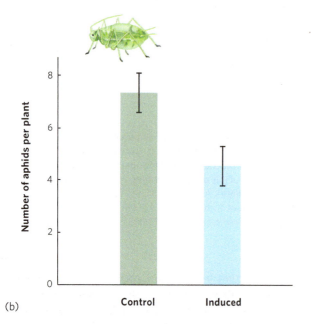

(b)

Figure 4.4 Herbivore-induced responses by plants.
(a) Virginia pepperweed (*Lepidium virginicum*) responds to herbivore attacks by growing more leaf hairs, known as trichomes, and more types of defensive chemicals, called glucosinolates.
(b) Plants attacked by herbivores in the past developed more trichomes and glucosinolates and have fewer aphids on them than plants that were not previously attacked by herbivores. Error bars are standard errors. After A. Agrawal, Benefits and costs of induced plant defense for *Lepidium virginicum* (*Brassicaceae*), Ecology 81 (2000): 1804–1813.

A striking example of a plastic response to variation in food availability can be found in the digestive tracts of snakes. The Burmese python (*Python bivittatus*), for example, can consume a large rodent that is 25 percent of its own body weight, but it may find such a prey item only once a

Figure 4.5 Jewelweed. When jewelweed is shaded by other plants, it elongates its stems to become taller and rise above the competing plants. Elongated jewelweeds experience greater fitness in high-competition environments, whereas short jewelweeds experience greater fitness in low-competition environments. AdamLongSculpture/Getty Images.

month (**Figure 4.6**). Between meals, the snake's stomach and intestine shrink a great deal, which reduces its weight and the associated carrying costs. Once a prey is consumed, however, the python can enlarge its cells and double the length of its digestive tract in just 24 hours. This dramatically increases the surface area of the intestine, which allows the snake to absorb more energy from the digested prey. The snake also sends 10 times more blood to the intestines to assist in the absorption of nutrients into the bloodstream. To handle this increase in blood flow, the snake increases the size of its heart by 40 percent, a remarkable achievement given that it occurs in 2 days.

MATES

When mates are rare, reproduction can be a challenge. Consider the situation of flowering plants that depend on pollinators to deliver the pollen that contain a male's sperm. In this situation, the probability of being pollinated can be highly variable, which means that the probability of finding a mate is also highly variable. There is a solution for flowering plants that are **hermaphrodites,** which are individuals that produce both male and female gametes. Such individuals are often self-compatible, that is, they are capable of fertilizing their eggs with their own sperm. This process, known as self-fertilization, often carries the potential cost of **inbreeding depression,** in which offspring can experience reduced fitness when deleterious alleles are inherited from both the egg and the sperm. Because mate availability can be quite

Hermaphrodite An individual that produces both male and female gametes.

Inbreeding depression The decrease in fitness caused by matings between close relatives due to offspring inheriting deleterious alleles from both the egg and the sperm.

Figure 4.6 The plasticity of a python. (a) A Burmese python can consume large prey, but may do so only once a month. **(b)** A fasting animal has a shrunken intestine. **(c)** Within 2 days of eating, the intestine doubles in length and swells in diameter. (Note that DPF means days post feeding.) **(d)** Ten days later, digestion is complete and the intestine shrinks again. (a) Bryan Rourke, California State University; (b through d) Reproduced/adapted with permission Secor S M J Exp Biol 2008;211:3767-3774 © 2008 by The Company of Biologists Ltd.

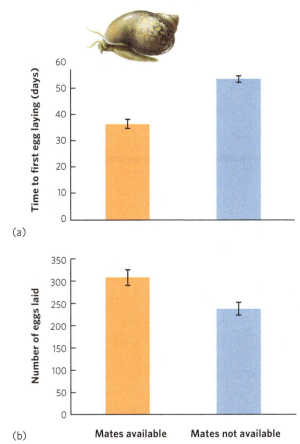

Figure 4.7 nail responses to variation in mates. In the common pond snail, mate availability can vary a great deal. **(a)** Snails raised without mates wait nearly 2 weeks before self-fertilizing their eggs. **(b)** Snails without mates produce fewer progeny compared to snails with mates, but this strategy is more fit than forgoing all reproduction. Error bars are standard errors. After A. Tsitrone et al., Delayed selfing and resource reallocations in relation to mate availability in the freshwater snail *Physa acuta, American Naturalist* 162 (2003): 474–488.

variable and self-fertilization can have a substantial cost, the organism must develop one of two life history choices: wait to find a mate and enjoy higher fitness, or self-fertilize. In some cases, this choice is made at the level of the entire organism. In other cases, as in many hermaphroditic species of flowering plants, the choice is made by each flower. If a flower is not pollinated within a given period of time, the flower will self-fertilize. Although self-fertilizing may result in lower fitness than breeding with a mate, low fitness is better than the zero fitness the organism would obtain by avoiding self-fertilization altogether.

Reproductive adaptations to variation in mate availability have been studied in a variety of hermaphroditic plants

and animals. For example, the common pond snail (*Physa acuta*) is a hermaphrodite that experiences large variation in population sizes, which means that it also experiences large variation in the availability of mates. When potential mates are abundant, a snail will normally mate with another individual; when potential mates are rare, it can fertilize its eggs with its own sperm. In an experiment designed to measure the fitness effects of the two alternative strategies, researchers assigned snails to one of two groups: snails living with mates or snails living without mates. The researchers then observed the time it took for each group to begin reproduction and the total number of eggs that each group of snails laid. As you can see in **Figure 4.7a**, snails living without mates delayed their reproduction for 2 weeks before they used the alternative strategy of self-fertilization. As predicted, this choice came at a cost. Looking at Figure 4.7b, we see that self-fertilizing snails laid many fewer eggs than snails that had mates available. However, the self-fertilizing snails did gain some fitness, which is better than forgoing reproduction. The good news for the pond snail, and many other hermaphroditic organisms, is that by evolving multiple strategies they can experience higher fitness over time than would be possible without a phenotypically plastic strategy.

CONCEPT CHECK

1. How might predators and prey both evolve phenotypically plastic strategies to improve their fitness?
2. How is the phenotypic plasticity of intestines an adaptation to variable food availability?
3. If self-fertilization causes lower fitness than cross-fertilization, under what environmental conditions would self-fertilization be the superior strategy?

4.3 Many organisms have evolved adaptations to variable abiotic conditions

We have seen that variation in biotic conditions, including enemies, competitors, and mates, can be quite high. Abiotic conditions, including temperature, water availability, salinity, and oxygen, also vary. Faced with this abiotic variation, many species have evolved phenotypically plastic traits that allow them to improve their fitness.

TEMPERATURE

Organisms have evolved a number of plastic responses to temperature variation. As we saw in Chapter 2, isozymes in rainbow trout allow the fish to have proper nerve transmission in cold winter water and warm summer water. Isozymes are actually a form of phenotypic plasticity with a rapid response time, in some cases a matter of hours or days. For example, goldfish (*Carassius auratus*) can be held at either 5°C or 25°C for a few days and then tested to determine how fast they can swim in a variety of temperatures. As shown in **Figure 4.8**, fish acclimated to 5°C swim faster at low temperatures but swim poorly at high temperatures. In contrast, fish acclimated to 25°C swim faster at warm temperatures but swim poorly at low temperatures. This demonstrates that goldfish can adjust their physiology to maintain relatively high swimming speeds across different environmental temperatures.

Many animals respond to changing temperatures by moving to habitats with more favorable temperatures. Migrating birds present an extreme example; they fly every autumn to warmer latitudes. Not all animals travel a long distance. Some move to a specific location within a habitat, called a **microhabitat,** that contains more favorable abiotic conditions.

The use of microhabitats can be illustrated by the daily behavior cycle of the desert iguana (*Dipsosaurus dorsalis*), shown in **Figure 4.9**. Although the desert iguana cannot regulate its temperature by generating heat metabolically, it can take advantage of sunny and shady microhabitats to alter its

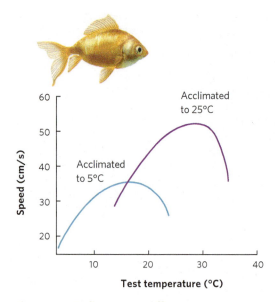

Figure 4.8 Acclimation to different temperatures. Goldfish raised at low temperatures swim faster in cold water and more slowly in warmer water. Individuals raised at high temperatures swim faster in warmer water than in cold water. After F. E. J. Fry and J. S. Hart, Cruising speed of goldfish in relation to water temperature, *Journal of the Fisheries Research Board of Canada* 7 (1948): 169–174.

temperature. The iguana has a preferred temperature range of 39°C to 43°C. In the southwestern United States, the air temperature can reach 45°C. As temperatures rise during the day, the lizard first moves to the shade of plants or rocks and then to a cooler underground burrow. If temperatures begin to cool, the iguana can move out of its burrow and bask in the sun to raise its temperature to its preferred range. Such behavioral plasticity allows the iguana to remain within its preferred temperature range for a greater proportion of the day. In effect, the plasticity of the iguana's behavior permits homeostasis in its body temperature.

WATER AVAILABILITY

When faced with changes in water availability, most animals can move among different microhabitats. However, plants are typically rooted in a single location and therefore face a tremendous challenge in locating water. As a result, plants have a number of phenotypically plastic adaptations for coping with water variability. Closing the stomata is one of the most common adaptations. As we saw in Chapter 3, when water is plentiful, the guard cells in a plant's leaves open and transpiration occurs through the stomata. However, this process causes water loss. When water is scarce, these cells change shape and the stomata close to conserve water. In this way, the plant can transpire when water is abundant but stop transpiring when water is scarce.

Other plant strategies in response to a lack of water are even more dramatic. For example, plants living on coastal

---

Microhabitat A specific location within a habitat that typically differs in environmental conditions from other parts of the habitat.

Figure 4.9 Selecting microhabitats. The desert iguana regulates its body temperature by selecting microhabitats that contain favorable abiotic conditions. When the lizard is cool, it can bask in the sun to increase its internal temperature. As the temperature becomes hot during the day, the lizard can seek shade or move into a burrow to lower its temperature.

dunes in Europe commonly experience drought because water drains quickly through the sandy substrate. Three common plants of these coastal dunes—gray hair-grass (*Corynephorus canescens*), mouse-ear hawkweed (*Hieracium pilosella*), and sand sedge (*Carex arenaria*)—demonstrate how plants adjust their relative allocations of energy and materials to grow either roots or shoots. To demonstrate

this phenomenon, researchers grew each of the three plants under conditions of abundant versus scarce water. After 5 months of growth, the researchers measured the ratio of root growth to shoot growth and the results are shown in **Figure 4.10**. When water was abundant, the plants devoted more energy to the growth of shoots, which function primarily to photosynthesize. When water was scarce, the

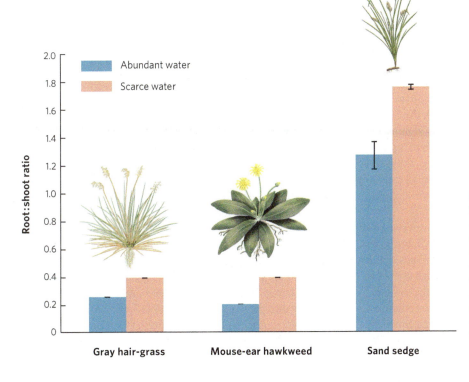

Figure 4.10 Morphological plasticity in response to water. Plants living under scarce water conditions have a higher root-to-shoot ratio. By devoting more growth into the roots than into the shoots, these plants can better obtain water when it is scarce. Error bars are standard errors. After A. Weigelt et al., Competition among three dune species: The impact of water availability on below-ground processes, *Plant Ecology* 176 (2005): 57–68.

(a)

Figure 4.11 Adaptations to fluctuating salt environments.
(a) Pools of water on the rocky coasts of La Jolla Cove in San Diego, California, are filled with a mixture of rain water and sea water from waves. Evaporation can cause high concentrations of salts and other solutes. **(b)** The tiny copepod *Tigriopus*, shown here carrying eggs, is able to handle the widely fluctuating solute concentrations in its environment by adjusting its solute concentrations through the production of amino acids. Peter Bennett/Green Stock Photos Peter Bennett/Citizen of the Planet; Morgan kelley.

(b)

plants devoted more energy to the growth of roots, which expanded their ability to capture what little water was available. Given that these plants experience variation in water availability, it is clear that no single allocation strategy would be as beneficial as the phenotypically plastic strategy that they exhibit.

SALINITY

In Chapter 2, we learned that freshwater and saltwater organisms have evolved numerous adaptations to handle their aquatic environments. However, some organisms live in aquatic environments characterized by solute concentrations that fluctuate widely over short periods of time. To survive, these organisms must have the ability to make rapid physiological adjustments. For example, the copepod *Tigriopus* lives along rocky Pacific coasts in pools that receive sea water infrequently, from the splash of high waves (**Figure 4.11**). As water in the pools evaporates, the salt concentration rises to high levels, but a heavy rainfall can lower the salt concentration—a rapid reversal of environmental conditions for the copepod.

Like sharks and rays, *Tigriopus* manages its water balance by changing the osmotic potential of its body fluids. When the salt concentration in a pool is high, individuals synthesize large quantities of certain amino acids such as alanine and proline. These small molecules increase the osmotic potential of the body fluids to match that of the environment without the deleterious physiological consequences that come with high levels of salts or urea. However, this

response to excess salts in the environment is costly in terms of the energy it requires. When individual *Tigriopus* are switched from 50 percent sea water to 100 percent sea water, the respiration rate of the copepods initially declines, owing to salt stress, and then increases as they synthesize alanine and proline to restore their water balance. When switched from 100 percent sea water back to 50 percent sea water, the copepods' respiration rate immediately increases as they rapidly degrade and metabolize excess free amino acids to reduce their osmotic difference to be more in line with their new environment.

OXYGEN

If you have ever been to a location at a high elevation, you probably experienced the challenge of low oxygen. As you continue to rise above sea level, the air pressure becomes lower, which reduces the amount of available oxygen. At thousands of meters above sea level, breathing becomes labored and physical activity is very difficult. Animals that travel up and down mountains as part of their daily or seasonal movements, for example, the llamas of South America, are able to adjust their physiology to this variation in oxygen concentration. Similarly, human mountaineers, who tackle Mount Everest, experience an extreme oxygen challenge for which their bodies can only partially adjust. At its peak, the mountain is 8,848 m above sea level and the oxygen pressure is only one-third of that found at sea level. To acclimate to the low oxygen conditions, climbers stop periodically for several days along the way. Initial changes include more rapid breathing and an increased heart rate. After a week or two, additional changes improve the body's ability to carry oxygen, including an increase in the number of red blood cells and an increased concentration of hemoglobin in the red blood cells. At the highest altitudes, where humans rarely spend much time, the body is not able to fully adjust to such low oxygen concentrations; this has contributed to the deaths of many climbers at high altitudes. When climbers later return to low altitudes, their physiological changes slowly revert to their original state.

CONCEPT CHECK

1. How does the function of isozymes represent an example of phenotypic plasticity?
2. Why do plants alter their root-to-shoot ratios?
3. What are three adaptations humans undergo in response to decreased oxygen at high elevations?

4.4 Migration, storage, and dormancy are strategies to survive extreme environmental variation

In many parts of the world, extremes of temperature, drought, darkness, and other adverse conditions are so severe that individuals either cannot change enough to maintain their normal activities, or the change required would not be worth the cost. Under such conditions, organisms resort to a number of extreme phenotypically plastic responses. These include migration, storage, and dormancy.

MIGRATION

Migration is the seasonal movement of animals from one region to another. In this case, the phenotype is the behavior of living in a particular location and the plasticity is displayed in the act of migrating, which allows the animal to express the alternative phenotypes of living in multiple locations. Each fall, hundreds of species of land birds leave temperate and Arctic North America, Europe, and Asia for the south, in anticipation of cold winter weather and dwindling food supplies. In east Africa, many large herbivores, such as wildebeests (*Connochaetes taurinus*), migrate long distances, following the geographic pattern of seasonal rainfall and fresh vegetation. Some insects also migrate. Monarch butterflies offer a fascinating example of insect migration, as shown in **Figure 4.12**. The adult butterflies living in the northern United States and southern Canada migrate to wintering areas in the southern United States and Mexico. Here, they hibernate through the winter and then begin heading back north. On their way north, they breed and produce a second generation of butterflies. This second generation completes the migration back to the summer breeding areas. In all cases, the decision to migrate is a plastic behavior in response to environmental changes including lower temperature and shorter day length.

Some migratory movements occur in response to reductions in local food supplies. For example, consider locusts, which are species of short-horned grasshoppers. Locust migrations occur when the insects leave an area where they have a large population and a depleted food supply (**Figure 4.13**). The migrations can include billions of locusts and, when they

---

Migration The seasonal movement of animals from one region to another.

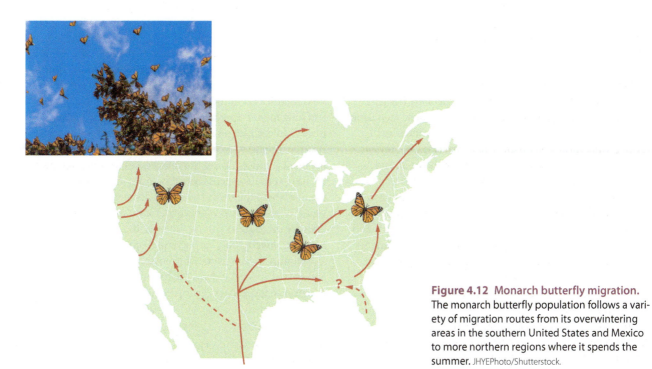

Figure 4.12 Monarch butterfly migration. The monarch butterfly population follows a variety of migration routes from its overwintering areas in the southern United States and Mexico to more northern regions where it spends the summer. JHYEPhoto/Shutterstock.

Figure 4.13 Locust migration. When locusts experience a reduced food supply, the normally solitary grasshoppers produce offspring that are highly gregarious and mobile. These offspring move across the landscape in huge swarms and cause major damage by consuming crops, such as this example in North Africa. Avalon/Photoshot License/Alamy.

pass through an area, these herbivores can cause extensive crop damage over wide areas. As a result, their arrival can cause major damage to human food supplies. This migratory behavior in locusts is the result of a number of behavioral changes. In sparse populations, adult locusts are solitary and sedentary. In dense populations, however, frequent contact with other locusts stimulates young individuals to develop gregarious, highly mobile behavior, which can grow into a mass migration.

STORAGE

Where environmental variation shifts the food supply from feast to famine and migration is not a possibility, storing resources can be an adaptive strategy. For example, during infrequent rainy periods, desert cacti swell with water stored in their succulent stems, as discussed in Chapter 3. In habitats that frequently burn—such as the chaparral of southern California—perennial plants store food reserves in fire-resistant root crowns, shown in **Figure 4.14**. The surviving root crowns send up new shoots shortly after a fire has passed.

Many temperate and Arctic animals accumulate fat during mild weather as a reserve of energy for periods of harsher weather when snow and ice make food inaccessible. However, fat reserves can make an animal slower and less agile, and therefore more likely to be caught by predators. One way to avoid this problem is to store the food before consuming it. Some mammals and birds that are active during the winter—for example, beavers, squirrels, pikas, acorn woodpeckers, and jays—cache food supplies underground or under the bark of trees for later retrieval. These caches can be large enough to sustain individuals for long periods.

DORMANCY

Environments sometimes become so cold, dry, or depleted of nutrients that organisms can no longer function normally. Some species that do not migrate have evolved a strategy of

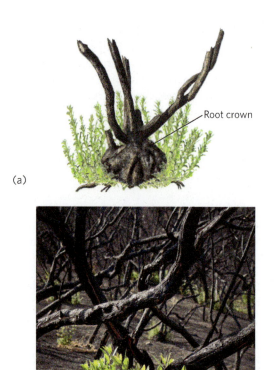

Root crown

(a)

(b)

Figure 4.14 Energy storage. (a) Some species of plants, including one known as death camas due to the toxic chemicals it produces, possess root crowns that are resistant to fire and store large amounts of energy. **(b)** This stored energy can be used by the plant to quickly re-sprout after a fire. When fire burns an area, such as this site in Angeles National Forest, California, the death camas plant can quickly re-sprout. Rob Sheppard/Danita Delimont Stock Photography.

dormancy, a condition in which they dramatically reduce their metabolic processes. One of the most obvious forms of dormancy occurs when many temperate and Arctic trees shed their leaves in the fall before the onset of winter frost and long nights. Similarly, many tropical and subtropical trees shed their leaves during seasonal periods of drought. Plant seeds and spores of bacteria and fungi also exhibit dormancy. Indeed, there are many cases in which researchers have sprouted seeds recovered from archeological excavations where the seeds have been dormant for hundreds of years. Whereas many plants experience dormancy, in animals we distinguish among four specific types of dormancy: *diapause, hibernation, torpor,* and *aestivation*.

In most species, worsening environmental conditions are anticipated and individuals proceed through a series of physiological changes that prepare them for a partial or complete

---

Dormancy A condition in which organisms dramatically reduce their metabolic processes.

Figure 4.15 Hibernation. Some mammals, such as this chipmunk, spend the winter in a deep sleep. During this time, the chipmunk slows its breathing and heart rate and reduces its temperature close to 0°C to help conserve energy through the winter. Leonard Lee Rue III/Getty Images.

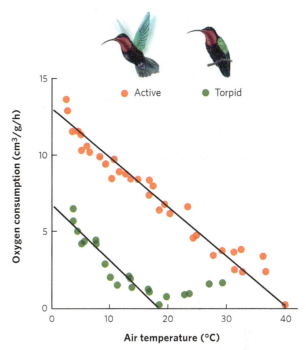

Figure 4.16 Torpor. The tiny West Indian hummingbird has a high surface-to-volume ratio, causing it to lose much of the heat it generates during periods of cold air temperatures. To save energy, measured as the amount of energy consumed to fuel the metabolism of its food, the bird enters torpor when it is resting at night. Data from F. R. Hainsworth and L. L. Wolf, Regulation of oxygen consumption and body temperature during torpor in a hummingbird, *Eulampis jugularis, Science* 168 (1970): 368–369.

physiological shutdown. **Diapause** is a type of dormancy that is common in insects in response to unfavorable environmental conditions. For example, as winter approaches, insects shut down their metabolism to barely detectable levels. In doing so, the insects need to either reduce the quantity of water in their bodies or chemically bind the water in their bodies to prevent freezing. Similarly, insects facing drought conditions can enter a summer diapause by dehydrating themselves. They either tolerate the desiccated condition of their bodies or secrete an impermeable outer covering to prevent further drying.

During **hibernation**, a less extreme type of dormancy that occurs in mammals, animals reduce the energetic costs of being active by lowering their heart rate and decreasing their body temperature. Many mammals, including ground squirrels and bats, hibernate during seasons when they are unable to find food (**Figure 4.15**). Prior to hibernation, the animal consumes enough food to produce a thick layer of fat that provides the energy necessary to survive the hibernation period without eating.

Some types of dormancy occur over short periods of time to deal with cold temperatures. At low air temperatures, some birds and mammals are not able to maintain a high body temperature. Doing so would require that the animal burn up its stored energy faster than it can consume and digest the food needed to replace the energy lost in creating body heat. In this situation, the animal might go into a brief period of dormancy, known as **torpor,** during which the animal reduces its activity and its body temperature decreases. During torpor, reduced activity and reduced body temperature help conserve energy. Torpor may last as little as a few hours or extend for a few days and is a voluntary, reversible condition.

Many small birds and mammals use torpor. Hummingbirds, a group of tiny birds with body lengths of 7.5 to 13 cm, provide a good example. These small birds have a high surface-to-volume ratio. This causes a rapid

loss of heat across the body surface relative to the volume of body that can produce heat. As the air temperature declines, hummingbirds must metabolize increasing amounts of energy to maintain a resting body temperature near 40°C. **Figure 4.16** shows the relationship between air temperature and energy needs for the West Indian hummingbird (*Eulampis jugularis*). Its metabolic rate is measured as the amount of oxygen consumed as it converts its stored energy into body heat. When the bird enters torpor, it reduces its resting body temperature by 18°C to 20°C. If the air temperature drops to 20°C, torpor allows the bird to stop burning extra energy to generate body heat and thereby conserve its energy reserves. Torpor does not mean that the animal ceases to regulate its body temperature; it merely changes the set point on its thermostat to reduce the difference between ambient and body temperature, and thereby reduces the energy expenditure needed to maintain its temperature at the set point.

Diapause A type of dormancy in insects that is associated with a period of unfavorable environmental conditions.

Hibernation A type of dormancy that occurs in mammals in which individuals reduce the energetic costs of being active by lowering their heart rate and decreasing their body temperature.

Torpor A brief period of dormancy that occurs in birds and mammals in which individuals reduce their activity and their body temperature.

ANALYZING ECOLOGY

Correlations

Eco TV

macmillanlearning.com/ ricklefsvideo

In the hummingbird example, we saw that as air temperature decreased, the bird's consumption of oxygen increased (Figure 4.16). This is an example of a statistical *correlation*. A **correlation** is a statistical description of how one variable changes in relation to another variable. For example, at the beginning of this chapter, we observed that when tadpoles faced more dangerous predators, they exhibited larger phenotypic changes.

Two variables can be related to each other in a variety of ways, shown in the graphs below. A *positive correlation* (a) indicates that as one variable increases in value, the second variable also increases. A *negative correlation* (b) indicates that as one variable increases in value, the second variable decreases. Such increases or decreases may be linear, meaning that the data fall along a straight line, as shown in the example. They may also be curvilinear, as shown in (c) and (d), meaning that the data follow a curved line.

Correlations do not tell us anything about causation. For example, the positive correlation between variable A and variable B in panel (a) might occur because a change in variable A causes the change in variable B. But it could also be that a change in variable B causes the change in variable A. Alternatively, a third, unmeasured variable might cause both A and B to change.

Consider the case of humans climbing Mount Everest. As a person ascends the mountain over several weeks, the temperature continually decreases and the climber's efficiency in obtaining oxygen from the air increases. This is a correlation, but it is not causation. Lower temperatures do not cause a person to acquire oxygen more efficiently. We know that a third variable, the declining oxygen pressure at high altitudes, is the real cause of the increased efficiency in climbers obtaining oxygen from the air during the time that they spend climbing.

YOUR TURN For the following set of data, construct a graph that demonstrates the relationship between the number of predators in a pond and the activity level of gray treefrog tadpoles (i.e., the proportion of their time spent moving). After constructing the graph, determine (a) whether it is a positive or negative correlation and (b) whether it is linear or curvilinear.

Number of predators	Activity level
0	40
1	20
2	10
3	5
4	3

Correlation A statistical description of how one variable changes in relation to another variable.

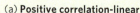

(a) **Positive correlation-linear**

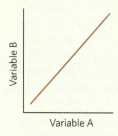

Variable B

Variable A

(b) **Negative correlation-linear**

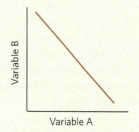

Variable B

Variable A

(c) **Positive correlation-curvilinear**

Variable B

Variable A

(d) **Negative correlation-curvilinear**

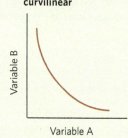

Variable B

Variable A

Correlations. (a) When an increase in one variable is associated with a linear increase in another variable, it is a positive linear correlation. **(b)** When an increase in one variable is associated with a linear decrease in another variable, it is a negative linear correlation. Correlations also can be **(c)** positive and curvilinear or **(d)** negative and curvilinear.

A fourth type of dormancy is **aestivation,** which is the shutting down of metabolic processes during the summer in response to hot or dry conditions. Well-known aestivating animals include snails, desert tortoises, and crocodiles.

By whatever mechanism it occurs, dormancy reduces exchange between organisms and their environments, enabling animals and plants to survive in unfavorable conditions.

ADAPTATIONS TO PREVENT FREEZING

In Chapter 2, we discussed the adaptations that some aquatic animals have, to avoid the damaging effects that freezing can cause to their tissues. In a similar fashion, some terrestrial animals survive very cold weather on land by using special adaptations, including the production of antifreeze chemicals that prevent or control the formation of ice crystals. For example, in cold northern climates, many insects spend the winter living just beneath the bark of trees. This bark helps insulate insects to some degree, but they still experience temperatures below the freezing point of water. Similarly, many species of amphibians spend the winter buried a short distance below the soil surface. These animals can freeze solid underground in a state that requires very little metabolic activity (**Figure 4.17**). Two strategies—the use of antifreeze chemicals and forming ice crystals between rather than within cells—help insects and amphibians avoid tissue damage. As temperatures warm in the spring, the animals slowly thaw out and resume their normal activities.

Migration, storage, and dormancy represent phenotypically plastic strategies that allow organisms to cope with extreme changes in their environment. Such behavioral and physiological flexibility provides a substantial selective advantage.

Figure 4.17 Freezing solid. Many species of frogs, such as this gray treefrog, can become dormant during cold winter months and then thaw out when spring arrives. By producing antifreeze and forming ice crystals between their cells rather than within their cells, the frogs can dramatically reduce their metabolic activity through the winter. Photo by K.B. Storey, Carleton University.

> ### CONCEPT CHECK
> 1. Explain how migration and dormancy are both examples of phenotypic plasticity.
> 2. Why would it be adaptive for subtropical trees to shed their leaves during seasonal periods of drought?
> 3. Explain the differences between the four types of dormancy in animals.

4.5 Variation in food quality and quantity is the basis of optimal foraging theory

As we discussed earlier in this chapter, animal behavior is a form of phenotypic plasticity. Foraging is one of many important behaviors for animals and a great deal of research has been conducted on how animals search for food and select from a diversity of food choices. Because abundance of food items varies over space and time, no single feeding strategy can maximize an animal's fitness. Hence, feeding decisions represent phenotypically plastic behavior because different feeding strategies represent different behavioral phenotypes. These phenotypes are induced by unique environmental cues and each alternative feeding behavior is well suited to a particular environment but not well suited to other environments. Therefore, the alternative behavioral phenotypes experience fitness trade-offs.

Animals must determine where to forage, how long to feed in a certain habitat, and which types of food to eat. Ecologists predict how animals should make foraging decisions based on estimated costs and benefits of feeding in particular environmental situations. They then compare these predictions to observations of foraging animals to see which strategy provides the highest fitness. Although it would be ideal to measure costs and benefits in terms of survival and reproduction, these components of evolutionary fitness can be difficult to measure. Consequently, ecologists usually look at factors correlated with fitness, such as foraging efficiency. This is based on an assumption that animals able to gather more food in less time should be more successful at survival and reproduction. The idea that animals should strive for the best balance between the costs and benefits of different foraging strategies is known as **optimal foraging theory.**

Animals have four responses to food variation in space and time: *central place foraging, risk-sensitive foraging, optimal diet composition,* and *diet mixing.*

CENTRAL PLACE FORAGING

When birds feed their offspring in a nest, the chicks are tied to a single location, while the parents are free to search for food at

---

Aestivation The shutting down of metabolic processes during the summer in response to hot or dry conditions.

Optimal foraging theory A model describing foraging behavior that provides the best balance between the costs and benefits of different foraging strategies.

a distance. This situation is known as **central place foraging** because acquired food is brought to a central place, such as a nest with young. As the parent ranges farther from the nest, it finds a greater amount of potential food sources. However, traveling a longer distance increases time, energy costs, and exposure to risk. The animal must choose the amount of time spent gathering food before returning to the nest as well as how much food to bring back with each trip.

Researchers have used these choices to investigate the feeding behavior of European starlings (*Sturnus vulgaris*). During the summer, starlings typically forage on lawns and pastures for the larvae of craneflies, called leatherjackets. Starlings feed by thrusting their bills into the soft ground and spreading their beak to expose the prey. When they are gathering food for their young, they hold captured leatherjackets in the base of their bill. Researchers predicted that as a starling continued to capture more leatherjackets and to hold them in its bill, grabbing the next one would become more difficult This is analogous to shopping for items in a grocery store without a cart or a basket; the more items you hold, the harder it is to add another. As a result, the number of prey caught by the starlings should slow down over time. The prediction, shown in **Figure 4.18**, was supported by the data. The shape of the curve shows that the rate of food gathering rises rapidly at first and then, as the starling fills its beak, begins to slow down. We say that the starling experiences diminishing benefits over time.

The rate at which the parent bird brings food back to its offspring is a function of how much food it obtains and how much total time it takes to obtain the food. The total time it takes to obtain the food depends on the time needed to fly round-trip to the site that contains the food, known as the *traveling time,* plus the time spent obtaining the food once the bird arrives at the site, known as the *searching time.* **Figure 4.19** shows a graphical model of how an animal should make decisions as an optimal forager. The diminishing benefits line (from Figure 4.18) is shown in orange. To this, we can add a fixed traveling time, which is the amount of time the bird needs to get to the feeding area. We can then

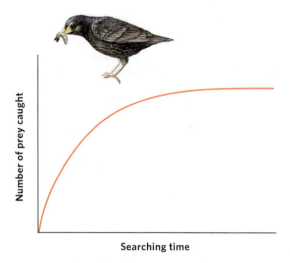

Figure 4.18 Diminishing benefits over time. The rate of food gathering for the European starling is rapid at first but, as time passes, it experiences diminishing benefits because the amount of prey gathered per unit of time decreases.

draw a red line from the origin of the trip to intersect the benefits curve. If we draw the red line at the steepest slope that intersects the orange benefits curve, the two lines cross at the red dot on the figure. This intersection point—which is drawn tangent to the orange benefits curve—represents the highest rate of food capture the bird can obtain, including traveling time. If the starling expressed any alternative behavioral phenotypes—for example, if it stayed at the feeding site for longer or shorter periods indicated by the black dots in the figure—it would have a lower rate of food acquisition. You can see this lower rate of food acquisition as the dashed black line in the figure.

Given our understanding of how the starling should forage when the feeding location is at a fixed distance from its nest, how should the bird's behavior change when the food source is closer or farther away? At sites that are farther

---

Central place foraging Foraging behavior in which acquired food is brought to a central place, such as a nest with young.

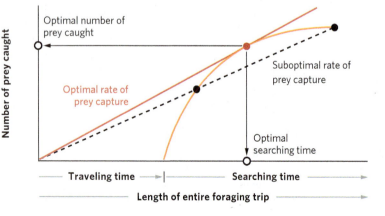

Figure 4.19 Optimal foraging. The optimal rate of foraging for an animal that leaves its nest to find food depends on the time needed to travel to a location that contains food and the time spent feeding once it has arrived. For a given benefits curve (the orange line), the optimal rate of prey capture is found by drawing a straight line from the origin of the trip tangent to the benefits curve. The point of tangency indicates the optimal time that the animal should spend searching and the optimal amount of food it should bring back. Spending more or less time feeding in the location, as indicated by black dots, results in suboptimal amounts of food obtained per unit of time.

away, the bird should spend more time searching for food and bring back more food to help offset the extra travel time. In contrast, as travel time decreases for sites that are closer to the nest, the bird should spend less time searching the site for food and bring back less food. Recall the example of the grocery store. If the store happened to be across the street from your house, you would probably make frequent trips, spend a short amount of time searching for food, and bring back a few items on each trip. If the store were an hour's drive away, you would likely make fewer trips, spend a longer time searching for food, and bring back an armload of items on each trip. As with the starling, these decisions improve your efficiency in bringing back food.

To what extent do organisms actually forage optimally? Researchers addressed this question in a clever experiment. They trained starlings to visit feeding tables that offered mealworms through a plastic tube at precisely timed intervals. A starling would arrive at the table, pick up the first mealworm, and then wait for the next mealworm to be delivered. Each successive mealworm was presented at progressively longer intervals, mimicking the longer intervals at which a starling would catch leatherjackets as its beak became increasingly full. The researchers set up feeding tables at different distances from the starling nests and observed how many mealworms a starling picked up before it departed back to its nest. The predicted number of prey brought back to the bird's nest, shown as a blue line in **Figure 4.20**, agrees with the actual number observed in the experiment, represented by the red data points.

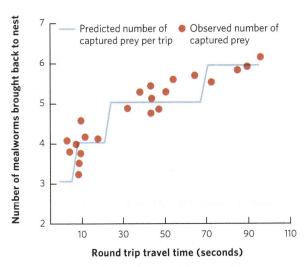

Figure 4.20 Predicted versus observed prey capture for a central-place forager. Based on an optimal foraging model, researchers predicted that longer travel times would cause starlings to return to their nests with a larger number of mealworms. The researchers offered mealworms to starlings on tables that were located at different distances from their nests. The observed number of mealworms brought back to the nests show agreement with the predictions. After A. Kacelnik, Central place foraging in starlings (*Sturnus vulgaris*), I. Patch residence time, *Journal of Animal Ecology* 53 (1984): 283–299.

RISK-SENSITIVE FORAGING

Our predictions of how animals should forage have assumed that animals are simply maximizing their rate of energy gain. However, most animals have other considerations, including predators. Because the act of feeding puts most animals at risk for predation, they must consider this danger when making their foraging decisions. Animals that incorporate the risk of predation into their foraging decisions are said to practice **risk-sensitive foraging.**

The creek chub is a fish that faces the common challenge of finding its lunch rather than becoming lunch for a predator. Small creek chubs feed on sludge worms (*Tubifex* spp.) that live in the mud of streams; understandably, the fish prefer to feed in locations that have more worms. But what if locations containing more worms also contain more predators including cannibalistic larger creek chubs? How much food would it take to entice the small creek chubs to feed in the riskier location? To address this question, researchers placed small creek chubs into an artificial stream that contained a refuge from predators in the middle section of the stream. On one end of the stream, the researchers placed a large creek chub and a low density of worms. On the other end of the stream, they placed two large creek chubs and manipulated different densities of worms. As you can see in **Figure 4.21**, when the opposite end of the stream had two large creek chubs, the small creek chubs would not move to the side containing two predators until that side contained at least three times as many worms to feed on. In short, the small creek chubs balance the risk of predation against the benefits of food availability and adjust their behavioral phenotype accordingly.

OPTIMAL DIET COMPOSITION

Most animals do not consume a single food item but choose from a range of food items. For example, consider the food choices of coyotes living in the western United States. In Idaho, the coyote can consume a variety of prey species, including small prey such as voles (*Microtus montanus*), medium prey such as cottontail rabbits (*Sylvilagus nuttallii*), and large prey such as jackrabbits (*Lepus californicus*). Larger prey species provide larger energy benefits to the coyote, but require more time and energy to subdue and consume. Given these options, the coyote has to decide which prey species it should pursue and which prey species it should ignore.

To determine the optimal food decision, we must balance the energy obtained from the prey and the **handling time**—the time required to subdue and consume the prey. Assuming the same amount of handling time for each choice, the optimal decision for the predator will depend on the energy obtained from each prey and the abundance of the prey. The optimal

Risk-sensitive foraging Foraging behavior that is influenced by the presence of predators.

Handling time The amount of time that a predator takes to consume a captured prey.

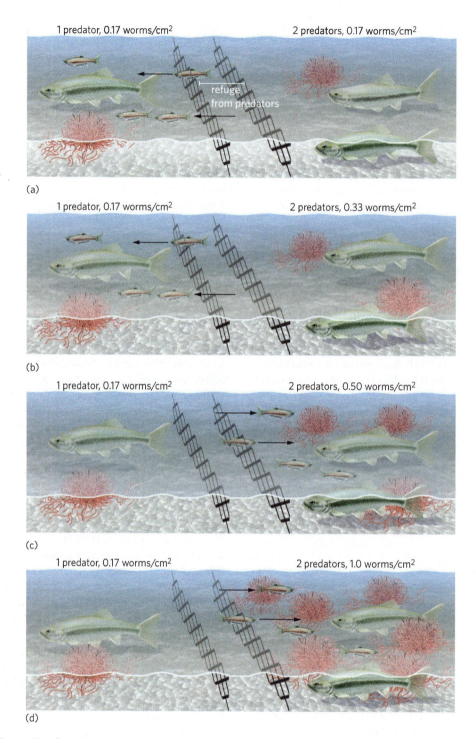

(a)

(b)

(c)

(d)

Figure 4.21 Risk-sensitive foraging. The sensitivity of young creek chubs (*Semotilus atromaculatus*) to food density and predators was tested in artificial streams. All streams had one side that contained one predatory adult creek chub and a low density of food (0.17 worms/cm²). **(a)** When the right end of the stream contained two predators and the same low density of food, the young chubs moved to the left side. **(b)** When the right end contained two predators and a medium food density (0.17 worms/cm²), the young chubs still moved to the left side. Only when the side with two predators contained either a **(c)** high food density (0.50 worms/cm²) or **(d)** very high food density (1.0 worms/cm²) did the young chubs move to the right side of the stream. After J. F. Gilliam and D. F. Fraser, Habitat selection under predation hazard: Test of a model with foraging minnows, *Ecology* 68 (1987): 1856–1862.

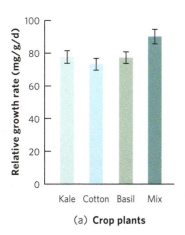

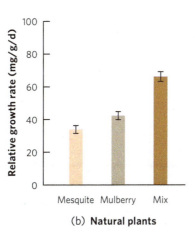

(a) **Crop plants**

(b) **Natural plants**

Figure 4.22 Mixed diets. Young grasshoppers grow faster on mixed diets than on any single diet, regardless of whether comparisons were made using **(a)** crop plants or **(b)** natural plants. In general, mixed diets supply a more complete range of nutrients needed by animals than single diets. Error bars are standard errors. After E. A. Bernays et al., Dietary mixing in a generalist herbivore: Tests of two hypotheses, *Ecology* 75 (1994): 1997–2006.

decision can change if handling time is not equal. In this situation, we need to consider the amount of energy gained per unit time for each prey species. We can do this by dividing the energy benefit of a prey item by its handling time. When we do this, we sometimes find that the smallest prey should be consumed because, although they provide a smaller energy benefit than larger prey, their low handling time can provide the predator with the highest energy gain per unit time. In the case of the coyotes, researchers have found that although jackrabbits require more effort to catch and consume than cottontail rabbits or voles, jackrabbits also provide a larger energy benefit, so the coyotes should rank jackrabbits as the most profitable food item, followed by cottontail rabbits and then voles.

Once we know how different food items compare in terms of energy gained per unit of handling time, we can make a number of predictions. For example, the predator should always eat the prey species that provides the highest energy benefit; if this prey is abundant, it is the only prey that the predator should consume. This strategy maximizes the animal's energy gain. However, if this highest-energy prey is rare and the predator's energy needs are not met, the animal should include less profitable items in its diet. Prey species of very low energy value should never be included in the diet unless all higher-energy prey are scarce. In the case of the coyote, researchers have found that the coyotes appear to be making optimal diet choices. The coyotes always consumed the jackrabbits regardless of their abundance. However, when jackrabbits became less abundant, the coyotes would increase their consumption of cottontail rabbits and voles.

DIET MIXING

Some foragers consume a varied diet because one type of food might not provide all the necessary nutrients. Humans, for example, can synthesize many amino acids, but essential amino acids can only be obtained from one's diet. A diet of only rice or only beans does not possess the complete set of essential amino acids needed by humans. However, a diet that combines rice and beans contains all the required essential amino acids because each contains the essential amino acids that are missing in the other.

The benefits of diet mixing have been demonstrated using nymphs (immature stages) of the American grasshopper (*Schistocerca americana*). As you can see in **Figure 4.22a**, grasshopper nymphs grew faster when fed a mixture of kale, cotton, and basil than when they were offered any one of these food plants alone. The effect was even more pronounced on lower-quality, natural food plants, such as mesquite and mulberry: Nymphs with mixed diets grew almost twice as fast as those feeding on either one of these plant species alone, shown in Figure 4.22b. Based on these data, we might predict that when given a choice, these grasshoppers would decide to forage on a mixed diet to improve their fitness.

Throughout this chapter, we have seen that organisms commonly experience spatial and temporal variation in their environment. In response to this variation, many have evolved the ability to produce multiple phenotypes from a single genotype. The strategy of using multiple phenotypes—including changes in morphology, physiology, or behavior—is effective when there are trade-offs such that no single phenotype performs well in all environments. The evolution of phenotypic plasticity is common among all groups of organisms on Earth, wherever there are reliable environmental cues.

CONCEPT CHECK

1. Why is optimal foraging an example of phenotypic plasticity?
2. Why does central place foraging cause animals that travel farther to bring back larger amounts of food?
3. What are the costs and benefits that animals must consider during risk-sensitive foraging?

ECOLOGY TODAY   CONNECTING THE CONCEPTS

Responding to Novel Environmental Variation

Elevated CO$_2$ experiment. Tall towers at the Duke University Forest have been emitting CO$_2$ into the atmosphere for several years and researchers have been tracking the effects on the plants

Jeffery S. Pippen, http:people. duke.edu/~jspippen/nature.htm.

Ecologists have a good understanding of phenotypically plastic adaptations to environmental variation that has been present for hundreds of thousands of generations, long enough to evolve an appropriate phenotypic response mechanism. But how do organisms respond to more recent environmental variation?

One of the most profound changes in our environment has been the global increase in atmospheric CO$_2$. In 1958, Charles Keeling began recording atmospheric CO$_2$ concentrations atop 3,400-m-high Mauna Loa on the island of Hawaii. Keeling wanted to determine whether anthropogenic emissions were increasing the concentration of CO$_2$ in the atmosphere. At the time he began his study, scientists had no accurate long-term measurements of atmospheric CO$_2$ concentrations. In 1958, the CO$_2$ concentration was about 316 parts per million (ppm; 316 CO$_2$ molecules per million molecules of air, mostly nitrogen and oxygen). During the subsequent decades, the concentration of CO$_2$ in the atmosphere has increased dramatically, rising to 354 ppm by 1990 and 404 ppm by the end of 2016, with no sign of leveling off. Other research indicates that this concentration of CO$_2$ has not been present on Earth for at least the past 10,000 years. As demand for energy and agricultural land increases, the concentration of CO$_2$ is expected to reach 500 to 1,000 ppm by the year 2100.

How will organisms respond to such a change in their environment given that they have not experienced such high concentrations of CO$_2$ in the past 10,000 years? To address this question, scientists began conducting large outdoor experiments in the1990s in which tall towers emit CO$_2$ gas over forests, deserts, wetlands, and croplands to cause an elevation in CO$_2$ ranging from 475 to 600 ppm. Many of these experiments continue today and, averaged across several such experiments, researchers have found that elevated CO$_2$ causes an increase in the rate of photosynthesis by 40 percent. In addition, because plants open their stomata to obtain CO_2, the elevated CO_2 concentrations allow plants to keep their stomata closed more often, resulting in a 22 percent reduction in transpiration of water. All this translates into improved plant growth. Plants experiencing elevated CO_2 had a 17 percent increase in the growth of their shoots and a 30 percent increase in the growth of their roots.

These growth responses represent averages across a variety of species, but not all species responded in the same manner. For example, growth improved in C_3 plants but not in C_4 plants. Researchers hypothesize that because the C_4 pathway of photosynthesis already pumps high concentrations of CO_2 into the bundle sheath cells of the leaf, higher atmospheric concentration of CO_2 has little additional effect. Given that most plant species use the C_3 pathway, most plants will experience higher growth unless other nutrients become limiting or if herbivory of the plants also increases and causes an increased loss of plant tissues. On the other hand, C_4 plants, which include corn, sugarcane, and many other important crops, are not expected to grow any faster as humans continue to elevate the concentration of CO_2 in the atmosphere.

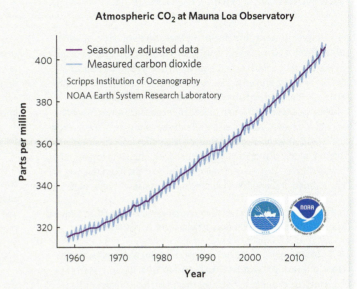

Atmospheric CO_2 at Mauna Loa Observatory

Scripps Institution of Oceanography
NOAA Earth System Research Laboratory

Changes in atmospheric CO_2 over time. Measurements on the island of Hawaii have shown that CO_2 concentrations have continuously increased during the past 50 years. After http://www.esrl.noaa.gov/gmd/ccgg/trends/.

In 2016, researchers announced that they were planning to begin the first CO_2 emission experiments in a tropical rainforest in which they will expose small areas of the forest to 600 ppb of CO_2. This is important because it has been hypothesized that tropical rainforests may absorb a lot of atmospheric CO_2. The project, which will be based in Manaus, Brazil, faces a number of challenges, including raising tens of millions of dollars to fund the research project and the practical problem of how to transport thousands of metric tons of CO_2 along a 34-km treacherous dirt road in the Amazon. The project is expected to last at least 12 years and provide a number of insights regarding how much of the elevated CO_2 will be absorbed by the plants and whether the forest will respond with increased growth rates.

The change in CO_2 concentrations is just one example of the many anthropogenic changes occurring on Earth today. Most organisms have flexible phenotypes that have been shaped by natural selection in response to past environmental variation. Organisms facing novel environmental variation from anthropogenic causes may be able to make use of these existing adaptations and they may also experience continued evolution for new types of flexible phenotypes. However, many other types of anthropogenic impacts—such as air and water pollution—may be far outside the range of historic environmental variation for a population. As a result, populations may not possess phenotypically plastic strategies that will allow them to perform well when facing these types of anthropogenic impacts.

SOURCES:
Grossman, D. 2016. Amazon rainforest to get a growth check. *Science* 353:635–636.
Jaub, D. 2010. Effects of rising atmospheric concentrations of carbon dioxide on plants. *Nature Education Knowledge* 1:21.
Ainsworth, E. A., and S. P. Long. 2005. What have we learned from 15 years of free-air CO_2 enrichment (FACE)? A meta-analytic review of the responses of photosynthesis, canopy properties and plant production to rising CO_2. *New Phytologist* 165:351–372.

SUMMARY OF CHAPTER CONCEPTS

4.1 Environmental variation favors the evolution of variable phenotypes. Temporal variation occurs across a range of hours to years, with the most extreme variation being the least frequent. Spatial variation also exists due to differences in climate, topography, and soils. Phenotypic plasticity, the ability to produce alternative phenotypes, is favored when organisms experience environmental variation, when reliable cues indicate the current state of the environment, and when no single phenotype is superior in all environments. Phenotypically plastic traits include behavior, physiology, morphology, and life history. Each type of trait differs in how fast it can respond to environmental change and whether the responses are reversible.

Key Terms: Temporal environmental variation, Weather, Climate, Phenotypic trade-off, Phenotypic plasticity, Acclimation

4.2 Many organisms have evolved adaptations to variation in enemies, competitors, and mates. Responses to enemies include changes in behavior that make individuals harder to detect, morphological defense that makes prey harder to capture, and chemical defense that makes prey less palatable. Responses to competitors include morphological changes in plants to make them better able to obtain resources, morphological changes in animals that make them better able to consume and digest scarce food, and behavioral strategies in animals that improve their ability to find scarce food. Organisms generally favor breeding with another individual, but a scarcity of mates can make self-fertilization a viable alternative for some species.

Key Terms: Hermaphrodite, Inbreeding depression

4.3 Many organisms have evolved adaptations to variable abiotic conditions. Variation in temperature has favored the evolution of isozymes and switching between microhabitats. Variation in water availability has favored plants that can open and close their stomata and alter their root-to-shoot ratios. Variation in salinity has favored the evolution of novel ways of adjusting solute concentrations to minimize the cost of osmoregulation. Variation in oxygen can cause adaptive increases in red blood cells and hemoglobin to improve the uptake of oxygen at high altitudes.

Key Term: Microhabitat

4.4 Migration, storage, and dormancy are strategies used to survive extreme environmental variation. Migration allows organisms to leave areas with degrading environments, storage allows organisms to have an extra supply of energy to make it through periods of degrading environments, and dormancy allows organisms to shut down their metabolism until the harmful environmental conditions have passed.

Key Terms: Migration, Dormancy, Diapause, Hibernation, Torpor, Correlation, Aestivation

4.5 Variation in food quality and quantity is the basis of optimal foraging theory. Central place foraging predicts that the amount of time spent foraging at a site and the amount of food brought back to a central nest will depend on the benefits gained over time at the site and the round-trip travel time. Risk-sensitive foragers consider not only the energy to be gained, but also the predation risk posed. Many animals must also consider a range of alternative food items, the energy and abundance of each food item, and whether they should consume a mixture of food items to meet all their nutritional needs.

Key Terms: Optimal foraging theory, Central place foraging, Risk-sensitive foraging, Handling time

CRITICAL THINKING QUESTIONS

1. What is the difference between weather and climate in terms of temporal variation?

2. How could spatial heterogeneity be perceived by an organism as temporal heterogeneity?

3. Why do we have to consider the mean fitness across all environments when evaluating whether evolving a phenotypically plastic genotype will be favored over a non-plastic genotype?

4. How would the presence of unreliable environmental cues affect the evolution of phenotypically plastic responses to environmental variation?

5. Would the frequency of experiencing predator versus no-predator environments affect the evolution of phenotypically plastic traits?

6. If predators reduce prey abundance to very low levels, how might variable predator environments affect the evolution of flexible mating strategies?

7. If a plant can improve its ability to obtain water by growing more roots, why shouldn't the plant always grow more roots?

8. How could you experimentally determine whether migrating birds use day length or temperature as an environmental cue for migration?

9. What is the relationship between correlation and causation?

10. When determining whether to feed on small, medium, or large prey, why should predators evaluate both the energy obtained from each prey item, the abundance of each prey item, and the handling time of each prey item?

GRAPHING THE DATA The Foraging Behavior of American Robins

The following data were collected by a scientist observing the number of earthworms that American robins (*Turdus migratorius*) can hold in their beaks as they searched a lawn after a summer rain. Graph the data and describe the relationship between time and the number of collected earthworms. Does this correlation between time and the number of worms collected represent both a correlation and causation?

Time (min)	Average number of earthworms collected
0	0
1	1.0
2	2.0
3	2.8
4	3.4
5	3.7
6	3.9
7	4.0
8	4.0

5 Climates and Soils

Where Does Your Garden Grow?

If you have ever planted a garden, you know that you have a lot of decisions to make. You can choose from a vast selection of fruits and vegetables, not to mention a dizzying array of flowers, shrubs, and trees. Although you have a large number of choices, not all plants grow well in all locations. To help gardeners determine what plants can survive and flourish in their location, the U.S. Department of Agriculture has developed a map of plant hardiness zones.

Plant hardiness zones take into account the coldest temperatures that occur during the winter. The hardiest plants tolerate very cold temperatures, whereas others are too sensitive to survive harsh winters. The plant hardiness zones follow the minimum temperature typically reached in locations throughout North America. Zone 10, for example, is found in southern Florida, where the average minimum temperature in winter is between −1°C and 4°C. In contrast, zone 1 is found in Alaska, with an average minimum winter temperature below −45°C.

In looking at the map of plant hardiness zones, several patterns emerge. First, there appears to be an ordering of zones that is related to latitude, particularly through the middle of the continent. Higher latitudes receive less sunlight in the winter and have lower winter temperatures. However, a second pattern of plant hardiness zones curves up along the coastlines. For instance, in the interior of the continent, zone 8 spans states such as Louisiana, Alabama, and Georgia. Along the East Coast, however, zone 8 extends all the way up to Virginia. Along the West Coast, zone 8 extends north, all the way through the state of Washington. These patterns occur because the two coasts are adjacent to oceans, which contain warm tropical waters that circulate up from the equator. These warm waters heat the air during winter and make the land along the coasts warmer than the interior of the continent at the same latitude. A third pattern can be seen in elevation; mountain tops have colder temperatures than lower elevations.

> "The map of plant hardiness zones illustrates how climates around the world are the result of a complex combination of sunlight, latitude, elevation, air currents, and ocean currents."

The hardiness zone map also shows that the West Coast has warmer winter temperatures than the East Coast. This allows farmers in California to grow fruits and vegetables during the winter. The difference in temperatures between the East and West Coasts is caused by prevailing winds that blow from west to east. In the winter along the West Coast, the winds carry the warmer ocean air toward the coast, making it warm. Along the East Coast, however, winds carry cold air from the middle of the continent toward the coast and push the warm ocean air away from the coast. As a result, the East Coast remains colder than the West Coast during the winter.

The map of plant hardiness zones illustrates how climates around the world are the result of a complex combination of sunlight, latitude, elevation, air currents, and ocean currents. In this chapter, we will explore how global processes affect the distribution of climates on Earth and how climates affect the types of soils that form.

A beautiful garden. This garden is located at the Sequim Gardens, in Washington State. Mitch Diamond/Getty Images

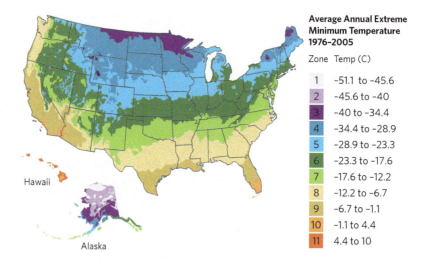

Average Annual Extreme Minimum Temperature 1976–2005

Zone	Temp (C)
1	–51.1 to –45.6
2	–45.6 to –40
3	–40 to –34.4
4	–34.4 to –28.9
5	–28.9 to –23.3
6	–23.3 to –17.6
7	–17.6 to –12.2
8	–12.2 to –6.7
9	–6.7 to –1.1
10	–1.1 to 4.4
11	4.4 to 10

SOURCE: USDA Plant Hardiness Zone Map, http://planthardiness.ars.usda.gov/PHZMWeb/.

Plant hardiness zones for North America. The warmest zones occur in the southern United States and along the coasts.

LEARNING OBJECTIVES

After reading this chapter, you should be able to:

5.1 Describe how Earth is warmed by the greenhouse effect.

5.2 Explain the unequal heating of Earth by the Sun.

5.3 Illustrate how the unequal heating of Earth drives air currents in the atmosphere.

5.4 Demonstrate how ocean currents also affect the distribution of climates.

5.5 Describe the role of smaller-scale geographic features in affecting regional and local climates.

5.6 Explain how climate and the underlying bedrock interact to create a diversity of soils.

As we saw in Chapter 3, the climate of a region on Earth refers to the average atmospheric conditions measured over many years. Climates can range widely, from very cold areas near the North and South Poles, to hot and dry deserts at approximately 30° N and 30° S latitudes, to hot and wet areas near the equator. In this chapter, we will examine factors that determine the location of climates around the world. With an understanding of climates, we will move on to look at how soils are formed. As we will see in subsequent chapters, these differences in climates and soils help determine the distribution of organisms around the globe.

A number of factors contribute to the different climate patterns. Some of the most important are the greenhouse effect, the unequal heating of Earth by solar energy, atmospheric convection currents, the rotation of Earth, ocean currents, and a variety of smaller-scale topographic features.

5.1 Earth is warmed by the greenhouse effect

Solar radiation provides the vast majority of the energy that warms Earth and that organisms use. However, solar radiation alone is not sufficient to warm the planet; gases in the atmosphere also play a critical role.

THE GREENHOUSE EFFECT

Solar radiation warms Earth through a series of steps, illustrated in **Figure 5.1**. About one-third of the solar radiation emitted toward Earth is reflected by the **atmosphere**—the 600-km thick layer of air that surrounds the planet—and heads back into space. The remaining solar radiation penetrates the atmosphere. Much of the high-energy radiation—including ultraviolet radiation—is absorbed in the atmosphere. The rest passes with most of the visible light through the atmosphere. When this radiation subsequently strikes clouds and the surface of Earth, a portion is reflected back into space and the rest is absorbed. As clouds and Earth's surface absorb this radiation, they begin to warm and emit lower-energy infrared radiation. The heat you feel radiating into the air if you stand on a hot asphalt road is an example of this infrared radiation.

Atmosphere The 600-km thick layer of air that surrounds the planet.

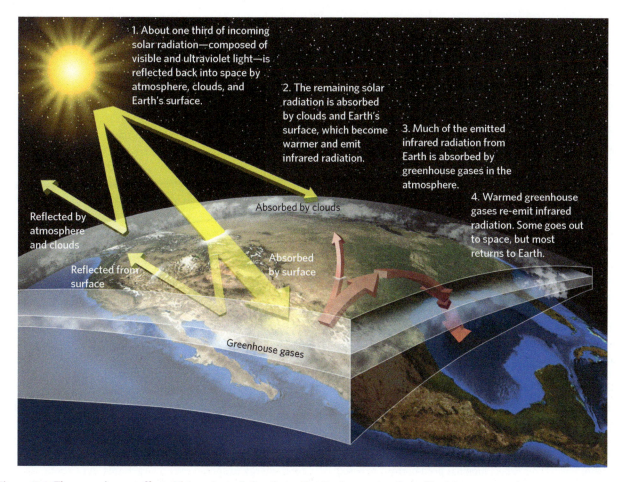

1. About one third of incoming solar radiation—composed of visible and ultraviolet light—is reflected back into space by atmosphere, clouds, and Earth's surface.

2. The remaining solar radiation is absorbed by clouds and Earth's surface, which become warmer and emit infrared radiation.

3. Much of the emitted infrared radiation from Earth is absorbed by greenhouse gases in the atmosphere.

4. Warmed greenhouse gases re-emit infrared radiation. Some goes out to space, but most returns to Earth.

Reflected by atmosphere and clouds

Absorbed by clouds

Reflected from surface

Absorbed by surface

Greenhouse gases

Figure 5.1 The greenhouse effect. Of the solar radiation that strikes Earth, some is reflected back into space and the rest penetrates the atmosphere, where much of it warms clouds and the planet's surface. These warmed objects emit infrared radiation back toward the atmosphere, where it is absorbed by greenhouse gases. The warmed greenhouse gases re-emit infrared radiation back toward Earth, which causes the surface to warm further.

Infrared radiation is readily absorbed by gases in the atmosphere. The gases are warmed by the infrared radiation and then re-emit infrared radiation in all directions. Some of this energy goes out into space and some goes back toward the planet's surface. This process of solar radiation striking Earth, being converted to infrared radiation, and then being absorbed and reradiated by atmospheric gases back to Earth is known as the **greenhouse effect.** The name comes from the fact that the effect resembles a gardener's greenhouse with windows that hold in the heat from solar radiation.

GREENHOUSE GASES

There are many different gases in the atmosphere, but only those that absorb and re-emit infrared radiation and contribute to the greenhouse effect are known as greenhouse gases. In fact, if we exclude water vapor, 99 percent of the gases in the atmosphere are oxygen (O_2) and nitrogen (N_2), and neither gas functions as a greenhouse gas. This means that greenhouse gases, which play such a major role in keeping our planet warm, compose only a small fraction of the atmosphere. This also means that even small changes in the concentrations of greenhouse gases can have large impacts on Earth's temperature.

The two most prevalent greenhouse gases are water vapor (H_2O) and carbon dioxide (CO_2). Other naturally occurring greenhouse gases include methane (CH_4), nitrous oxide (N_2O), and ozone (O_3). These gases have a variety of natural sources: H_2O comes from large bodies of water and the transpiration of plants; CO_2 comes from decomposition, respiration of organisms, and volcanic eruptions; CH_4 comes from anaerobic decomposition; N_2O comes from wet soils and low-oxygen regions of water bodies; and O_3 comes from ultraviolet radiation breaking apart O_2 molecules in the atmosphere and causing each molecule to combine with

Greenhouse effect The process of solar radiation striking Earth, being converted to infrared radiation, and being absorbed and re-emitted by atmospheric gases.

another O_2 molecule. The naturally occurring greenhouse effect is quite beneficial to organisms on Earth. Without this phenomenon, the average temperature on Earth, which is currently 14°C, would be much colder, about −18°C.

 The concentrations of greenhouse gases in the atmosphere are increasing. As we noted in Chapter 4, the concentration of CO_2 in the atmosphere has substantially increased over the past two centuries due to increased combustion of fossil fuels by automobiles, electric generating plants, and other industrial processes. At the same time, there have also been increases in methane and nitrous oxide from a variety of anthropogenic sources that include agriculture, landfills, and the combustion of fossil fuels. Finally, there are greenhouse gases that are not naturally produced, such as chlorofluorocarbons that have been manufactured to serve as propellants in aerosol cans and as refrigerants in freezers, refrigerators, and air conditioners. Although these human-created compounds exist at much lower concentrations than water vapor or CO_2, each molecule can absorb much more infrared radiation than water vapor and CO_2. In addition, each of these human-created molecules persists in the atmosphere for hundreds of years.

A steady increase of these manufactured gases has raised concerns among scientists, environmentalists, and policy makers. Because greenhouse gases absorb and re-emit infrared radiation to Earth and its atmosphere, it makes logical sense that an increase in the concentration of greenhouse gases in the atmosphere would cause an increase in the average temperature of Earth. This expectation has been borne out. Based on thousands of measurements made throughout the world, the average air temperature of the planet's surface increased by approximately 1°C from 1880 to 2016. While some regions, such as Antarctica, have become 1°C to 2°C cooler, other regions, such as Northern Canada, have become up to 4°C warmer. In fact, over the 136-year period of monitoring temperatures around the world, 16 of the 17 warmest years occurred from 2001 to 2016. As we will see at the end of this chapter, these temperature changes are altering global climates.

of the globe, the depth of the atmosphere that the energy passes through, and seasonal changes in the position of Earth relative to the Sun.

THE PATH AND ANGLE OF THE SUN

Consider the position of the Sun during the March and September equinoxes when the Sun is positioned directly over the equator. At these times of the year, the equator receives the greatest amount of solar radiation and the poles receive the least. Three factors dictate this pattern: the distance that sunlight must pass through Earth's atmosphere, the angle at which the Sun's rays hit Earth, and the reflectivity of Earth's surface.

As illustrated in **Figure 5.2**, before the rays of the Sun reach Earth, they must travel through Earth's atmosphere. When they pass through the atmosphere, gases absorb some of the solar energy. Following the path of the rays, you can see that the distance traveled through the atmosphere is shorter at the equator than at the poles. This means that less solar energy is removed by the atmosphere before it strikes Earth at the equator.

The intensity of solar radiation that strikes an area also depends on the angle of the Sun's rays. Looking again at Figure 5.2, you can see that when the Sun is positioned directly above the equator, the rays of the Sun strike Earth at a right angle. This causes a large quantity of solar energy to strike a small area. In contrast, near the poles the rays of the Sun strike Earth at an oblique angle, which causes the solar energy to spread over a larger area. As a result, Earth's surface receives more solar energy per square meter near the equator than near the poles. These differences in solar intensity translate into differences in temperatures with latitude, as we saw in our opening story on plant hardiness zones.

Finally, some surfaces of the globe reflect solar energy more than others. Light-colored objects reflect a higher percentage of solar energy than dark-colored objects, which

---

CONCEPT CHECK

1. What are the steps involved in the greenhouse effect?
2. How does the human production of greenhouse gases lead to global warming?
3. If 99% of all gases in the atmosphere (excluding water vapor) are nitrogen and oxygen, why aren't climate researchers focused on changes in the concentrations of these gases?

5.2 There is an unequal heating of Earth by the Sun

The differences in temperature around the globe are the result of how much solar radiation strikes the surface of Earth at a given location. Differences in solar radiation are determined by the angle of the Sun striking different regions

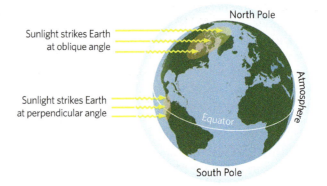

Figure 5.2 Unequal heating of Earth. When the Sun is directly over the equator, its rays travel through less atmosphere and are spread over a smaller area. Near the poles, however, the rays of the Sun must travel through more atmosphere and are spread over a larger area.

Figure 5.3 Albedo effect. Light-colored objects such as fresh snow reflect a high percentage of incoming solar energy, and dark-colored objects reflect very little. The average albedo of Earth is 30 percent.

absorb most incoming solar energy. For example, asphalt absorbs 90 to 95 percent of the total solar energy that strikes its surface, which explains why asphalt pavement becomes so hot on a sunny summer afternoon. On the other hand, cropland reflects 10 to 25 percent of the total solar energy that strikes its surface and fresh snow reflects 80 to 95 percent. The fraction of solar energy reflected by an object is its **albedo.** As you can see in **Figure 5.3**, the more solar energy reflected, the higher the albedo.

The unequal heating of Earth explains the general pattern of declining temperatures as we move from the equator to the poles. At the equator, the Sun's rays lose less energy to the atmosphere, solar energy is spread over a smaller area, and the low albedo of dark-colored forests causes much of this energy to be absorbed. Near the poles, however, the Sun's rays lose much more of their energy to the atmosphere, solar energy is spread over a larger area, and the high albedo of the snow-covered land causes much of this solar energy to be reflected. This helps explain why the plant hardiness zone numbers we discussed at the beginning of this chapter generally decrease as you move to higher latitudes.

SEASONAL HEATING OF EARTH

The relationship between the Sun and Earth also causes seasonal differences in temperatures on Earth. The axis of Earth is tilted 23.5° with respect to the path Earth follows in its orbit around the Sun. **Figure 5.4** illustrates how this tilt affects the seasonal heating of Earth. During the March equinox, the Sun is directly over the equator. As we approach the June solstice, the orbit and tilt of Earth cause the Sun to be directly over 23.5° N latitude, which is also known as the Tropic of Cancer. In September, the Sun is back to being directly over the equator, and in December, the Sun is directly over 23.5° S latitude, which is also known as the Tropic of Capricorn.

The tilt of Earth as it orbits around the Sun causes the Northern Hemisphere to receive more solar energy between March and September than the Southern Hemisphere. During this time the daylight period in the Northern Hemisphere is greater than the nighttime period, and the Sun's angle is 90° somewhere over the Northern Hemisphere. This means that more intense solar radiation is produced per unit area and for a longer period of time. Between the fall equinox in September and the spring equinox in March, the situation reverses and the Southern Hemisphere has longer days and receives more direct solar energy than the Northern Hemisphere. The latitude that receives the most direct rays of the Sun, known as the **solar equator,** shifts throughout the year—from 23.5° N latitude in June to 23.5° S latitude in December. These are the warmest latitudes on Earth and are known as the tropical latitudes.

Albedo The fraction of solar energy reflected by an object.

Solar equator The latitude receiving the most direct rays of the Sun.

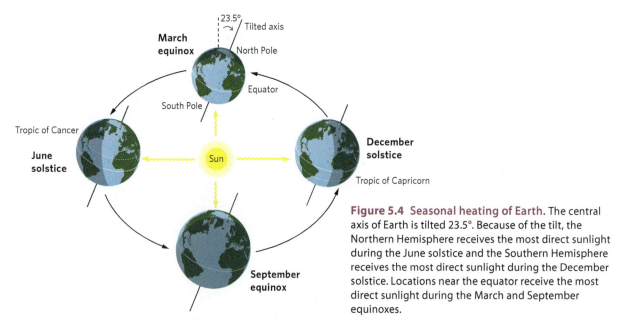

Figure 5.4 Seasonal heating of Earth. The central axis of Earth is tilted 23.5°. Because of the tilt, the Northern Hemisphere receives the most direct sunlight during the June solstice and the Southern Hemisphere receives the most direct sunlight during the December solstice. Locations near the equator receive the most direct sunlight during the March and September equinoxes.

Seasonal changes in temperature vary as Earth traces its annual path around the Sun. While the average temperatures of the warmest and coldest months in the tropics differ by as little as 2°C to 3°C, at higher latitudes in the Northern Hemisphere, average monthly temperatures vary by an average of 30°C over the year and extreme temperatures vary by more than 50°C annually.

CONCEPT CHECK

1. Why is solar energy per unit more intense near the equator than near the poles?
2. What is the albedo effect?
3. What is the solar equator?

ANALYZING ECOLOGY

Regressions

Eco TV
macmillanlearning.com/ricklefsvideo

As we have discussed, latitudes closer to the equator receive more solar radiation than latitudes closer to the poles. Given this observation, lower latitudes should also have warmer temperatures than higher latitudes. In fact, understanding the nature of this relationship would help us determine exactly how much temperature changes with latitude. When we want to know how one variable changes in relation to another, we use a statistical tool called **regression**. In Chapter 4, we saw that a correlation determines if there is a relationship between two variables. A regression determines whether there is a relationship and also describes the nature of that relationship.

To help illustrate this idea, we can use data on the average January temperature from 56 cities around the United States, spanning the latitudes of the contiguous 48 states. If we plot the relationship between city latitude

and average city temperature in January, we obtain the following graph:

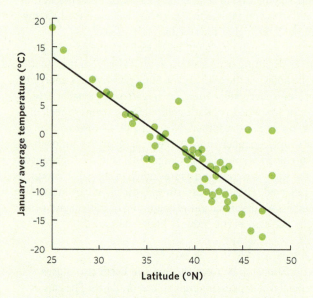

In this case, the relationship between the two variables follows a straight line, and we have drawn a line of best fit

Regression A statistical tool that determines whether there is a relationship between two variables and that also describes the nature of that relationship.

5.3 The unequal heating of Earth drives air currents in the atmosphere

Earth's uneven heating helps determine **atmospheric convection currents,** which are the circulations of air between the surface of Earth and the atmosphere. The patterns of air circulation play a major role in the location of tropical rainforests, deserts, and grasslands throughout the world. In this section, we will explore how the interaction of Earth's unequal heating and the properties of air creates atmospheric convection currents.

PROPERTIES OF AIR

Four properties of air influence atmospheric convection currents: density, water vapor saturation point, latent heat release, and adiabatic heating or cooling.

Air density

In regard to density, when air warms, it expands and becomes less dense. As a result, when air becomes warmer next to the surface of Earth, it becomes less dense than the air above it. This causes the warm air to rise. This is the initial step in creating convection currents.

Water vapor saturation point

As air temperature increases, its capacity to contain water vapor—the gaseous form of water—increases. The graph in **Figure 5.5** shows the relationship between air temperature and the maximum amount of water vapor that the air can contain. Although the capacity to contain water increases at higher temperatures, there is always a limit, known as the **saturation point.** When the water vapor content of air exceeds the saturation point at any given temperature, the excess water vapor condenses and changes into either liquid water or ice. When the water vapor content is below the saturation point, liquid water or ice can be converted to water vapor. For example, at 30°C, air can contain up to 30 g of water vapor per m³. Air that contains the maximum amount of water vapor has reached its saturation point. If the air at 30°C cools to 10°C, the saturation point of the air will decrease to 10 g of water vapor per m³. As a result, the excess vapor changes phases to liquid water and produces clouds and precipitation. The relationship between temperature and water vapor saturation affects patterns of evaporation and precipitation around the world. This, in combination with air currents, determines the distribution of wet and dry environments around the world.

---

through the distribution of the data points. This is a regression line because it represents the relationship between the two variables. It informs us about the nature of the relationship through the slope and intercept of the line. For these data, the regression can be described using the equation of a straight line, where Y is the dependent variable, X is the independent variable, m is the slope of the line, and b is the Y-intercept of the line at the point where $X = 0$. In this example, the slope is -1.2 and the intercept is 43:

$$Y = mX + b$$

$$\text{Temperature} = -1.2 \times \text{latitude} + 43$$

This regression equation tells us that for every 1 degree increase in latitude, the average temperature in January decreases by 1.2°C. Note that while the simplest form of a regression is a straight line, regression lines can also be curvilinear.

YOUR TURN Based on the relationship between latitude and temperature, use the regression equation to estimate the average January temperature at 10, 20, and 30 degrees of latitude.

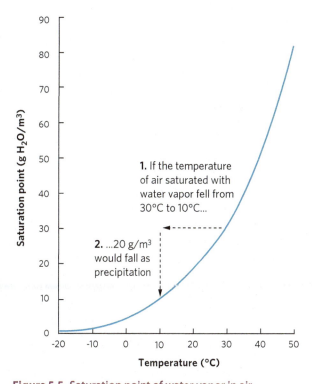

1. If the temperature of air saturated with water vapor fell from 30°C to 10°C...

2. ...20 g/m³ would fall as precipitation

Figure 5.5 Saturation point of water vapor in air. As the temperature of the air increases, the air is able to contain greater amounts of water vapor.

Atmospheric convection currents The circulations of air between the surface of Earth and the atmosphere.

Saturation point The limit of the amount of water vapor the air can contain.

Latent heat release

Another property of air to consider when contemplating atmospheric convection currents is the release of heat. As you might recall from our discussion of the thermal properties of water in Chapter 2, converting liquid water to water vapor requires a great deal of energy. In the reverse process, known as **latent heat release,** water vapor converted back to liquid water releases energy in the form of heat. Latent heat release is significant because whenever water vapor exceeds its saturation point, condensation will cause a release of heat that warms the surrounding air.

Adiabatic heating and cooling

The final factor to consider in regard to convection currents is the movement of air in response to changes in pressure. Near the surface of Earth, the gravitational pull on all molecules in the atmosphere brings many of the molecules close to Earth's surface. An increase in the number of molecules causes an increase in air pressure near Earth's surface. As one moves higher into the atmosphere, the air contains fewer total molecules, which lowers the air pressure. As a result, when air moves between the surface of Earth and the atmosphere, it experiences changes in pressure.

Air pressure is related to the frequency of collisions among air molecules, which also influences temperature. Lower rates of collision cause lower temperatures. As a result, when air moves higher up into the atmosphere and experiences lower pressure, the air expands and the temperature decreases—a process known as **adiabatic cooling.** Conversely, when air moves down to Earth's surface and experiences higher pressure, the air compresses and the temperature increases in a process known as **adiabatic heating.**

FORMATION OF ATMOSPHERIC CONVECTION CURRENTS

Understanding the above four properties of air helps us understand how atmospheric convection currents are formed. Let's begin by looking at the equator during the March or September equinoxes when the Sun is directly over the equator. As you can see in **Figure 5.6,** solar

---

Latent heat release When water vapor is converted back to liquid, water releases energy in the form of heat.

Adiabatic cooling The cooling effect of reduced pressure on air as it rises higher in the atmosphere and expands.

Adiabatic heating The heating effect of increased pressure on air as it sinks toward the surface of Earth and decreases in volume.

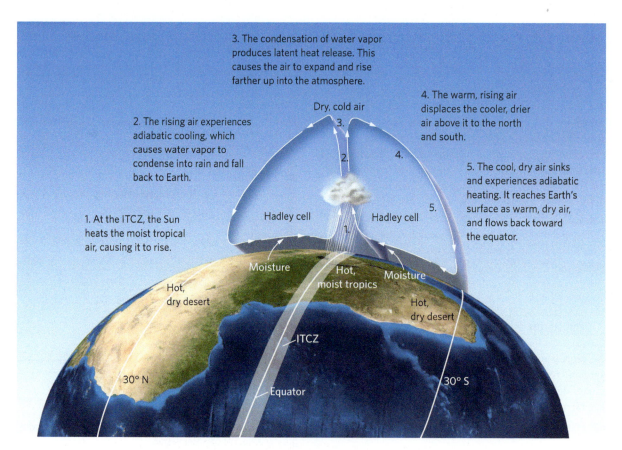

Figure 5.6 The circulation of air in Hadley cells. In this example, the Sun is directly over the equator, as happens during the March and September equinoxes. At the latitude receiving the most direct sun, a column of warm air rises and the intertropical convergence zone (ITCZ) drops its precipitation. After rising more than 10 km into the atmosphere, the now cool, dry air circulates back to Earth at approximately 30° N and 30° S latitudes.

energy warms the air at the surface of Earth. This warming causes the air to expand and rise. As air rises into regions of decreased atmospheric pressure, it expands. As the air expands, the temperature of the air cools through the mechanism of adiabatic cooling. This cooled air has a reduced capacity to contain water vapor, so the excess water vapor condenses and falls back to Earth as rain. This process, in which the surface air heats, rises, and releases excess water vapor in the form of rain, is the primary reason that latitudes near the equator experience high amounts of rainfall.

Returning to Figure 5.6, we can see that when the water vapor condenses, it causes latent heat release, which further warms the air. This enhances the upward motion of the air, condensation, and rainfall. As the air pressure continues to fall with rising altitude, the air temperature continues to drop. At high altitudes, the cool, dry air is pushed from below by more rising air and begins to move horizontally toward the poles.

The upward movement of air is the driving force behind atmospheric convection currents, but it is just the first of a series of steps in the process. Once the cool, dry air is displaced horizontally toward the poles, it begins to sink back toward Earth at approximately 30° N and 30° S latitudes. As Figure 5.6 illustrates, this dry air sinks toward Earth where increased pressure causes it to compress. As the air compresses, it experiences adiabatic heating. By the time the air falls back to the surface of Earth, it is hot and dry. This explains why many of the major deserts of the world—which are characterized by hot, dry air—are located at approximately 30° N and 30° S latitudes. Once this hot, dry air reaches the ground, it flows back toward the equator, completing the air circulation cycle. The two circulation cells of air between the equator and 30° N and 30° S latitudes are known as **Hadley cells.** The area where the two Hadley cells converge and cause large amounts of precipitation is known as the **intertropical convergence zone (ITCZ).**

The intense sunlight at the solar equator drives the Hadley cells and the ITCZ, causing the warmed air to rise and precipitation to be released in the form of rain. As we saw in our earlier discussion about unequal seasonal heating of Earth, we know that the solar equator shifts throughout the year, from 23.5° N latitude in June to 23.5° S latitude in December. Because the latitude of the solar equator moves throughout the year and the latitude of the solar equator determines the latitude of the ITCZ, the latitude of the ITCZ also moves throughout the year. This also means that the seasonal movement of the solar equator influences seasonal patterns of rainfall.

You can see the effect of the ITCZ movement in **Figure 5.7** by examining the patterns of rainfall across

---

Hadley cells The two circulation cells of air between the equator and 30° N and 30° S latitudes.

Intertropical convergence zone (ITCZ) The area where the two Hadley cells converge and cause large amounts of precipitation.

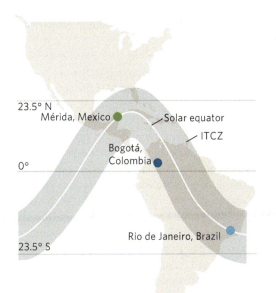

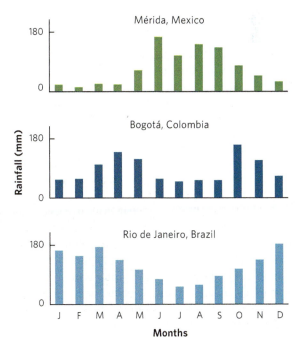

Figure 5.7 Rainy seasons and the ITCZ. As the solar equator moves throughout the year, so does the ITCZ. As a result, latitudes north and south of the equator have a single distinct rainy season, whereas latitudes close to the equator have two rainy seasons.

three locations in the Western Hemisphere. The city of Mérida, Mexico, lies about 20° N of the equator. The intertropical convergence reaches Mérida only during June, which is why June is the rainy season for Mérida. In comparison, Rio de Janeiro, Brazil, lies about 20° S latitude. The intertropical convergence reaches Rio de Janeiro in December, which is the middle of the rainy season for that city. Close to the equator, in Bogotá, Colombia, the intertropical convergence zone passes overhead twice each year, during the March and September equinoxes. As a result, Bogotá experiences two rainy seasons.

The pattern of air circulation near the equator also exists near the two poles. At approximately 60° N and 60° S latitudes, air rises up into the atmosphere and drops moisture. The cold, dry air then moves toward the poles and sinks back to Earth at approximately 90° N and 90° S latitudes. This air then moves along the ground back to 60° N and 60° S latitudes where it rises again. The atmospheric convection currents that move air between 60° and 90° latitudes are called **polar cells.**

Between the Hadley cells and polar cells, from latitudes of approximately 30° to 60°, is an area of air circulation that lacks distinct convection currents. In this range of latitudes in the Northern Hemisphere—which includes much of the United States, Canada, Europe, and Central Asia—some of the warm air from the Hadley cells that descends at 30° latitude moves toward the North Pole, while some of the cold air from the polar cells that is traveling toward 60° moves toward the equator. The movement of air in this region also helps redistribute the warm air of the tropics and the cold air of the polar regions toward the middle latitudes.

The region between Hadley cells and polar cells can have dramatic changes in wind direction and therefore can experience large fluctuations in temperature and precipitation. However, winds generally move from west to east. This wind direction contributes to warmer conditions on the west coast of North America than on the east coast, as you can see from the plant hardiness zones.

EARTH'S ROTATION AND THE CORIOLIS EFFECT

Hadley cells and polar cells are important drivers of wind direction on Earth. However, wind direction is also affected by the speed of Earth's rotation, which changes with latitude. Earth completes a single rotation in 24 hours. Because the circumference of the planet at the equator is much larger than its circumference near the poles, the speed of rotation is faster at the equator. As **Figure 5.8** shows, an object at the equator rotates at 1,670 km/hr, an object at 30° N rotates at 1,445 km/hr, and an object at 80° N rotates at 291 km/hr.

The different rotation speeds deflect the direction of surface air circulation in the Hadley and polar cells.

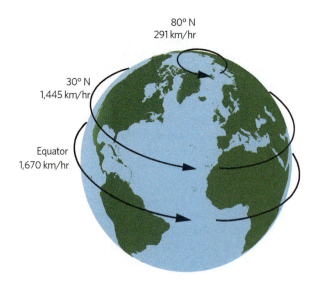

Figure 5.8 Earth's speed of rotation. An object sitting at the equator travels at a much higher speed to complete a rotation in 24 hours than do objects at higher latitudes.

Imagine standing at the North Pole and throwing a baseball straight south to the equator, as shown in **Figure 5.9**. While the ball is flying through the air, the planet continues to rotate. As a result, the ball does not land straight south at the equator. Instead, it travels along a path that appears to deflect to the west. In fact, the ball is traveling straight, but because the planet rotates while the ball is in motion, the ball lands west of its intended target. With respect to the planet, the ball's path appears to be deflected. The deflection of an object's path due to the rotation of Earth is known as the **Coriolis effect.**

The deflected path of the ball in our example mimics the Coriolis effect on air moving to the north or south. For example, Hadley cells north of the equator move air along the surface from north to south. As we can see in **Figure 5.10**, the Coriolis effect causes this path to deflect so that it moves from the northeast to the southwest. These winds are known as the *northeast trade winds.* Below the equator, the Hadley cells are moving air along the ground from the south to the equator. The Coriolis effect causes this path to deflect so that it moves from the southeast to the northwest. These winds are known as the *southeast trade winds.* A similar phenomenon occurs in the polar cells. In the latitudes between Hadley cells and polar cells, wind direction can be quite variable. However, these winds often move away from the equator and toward the poles, only to be

---

Polar cells The atmospheric convection currents that move air between 60° and 90° latitudes in the Northern and Southern Hemispheres.

Coriolis effect The deflection of an object's path due to the rotation of Earth.

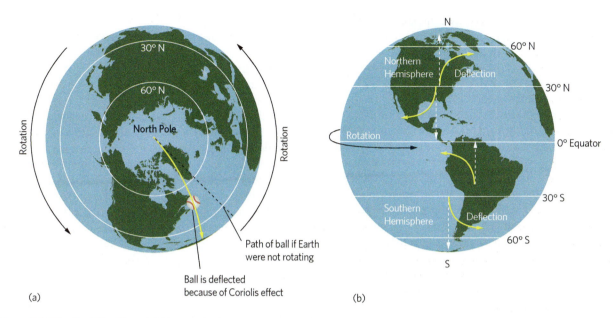

Figure 5.9 The Coriolis effect. (a) Because Earth rotates, the path of any object that travels north or south is deflected. **(b)** This deflection causes the predominant air circulation currents along Earth's surface to be deflected.

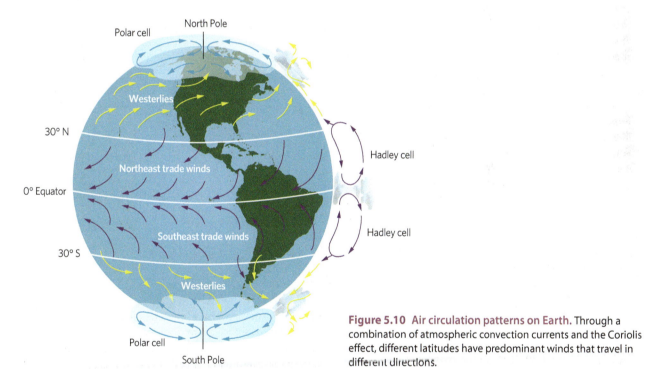

Figure 5.10 Air circulation patterns on Earth. Through a combination of atmospheric convection currents and the Coriolis effect, different latitudes have predominant winds that travel in different directions.

deflected by the Coriolis effect. This causes winds known as *westerlies*. Thus, weather in the middle latitudes tends to move from west to east. When considering the Coriolis effect, the general rule is that surface winds are deflected to the right in the Northern Hemisphere and to the left in the Southern Hemisphere.

CONCEPT CHECK

1. What are the four properties of air that are important in driving atmospheric convection currents?
2. Given that the position of the solar equator moves throughout the year, what does its changing position suggest about the location of the intertropical convergence zone throughout the year?
3. What is the Coriolis effect and how does it affect atmospheric convection currents?

5.4 Ocean currents also affect the distribution of climates

Like air currents, ocean currents distribute unequal heating of Earth's water and therefore influence the location of different climates. **Figure 5.11** shows some general patterns of circulation: Warm tropical water circulates up through the western reaches of ocean basins toward the poles and the cold polar water circulates down the eastern edges of ocean basins toward the tropics. Many factors create these currents, including unequal heating, Coriolis effects, predominant wind directions, the topography of the ocean basins, and differences in salinity. In this section, we will examine the drivers of major ocean currents, including *gyres* and *upwelling*. We will then investigate how natural changes in ocean currents can have large effects on global climates through a process known as the *El Niño–Southern Oscillation*. Finally, we will explore the thermohaline ocean circulation, a deep ocean current that can take hundreds of years to complete a single path around the globe.

GYRES

We have noted how tropical regions of Earth receive more direct sunlight than regions at higher latitudes. This causes ocean waters near the equator generally to be warmer than ocean waters at higher latitudes. Because of the unequal heating, the tropical waters expand as they warm. This expansion causes the water near the equator to be approximately 8 cm higher in elevation than the water in mid-latitudes. Although the difference may seem small, it is enough for the force of gravity to cause movement of water away from the equator.

Around the globe, ocean circulation patterns are also affected by the dominant wind directions and Coriolis effects. North of the equator, for example, the northeast trade winds push surface water from the northeast to the southwest. At the same time, the Coriolis forces deflect ocean currents to the right. The combination of the two forces causes tropical water above the equator to move from east to west. The topography of ocean basins, particularly the locations of continents, forces these currents to change

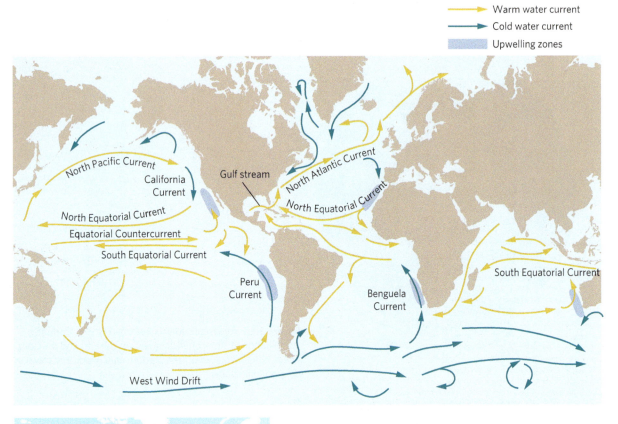

Figure 5.11 Ocean currents. Ocean currents circulate as a result of unequal heating, Coriolis effects, predominant wind directions, and the topography of the ocean basins. Each of the five major ocean basins contains a gyre. These gyres are driven by the trade winds in the tropics and the westerlies at mid-latitudes. This produces a clockwise circulation pattern in the Northern Hemisphere and a counterclockwise circulation pattern in the Southern Hemisphere. Along the west coasts of many continents, currents diverge and cause the upwelling of deeper and more fertile water.

their direction. At mid-latitudes, the westerlies push surface waters to the northeast. As this occurs, Coriolis forces deflect the ocean currents to the right, which causes the ocean currents to move from west to east at mid-latitudes in the Northern Hemisphere. These large-scale water circulation patterns between continents are called **gyres.** The direction of the deflections caused by Coriolis forces depends on latitude; as you can see in Figure 5.11, gyres move in a clockwise direction in the Northern Hemisphere and in a counterclockwise direction in the Southern Hemisphere.

Gyres redistribute energy by transporting both warm and cold ocean water around the globe. The proximity of these ocean waters to continents can make the continents considerably warmer or colder and therefore influence terrestrial climates. You can see this impact on the coastal patterns of the plant hardiness zones we discussed in the beginning of the chapter. For example, England and Newfoundland, Canada, are at similar latitudes. However, England is adjacent to a warm-water current, the Gulf Stream, that comes out of the Gulf of Mexico, whereas Newfoundland is adjacent to a cold-water current that comes down from the west side of Greenland and pushes the warmer Gulf Stream water offshore. As a result, England experiences winter temperatures that are, on average, 20°C warmer than Newfoundland.

UPWELLING

Any upward movement of ocean water is referred to as **upwelling.** Illustrated as dark blue areas in Figure 5.11, upwelling occurs in locations along continents where surface currents move away from the coastline. As surface water moves away from land, cold water from beneath is drawn upward. Strong upwelling zones occur on the western coasts of continents where gyres move surface currents toward the equator and then veer from the continents. As surface water moves away from the continents, it is replaced by water rising from greater depths. Because deep water tends to be rich in nutrients, upwelling zones are often regions of high biological productivity. Major commercial fisheries are often located in these zones.

THE EL NIÑO–SOUTHERN OSCILLATION

Sometimes ocean currents are substantially altered and this can affect climatic conditions. One of the best-known examples is the **El Niño–Southern Oscillation (ENSO),** in which periodic changes in winds and ocean currents in the South Pacific cause weather changes throughout much of the world. This is illustrated in **Figure 5.12.** During most years in the South Pacific Ocean, southeast trade winds and Coriolis forces push the surface waters of the Peru Current, which causes them to flow northwest along the west coast of South America, with upwelling of cold water along the coast. Equatorial winds—powered by high air pressures in the eastern Pacific and low air pressures in the western Pacific—then push these surface waters offshore at Ecuador and west across the Pacific Ocean. As this water moves west, it warms. This warm water drives thunderstorm activity in the western Pacific, which results in high amounts of precipitation.

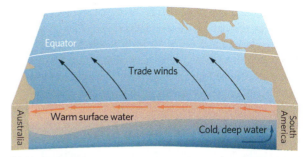

(a) **Normal year**

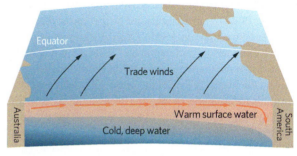

(b) **El Niño year**

Figure 5.12 The El-Niño–Southern Oscillation (ENSO). Changes in the strength of trade winds near the equator can have major impacts on the climates of the world. **(a)** In most years, strong trade winds push warm surface waters away from the west coast of South America. This causes cold, deep waters to upwell along the coast. **(b)** During an ENSO year, the trade winds weaken or reverse and the warm surface water moves from west to east. As a result, warm water builds up along the west coast of South America and prevents the upwelling of the cold, deep water. This change in ocean circulation alters climates around the world.

Every 3 to 7 years, however, this series of events changes. In the atmosphere, the normal difference in air pressures reverses and the equatorial winds weaken. In some years, these winds can even reverse direction. This change in air pressures in the Southern Hemisphere is the Southern Oscillation element of the ENSO. With either weakened or reversed equatorial winds, the warm surface waters of the western Pacific move east toward South America. As a result, the upwelling of nutrients shuts down and the normally productive fisheries in the area become much less productive. The accumulating warm water also serves as a source of increased precipitation for this region. The unusually warm water is the El Niño ("the baby boy") element of the ENSO, so named because it typically occurs around Christmas time.

Gyre A large-scale water circulation pattern between continents.

Upwelling An upward movement of ocean water.

El Niño–Southern Oscillation (ENSO) The periodic changes in winds and ocean currents in the South Pacific, causing weather changes throughout much of the world.

Because air and water currents are responsible for distributing energy throughout the world, the effects of an ENSO event extend over much of the world. In North America, ENSO events bring cooler, wetter, and often stormy weather to the southern United States and northern Mexico, and warm, dry conditions to the northern United States and southern Canada. Some ENSO events have been particularly strong. For example, a strong ENSO event in 1982-83 disrupted fisheries and destroyed kelp beds in California, caused reproductive failure of seabirds in the central Pacific Ocean, and killed off large areas of coral in Panama. Precipitation was also dramatically affected in many terrestrial ecosystems. Another ENSO event from 1991-92 was accompanied by the worst drought of the twentieth century in Africa, which caused poor crop production and widespread starvation. The event also brought extreme dryness to many areas of tropical South America, Australia, and the islands of the South Pacific. Heat and drought in Australia reduced populations of red kangaroos to less than half their pre-ENSO levels. The El Niño event of 1997-98 was blamed for 23,000 human deaths—mostly from famine—and $33 billion in damages to crops and property worldwide. The most recent ENSO event happened in 2015-16 and was one of the strongest ENSO events in recent decades. It caused 1 to 5 °C warmer winter temperatures in Canada and increased precipitation in Northern California. The impacts around the Pacific were even larger, including record-setting heat and drought in Thailand, Malaysia, and India.

The El Niño portion of an ENSO event is typically followed by a La Niña portion in which conditions in the southern Pacific Ocean oscillate strongly in the opposite direction. During La Niña, equatorial winds blow much stronger to the west and all the effects of El Niño event are reversed. Regions that become hotter and drier during the El Niño become cooler and wetter during La Niña. After the cycle of El Niño and La Niña, we typically experience several years of more normal weather conditions.

THERMOHALINE CIRCULATION

Ocean currents are also driven by the **thermohaline circulation,** a global pattern of surface- and deep-water currents that flow as a result of variations in temperature and salinity that change the density of water. The thermohaline circulation, shown in **Figure 5.13**, is responsible for the global movement of great masses of water between the major ocean basins. As wind-driven surface currents—for example, the Gulf Stream—move toward higher latitudes, the water

---

Thermohaline circulation A global pattern of surface- and deep-water currents that flow as a result of variations in temperature and salinity that change the density of water.

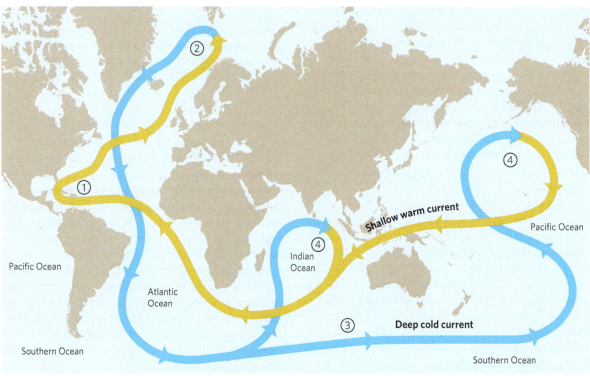

① Warm water flows from the Gulf of Mexico to the North Atlantic, where some of it freezes and evaporates.

② The remaining water, now saltier and denser, sinks to the ocean bottom.

③ The cold water travels along the ocean floor, connecting the world's oceans.

④ The cold, deep water eventually rises to the surface and circulates back to the North Atlantic.

Figure 5.13 Thermohaline circulation. This slow circulation of deep water and surface waters is driven by the sinking of cold, high-salinity water near Greenland and Iceland.

cools and becomes denser. In the far north, toward Iceland and Greenland, the surface of the ocean freezes in winter. Because ice does not contain salts, the salt concentration of the underlying water rises, which causes the cold water to become even denser. This high-density water begins to sink and acts as the driving force behind a deep-water current in the Atlantic Ocean known as the North Atlantic Deep Water. Similar sinking currents are formed around the edges of Antarctica in the Southern Ocean. These cold, dense waters then flow through the deep ocean basins and back into equatorial regions, where they eventually surface as upwelling currents. These upwelling currents become warm and begin to make their way back to the North Atlantic. Like a giant conveyor belt, the thermohaline circulation slowly redistributes energy and nutrients among the oceans of the world in a trip that can take hundreds of years.

CONCEPT CHECK

1. Where are the five major gyres in the world?
2. Why are areas of ocean upwelling important to commercial fishing?
3. For El Niño–Southern Oscillation events, what causes the oscillation?

5.5 Smaller-scale geographic features can affect regional and local climates

As we have seen, the primary global patterns in Earth's climate are the result of the unequal solar heating of Earth's surface. However, a number of other factors have secondary effects on local temperature and precipitation. These include continental land area, proximity to coasts, and rain shadows.

CONTINENTAL LAND AREA

The positions of continents exert important secondary effects on temperature and precipitation. For example, oceans and lakes, the sources of most atmospheric water vapor, cover 81 percent of the Southern Hemisphere but only 61 percent of the Northern Hemisphere. Because of this difference, more rain falls at any given latitude in the Southern Hemisphere than in the Northern Hemisphere. The presence of water has a moderating influence on land temperatures, so temperatures in the Northern Hemisphere vary more than they do in the Southern Hemisphere.

PROXIMITY TO COASTS

The interior of a continent usually experiences less precipitation than its coasts, simply because the interior lies farther from the oceans, which are the major sources of atmospheric water. Furthermore, as we saw in our discussion of plant hardiness zones, coastal climates vary less than interior climates because the heat storage capacity of ocean water reduces temperature fluctuations along the coasts. The ocean warms the air near the coasts during the winter and cools the air near the coasts during the summer. For example, the hottest and coldest average monthly temperatures near the Pacific coast of North America at Portland, Oregon, differ by only 16°C. As we move farther inland, we observed that this range increases to 23°C at Spokane, Washington; 26°C at Helena, Montana; and 33°C at Bismarck, North Dakota.

RAIN SHADOWS

Mountains also play a secondary role in determining climates, as we can see in **Figure 5.14**. When winds blowing inland from the ocean encounter coastal mountains, the mountains force the air upward, which causes adiabatic cooling, condensation, and precipitation. The air, which is now dry and warmed by latent heat release, descends the

Figure 5.14 Rain shadows. When winds carry warm moist air up over a mountain, the air cools and releases much of its moisture as precipitation. After crossing the mountain, the now dry air descends down the mountain, which causes the environment on this side of the mountain to be very dry.

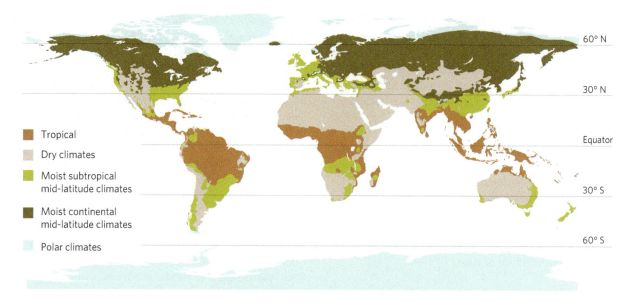

Figure 5.15 Broad climate patterns around the world. Near the tropics, the climate is warm with high amounts of precipitation. The world's great deserts are near 30° N and 30° S latitudes. Cold and snowy polar regions are located at even higher latitudes. In addition, we can see that regions of high precipitation sometimes occur on the western side of mountains, as in western Canada, and that deserts occur in the rain shadows of mountains, such as on the eastern sides of the Cascade Mountains and Sierra Nevadas in North America and the Andes Mountains in South America.

other side of the mountain, warms adiabatically, and travels across the lowlands beyond, where it creates relatively warm, arid environments called **rain shadows.** The Great Basin Desert of the western United States, for example, lies in the rain shadow of the Sierra Nevadas and the Cascade Mountains, and covers a large area that includes nearly all of Nevada and much of western Utah. The processes involved in creating rain shadows have much in common with the processes we saw occurring in Hadley cells, including adiabatic cooling, adiabatic heating, and release of latent heat.

We can now use what we have learned to draw a complete picture of worldwide climate distribution, which can be categorized by temperature and precipitation. Looking at **Figure 5.15**, we see repeated patterns that show where different climates exist. **Tropical climates,** characterized by warm temperatures and high precipitation, occur in regions near the equator. At approximately 30° N and 30° S latitudes, we commonly find the **dry climates** that experience a wide range of temperatures. Dry climates are not only affected by latitude, however. Many dry climates are caused by rain shadows, for example, the extensive regions that lie just east of the Andes Mountains in western South America. **Moist subtropical mid-latitude climates** are characterized by warm, dry summers and cold, wet winters. **Moist continental mid-latitude climates** exist at the interior of continents and typically have warm summers, cold winters, and moderate amounts of precipitation. Finally, closest to the poles, we find **polar climates** that experience very cold temperatures and relatively little precipitation.

We have seen the processes that account for different climates around the world. Before we look more closely at the individual climate types and the plants that they support in the next chapter, we need to know something about the formation of soil, which supports all life.

CONCEPT CHECK

1. What is the impact of continental land masses on the amount of precipitation falling in the Northern versus Southern Hemisphere?
2. Why do coastal continental temperatures typically fluctuate less than inland temperatures?
3. How do rain shadows cause desert formation?

Rain shadow A region with dry conditions found on the leeward side of a mountain range as a result of humid winds from the ocean, causing precipitation on the windward side.

Tropical climate A climate characterized by warm temperatures and high precipitation, occurring in regions near the equator.

Dry climate A climate characterized by low precipitation and a wide range of temperatures, commonly found at approximately 30° N and 30° S latitudes.

Moist subtropical mid-latitude climate A climate characterized by warm, dry summers and cold, wet winters.

Moist continental mid-latitude climate A climate that exists at the interior of continents and is typically characterized by warm summers, cold winters, and moderate amounts of precipitation.

Polar climate A climate that experiences very cold temperatures and relatively little precipitation.

5.6 Climate and the underlying bedrock interact to create a diversity of soils

Climate indirectly affects the distributions of plants and animals through its influence on the development of *soil*, which provides the substratum for plant roots to grow and in which many animals burrow. **Soil** defies a simple definition, but we can describe it as the layer of chemically and biologically altered material that overlies bedrock or other unaltered material at Earth's surface. Because the layer of bedrock that underlies soils plays a major role in determining the type of soil that will form above it, soil scientists call bedrock the **parent material.**

SOIL FORMATION

Soil includes minerals derived from the parent material; modified minerals formed within the soil; organic material contributed by plants, air, and water within the pores of the soil; living roots of plants; microorganisms; and the larger worms and arthropods that make the soil their home. For example, if you have ever seen a recently excavated road cutting through a hillside, you may have observed that soils have

distinct layers, called **horizons,** as shown in **Figure 5.16**. Soil horizons are categorized by the components and processes that occur at each level.

Soils exist in a dynamic state and their characteristics are determined by climate, parent material, vegetation, local topography, and, to some extent, age. Groundwater removes some substances by dissolving them and moving them down through the soil to lower layers, a process known as **leaching.** Other materials enter the soil from vegetation, in precipitation, as dust from above, and from the parent rock below. Where little rain falls, the parent material breaks down slowly, and sparse plant production means that little organic material is added to the soil. Thus, dry climates

---

Soil The layer of chemically and biologically altered material that overlies bedrock or other unaltered material at Earth's surface.

Parent material The layer of bedrock that underlies soil and plays a major role in determining the type of soil that will form above it.

Horizon A distinct layer of soil.

Leaching A process in which groundwater removes some substances by dissolving them and moving them down through the soil to lower layers.

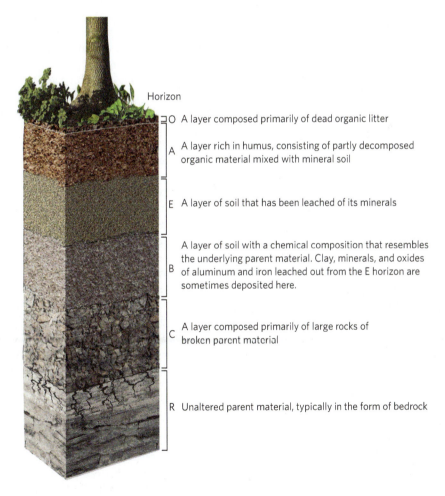

Horizon

O A layer composed primarily of dead organic litter

A A layer rich in humus, consisting of partly decomposed organic material mixed with mineral soil

E A layer of soil that has been leached of its minerals

B A layer of soil with a chemical composition that resembles the underlying parent material. Clay, minerals, and oxides of aluminum and iron leached out from the E horizon are sometimes deposited here.

C A layer composed primarily of large rocks of broken parent material

R Unaltered parent material, typically in the form of bedrock

Figure 5.16 Soil horizons. Soils develop distinct horizons that differ in thickness, depending on climates and parent material.

typically have shallow soils, with bedrock lying close to the surface. In places where decomposed bedrock and organic material erode as rapidly as they form, soils may not form at all. Soil development also stops short on alluvial deposits, where fresh layers of silt deposited each year by floodwaters bury older material. At the other extreme, soil formation proceeds rapidly in tropical climates, where chemical alteration of parent material may extend to depths of 100 m. Most soils of mid-latitude climates are intermediate in depth, extending to an average of about 1 m.

WEATHERING

Weathering is the physical and chemical alteration of rock material near Earth's surface. It occurs whenever surface water penetrates the parent material. In cold climates, for example, the repeated freezing and thawing of water in crevices cause the rock to break into smaller pieces and expose a greater surface area of the rock to chemical reactions. The initial chemical alteration of the rock occurs when water dissolves some of the more soluble minerals, such as sodium chloride (NaCl) and calcium sulfate ($CaSO_4$). Further chemical reactions continue the soil building process.

The weathering of granite illustrates some basic processes of soil formation. The minerals responsible for the grainy texture of granite—feldspar, mica, and quartz—consist of various combinations of oxides of aluminum, iron, silicon, magnesium, calcium, and potassium. The key aspect of the process of weathering is the displacement of many of these elements by hydrogen ions, followed by the reorganization of the remaining minerals into new types of minerals. The hydrogen ions involved in weathering are derived from two sources. One source is the carbonic acid that forms when carbon dioxide dissolves in rainwater, as discussed in Chapter 2. The other source of hydrogen ions is the decomposition of organic material in the soil itself. The metabolism of carbohydrates, for example, produces carbon dioxide. This carbon dioxide is converted into carbonic acid in water, which generates additional hydrogen ions.

As granite weathers, many of the positively charged elements—such as iron (Fe^{3+}) and calcium (Ca^{2+})—are replaced by hydrogen ions to form new, insoluble materials, such as the clay particles we discussed in Chapter 3. Clay particles are important to the water-holding capacity of soils. They accumulate negative charges on their surfaces that attract positively charged ions called *cations*. Cations—including calcium (Ca^{2+}), magnesium (Mg^{2+}), potassium (K^+), and sodium (Na^+)—are important nutrients for plants. The ability of a soil to retain these cations, called its **cation exchange capacity,** provides an index to the fertility of that soil. Young soils have relatively few clay particles and little added organic material; this low cation exchange capacity leads to relatively low fertility. Older soils generally have a higher cation exchange capacity and therefore relatively high fertility. Soil fertility improves with time, up to a point.

Eventually, weathering breaks down clay particles, cation exchange capacity decreases, and soil fertility drops.

Podsolization

Under mild temperatures and moderate precipitation, sand grains and clay particles resist weathering and become stable components of the soil. This allows soils to retain relatively high fertility. However, in acidic soils typical of cool, moist regions, clay particles break down in the E horizon, and their soluble ions are transported down to the lower B horizon. This process, known as **podsolization,** reduces the fertility of the soil's upper layers.

Acidic soils occur primarily in cool regions where needle-leaved trees such as spruces and firs dominate the forests. The slow decomposition of the spruce and fir needles produces organic acids that promote high concentrations of hydrogen ions. In the moist regions where podsolization occurs, rainfall usually exceeds evaporation. Because water continuously moves downward through the soil profile, little clay-forming material is transported upward from the weathered bedrock.

In North America, podsolization is most advanced under spruce and fir forests of New England, in the Great Lakes region, and across a wide belt of southern and western Canada. A typical profile of a highly podsolized soil, shown in **Figure 5.17**, reveals striking bands corresponding to the regions of leaching and redeposition. The A horizon is dark and rich in organic matter. It is underlain by a light-colored E horizon that has been leached of most of its clay, leaving behind sandy material that holds neither water nor nutrients well. One usually finds a dark band immediately below the E horizon. This is the uppermost layer of the B horizon, where iron and aluminum oxides are redeposited. Other, more mobile minerals may accumulate to some extent in lower parts of the B horizon, which then grades almost imperceptibly into a C horizon and the parent material.

Laterization

In the warm, humid climates of many tropical and subtropical regions, soils weather to great depths. One of the most conspicuous features of weathering under these conditions is the breakdown of clay particles, which causes silicon to leach from the soil and leaves oxides of iron and aluminum to dominate throughout the soil profile—a process that is called **laterization.** The iron and aluminum oxides give such soils a

---

Weathering The physical and chemical alteration of rock material near Earth's surface.

Cation exchange capacity The ability of a soil to retain cations.

Podsolization A process occurring in acidic soils typical of cool, moist regions, where clay particles break down in the E horizon, and their soluble ions are transported down to the lower B horizon.

Laterization The breakdown of clay particles, which results in the leaching of silicon from the soil, leaving oxides of iron and aluminum to predominate throughout the soil profile.

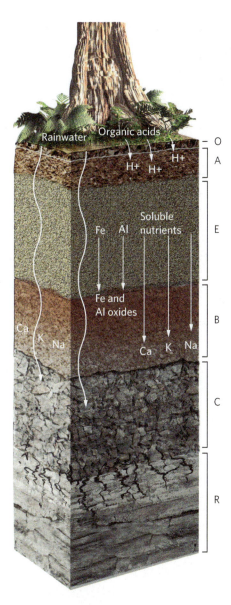

Figure 5.17 Podsolization. In cool, moist conditions with highly acidic soils, clay particles normally found in the E horizon are weathered and leached down, leaving a very sandy layer with little capacity to retain nutrients for plants.

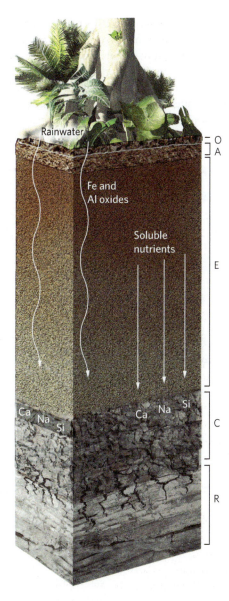

Figure 5.18 Laterization. Under conditions of warm temperatures and high precipitation, clay particles are broken down and leave behind a soil that has a low cation exchange capacity and low fertility.

characteristic reddish coloration, as illustrated in **Figure 5.18**. Even though the rapid decomposition of organic material in tropical soils contributes an abundance of hydrogen ions, bases formed by the breakdown of clay particles neutralize them. Consequently, lateritic soils are not usually acidic, even though they may be deeply weathered. Regardless of the parent material, weathering reaches deepest, and laterization proceeds farthest, on low-lying soils, such as those of the Amazon basin, where highly weathered surface layers are not eroded away and the soil profiles are very old.

Laterization causes many tropical soils to have a low cation exchange capacity. In the absence of clay and organic matter, mineral nutrients are readily leached from the soil. Where soils are deeply weathered, new minerals formed

by the decomposition of the parent material are simply too far from the surface to contribute to soil fertility. In addition, heavy rainfall in the tropics keeps water moving down through the soil profile, preventing the upward movement of nutrients. In general, the deeper the ultimate sources of nutrients in the unaltered bedrock, the lower the fertility of the surface layers. The high productivity of tropical rainforests depends more on rapid cycling of nutrients close to the surface of the soil than on the nutrient content of the soil itself. Rich soils do, however, develop in many tropical regions, particularly in mountainous areas where erosion continually removes nutrient-depleted surface layers of soil, and in volcanic areas where the parent material of ash and lava is often rich in nutrients such as potassium.

From our discussion of soils, you can see that the composition of the soils present in various parts of the world depends on differences in climate, underlying parent material, and vegetation. In the next chapter, we will discuss how these regional differences in climate and the associated effects on soils affect the types of plants and animals that can live in each region.

CONCEPT CHECK

1. What are the different soil horizons and of what are they composed?
2. Why is cation exchange capacity important for determining soil fertility?
3. Why are tropical soils weathered to greater depths than soils in the northern United States?

ECOLOGY TODAY   CONNECTING THE CONCEPTS

Global Climate Change

A polar bear hunting for seals on the Arctic sea ice of Norway. Warming trends over the last few decades have caused the Arctic ice to melt earlier in the year; this means that polar bears have less time to hunt for seals, which make up a large part of their diet. Steven Kazlowski/naturepl.com

As we have seen in this chapter, a substantial number of interacting factors determine the different climates of the world. For example, the differential heating of Earth drives the movements of air and water, which are further modified by the Coriolis effect and the position of continents. Because these interactions are complex, any changes in these factors can have far-reaching effects on the entire system. Such is the case for global warming and *global climate change*. Global warming is the increase in the average temperature of the planet due to an increased concentration of greenhouse gases in the atmosphere. **Global climate change** is a much broader phenomenon that refers to changes in Earth's climates and includes global warming, changes in the global distribution of precipitation and temperature, changes in the intensity of storms, and altered ocean circulation. Throughout the history of Earth, long periods of gradual global warming and cooling have been associated with substantial global climate change. During the past two centuries, however, human activities have caused a rapid change in conditions that have led to global warming and global climate change.

Global warming is a major driver of current changes in global climates. One direct impact is the increase in temperatures in many parts of the world, particularly at high latitudes in the Northern Hemisphere. The rise in temperature in these regions has had a wide range of effects. For example, in high latitudes and at high altitudes, the lower layers of soil may be permanently frozen, a phenomenon that is known as **permafrost.** Warmer temperatures cause these highly organic soils to thaw and begin decomposing. Because these soils are waterlogged and anaerobic, the decomposition produces methane, a greenhouse gas that can further contribute to global warming.

Increased global temperatures, which affect high latitudes more than lower latitudes, also affect the amount of ice melting around the world. From 1979 to 2016, the polar ice cap that exists between the United States, Canada, Europe, and Russia has declined at a rate of 13 percent per year. The remaining ice has also become thinner. In 2016, the warmest year around the world on record since 1880, the Arctic ice cap melted to the second-smallest area since records began in 1979. The ice of Greenland and Antarctica is also melting. NASA scientists found, from 2003 to 2016 that the two regions lost an average of 412 gigatonnes (Gt) of ice per year and that the annual rate of ice loss is accelerating. By 2016, Greenland was losing 350 Gt of ice per year. Similarly, glaciers are melting in many regions of the world. In Montana's Glacier National

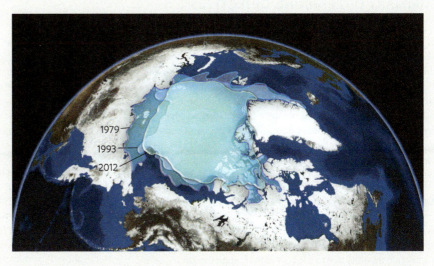

The melting polar ice cap. The area covered by the Arctic ice cap has declined at a rate of 13 percent annually.

Park, for example, there were 150 glaciers in 1850. Today, only 25 glaciers remain. All this ice melting, combined with warmer oceans expanding in volume, has caused sea levels to rise 200 mm since 1870, and scientists predict that continued melting could raise sea levels an additional 280 to 430 mm by the year 2100. In 2016, researchers reported that based on the close correlation between increased CO_2 emissions and ice loss in the polar ice cap, a complete loss of ice could occur in summer by approximately 2050.

Because the complex nature of the global climate system can make it difficult to predict how climate will change in the coming decades, scientists have developed computer models that incorporate our best understanding of the processes that govern climate along with the changes being caused by increased atmospheric greenhouse gas concentrations. Although different models make somewhat different predictions, there is general agreement regarding several aspects of predicted change. For example, the increase in temperatures is expected to cause longer periods of hot weather and fewer days of extremely cold weather. Because heat is the driver of evaporation and air circulation that determine precipitation, precipitation patterns are also predicted to change across the globe, with some regions of the world receiving increased amounts of rain and snow while other regions receive less. The intensity of storms, such as hurricanes, is also predicted to increase due to an increased warming of the world's oceans.

Ocean currents may also be affected by global warming. As we discussed earlier in this chapter, ocean currents are driven by the differential heating of Earth and, in turn, play a major role in determining the temperature of nearby continents. Of particular concern is the potential impact on thermohaline circulation. As you may recall, thermohaline circulation is the slow, deep circulation of ocean water around the globe that is driven by the dense, salty water that sinks near Greenland. With the increased melting of the polar ice cap and the ice sheets of Greenland, climate scientists are concerned that the water in the North Atlantic may not be dense enough to sink and therefore may cause the thermohaline circulation to shut down. While researchers do not have a prediction for when the thermohaline circulation might stop, the disappearance of this deep-water current would effectively stop the circulation of warm water from the Gulf of Mexico to Europe and cause a substantial cooling of Europe—with potentially devastating consequences for the people and environment of that region.

Global climate change
A phenomenon that refers to changes in Earth's climates, including global warming, changes in the global distribution of precipitation and temperature, changes in the intensity of storms, and altered ocean circulation.

Permafrost A phenomenon whereby layers of soil are permanently frozen

SOURCES:

Notz, D., and J. Stroeve. 2016, November 11. Observed Arctic sea-ice loss directly follows anthropogenic CO_2 emission. *Science* 354, no. 6313:747-750. doi:10.1126/science.aag2345.

Climate Change 2007: Synthesis Report. *Fourth Assessment Report of the Intergovernmental Panel on Climate Change.* http://www.ipcc.ch/pdf/assessment-report/ar4/syr/ar4_syr.pdf.

Global Climate Change Impacts in the United States. 2009. *U.S. Global Change Research Program.* http://downloads.globalchange.gov/usimpacts/pdfs/climate-impacts-report.pdf.

SUMMARY OF LEARNING OBJECTIVES

5.1 Earth is warmed by the greenhouse effect. Much of the ultraviolet and visible light emitted by the Sun passes through the atmosphere and strikes clouds and the surface of Earth. Clouds and the planet begin to warm and emit infrared radiation back toward the atmosphere. The gases in the atmosphere absorb the infrared radiation, become warmer, and re-emit infrared radiation back toward Earth. These greenhouse gases allow the planet to become warmer than would otherwise be possible. The increase in production of greenhouse gases due to human activities increases the greenhouse effect and leads to global warming.

Key Terms: Atmosphere, Greenhouse effect

5.2 There is an unequal heating of Earth by the Sun. Each year, high-latitude regions of the world receive solar radiation that is weaker in intensity due to a longer path through the atmosphere with a less direct angle, which causes the energy of the Sun to be spread over a larger area. In addition, the axis of Earth is tilted 23.5° and this causes seasonal changes in temperature.

Key Terms: Albedo, Solar equator, Regression

5.3 The unequal heating of Earth drives air currents in the atmosphere. Due to the properties of air, the warmer temperatures near the equator drive atmospheric convection currents known as Hadley cells between approximately 0° to 30° latitude in the Northern and Southern Hemispheres. Polar cells are at higher latitudes, between approximately 60° to 90°. These air convection currents cause the distribution of heat and precipitation around the globe. Their path is also influenced by Coriolis forces that are created by the rotation of Earth.

Key Terms: Atmospheric convection currents, Saturation point, Latent heat release, Adiabatic cooling, Adiabatic heating, Hadley cells, Intertropical convergence zone (ITCZ), Polar cells, Coriolis effect

5.4 Ocean currents also affect the distribution of climates. Ocean currents are driven by the unequal warming of Earth combined with Coriolis effects, atmospheric convection currents, and differences in salinity. Gyres exist on both sides of the equator and help distribute heat and nutrients to higher latitudes. El Niño–Southern Oscillation (ENSO) events represent a disruption in normal ocean currents in the South Pacific, and the impacts on climates can be felt around the world. Thermohaline circulation is a deep and slow circulation of the world's oceans driven by changes in salt concentration in the waters of the North Atlantic.

Key Terms: Gyre, Upwelling, El Niño–Southern Oscillation (ENSO), Thermohaline circulation

5.5 Smaller-scale geographic features can affect regional and local climates. Increased land area of continents reduces the amount of evaporation possible, which causes the Northern Hemisphere to experience less precipitation than the Southern Hemisphere. Proximity to the coast can also affect climates; regions that are more distant from coastlines typically have lower precipitation and higher variation in temperature. Mountain ranges force air to rise over them, causing higher precipitation on one side of the mountain range and rain shadows on the opposite side.

Key Terms: Rain shadow, Tropical climate, Dry climate, Moist subtropical mid-latitude climate, Moist continental mid-latitude climate, Polar climate

5.6 Climate and the underlying bedrock interact to create a diversity of soils. Soils are made up of horizons that contain different amounts of organic matter, nutrients, and minerals. Soils can be weathered by processes including freezing, thawing, and leaching. In acidic soils of cool, moist regions, soils can experience podsolization, a process that breaks down clay particles and reduces fertility. In warm, humid climates, soils can experience laterization, a process that breaks down clay particles and leaches nutrients from the soil.

Key Terms: Soil, Parent material, Horizon, Leaching, Weathering, Cation exchange capacity, Podsolization, Laterization, Global climate change, Permafrost

CRITICAL THINKING QUESTIONS

1. Of the many different types of greenhouse gases, which ones will likely decline faster if we reduce their production?

2. Some sources of air pollution produce tiny black particles that can be transported around the world and settle on regions covered in snow and ice. Based on the albedo effect, how might this air pollution contribute to warming global temperatures, melting polar ice caps, and rising sea levels?

3. If Earth was not tilted on its axis, how would it affect the seasonality of rainfall at the equator?

4. What parallels can you draw between the processes that drive Hadley cells versus rain shadows?

5. If there was no Coriolis effect, how would it affect atmospheric convection currents and ocean currents?

6. How do ocean gyres affect the pattern of plant hardiness zones in North America?

7. Why do El Niño and La Niña events cause opposite weather climates around the world?

8. Based on your knowledge of the thermohaline circulations, how might melting of the ice in the Arctic Ocean affect the climate of Europe?

9. What are the processes that are responsible for the locations of the world's major deserts?

10. Compare and contrast podsolization and laterization.

GRAPHING THE DATA Precipitation in Mexico City, Quito, and La Paz

As we have seen in this chapter, cities around the world often differ in their pattern of monthly precipitation. Using the data provided in the table, create a bar graph for each of the three cities.

(a) Based on these graphs, how many peaks in precipitation does each city receive? (b) Based on their geographic locations, why does the number of peaks in precipitation in these cities differ?

Average Monthly Precipitation (mm) in Three Cities

Month	Mexico City, Mexico	Quito, Ecuador	La Paz, Bolivia
January	10.2	114.3	129.5
February	10.2	129.5	104.1
March	12.7	152.4	71.1
April	27.9	175.3	35.6
May	58.4	124.5	12.7
June	157.5	48.3	5.1
July	182.9	20.3	7.6
August	172.7	25.4	15.2
September	144.8	78.7	30.5
October	61.0	127.0	40.6
November	5.1	109.2	50.8
December	0.8	104.1	94.0

6 Terrestrial and Aquatic Biomes

The World of Wine

The fascinating history of winemaking dates back thousands of years. Archaeologists have found signs of winemaking in many cultures around the Mediterranean Sea, including those of the ancient Egyptians, Romans, and Greeks. Indeed, the entire Mediterranean region has a long tradition of cultivating wine grapes, and the production of wine has played an important role in the economic development of many societies and in religious rituals.

European explorers spread winemaking to other parts of the world. For example, in the sixteenth century, Spanish explorers brought grape vines to Chile, Argentina, and California. Grape vines accompanied the Dutch to South Africa in the seventeenth century, and the British to Australia in the nineteenth century.

Although grapes can grow in many parts of the world, specific growing conditions are required to produce grapes for the best wines. The ideal climate is a combination of hot, dry summers and wet, mild winters. The hot and dry summer climate allows the grapes to develop the right balance of sugar and acidity that provides the complex flavors of a fine wine. The dry summers also prevent various plant diseases that flourish under moist conditions. Domesticated grapes have deep roots, so they are well adapted to dry summer landscapes. Because below-freezing temperatures can harm the vines, the presence of mild, wet winters is equally important. While climate is critical, the flavor of a fine wine is also influenced by the pH, fertility, and mineral content of the soils in which the vines grow. The composition of a soil affects how well the grape vines grow and gives the grapes a distinctive flavor that characterizes the wine made from them. In short, distinctive tasting wines from around the world are the result of unique combinations of climate and soil.

Given the conditions required to make a great wine, it is perhaps not surprising that most of the major wine-producing locations around the world have the same climate—hot, dry summers followed by cool, moist winters. This is the climate of the countries surrounding most of the Mediterranean Sea. It is also the climate of most regions where wine grapes have been introduced, including Chile, Argentina, California, South Africa, and the southwestern coast of Australia. Interestingly, these regions all lie on the west side of continents and are located at latitudes between 30° and 50° in the Northern and Southern Hemispheres. Superior winemaking regions not only have similar climates, but their landscapes also contain similar-looking plants, despite being separated by thousands of

> "Superior wine-making regions not only have similar climates, but their landscapes also contain similar looking plants, despite being separated by thousands of kilometers."

Growing grapes for wine during the hot, dry summer. At the Chateau Corcelles in southern France, the climate is ideal for growing the grapes that are used for wine. robert paul van beets/Shutterstock.

SOURCES:

Retallack, G. J., and S. F. Burns. 2016. The effects of soil on the taste of wine. *Geological Society of America Today* 26:4-9. http://www.geosociety.org/gsatoday/archive/26/5/article/i1052-5173-26-5-4.htm.

A brief history of wine. 2007. *New York Times*, November 5. http://www.nytimes.com/2007/11/05/timestopics/topics-winehistory.html.

kilometers. For example, while each winemaking region contains a large number of unique plant species, the plants are similar in their growth form. Whether in France, California, Chile, or South Africa, the plant communities are dominated by drought-adapted grasses, wildflowers, and shrubs.

In this chapter, we will explore how particular climates that are found in different locations around the world are associated with very similar-looking plants and how scientists use these patterns to categorize terrestrial ecosystems. We will also examine why scientists categorize aquatic ecosystems in a different way, based on differences in salinity, flow, and depth.

LEARNING OBJECTIVES

After reading this chapter, you should be able to:

6.1 Explain how terrestrial biomes are categorized by their major plant growth forms.

6.2 Describe the nine categories of terrestrial biomes.

6.3 Describe the many aquatic biomes, which are categorized by their flow, depth, and salinity.

As we saw in Chapter 5, climate patterns around the globe are determined by a range of factors, including air currents, water currents, Coriolis forces, and local geographic features. Together, these factors are responsible for the climates that occur in different regions of the world. Different climates provide unique seasonal temperature and precipitation conditions, and these unique conditions favor different types of plants.

6.1 Terrestrial biomes are categorized by their major plant growth forms

Successful survival strategies vary with climate. In the world's deserts, for example, we find plants that are well adapted to scarce water availability. In North American deserts, many species of cacti have thick, waxy outer layers covered with hairs and spines to help reduce water loss. In Africa, we find a group of plants called euphorbs that are not closely related to the cacti of North America yet have many similar features (**Figure 6.1**). Though the two groups of desert-adapted plants are descended from unrelated ancestors, they look similar because they have evolved under similar selective forces, a phenomenon known as **convergent evolution.** Convergent evolution can be observed in many organisms. For example, sharks and dolphins are not closely related to each other—one is a fish and the other is

Convergent evolution A phenomenon in which two species descended from unrelated ancestors look similar because they have evolved under similar selective forces.

(a)

(b)

Figure 6.1 Convergent evolution. Similar conditions in the deserts of the world have selected for similar water-conserving life forms in two groups of unrelated plants: **(a)** an organ pipe cactus (*Stenocereus thurberi*) in Organ Pipe Cactus National Monument, Arizona, and **(b)** a euphorb (*Euphorbia virosa*) in Namibia, Africa.

All Canada Photos/Alamy; Alessandra Sarti/imageBROKER/Newscom.

a mammal—yet both have evolved fins, powerful tails, and streamlined bodies. To perform well in an aquatic environment, natural selection has favored this set of traits because it allows both groups of animals to swim rapidly.

Convergent evolution explains why we can recognize an association between the forms of organisms and the environments in which they live. Trees found in tropical rainforests have the same general appearance no matter where they are located on Earth or their evolutionary lineage. The same can be said of shrubs inhabiting seasonally dry environments; they tend to have small, deciduous leaves and often have stems with spines to discourage herbivores from eating them.

Geographic regions that contain communities composed of organisms with similar adaptations are called **biomes.** Because of convergent evolution, we can categorize terrestrial ecosystems by dominant plant forms that are associated with distinct patterns of seasonal temperatures and precipitation. In aquatic ecosystems, the major producers are often not plants but algae. As a result, aquatic biomes are not easily characterized by the dominant growth forms of the producers. Instead, aquatic biomes are characterized by distinct patterns of depth, flow, and salinity.

Biomes provide convenient reference points for comparing ecological processes around the globe, which makes the biome concept a useful tool that enables ecologists to understand the structure and functioning of large ecological systems. As in all classification systems, exceptions occur. Boundaries between biomes can be unclear and not all plant growth forms correspond to climate in the same way. Australian eucalyptus trees, for example, form forests under climatic conditions that support only scrubland or grassland on other continents. Finally, plant communities reflect factors other than temperature and rainfall. Topography, soils, fire, seasonal variations in climate, and herbivory all affect which species can live in different plant communities.

The overview of the major terrestrial biomes in this chapter emphasizes the distinguishing features of the physical environment and how these features are reflected in the form of the dominant plants. As a final note, although ecologists use plant forms to categorize biomes, there is generally a good association between the plant forms in a biome and the animal forms that live there. For example, deserts contain plants and animals that are adapted to dry conditions.

We will use a classification system that recognizes nine major terrestrial biomes, which are listed in **Figure 6.2**. If we consider all combinations of average annual temperatures and average annual precipitation, as shown in the figure, we see that most places on Earth are located inside a triangular area with corners representing warm moist, warm dry, and cool dry climates. Cold regions with high rainfall are rare because water does not evaporate rapidly at low temperatures and because the atmosphere in cold regions contains very little water vapor.

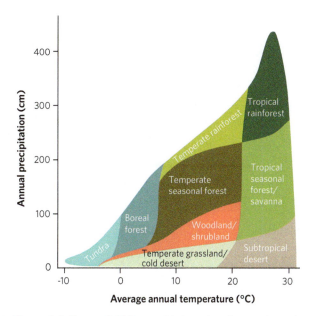

Figure 6.2 Terrestrial biomes. Distinct plant forms exist under different combinations of average annual precipitation and average annual temperature.

The nine biomes fall within three temperature ranges that we refer to often throughout this book. Boreal forest and tundra biomes have average annual temperatures that are generally below 5°C. Temperate biomes—temperate rainforest, temperate seasonal forest, woodland/shrubland, and temperate grassland/cold desert—are warmer, with average annual temperatures generally between 5°C and 20°C. Finally, tropical biomes—tropical rainforest, tropical seasonal forest/savanna, and subtropical desert—are the warmest biomes, with average annual temperatures greater than 20°C. The global distribution of these biomes is illustrated in **Figure 6.3**. As we will see, the average annual precipitation within each of these temperature categories can vary widely.

CLIMATE DIAGRAMS

To visualize the patterns of temperature and precipitation that are associated with particular biomes, scientists use **climate diagrams,** which are graphs that plot the average monthly temperature and precipitation of a specific location on Earth. **Figure 6.4** provides two sample climate diagrams. As you can see, the shaded area on the *x*-axis indicates the months in which the average temperature exceeds 0°C. These months are warm enough to allow plant growth and therefore represent the **growing season** of the

Biome A geographic region that contains communities composed of organisms with similar adaptations.

Climate diagram A graph that plots the average monthly temperature and precipitation of a specific location on Earth.

Growing season The months in a location that are warm enough to allow plant growth.

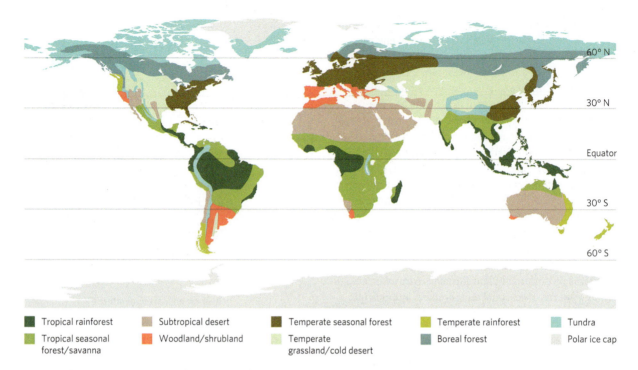

Tropical rainforest   Subtropical desert   Temperate seasonal forest   Temperate rainforest   Tundra

Tropical seasonal   Woodland/shrubland   Temperate   Boreal forest   Polar ice cap
forest/savanna   grassland/cold desert

Figure 6.3 The global distribution of biomes. The nine terrestrial biomes represent locations with similar average annual temperatures and precipitations and similar plant growth forms. Also shown are polar ice caps, which lack plants and therefore are not part of the biome classification system.

biome. Climate diagrams can also indicate whether plant growth is more limited by temperature or by precipitation. For every 10°C increase in temperature, plants require an additional 2 cm of monthly precipitation to meet the increased water needs. Climate diagrams adjust their temperature and precipitation axes such that every 10°C increase in average monthly temperature corresponds to a 2 cm increase in monthly precipitation. This means that

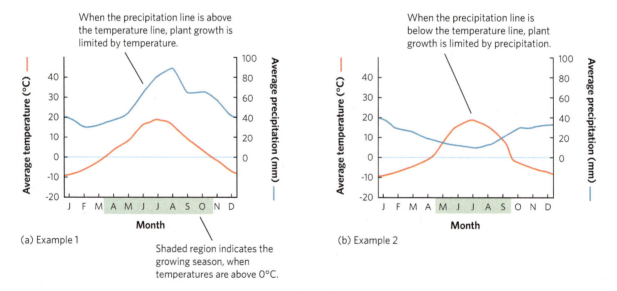

Figure 6.4 Climate diagrams. By plotting the average monthly temperature and precipitation values over time for a particular location on Earth, we can determine how climates vary throughout the year and the length of the growing season. **(a)** In this hypothetical climate diagram, there is a seven-month growing season and plant growth is limited by temperature throughout the year. **(b)** In this example, there is a five-month growing season and plant growth is limited by precipitation.

ANALYZING ECOLOGY

Mean, Median, and Mode

Eco TV

 macmillanlearning.com/ricklefsvideo

Climate diagrams are a useful way of conveying a good deal of information about the average monthly changes in temperature and precipitation. While the climate of a particular location can vary from year to year, the climate diagrams display the typical conditions based on several years of collecting data. Using these data, one can then determine the *mean* temperature and precipitation for a particular month. The mean, or average, is calculated by summing all the data and dividing by the total number of data points. The mean value gives you a sense of where the middle value lies in a set of data. However, this assumes that the data have a symmetrical distribution such that half of the values fall above the mean and half of the values fall below the mean.

In some sets of data, the values are not symmetrically distributed around a middle value. In such cases, a better estimate of the middle value is the *median*. The median is found by placing the data in order, from lowest to highest, and finding the number that occurs in the middle. If there is an even number of values, then there are two numbers in the middle and the median is found by taking the average of these two middle numbers. For example, consider the values

95, 93, 90, 85, 81, 75, 63, 42, 21:

the mean = $(95 + 93 + 90 + 85 + 81 + 75 + 63 + 42 + 21) \div 9 = 71.7$

In contrast, the median = 81.

For a set of data that contains an even number of values, the median is calculated as the mean of the middle two values. For example, consider the following values:

95, 93, 90, 85, 81, 79, 75, 63, 42, 21

In this case, there is an even number of values and two numbers, 79 and 81, are the middle values. As a result, the median is the average of these two numbers = 80.

Sometimes scientists are not as interested in the mean or median from a set of data, but rather want to know which values occur most frequently. In this case, they determine how often each value occurs; the *mode* is the value that occurs most frequently. For example, consider the following values:

95, 93, 90, 85, 81, 81, 75, 63, 42, 21

In this list, 81 appears more frequently than the other values, so the mode is 81. Generally, the mode is useful only for large samples, where sampling of each possible value is reasonably good.

YOUR TURN For the following set of data, determine the mean, median, and mode:

12, 13, 15, 18, 17, 19, 18, 17, 12, 14, 10, 17, 19, 16, 17

Why is the mean of these values different from the median and the mode?

---

in any month in which the precipitation line goes below the temperature line, plant growth is constrained by a lack of sufficient precipitation. In contrast, any month in which the temperature line goes below the precipitation line, plant growth is constrained by a lack of sufficient temperature. Since climate is the primary force determining the plant forms of different biomes, locations around the world that are from a particular biome have similar climate diagrams. With this background, in the next section, we will take a closer look at the nine terrestrial biomes and their associated climate diagrams.

CONCEPT CHECK

1. Why do unrelated plants often assume the same growth form in different parts of the world?
2. How do we break down biomes into three broad temperature categories?
3. What information about a biome can you obtain from a climate diagram?

6.2 There are nine categories of terrestrial biomes

Terrestrial biomes are traditionally placed into nine categories. In this section, we will take a tour of the biomes. We start with tundras and boreal forests, which have average annual temperatures of less than 5°C. We will then examine the biomes in temperate regions, with average temperatures between 5°C and 20°C. Finally, we will explore the biomes of tropical regions, which have average annual temperatures of more than 20°C. As we will see, seasonal patterns and amounts of precipitation can differ a great deal within a given temperature range, producing different types of biomes.

TUNDRAS

The **tundra,** shown in **Figure 6.5**, is the coldest biome and is characterized by a treeless expanse above permanently

---

Tundra The coldest biome, characterized by a treeless expanse above permanently frozen soil.

Denali National Park, Alaska, United States
Average annual precipitation: 343 mm

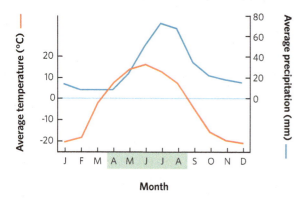

Figure 6.5 Tundra biome. The Denali National Park in Alaska is an example of a tundra biome, which is characterized by a lack of trees and by soil that is permanently frozen.

Noppawat Tom Charoensinphon/Getty Images.

frozen soil, or permafrost. The soils thaw to a depth of 0.5 to 1 m during the brief summer growing season. Annual precipitation is generally less, and often much less, than 600 mm, but in low-lying areas where permafrost prevents drainage, soils may remain saturated with water throughout most of the growing season. Tundra soils contain few nutrients. They also tend to be acidic because of their high content of organic matter, which is the result of cold conditions dramatically slowing the decomposition of organic matter. In this nutrient-poor environment, plants hold their foliage for years. Most plants are dwarf, prostrate woody shrubs, which grow low to the ground to gain protection under the winter blanket of snow and ice, since anything protruding above the surface of the snow is sheared off by blowing ice crystals. For most of the year, the tundra is an exceedingly

harsh environment, but during summer days with 24 hours of sunlight, there is a rush of biological activity.

The tundra is found in the Arctic regions of Russia, Canada, Scandinavia, and Alaska, and in the Antarctic regions along the edge of Antarctica and nearby islands. At high elevations within temperate latitudes, even within the tropics, one finds vegetation resembling that of Arctic tundra, including some of the same species or their close relatives. These areas of *alpine tundra* above the tree line occur widely in the Rocky Mountains of North America, the Alps of Europe, and, especially, on the Plateau of Tibet in central Asia. In spite of their similarities, alpine and Arctic tundra have important differences. Areas of alpine tundra generally have warmer and longer growing seasons, higher precipitation, less severe winters, greater productivity, better-drained soils, and higher species diversity than Arctic tundra. However, harsh winter conditions ultimately prevent the growth of trees in both Arctic tundra and alpine tundra.

BOREAL FORESTS

Stretching in a broad belt centered at about 50° N in North America and about 60° N in Europe and Asia lies the *boreal forest*. As shown in **Figure 6.6,** the **boreal forest,** sometimes called **taiga,** is a biome densely populated by evergreen needle-leaved trees, with a short growing season and severe winters. The average annual temperature is generally below 5°C and annual precipitation generally ranges between 40 and 1,000 mm. Since evaporation is low, soils are moist throughout most of the growing season. The vegetation consists of dense, seemingly endless stands of 10- to 20-m tall evergreen needle-leaved trees, mostly spruces and firs. Because of the low temperatures, plant litter decomposes very slowly and accumulates at the soil surface, forming one of the largest reservoirs of organic carbon on Earth. The needle litter produces high levels of organic acids, so the soils are acidic, strongly podsolized (as discussed in Chapter 5), and generally of low fertility. Growing seasons rarely exceed 100 days, and are often half that long. The vegetation is extremely frost-tolerant; temperatures may reach −60°C during the winter. Since few species can survive in such harsh conditions, species diversity is very low. The boreal forest is not well suited for agriculture, but it serves as a source of timber products that include lumber and paper.

TEMPERATE RAINFORESTS

As we begin to move closer to the equator, we find the four temperate biomes: temperate rainforest, temperate seasonal forest, woodland/shrubland, and temperate grassland/cold desert. The **temperate rainforest** biome, shown in **Figure 6.7,** is known for mild temperatures and abundant precipitation and is dominated by evergreen forests. These conditions are due to nearby warm ocean currents. This biome is most extensive near the Pacific coast

---

Boreal forest A biome densely populated by evergreen needle-leaved trees, with a short growing season and severe winters. *Also known as* **Taiga.**

Temperate rainforest A biome known for mild temperatures and abundant precipitation, dominated by evergreen forests.

Winton, Minnesota, United States
Average annual precipitation: 710 mm

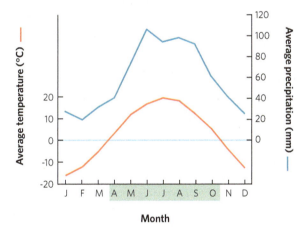

Figure 6.6 Boreal forest biome. Boreal forests, such as this one in the Boundary Waters Canoe Area Wilderness of the Superior National Forest, Minnesota, typically have cold temperatures and are dominated by evergreen trees, including spruces and firs. Gary Cook/Alamy.

in northwestern North America and in southern Chile, New Zealand, and Tasmania. The mild, rainy winters and foggy summers create conditions that support evergreen forests. In North America, these forests are dominated toward the south by coast redwood (*Sequoia sempervirens*) and toward the north by Douglas fir. These trees are typically 60 to 70 m tall and may grow to over 100 m, making them very attractive for harvesting as lumber. The fossil record shows that these plant communities are very old and they are remnants of forests that were vastly more extensive 70 million years ago. In contrast to tropical rainforests, temperate rainforests typically support few species.

Carmanah Point, British Columbia, Canada
Average annual precipitation: 2991 mm

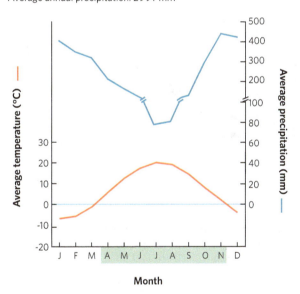

Figure 6.7 Temperate rainforest biome. Temperate rainforests exist along the coasts of several continents, including this forest of giant Sitka spruce trees (*Picea sitchensis*) in British Columbia, Canada. They have mild temperatures and high amounts of precipitation. Radius Images/Alamy.

TEMPERATE SEASONAL FORESTS

The **temperate seasonal forest** biome, shown in **Figure 6.8**, occurs under moderate temperature and precipitation conditions, and is dominated by deciduous trees. Winter temperatures can drop below freezing in this biome. The

Temperate seasonal forest A biome with moderate temperature and precipitation conditions, dominated by deciduous trees.

Bialystok, Podlaskie, Poland
Average annual precipitation: 580 mm

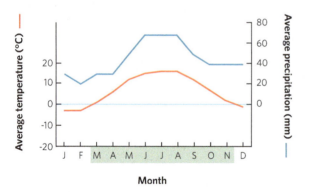

Figure 6.8 Temperate seasonal forest biome. Temperate seasonal forests have warm summers, cold winters, and a moderate amount of precipitation that favors the growth of deciduous trees. Pictured here is the Bialowieza Forest in Poland. Aleksander Bolbot/AGE Fotostock.

environmental conditions in this biome fluctuate much more than they do in the temperate rainforests because they do not benefit from the moderating effects of nearby warm ocean waters. In North America, the dominant plant growth form is deciduous trees, including maple, beech, and oak, which lose their leaves each fall. In North America, this biome stretches across the eastern United States and southeastern Canada; it is also widely distributed in Europe and eastern Asia. This biome is not common in the Southern Hemisphere, where the larger ratio of ocean surface to land moderates winter temperatures at many high altitudes and prevents frost.

In the Northern Hemisphere, the length of the growing season in this biome varies from 130 days at higher latitudes

to 180 days at lower latitudes. Precipitation usually exceeds evaporation and transpiration; consequently, water tends to move downward through soils and to drain from the landscape as groundwater and as surface streams and rivers. Soils are often podsolized, tend to be slightly acidic and moderately leached, and contain abundant organic matter. The vegetation often includes a layer of smaller tree species and shrubs beneath the dominant trees, as well as herbaceous plants on the forest floor. Many of these herbaceous plants complete their growth and flower in early spring before the trees have fully leafed out.

Warmer and drier parts of the temperate seasonal forest biome, especially where soils are sandy and nutrient poor, tend to develop needle-leaved forests dominated by pines. This includes the pine forests of the coastal plains of the Atlantic and Gulf coasts of the United States; pine forests also exist at higher elevations in the western United States. Because of the warm climate in the southeastern United States, decomposition is rapid. The rapid decomposition rates and sandy soils lead to the low availability of nutrients. The low nutrients and relatively dry conditions, in turn, favor the evergreen, needle-leaved trees, which resist desiccation and give up nutrients slowly because they retain their needles for several years. Since soils in this biome tend to be dry, fires are frequent in the pine forests, although most species are able to resist fire damage. The temperate seasonal forest was one of the first biomes that European settlers in North America used for agriculture.

WOODLANDS/SHRUBLANDS

The **woodland/shrubland** biome, shown in **Figure 6.9**, is characterized by hot, dry summers and mild, wet winters, a combination that favors the growth of drought-tolerant grasses and shrubs. Because this type of climate is found around much of the Mediterranean Sea, it is often referred to as a Mediterranean climate regardless of where it is actually found. The woodland/shrubland biome has many different regional names, including *chaparral* in southern California, *matorral* in southern South America, *fynbos* in southern Africa, and *maquis* in the area surrounding the Mediterranean Sea.

As you can see in the climate diagram, although there is a 12-month growing season, plant growth is limited by dry conditions in the summer and by cold temperatures in the winter. This biome supports thick evergreen shrubby vegetation 1 to 3 m in height, with deep roots and drought-resistant foliage. The small, durable leaves of typical Mediterranean-climate plants have earned the label of **sclerophyllous** ("hard-leaved") vegetation. Fires are frequent in the woodland/shrubland biome, and most plants have either fire-resistant seeds or root crowns that resprout soon after a fire. Traditional human use of this biome has been for grazing animals and growing deep-rooted crops such as wine grapes, as we discussed at the beginning of the chapter.

Woodland/shrubland A biome characterized by hot, dry summers and mild, wet winters, a combination that favors the growth of drought-tolerant grasses and shrubs.

Sclerophyllous Vegetation that has small, durable leaves.

Paso Robles, California, United States
Average annual precipitation: 280 mm

Medora, North Dakota, United States
Average annual precipitation: 374 mm

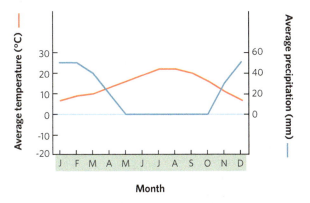

Figure 6.9 Woodland/shrubland biome. This biome is characterized by hot, dry summers and mild, wet winters, a combination that favors the growth of drought-tolerant grasses and shrubs. An example of this biome can be found in Cape Town, South Africa. The climate diagram is from Paso Robles, California. Rodger Shagam/ Getty Images.

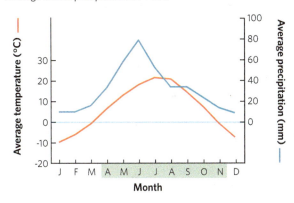

Figure 6.10 Temperate grassland/cold desert biome. Grasslands such as this one at the Theodore Roosevelt National Park in North Dakota are characterized by hot, dry summers and very cold winters. Where moisture is more abundant, the dominant vegetation is grass. Where moisture is less abundant, in areas known as cold deserts, the dominant vegetation is composed of widely scattered shrubs. Thomas & Pat Leeson/Science Source.

TEMPERATE GRASSLANDS/COLD DESERTS

The **temperate grassland/cold desert** biome, shown in **Figure 6.10**, is characterized by hot, dry summers and cold, harsh winters, and is dominated by grasses, nonwoody flowering plants, and drought-adapted shrubs. Plant growth is constrained by a lack of precipitation in the summer and by cold temperatures in the winter. The biome is also known by a variety of different names around the world, including *prairies* in North America, *pampas* in South America, and *steppes* in eastern Europe and central Asia.

As the biome name suggests, the dominant plant forms in temperate grasslands are grasses and nonwoody flowering plants that are well adapted to the frequent fires. Precipitation

varies widely across this biome. For example, on the eastern edge of North American prairies, annual precipitation can be 1,000 mm. In such areas, grasses can grow to more than 2 m high and are referred to as *tallgrass prairies*. There is even enough moisture in such areas to support the growth of trees, but the frequent fires prevent the trees from becoming a dominant component of this biome. As one moves west, annual precipitation declines to 500 mm or less. In these areas, grasses generally do not grow taller than 0.5 m and

---

Temperate grassland/cold desert A biome characterized by hot, dry summers and cold, harsh winters and dominated by grasses, nonwoody flowering plants, and drought-adapted shrubs.

are referred to as *shortgrass prairies.* Because precipitation is infrequent, organic detritus does not decompose rapidly, and this makes the soils rich in organic matter. In addition, the weak acidity of the soils means that they are not heavily leached, and they tend to be rich in nutrients.

Even farther west in North America, annual precipitation drops below 250 mm and the temperate grasslands grade into cold deserts, also known as temperate deserts. In the United States, the cold desert extends across most of the Great Basin, which sits in the rain shadow of the Sierra Nevada and Cascade Mountains. In the northern part of the region, the dominant plant is sagebrush, whereas toward the south and on somewhat moister soils, widely spaced juniper and piñon trees predominate, forming open woodlands with trees less than 10 m in stature and sparse coverings of grass. In these cold deserts, evaporation and transpiration exceed precipitation during most of the year, resulting in dry soils. Fires are infrequent in cold deserts because the habitat produces so little plant material to burn. However, because of the low productivity of the plant community, grazing can exert strong pressure on the vegetation and may even favor the persistence of shrubs, which are not good forage. Indeed, many dry grasslands in the western United States and elsewhere in the world have been converted to deserts by overgrazing.

TROPICAL RAINFORESTS

Our final group of biomes is found in areas of tropical temperatures and includes tropical rainforests, tropical seasonal forests/savannas, and subtropical deserts. **Tropical rainforests,** shown in **Figure 6.11,** generally fall within 20° N and 20° S of the equator, are warm and rainy, and are characterized by multiple layers of lush vegetation. Tropical rainforests have a continuous canopy of 30 to 40 m trees with emerging trees that occasionally reach 55 m. Shorter trees and shrubs form a layer known as the *understory* below the canopy. The understory also contains an abundance of epiphytes and vines. Species diversity is higher in tropical rainforests than anywhere else on Earth. This biome occurs in much of Central America, the Amazon Basin, the Congo in southern West Africa, the eastern side of Madagascar, Southeast Asia, and the northeast coast of Australia. In many of these locations, however, much of the rainforest has been destroyed to harvest lumber and to make room for agriculture.

Climates that support tropical rainforests are always warm and receive at least 2,000 mm of precipitation throughout the year, with rarely less than 100 mm during any single month. The tropical rainforest climate often exhibits two peaks of rainfall centered on the equinoxes, corresponding to the periods when the intertropical convergence zone passes over the equator (as discussed in Chapter 5). Rainforest soils are typically old and deeply weathered from the high amounts of precipitation. Because they are relatively devoid of organic matter and clay, they take on the reddish color of aluminum and iron oxides and they retain nutrients poorly. This is the process of laterization that we also discussed in Chapter 5. Despite the

Sandakan, Sabah, Malaysia
Average annual precipitation: 3060 mm

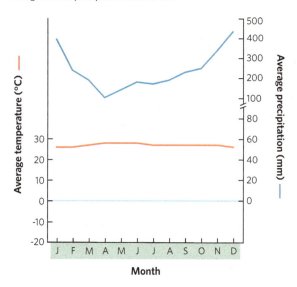

Figure 6.11 Tropical rainforest biome. Tropical rainforests, such as this site in Borneo, have very warm temperatures and very high amounts of precipitation. As a result, this biome has multiple layers of lush vegetation. Nick Garbutt/naturepl.com.

poor ability of these soils to hold nutrients, the biological productivity of tropical rainforests per unit area exceeds that of any other terrestrial biome. Moreover, the standing biomass exceeds that of all other biomes except temperate rainforests. This tremendous growth is possible because the continuously high temperatures and abundant moisture cause organic matter to decompose quickly, and vegetation immediately takes up the released nutrients. While rapid nutrient cycling supports the high productivity of the

Tropical rainforest A warm and rainy biome, characterized by multiple layers of lush vegetation.

rainforest, it also makes the rainforest ecosystem extremely vulnerable to disturbance. When tropical rainforests are cut and burned, many of the nutrients are carted off in logs or go up in smoke. The vulnerable soils erode rapidly and fill the streams with silt. In many cases, the environment degrades rapidly and the landscape becomes unproductive.

TROPICAL SEASONAL FORESTS/SAVANNAS

Tropical seasonal forests, shown in **Figure 6.12**, are located mostly beyond 10° N and 10° S of the equator. These regions

Nairobi, Nairobi, Kenya
Average annual precipitation: 750 mm

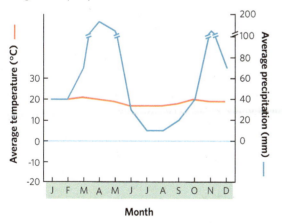

Figure 6.12 Tropical seasonal forest/savanna biome. Tropical seasonal forests and savannas have warm temperatures like the tropical rainforests, but also distinct wet and dry seasons due to the movement of the intertropical convergence zone. As a result, this biome has trees that shed their leaves during the dry season. An example of this biome can be found in the Masai Mara National Reserve of Kenya. Denis-Huot/Newscom/ZUMA Press/Kenya.

experience warm temperatures and, as the intertropical convergence zone moves during the year, pronounced wet and dry seasons. Because tropical seasonal forests have a preponderance of deciduous trees that shed their leaves during the dry season, this biome is sometimes referred to as a *tropical deciduous forest*. In areas where the dry season is longer and more severe, the vegetation becomes shorter and develops thorns to protect leaves from grazing animals. With even longer dry periods, the vegetation grades from dry forest into thorn forest and finally into *savannas*, which are open landscapes containing grasses and occasional trees, including acacia and baobab trees.

The tropical seasonal forest/savanna biome occurs in Central America, the Atlantic coast of South America, sub-Saharan Africa, southern Asia, and northwestern Australia. Fire and grazing play important roles in maintaining the character of the savanna biome. Under these conditions, grasses can persist better than other forms of vegetation. When grazing and fire are prevented within a savanna habitat, dry forest often begins to develop. As in more humid tropical environments, the soils tend to hold nutrients poorly but the warm temperatures favor rapid decomposition. Rapid decomposition provides a rapid recycling of nutrients into the soil, which trees can quickly take up and use for growth and reproduction. Such a rapid cycling of nutrients also makes this biome an attractive place for agriculture, including raising cattle. On the Pacific coast of Central America and on the Atlantic coast of South America, for example, over 99 percent of this biome has been converted to agriculture.

SUBTROPICAL DESERTS

Subtropical deserts, shown in **Figure 6.13**, are characterized by hot temperatures, scarce rainfall, long growing seasons, and sparse vegetation. Also known as *hot deserts,* subtropical deserts develop at 20° to 30° north and south of the equator, in areas associated with the dry, descending air of Hadley cells that we discussed in Chapter 5. Subtropical deserts include the Mojave Desert in North America, the Sahara Desert in Africa, the Arabian Desert in Asia, and the Great Victoria Desert in Australia.

Because of the low rainfall, the soils of subtropical deserts are shallow, virtually devoid of organic matter, and neutral in pH. Whereas sagebrush dominates the cold deserts of the Great Basin, creosote bush (*Larrea tridentata*) dominates the subtropical deserts of the Americas. Moister sites support a profusion of succulent cacti, shrubs, and small trees, such as mesquite and paloverde (*Cercidium microphyllum*). Most subtropical deserts receive summer rainfall. After summer rains, many herbaceous plants sprout from dormant seeds,

Tropical seasonal forest A biome with warm temperatures and pronounced wet and dry seasons, dominated by deciduous trees that shed their leaves during the dry season.

Subtropical desert A biome characterized by hot temperatures, scarce rainfall, long growing seasons, and sparse vegetation.

Antofagasta, Antofagasta, Chile
Average annual precipitation: 0 mm

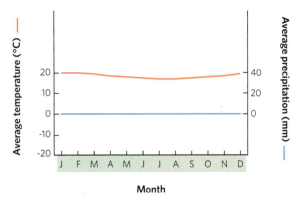

Figure 6.13 Subtropical desert biome. Subtropical deserts, such as this site in the Atacama Desert of Chile, have hot temperatures and scarce rainfall. This favors drought-resistant plants such as cactus, creosote bush, mesquite, and euphorbs. imageBROKER/Alamy.

grow quickly, and reproduce before the soils dry out again. Few plants in subtropical deserts are frost-tolerant. Species diversity is usually much higher than in temperate arid lands.

CONCEPT CHECK

1. Why is the boreal forest biome found on several different continents, including North America, Europe, and Asia?
2. What types of terrestrial plants are found in each of the four biomes situated at temperate latitudes?
3. Why do tropical rainforests experience two peaks of rainfall?

6.3 Aquatic biomes are categorized by their flow, depth, and salinity

As we discussed earlier in this chapter, ecologists categorize aquatic biomes using a range of physical factors, including water depth, water flow, and salinity. The major types of aquatic biomes include streams and rivers, lakes and ponds, freshwater wetlands, salt marshes, mangrove swamps, intertidal zones, coral reefs, and the open ocean.

STREAMS AND RIVERS

Because streams and rivers are characterized by flowing fresh water, they are often referred to as **lotic** systems. Although there is no exact specification to determine classification differences between a stream and a river, in general, **streams,** also called **creeks,** are narrow channels of fast-flowing fresh water, whereas **rivers** are wide channels of slow-flowing fresh water (**Figure 6.14**). As streams flow down from their headwaters, they join together with other streams and eventually grow large enough to be considered a river. Streams and some rivers are usually bordered by a **riparian zone,** a band of terrestrial vegetation influenced by seasonal flooding and elevated water tables.

Downstream, water flows more slowly and becomes warmer and richer in nutrients. Under these conditions, ecosystems generally become more complex and more productive. In general, streams support fewer species than other aquatic biomes. Small streams are often shaded and nutrient poor, which limits the productivity of algae and other photosynthetic organisms. Much of the organic content of stream ecosystems depends on **allochthonous** inputs of organic matter, such as leaves, that come from outside the ecosystem. In large rivers, a higher proportion of the organic inputs are **autochthonous,** meaning that they are produced from inside the ecosystem by algae and aquatic plants. As rivers progress from their source, they typically become wider, slower-moving, more heavily laden with nutrients, and more exposed to direct sunlight. They also accumulate sediments that are washed in from the land and carried downstream. High turbidity caused by suspended sediments in the lower reaches of silt-laden rivers can block light and reduce production.

Lotic systems are extremely sensitive to modification of their water flow by dams. In the United States, tens of thousands of dams—built to control flooding, to provide water for irrigation, or to generate electricity—interrupt stream flow. Dams also alter water temperature and rates of sedimentation. Typically, water behind dams becomes warmer,

Lotic Characterized by flowing fresh water.

Stream A narrow channel of fast-flowing fresh water. *Also known as* **Creek**.

River A wide channel of slow-flowing fresh water.

Riparian zone A band of terrestrial vegetation alongside rivers and streams that is influenced by seasonal flooding and elevated water tables.

Allochthonous Inputs of organic matter, such as leaves, that come from outside of an ecosystem.

Autochthonous Inputs of organic matter that are produced by algae and aquatic plants inside an ecosystem.

Figure 6.14 Streams and rivers. Streams and rivers are characterized by flowing fresh water. This example is on the Vefsna River in Norway. © Erwan Balança/Biosphoto.

and the original stream bottoms become filled with silt that destroys habitat for fish and other aquatic organisms. Water released downstream from large dams often has low concentrations of dissolved oxygen. Using dams for flood control changes the natural seasonal cycles of flooding that are necessary for maintaining many kinds of riparian habitats on floodplains. Dams also disrupt the natural movement of aquatic organisms upstream and downstream, fragmenting river systems and isolating populations.

PONDS AND LAKES

Ponds and **lakes** are aquatic biomes characterized by nonflowing fresh water with at least some area of water that is too deep for plants to rise above the water's surface (**Figure 6.15a**). Although there is no clear-cut distinction between ponds and lakes, ponds are smaller. Many lakes and ponds were formed as glaciers retreated, gouging out basins and leaving behind glacial deposits containing blocks of ice that eventually melted. The Great Lakes of North America formed in glacial basins, overlain until 10,000 years ago by thick ice. Lakes are also formed in geologically active regions, such as

the Great Rift Valley of Africa, where vertical shifting of blocks of the Earth's crust created basins in which water accumulates. Broad river valleys, such as those of the Mississippi and Amazon rivers, have *oxbow lakes,* which are broad bends of what was once the river, cut off by shifts in the main channel.

As shown in Figure 6.15b, lakes can be subdivided into several ecological zones, each with distinct physical conditions. The **littoral zone** is the shallow area around the edge of a lake or pond containing rooted vegetation, such as water lilies and pickerelweed. The open water beyond the littoral zone is the **limnetic** (or **pelagic**) **zone,** where the dominant photosynthetic organisms are floating algae, or *phytoplankton.* Very deep lakes also have a **profundal zone** that does not receive sunlight because of its depth. The absence of photosynthesis, as well as the presence of bacteria that decompose the detritus at the bottom of the lake, causes the profundal zone to have very low concentrations of oxygen. The sediments at the bottoms of lakes and ponds constitute the **benthic zone,** which provides habitat for burrowing animals and microorganisms.

Circulation in Ponds and Lakes

While lakes and ponds can be separated into four zones based on the proximity to the shore and the amount of light penetration, the water depths can also be classified by temperature. In most lakes and ponds in temperate and polar regions, the temperature of the water forms layers. The surface water, known as the **epilimnion,** can have a different temperature than the deeper water, known as the **hypolimnion.** Between these two temperature regions is the **thermocline,** which is a middle depth of water that experiences a rapid change in temperature over a relatively short distance in depth. The thermocline serves as a barrier to a mixing between the epilimnion and hypolimnion.

Most production in a lake occurs in the epilimnion, where sunlight is most intense. Oxygen produced by photosynthesis and oxygen entering the lake at the interface of the water and atmosphere keep the epilimnion well aerated and thus suitable for animal life. Throughout a growing

Pond An aquatic biome that is smaller than a lake and is characterized by nonflowing fresh water with some area of water that is too deep for plants to rise above the water's surface.

Lake An aquatic biome that is larger than a pond and is characterized by nonflowing fresh water with some area of water that is too deep for plants to rise above the water's surface.

Littoral zone The shallow area around the edge of a lake or pond containing rooted vegetation.

Limnetic zone The open water beyond the littoral zone, where the dominant photosynthetic organisms are floating algae. *Also known as* **Pelagic zone**.

Profundal zone The area in a lake that is too deep to receive sunlight.

Benthic zone The area consisting of the sediments at the bottoms of lakes, ponds, and oceans.

Epilimnion The surface layer of the water in a lake or pond.

Hypolimnion The deeper layer of water in a lake or pond.

Thermocline A middle depth of water in a lake or pond that experiences a rapid change in temperature over a relatively short distance in depth.

Figure 6.15 Ponds and lakes. Ponds and lakes are characterized by nonflowing fresh water with areas of water that are too deep for emergent vegetation. **(a)** Red Rock Lake, Colorado. **(b)** Lakes contain a variety of zones. The littoral zone exists around the edge of the lake and contains rooted, emergent plants. The limnetic zone consists of the open water in the middle of the lake, where the dominant photosynthetic organisms are floating algae. Below the limnetic zone is the profundal zone, which is too deep for sunlight to penetrate enough to permit photosynthesis. The layer of sediments at the bottom of the lake is the benthic zone. Lee Wilcox.

(a)

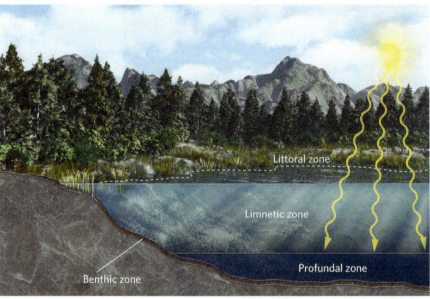

(b)

season, however, plants and algae often deplete the supply of dissolved mineral nutrients in the epilimnion and this curtails their growth. In the hypolimnion, which can include the lower limnetic zone and the profundal zone, bacteria continue to decompose organic material, but the reduced intensity of light causes a reduction in photosynthesis. The result is that oxygen is used up faster than it is produced, and this leads to anaerobic conditions. Oxygen is in particularly short supply deep in productive lakes that generate abundant organic matter in the epilimnion.

Lakes in the temperate zone experience changing temperatures across the seasons. These temperature changes drive changes in water density, which, in turn, causes the shallow and deep water to mix. **Figure 6.16** shows this process. As you may recall from Chapter 2, water becomes more dense as it cools to 4°C and then less dense as it cools below 4°C. During the winter in cold climates, the coldest lake water (0°C) lies at the surface just beneath the ice, while the slightly warmer, denser water (4°C) sinks to the bottom of the lake. In early spring, the sun gradually warms the lake. As the surface temperature increases toward 4°C, the sun-warmed water sinks into the cooler layers immediately below and the

water begins to mix. At the same time, winds drive surface currents that can cause deep water to rise in a manner similar to upwelling currents in the oceans. The vertical mixing of the lake water that occurs in early spring and is assisted by winds that drive the surface currents is known as the **spring turnover.** The spring turnover brings nutrients from sediments on the bottom to the surface and oxygen from the surface to the depths. This mixing results in the rapid growth of phytoplankton, the algae the float throughout the water column and serve as a major food source for herbivores.

In late spring and early summer, surface layers of water gain heat faster than deeper layers. At this point, the thermocline is created. Once the thermocline is well established, the surface and deep waters no longer mix because the warmer, less dense surface water floats on the cooler, denser water below, a condition known as **stratification.** During the fall, the temperature of the surface layers of the lake drops. As this water becomes

---

Spring turnover The vertical mixing of lake water that occurs in early spring, assisted by winds that drive the surface currents.

Stratification The condition of a lake or pond when the warmer, less dense surface water floats on the cooler, denser water below.

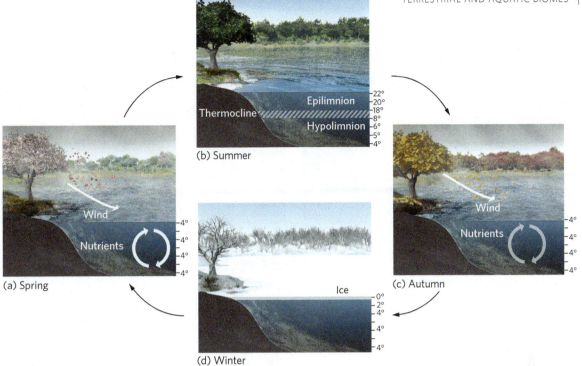

Figure 6.16 Circulation and turnover in temperate lakes. (a) In the spring, seasonal winds cause the lake water to mix, which brings nutrients from the sediments to the surface water and oxygen from the surface water down to the deeper water. **(b)** During the summer, the surface water warms faster than the deep water, so the lake experiences thermal stratification. The zone at which water temperature changes rapidly with depth is known as the thermocline. **(c)** In autumn, the surface water cools, stratification breaks down, and autumn winds cause the surface water and deep waters to mix once again. **(d)** In winter, the surface waters are exposed to freezing temperatures, so ice forms at the surface. Because 4°C water is the most dense, the bottom of the lake does not freeze.

denser than the underlying water, it begins to sink. The vertical mixing that occurs in the fall and is assisted by winds that drive surface currents is called **fall turnover.** Similar to the spring turnover, fall turnover brings oxygen to deep waters and nutrients to the surface. The infusion of nutrients into surface waters in the fall may cause a second phytoplankton bloom. This mixing persists into late fall, until the temperature at the lake surface drops below 4°C and winter stratification ensues.

The spring and fall turnover are typical of lakes that exist in temperate climates because they experience cold winters and warm summers. The seasonality of vertical mixing is much less dramatic in lakes that are not exposed to such dramatic temperature changes. In warmer climates, water temperatures do not fall below 4°C. As a result, such lakes do not stratify in the winter, and many have only one mixing event each year following summer stratification.

FRESHWATER WETLANDS
Freshwater wetlands are aquatic biomes that contain standing fresh water, or soils saturated with fresh water for at least part of the year, and are shallow enough to have emergent vegetation throughout all depths. Most of the plants that grow in wetlands can tolerate low oxygen concentrations in the soil; many are specialized for these anoxic conditions and grow nowhere else.

Freshwater wetlands include swamps, marshes, and bogs (**Figure 6.17**). *Swamps* contain emergent trees. Some of the best-known swamps are the Okefenokee Swamp in Georgia and Florida and the Great Dismal Swamp in Virginia and

North Carolina. *Marshes* contain emergent nonwoody vegetation such as cattails. Some of the largest marshes in the world include the Everglades in Florida and the Pantanal of Brazil, Bolivia, and Paraguay. In contrast to swamps and marshes, *bogs* are characterized by acidic waters and contain a variety of plants, including sphagnum mosses and stunted trees that are specially adapted to these conditions. Some of the largest bogs are found in Canada, northern Europe, and Russia.

Freshwater wetlands provide important habitats for a wide variety of animals, notably waterfowl and the larval stages of many species of fish and invertebrates that are characteristic of open waters. Wetland sediments immobilize potentially toxic or polluting substances dissolved in water and therefore function as a water purification system.

SALT MARSHES/ESTUARIES
Salt marshes are a saltwater biome that contains nonwoody emergent vegetation. Salt marshes are found along the coasts of continents in temperate climates, often within **estuaries,**

Fall turnover The vertical mixing of lake water that occurs in fall, assisted by winds that drive the surface currents.

Freshwater wetland An aquatic biome that contains standing fresh water, or soils saturated with fresh water for at least part of the year, and which is shallow enough to have emergent vegetation throughout all depths.

Salt marsh A saltwater biome that contains nonwoody emergent vegetation.

Estuary An area along the coast where the mouths of freshwater rivers mix with the salt water from oceans.

(a)

(b)

(c)

Figure 6.17 Freshwater wetlands. This biome includes a variety of aquatic habitats. **(a)** Swamps contain emergent trees, such as this bald cypress swamp (*Taxodium distichum*) in Reelfoot Lake State Park, Tennessee. **(b)** Marshes contain emergent nonwoody vegetation that includes cattails, such as in this location near Fairfax, Virginia. **(c)** Bogs are characterized by acidic waters and plants that are well adapted to these conditions, such as this bog in northern Wisconsin. (a) Byron Jorjorian/Science Source; (b) Corey Hilz/Danita Delimont.com/Newscom; (c) Lee Wilcox.

which are areas along the coast where the mouths of freshwater rivers mix with the salt water from oceans (**Figure 6.18**). Estuaries are unique because of their mix of fresh and salt water. In addition, they contain an abundant supply of nutrients and sediments carried downstream by rivers. The rapid exchange of nutrients between the sediments and the surface in the shallow waters of an estuary supports extremely high

growth of plants and algae. Because estuaries tend to be areas of sediment deposition, they are often edged by extensive saltwater marshes at temperate latitudes and by mangrove swamps in the tropics. With a combination of high nutrient levels and freedom from water stress, tidal marshes are among the most productive habitats on earth. They contribute organic matter to estuarine ecosystems, which, in turn, support large populations of oysters, crabs, fish, and the animals that feed on them.

MANGROVE SWAMPS

Mangrove swamps are a biome that exists in salt water along tropical and subtropical coasts and contains salt-tolerant trees with roots submerged in water. This biome can also occur in estuaries where fresh water and salt water mix (**Figure 6.19**). By living along the coasts, these salt-tolerant trees play important roles in preventing the erosion of coastal shorelines from constant incoming waves. The swamps also provide critical habitats to many species of fish and shellfish.

INTERTIDAL ZONES

The **intertidal zone** is a biome consisting of the narrow band of coastline between the levels of high tide and low tide. As the tides come in and go out, the intertidal zone experiences

Figure 6.18 Salt marsh. Salt marshes occur in salt water and estuaries and contain nonwoody emergent vegetation. An example of a salt marsh can be found in Plum Island Sound off the coast of Massachusetts. Jerry Monkman/naturepl.com.

Mangrove swamp A biome that occurs along tropical and subtropical coasts and contains salt-tolerant trees with roots submerged in water.

Intertidal zone A biome consisting of the narrow band of coastline between the levels of high tide and low tide.

Figure 6.19 Mangrove swamps. Mangrove swamps, including this location off the coast of Australia, are saltwater biomes that contain salt-tolerant trees along tropical and subtropical coastlines. © T. & S. Allofs/Biosphoto.

widely fluctuating temperatures and salt concentrations. The species living in this biome—including crabs, barnacles, sponges, mussels, and algae—must therefore possess adaptations that allow them to tolerate such harsh conditions. Intertidal zones can occur along steep rocky coastlines, as one might find in Maine, or gently sloping mudflats, as one might find in Cape Cod Bay in Massachusetts (**Figure 6.20**).

(a)

CORAL REEFS

Coral reefs are a marine biome found in warm, shallow waters that remain above 20°C year-round. Coral reefs often surround volcanic islands, where they are fed by nutrients eroding from the rich volcanic soil and by deep-water currents forced upward by the profile of the island.

Corals are tiny animals—related to hydra and other cnidarians—that live in a mutualistic relationship with algae. An individual coral is a hollow tube that secretes a hard exoskeleton made of calcium carbonate. It also has tentacles that sweep food particles of detritus and plankton into the tube. As it digests these particles, the coral produces CO_2 that can be used by their symbiotic algae in photosynthesis. Some of the sugars and other organic compounds the algae produce leak into the coral tissues and further support coral growth. Although an individual coral is tiny, corals live in huge colonies. As an individual coral dies, the soft tissues decompose but the hard outer skeletons remain behind. Over time, these skeletons accumulate to form massive coral reefs.

The complex structure that corals build provides a wide variety of substrates and hiding places for algae and animals. This helps to make coral reefs among the most diverse biomes on Earth (**Figure 6.21**). As you may recall from our discussion of coral reefs in Chapter 2, rising sea surface temperatures in the tropics are causing the departure of the algal symbionts of corals over large areas—a phenomenon known as *coral bleaching*. Because the algal symbionts are critical to the survival of the coral, the stability of these biomes is now at risk.

Coral reef A marine biome found in warm, shallow waters that remain 20°C year-round.

Figure 6.20 Intertidal zone. Intertidal biomes are the coastal regions around the world that exist between the high tide and the low tide of the oceans. **(a)** Rocky coasts produce rocky intertidal habitats, such as this one along the Alaskan coast. **(b)** Muddy coasts produce mudflat habitats around the world, including this site at Los Llanos, Venezuela. (a) Mark Conlin/V&W/imagequestmarine.com; (b) Lee Dalton/Alamy.

(b)

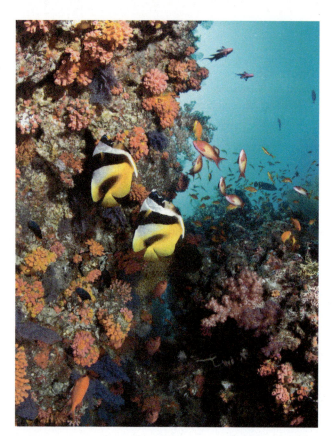

Figure 6.21 Coral reefs. The hard exoskeletons of millions of tiny corals form massive coral reefs in the ocean, which serve as home to an incredible diversity of organisms. Coral reefs can be found in shallow, warm ocean waters, such as this reef off the coast of the Maldives. Lea Lee/Getty Images.

THE OPEN OCEAN

The open ocean is characterized as the part of the ocean that is away from the shoreline and coral reefs. Open oceans cover the largest portion of the surface of Earth. Beneath the surface lies an immensely complex realm with large variations in temperature, salinity, light, pressure, and currents. Ecologists recognize a number of zones in the open ocean, shown in **Figure 6.22**. Beyond the range of the lowest tidal level, the **neritic zone** extends to depths of about 200 m, which corresponds to the edge of the continental shelf. Because strong waves move nutrients to the sunlit surface layers from sediments below, the neritic zone is generally a region of high productivity. Beyond the neritic zone, the

Neritic zone The ocean zone beyond the range of the lowest tidal level, and which extends to depths of about 200 m.

Figure 6.22 Open ocean. The open ocean is represented by water that is offshore and away from coral reefs. This biome can be broken up into several zones.

seafloor drops rapidly to the great depths of the **oceanic zone.** Here, nutrients are sparse, and production is strictly limited. Finally, the benthic zone consists of the seafloor underlying the neritic and oceanic zones.

The neritic and the oceanic zones may be subdivided vertically into a *photic zone* and an *aphotic zone.* The **photic zone** is the area of the neritic and oceanic zones that contains sufficient light for photosynthesis by algae. The **aphotic zone** is the area of the neritic and oceanic zones where the water is so deep that sunlight cannot penetrate. However, as we saw in Chapter 1, bacteria in the aphotic zone use chemosynthesis to convert inorganic carbon into simple sugars. Other organisms in the aphotic zone depend on the organic material that falls from the photic zone. One of the fascinating adaptations of many organisms in the aphotic zone is the ability to generate their own source of light, known as *bioluminescence,* to help them find and consume prey. A number of jellyfish, crustaceans, squid, and fish species have independently evolved this ability.

In this chapter, we have explored how differences in climates determine the types of dominant plant forms that can persist in different parts of the world, forming the basis of categorizing terrestrial biomes. In contrast, the aquatic biomes are categorized by differences in water flow, depth, and salinity. In all cases, there is a close association between the environmental conditions and the species that have evolved adaptations to live under these conditions. Of course, adaptations reflect not only the physical factors in the environment but also the many interactions with other organisms. In the next chapter, we examine the process of evolutionary adaptation and see how it has created the tremendous diversity of life on Earth.

CONCEPT CHECK

1. How do headwater streams and larger rivers differ in their major source of organic material?
2. Why does productivity in the ocean differ between the photic and aphotic zones?
3. What are the five types of saltwater biomes?

Oceanic zone The ocean zone beyond the neritic zone.

Photic zone The area of the neritic and oceanic zones that contains sufficient light for photosynthesis by algae.

Aphotic zone The area of the neritic and oceanic zones where the water is so deep that sunlight cannot penetrate.

ECOLOGY TODAY   CONNECTING THE CONCEPTS

Changing Biome Boundaries

Climate change. The changing climates around the world are predicted to alter the distribution of many organisms, including plants in the genus *Banksia* in the shrublands of southwestern Australia. These changes are also affecting human agriculture, including the vineyards that are planted in the shrubland biomes around the world. Phil Morley/AGE Fotostock.

Throughout this chapter, we have seen that climate largely determines the location of the terrestrial biomes. Climatic conditions, combined with species interactions, set the edges of biome boundaries. Given our understanding of how biome boundaries form, what would happen to the biomes, and the species living in them, if the climate changed?

 Records show us that during the past 130 years, temperatures of the surface of Earth have increased an average of 1°C. In fact, according to NASA, 15 of the 16 warmest years recorded since 1880 have occurred since 2001. This small average increase in Earth's temperature conceals the fact that some regions have become 1°C to 2°C cooler during this time, while others have become up to 4°C warmer. Scientists predict even greater increases in temperature and large changes in precipitation patterns for the twenty-first century. If climate determines the location of biomes and the climate is changing, it seems reasonable to predict that the boundaries of biomes are also going to change.

In some cases, scientists believe that shifting biome boundaries might occur relatively easily. Where no barriers prevent movement, plant and animal populations will be able to shift north or south over several decades without much difficulty. However, when movement is blocked, for example, by mountain ranges or large highways, the plants and animals may not be able to survive in their current locations given the changing conditions. Consider the woodland/shrubland biome on the southwestern coast of Australia. This small biome is located on a relatively small area of coastal land, with an ocean to the south and west, and a desert to the north and east. Scientists predict that this biome will become hotter and drier during the current century. If this prediction is correct, organisms that cannot tolerate the increase in temperature will have nowhere hospitable to go since the neighboring desert biome is already too dry for them to survive. Scientists who examined one group of plants from the genus *Banksia*, which is composed of 100 species, concluded that over the next 70 years, 66 percent of these species will decline in abundance and 25 percent will go extinct.

Climate changes also affect human agriculture. Recall from the beginning of this chapter that most of the world's wine is produced in the shrubland/woodland biome. However, changes in climate have already affected the cultivation of grapes used in winemaking. In France, for example, over the past 30 years warmer growing seasons have caused grapes to ripen 16 days earlier. This warmer climate alters the sugar content and acidity of the grapes—two components that must be in balance to make a fine-tasting wine. This problem is so serious for winemakers and wine consumers that in 2009 French wine growers called upon world leaders to take immediate action to try to reverse global climate change. On the other hand, locations at slightly higher latitudes that had temperatures historically too cool for growing quality wine grapes are now reporting increased summer temperatures, which have allowed them to produce some of the best wine grapes in years. In England, for example, the amount of land dedicated to growing grapes for wine has increased by 148 percent during the past decade. In fact, researchers in 2016 forecasted an additional increase of 2.2 °C and a 6 percent increase in precipitation by 2100, which will favor the production of wine grapes even more. This is very good news for winemakers in England, but devastating for the French who have a long history of making some of the world's finest wines.

SOURCES:

Chaudhuri, S. 2016. Climate change uncorks British wine production. *Wall Street Journal*, December 1. http://www.wsj.com/articles/britains-wine-production-could-be-boosted-by-climate-change-1480619748.

Iverson, J. T. 2009. How global warming could change the winemaking map. *Time Magazine*, December 3.

Fitzpatrick, M. C., A. D. Grove, N. J. Sanders, and R. R. Dunn. 2008. Climate change, plant migration, and range collapse in a global biodiversity hotspot: The *Banksia* (*Proteaceae*) of Western Australia. *Global Change Biology* 14:1337–1352.

SUMMARY OF LEARNING OBJECTIVES

6.1 Terrestrial biomes are categorized by their major plant growth forms. Ecologists use the dominant plant forms to categorize ecosystems into terrestrial biomes because many plants have evolved convergent forms in response to similar climate conditions. Climate and dominant plant forms are similar within biomes.

Key Terms: Convergent evolution, Biome, Climate diagram, Growing season

6.2 There are nine categories of terrestrial biomes.
The coldest biomes are the tundra and boreal forests. In temperate regions, we can find temperate rainforests, temperate seasonal forests, woodlands/shrublands, and temperate grasslands/cold deserts. In tropical latitudes, the biomes can be categorized as tropical rainforests, tropical seasonal forests/savannas, and subtropical deserts.

Key Terms: Tundra, Boreal forest, Temperate rainforest, Temperate seasonal forest, Woodland/shrubland, Sclerophyllous, Temperate grassland/cold desert, Tropical rainforest, Tropical seasonal forest, Subtropical desert

6.3 Aquatic biomes are categorized by their flow, depth, and salinity. Freshwater biomes include streams and rivers, ponds and lakes, and freshwater wetlands. Saltwater biomes include salt marshes/estuaries, mangrove swamps, intertidal zones, coral reefs, and the open ocean.

Key Terms: Lotic, Stream, River, Riparian zone, Allochthonous, Autochthonous, Pond, Lake, Littoral zone, Limnetic zone, Profundal zone, Benthic zone, Epilimnion, Hypolimnion, Thermocline, Spring turnover, Stratification, Fall turnover, Freshwater wetland, Salt marsh, Estuaries, Mangrove swamp, Intertidal zone, Coral reef, Neritic zone, Photic zone, Aphotic zone

CRITICAL THINKING QUESTIONS

1. Compare and contrast the factors used to categorize terrestrial biomes with those used to categorize aquatic biomes.

2. How do atmospheric convection currents discussed in Chapter 5 help to determine the locations of tropical seasonal forests?

3. In considering all the terrestrial biomes, what is the general effect of precipitation on soil fertility?

4. Given your knowledge of terrestrial biomes, why can temperate seasonal forests retain more of their soil fertility than tropical rainforests after they are logged?

5. How do environmental conditions differ among the four temperate biomes?

6. Compare and contrast allochthonous and autochthonous inputs to streams and rivers.

7. What parallels can you draw between the zones in a lake and zones in the ocean?

8. If northern latitudes continue to become warmer over the centuries, what effect might this have on lake circulation during the spring and fall?

9. Compare and contrast swamps, marshes, and bogs.

10. How will increasing global temperatures over the coming centuries likely affect the future distributions of biomes?

GRAPHING THE DATA   Creating a Climate Diagram

Scientists have collected climate data for locations all over the world. Using the monthly temperature and precipitation data for Miami, Florida (provided in the table), create a climate diagram. Remember to make each 10°C increase in temperature correspond to a 20 mm increase in precipitation.

Month	Temperature (°C)	Precipitation (mm)
January	2	45
February	5	50
March	9	104
April	15	100
May	18	120
June	23	110
July	28	88
August	25	100
September	21	140
October	15	98
November	8	100
December	3	65

7 Evolution and Adaptation

Favoring Flightless Birds

One of the distinguishing characteristics of birds is that they possess feathers that help them fly. However, not all birds have the ability to fly, including penguins and a group of birds known as ratites. Ratites are a group of well-known species, such as the ostriches of Africa (**Struthio** spp.), the emus and cassowaries of Australia (*Dromaius* spp., *Casuarius* spp.), the kiwis of New Zealand (*Apteryx* spp.), and the rheas of South America (*Rhea* spp.). For decades, scientists have assumed that the ratites all shared a common ancestor that was flightless and, following the breakup of the continents, the isolated populations of this ancestor gradually evolved into the distinct species that we see today. While this interpretation is an attractive story due to the simple explanation, research during the past decade has demonstrated that this evolution of flightless ratites was much more complex.

Using DNA from the various species of ratites and other birds has allowed researchers to build a family tree to determine which species are closely versus distantly related to each other. It turns out that some species of ratites are more closely related to other species of flying birds than to other species of ratites. This means that the ancestor of the ratites was likely a species that could fly and that today's flightless ratites actually became flightless through many independent evolutionary events rather than flightlessness evolving first and then the ratites evolving into separate species.

In many cases, the evolution of flightless birds occurs on islands where predators are few or completely lacking. Predator-free islands should reduce the benefits of flying, since the birds would not need flight as a means of escape. Although flight certainly serves many other purposes, it leads one to hypothesize that birds living on islands, even if they have not evolved to be flightless, may evolve a reduction in the investment of their flight machinery. To test this hypothesis, researchers examined the size of the flight muscles and the length of the legs in bird species living on islands versus related species that live on the mainland continents.

> **"Flying species of birds that live on islands have evolved smaller flight muscles and longer legs."**

In 2016, they reported their discovery that flying species of birds that live on islands have evolved smaller flight muscles and longer legs. Moreover, the magnitude of this effect was greatest on those islands containing the fewest predators. The researchers interpreted the evolution of longer legs as permitting a slower, more energy-efficient takeoff for the birds by jumping higher, since the birds no longer needed to take off to avoid predators rapidly by primarily using their flight muscles. Such trait changes may represent the intermediate evolutionary steps that the ratite birds went through during their evolution.

The evolution of reduced flight muscles on predator-free islands also means that these birds are more susceptible to any predators that are introduced to the islands. As we will see in later chapters, the unintentional introduction of non-native bird predators, such as snakes, has decimated the bird fauna of many islands.

Ratite birds. Two-wattled Cassowary (*Casuarius casuarius*) female on a forest edge in Australia, Queensland, Moresby Range National Park. Photo by blickwinkel/Alamy.

SOURCES:
Wright N. A., et al. 2016. Predictable evolution toward flightlessness in Volant island birds. *PNAS* 113: 4765–4770.

Harshman J., et al. 2008. Phylogenomic evidence for multiple losses of flight in ratite birds. *PNAS* 105: 13462–13467.

The research on bird flight is just one example of how scientists can use the theory of evolution by natural selection to obtain tremendous insight into how natural selection operates in nature. By studying how evolution occurs in wild populations, we can see how natural selection can alter the traits of species and populations over time and, in turn, affects species interactions. In this chapter, we will explore the ways in which evolution causes populations to become genetically distinct and how this leads to the origin of new species.

LEARNING OBJECTIVES

After reading this chapter, you should be able to:

7.1 Describe how the process of evolution depends on genetic variation.

7.2 Explain how evolution can occur through random processes or through selection.

7.3 Illustrate how microevolution operates at the population level.

7.4 Describe the way that macroevolution operates at the species level and higher levels of taxonomic organization.

The story of evolving flightlessness demonstrates that evolution shapes the form and function of organisms according to properties of their environments. Evolution depends on genetic variation, and this variation can arise from a number of processes. Over time, populations and species can evolve changes in traits, such as the size of bird flight muscles. These changes can evolve due to random processes or due to the nonrandom process of selection. Some of the most important sources of natural selection include differences in physical conditions, food resources, and interactions with competitors, predators, pathogens, and individuals of the same species. In this chapter, we will examine such processes and explore how genes and the environment come together to cause the evolution of populations and new species.

7.1 The process of evolution depends on genetic variation

In Chapter 4, we discussed how the traits expressed by an individual are the outcome of genotypes and the environment interacting. When genetic variation is present, it allows evolution by natural selection. In this section, we review the structure of DNA, the process of how genes help determine the phenotypes of organisms, and the process by which variation in genes is produced.

THE STRUCTURE OF DNA

Genetic information is contained in the molecule **deoxyribonucleic acid,** also known as **DNA**—a molecule composed of two strands that are wound together into a shape known as a double helix. Each strand is composed of subunits called *nucleotides* and each nucleotide is composed of a sugar, a phosphate group, and one of four different nitrogenous bases: adenine (A), thymine (T), cytosine (C), and guanine (G). Just as a sequence of letters signifies a particular meaning, or word, genetic information is coded in the particular order of the different nitrogenous bases. Long strands of DNA are wound around proteins into compact structures called **chromosomes.**

GENES AND ALLELES

Genes are regions of DNA that code for particular proteins, which in turn affect particular traits. Different forms of a particular gene are referred to as **alleles.** In diploid organisms—those having two sets of chromosomes—one allele comes from the mother's gamete and the other comes from the father's gamete. As you may recall, each gamete is haploid, meaning that it has just one set of chromosomes.

In many cases, a change in alleles can create differences in an organism's phenotype. The ABO blood groups in humans, for example, are determined by which allele a person inherits from each parent—A, B, or O. The allele is responsible for production of the *antigens* A and B, which are molecules on the surface of our red blood cells that interact with the immune system. (Note that the O allele does not produce an antigen.) Individuals with blood type A have AA or AO genotypes; individuals with blood type B have BB or BO genotypes. All remaining individuals

---

Deoxyribonucleic acid (DNA) A molecule composed of two strands of nucleotides that are wound together into a shape known as a double helix.

Chromosomes Compact structures consisting of long strands of DNA that are wound around proteins.

Alleles Different forms of a particular gene.

have either AB or OO genotypes. In this case, the link between the genotype and the phenotype is direct, and the pattern of inheritance is straightforward. For example, children of an AA father and a BB mother will all have the AB genotype.

Whereas blood type is determined by different alleles on a single gene, **polygenic** traits reflect the effects of alleles from several genes. For example, eye color in humans is determined by at least three genes that control pigments in different parts of the iris of the eye. Patterns of phenotype inheritance that depend on interactions among multiple alleles can be quite complex.

Many phenotypes in a population can span a range of values because they are polygenic traits. Body size is a good example. Most populations exhibit a normal, or bell-shaped, distribution of body sizes, as shown in **Figure 7.1**. In this distribution, most individuals cluster near the middle of the range with progressively fewer located toward the upper and lower extremes. Part of this continuous variation might be due to environmental differences, such as the amount of resources available. However, much of the variation can be attributed to the actions of many genes, each with a relatively small influence on the value of the trait. If several genes influence body size, an individual's size will depend on the mix of alleles for all of these genes. The tendency for individuals to be concentrated toward the center of the distribution reflects the relative improbability of an individual inheriting mostly alleles that code for large body size or mostly alleles that code for small body size. Think of this in terms of flipping a coin. The chance of getting 10 tails in a row (about 1 in 1,000) is much more remote than the chance of getting 5 heads and 5 tails (about 1 in 4).

Whereas some genes only affect a single trait, such as size, other genes affect multiple traits, an effect referred to as **pleiotropy.** For example, chickens have a gene—known as the frizzle gene—which causes feathers to curl outward rather than to lay flat along the body. However, this gene causes other variations, including faster metabolism, slower digestion, and less frequent egg laying. When a gene has pleiotropic effects, any changes in the gene can have far-reaching effects on the traits of organisms.

In some cases, the expression of one gene can be controlled by other genes. This is known as **epistasis.** In the case of mouse hair color, for example, there is a gene that determines whether a mouse will make black or brown hair pigments. However, there is a second gene that determines whether the hair will receive any pigments at all. If alleles in the second gene prevent pigments from being deposited in the hair, the alleles of the first gene become irrelevant and the mouse will have a white coat.

DOMINANT AND RECESSIVE ALLELES

Every individual has two copies of each gene, one inherited from its mother and one from its father. Exceptions to this rule include genes located on sex chromosomes, the genes

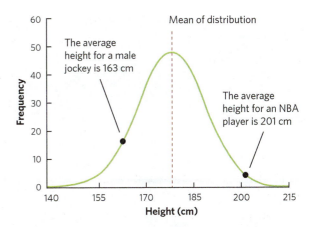

Figure 7.1 Frequency distribution of polygenic traits. When a continuous trait is determined by many genes, the distribution follows a bell-shaped curve. For the heights of adult males in the United States, for example, there is a symmetrical distribution around the mean. Most individuals express an intermediate trait value, whereas only a few individuals, such as professional basketball players and jockeys, express extreme trait values. Data from U.S. Census 2000.

of organisms that reproduce by self-fertilization, haploid organisms, and organisms such as plants that alternate between haploid and diploid generations as part of their normal life cycle. An individual with two different alleles of a particular gene is said to be **heterozygous** for that gene, as in the case of a person with the AB blood type. An individual with two identical alleles is said to be **homozygous,** for example, a person with AA blood type. When an individual is heterozygous, the two different alleles may produce an intermediate phenotype, as in the case of a person with the AB blood type that expresses both alleles. When both alleles contribute to the phenotype, the alleles are said to be **codominant.** Codominance is also found in the flower color of several species of plants (**Figure 7.2**). Alternatively, one allele may mask the expression of the other. In this case, the allele that is expressed is called **dominant** and the allele that is not expressed is called **recessive.** In domestic pigs, for example, the allele for white coat color is dominant and the allele for black coat color is recessive.

Polygenic When a single trait is affected by several genes.

Pleiotropy When a single gene affects multiple traits.

Epistasis When the expression of one gene is controlled by another gene.

Heterozygous When an individual has two different alleles of a particular gene.

Homozygous When an individual has two identical alleles of a particular gene.

Codominant When two alleles both contribute to the phenotype.

Dominant allele An allele that masks the expression of the other allele of a given gene.

Recessive An allele whose expression is masked by the presence of another allele.

Figure 7.2 Codominance. In snapdragons, the red and white flowers are homozygous genotypes. The pink flower gets its color from one red gene and one white gene that are codominant. *Photo by John Kaprielian/Science Source.*

Fortunately, most harmful alleles are recessive and are therefore not expressed in a heterozygous individual. Any dominant harmful alleles that might arise are expressed when homozygous or heterozygous. Because they reduce fitness, dominant harmful alleles are strongly selected against and removed from the population over time. In contrast, recessive harmful alleles are expressed when homozygous but not when heterozygous, so they can persist in a population because they are not selected against when they occur in heterozygous individuals. Examples of recessive harmful alleles in humans include those that cause sickle cell anemia and cystic fibrosis.

A **gene pool** consists of the alleles from all of the genes of every individual in a population. The gene pools of most populations that reproduce sexually contain substantial genetic variation. With the ABO blood type gene, for example, the human population of the United States includes 61 percent O alleles, 30 percent A alleles, and 9 percent B alleles. The proportions of these alleles vary among populations. Populations of Asian descent, for instance, tend to have higher frequencies of B alleles, while populations of Irish descent have higher frequencies of O alleles.

SOURCES OF GENETIC VARIATION

Now that we understand the role of genes and alleles, we need to review how we obtain genetic variation in the traits of organisms. One of the most common ways to generate variation is through sexual reproduction. By combining a haploid sex cell of one parent with a haploid sex cell of another parent, new combinations of alleles can be produced in the offspring across many different chromosomes. The chromosomes in a haploid gamete are a **random assortment** of the chromosomes in the parent's diploid cells, meaning that they can be any combination of chromosomes that the parent received from its mother and father. When an individual produces an egg cell, for example,

some chromosomes in the egg will have come from the individual's father, whereas other chromosomes in the egg will have come from the individual's mother. As we will see in Chapter 9, the creation of new gene combinations through sexual reproduction represents a major strategy for species to create offspring that are resistant to rapidly evolving parasites and pathogens.

Two additional ways genetic variation arises are through *mutation* and *recombination*. **Mutation** is a random change in the sequence of nucleotides in regions of DNA that either comprise a gene or control the expression of a gene. Mutations can occur anywhere along the chromosomes, although some regions of the chromosome can experience higher frequencies of mutation than others. Many mutations have no detectable effect and are referred to as *silent,* or *synonymous,* mutations. Other mutations may simply alter the appearance, physiology, or behavior of the individual. When phenotypic changes are better suited to the environment, these phenotypes will be favored by natural selection. Some mutations, however, can cause drastic, often lethal, changes in the phenotype. Many human genetic disorders, such as sickle cell anemia, Tay-Sachs disease, cystic fibrosis, and albinism, as well as tendencies to develop certain cancers and Alzheimer's disease, are caused by single-nucleotide mutations of individual genes.

Genetic **recombination** is the reshuffling of genes that can occur as DNA is copied during *meiosis,* the process

---

Gene pool The collection of alleles from all individuals in a population.

Random assortment The process of making haploid gametes in which the combination of alleles that are placed into a given gamete could be any combination of those possessed by the diploid parent.

Mutation A random change in the sequence of nucleotides in regions of DNA that either comprise a gene or control the expression of a gene.

Recombination The reshuffling of genes that can occur as DNA is copied during meiosis and chromosomes exchange genetic material.

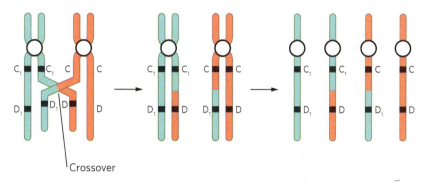

Figure 7.3 Genetic variation through recombination. During meiosis in eukaryotic organisms, pairs of homologous chromosomes line up together. When the chromosomes exhibit crossing over, they exchange DNA and each chromosome contains a new combination of genes.

that creates haploid gametes from diploid parent cells. During meiosis, pairs of *homologous chromosomes*—one member of which is inherited from each parent—line up next to each other. When the two chromosomes in the pair do not exchange any DNA, we end up with haploid cells that contain unaltered chromosomes. However, sometimes the two chromosomes in the pair do exchange DNA, in a process known as *crossing over*, as shown in **Figure 7.3**. In some cases, crossing over can also occur between *nonhomologous chromosomes*. In either case, new genes are not being created, but new combinations of alleles are produced that have the potential to produce new phenotypes.

One of the best-known examples of recombination involves the immune system of vertebrates. Vertebrates face a diversity of pathogens that continually evolve so that they become better at attacking their hosts. To combat these ever-changing pathogens, vertebrates need an ever-changing immune system that can identify and destroy the pathogens. Recombination provides the mechanism to create the high genetic variation in the immune system that vertebrates need to match the rapid evolution of their parasites. We will talk much more about this issue in Chapter 9.

CONCEPT CHECK

1. What is the difference between genes and alleles?
2. Why is it essential that traits be inherited for evolution to occur?
3. What are the three primary sources of genetic variation?

7.2 Evolution can occur through random processes or through selection

In western New York State, there is a herd of white-tailed deer that looks very different from most white-tailed deer. Many of the deer living on the 4,300-ha Seneca Army Depot do not have typical reddish brown coats. Rather, their coats are white (**Figure 7.4**). The white coat phenotype is due to a rare mutation that occurs in

Figure 7.4 A mutant white-tailed deer. At the Seneca Army Depot in western New York State, a mutation for white hair appeared in the 1940s. Since that time, the white phenotype was protected while the normal brown phenotype was hunted. Over the subsequent 70 years, the white phenotype has come to compose about 25 percent of the population.
Photo by Syracuse Newspapers/Dick Blume/The Image Works.

white-tailed deer populations. Because a white coat can frequently make a deer more visible to predators, it does not provide any fitness benefits and we would not expect it to persist. So why is there a high frequency of white deer on the Seneca Army Depot? It turns out that when the army depot was built in 1941, the 4,300-ha area was fenced and several dozen deer were trapped inside. A few years later, 2 white deer were observed. Given that white deer were such an unusual sight, authorities at the depot banned hunters from killing the deer with the white phenotype. Over time, the deer population grew larger and the white phenotype became more frequent. Today, of the nearly 800 deer living on the property, about 200 are white. In recent years, the U.S. Corps of Engineers has been trying to sell the land since it was no longer needed by the federal government. This made the future of the white deer uncertain, particularly if the ban on hunting white deer was lifted. In 2016, the land was sold to a company that pledged to develop some of the land for industry while still protecting the deer herd with the rare white phenotype. As a result, the rare phenotype that emerged due to random genetic processes will continue to persist.

The story of the white deer demonstrates how evolution often happens through multiple processes. Random events, such as mutations, may confer no fitness advantage when they first appear. Such is the case for most deer populations in which the white mutation occasionally arises but rarely becomes frequent in a population. However, if selection for the mutant phenotype occurs after the mutant appears, as happened at the Seneca Army Depot with reduced hunting pressure on deer with the white phenotype, this nonrandom process can cause the mutant to become more frequent in the population.

EVOLUTION THROUGH RANDOM PROCESSES
Random processes can facilitate evolutionary change in a population. In addition to mutation, random processes include *genetic drift, bottleneck effects,* and *founder effects.*

Mutation
We have seen that mutation is an important way in which genetic variation can arise in a population. Because genes often code for functions that are vital to performance and fitness, mutations that negatively impact these functions are not favored by selection. However, a small fraction of mutations can be beneficial. For example, **Figure 7.5** illustrates what happened in the deer herd of the Seneca Army Depot. In a herd of deer, a mutation for the white coat appeared, which added genetic variation to the population. After the mutation appeared, the ban on hunting white deer caused selection for the white phenotype and the frequency of white deer increased.

Mutation rates vary a great deal in different groups of organisms, but among genes that are expressed and can be observed as altered phenotypes, mutation rates range from 1 in 100 to 1 in 1,000,000 per gene per generation. The more genes that a species carries, the higher the probability that at least one gene will experience a

mutation. Similarly, the larger the size of the population, the higher the probability that an individual in the population will carry a mutation.

Genetic Drift
Genetic drift, another random process, occurs when genetic variation is lost because of random variation in mating, mortality, fecundity, or inheritance. Genetic drift is more common in small populations because random events can have a disproportionately large effect on the frequencies of genes in the population. But how does one determine whether an evolved phenotype is the outcome of drift versus some other process, such as natural selection? Research on the Mexican cavefish (*Astyanax mexicanus*) provides an answer.

The Mexican cavefish is a species composed of some populations that live in cave streams and other populations that live in surface streams. Although the populations can interbreed, they look quite different. As is the case with many cave-adapted animals, the cave populations have very reduced eyes and reduced pigmentation (**Figure 7.6a**). In contrast, populations living in surface streams have normal eyes and dark pigmentation. To determine whether these changes were due to natural selection or genetic drift, researchers raised individuals from the cave population, individuals from the surface population, and hybrid offspring made by interbreeding the cave and surface populations. The researchers then examined regions of the fish DNA that coded for eye size and pigmentation, which could contain one or more genes. In 2007, they reported that the 12 regions of DNA that coded for large eyes in the surface population all coded for small eyes in the cave population. The results are shown in Figure 7.6b. This suggests that natural selection favored all of the eye genes to evolve in a similar direction to produce smaller eyes. In contrast, when they examined the 13 regions of DNA that coded for pigmentation, they found that 5 of the regions

---

Genetic drift A process that occurs when genetic variation is lost because of random variation in mating, mortality, fecundity, and inheritance.

Mutation ⟶   Selection for the mutant phenotype

Figure 7.5 Evolution by mutation. Mutations, such as white coat in white-tailed deer, can arise in populations. If the mutation confers a fitness benefit, the mutation can increase in frequency in the population over multiple generations.

(a)

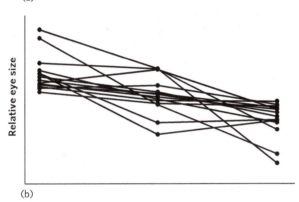

(b)

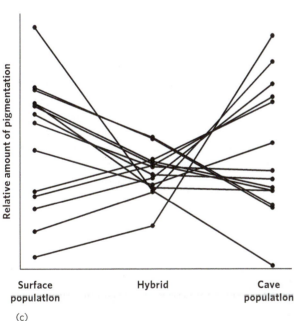

Surface population   Hybrid   Cave population

(c)

Figure 7.6 Evolution by genetic drift. (a) Mexican cavefish populations that live in surface streams have large eyes and dark pigmentation, whereas populations that live in caves have small eyes and reduced pigmentation. **(b)** When researchers compared how different regions of DNA that code for eye size changed between surface and cave populations, they found that all 12 regions coded for smaller eyes in the cave population than in the surface population. Because all 12 regions changed in the same direction, this suggests that natural selection selected for smaller eyes. **(c)** When researchers looked at changes in regions of DNA that code for pigmentation, they found 5 regions coded for increased pigmentation, while 8 regions coded for decreased pigmentation. Because the 13 regions did not change in the same direction, this suggests that changes in pigmentation were due to genetic drift. Data from M. Protas et al., Regressive evolution in the Mexican cave tetra, *Astyanax mexicanus, Current Biology* 17 (2007): 452–454. Image courtesy of Dr. Richard Borowsky.

populations, it may be the case that the cave population was initially very small.

Bottleneck Effects

A reduction in genetic variation can also occur because of a severe reduction in population size, known as the **bottleneck effect.** When a population experiences a large reduction in the number of individuals, the survivors carry only a fraction of the genetic diversity that was present in the original, larger population. Moreover, after being reduced to a small population by the bottleneck effect, the small population can then experience genetic drift. Population reductions can result from natural causes, for example, a drought that reduces the abundance of food, or anthropogenic causes, such as loss of habitat due to the construction of homes or factories.

One example of a bottleneck effect is the greater prairie chicken (*Tympanuchus cupido*), which is a grassland bird that historically lived throughout much of the middle United States, including Minnesota, Kansas, Nebraska, and Illinois. While prairie chickens have remained abundant in many neighboring states, the population in Illinois declined from approximately 12 million in 1860 to only 72 birds by 1990, as illustrated in **Figure 7.7a**. To determine whether this dramatic decline in population size was associated with a decline in genetic diversity, researchers collected DNA samples from museum specimens of prairie chickens from the 1930s, when the population was 25,000, and from the 1960s, when the population was 2,000. They defined this time period from the 1930s to the 1960s as "pre-bottleneck." They also compared the genetic diversity of the Illinois birds before and after the bottleneck to the genetic diversity in the present-day populations of prairie chickens in Minnesota, Kansas, and Nebraska. In all cases, they examined the number of alleles that a population contained for each of six different genes. As shown in Figure 7.7b, the large populations of the neighboring

coded for increased pigmentation in the cave populations and 8 of the regions coded for decreased pigmentation, as shown in Figure 7.6c. The lack of a consistent pattern among the 13 regions of DNA suggests that natural selection was not involved. Instead, the differences in pigmentation in the cavefish populations were likely produced by genetic drift. Given that small populations tend to experience more genetic drift than large

Bottleneck effect A reduction of genetic diversity in a population due to a large reduction in population size.

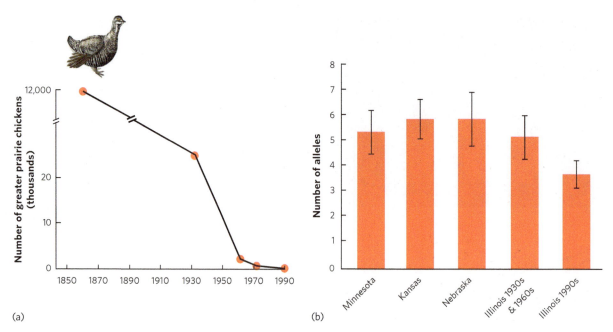

(a)

(b)

Figure 7.7 Evolution by the bottleneck effect. (a) The greater prairie chicken population in Illinois declined from approximately 12 million in the 1860s to 72 birds in 1990. **(b)** Averaged across six different genes, the mean number of alleles is high for birds in neighboring states that still have large populations and for the larger, historic population in Illinois that existed in the 1930s and 1960s. In the current population in Illinois, however, the small population is experiencing a population bottleneck and has a lower mean number of alleles. Error bars are standard errors. Data from J. L. Bouzat et al., The ghost of genetic diversity past: Historical DNA analysis of the greater prairie chicken, *American Naturalist* 152 (1998): 1–6.

states and the large historic population in Illinois all contained a high number of alleles. The current population in Illinois, however, has a lower number of alleles, which reflects the genetic bottleneck effect. The state of Illinois has since purchased more prairie habitat and introduced hundreds of prairie chickens from neighboring states to help bolster the Illinois population and increase its genetic diversity.

The bottleneck effect is of particular interest because the subsequent reduction in genetic diversity may prevent the population from adapting to future environmental changes. This is especially true for organisms that face deadly pathogens. An inability to evolve against new strains of a pathogen could lead to the extinction of the host organism. For example, the African cheetah (*Acinonyx jubatus*) experienced a population bottleneck approximately 10,000 years ago. Although the cause of that bottleneck is unknown, the current population contains very little genetic variation. This low genetic variation makes cheetahs more vulnerable to pathogens, including a deadly pathogen that causes the disorder known as AA amyloidosis and kills up to 70 percent of cheetahs held in captivity.

Founder Effects

The **founder effect** occurs when a small number of individuals leave a large population to colonize a new area and bring with them only a small amount of genetic variation. Following the founding by this small population, genetic drift can cause additional reductions in genetic

variation. The genetic variation remains low until enough time has passed to accumulate new mutations. The water hyacinth (*Eichhornia crassipes*), which has been introduced by humans to many parts of the world, provides an example of the founder effect. Water hyacinth is an aquatic plant that is native to South America. During the past 150 years, the plant has been either accidentally or intentionally introduced to many other parts of the world. Once introduced, it grows and spreads quite rapidly, dominating shallow-water areas and displacing native plants. Today, water hyacinth has become one of the most invasive plants in the world.

Because most water hyacinth introductions are thought to have happened with just a few individuals, researchers wondered if the plant would show indications of founder effects in those parts of the world where it was not historically present. They sampled 1,140 plants from around the world and determined their genotypes. In 2010, they reported a single widespread genotype occurred in 71 percent of the sampled plants, and this genotype dominated 75 percent of all populations that were outside the plant's native range, as shown in **Figure 7.8**. Moreover, 80 percent of all populations outside the native range were composed of a single genotype, whereas populations in the native range of South America

Founder effect When a small number of individuals leave a large population to colonize a new area and bring with them only a small amount of genetic variation.

Native range

Figure 7.8 Evolution by the founder effect. Water hyacinth is an aquatic plant that was native to South America, where many different genotypes exist, as indicated by the different-colored dots. Introductions around the world are thought to have occurred with low numbers of founders. Today, most populations existing outside of South America are represented by a single genotype. Data from Y.-Y Zhang, D.-Y. Zhang, and S. C. H. Barrett, Genetic uniformity characterizes the invasive spread of water hyacinth (*Eichhornia crassipes*), a clonal aquatic plant, *Molecular Ecology* 19 (2010): 1774–1786. Photo by National Geographic Creative/Alamy.

had up to five different genotypes. This pattern suggests that there were few founders in the invaded regions of the world and that they carried with them only a small proportion of the genetic diversity of native populations in South America.

EVOLUTION THROUGH SELECTION, A NONRANDOM PROCESS

The nonrandom process of *selection* also plays a substantial role in evolution. **Selection** is the process by which certain phenotypes are favored to survive and reproduce over other phenotypes. As we saw in the story of island birds at the beginning of this chapter, selection is a powerful force that can change the phenotypes (and therefore the gene frequencies) of a population. Depending on how the environment varies over time and space, selection can influence the distribution of traits in a population in three ways: *stabilizing selection, directional selection,* and *disruptive selection.*

Stabilizing Selection

When individuals with intermediate phenotypes have higher survival and reproductive success than those with extreme phenotypes, we call it **stabilizing selection.** As shown in **Figure 7.9a**, stabilizing selection begins with a

relatively wide distribution of phenotypes, as illustrated by the orange line. After stabilizing selects for parents who possess intermediate phenotypes, their progeny have a more narrow distribution of phenotypes, as illustrated by the blue line. In doing so, it performs genetic housekeeping for a population, sweeping away harmful genetic variation. An example of stabilizing selection can be seen in selection for body mass in a species of bird from South Africa called the sociable weaver (*Philetairus socius*). Over an 8-year period, researchers marked nearly 1,000 adult birds and examined how a bird's mass was related to its subsequent survival. As you can see in Figure 7.9b, the mass of the adults in the study follows a normal distribution with a mean of approximately 29 g. The researchers then asked how well birds of different mass survive. When mass was graphed against survival, as shown in Figure 7.9c, they found that the smallest and largest birds survived poorly, whereas birds of intermediate mass

Selection The process by which certain phenotypes are favored to survive and reproduce over other phenotypes.

Stabilizing selection When individuals with intermediate phenotypes have higher survival and reproductive success than those with extreme phenotypes.

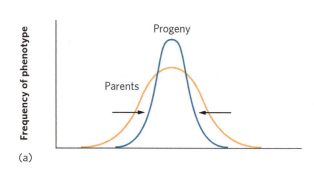

(a)

Figure 7.9 Stabilizing selection. (a) Stabilizing selection favors intermediate phenotypes and selects against both extremes. **(b)** In the sociable weaver bird, body size has a normal distribution. **(c)** The bird experiences stabilizing selection on body size because birds with intermediate body sizes experience high survival, while birds with low and high body sizes have low survival. This selection for intermediate phenotypes would cause the next generation of birds to have a narrower distribution of phenotypes. Data from R. Covas et al., Stabilizing selection on body mass in the sociable weaver *Philetairus sociu.*, *Proceedings of the Royal Society of London Series B 269* (2002): 1905–1909.

survived the best. That is, selection favored the intermediate phenotype. When the environment of a population is relatively unchanging, stabilizing selection is the dominant type of selection. Because the average phenotype does not change, little evolutionary change takes place.

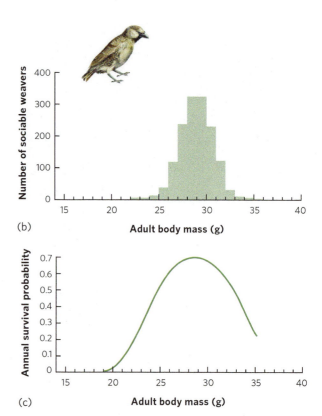

(b)

(c)

Directional Selection

Directional selection, which occurs when an extreme phenotype experiences higher fitness than the average phenotype of the population, as shown in **Figure 7.10a**. In the medium ground finch (*Geospiza fortis*), for example,

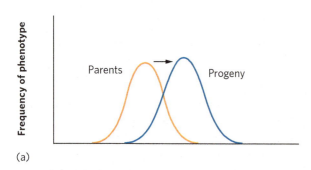

(a)

Figure 7.10 Directional selection. (a) Directional selection favors phenotypes of one extreme and selects against phenotypes of the other extreme. **(b)** Prior to a drought in 1976, the beak size in offspring of the medium ground finch had a mean depth of 8.9 mm, as indicated by the dashed red line. During the drought, when mostly large seeds were available, birds with larger beaks survived better. **(c)** Two years later, finch offspring had a mean beak depth of 9.7 mm, confirming that larger seeds caused directional selection for larger beaks. Data from R. Grant and P. Grant, What Darwin's finches can teach us about the evolutionary origin and regulation of biodiversity, *BioScience* 53 (2003): 965–975.

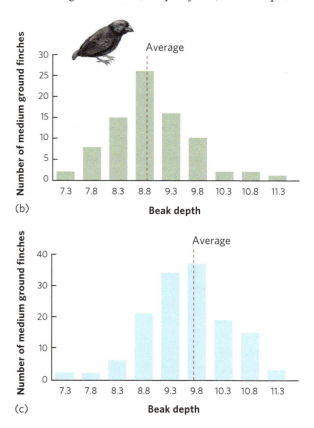

(b)

(c)

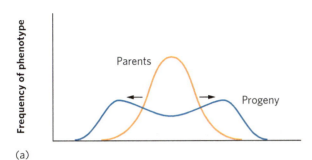

(a)

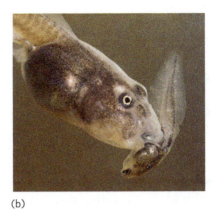

(b)

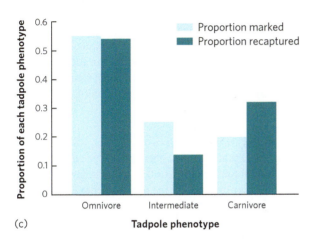

(c)

Figure 7.11 Disruptive selection. (a) Disruptive selection favors both extreme phenotypes and selects against intermediate phenotypes. **(b)** In the tadpoles of the spadefoot toad, an individual can have specialized mouthparts for carnivory and cannibalism (shown in this photo), omnivory, or an intermediate phenotype. **(c)** When more than 500 tadpoles of each phenotype were marked and released, about 10 percent were recovered. Of those recovered, the omnivores and carnivores experienced relatively high survival, while the intermediate phenotypes experienced lower survival. Data from R. A. Martin and D. W. Pfennig, Disruptive selection in natural populations: The roles of ecological specialization and resource competition, *American Naturalist* 174 (2009): 268–281. Photo by Thomas Wiewandt/Danita Delimont/Alamy.

Peter and Rosemary Grant quantified the distribution of beak sizes in offspring hatched in 1976, which was just prior to a drought. As you can see in Figure 7.10b, the beak sizes of these offspring followed a normal distribution with a mean beak depth of 8.9 mm. Individuals with large beaks can generate the force needed to crack the largest seeds, whereas individuals with small beaks are better at handling the smallest seeds. When a drought came, although all seeds became less abundant, there were proportionately more large seeds remaining. The large seeds are harder to crack, so birds with deeper beaks were better able to feed on hard seeds and they had better survival. Because beak depth is a heritable trait, the offspring that were hatched in 1978 had deeper beaks, as shown in Figure 7.10c.

Disruptive Selection

Under some circumstances, we see another type of selection known as *disruptive selection*. In **disruptive selection,** individuals with extreme phenotypes at either end of the distribution can have higher fitness than individuals with intermediate phenotypes. Disruptive selection is illustrated in **Figure 7.11a**. For example, tadpoles of the Mexican spadefoot toad (*Spea multiplicata*) can express a range of possible phenotypes that are related to what they eat. At one extreme is the omnivorous phenotype that has small jaw muscles, numerous little teeth, and a long intestine that makes it

well suited for feeding on detritus. At the other extreme is the carnivorous phenotype that has large jaw muscles, a notched mouthpart, and a short intestine that makes it well suited for feeding on freshwater shrimp and cannibalizing conspecifics. Intermediate phenotypes are not well suited to feed on either of the two food types. To test whether the tadpoles experienced disruptive selection, researchers collected more than 500 tadpoles from a desert pond, marked them to indicate their phenotype, and then returned them back to their pond. They sampled the pond again 8 days later to determine the survival of the three phenotypes. As you can see in Figure 7.11b, the omnivorous and carnivorous phenotypes survived relatively well, but the intermediate phenotypes—which had intermediate jaw muscles, tooth number, and intestine length—survived poorly. Because disruptive selection removes the intermediate phenotypes, it increases genetic and phenotypic variation within a population. In doing so, it creates a distribution of phenotypes with peaks toward both ends of the original distribution.

Directional selection When individuals with an extreme phenotype experience higher fitness than the average phenotype of the population.

Disruptive selection When individuals with either extreme phenotype experience higher fitness than individuals with an intermediate phenotype.

Strength of Selection, Heritability, and Response to Selection

Eco TV

macmillanlearning.com/
ricklefsvideo

Researchers often wish to know exactly how much selection will move the mean phenotype in a population. For example, if a plant breeder selects for larger tomato plants, she might want to know how much larger the next generation will be. Similarly, a government agency that regulates fishing might want to know if harvesting only the largest individuals might cause the population to evolve to be smaller in the next generation.

Let's consider the case of directional selection in which one end of the phenotypic distribution is favored. If there is selection for more extreme phenotypes and the phenotype has a genetic basis, directional selection will cause the mean phenotype to change. Can we determine exactly how much the mean phenotype will change in the next generation? To answer this question, we need to know both the *strength of selection* on the phenotype and the *heritability* of the phenotype.

The **strength of selection** is the difference between the mean of the phenotypic distribution before selection and the mean after selection, measured in units of standard deviations (see Chapter 2). For example, imagine that we wanted to select for larger tomatoes. The phenotype (tomato mass) follows a normal distribution with a mean of 100 g and a standard deviation of 10 g. Now imagine that we select the upper end of the distribution and use these individuals to breed the next generation of tomatoes. If this selected group has a mean of 115 g, our selected group has a mean that is 1.5 standard deviations away from the mean of the entire population. Thus, the strength of selection is 1.5.

We also know that phenotypes are the products of genes and the environment. If we wish to know how much the mean phenotype will change, we have to determine the proportion of the total phenotypic variation caused by genes, which is called the **heritability**. Heritability values can range between 0 and 1. If all of the phenotypic variation that we see in a normal distribution is due to the environment, the heritability is 0. If all of the phenotypic variation is due to genetic variation, the heritability is 1. By convention, the symbol for heritability is h^2. (This notation can be confusing because nothing is being squared.)

Using the concepts of strength of selection and heritability, we can build an equation that describes how much a population will respond to selection in the next generation. Because a population's response to selection is a function of the strength of selection and the heritability of the phenotype,

$$R = S \times h^2$$

where R is the response to selection, S is the strength of selection, and h^2 is the heritability.

Using our tomato example, we can calculate how much larger the tomatoes should be in the next generation. If we select parents that are 1.5 standard deviations above the population mean and if the heritability is 0.33, then

$$R = 1.5 \times 0.33 = 0.5$$

which means that the mean phenotype of the next generation of tomatoes will be 0.5 standard deviations— or 5 g—greater than that of the parent's generation.

YOUR TURN Given the following values for the strength of selection and heritability on tomato mass, calculate the expected response to selection in units of standard deviations and grams:

S	h^2
0.5	0.7
1.0	0.7
1.5	0.7
2.0	0.9
2.0	0.6
2.0	0.3
2.0	0.0

Based on your calculations, how is the response to selection affected by the strength of selection and heritability?

Strength of selection The difference between the mean of the phenotypic distribution before selection and the mean after selection, measured in units of standard deviations.

Heritability The proportion of the total phenotypic variation that is caused by genetic variation.

CONCEPT CHECK

1. Compare and contrast the processes of bottleneck effects and founder effects.
2. How should stabilizing selection and disruptive selection affect the magnitude of phenotypic variation from one generation to the next?
3. How does heritability affect the response to selection?

7.3 Microevolution operates at the population level

The random and nonrandom processes that cause evolution can operate at a variety of levels. The evolution of populations is known as **microevolution**, and it is pervasive. It is the process responsible for producing different breeds of cats, cattle, and dogs, and for producing distinct populations of wild organisms, including salmon, bears, and the flu virus. Microevolution is affected by both random process and selection. Selection at the microevolution level can be further divided into *artificial selection* and *natural selection*.

ARTIFICIAL SELECTION

In his book *On the Origin of Species,* Charles Darwin discussed the wide variety of domesticated animals that humans have bred to produce particular suites of traits. In the case of dogs, for example, humans began by domesticating gray wolves. Over time they bred individuals that possessed particular characteristics, such as body size, coat color, and hunting ability. As shown in **Figure 7.12**, just a few centuries of breeding have created dog breeds with widely divergent phenotypes—from Saint Bernards to Chihuahuas. All of these dogs belong to the same species as the wolf and could potentially interbreed. Thanks to dog breeders, there is an excellent paper trail of exactly how the various breeds have been produced. This is an example of **artificial selection,** in which humans decide which individuals will breed, and the breeding is done with a preconceived goal for the traits desired in the population. Similar artificial selection has occurred to create numerous breeds of other domesticated animals, including cattle, sheep, pigs, and chickens.

Microevolution Evolution at the level of populations.

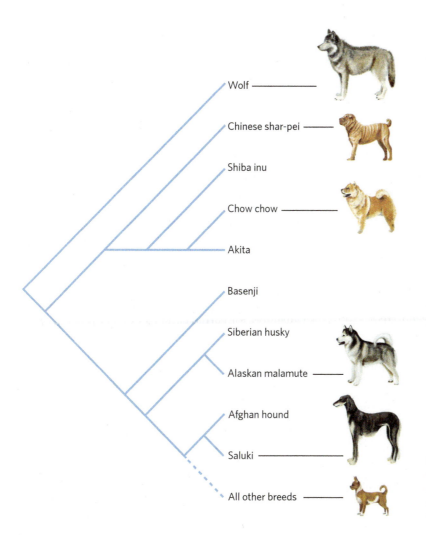

Figure 7.12 Breeds of domestic dogs. Beginning with domesticated individuals of the gray wolf, humans have created a wide diversity of dog breeds through the process of artificial selection. Data from H. G. Parker et al., Genetic structure of the purebred domestic dog, *Science* 304 (2004): 1160–1164.

Figure 7.13 Artificial selection on wild mustard. Over the years, plant breeders have produced a variety of common vegetables through artificially selecting for different traits of the wild mustard.

Artificial selection also has been applied to plants. One of the best-known examples is the artificial breeding of wild mustard (*Brassica oleracea*). As you can see in **Figure 7.13**, the wild mustard has been bred into a diversity of vegetables by selecting for unique stem, leaf, and flower traits. Today, the wild mustard can be consumed as cabbage, brussels sprouts, cauliflower, broccoli, kale, and kohlrabi.

Human practices can also cause artificial selection unintentionally, sometimes with detrimental consequences. For example, the widespread application of pesticides has caused resistance in more than 500 species of pests that harm food production and human health. Similarly, extensive use of antibiotics has caused many harmful human pathogens to evolve resistance to antibiotics, as described in "Drug-Resistant Tuberculosis" at the end of this chapter. In these cases, the role of evolutionary mechanisms is clear. When pesticides or antibiotics are targeted at millions of organisms, a small number of individuals commonly carry a mutation that confers resistance. Since only the mutants survive and the mutation is heritable, the next generation becomes more resistant.

Darwin's case for evolution by natural selection was strengthened by his observations of how artificial selection worked. He argued that if humans could produce such a wide variety of animal and plant breeds in only a few centuries through artificial selection, natural selection could certainly have had similar effects over millions of years.

NATURAL SELECTION

A person conducting artificial selection typically has a particular suite of traits in mind, for example, increased milk production in cattle. This is not the case for natural selection, which favors any trait combination that provides higher fitness to an individual. Artificial and natural selection both operate by favoring certain traits over others. Both select on traits that are heritable; the difference lies in how traits are selected. There can be multiple ways to improve an individual's fitness, and all of them are favored by natural selection, regardless of the resulting phenotype. For example, a prey organism might reduce its probability of being eaten either by hiding from predators so it is not detected or by growing spines so that it cannot be eaten. Both strategies are effective at improving the prey's fitness, and both traits could be favored by natural selection. In artificial selection, humans determine the fitness of traits, and they often select attributes for particular purposes that would actually reduce the fitness of individuals if they lived in the natural environment.

Most evolutionary biologists agree that the diversification of organisms over the long history of life on Earth has occurred primarily by natural selection. Natural selection is an ecological process: It occurs because of differences in reproductive success among individuals endowed with

Artificial selection Selection in which humans decide which individuals will breed and the breeding is done with a preconceived goal for the traits of the population.

different form or function in a particular environment. That is, as individuals interact with their environment—including physical conditions, food resources, predators, other individuals of the same species, and so on—traits that lead to greater fitness in that environment are passed on.

Evolution by natural selection is a common phenomenon in populations. For example, many predators cause selection on the traits of their prey. We can see this process in **Figure 7.14** for fish that prey on amphipods, a tiny crustacean species. The parent amphipods produce an abundance of offspring that vary in size. However, fish prefer to consume the largest amphipods because they offer the highest amount of energy per unit effort. Smaller amphipods are more likely to survive, and since body size is a heritable trait in amphipods, subsequent generations evolve smaller body sizes.

One of the most striking demonstrations of microevolution is the case of the peppered moth (*Biston betularia*). During the early nineteenth century in England, most individuals of this moth were white with black spots, but occasionally there would be a dark, or *melanistic*, individual (**Figure 7.15a**). Over the next hundred years, dark individuals became more common in forests near heavily industrialized regions, a phenomenon often referred to as **industrial melanism.** In regions that were not industrialized, the light phenotype still prevailed.

Since melanism is an inherited trait, it seemed reasonable to suppose that the environment must have been altered to give dark forms a survival advantage over light forms. The specific agent of selection was easily identified. Peppered moths rest on trees during the day. Scientists observed that air pollution in industrial areas had darkened the trees with soot, as shown in Figure 7.15b, so they suspected that the predatory birds could see the lighter moths more easily. Because trees in nonpolluted regions were much lighter in color, the dark moths in these regions would be more visible. To test these hypotheses, equal numbers of light and dark moths were placed on tree trunks in polluted and unpolluted woods. As you can see in Figure 7.15c, when both types of moths were attached to the light-colored trees in unpolluted regions, birds consumed more dark moths. When both types of moths were attached to the dark-colored trees in polluted regions, birds consumed more light moths. This confirmed that the change in phenotypes observed over time in England reflected evolution of the population in response to changing environmental conditions.

In recent years, as pollution control programs reduced the amount of soot in the air and improved the conditions of the forests, the frequencies of melanistic moths have decreased, as we would expect. **Figure 7.16** shows data for the area around the industrial center of Kirby in northwestern England. As the amount of pollution declined—measured in terms of sulfur dioxide and shown as a blue line—the bark of the trees began to turn lighter. After two decades of decreasing pollution, the trees became lighter and the frequency of the dark form of the moth decreased from more than 90 percent of the population in 1970 to about 30 percent by 1990, as shown by the black line. The story of the melanistic moths demonstrates that microevolution can occur in a relatively short time.

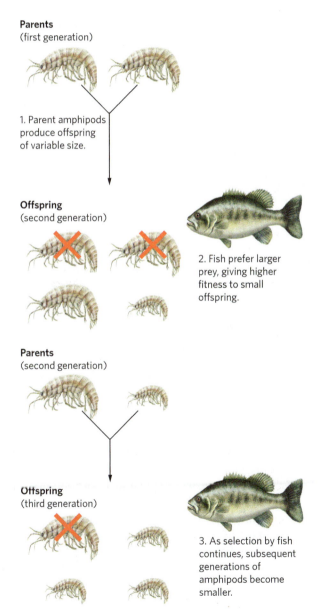

Parents
(first generation)

1. Parent amphipods produce offspring of variable size.

Offspring
(second generation)

2. Fish prefer larger prey, giving higher fitness to small offspring.

Parents
(second generation)

Offspring
(third generation)

3. As selection by fish continues, subsequent generations of amphipods become smaller.

Figure 7.14 Natural selection by predators on prey.
The amphipod, a tiny crustacean, produces an abundance of offspring that vary in size. Predatory fish prefer to eat the largest amphipods, and this causes selection for small body size.

Industrial melanism A phenomenon in which industrial activities cause habitats to become darker due to pollution and, as a result, individuals possessing darker phenotypes are favored by selection.

(a) **Unpolluted forests**

(b) **Polluted forests**

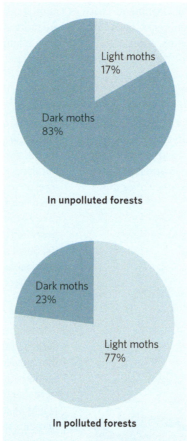

In unpolluted forests

In polluted forests

(c) **Moths consumed**

Figure 7.15 Selection by bird predation for different moth phenotypes. (a) In unpolluted forests, trees have light-colored bark and moths with the light phenotype are better camouflaged. **(b)** In polluted forests, trees have dark-colored bark and moths with the dark phenotype are better camouflaged. **(c)** When researchers placed both moth phenotypes on trees in polluted and unpolluted forests fewer light-colored moths were consumed by birds on unpolluted trees but fewer dark moths were consumed on polluted trees. Data from B. Kettlewell, Further selection experiments on industrial melanism in the Lepidoptera, *Heredity* 10 (1956): 287–301. Photos by Michael Willmer Forbes Tweedie/Science Source

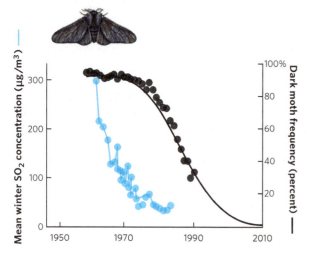

Figure 7.16 Reversing the effects of pollution. As industries around Kirby, England, reduced the amount of sulfur dioxide pollution they emitted into the air, the color of the trees became lighter. After a decade of declining pollution, the frequency of the dark form of the peppered moth began to decline rapidly. Data from C. A. Clarke et al., Evolution in reverse: Clean air and the peppered moth, *Biological Journal of the Linnean Society* 26 (1985): 189–199; G. S. Mani and M. E. N. Majerus, Peppered moth revisited: Analysis of recent decreases in melanic frequency and predictions for the future, *Biological Journal of the Linnean Society* 48 (1993): 157–165. Photos by Michael Willmer Forbes Tweedie/Science Source.

CONCEPT CHECK

1. How is the domestication of wild mustard an example of artificial selection?
2. What traits are favored by natural selection?
3. What are the four required conditions for natural selection to occur?

7.4 Macroevolution operates at the species level and higher levels of taxonomic organization

Whereas microevolution is a process that occurs at the level of the population, **macroevolution** is a process that occurs at higher levels of organization, including species, genera, families, orders, and phyla. For our purposes, we will restrict our discussion of macroevolution to the evolution of new species, a process known as **speciation**. The pattern of speciation over time can be illustrated using *phylogenetic* trees and speciation can occur in one of two ways: *allopatric speciation* or *sympatric speciation*.

Macroevolution Evolution at higher levels of organization, including species, genera, families, orders, and phyla.

Speciation The evolution of new species.

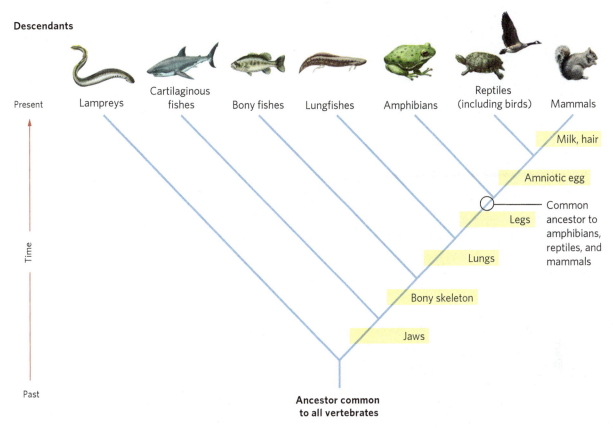

Figure 7.17 Phylogenetic tree. By looking for similarity in phenotypes and DNA, scientists can develop hypotheses about the relatedness of different groups of organisms. In this phylogenetic tree of the major vertebrate groups, the yellow boxes indicate points in time at which important new traits evolved.

PHYLOGENETIC TREES

Scientists can often document microevolution because it can happen in a relatively short time. In some cases, scientists have monitored wild populations over time to track the evolutionary process. In other cases, we have historical documents that trace the development of domesticated plants and animals. For example, most modern dog breeds are the result of artificial selection over the last three centuries and records show the older breeds that gave rise to newer breeds. Understanding how macroevolution has occurred, however, is a much more daunting challenge. Because we cannot travel back in time and there are no written records from millions of years ago, the true patterns of evolution can never be known for certain, although fossils can be helpful when examining the evolution of morphological traits. In the absence of more direct evidence, scientists work from the premise that species with the greatest number of traits in common are the most closely related. These traits can include shapes and sizes of structures of living and fossilized organisms as well as the ordering of nitrogenous bases in the DNA of different organisms. To map these relationships, scientists use **phylogenetic trees,** which are hypothesized patterns of relatedness among different groups such as populations, species, or genera. In essence, phylogenetic trees are attempts to understand the order in which groups evolved from other groups. **Figure 7.17** shows a phylogenetic tree for several major groups of vertebrates. From this tree, you can see that all vertebrates share a common ancestor. Over time, this ancestor has given rise to fish, amphibians, mammals, and reptiles (including birds).

ALLOPATRIC SPECIATION

Allopatric speciation is the evolution of new species through the process of geographic isolation. Imagine that we start with a single large population of a field mouse, shown as the first stage in **Figure 7.18**. At some point in time, a portion of the population is separated from the rest. This could occur because some individuals colonize a new island, such as the first finches to arrive in the Galápagos Islands from South America, which Charles Darwin described in his theory of evolution by natural selection. Alternatively, the population could be divided by a geographic barrier, such as a new river that cuts through the middle of the terrestrial habitat, a rising mountain range that cannot be crossed, or a large lake that becomes two smaller lakes. In each case, the two populations are isolated from each other, as shown in Step 2 of the figure. They are no longer able to interbreed because of physical separation, so each population evolves independently. If one or both of the populations has few individuals, founder effects and genetic drift can strongly influence the direction in which that population

Phylogenetic trees Hypothesized patterns of relatedness among different groups such as populations, species, or genera.

Allopatric speciation The evolution of new species through the process of geographic isolation.

1. Original field mouse population

2. River arises, splitting the population.

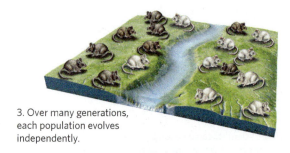

3. Over many generations, each population evolves independently.

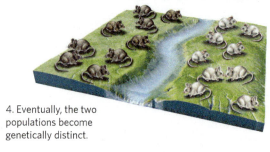

4. Eventually, the two populations become genetically distinct.

5. Later, even if the river dries up and the two populations come into contact, they may no longer be able to interbreed.

Figure 7.18 Allopatric speciation. When geographic barriers divide populations, each evolves independently. Over time, the two populations can become so different that they are not able to interbreed. At this point, they have become two distinct species.

evolves. When ecological conditions differ in the two isolated locations, natural selection will cause each population to evolve adaptations that improve fitness under local environmental

conditions. Over time, as shown in Steps 3–5, the populations can become so different that they are no longer capable of interbreeding, even if they are brought back together. At this point, the two populations have evolved into different species.

Allopatric speciation is thought to be the most common mechanism of speciation. **Figure 7.19** shows the process for Darwin's finches using a phylogenetic tree. Darwin hypothesized that the ancestor species of the finches he saw probably came from the mainland of South America. Once the ancestral finch species arrived in the Galápagos Islands, the population grew and eventually colonized many of the islands in the archipelago. The isolation and unique ecological conditions present on each island favored the process of allopatric speciation. Researchers hypothesize that these conditions gave rise to some of the 14 currently recognized species of finches in the Galápagos Islands, although other species of finches on the islands appear to have evolved through a different process known as *sympatric speciation*.

SYMPATRIC SPECIATION

In contrast to allopatric speciation, **sympatric speciation** gives rise to new species without geographic isolation. In some cases, species evolve into a diversity of new species within a given location. An example of this is the group of cichlid fish species that live in Lake Tanganyika in eastern Africa. Over millions of years, a single ancestor fish has given rise to more than 200 unique species of fish, including insect eaters, fish eaters, and mollusk eaters (**Figure 7.20**). This massive amount of speciation appears to have been facilitated by the presence of many distinct habitats throughout the lake, such as rocky versus sandy shorelines. This small-scale habitat variation may have favored the evolution of different phenotypes that then led to the evolution of new species.

A common way that sympatric speciation can occur in some types of organisms is through *polyploidy*. **Polyploid** species, which contain three or more sets of chromosomes, arise when homologous chromosomes fail to separate properly during meiosis, with the result that gametes are diploid rather than haploid. If a diploid egg, for example, is fertilized by a haploid sperm, the resulting zygote will contain three sets of chromosomes. At this point, the organism is a polyploid. Because it now has more than two sets of chromosomes, it is unable to breed with any diploid individuals. Thus, when a polyploid is formed, it instantly becomes a species that is genetically distinct from its parents. Several species of insects, snails, and salamanders are polyploid, as are 15 percent of all flowering plant species.

An interesting example of polyploidy can be found in a group of salamanders. The blue-spotted salamander (*Ambystoma laterale*) and the Jefferson salamander (*A. jeffersonianum*) are both diploid species. As **Figure 7.21** illustrates, at some point in the past, a blue-spotted salamander experienced incomplete meiosis and accidentally

Sympatric speciation The evolution of new species without geographic isolation.

Polyploid A species that contains three or more sets of chromosomes.

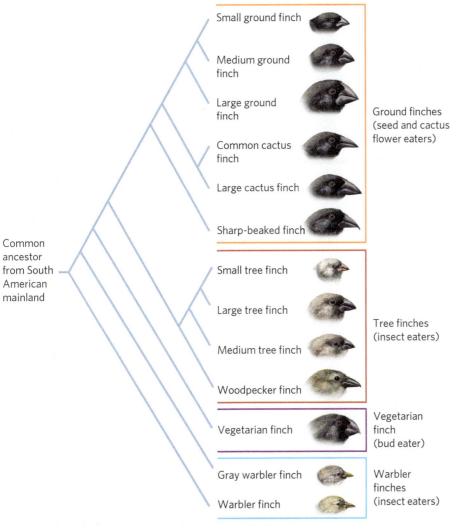

Figure 7.19 Allopatric speciation in Darwin's finches. Through allopatric speciation, a single ancestor species from the South American mainland has evolved into 14 different species of finches on the Galápagos Islands. (Note that the phylogenetic study was done on only 13 of the 14 species.)

Small ground finch

Medium ground finch

Large ground finch

Ground finches (seed and cactus flower eaters)

Common cactus finch

Large cactus finch

Sharp-beaked finch

Small tree finch

Large tree finch

Tree finches (insect eaters)

Medium tree finch

Woodpecker finch

Vegetarian finch

Vegetarian finch (bud eater)

Gray warbler finch

Warbler finch

Warbler finches (insect eaters)

Common ancestor from South American mainland

Figure 7.20 Sympatric speciation. More than 200 species of cichlid fishes of Lake Tanganyika have evolved from a single ancestor. From http://www.uni-graz.at/~sefck/.

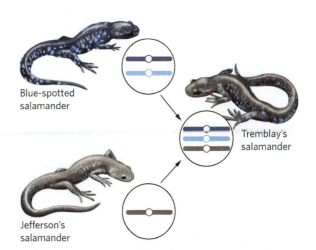

Blue-spotted salamander

Jefferson's salamander

Tremblay's salamander

Figure 7.21 Polyploidy in salamanders. Triploid species can occur when one individual that experiences incomplete meiosis and produces a diploid gamete mates with another individual that experiences normal meiosis and produces a haploid gamete. The Tremblay's salamander is an all-female, triploid species that arose through sympatric speciation from the mating of the blue-spotted salamander and the Jefferson salamander.

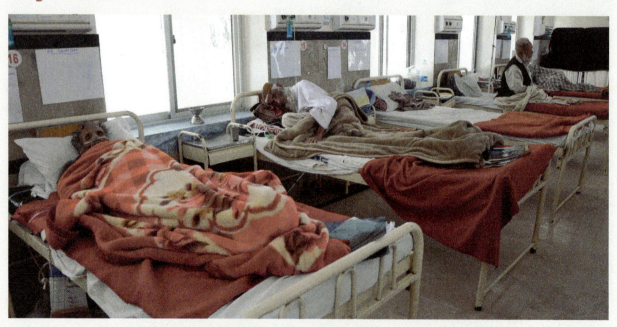

(a) Einkorn wheat   (b) Durum wheat   (c) Common wheat

Figure 7.22 Polyploidy in wheat. Starting with a diploid species of wheat, plant breeders have bred new species that contain four or six sets of chromosomes. **(a)** The ancestral einkorn wheat (*Triticum boeoticum*) has two sets of chromosomes and small seeds. **(b)** Durum wheat (*Triticum durum*), which is used to make pasta, was bred to have four sets of chromosomes. It has medium-sized seeds. **(c)** Common wheat (*Triticum aestivum*), which is used for bread and other baked goods, was bred to have six sets of chromosomes. It has the largest seeds.

species that produces clonal daughters. These salamanders will breed with another salamander species to stimulate their own reproduction, but they can make daughters without incorporating DNA from any of the other species. If they do incorporate the haploid sperm from a male of the other species, their offspring can then carry four sets of chromosomes, which would make them *tetraploids*.

Plant breeders have developed techniques to intentionally cause polyploidy to produce more desirable characteristics in plants. This is a form of artificial selection at the species level. By exposing plants to sudden cold temperatures at the time of reproduction, they can increase the chances of a plant producing gametes that are diploid rather than haploid. Polyploid plants tend to be larger and produce bigger flowers and fruits. Many beautiful flowers you see in a florist shop are the product of human-induced polyploidy. Many crops are also polyploids, including watermelons, bananas, strawberries, and wheat. As you can see in **Figure 7.22**, plant breeders used a species of wheat that has two sets of chromosomes to develop new species that have four or six chromosomes. The more sets of chromosomes in wheat, the larger the plant and the larger the seeds.

produced a diploid gamete. It then mated with a Jefferson salamander, which produced a normal haploid gamete. The resulting offspring was a triploid salamander that instantly became a distinct species known as Tremblay's salamander (*A. tremblayi*). The Tremblay's salamander is an all-female

CONCEPT CHECK

1. What is the assumption used when arranging species on a phylogenetic tree?
2. What is one requirement of allopatric speciation?
3. How does a polyploidy event immediately give rise to a new species?

ECOLOGY TODAY   **CONNECTING THE CONCEPTS**

Drug-Resistant Tuberculosis

Tuberculosis, or TB, is a highly infectious disease caused by a bacterium (*Mycobacterium tuberculosis*). TB causes extensive tissue damage, weakness, night sweats, and bloody sputum. It is highly contagious; when an infected individual coughs or talks, bacteria are expelled and can survive in the air for several hours and infect other people. TB has killed people for thousands of years. In 2009, for example, researchers discovered that the preserved tissues of a woman who died 2,600 years ago and was mummified have genetic markers of the TB bacterium. Today, experts estimate that nearly one-third of the world's human population is infected by the bacterium, although the bacterium remains inactive and does not cause any problems for most of these people. However, the bacterium does become active in about 9 million people every year.

Around the world each year, 2 million people die of TB, which makes it the number one cause of death by infectious disease in the world. Fortunately, medical researchers have developed an inexpensive drug to combat tuberculosis. While the drug has been highly effective in reducing the number of people infected with TB, the bacterium has started to evolve resistance to this drug.

Drug-resistant tuberculosis is a rapidly growing problem around the world, particularly in Africa, Russia, and China. The reason is not a mystery. Bacteria can quickly grow to incredibly high numbers and, as you may recall from our discussion of evolution, very large populations are likely to have a few individuals that possess mutations. Occasionally, a mutation makes a bacterium more resistant and these mutations are heritable. Antibiotics represent a strong selective force that can quickly kill the vast majority of sensitive bacteria, thereby leaving resistant bacteria to flourish.

One of the biggest contributors to the evolution of TB resistance is thought to be the behavior of TB patients. The typical drug treatment of TB requires taking pills daily for 1 year. Although many bacteria are killed early in the treatment, continued treatment helps eliminate every last pathogen. Sometimes patients stop taking their pills because they feel better after a few months or they simply lack the money to pay for an entire year of treatment. In either case, the most resistant bacteria remain in their bodies and they eventually become abundant, making the patient no longer responsive to the inexpensive TB drug.

Drug-resistant TB is becoming a major problem. Researchers have developed new types of TB drugs to try to select different TB traits with the hope that evolving resistance to one drug will still make the pathogen susceptible to other drugs. However, there is now an increase in "Multiple Drug Resistant TB" or MDR-TB, a strain of the bacterium that has evolved resistance to several different drugs. In Russia, for example, nearly 20 percent of all people infected with TB are carrying the MDR-TB strain. These strains are much harder for doctors to kill; the drugs required are 10 times more expensive than traditional ones, they may need to be taken for two years, and there are major side effects, including hearing loss. In 2016, researchers announced that a new regimen of seven drugs administered over nine months is turning out to be highly effective against MDR-TB.

Even more sobering is the discovery of what is being called "Extensively Drug Resistant TB." This version of TB has been detected in 45 countries, including Russia. Currently, drug regimens are only effective on 30 to 50 percent of patients, depending on the particular strain of bacteria and the state of the patient's immune system.

The evolution of TB resistance is an excellent example of why we need to understand the process of evolution. Knowing the sources of genetic variation and how selection operates on this variation helps us develop drug treatment programs that are better able to control pathogens without producing strains that have multiple drug resistance.

Death from tuberculosis.
Tuberculosis patients being treated in a hospital in Tomsk, Russia.
Photo by Xinhua/Alamy.

SOURCES:

Coghlan, A. 2016. Superfast therapy cracks multidrug-resistant tuberculosis. *New Scientist*, October 26. https://www.newscientist.com/article/2110555-superfast-therapy-cracks-multidrug-resistant-tuberculosis/.

Altman, L. K. 2008. Drug-resistant TB rates soar in former Soviet regions. *New York Times*, February 27. http://www.nytimes.com/2008/02/27/health/27tb.html.

Goozner, M. 2008. A report from the Russian front in the global fight against drug-resistant tuberculosis. *Scientific American*, August 25. http://www.scientificamerican.com/article.cfm?id=siberia-drug-resistant-tuberculosis.

SUMMARY OF CHAPTER CONCEPTS

7.1 The process of evolution depends on genetic variation. Among and within populations, genetic variation is caused by the presence of different alleles, which can be dominant, codominant, or recessive. Genetic variation can be generated through mutation or recombination.

Key Terms: Deoxyribonucleic acid (DNA), Chromosomes, Alleles, Polygenic, Pleiotropy, Epistasis, Heterozygous, Homozygous, Codominant, Dominant, Recessive, Gene pool, Random assortment, Mutation, Recombination

7.2 Evolution can occur through random processes or through selection. The four random processes that cause evolution are mutation, genetic drift, bottleneck effects, and founder effects. Evolution can also occur by selection, which can be stabilizing, directional, or disruptive. Whether evolution occurs by random processes or by selection, scientists can use similarities in traits to arrange hypothesized patterns of relatedness among different groups on phylogenetic trees.

Key Terms: Genetic drift, Bottleneck effect, Founder effect, Selection, Stabilizing selection, Directional selection, Disruptive selection, Strength of selection, Heritability

7.3 Microevolution operates at the population level. Populations can evolve due to artificial selection to produce breeds of domesticated animals and plants. Populations can also evolve due to natural selection, such as when predators selectively consume prey and when pesticides and antibiotics selectively kill the most sensitive individuals, allowing the most resistant individuals to survive and reproduce.

Key Terms: Microevolution, Artificial selection, Industrial melanism

7.4 Macroevolution operates at the species level and higher levels of taxonomic organization. The most common process causing macroevolution is allopatric speciation in which populations become geographically isolated and independently evolve to become distinct species over time. The less common process is sympatric speciation in which species become reproductively isolated without being geographically isolated, often by forming polyploids.

Key Terms: Macroevolution, Speciation, Phylogenetic trees, Allopatric speciation, Sympatric speciation, Polyploid

CRITICAL THINKING QUESTIONS

1. Given that mutations are rare in populations, how does a mutation spread through a population and become common?

2. Compare and contrast genetic variation caused by random assortment versus recombination.

3. The insecticide DDT has been widely used to control the mosquitoes that carry malaria. How would you explain the fact that many mosquito populations are now resistant to DDT?

4. How does the introduction of new individuals to a population help offset problems associated with genetic drift?

5. Compare and contrast evolution by artificial selection with evolution by natural selection.

6. Compare and contrast stabilizing, directional, and disruptive selection with regard to how each affects the mean phenotype of the population as well as the variance in the phenotype.

7. How does the breeding of domesticated animals provide evidence for the power of evolution on diverse phenotypes?

8. How is it that polyploidy can allow us to observe the evolution of a new species within one generation?

9. Distinguish between microevolution and macroevolution.

10. What is the difference between the processes involved in allopatric speciation and sympatric speciation?

GRAPHING THE DATA Natural Selection of Finch Beaks

The following table lists the frequency distributions of finch beak sizes, both before and after selection. Using a bar graph, plot the relationships between beak size and frequency. Then determine how much the mean beak size changed due to selection and decide which type of selection has occurred.

Beak size (mm)	Frequency before selection	Frequency after selection	R
10.00	0.00	0.00	
10.20	0.00	0.00	
10.40	0.02	0.00	
10.60	0.04	0.00	
10.80	0.08	0.00	
11.00	0.16	0.00	
11.20	0.20	0.00	
11.40	0.20	0.00	
11.60	0.16	0.02	
11.80	0.08	0.04	
12.00	0.04	0.08	
12.20	0.02	0.16	
12.40	0.00	0.20	
12.60	0.00	0.20	
12.80	0.00	0.16	
13.00	0.00	0.08	
13.20	0.00	0.04	
13.40	0.00	0.02	
13.60	0.00	0.00	
13.80	0.00	0.00	

The Many Ways to Make a Frog

When we think of a frog's life, we often think of an egg hatching into a tadpole, the tadpole growing larger, and ultimately metamorphosing into a frog. Within this lifestyle, there is tremendous variation; some species of frogs spend up to 2 years as a tadpole, whereas other species spend as few as 10 days. In addition, the number of eggs laid can range from dozens to tens of thousands. However, this is only the beginning of the story of how different species of frogs live their lives.

It turns out that the schedule of a frog's life can take many different paths. While about half of all species lay eggs in the water and the eggs hatch into an aquatic tadpole, other species of frogs have evolved very different solutions. For example, some frogs hold their eggs (and hatched tadpoles) on their back. The gastric-brooding frogs (*Rheobatrachus* spp.) hold their eggs in their stomach, which requires specialized adaptations, including the shutdown of acids in the stomach. Other species lay their eggs in foam nests on branches that hang over water; once the eggs hatch, the tadpoles drop into the water to begin an aquatic tadpole stage. Among all these species, tadpoles either feed themselves or they are given sufficient yolk from their mother to complete the tadpole stage without feeding.

Still other species of frogs lay their eggs on land and completely skip the aquatic tadpole stage. In a process known as direct development, the embryo changes from an egg into a fully formed—albeit very tiny—frog (referred to as a froglet). For decades, scientists hypothesized that frogs with direct development likely had ancestors that began with the common aquatic tadpole stage, but over time they evolved to have terrestrial eggs, larvae that lived only on yolk, and then terrestrial eggs that could directly develop into froglets.

To test this hypothesis, researchers in 2012 examined the reproductive patterns in 720 species of frogs to determine whether this hypothesis of gradual transitions was supported. Using genetic data, they could determine how the species were related to each other and the paths that frog evolution has taken over more than 200 million years. When they examined the patterns in the data, they found that many species of direct-developing frogs evolved by transitioning through the step of having terrestrial eggs. However, it was nearly as common to evolve from ancestors with fully aquatic eggs and tadpoles, thereby skipping the intermediate step of having terrestrial eggs. They also found that frog species with terrestrial eggs commonly produce fewer eggs, but the eggs are larger in diameter, which may provide space for more water and greater provisioning of energy in the form of yolk.

> **"The schedule of a frog's life can take many different paths."**

Reproduction in frogs. While many frogs have a typical life of hatching into a tadpole that later metamorphoses into a frog, other species skip the entire tadpole stage by hatching into a tiny froglet. Photo by Danté Fenolio/Science Source.

SOURCES:
Gomez-Mestre, et al. 2012. Phylogenetic analyses reveal unexpected
patterns in the evolution of reproductive modes in frogs. *Evolution*
66: 3687–3700.

The diversity of frog reproduction highlights the notion that species on Earth have evolved a tremendous variety of reproductive strategies. As we shall see in this chapter, organisms have evolved a wide range of alternative strategies for growth, development, and reproduction, and these commonly reflect important fitness trade-offs.

LEARNING OBJECTIVES

After reading this chapter, you should be able to:

8.1 Describe life history traits as the schedule of an organism's life.

8.2 Explain how life history traits are shaped by trade-offs.

8.3 Recognize that organisms differ in the number of times that they reproduce, but they all eventually become senescent.

8.4 Explain how life histories are sensitive to environmental conditions.

As we have seen in previous chapters, organisms are generally well adapted to the conditions of their environments. Their form and function are influenced by physical and biological factors. Similarly, the strategies that organisms have evolved for sexual maturation, reproduction, and longevity are also shaped by natural selection. In this chapter, we will explore the wide array of strategies that species have evolved and the trade-offs among different traits.

8.1 Life history traits represent the schedule of an organism's life

The schedule of an organism's growth, development, reproduction, and survival makes up what ecologists call the **life history** of the organism. As you can see in **Figure 8.1**, the life history of an organism includes the traits connected to the birth or hatching of offspring. These include the time required to reach sexual maturity; **fecundity,** which is the number of offspring produced per reproductive episode; **parity,** which is the number of episodes of reproduction; **parental investment,** which is the amount of time and energy given to offspring; and **longevity,** or **life expectancy,** which is the life span of an organism. In essence, life history traits describe an organism's strategy for obtaining evolutionary fitness throughout its lifetime. In addition, life history traits represent the combined effect of many morphological, behavioral, and physiological adaptations of organisms, all of which interact with environmental conditions to affect survival, growth, and reproduction. In this section, we will explore the wide range of life traits that exist in nature and how these traits are often organized into strategies that allow organisms to persist under different ecological conditions.

THE SLOW-TO-FAST LIFE HISTORY CONTINUUM

Life history traits vary widely among species and among populations within a species. An organism's life history represents a solution to the problem of allocating limited time and resources to achieve maximum reproductive success. A remarkable fact about reproductive success is that the result is always nearly the same. On average, only one of the offspring that an individual produces lives to reproduce. In short, each individual replaces only itself. If this were not the case, populations would either dwindle to extinction because individuals fail to replace themselves or populations would continually expand.

How organisms grow and produce offspring varies in all imaginable ways. A female sockeye salmon (*Oncorhynchus nerka*), after swimming up to 5,000 km from her Pacific Ocean feeding ground to the mouth of a coastal river in British Columbia, still faces an upriver journey of 1,000 km to her spawning ground. There, she lays thousands of eggs, and promptly dies, her body wasted from the exertion. A female African elephant gives birth to a single offspring at intervals of several years, lavishing intense care on her baby until it is old enough and large enough to fend for itself in the world of elephants. Thrushes, a group of birds that includes the American robin (*Turdus migratorius*), start to reproduce when they are 1 year old, and may produce several *clutches*—which are sets of eggs—each year, with each clutch containing three or four chicks. Adult thrushes rarely live beyond 3 or 4 years. In contrast, storm petrels, which are seabirds about the size of thrushes, do not begin to reproduce until they are 4 or 5 years old and then only rear

---

Life history The schedule of an organism's growth, development, reproduction, and survival.

Fecundity The number of offspring produced by an organism per reproductive episode.

Parity The number of reproductive episodes an organism experiences.

Parental investment The amount of time and energy given to an offspring by its parents.

Longevity The life span of an organism. *Also known as* **Life expectancy.**

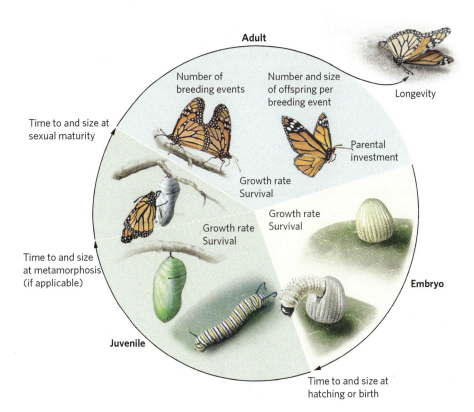

Adult

Number of breeding events

Number and size of offspring per breeding event

Time to and size at sexual maturity

Longevity

Parental investment

Growth rate Survival

Growth rate Survival

Growth rate Survival

Time to and size at metamorphosis (if applicable)

Embryo

Juvenile

Time to and size at hatching or birth

Figure 8.1 Life history traits. The schedule of an organism's life begins as an embryo. It either hatches or is born at a particular size and time. During the subsequent juvenile stage, the organism grows and eventually becomes a sexually mature adult. To reach this stage, many species must first pass through metamorphosis. Adults can then reproduce in one or more breeding events with a particular level of fecundity, parental investment, and longevity. At all stages, species experience a characteristic amount of growth and some probability of surviving to the next stage.

a single chick each year. They may live for 30 or 40 years. This wide variation in life history traits among species has captured the interest of researchers who wish to understand ecological conditions that favor such disparate evolutionary outcomes.

Two points can be made about this variation in life histories. First, life history traits often vary consistently with respect to life form, habitat, or conditions in the environment. Seed size, for example, is generally larger for trees than for grasses. Second, variation in one life history trait is often correlated with variation in other life history traits. For example, the number of offspring produced during a single breeding event is often negatively correlated with the size of the offspring. As a result, variations in many life history traits can be organized along a continuum of values.

We can refer to one extreme as the "slow" end of the spectrum. At this extreme, organisms such as elephants, albatrosses, giant tortoises, and oak trees require a long time to reach sexual maturity. They commonly have long life spans, low numbers of offspring, and a high parental investment in the energy given to their offspring, such as parental care, the amount of yolk in an egg, or the amount of energy stored in a seed. At the fast end of the spectrum are organisms such as fruit flies and herbaceous weedy plants, which exhibit short times to sexual maturity, high numbers of offspring, little parental investment, and short life spans.

COMBINATIONS OF LIFE HISTORY TRAITS IN PLANTS

The English ecologist J. Philip Grime conceptualized the relationship between the life history traits and environmental conditions as a triangle, with each of the three points representing an extreme environmental condition: abiotic stress, competition, and the frequency of disturbances (see **Figure 8.2**). Grime proposed that plants functioning at the extremes of these three axes possessed combinations of traits that could be categorized as *stress tolerators, competitors,* or *ruderals.* **Table 8.1** lists some of the major differences in these three plant strategies.

As their name implies, stress tolerators can survive and reproduce under extreme environmental conditions, such as very low water availability, very cold temperatures, or high salt concentrations. For example, plants living in the tundra biome, such as the wooly lousewort (*Pedicularis dasyantha*), are typically small herbs that live for many years, grow very slowly, and achieve sexual maturity relatively late in life. Similarly, many plants that live in the desert, such as cactus, are stress tolerators because they can survive long periods of hot temperatures with no precipitation. Because growing from a seed is very difficult in such stressful environments, stress tolerators put little of their energy into seeds. Instead, they rely on *vegetative reproduction,* a form of asexual reproduction in which new plants develop from the roots and stems of existing plants.

Where conditions for plant growth are less stressful, plants can evolve life history traits that fall along a continuum from competitors to ruderals. Without the abiotic stressors of extreme temperatures or an extreme lack of water and without frequent disturbances, plants can grow rapidly for long periods of time, and this creates

Goldenrod

Figure 8.2 Combinations of life history traits in plants.
Plants face the environmental challenges of competition, disturbance, and stress. Plant species living at each extreme environmental condition have evolved suites of life history traits that make the plants well suited to these environments.
Data from J. P. Grime, *Plant Strategies and Vegetation Processes* (Wiley, 1979).

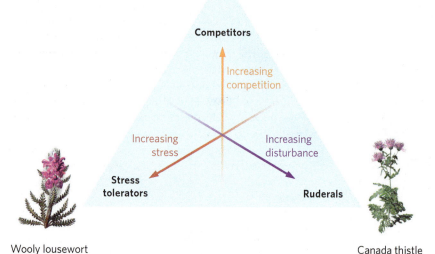

Wooly lousewort

Canada thistle

more competition among plants for soil nutrients and light. Competitors can grow relatively quickly, achieve sexual maturity early in life, and still devote only a small proportion of their energy to seed production because they often spread by vegetative reproduction. Competitors also tend to grow to larger sizes and exhibit long life spans. In forests, most species of trees fit the competitor category. In abandoned farm fields, a strong competitor is the group of tall herbaceous plants known as goldenrods.

At the other point of the triangle, with low stress and high frequency of disturbance, we find ruderals. These plants colonize disturbed patches of habitat, exhibit fast growth, early maturation, and use a high proportion of their energy to make seeds. Ruderals include many plants that we might call "weeds" in a garden, including dandelions (*Taraxacum officinale*), common ragweed

(*Ambrosia artemisiifolia*), and Canada thistle (*Cirsium arvense*). Ruderals typically have seeds that are easily dispersed and that can persist in the environment for many years as they wait for favorable environmental conditions. This collection of traits allows ruderal plants to reproduce quickly and to disperse their seeds to other disturbed sites.

CONCEPT CHECK

1. How would you differentiate fecundity and parity?
2. Using Grime's categorization of plant life history traits, why might ruderals spread via easily dispersed seeds, whereas stress tolerators spread vegetatively?
3. Why do we expect that, on average, only one of the offspring that an individual produces survives to reproduce?

TABLE 8.1

The Life History Traits of Plants That Are Positioned at the Environmental Extremes of Stress, Competition, and Disturbance

Life history traits	Stress tolerators	Competitors	Ruderals
Potential growth rate	Slow	Fast	Fast
Age of sexual maturity	Late	Early	Early
Proportion of energy used to make seeds	Low	Low	High
Importance of vegetative reproduction	Frequently important	Often important	Rarely important

8.2 Life history traits are shaped by trade-offs

If we consider the many types of life history traits, it would seem that an organism could have very high fitness if it could grow fast, achieve sexual maturity at an early age, reproduce at a high rate, and have a long life span. However, no organism has the best of all such life history traits, which highlights the fact that organisms face trade-offs. When one life history trait is favored, it often prevents the adoption of another advantageous life history trait. In some cases, there are physical constraints, such as the size of a mammal's uterus, which places a limit on the total volume of offspring that can be produced at one time. Thus, a female can produce several small offspring, or a few large offspring, but not several large offspring. In other cases, the trade-off reflects the genetic makeup of the organism. Because some genes have multiple effects, selection that favors genes for one life history trait can cause other life history traits to change. For instance, in the plant known as the mouse-ear cress (*Arabidopsis thaliana*), artificial selection on genes that cause earlier flowering also results in reduced seed production because the genes have multiple effects. In still other cases, the trade-offs are the result of how an organism allocates a finite amount of time, energy, or nutrients. For example, the more time that a gazelle spends looking around for predators, the less time it has to spend looking for and eating food. In this section, we will discuss the *principle of allocation* and highlight some of the most common trade-offs that have been observed.

THE PRINCIPLE OF ALLOCATION

Organisms often have limited time, energy, and nutrients at their disposal. According to the **principle of allocation,** when these resources are devoted to one body structure, physiological function, or behavior, they cannot be allotted to another. As a result, natural selection will favor those individuals that allocate their resources in a way that achieves maximum fitness.

Selection on life history traits can be complex because when one trait is altered, it often influences several components of survival and reproduction. As a result, the evolution of a particular life history trait can be understood only by considering the entire set of consequences. For example, an increase in the number of seeds an oak tree produces may contribute to higher fitness. However, if a higher number of seeds is achieved by making each seed smaller, and if smaller seeds experience lower survival, then producing more seeds could negatively affect the tree's overall fitness. In this case, to achieve an outcome that maximizes overall fitness, evolution should favor a strategy that balances the trade-off between seed number and seed survival.

From an evolutionary point of view, individuals exist to produce as many successful progeny as possible. Doing so, however, involves many allocation problems, including the timing of sexual maturity, the number of offspring to have at any one time, and the amount of parental care to bestow on the offspring. An optimized life history is one that resolves conflicts between the competing demands of survival and reproduction to the best advantage of the individual in terms of its fitness. Although it is widely believed that trade-offs constrain the specific combination of life history traits that a species can evolve, demonstrating this has proved difficult. In some cases, trade-offs can only be exposed by using experimental manipulations.

OFFSPRING NUMBER VERSUS OFFSPRING SIZE

Most organisms face a trade-off between the number of offspring they can produce and the size of those offspring in a single reproductive event. As we noted for mammals, the number of offspring in any given pregnancy can only increase if the size of each individual offspring decreases. The trade-off between offspring number and offspring size for a given reproductive event can also be limited by energy and nutrients. An example of this can be seen in **Figure 8.3**, which illustrates the relationship between seed size and seed number in plants of the goldenrod genus (*Solidago*). Across

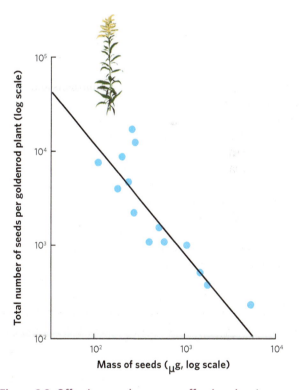

Figure 8.3 Offspring number versus offspring size. Among 14 populations and species of goldenrod plants, individuals that produce more seeds also produce smaller seeds, demonstrating that a trade-off exists between the two life history traits. Data from P. A. Werner and W. J. Platt, Ecological relationships of co-occurring goldenrods (*Solidago*: Compositae), The *American Naturalist* 110 (1976): 959–971.

Principle of allocation The observation that when resources are devoted to one body structure, physiological function, or behavior, they cannot be allotted to another.

populations and species, there is a negative correlation, demonstrating that goldenrod plants that produce more seeds also produce smaller seeds. While the trade-off between offspring number and offspring size can be seen in a number of species, the expected trade-off is often not observed. In some species, the number of offspring can be quite variable among individuals, but the size of offspring can be relatively constant. This suggests that selection often favors a uniform, perhaps even optimal, offspring size and that an individual able to acquire additional energy can only use it to make greater numbers of offspring.

OFFSPRING NUMBER VERSUS PARENTAL CARE

The number of offspring produced in a single reproductive event can also cause a trade-off with the amount of parental care that can be provided. As the number of offspring increases, the efforts of the parents to provide food and protection will be increasingly spread thin.

In a classic study of life history evolution, David Lack of Oxford University considered the number of offspring produced by songbirds. Lack observed that songbirds breeding in the tropics lay fewer eggs at a time—an average of 2 or 3 per nest—than songbirds breeding at higher latitudes, which, depending on the species, typically lay 4 to 10 eggs. In 1947, he proposed that these different reproductive strategies evolved in response to differences between tropical and temperate environments. Lack recognized that birds could improve their overall reproductive success by increasing the number of eggs in each reproductive event, as long as a larger number of offspring did not cause a decrease in offspring survival. He hypothesized that the ability of parents to gather food for their young was limited and, if they could not provide enough, the offspring would be undernourished and therefore less likely to survive. We should therefore expect that parents produce the number of offspring that they can successfully feed. One difference between the low latitudes

---

ANALYZING ECOLOGY

Coefficients of Determination

Eco TV

macmillanlearning.com/ricklefsvideo

When ecologists want to look for life history trade-offs, they commonly plot one life history variable against another life history variable and look for a negative relationship. We saw an example of this in the case of the goldenrod seeds; the researchers used a regression analysis to demonstrate a trade-off between seed number and seed size. As we saw in Chapter 5, a regression analysis describes the mathematical relationship between two variables. For example, data points that tend to follow a straight line might be best described by using a linear regression described by the following equation:

$$Y = mX + b$$

where X and Y are the measured variables; m is the slope, which is negative in the case of the goldenrod example; and b is the intercept.

Although this equation reveals how one variable is associated with another variable, it does not tell us how strongly the two variables are related. For example, it would be valuable to know if the data points fit closely along the line or if the data points are highly variable around the line. We can answer this question using a statistic known as the *coefficient of determination*. The **coefficient of determination,** which is abbreviated

as R^2, is an index that tells us how well data fit to a line. Values can range from 0 to 1, with 0 indicating a very poor fit of the data and 1 indicating a perfect fit of the data. In terms of life history trade-offs, higher R^2 values indicate that a variation in one life history trait explains a large amount of the variation in the other life history trait.

Consider the following set of hypothetical data for plant seeds that all have the same linear relationship between seed mass and seed number, as given in

$$\text{Seed number} = (-4 \times \text{Seed mass}) + 24$$

Seed mass (g)	Seed number for population A	Seed number for population B	Seed number for population C
1	21	22	24
1	19	18	16
3	13	14	16
3	11	10	8
5	5	6	7
5	3	2	1
	$\overline{Y} = 12$	$\overline{Y} = 12$	$\overline{Y} = 12$

Coefficient of determination (R^2) An index that tells us how well data fit to a line.

of the tropics and higher, temperate, latitudes is the number of daylight hours. Lack noted that parents at higher latitudes had more hours to gather food to feed their young. Therefore, assuming that the rate of food gathering is similar at low and high latitudes, he hypothesized that birds breeding at high latitudes could rear more offspring than birds breeding in the low latitudes of the tropics.

Lack made three important points. First, he stated that life history traits, such as the number of eggs laid in a nest, not only contribute to reproductive success but also influence evolutionary fitness. Second, he demonstrated that life histories vary consistently with respect to factors in the environment, such as the number of daylight hours available for gathering food for their young. This observation suggested that life history traits are molded by natural selection. Third, he hypothesized that the number of offspring parents can successfully rear is limited by food supply. To test this idea, one could add eggs to nests to create unnaturally high numbers of hatchlings. According to Lack's hypothesis, parents should not be able to rear larger broods of chicks because they cannot gather the additional food that is required.

Lack's hypothesis has been supported by numerous experiments during the last several decades. For example, the European magpie (*Pica pica*) typically lays seven eggs in its nest. To determine whether this is the most fit strategy for the magpie, researchers manipulated the number of eggs in magpie nests by removing one or two eggs from several nests and adding these eggs to other nests. To control for their disturbance of the nests, they also switched eggs between nests without changing their number. The researchers then waited to see how many chicks could be raised to the fledgling stage, when the offspring can leave the nest. As shown in **Figure 8.4**, magpies that had fewer than seven eggs or more than seven eggs produced fewer fledgling birds. Nests that had eggs removed produced fewer fledglings because they started with fewer offspring. In contrast, nests with eggs added produced fewer total fledglings because the chicks had to share the food with more siblings. This competition with siblings caused the chicks to grow more slowly and experience higher mortality rates because the parents were unable to feed so many offspring. As

---

For each set of data, we can graph the relationships and include the regression line, as shown in the graphs below.

To calculate R^2, we need to first calculate the mean Y value, which for all populations is a seed number of 12. We also need to determine the expected seed numbers if the data were to fall perfectly along the line. Using the line equation and the above data for seed mass, the six expected seed numbers are 20, 20, 12, 12, 4, and 4.

Next we need to calculate the *total sums of squares,* which is the sum of the squared differences between each observed seed number (y_i) and the mean seed number ($\overline{Y}$):

$$\Sigma(y_i - \overline{Y})^2$$

Then we calculate the *error sums of squares,* which is the sum of the squared differences between each expected seed number (f_i) and the observed seed number (y_i):

$$\Sigma(y_i - f_i)^2$$

Finally, we can calculate the value of R^2 as follows:

$$R^2 = 1 - \left(\frac{\Sigma(y_i - f_i)^2}{\Sigma(y_i - \overline{Y})^2} \right)$$

For population A, we can calculate the value of R^2 as

$$R^2 =$$
$$1 - \left(\frac{(21-20)^2+(19-20)^2+(13-12)^2+(11-12)^2+(5-4)^2+(3-4)^2}{(21-12)^2+(19-12)^2+(13-12)^2+(11-12)^2+(5-12)^2+(3-12)^2} \right) =$$
$$1 - \left(\frac{6}{262} \right) = 0.98$$

YOUR TURN Using the above formulas, calculate R^2 for populations B and C. Based on your three R^2 values, which set of data best fit the regression line? Which population gives you the strongest confidence that there is a negative relationship between seed size and seed number?

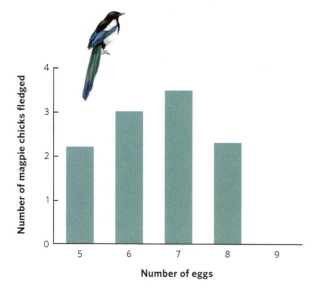

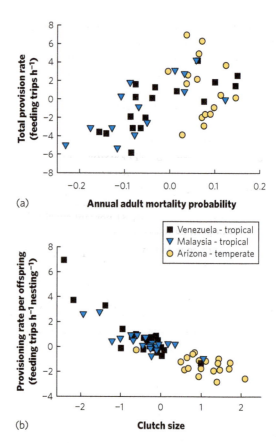

Figure 8.4 Manipulating the number of eggs in a nest. In the European magpie, the adults typically lay seven eggs. When researchers removed one or two eggs or added one or two eggs, the number of chicks that survived to fledge declined. This suggests that the typical number of eggs laid may be the optimal number for the magpie. Data from G. Högstedt, Evolution of clutch size in birds: Adaptive variation in relation to territory quality, *Science* 210 (1980): 1148–1150.

Figure 8.5 Alternative life history strategies of temperate versus tropical birds. Researchers examined the relative rates of mortality and relative feeding effort of 72 bird species in temperate and tropical regions of the world. **(a)** Temperate birds, which typically lay one or two more eggs than closely related birds in the tropics, experience higher rates of adult mortality and conduct a greater number of foraging trips to feed their total clutch of offspring. **(b)** However, tropical birds make more foraging trips for their offspring on a per-nestling basis, resulting in the higher survival of those nestling while in the nest and throughout their later adult life. In all cases, relative rates are calculated after controlling for numbers other factors including differences in bird mass. Data from T. Martin, Age-related mortality explains life history strategies of tropical and temperate songbirds, *Science* 349 (2015): 966–970.

Lack predicted, the number of eggs the magpie produces maximizes the number of offspring it can successfully raise.

Despite the support for Lack's hypothesis, research in 2015 revealed that the underlying premise that tropical birds lay fewer eggs than temperate birds because they are unable to provide as much food to their nestlings is not correct. By examining dozens of bird species across Arizona, Venezuela, and Malaysia, the study found that tropical birds had slower growth than temperate birds early in their nestling life, but they had faster growth later in their nestling life. Moreover, while temperate birds make more foraging trips to feed their total number of offspring, tropical birds make more foraging trips on a per-offspring basis, as shown in **Figure 8.5**. All this suggests that, contrary to Lack's hypothesis, the tropical parents are quite capable of finding abundant food for their offspring. In addition, the tropical nestlings grow longer wings, and this makes them less susceptible to predators compared to temperate nestlings. In short, tropical birds produce a smaller number of higher-quality offspring, not because they are unable to feed more offspring but because they have a different overall strategy. The tropical birds' strategy is to provide more food per nestling as a way to produce offspring with higher survival rates after they leave the nest. In contrast, the temperate birds' strategy is to produce more nestlings, at the cost of each nestling being of lower fitness.

FECUNDITY AND PARENTAL CARE VERSUS PARENTAL SURVIVAL

We saw how adding eggs to the nests of birds increases the effort for parents to supply food. Consequently, parents with small or intermediate size broods have the greatest fitness. Sometimes, however, having more mouths to feed stimulates parents to hunt harder for food for their chicks. In this case, an artificially enlarged brood might result in higher reproductive success in the short term. However, the additional parental effort can impose a cost on the parents that affects their subsequent fitness.

European kestrels (*Falco tinnunculus*) provide an example of the trade-off between fecundity and parental survival. Kestrels are small falcons that feed on voles and shrews caught in open fields. While foraging requires a high rate of energy expenditure, the small mammals are so abundant that kestrel pairs normally can catch enough prey to feed their brood in a few hours each day. Kestrels lay an average of five eggs per brood. In one study, when chicks were about a week old, broods in a sample of nests were subjected to one of three manipulations: Investigators removed two chicks from a nest, switched chicks between nests without changing their number, or added two chicks to a nest. Investigators expected that parents of the artificially reduced and enlarged clutches would alter the amount of energy they expended looking for food for the chicks.

While parents with fewer eggs ultimately had fewer chicks fledge than the control group, parents with additional eggs had more chicks fledge, as shown in **Figure 8.6**. However, in spite of the increased hunting efforts of their parents, chicks in the enlarged broods were somewhat undernourished and only 81 percent survived to fledge, compared with 98 percent survival in control nests and in nests with eggs removed. Consequently, the extra hunting effort to feed the additional two chicks only netted the parents an extra 0.8 chick, and this gain may have been diminished by the later deaths of some of the underweight fledglings. Moreover, the increased hunting efforts caused lower survival of these adults into the next breeding season. This means that increasing the number of offspring provides diminishing benefits to the parents in terms of the number of offspring that survive. Simultaneously, it causes greater adult mortality because parents must spend more energy securing food. At some point, the gains made by increasing current reproduction, which requires a large increase in parental care, are offset by the increased adult mortality, which reduces the chance of future reproduction.

GROWTH VERSUS AGE OF SEXUAL MATURITY AND LIFE SPAN

Organisms also commonly face a trade-off between putting their energy into growth or putting their energy into reproduction. In most species of birds and mammals, females grow to a particular size before they begin to reproduce. Once they initiate reproduction, females do not grow anymore—a phenomenon known as **determinate growth.** In contrast, many plants and invertebrates, as well as many fishes, reptiles, and amphibians, do not have a characteristic adult size. Rather, they continue to grow after initiating reproduction, a phenomenon known as **indeterminate growth.** Typically, indeterminate growth occurs at a decreasing rate over time.

Whether a species has determinate or indeterminate growth, the key feature shaping the trade-off between growth and reproduction is that larger females commonly produce more offspring. Because offspring production and growth draw on the same resources of assimilated energy and nutrients, increased fecundity during one year occurs at the cost of further growth that year. Moreover, for indeterminate growers, the reduced growth in one year can cause reduced fecundity in subsequent years. An organism with a long life expectancy should favor growth over fecundity during the early years of its life. In contrast, organisms with a short life expectancy should allocate

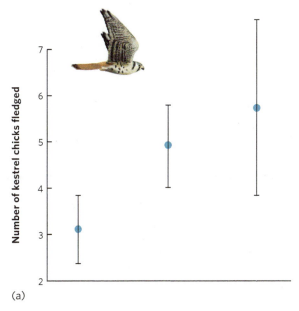

(a)

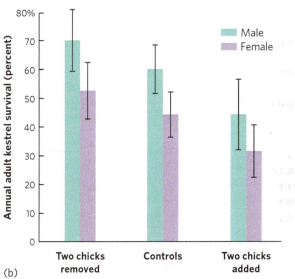

(b)

Figure 8.6 Parental care versus parental survival.
(a) When researchers removed eggs from kestrel nests, fewer chicks fledged, but when researchers added eggs, more chicks fledged. Thus, having more eggs leads to more fledgling offspring. Error bars are standard deviations. **(b)** However, adding more eggs causes a tradeoff. The parents work harder to feed the young, causing the adults to experience decreased survival. Error bars are standard errors. Data from C. Dijkstra et al., Brood size manipulations in the kestrel (*Falco tinnunculus*): Effects on offspring and parental survival, *Journal of Animal Ecology* 59 (1990): 269–286.

their resources to producing eggs early in life rather than delaying reproduction and growing more. These predictions can be tested by examining the relationships among these life history traits across many different species and by conducting manipulative experiments on species in nature.

Determinate growth A growth pattern in which an individual does not grow any more once it initiates reproduction.

Indeterminate growth A growth pattern in which an individual continues to grow after it initiates reproduction.

Comparisons Across Species

When we look across many different species, we see that long-lived organisms typically begin to reproduce later in life than short-lived ones. Why is this the case? If an organism has a long life span and if delaying maturity allows an organism to grow larger and produce more offspring per year once reproduction begins, natural selection will favor delayed age of maturity in these organisms. An analysis of hundreds of populations and species of animals illustrates this relationship. As you can see in **Figure 8.7**, as the age of sexual maturity increases, there is an associated increase in the number of years that an animal will survive after reaching sexual maturity. Moreover, different taxonomic groups fall along different regression lines. For species that are expected to live 2 years after sexual maturity, birds and mammals have the shortest times to sexual maturity, whereas reptiles, fish, and shrimp have the longest times. This reflects the fact that endothermic animals can grow more rapidly than ectothermic animals. Among species with an age of maturity of 1 year, birds have the longest expected life spans after sexual maturity. This reflects the generally lower predation risk for birds due to their ability to fly.

Manipulative Experiments

Another way to examine the trade-offs of growth, age of maturity, and life span is by conducting a manipulative experiment. The Trinidadian guppy (*Poecilia reticulata*), for instance, lives in the streams of Trinidad, a large island in the southern Caribbean Sea. In the lower reaches of these streams, the guppies live with a number of predatory fish species, including the pike cichlid (*Crenocichla alta*), which preys on adult guppies, and the smaller killifish (*Rivulus*

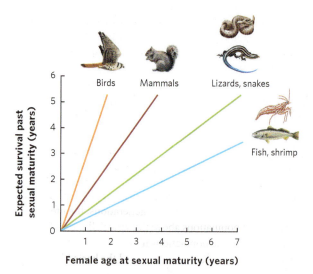

Figure 8.7 Age of sexual maturity versus life span. Using hundreds of different populations and species, we can see that different groups of animals have different relationships between these two life history traits. Data from E. L. Charnov and D. Berrigan, Dimensionless numbers and life history evolution: Age of maturity versus the adult life span, *Evolutionary Ecology* 4 (1990): 273–275.

hartii), which preys primarily on juvenile guppies. Because of this predation, these guppies have short life expectancies. However, at higher elevations, where guppies have been able to ascend numerous small waterfalls, they live in a relatively predator-free environment and have longer life expectancies. **Figure 8.8** shows the life history traits of guppy populations. In populations that face high predation risk and short life spans, males mature at smaller sizes. Females allocate a larger

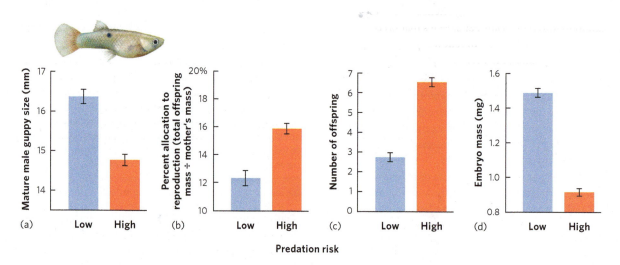

Figure 8.8 Alternate life history strategies of guppies in Trinidad. Guppy populations in streams with a high risk of predation have shorter life spans, and those in streams with low predation risk have longer life spans. In response to this difference in longevity, guppies living in predator environments have evolved to **(a)** mature as smaller males, **(b)** allocate a greater fraction of energy to offspring, which is quantified as the percentage of the mother's mass that is dedicated to the production of her offspring, **(c)** produce more offspring, and **(d)** produce smaller offspring. Error bars are standard deviations. Data from D. N. Reznick et al., Life history evolution in guppies (*Poecilia reticulata*: Poeciliidae). 4. Convergence in life history phenotypes, *American Naturalist* 147 (1996): 319–338. Photos by (a) Joi Ito; (b) Dusty Pixel photography/Getty Images.

percentage of their body mass to reproduction, they produce more offspring, and these offspring are smaller. As predicted, researchers found the opposite to be true for populations in the predator-free sections of the streams above the waterfalls: Males mature at larger sizes, and females allocate less of their body mass to reproduction and produce fewer, larger offspring.

Researchers then conducted a manipulative experiment to test the hypothesis that increased mortality from predation altered life history strategies of the local guppy populations. They transplanted predators from the lower reaches of the streams to the areas above the waterfalls where predators did not historically exist. Within a few generations, the life histories of the populations above the waterfalls came to resemble those of the populations in the lower reaches of the stream. This finding not only confirmed the basic ideas about the optimization of life history patterns but also demonstrated that predation is a strong selective force in evolution.

CONCEPT CHECK

1. Why are trade-offs among life history traits so commonly observed?
2. Why should organisms with low annual survival rates begin to reproduce at an early age?
3. Why do organisms face a fundamental trade-off between growth and fecundity?

8.3 Organisms differ in the number of times that they reproduce, but they eventually become senescent

The number of times an individual reproduces during its lifetime varies a great deal among species. But in almost all species, individuals eventually experience a decline in body condition followed by death. For species that breed once, physiological decline and death follow rapidly. For species that breed many times, the decline in physiology comes more gradually. An interesting exception is in the case of bacteria, in which individuals do not always die. Instead, they undergo fission whereby a single-celled organism divides into two single-celled organisms. In this section, we will examine the life history strategies of breeding once versus breeding several times and investigate the causes of a gradual decline in physiology.

SEMELPARITY AND ITEROPARITY

Organisms can evolve to reproduce only once, a phenomenon known as **semelparity,** or multiple times throughout their life, a phenomenon known as **iteroparity.** Semelparity and iteroparity do not tell us whether an organism's life history is **annual,** meaning a life span of one year, or **perennial,** referring to a life span of more than one year. For example, organisms that live for only one year can undertake more than one episode of reproduction, or even prolonged continuous reproduction, during that

time. Iteroparity is a common life history that occurs in most species of birds, mammals, reptiles, and amphibians. Semelparity is relatively rare in vertebrate animals, but it occurs in insects and many species of plants. For example, semelparity is a common life history of crop plants, including wheat and corn.

Bamboos and Agaves

Although most semelparous organisms are short-lived, the best-known cases of plants occur in the long-lived bamboos and agaves, two distinctly different groups. Most bamboos are plants of tropical or warm temperate climates, and they often form dense stands. Reproduction in bamboos does not appear to require substantial preparation or resources, as in the case of salmon. But bamboos probably have few opportunities for successful seed germination. Once a bamboo plant establishes itself in a disturbed habitat, it spreads for years by vegetative growth, continually sending up new stalks until the habitat in which it germinated is densely packed with bamboo shoots (**Figure 8.9a**).

In many species of bamboo, breeding is highly synchronous over large areas, such that every individual produces flowers and makes seeds in the same year. After reproduction, the future of the entire population rests with the crop of seeds. Synchronous breeding may facilitate fertilization in this wind-pollinated plant group and also may overwhelm seed predators, which cannot consume such a large crop of seeds. Some species of bamboo, such as Chinese bamboo (*Phyllostachys bambusoides*), have a 120-year cycle of germinating, growing, and then simultaneously flowering and producing seeds.

In contrast to bamboo plants, most species of agave plants live in arid climates with sparse and erratic rainfall. Ranging from the southwestern United States through Central America, agaves grow as a rosette of leaves over several years, with the duration of this growth varying among species. When the plant is ready to reproduce, it grows a gigantic flowering stalk that produces a large number of seeds (**Figure 8.9b**). The growth of the stalk is so rapid that it cannot be fully supported by photosynthesis in the stalk and uptake of water by the roots. Instead, the nutrients and water necessary for stalk growth are drawn from the leaves, which die soon after the seeds are produced.

When we consider all the plants and animals that practice semelparity, we see that semelparity appears to arise when there is a massive amount of energy required for reproduction, such as the long migrations of salmon and the production of giant flowering stalks of agaves. These large energy requirements make it difficult for the individuals to survive after the reproductive event.

Semelparity When organisms reproduce only once during their life.

Iteroparity When organisms reproduce multiple times during their life.

Annual An organism that has a life span of one year.

Perennial An organism that has a life span of more than one year.

(a)

(b)

Figure 8.9 Semelparous plants. (a) Bamboos and **(b)** agaves are two groups of plants that live for many years, reproduce one time, and then die. The flowering bamboos are from a location in Kyoto, Japan. The flowering agave is from Sodona, Arizona. Photos by (a) Joi Ito; (b) Dusty Pixel Photography/Getty Images

Salmon

Salmon are a group of species that vary a great deal in their parity. For example, Coho salmon (*O. kisutch*) lay eggs in rivers that empty into the North Pacific Ocean, from California to Alaska to eastern Russia. After growing in the river for a year, the fish swim out to the ocean, where they continue to eat and grow for 1 to 3 more years. When they are ready to reproduce, they migrate back to the same river where they hatched. The females make nests in the river bottom and deposit their eggs, which the males then fertilize with their sperm. Shortly after reproducing for the first time, both the male and female salmon rapidly lose strength and physiological abilities, and die. Several closely related species, including the chinook salmon (*O. tshawytscha*) and the sockeye salmon, also migrate to sea as juveniles, return for a single breeding event, and then die.

In contrast, other salmon species, such as the rainbow trout, breed several times during their life. Some populations—known as resident rainbow trout—do not migrate out to the ocean but remain in freshwater rivers. Other populations of rainbow trout—commonly known as steelheads—migrate out to sea and return to the rivers to breed, much like the coho salmon. Unlike the coho, however, steelheads migrate several times and breed each time they return to the freshwater rivers.

Cicadas

One of the most remarkable cases of semelparity in animals is the life cycle of periodical cicadas (**Figure 8.10**). Cicadas spend the first part of their life underground, where they obtain nutrients from the xylem tissue of plant roots and, after some time, emerge as adults. Their mating calls from the trees can be heard in the summer

days in many parts of the Northern Hemisphere. Some cicada species have annual life cycles, while other species spend several years underground, with a fraction of them emerging each summer. Periodical cicadas, however, are different. They live as nymphs underground for 13 or 17 years and then emerge from the ground in synchrony to mate. The emergence of the periodical cicada is marked by nearly deafening noise as the males call to attract females during their brief mating period. The long life cycle gives the larvae time to grow to adulthood on a diet of low nutritional quality. The synchrony is probably a mechanism to overwhelm potential predators. Most of the occasional individuals that fail to emerge in synchrony come out a year earlier or later and are grabbed by predators attracted to the loud mating calls.

Scientists long wondered how periodical cicadas know when to emerge from the ground. They speculated that the cicada might count the years by the warming and cooling of the soil, or by the physiological cycles of their hosts. Researchers performed a clever experiment by growing 17-year periodical cicadas on peach trees that had been artificially selected to drop their leaves and flower twice each year. The cicadas emerged after 17 fruiting seasons had passed, rather than 17 years, demonstrating that the cicadas are sensitive to the reproductive cycles of their hosts rather than annual physical changes in their environments. How they count to 17 remains a mystery.

SENESCENCE

A few long-lived semelparous organisms die immediately after reproduction. However, iteroparous organisms experience a gradual deterioration in their physiological function over their lifetime. This leads to a gradual decrease

Figure 8.10 Periodical cicadas.
The 13- and 17-year cicadas are semelparous insects. They spend many years underground, emerge as adults to breed, and die soon after. Pictured here is the pharaoh cicada (*Magicicada septendecim*), which is a species of 17-year cicada. Photo by ARS Information Staff.

in fecundity with an increase in the probability of mortality, a phenomenon known as **senescence.** Humans are an example of an organism that experiences senescence. Most physiological functions in humans decrease between the ages of 30 and 85 years; for example, the rate of nerve conduction and basal metabolism decrease by 15 to 20 percent, the volume of blood circulated through the kidneys decreases by 55 to 60 percent, and maximum breathing capacity decreases by 60 to 65 percent. Over time the function of the immune system and other repair mechanisms also declines. Using data from the United States population in 2014, **Figure 8.11** illustrates that the incidence of death from cancers and cardiovascular diseases rises sharply with age. Birth defects in offspring and

infertility also occur with increasing prevalence in women after 30 years of age, and fertility decreases dramatically in men after 60 years.

If maintaining high survival and reproduction would increase an individual's fitness at any age, why does senescence exist? Studies of aging in a variety of animals demonstrate that senescence is an inevitable consequence of natural wear and tear. It is impossible to build a body that will not wear out eventually, just as it is impossible to build an automobile that will not wear out. Senescence might simply reflect the accumulation of molecular defects

Senescence A gradual decrease in fecundity and an increase in the probability of mortality.

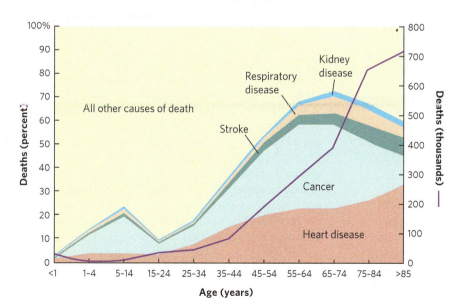

Figure 8.11 Senescence.
Throughout life, humans experience a gradual decrease in physiological function and an increase in cancer and cardiovascular diseases. These contribute to increased probability of mortality. Data from M. Heron, 2014, *National Vital Statistics Reports. Deaths: Leading Causes for 2014*, U.S. Department of Health and Human Services, Centers for Disease Control and Prevention, National Center for Health Statistics.

that fail to be repaired. For example, ultraviolet radiation and highly reactive forms of oxygen break chemical bonds, macromolecules become inactivated, and DNA accumulates mutations. However, this wear and tear cannot be the entire explanation for patterns of aging because maximum longevity varies widely even among species of similar size and physiology. For instance, many small insectivorous bats live 10 to 20 years in captivity, whereas mice of similar size rarely live beyond 3 to 5 years.

The rate of deterioration can be modified by a variety of physiological mechanisms that either prevent or repair damage. One major difference among groups that differ dramatically in life span is the cellular mechanisms for reducing the production of reactive forms of oxygen and for repairing damaged DNA and protein molecules. These appear to be better developed in long-lived animal species than in their short-lived relatives. Because these mechanisms are under genetic control, they can be modified by evolution. Mechanisms of prevention and repair require investments of time, energy, nutrients, and tissues. Therefore, the allocation of resources to these mechanisms depends on the expected life span of the individual. When a population has a low survival rate, selection should favor improvements in reproductive success at young ages and selection to delay senescence should be weak. In a population with a high survival rate, selection for delayed senescence should be strong.

This prediction is consistent with observations of natural populations. For example, because bats and birds can fly to escape predators, they lead safer lives than do rodents of a similar size. As a result, the maximum potential life spans of birds and bats are much longer than the life spans of rodents. Storm petrels, a small seabird, have a body size and metabolic rate that is similar to those of many rodents. However, the storm petrel can live more than 40 years, whereas a rodent may only live 1 or 2 years. Because of their long life spans, birds and bats age more slowly than rodents of a similar size.

CONCEPT CHECK

1. Compare and contrast semelparous and iteroparous life history strategies.
2. What is a cause of senescence?
3. Across species, what is the expected relationship between longevity and senescence?

8.4 Life histories are sensitive to environmental conditions

As we saw in the discussion of phenotypic plasticity in Chapter 4, many traits show flexibility in response to different environmental conditions, and life history traits are no exception. As a result, researchers have continued to discover a fascinating array of life history traits that can be altered by changing environmental conditions.

STIMULI FOR CHANGE

Many events in the life history of an organism are timed to match seasonal changes in the environment. The right timing is essential so behavior and physiology match changing environmental conditions. For instance, flowering plants must bloom when pollinators are present and most birds must breed when there is an abundance of food to feed their chicks. To get their timing right, organisms rely on various indirect cues in the environment.

Virtually all organisms sense the amount of light that occurs each day, known as the **photoperiod.** Many can distinguish whether the photoperiod is getting shorter or longer. Within a single species, populations may be exposed to a variety of environmental conditions. Each population develops a particular response to the photoperiod in its environment. For example, consider the grass known as sideoats grama (*Bouteloua curtipendula*). Southern populations living at 30° N flower in autumn in response to a photoperiod of 13 hours per day. In contrast, northern populations living at 47° N flower in summer in response to a photoperiod that exceeds 16 hours each day. Because organisms experience longer days at higher latitudes, as we saw in Chapter 5, each population responds to a photoperiod that is indicative of summer at a particular latitude.

Another example occurs in water fleas of the genus *Daphnia*. In Michigan, water fleas enter diapause in mid-September, when the photoperiod declines to fewer than 12 hours of sunlight, but related species in Alaska enter diapause in mid-August, when the photoperiod decreases to fewer than 20 hours of sunlight. Water fleas never see 20-hour days in Michigan, but Alaskan water fleas would perish from the cold if they waited for 12-hour days before entering diapause. From this we see that the critical stimulus for these organisms is the change in environmental conditions associated with a particular photoperiod that is relevant to a particular latitude. The sensitivity of individuals to these cues has been adjusted by natural selection so that the individual's response to an environmental cue is well matched to the environmental condition.

THE EFFECTS OF RESOURCES

Many types of organisms undergo dramatic life history changes during the course of their development. One of the most striking changes is the process of metamorphosis in which a larva changes into a juvenile or adult organism. Metamorphosis can be seen in many species of insects and amphibians, as in the transformation from tadpole to frog. Organisms that metamorphose have a wide range of timing options. Environmental conditions that influence timing include the amount of resources available, the temperature, and the presence of enemies.

Let's consider the different options for timing of metamorphosis by looking at the two growth curves in **Figure 8.12**. These curves represent the change in

Photoperiod The amount of light that occurs each day.

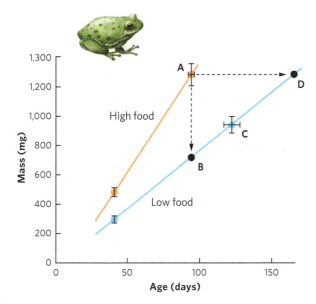

Figure 8.12 Alternative growth curves of a metamorphosing organism. The amount of food available can affect an organism's mass and age at metamorphosis. In the barking treefrog, an individual living under high food is able to metamorphose at a large mass and young age, represented by point A. An individual living under low food conditions could achieve the same age at metamorphosis if it emerges at a smaller mass, for example, at point B. It could achieve the same mass at metamorphosis if it took longer to metamorphose, for example, at point D. In reality, the tadpoles reach a compromise and metamorphose at a somewhat smaller mass and a somewhat later age, as indicated by point C. Error bars are standard deviations. Data from J. Travis, Anuran size at metamorphosis: Experimental test of a model based on intraspecific competition, *Ecology* 65 (1984): 1155–1160.

the mass of the barking treefrog (*Hyla gratiosa*) raised under conditions of high or low food availability On any particular day early in life, the individuals raised under high food availability have a larger mass than those raised under low food availability. As time progresses, an individual with high food availability is able to metamorphose at a relatively large mass and young age. An individual with low food availability cannot achieve the same combination of mass and age, but it can follow several alternative strategies. It might wait to metamorphose when it achieves the same mass as individuals raised under high food availability, as shown by point D in the figure, though it will take longer to achieve that mass and the delay in reproduction could reduce its fitness. Alternatively, it might metamorphose at the same age as individuals raised under high food availability, as shown by point B in the figure, even though it will be significantly smaller. The drawback of this strategy is that a smaller size at metamorphosis makes the organism more vulnerable to predation. For most metamorphosing organisms, the optimum solution is usually a compromise between these two strategies. So an organism with low food availability typically

metamorphoses at an older age and smaller mass, as shown by the decision of the barking treefrog at point C in the figure.

THE EFFECTS OF PREDATION

The risk of predation is also an important factor in affecting the life history of organisms. As we saw in Chapter 4, predation can affect a wide range of life history traits, including time to and size at hatching, time to and size at metamorphosis, and the time to and size at sexual maturity. One of the more remarkable effects of predators is their impact on the embryos of many species of aquatic organisms. In a variety of groups, including fish and amphibians, the embryo that develops inside an egg can detect the presence of an egg predator. Many embryos sense the chemical odors that predators emit, while other embryos can detect the vibrations produced by predators. When predators are detected, the embryos can speed up their time of hatching in an attempt to leave the egg before the predator eats it. For example, the red-eyed treefrog (*Agalychis callidryas*) lives in Central America and the adults lay their eggs on leaves that hang over water. When the embryos have developed sufficiently, they hatch and drop into the water. Should a cat-eyed snake (*Leptodeira septentrionalis*) appear, however, the frog embryos sense the vibrations of the approaching snake and begin to hatch earlier than usual and drop into the water to avoid the snake (**Figure 8.13**).

Figure 8.13 Hatching early in response to predators. As the cat-eyed snake begins to attack the eggs of the red-eyed treefrog, the embryos are stimulated to hatch early. Note the tadpole escaping the snake's attempt to eat the egg in which the tadpole lived. This photo was taken at the Corcovado National Park in Costa Rica. Photo by Karen M. Warkentin.

However, this response comes at the cost of hatching at a smaller size that can make the hatchling tadpoles more susceptible to predators living in the water. Therefore, when egg predators are not present, the embryo stays in the egg longer and hatches at a larger and safer size.

Studies of metamorphosing animals also find that predators commonly play an important role in affecting the size and time at which metamorphosis takes place. For example, many high-elevation streams of western Colorado contain trout, an important predator of mayfly larvae, whereas other streams lack trout. Larval mayflies living with trout metamorphose and leave the streams earlier and at smaller sizes than mayflies in comparable streams that lack trout. Growth rates in the two types of streams are similar, so the difference in the time to and size at metamorphosis is entirely the result of predation risk.

Predators can also affect when organisms achieve sexual maturity. Several species of freshwater snails, for example, face higher risks of predation when they are small. As a result, when predators are present, a snail is likely to do better if it delays reproduction and uses its energy to grow to a larger size before reproducing, as shown in **Figure 8.14 a–b**. Once it has grown to a safer size, it can reproduce. Although this strategy can improve the snail's probability of survival in the presence of predators, the cost of delaying sexual maturity can be reduced fecundity. However, once the predator-induced snails begin reproducing, they can produce more eggs in each clutch because they have larger bodies and they can live longer. In these cases, the snails can achieve the same lifetime fecundity as snails raised without predators, as shown in **Figure 8.14c**.

THE EFFECTS OF GLOBAL WARMING

We have seen how the life history of organisms responds to different environmental conditions found in nature. Over the past 100 years, human activity has caused a warming trend on Earth. In many regions, the difference in temperature is relatively small—a rise of 1°C or 2°C. However, even small changes in temperature can have substantial impact on an organism's physiological processes, as we saw in Chapter 2. During the past decade, researchers have begun to discover that the increase in global temperatures has caused changes in the breeding times of many animals and plants.

Animal Breeding

 Researchers interested in the effect of global climate change on the life histories of animals have focused on the breeding times of numerous species.

A change in breeding dates has been observed in several species of amphibians. In Britain, researchers

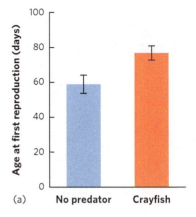

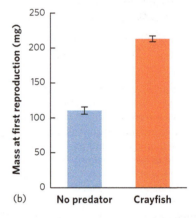

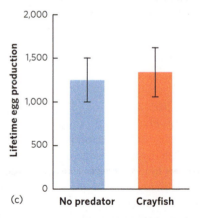

Figure 8.14 Snail life history strategies are altered in the presence of predators. When snails detect the smell of predatory crayfish in the water, **(a)** they delay their age at first production in favor of **(b)** increasing their mass at first reproduction. Because snails living with the scent of predators experience an extended longevity, their **(c)** lifetime production of offspring is similar to that of snails not living with predators. Data from J. Auld and R. Relyea, Are there interactive effects of mate availability and predation risk on life history and defence in a simultaneous hermaphrodite?, *Journal of Evolutionary Biology* 21 (2008): 1371–1378.

monitored three species of frogs and three species of salamanders for 17 years. At the end of this period, they found that two of the three frog species monitored were breeding 2 to 3 weeks earlier and all three species of salamanders were breeding 5 to 7 weeks earlier. These changes in breeding times were correlated with the average maximum temperatures that occurred just prior to breeding, which generally increased over the 17 years. A similar study of North American amphibians, however, failed to find a relationship between changes in mean maximum air temperature over time and the initiation of breeding. At the present time, researchers do not know why amphibians in different regions of the world show different responses to global warming.

While global climate change includes warming temperatures, it also includes changes in the pattern of precipitation. Researchers studying leatherback sea turtle (*Dermochelys coriacea*) breeding examined how the hatchling survive in light of changing precipitation patterns that have been observed at four sites around the world during the past several decades (1982–2010). In some sites, no relationship existed between precipitation and the proportion of turtle eggs that hatched. At other sites, however, there was a clear decline in precipitation over the decades as well as a decline in the proportion of turtle eggs that hatched (**Figure 8.15**). Such data suggest that the impacts of global climate change can vary among different geographic locations on Earth.

Plant Flowering

Plants are also susceptible to climate change, which has the potential to alter the initiation of flower production. One of the longest studies began in the nineteenth century with the writer Henry David Thoreau, who is best known for having spent a year in a small cabin at Walden Pond in Concord, Massachusetts, and for his numerous essays about the natural world. Thoreau kept data on more than 500 species of flowering plants in Concord. From 1852 to 1858, he took notes on the dates when each plant species first began flowering in Concord.

After Thoreau's death, a local shopkeeper continued his work by observing the first flowering times of more than 700 plant species. More recently, two ecologists realized that the data could help them determine whether long-term changes in global temperatures might be associated with changes in the initial flowering times of plants. Because flowering time is sensitive to temperature as well as to photoperiod, they predicted that warmer global temperatures would cause plants to flower earlier today than in Thoreau's time. To test their hypothesis, they collected data for plant flowering times in Concord from 2003 to 2006.

The researchers reported that over the 154-year period from 1852 to 2006, local temperatures in Concord had increased by 2.4°C. You can see these data in **Figure 8.16**. They also found that for the 43 most common species of plants, flowering time today is an average of 7 days earlier than in Thoreau's time. Interestingly, not all plants responded to the temperature change in the same way. In some species, initial flowering time remained unchanged, perhaps because these species use day length as their cue for flowering and day length has not changed. Other species, such as highbush blueberry (*Vaccinium corymbosum*) and yellow wood sorrel (*Oxalis europaea*), flower 3 to 4 weeks earlier now than they did in 1852. These unique data collected over a century and a half indicate that a seemingly small change in average annual temperature has been associated with dramatic changes in initial flowering time.

Consequences of Altered Breeding Events

The changing breeding seasons of plants and animals in response to global warming do not by themselves cause any problems to the species that are responding. Problems can arise, however, when a species depends on the environment to provide the necessary resources with an altered breeding season. The pied flycatcher (*Ficedula hypoleuca*), for example, is a bird that breeds in Europe each spring. In 1980, researchers in the Netherlands found that the date of egg hatching of the flycatcher began just a few days before the peak abundance of caterpillars, which are a major prey item for the flycatcher chicks. As spring temperatures warmed over the next 2 decades, however, tree leaves appeared 2 weeks earlier and the caterpillars reached their peak abundance 2 weeks earlier. The pied flycatcher, however, retained its normal date of egg hatching, which was 2 weeks later than the new time for peak caterpillar abundance. As a result, the chicks of the flycatcher no longer have a major source of food and the pied flycatcher population has declined by 90 percent.

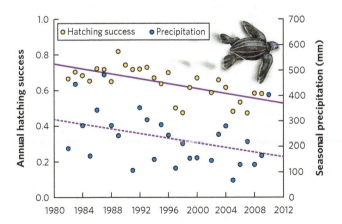

Figure 8.15 Hatching survival in leatherback sea turtles.
At one of four sites where sea turtles breed, researchers monitored precipitation and the proportion of turtle eggs that hatched from 1982 to 2010. Data from P. S. Tomillo et al., Global analysis of the effect of local climate on the hatchling output of leatherback turtles, *Nature Scientific Reports* 5 (2015). doi:10.1038/srep16789

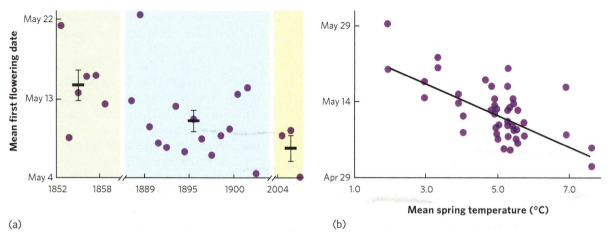

Figure 8.16 First flowering dates for plants in Concord, Massachusetts. (a) The mean flowering time today is 7 days earlier than in the 1850s. Error bars are standard errors. **(b)** The variation in first flowering time is associated with the mean temperature of the 1 or 2 months preceding each species' flowering time. Data from A. J. Miller-Rushing and R. B. Primack, Global warming and flowering times in Thoreau's Concord: A community perspective, *Ecology* 89 (2008): 332–341.

A similar phenomenon was observed by researchers who revisited the plants that Henry David Thoreau studied. From a series of studies conducted in the Boston area, these researchers noted a variety of organismal responses to warming temperatures. As you can see in **Figure 8.17**, the mean time for trees to leaf out is currently about 18 days earlier than Thoreau observed in the 1850s. The same team examined several other organisms' responses to warming temperatures as well. Butterflies are being seen about 12 days earlier, but the arrival of migratory birds is only about 3 days earlier. As a result, changes in the historic timing of birds, insects, and plants beginning their spring activities have occurred, and this has the potential to dramatically alter how these species interact in regard to predation, herbivory, and pollination.

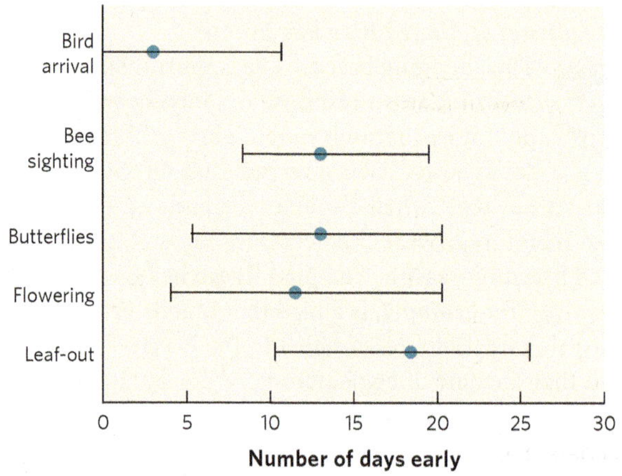

From our discussion of life history traits, you can see that natural selection has favored a wide variety of strategies for the life history of different species. Different life histories evolve as a result of different selection pressures on traits such as mortality, fecundity, and longevity combined with a considerable number of potential trade-offs among life history traits. As is true for other traits, the genes that code for these traits interact with the environments that organism's experience, ultimately to produce the life history traits of individuals.

Figure 8.17 Effects of global warming on birds, plants, and insects. Since the 1850s in the Boston area, the dates of birds arriving, bees and butterflies being first observed, and plants flowering and leafing out from their winter buds are now all earlier than in the past. Error bars are 95% confidence intervals. Data compiled in R. Primack, Spring budburst in a changing climate, *American Scientist* 104 (2016): 102–109.

CONCEPT CHECK

1. Why is it important for organisms to adjust the timing of their life history to seasonal changes in the environment?
2. How do different resource levels affect the decision of when, and at what mass, an animal should metamorphose?
3. Why do prey species often delay their time to first reproduction in favor of increasing their mass at first reproduction?

Selecting on Life Histories with Commercial Fishing

Human selection on the life history of fish. For several decades, commercial fishing vessels, like this boat in Alaska, have harvested the largest individuals and thereby caused unintentional selection for smaller fish. Because of this selection, some species of fish now achieve sexual maturity earlier. Photo by Design Pics Inc/Alamy.

Throughout this chapter, we have seen how natural selection has shaped the evolution of life histories by favoring individuals that are best suited to their environment. But what if a change in the selection process favors individuals with different suites of traits? This is precisely what happens in many commercial fisheries because they harvest only the largest individuals. For many years, this seemed like a wise way to manage the exploitation of wild populations because it protected the small individuals, allowing them a chance to grow. It is also a common practice for state agencies to set a minimum size for fish, such as bass and salmon, that anglers are permitted to keep.

Based on our discussion in this chapter, you should be able to predict what will happen to the life history traits of species that experience a great deal of fishing pressure, particularly from large commercial fishing boats that can collect thousands of fish. When the smallest fish are either not caught or are thrown back into the water, we impose a high mortality rate on large adults and leave the smaller, younger fish behind for breeding. As we have seen in this chapter, high adult mortality favors the evolution of smaller adult sizes, earlier times to maturity, higher fecundity, and shorter life span.

During the past two decades, researchers investigated whether large-scale fishing can cause unintended evolution of fish life histories, and they have confirmed that commercial fishing imposes substantial selection on fished populations. In keeping with the requirements for evolution, there is sufficient heritability present in fish populations for the selection to cause a change in subsequent generations. For example, in the 1930s and 1940s, the northeast Atlantic Cod (*Gadus morhua*) had a median age at maturity that ranged between

9 and 11 years. By the 1960s and 1970s, age at maturity ranged between 7 and 9 years. The data collected by commercial fishing boats do not typically contain information on other life history traits such as fecundity and longevity. However, based on our knowledge of common life history trade-offs, it is reasonable to assume that an increase in adult mortality and a decline in age at maturity imposed by fishing practices coincide with increases in size-adjusted fecundity and declines in longevity.

One challenge in determining the effect of fishing pressure on life history is that it can cause other significant changes. As we noted earlier in this chapter, environmental changes—such as resource levels—can also affect life history traits. For example, fishing from commercial fishing boats could reduce competition among the remaining fish, which allows for a faster time to sexual maturity. In some cases, researchers cannot differentiate between the environmental induction of life history changes from reduced competition and the evolution of life history changes. In other cases, however, scientists have been able to document that a fish population held at low numbers for several decades continues to exhibit life history changes over time. In such cases, the changes in life history are more likely to be the result of evolution by artificial selection.

To help clarify the role of selection on fish life histories without the complications of other environmental changes, including reduced competition, researchers in 2013 reported the results of an experiment in which they selected guppies by culling large individuals, culling small individuals, and culling the population randomly. After only three generations, the researchers found an evolutionary response to the selection on body size. Fish populations that experienced a culling of the largest individuals reached a smaller size at maturity and a younger age at maturity. In contrast, populations that experienced a culling of the smallest individuals reached a larger size at maturity and an older age at maturity.

Experimental selection on the life history of fish. Over six generations, populations selected for larger size evolved to become larger than the unselected control population, whereas populations selected for smaller size evolved to become smaller. Error bars are standard errors of the mean. Data from S. J. van Wijk et al. Experimental harvesting of fish populations drives genetically based shifts in body size and maturation, *Frontiers in Ecology and the Environment* 11 (2013): 181–187.

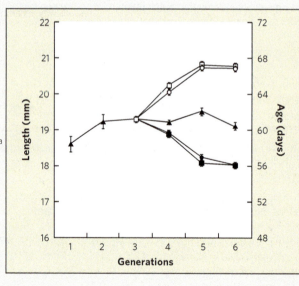

The impact of human selection on natural populations is not limited to fish; similar impacts have been found in hunted mammals and some plants. In all these cases, identifying the factors that naturally cause the evolution of life histories and the ways in which various life history traits trade off with each other has helped fisheries managers understand how the human harvest of wild populations can have unintended consequences.

SOURCES:

van Wijk, S. J., et al. 2013. Experimental harvesting of fish populations drives genetically based shifts in body size and maturation. *Frontiers in Ecology and the Environment* 11: 181–187.

Law, R. 2000. Fishing, selection, and phenotypic evolution. *CIES Journal of Marine Science* 57: 659–668.

Darimont, C. T., et al. 2009. Human predators outpace other agents of trait change in the wild. *Proceedings of the National Academy of Science* 106: 952–954.

SUMMARY OF CHAPTER CONCEPTS

8.1 Life history traits represent the schedule of an organism's life. Species differ in a wide range of traits that help determine their fitness throughout their life. These traits are under the influence of natural selection and often evolve in particular combinations.

Key Terms: Life history, Fecundity, Parity, Parental investment, Longevity

8.2 Life history traits are shaped by trade-offs. Trade-offs can occur because of physical constraints, time or energetic constraints that affect allocation, or genetic correlations that cause selection favoring one trait to come at the cost of another trait. Common trade-offs include offspring number versus offspring size, and growth versus reproduction.

Key Terms: Principle of allocation, Coefficient of determination (R^2), Determinate growth, Indeterminate growth

8.3 Organisms differ in the number of times that they reproduce, but they all eventually become senescent. Semelparous organisms breed once, whereas iteroparous organisms breed more than once. All organisms ultimately experience a decay in physiological function followed by death. In semelparous organisms, this decay in function occurs rapidly after reproduction. In iteroparous organisms, the decay in function can be very gradual.

Key Terms: Semelparity, Iteroparity, Annual, Perennial, Senescence

8.4 Life histories are sensitive to environmental conditions. Life history traits are the product of genes and environments interacting. Some of the most common environmental influences on life history traits include variation in resources and predators, both of which can induce substantial changes in the life history of organisms. Current anthropogenic changes in the environment can also induce life history changes.

Key Term: Photoperiod

CRITICAL THINKING QUESTIONS

1. How should selection for greater longevity affect time to maturity and size at maturity?

2. Compare and contrast the concept of a regression versus a coefficient of determination.

3. What factors might favor the evolution of semelparity versus iteroparity in different species of salmon?

4. Compare and contrast the two leading arguments regarding why temperate bird species lay one or two more eggs than closely related tropical bird species.

5. Given your knowledge of how predators and resources can affect life history traits, hypothesize how the four combinations of high resources, low resources, predator presence, and predator absence would affect the size at maturity of a prey species.

6. Why might natural selection act more strongly on traits that improve reproductive success early in life rather than later in life?

7. Give two reasons why a mammal such as a gray wolf could face a trade-off between offspring number and offspring size.

8. What is the mechanism by which parental care can increase current fitness but decrease future fitness?

9. Why might an organism use cues such as photoperiod to predict the future state of its environment?

10. In terms of environmental cues, hypothesize why some, but not all, species of plants flower earlier in association with warmer spring temperatures.

GRAPHING THE DATA Lizard Offspring Number Versus Offspring Mass

The common lizard (*Lacerta vivipara*) can produce 2 to 15 offspring in a single breeding event. Using the following data, create a scatter graph to illustrate the relationship between the number of lizard offspring and the average mass of a lizard's offspring. (You can review scatterplots in the graphing appendix.) After graphing the relationship, describe the relationship in words.

Offspring number	Offspring mass (g)
2	242
3	238
4	230
5	223
6	207
7	200
8	189
9	180
10	173
11	157
12	150
13	142
14	138
15	130

9 Reproductive Strategies

The Sex Life of Honeybees

Honeybees (*Apis mellifera*) have a complicated sex life. They live in hives that may contain tens of thousands of bees, usually progeny of the same mother, known as the *queen*. Like many organisms, the queen bee produces sons and daughters, but she does so in a rather unique way. Early in her life, the queen bee flies out of the hive and mates in the air with a group of male bees. Male bees, known as *drones*, are smaller than the queen. A queen mates with several drones, but the larger drones contribute more sperm than the smaller drones. The queen stores sperm from this single mating event in a special organ in her body, known as the *spermatheca*, where it remains viable for several years. She uses these sperm to fertilize her eggs and make diploid daughters, known as *workers*. In contrast, she creates drones by laying unfertilized haploid eggs. The drones a queen produces rarely mate with her, but instead mate with other queens outside the hive. After mating, the drones die.

A key issue for a successful hive is the proper ratio of drones and workers. In a typical beehive, the queen bee may produce a few dozen drones but tens of thousands of workers. Since the workers do the vast majority of the work in the hive, producing many more workers than drones is in the best interest of the queen. Since workers only live for 4 to 7 weeks, the queen must constantly produce more of them.

Genetically, the workers and the queen are quite similar. They are both female and they both arise from a fertilized, diploid egg. What makes them different is the food they are given as larvae. For the first few days of life, all larvae are fed royal jelly, a liquid produced by the worker bees, but after this, the larvae destined to be workers are switched to a diet of honey and pollen. These workers are not capable of mating with a drone, but they can lay unfertilized eggs. Larvae destined to be queens continue to be fed royal jelly, which allows the future queen to grow very large. The queen's size allows her to eventually lay as many as 2,000 eggs per day.

> "In a typical beehive, the queen bee may produce a few dozen drones but tens of thousands of workers."

When a beehive experiences the decline of its queen, new larval queens are often already being produced as replacements. Sometimes, however, the death or departure of a queen happens unexpectedly and there are no replacements. In such cases, some of the workers lay eggs but, since the workers cannot mate, their eggs are haploid and destined to turn into drones. Without a queen to lay fertilized eggs, the colony will eventually die.

Scientists have recently discovered an exception to this scenario in a subspecies of the honeybee known as the Cape honeybee (*Apis mellifera capensis*) that is found in southern Africa. Cape honeybee workers can lay diploid eggs without ever mating by using an unusual form of meiosis, in which they produce haploid egg cells and then merge two haploid cells to form a diploid cell. Researchers recently discovered that multiple genes control this ability. As a result, depending on which alleles they carry, individual workers can lay either haploid eggs that give rise to drones or diploid eggs that give rise to workers. Although laying diploid is normally the queen's job, evolving this behavior in workers can ensure the persistence of a beehive

A honeybee hive. In most populations of honeybees, a queen can reproduce by laying either haploid eggs that produce sons or fertilized diploid eggs that produce daughters. Photo by StockMediaSeller/Shutterstock.

SOURCES:

Wallberg, A., C. W. Pirk, M. H. Allsopp, and M. T. Webster. 2016. Identification of multiple loci associated with social parasitism in honeybees. *PLoS Genetics* 12: e1006097.

Lattorff, H. M. G., R. F. A. Moritz, and S. Fuchs. 2005. A single locus determines thelytokous parthenogenesis of laying honeybee workers (*Apis mellifera capensis*). *Heredity* 94: 533–537.

that has lost its queen. Even more interesting is that these workers can sneak into other beehives, eat the stored food of the hive, and lay their eggs. In doing so, a worker can have even greater fitness by having other hives rear her offspring.

The complexity of honeybee reproduction is an excellent example of the variety of reproductive strategies that have evolved. These options include reproducing with or without a sexual partner, choosing the number of mating partners, selecting the best traits of a sexual partner, varying the way the sex of offspring is determined, and controlling the number of sons versus daughters. In this chapter, we will explore the wide range of reproductive strategies in a diversity of organisms.

LEARNING OBJECTIVES

After reading this chapter, you should be able to:

9.1 Describe both sexual and asexual reproduction.

9.2 Explain how organisms can evolve as separate sexes or as hermaphrodites.

9.3 Describe why sex ratios of offspring are typically balanced, but can be modified by natural selection.

9.4 Explain how mating systems describe the pattern of mating between males and females.

9.5 Explain how sexual selection favors traits that facilitate reproduction.

The evolution of reproductive strategies involves a large number of different factors, many of which are influenced by ecological conditions. For example, organisms can evolve to reproduce sexually or asexually; each strategy has unique costs and benefits, particularly when species interact with parasites and pathogens. Organisms can evolve to reproduce as separate sexes or as hermaphrodites that possess both female and male sexual organs. If they reproduce as hermaphrodites, they must also evolve solutions to the problems associated with self-fertilization. In many species, the sex ratio of offspring can be altered to respond to changing ecological conditions. Finally, we see many different mating strategies to improve fitness, including the number of mates and the preference for certain traits in the opposite sex. This chapter explores how ecological conditions affect the evolution of sex and the strategies that organisms have evolved to increase their fitness.

9.1 Reproduction can be sexual or asexual

All organisms reproduce, but they have a variety of ways in which they accomplish this task. In plants, animals, fungi, and protists, reproduction may be accomplished through either *sexual reproduction* or *asexual reproduction*. In this section, we will look at the two processes and compare their costs and benefits.

SEXUAL REPRODUCTION

As we discussed in **Chapter 7**, reproductive function in most animals and plants is divided between two sexes. When progeny inherit DNA from two parents, we say they are the result of **sexual reproduction**. Haploid germ cells, called gametes, are produced through meiosis within sex organs called gonads. Each gamete contains a single full set of chromosomes. In animals, these haploid cells can immediately act as gametes. In plants and many protists, the haploid cells develop into multicellular, haploid stages of the life cycle that eventually produce gametes. When two gametes combine to form an offspring, each parent contributes one set of chromosomes. During meiosis, the distribution of the chromosomes into the gametes is generally random and the subsequent mixing of chromosomes from the two parents results in new combinations of genes in the offspring. Ultimately, two gametes join together in an act of fertilization to produce a diploid zygote.

ASEXUAL REPRODUCTION

In contrast to sexual reproduction, progeny produced by **asexual reproduction** inherit DNA from a single parent. This can be accomplished through either *vegetative reproduction* or *parthenogenesis*.

Vegetative Reproduction

Vegetative reproduction occurs when an individual is produced from the nonsexual tissues of a parent. Many plants can reproduce by developing new shoots that sprout

---

Sexual reproduction A reproduction mechanism in which progeny inherit DNA from two parents.

Asexual reproduction A reproduction mechanism in which progeny inherit DNA from a single parent.

Vegetative reproduction A form of asexual reproduction in which an individual is produced from the nonsexual tissues of a parent.

Figure 9.1 Vegetative reproduction. Organisms that use vegetative reproduction produce offspring from nonsexual tissues. The walking fern shown here produces offspring when the tips of its leaves touch the soil.

from leaves, roots, or rhizomes (i.e., underground shoots). **Figure 9.1** shows an example of this in a walking fern (*Asplenium rhizophyllum*), which produces offspring when the tips of its leaves touch the soil. If you have ever placed a plant clipping into a glass of water and watched it grow roots to become a new plant, you have witnessed this type of reproduction. Individuals that descend asexually from the same parent and bear the same genotype are known as **clones**. Many simple animals, such as hydras, corals, and their relatives, also reproduce in this way; they produce buds along their body that develop into new individuals. Bacteria and some species of protists reproduce by duplicating their genes and then dividing the cell into two identical cells, a process known as **binary fission**.

Parthenogenesis

In contrast to vegetative reproduction, some organisms reproduce asexually by producing an embryo without fertilization, a process known as **parthenogenesis**. In most cases, parthenogenetically produced offspring arise from diploid eggs, which do not require any genetic contribution from sperm. Parthenogenesis has evolved in plants and several groups of invertebrates, including water fleas, aphids, and the Cape honeybees mentioned at the beginning of this chapter. Animal species that reproduce only by parthenogenesis are typically composed entirely of females.

Parthenogenesis is relatively rare in vertebrates. It has never been observed as a natural occurrence in mammals but has been confirmed in a few species of lizards, amphibians, birds, and fishes. For a long time, it was thought that snakes and sharks were not capable of parthenogenesis. In 2007, however, researchers confirmed that a virgin female hammerhead shark (*Sphyrna tiburo*) gave birth to daughters that were genetically identical to the mother. In 2011, researchers discovered that an isolated female boa constrictor gave birth to two litters of daughters through parthenogenesis (**Figure 9.2**). The growing evidence

Clones Individuals that descend asexually from the same parent and bear the same genotype.

Binary fission Reproduction through duplication of genes followed by division of the cell into two identical cells.

Parthenogenesis A form of asexual reproduction in which an embryo is produced without fertilization.

Figure 9.2 A fatherless boa constrictor. This female boa constrictor was the product of parthenogenesis. As a result, the recessive caramel coloration of the mother was passed on to all her clonal daughters.
Photo by Warren Booth.

suggests that parthenogenesis may be more common than we currently appreciate and that many species can reproduce through both sexual and asexual reproduction.

Parthenogenesis can produce offspring that are either clones of the parent or offspring that are genetically variable. Clones are produced when germ cells develop directly into egg cells without going through meiosis. In contrast, genetically variable offspring are produced when germ cells proceed partially or entirely through meiosis. In partial meiosis, germ cells pass through the first meiotic division, but suppression of the second meiotic division results in diploid egg cells. Although a sexual union is not involved, these eggs differ from one another genetically because of recombination between pairs of homologous chromosomes and the independent assortment of chromosomes during the first meiotic division. When the germ cells experience complete meiosis, the gamete-forming cells of the female are haploid and then fuse to form a diploid embryo. We saw an example of this in the Cape honeybee.

COSTS OF SEXUAL REPRODUCTION

Sexual and asexual reproduction are both viable strategies, but sexual reproduction comes with a number of costs to the organism. For example, sexual organs require considerable energy and use resources that could be devoted to other purposes. In addition, mating itself can be a substantial task. Many plants must produce floral displays to attract pollinators and most animals conduct elaborate courtship rituals to attract mates. These activities require time and resources. They can also elevate the risk of herbivory, predation, and parasitism.

For organisms in which an individual is either male or female, sexual reproduction has an additional cost of reduced fitness. To understand this cost, we need to remember that every parent's goal is to maximize their fitness by contributing as many copies of its genes as possible to the next generation. In the case of asexual reproduction, as shown in **Figure 9.3a**, a parent contributes two sets of chromosomes to each of its offspring. In the case of sexual reproduction, each parent contributes only one set of chromosomes to each offspring because the gametes produced by meiosis are haploid, as shown in Figure 9.3b. Females using either mode of reproduction produce the same number of offspring, but the female parent using sexual reproduction leaves behind half as many copies of its genes as a female using asexual reproduction. This 50 percent reduction due to sexual versus asexual reproduction is known as the **cost of meiosis**.

The cost of meiosis is counterbalanced in hermaphrodites, individuals that both possess both male and female function, a situation found in most plants and in many invertebrates. Such an individual can contribute one set of its genes to offspring produced through female function and another set to offspring produced through male function. As shown in Figure 9.3c, this allows a hermaphrodite to contribute twice as many copies of its genes to its offspring than is possible for an individual that can be only male or female.

Cost of meiosis The 50 percent reduction in the number of a parent's genes passed on to the next generation via sexual reproduction versus asexual reproduction.

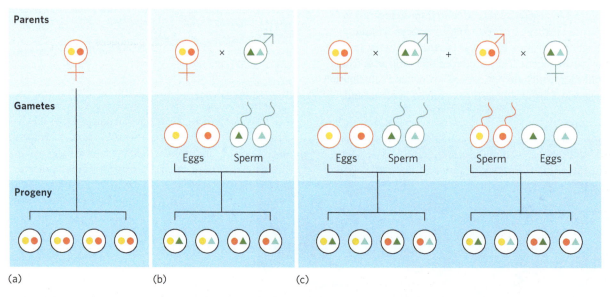

(a) (b) (c)

Figure 9.3 The cost of meiosis. If two hypothetical female organisms can only produce four eggs, **(a)** the female that uses asexual reproduction contributes eight copies of her genes. **(b)** In contrast, the female that uses sexual reproduction only contributes four copies of her genes. **(c)** A hermaphrodite can pass on eight copies of its genes by producing four eggs that are fertilized by another individual's sperm and four copies of its genes when it uses its sperm to fertilize another individual's eggs.

The cost of meiosis can also be offset when an individual is either male or female and the male helps the female take care of the offspring. For example, if a male's parental assistance allows the female to raise twice as many offspring than she could care for alone, then the cost of meiosis is eliminated.

BENEFITS OF SEXUAL REPRODUCTION

If sexual reproduction is so costly, then it must persist because it provides substantial benefits. These benefits include purging harmful mutations and creating genetic variation that helps offspring deal with future environmental variation, including the existence of rapidly evolving parasites and pathogens.

Purging Mutations

Mutations occur in all organisms, and some are harmful. In asexually reproducing organisms, mutations are transmitted from one generation to the next, and so accumulate over generations, especially if the asexual parents produce clonal offspring. In contrast, during the random assortment of genes during meiosis in sexually reproducing organisms, deleterious mutations may not be transmitted to gametes. Further, of all the gametes that are produced, those that form zygotes may not contain the mutation. Alternatively, if both parents carry a copy of a gene with a harmful mutation, and some of the gametes produced do contain the mutation, then a union of two such gametes will be homozygous for the harmful mutation. An offspring that is homozygous for the harmful mutation is likely not viable and will fail to breed and pass the mutation to the next generation.

Because species that reproduce only asexually do not have any means of purging mutations, deleterious mutations slowly accumulate over many generations. In time we would expect individuals of such a species to experience poor growth, survival, and reproduction, which would lead to eventual extinction. If this hypothesis is correct, then species that use asexual reproduction would probably have gone extinct by now and species that use asexual reproduction would have adopted this mode of reproduction only recently.

To test the hypothesis that asexual reproducing species do not persist in nature as long as sexual reproducing species, we can look at the patterns of asexual reproduction within a phylogeny. If the hypothesis is correct, we should observe that asexual reproduction has evolved relatively recently. For example, most vertebrate species that reproduce asexually belong to genera that have a sexual ancestor and contain mostly sexual species, with asexual species that are relatively recently evolved. We observe this pattern with salamanders in the genus *Ambystoma,* fishes in the genus *Poeciliopsis,* and lizards in the genus *Cnemidophorus.* These observations suggest that purely asexual species typically do not have long evolutionary histories. If they did, we would expect to see large groups of related species—such as most species within a genus—all using asexual reproduction. Such a pattern would suggest that their common ancestor used asexual reproduction. However, the long-term evolutionary persistence of asexual populations appears to be low, which supports the explanation that the accumulation of mutations and lack of genetic variation cause species that reproduce asexually to go extinct.

However, not all asexual species fit this pattern. For example in the bdelloid rotifers, an ancient group of more than 300 species of freshwater and soil organisms, all species are asexual and entirely female. Similarly, some groups of protists have existed for hundreds of millions of years and appear to lack sexual reproduction. One way that such species could avoid extinction is by producing offspring more rapidly than new deleterious mutations arise. In this case, some individuals would always retain the nonmutated parental genotype and produce the next generation, a process known as clonal selection. However, groups such as these continue to challenge our efforts to understand the full range of costs and benefits that favor the evolution of sexual or asexual reproduction.

Genetic Variation and Future Environmental Variation

A second benefit of sexual reproduction is the production of offspring with greater genetic variation. If the environment were homogeneous across time and space, parents that are well adapted to the environment could use asexual reproduction to produce clonal offspring that are also well adapted. However, as we have discussed in earlier chapters, environmental conditions typically change across time and space. As a result, offspring are likely to encounter different environmental conditions than their parents did. Because environmental conditions vary, offspring with genetic variation have an increased probability of possessing gene combinations that will help them adapt to different environmental conditions. However, most theoretical models conclude that a greater ability to adapt to temporal and spatial variation in the physical environment does not produce a large enough advantage to offset the cost of meiosis. However, temporal and spatial variation in the biotic environment—particularly variation in the occurrence of different pathogens—could provide a large advantage to the genetic variation created by sexual reproduction.

Genetic Variation and Evolving Parasites and Pathogens

To understand why sexual reproduction provides an evolutionary benefit when species experience variation in both types and abundance of pathogens, we first must realize that pathogens have much shorter generation times and much larger population sizes than most of the host species they infect. Because pathogens have the potential to evolve at a much faster rate than their hosts, they can evolve ways to counteract host defenses. Without rapid host evolution, pathogens could drive their hosts to low numbers or even to extinction. For example, in 1998 researchers described a newly discovered species of pathogen that was causing widespread amphibian deaths in Central America. The pathogen is a type of chytrid fungus (*Batrachochytrium dendrobatidis*) that can infect a wide variety of amphibian species. By 2012, the fungus had been detected on every continent inhabited by amphibians. In some parts of the world, including Central

America, it appears that this deadly pathogen may have been recently introduced and that many species of frogs in the region have no adaptations to combat it. As a result, it appears that the pathogen has caused dozens of species to go extinct.

The harmful effects of pathogens place a premium on hosts to evolve new defenses rapidly. As we have seen, sexual reproduction produces offspring with a greater range of genetic combinations, and some of these combinations might be better suited to combat the pathogen. In short, there is an evolutionary race between hosts, which are trying to evolve adaptations rapidly enough to combat the pathogen, and pathogens, which are trying to evolve adaptations rapidly enough to counter host defenses. The hypothesis that sexual selection allows hosts to evolve at a rate sufficient to counter the rapid evolution of parasites is called the **Red Queen hypothesis**, after the famous passage in Lewis Carroll's *Through the Looking-Glass, and What Alice Found There*, in which the Red Queen tells Alice, "Now, here, you see, it takes all the running you can do, to keep in the same place."

Testing the Red Queen Hypothesis

One of the most compelling tests of the Red Queen hypothesis focuses on a species of freshwater snail (*Potamopyrgus antipodarum*) that is a common inhabitant of lakes and streams in New Zealand. The snails can become infected by parasites, including trematode worms in the genus *Microphallus*. The life cycle of the pathogen is shown in **Figure 9.4**. The worm's life cycle begins when the snails ingest worm eggs. The eggs hatch into larvae that form cysts in the snail gonads, causing the snails to become sterile. Ducks then eat the infected snails and the pathogens mature sexually inside the intestines of the ducks, where they produce eggs asexually. These eggs exit the duck when it defecates in the water, thereby completing the cycle. Not surprisingly, *Microphallus* is most abundant in shallow waters of lakes where ducks feed.

The snail's mode of reproduction depends on the depth of water in which it lives. In the shallower regions of the lake, where the parasitic worm is more common, a higher proportion of the snails use sexual reproduction. Such populations contain about 13 percent males—enough to maintain some genetic diversity through sexual reproduction. In deeper regions, where the parasite is rare, a higher proportion of the snails use asexual reproduction. Although asexual populations reproduce faster than sexual populations, asexual clones cannot persist in the face of high rates of parasitism. As a result, the asexual snails do not survive well in the shallower regions of the lake, where they are more likely to encounter the parasitic worm.

If the parasitic worms evolve to specialize on the snails with which they coexist, then parasites living in the shallow water should be good at infecting snail populations that live in shallow water. Similarly, other parasites that live in deep water should be good at infecting snail populations in that region. Researchers tested this hypothesis with parasites and snails from several different lakes in New Zealand. As shown in **Figure 9.5**, shallow-water snails from several different lakes were infected most readily by shallow-water parasites, and

---

Red Queen hypothesis The hypothesis that sexual selection allows hosts to evolve at a rate that can counter the rapid evolution of parasites.

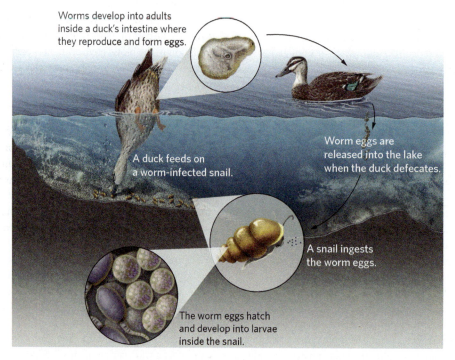

Worms develop into adults inside a duck's intestine where they reproduce and form **eggs**.

A duck feeds on a worm-infected snail.

Worm eggs are released into the lake when the duck defecates.

A snail ingests the worm eggs.

The worm eggs hatch and develop into larvae inside the snail.

Figure 9.4 A pathogen's life cycle through snails and ducks. Eggs of the pathogenic worm *Microphallus* are inadvertently consumed by snails. The eggs hatch into larvae and form cysts in the snails, causing the snails to become sterile. When the snails are consumed by ducks in the shallow waters, the worm turns into an adult and reproduces in the ducks' intestines. When the ducks defecate, the eggs are deposited back into the water.

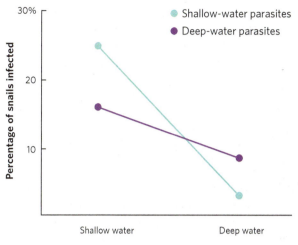

Figure 9.5 Snail infection by a pathogen. Shallow water holds more parasites than deep water because ducks defecate and release pathogen eggs more frequently in shallow water. The large number of pathogens allows them to evolve rapidly in response to any changes in the snails. Snails in shallow water are challenged by a need to evolve defenses rapidly against the pathogens that respond quickly. As a result, a much higher percentage of shallow-water snails than deep-water snails become infected. The pathogens also show an increased ability to infect the populations of snails with which they coexist. Pathogens from shallow water are better able to infect shallow-water snails than deep-water snails. Similarly, pathogens from deep water are better able to infect deep-water snails than shallow-water snails. Data from C. M. Lively and J. Jokela, Clinal variation for local adaptation in a host-parasite interaction, *Proceedings of the Royal Society of London B* 263 (1996): 891–897.

deep-water snails were infected most readily by deep-water parasites. Averaged across all parasite sources, infection rates were relatively low in deep-water snails because few parasites live in this region and, therefore, have had less opportunity to evolve an ability to infect the deep-water snails. However, because deep-water habitats contain fewer parasites, asexual lineages of snails have a reproductive advantage over sexual lineages because of their more rapid reproduction.

Recent studies have continued to support the Red Queen hypothesis. In the roundworm (*Caenorhabditis elegans*), for example, researchers raised individuals in the laboratory that were genetically destined to reproduce either sexually or asexually and then exposed populations containing two different types of worms to a bacterial parasite. When researchers allowed the bacteria to evolve to infect the worms, the parasite quickly drove the asexual worms to extinction. In contrast, the sexual worms continually evolved resistance against the parasite and persisted. When the researchers prevented the bacteria from evolving, the asexual worms came to dominate the population.

CONCEPT CHECK

1. What are the benefits of asexual reproduction?
2. What are the costs of sexual reproduction?
3. How does genetic variation help a population survive pathogens?

9.2 Organisms can evolve as separate sexes or as hermaphrodites

Species in nature have evolved an incredible array of sexual strategies for male and female function, as you can see in **Figure 9.6**. Most vertebrates have separate sexes, whereas most plants are hermaphrodites. Hermaphroditic plants, such as the wildflower known as Saint-John's-wort (*Hypericum perforatum*), possess male and female functions in the same flower. Flowers that have both male and female parts within each flower are known as **perfect flowers**. When the two functions are produced at the same time, we call it a **simultaneous hermaphrodite**. Examples of simultaneous hermaphrodites include many species of mollusks, worms, and plants. When an individual possesses one sexual function and then switches to the other, we call it a **sequential hermaphrodite**. Some species of plants are sequential hermaphrodites, as are some mollusks, echinoderms, and fishes.

Some plants produce separate male and female flowers. Plants that have separate male and female flowers on the same individual plant are known as **monoecious**. For example, every hazelnut tree (*Corylus americana*) has separate male flowers and female flowers. When an individual plant contains only male flowers or only female flowers, the species is known as **dioecious**. For example, the white campion (*Silene latifolia*) is a wildflower that includes some individuals that produce only male flowers and other individuals that produce only female flowers. Although perfect-flowered hermaphrodites account for more than two-thirds of flowering plant species, nearly all-imaginable sexual patterns are known. Populations of some plant species can be composed of a complex mixture of hermaphrodites, males, females, and monoecious individuals. In other species, individual plants produce both perfect flowers and flowers that are only male or only female.

COMPARING STRATEGIES

We would expect natural selection to favor the reproductive strategy with the highest fitness. For instance, in organisms such as flowering plants, a plant could evolve to make male flowers, female flowers, or hermaphrodite flowers. To determine whether evolution should favor separate sexes or hermaphrodites, we need to compare the amount of fitness an individual would gain by investing in only male or female reproduction versus the amount of fitness it would

---

Perfect flowers Flowers that contain both male and female parts.

Simultaneous hermaphrodites Individuals that possess male and female reproductive functions at the same time.

Sequential hermaphrodites Individuals that possess male or female reproductive function and then switch to possess the other function.

Monoecious Plants that have separate male and female flowers on the same individual.

Dioecious Plants that contain either only male flowers or only female flowers on a single individual.

Cluster of female flowers

Monoecious plants, such as this hazelnut shrub, have separate male and female flowers on a single individual.

Cluster of male flowers

Female

Male

Hermaphroditic plants, such as this St.-John's-wort wildflower, have flowers with both male and female parts.

Male

Female

Dioecious plants, such as this white campion wildflower, have either male or female flowers on different individuals.

Figure 9.6 Plant mating strategies. Hermaphroditic plants, such as the St.-John's-wort wildflower, possess perfect flowers that contain male and female structures within a single flower. Monoecious plants, such as hazelnut shrubs, have separate male and female flowers, but a single plant possesses both types of flowers. Dioecious plants, such as the white campion wildflower, include some individuals that only contain male flowers and other individuals that only contain female flowers.

gain by investing in both male and female reproduction. As depicted in **Figure 9.7**, if a male individual can invest in female function and gain a lot of female fitness while only giving up a small amount of male fitness, then selection will favor the evolution of hermaphrodites. A similar scenario can be considered for a female individual that adds male function. This occurs because the total fitness as a hermaphrodite through male plus female function exceeds the fitness of being only a male or only a female. In the case of flowers, the basic flower structure and the floral display necessary to attract pollinators—for those species that rely on them—are already in place in male flowers and female flowers. This should make the fitness costs of adding a sexual function relatively small while providing large fitness benefits. As we noted earlier, about two-thirds of all flowering plant species are hermaphrodites.

In some cases, the cost of investing in a second sexual function is too high to be offset by the benefits of being a hermaphrodite. **Figure 9.8** illustrates how a reduction in female function can allow an investment in male function, but the total fitness is less than males and females would experience if they were to retain a single sexual function. For

instance, sexual function in complex animals requires gonads, ducts, and other structures for transmitting gametes. Moreover, in many animals, being male requires large expenditures of time and energy for attracting mates and fighting with other males, while being female requires specializations for egg production or time and energy needed to care for offspring. Because these costs can be quite high, we would predict that hermaphroditism should occur only rarely among animal species that actively seek mates and engage in brood care. In contrast, we would predict that hermaphroditism should occur commonly among sedentary aquatic animals that mate by simply shedding their gametes into the water. Researchers have found evidence that supports both of these predictions.

SELFING VERSUS OUTCROSSING OF HERMAPHRODITES

One of the challenges for individuals that have both male and female function is the problem of *self-fertilization*. Self-fertilization, also known as selfing, occurs when an individual uses its male gametes to fertilize its own female gametes. As we discussed in **Chapter 4**, self-fertilization poses a fitness cost due to inbreeding depression. Therefore, selection should favor individuals that do not

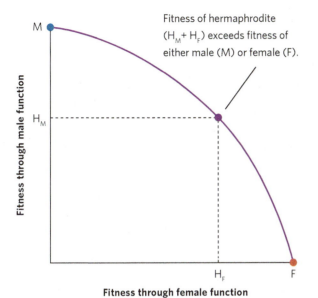

Figure 9.7 When hermaphrodites have a fitness advantage over separate sexes. When male or female individuals can add the other sexual function with little decline in original sexual function, they can achieve a higher total fitness by being a hermaphrodite than by being only male or female. In this example, the fitness of a hermaphrodite equals the fitness derived from male function (H_M) plus the fitness derived from female function (H_F).

use selfing when they have an opportunity to breed with other individuals, a strategy known as *outcrossing*. Some species avoid the problems of selfing by being sequential hermaphrodites. For example, the blue-headed wrasse (*Thalassoma bifasciatum*), a fish species common in coral reefs, can be functionally female when it is a small adult but then becomes functionally male later in life when it has grown larger. Similarly, if a plant releases the pollen in its anthers before making its stigma receptive to pollen, the flower will not be able to pollinate itself. Other species have self-incompatibility genes. Individuals with the same self-incompatibility genotype—including an individual mating with itself—cannot produce offspring.

MIXED MATING STRATEGIES

As we discussed in **Chapter 4**, some hermaphrodites use a mixture of mating strategies. When a mate can be found, the individual prefers to breed by outcrossing to avoid the costs of inbreeding. When a willing mate cannot be found, the individual self-fertilizes. Self-fertilizing will not provide as many viable offspring as outcrossing, but it might be better than not reproducing at all.

In some cases, using a mixture of outcrossing and selfing is a response to a lack of resources. Attracting mates can be energetically expensive, as is the case with plants that must produce nectar to attract pollinators. For example, in orange jewelweed, the production of outcrossing flowers is more energetically expensive than the production of selfing flowers, which do not need to invest in nectar to attract pollinators. Plants that experience herbivory of their leaves have less energy to make outcrossing flowers. As a result, these plants produce a higher proportion of selfing flowers, as shown by the data for orange jewelweed in **Figure 9.9**.

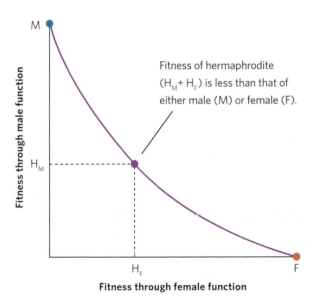

Figure 9.8 When separate sexes have a fitness advantage over hermaphrodites. When male or female individuals add the other sexual function and experience a large decline in original sexual function, they can achieve a higher total fitness by staying as separate sexes rather than being a hermaphrodite. In this example, the fitness of a hermaphrodite equals the fitness derived from male function (H_M) plus the fitness derived from female function (H_F).

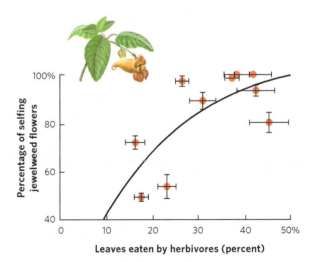

Figure 9.9 Mixed mating strategies in orange jewelweed. In a survey of populations, those populations receiving higher amounts of herbivory also produced higher proportions of selfing flowers and, therefore, a lower proportion of outcrossing flowers. **Error bars are standard errors.** Data from A. A. Steets and T.-L. Ashman, Herbivory alters the expression of a mixed-mating system, *American Journal of Botany* 91 (2004): 1046–1051.

9.3 Sex ratios of offspring are typically balanced, but they can be modified by natural selection

In organisms with separate sexes, the sex ratio of male to female offspring is often one to one. The exceptions offer interesting insights into evolutionary forces that favor particular sex ratios in an individual's offspring. In this section, we will examine the mechanisms that determine whether an offspring will be male or female. We will then examine the underlying reasons for the wide range of sex ratios observed in nature.

MECHANISMS OF SEX DETERMINATION

In previous chapters, we learned that the phenotypes of organisms are determined by both genetic and environmental factors. The sex of offspring is no different, although the influence of genetics and the environment on the sex of offspring differs among species.

Genetic Sex Determination

In mammals, birds, and many other organisms, sex is determined by inheritance of sex-specific chromosomes. In most mammals, females have two X chromosomes, whereas males have one X chromosome and one Y chromosome. Birds have the opposite pattern of genetic sex determination: males have two copies of the Z chromosome, whereas females have one Z chromosome and one W chromosome. In both cases, the sex that has two different chromosomes—male mammals and female birds—produces an approximately equal number of gametes containing each sex chromosome. On average, half the progeny in these populations will be female and half will be male.

In insects, genetic sex determination is accomplished in a variety of ways. In grasshoppers and crickets, for instance, all individuals are diploid, but females have two sex chromosomes, whereas males have only one. In honeybees and other members of their order, including other bees, ants, and wasps, sex is determined by whether or not an egg is fertilized. Fertilized eggs, which receive two sets of chromosomes, become females, whereas unfertilized eggs become males.

Environmental Sex Determination

In some species, sex is determined largely by the environment, in a process known as **environmental sex determination**. In reptiles, including several species of turtles, lizards, and alligators, the sex of an individual is determined by the temperature at which the egg develops. In turtles, embryos incubated at lower temperatures typically produce males, whereas those incubated at higher temperatures produce females. The reverse is generally true in alligators and lizards. This type of environmental sex determination is known as *temperature-dependent sex determination*. Because the genotype has the ability to produce multiple phenotypes, temperature-dependent sex determination is a type of phenotypic plasticity.

For decades, biologists have wondered whether temperature-dependent sex determination is adaptive. For example, if the temperatures that cause eggs to become male produce the most fit males, and the temperatures that cause eggs to become female produce the most fit females, temperature-dependent sex determination would be adaptive. To test this hypothesis, one would need to produce male and female offspring from eggs incubated across a range of different temperatures. However, since temperature is the very factor that determines their sex, we cannot naturally produce males and females at each temperature and compare their performance. This problem was solved in a study of the Jacky dragon (*Amphibolurus muricatus*). In this lizard species from Australia, females are produced when incubated at low and high temperatures, whereas both males and females are produced when incubated at intermediate temperatures. Because of this pattern, researchers could easily produce females at all three temperatures. To produce males at the highest and lowest temperatures, the researchers injected a hormone inhibitor that prevented the embryos from becoming female, thereby overriding the normal response to temperature effects. This manipulation allowed the researchers to produce male and female lizards at all three temperatures. Once the lizards hatched, the animals were put outside in field enclosures for 3 years. At the end of 3 years, the researchers determined the number of offspring produced by adults that had been incubated as eggs at different temperatures. As shown in **Figure 9.10**, males incubated at the intermediate temperature subsequently fathered more offspring than males incubated at the low and high temperatures. Females that incubated at the high temperature subsequently laid more eggs than females incubated at intermediate temperatures, although females incubated at low temperatures laid a similar number of eggs as females incubated at intermediate temperatures. This was one of the first studies in reptiles to demonstrate that temperature-dependent sex determination appeared to be adaptive.

Environmental sex determination A process in which sex is determined largely by the environment.

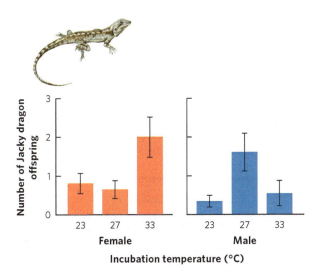

Figure 9.10 Environmental sex determination in the Jacky dragon. Under natural conditions, the eggs of the Jacky dragon become female under the low and high temperatures. Eggs become both male and female under intermediate temperatures. Female fitness is highest for females raised at the high temperature. Male fitness is highest for males raised at the intermediate temperature. Error bars are standard errors. Data from D. A. Warner and R. Shine, The adaptive significance of temperature-dependent sex determination in a reptile, *Nature* 451 (2008): 566–569.

Not all cases of environmental sex determination are driven by temperature. In some species, sex determination is driven by the social environment in which an individual lives. As mentioned in our discussion of sequential hermaphrodites, the blue-headed wrasse is a coral reef fish that is able to change from female to male as it ages (**Figure 9.11**). Blue-headed wrasses typically begin life as females and live in large schools along with one or a few dominant males. If the dominant male leaves or dies, the largest female will then change sex and become the new dominant male. Being large is important for males, because they must defend a territory from other males that attempt to breed with the females in the school.

OFFSPRING SEX RATIO

Now that we understand the mechanisms that help determine the sex of offspring, we can consider the factors that favor particular sex ratios of offspring. Biologists increasingly appreciate that a female can have a large influence on the sex ratios of her offspring. In species where the males have two different types of sex chromosomes, as in mammals, researchers have found that the females of some species can control whether X-chromosome or Y-chromosome sperm are allowed to fertilize their eggs. In species where the females contain two different types of sex chromosomes, as in birds, the females of some species can determine the sex ratio of their offspring by controlling the fraction of their eggs that receive the Z versus the W chromosome during meiosis. In the hymenopteran

Figure 9.11 Blue-headed wrasse. Young fish are typically female and live in schools with a dominant male. If the male leaves or dies, the largest female changes into a new dominant male. Photo by Steve Simonsen/Getty Images.

insects—bees, wasps, and ants—the female determines the sex of her offspring by whether or not she fertilizes the eggs.

A different approach to controlling offspring sex ratio is through selective abortion. In red deer (*Cervus elaphus*), for example, adult females breed in early autumn and give birth the following spring. Sons are often larger at birth and require more of the mother's milk than daughters. As a result, sons require a larger investment from the mother than daughters. Researchers in Spain examined the fetuses of 221 harvested red deer to determine whether the sex ratio of the fetuses was affected by the mother's age—categorized as adult, subadult, or yearling—and whether the sex ratio changed through the gestation period. As shown in **Figure 9.12a**, adult mothers produced a nearly even offspring sex ratio. In contrast, yearling mothers, which are smaller and have less energy, were much more likely to produce the less energetically expensive daughters. Averaged across all months of the deer harvest, yearling mothers carried about 25 percent male fetuses and 75 percent female fetuses. To determine how the yearling mothers accomplish this uneven sex ratio, researchers examined yearling mothers that were harvested at different months throughout the winter, illustrated in Figure 9.12b. They found that the yearling mothers initially had a nearly even sex ratio of offspring. As winter progressed, however, the proportion of male fetuses declined sharply, suggesting that the yearling mothers were capable of selectively aborting the more expensive male fetuses as gestation progressed. This phenomenon is not restricted to red deer; it has also been observed in other mammals, including rats and mice.

Regardless of how the sex ratio is controlled, in most species the sex ratio of male to female offspring is nearly even. Is this a ratio produced by default, or are there adaptive reasons for it? To answer this question, we can compare the conditions that favor a one-to-one sex ratio with the conditions that favor a deviation from that ratio.

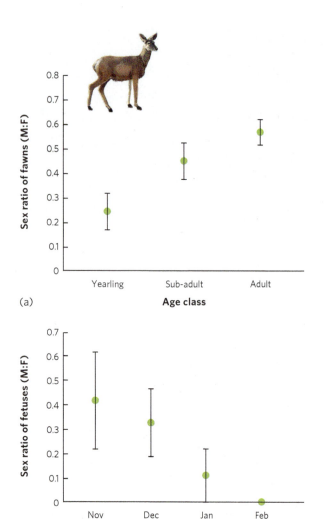

(a)

(b)

Figure 9.12 Offspring sex ratios in red deer. (a) Yearling deer produce a low proportion of male fawns, but this proportion increases in older deer. **(b)** Among yearling deer, the proportion of male fetuses is initially high but then declines as gestation continues throughout the winter as female deer selectively abort male fetuses. Error bars are standard errors. Data from T. Landete-Castillejos et al., Age-related foetal sex ratio bias in Iberian red deer (*Cervus elaphus hispanicus*): Are male calves too expensive for growing mothers?, *Behavioral Ecology and Sociobiology* 56 (2004): 1–8.

Frequency-Dependent Selection

To understand how selection commonly favors a one-to-one sex ratio, we have to consider how natural selection might favor the offspring sex ratio that an individual produces when a population deviates from the one-to-one ratio. Let's imagine, for instance, that a population has more females than males and each male breeds with only one female. In this situation, some females would remain unbred. A parent that produces all sons would have higher fitness than a parent that produces an even number of sons and daughters, because some of the daughters might not find mates. Similarly, if a population has

a surplus of males and one male breeds with only one female, a parent that produces all daughters would have higher fitness than a parent that produces an even number of sons and daughters, because some of the sons might not find mates. As you can see in this example, individuals of the less abundant sex enjoy greater reproductive success because they compete with fewer individuals of the same sex for breeding. Thus, whenever the population has an abundance of one sex, natural selection will favor any parents that produce offspring of the less abundant sex. Over time, as the less abundant sex becomes more common and the common sex becomes less abundant, populations tend to be balanced at an approximate one-to-one ratio. Therefore, the best sex ratio strategy for parents depends on the frequencies of males and females in a population. The evolution of the sex ratio is said to be the product of **frequency-dependent selection**, which occurs when natural selection favors the rarer phenotype in a population.

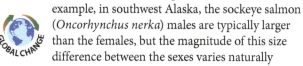

Researchers are beginning to appreciate that human activities can cause unnatural changes in sex ratios. For example, in southwest Alaska, the sockeye salmon (*Oncorhynchus nerka*) males are typically larger than the females, but the magnitude of this size difference between the sexes varies naturally among the populations. Now consider that the regulations for salmon fishing require that anglers only harvest the larger fish. This causes anglers to inadvertently harvest more males than females. In 2012, researchers reported that the sex ratios among the different populations range from 36 to 47 percent males due to the long-term effects of fishing, as illustrated in **Figure 9.13**.

Highly Skewed Sex Ratios

We have seen that natural selection favors a one-to-one sex ratio. In some cases, however, we observe highly skewed sex ratios, such as in Fig wasps (*Pegoscapus assuetus*). The female fig wasp arrives at a flower of a fig tree carrying pollen from another fig tree. She crawls in through a small hole in the flower and pollinates the flower. Once inside, the female lays her eggs in the developing fruit and then dies. Her eggs can be composed of up to 90 percent daughters! Similar to the bees that we discussed at the beginning of this chapter, wasps can easily adjust the sex ratio of their offspring because fertilized eggs become daughters, whereas unfertilized eggs become sons. Once the eggs hatch, the larvae feed on the fig fruit and seeds. The larvae metamorphose into adult wasps and breed with their siblings while still inside the flower. The young males breed with the young females, chew a hole in the side of the flower, and then die. The fertilized young females then escape through the hole and fly off to pollinate a new fig flower.

Frequency-dependent selection When the rarer phenotype in a population is favored by natural selection.

ANALYZING ECOLOGY

Frequency-Dependent Selection

Eco TV

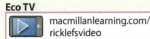 macmillanlearning.com/ricklefsvideo

We can better understand frequency-dependent selection if we work though an example with real numbers. The American black vulture (*Coragyps atratus*) is a large bird that feeds on animal carcasses throughout much of North and South America. The female vulture typically lays two eggs in its nest.

If a population is composed of five females and two males, and we assume that a male can breed with more than one female, how many copies of genes, on average, does each male and female contribute to the next generation?

Total number of eggs = 10

Average female fitness = 10 female gene copies ÷ 5 females = 2 gene copies/female

Average male fitness = 10 male gene copies ÷ 2 males = 5 gene copies/male

If a population is composed of five females and eight males, how many copies of genes, on average,

does each male and female contribute to the next generation?

Total number of eggs = 10

Average female fitness = 10 female gene copies ÷ 5 females = 2 gene copies/female

Average male fitness = 10 male gene copies ÷ 8 males = 1.25 gene copies/male

In this example, you can see that when males are the less abundant sex, males have the higher fitness. When females are the less abundant sex, females have the higher fitness.

YOUR TURN Using the same assumptions above, calculate the number of gene copies per male and per female under the following two scenarios:

1. Four males and five females
2. Six males and five females

Given the four scenarios that we have explored, what sex ratio will be favored by natural selection over the long term?

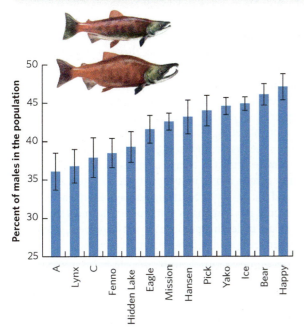

Figure 9.13 Sex ratios in sockeye salmon populations. Among populations of sockeye salmon in southwest Alaska, those with much large males than females are experiencing declines in the ratios of males to females due to fishing regulations that require harvesting the largest fish. Data from N. W. Kendall and T. P. Quinn, Size-selective fishing affects sex ratios and the opportunity for sexual selection in Alaskan sockeye salmon *Oncorhyncus nerka, Oikos* 122 (2012): 411–420.

The high proportion of daughters among fig wasps occurs because male competition for females is among brothers. This phenomenon, known as **local mate competition**, occurs when competition for mates takes place in a limited area and only a few males are required to fertilize multiple females. When only one fertilized mother enters a fig flower, the only males available to fertilize her daughters are her sons. From the mother's perspective, it does not matter which son passes along her genes to her grandoffspring. One son can fertilize multiple daughters, which means there is no fitness benefit to having multiple sons. The mother's fitness is therefore a function of how many eggs can be produced by her daughters, so it is in the mother's best interest to produce many daughters and just enough sons to fertilize those daughters. As you can see in **Figure 9.14**, when mating options are restricted such that the only mates available for daughters are their brothers, mothers that produce a higher proportion of daughters than sons will have more grandoffspring and therefore greater evolutionary fitness.

Sometimes, however, two or more females lay their eggs in the same flower. In this situation, a mother's male offspring can mate either with their sisters or with the daughters of the

Local mate competition When competition for mates occurs in a very limited area and only a few males are required to fertilize all the females.

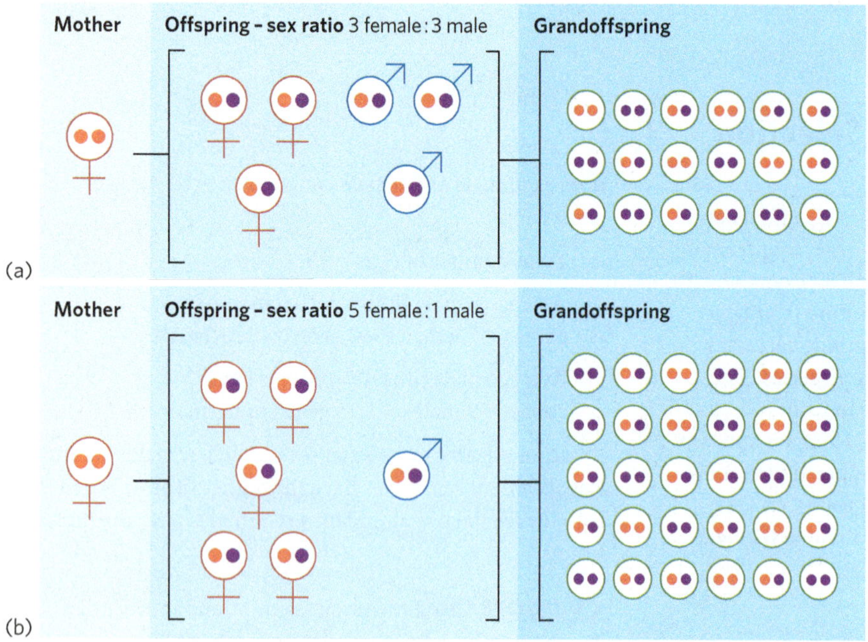

Figure 9.14 The effect of different sex ratios on a mother's fitness. When one son can breed with multiple daughters in an isolated population, the sex ratio of the mother's offspring will affect the number of grandoffspring that can be produced. In this example, we assume each female can produce six offspring. **(a)** If a mother produces three daughters and three sons, 18 grandoffspring can be produced. **(b)** If a mother produces five daughters and one son, 30 grandoffspring can be produced.

other mothers. When this occurs, a mother will obtain greater fitness if she produces extra sons so she has enough sons to fertilize all the females in the flower. As expected, researchers observe that when a fig flower contains more than one mother laying eggs, the mothers lay a higher proportion of male eggs.

CONCEPT CHECK

1. What is the difference between genetic and environmental sex determination?
2. When a population is composed of two sexes, why does the rarer sex have a fitness advantage?
3. How does local mate competition favor the production of female-biased sex ratios in offspring?

9.4 Mating systems describe the pattern of mating between males and females

Although many species of algae and fungi have gametes that are of similar size, animals and plants produce sperm that are a fraction of the size of eggs. This smaller sperm takes much less energy to produce than an egg. Because of this difference, a female's reproductive success depends both on the number of eggs she can make and on the quality of the mates that she finds. Because most males can produce millions of sperm, a male's reproductive success generally depends on how many females he can fertilize. In this section, we will discuss the **mating system** of species, which describes the number of mates each individual has and the permanence of the relationship among mates. Like sex ratios, the mating system of a population is subject to natural selection. Consequently, mating systems are often a product of the ecological conditions

under which species live. **Figure 9.15** illustrates the four mating systems: promiscuity, polyandry, polygyny, and monogamy.

PROMISCUITY

Promiscuity is a mating system in which individuals mate with multiple partners and do not create a lasting social bond. Among animal taxa as a whole, promiscuity is by far the most common mating system. Promiscuity is universal among outcrossing plants because they send pollen out to fertilize the eggs of multiple individuals and receive pollen from multiple individuals.

When animals release eggs and sperm directly into the water or when pollen is released into the wind, much of the variation in mating success is random. Whether a particular sperm is the first to find an egg is largely a matter of chance. When males attract or compete for mates, however, reproductive success can be influenced by factors such as body size and the quality of courtship displays. Even when fertilization is random, males that produce the most sperm or pollen and males that produce the most competitive sperm or pollen are bound to father the most offspring.

POLYGAMY

Polygamy is a mating system in which a single individual of one sex forms long-term social bonds with more than one individual of the opposite sex. Most often, a male mates

Mating system The number of mates each individual has and the permanence of the relationship with those mates.

Promiscuity A mating system in which males mate with multiple females and females mate with multiple males and do not create a lasting social bond.

Polygamy A mating system in which a single individual of one sex forms long-term social bonds with more than one individual of the opposite sex.

Figure 9.15 Mating systems. Promiscuity occurs in species of outcrossing plants such as the prairie sunflower (*Helianthus petiolaris*). Polygyny exists when a male mates with several females, as in elk (*Cervus canadensis*). Polyandry occurs when a female mates with multiple males, as is the case for the western toad (*Anaxyrus boreas*). Monogamy for life was thought to be the rule in 90 percent of all birds, as in the sandhill crane (*Grus canadensis*). However, recent genetic analyses have confirmed that most bird species actually participate in extra-pair copulations.

with more than one female, which is known as **polygyny**. In some species, polygyny evolves when males compete for females and the females all prefer only the few best males. In this case, the largest and healthiest males may mate with the vast majority of the females. Polygyny also can evolve when a male is able to defend a group of females from other males or when a male can defend a patchy resource that is attractive to multiple females. Consider the guanaco (shown in the opening to **Chapter 3**), a relative of the camel that lives in the Patagonian region of South America. The land in Patagonia is generally dry, with patches of wet habitat that grow nutritious plants that guanacos prefer to eat. When a male guanaco defends a wet patch from other males, he is able to mate with many of the females who come to the wet patch to eat the plants.

The opposite of polygyny is **polyandry**, a mating system in which a single female breeds with multiple males. The account of the queen honeybee at the beginning of this chapter is an example of polyandry because she mates with multiple drones. Polyandry commonly occurs when the female is in search of genetically superior sperm or has received material benefits from each male with whom she mates. For instance, in some species of insects, including some butterflies, the female receives a nutritional package of food—known as a *spermatophore*—from a male. The female uses the protein in the spermatophore to make her eggs. In polyandrous species, the more spermatophores that a female can collect from multiple males, the more eggs she can produce.

MONOGAMY

Monogamy is a mating system in which a social bond between one male and one female persists through the period that is required for them to rear their offspring. In some cases, the bond may endure until one of the pair dies. Monogamy is favored when males can make important contributions to raising the offspring. Monogamy occurs in about 90 percent of bird species because male birds can offer much of the same care to offspring as females, such as incubating eggs, gathering food for the chicks, and protecting the chicks from predators. A male bird's help enables his chicks to grow and survive much better than they would without his help. Therefore, his parenting improves

Polygyny A mating system in which a male mates with more than one female.

Polyandry A mating system in which a female mates with more than one male.

Monogamy A mating system in which a social bond between one male and one female persists through the period that is required for them to rear their offspring.

his fitness. In mammals, however, fewer than 10 percent of species have a social bond with only one other individual. Because male mammals cannot provide the same offspring care as females, particularly because females lactate, the growth and survival of mammal offspring are less reliant on a male's presence.

Although most female birds form a single social bond with a male partner, DNA analysis of offspring in a nest has revealed that the offspring often have different fathers. This means that the mother copulated with other males. While these females had a social bond with only one male, they were actually breeding with other males, a behavior known as **extra-pair copulation**. In some monogamous species, a third or more of the clutches contain offspring sired by another male—usually, a male in a neighboring territory. Since neighboring males also have a social bond with their own partner, both males and females use extra-pair copulations. Thanks to DNA analysis, we now know that 90 percent of bird species thought to be monogamous engage in extra-pair copulations.

Extra-pair copulations surely increase the fitness of the neighboring males, but how does this behavior increase the fitness of the female? One way would be if her reproductive success is improved by creating greater genetic variation among her offspring so at least one might be better suited to future environmental conditions. She could also benefit if the neighboring males have better genotypes than her mate. In a bird known as the bluethroat (*Luscinia svecica*), for example, extra-pair copulations are common. Researchers in Norway examined the immune response of chicks whose father was the social partner of the female versus the extra-pair partner of the female. To do this, the researchers inject a small amount of foreign material—which is extracted from kidney beans—into a bird's wing and then they measure the amount of swelling that occurs at the site of the injection; the greater the swelling, the stronger the bird's immune response. As you can see in **Figure 9.16**, the amount of swelling in the wing was greater in offspring produced by extra-pair copulations, which demonstrates that they had a stronger immune response. This suggests that females seek extra-pair copulations as a way to obtain superior genotypes from another male and produce offspring that have superior immune systems.

The male participating in a monogamous social bond does not benefit if his mate breeds with other males. The threat that she will participate in extra-pair copulations has selected for **mate guarding**, a behavior in which one partner prevents the other partner from participating in extra-pair copulations. A variety of mate guarding behaviors have evolved. In some species, an individual simply stays near its mate and keeps other potential suitors away, while in other cases, the mate makes future breeding physically impossible. For instance, researchers discovered that the golden orb-weaving spider (*Argiope aurantia*) inserts its sperm-transferring appendages into the two openings of

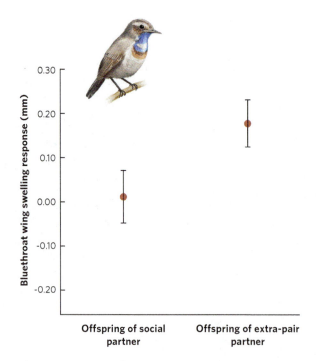

Figure 9.16 Female benefits of extra-pair copulations. In bluethroats, the strength of the immune system can be measured in terms of a wing-swelling response when the wing is injected with a chemical extracted from kidney beans. Larger wing-swelling responses indicate stronger immune systems. Error bars are standard deviations. Data from A. Johnsen et al., Female bluethroats enhance offspring immunocompetence through extra-pair copulations, *Nature* 406 (2000): 296–299.

the female's reproductive system. Within a few minutes of transferring his sperm, the male's heart stops and he dies with his appendages still stuck inside the female, making it impossible for the female to mate again. Such adaptations are effective at reducing the chance that an individual's partner will breed with another mate.

The various mating systems are the product of natural selection and are shaped by the ecological conditions under which each species lives. Because the mating systems often involve attracting a mate or defending against other members of the same sex, natural selection has also caused the evolution of many sex-specific traits, as we will see in the next section.

CONCEPT CHECK

1. Under what conditions would natural selection favor the evolution of polygyny?
2. Under what conditions would natural selection favor the evolution of polyandry?
3. What behavior can a mate use to enforce monogamy?

Extra-pair copulations When an individual that has a social bond with a mate also breeds with other individuals.

Mate guarding A behavior in which one partner prevents the other partner from participating in extra-pair copulations.

9.5 Sexual selection favors traits that facilitate reproduction

We have seen that a male's reproductive success is typically determined by the quantity of females he can breed, whereas a female's reproductive success is typically determined by the quality of males that can fertilize her limited number of eggs. This means that females should generally be the choosier sex. But what exactly should she be choosing? In broad terms, females should select those males that improve her fitness by the greatest amount. She might select the male with the best genotype or the male with the most resources for her and her offspring.

Because females are choosy, males should compete strongly with other males for the opportunity to breed. This intense competition among males for mates has resulted in the evolution of male traits that are either used to attract females or used in contests and combat among males. Natural selection for sex-specific traits related to reproduction is referred to as **sexual selection**. In this section, we will explore how males and females have evolved different traits as a result of sexual selection, what traits females prefer in their mates, and how the fitness interests of males and females can cause conflict between the sexes.

SEXUAL DIMORPHISM

One result of sexual selection is **sexual dimorphism**, which is a difference in the phenotype between males and females of the same species. Sexual dimorphism is seen in honeybees and sockeye salmon, which we discussed earlier in this chapter. Sexual dimorphism includes differences in body size, ornaments, color, and courtship behavior. Traits related to fertilization—such as gonads—are referred to as **primary sexual characteristics**, while traits related to differences in body size, ornaments, color, and courtship are known as **secondary sexual characteristics**.

Sexual dimorphism can evolve due to differences in life history between the sexes, contests between males, or mate choice by females. Body size differences are common between the sexes of many animals because there has been selection for an increased number of gametes produced or for an increase in parental care by one sex. In fish and spiders, for instance, egg production is directly related to body size. This selection for greater gamete production in females without concurrent selection for greater sperm production in males could be the underlying cause of the larger size of females than males in many species of spiders and fish (**Figure 9.17**).

Sexual dimorphism can also occur when males compete for mates. In this case, selection will favor the evolution of weapons for combat. Such weapons include the antlers of male elk, the horns of mountain sheep, and the leg spurs of chickens and turkeys. When fighting ability is also improved by having a large body, contests between males can also favor the evolution of larger male bodies (**Figure 9.18**). Males that win such contests are more likely to gain access to females. Sexual dimorphism also commonly arises when one sex, usually the female, is particularly choosy when selecting a mating partner. In these cases, female selection of males with particular traits can cause the sexual selection for those traits, such as male deer growing antlers.

 Human activities can also affect sexual dimorphism. In species that are hunted, for example, hunters either prefer or are required to harvest only the largest males. This can be

Sexual selection Natural selection for sex-specific traits related to reproduction.

Sexual dimorphism The difference in the phenotype between males and females of the same species.

Primary sexual characteristics Traits related to fertilization.

Secondary sexual characteristics Traits related to differences between the sexes in terms of body size, ornaments, color, and courtship.

Figure 9.17 Sexual size differences. In the golden silk spider (*Nephila clavipes*), females are much larger than males. Photo by Millard H. Sharp/Science Source.

Figure 9.18 Male weapons. In some species, such as these elk in Alberta, Canada, males fight with each other for the right to breed females. Repeated contests cause selection for large weapons, including large antlers. Photo by Robert Harding Picture Library/Newscom.

measured as the number of points on the antlers of deer or elk or the length of the curled horns in wild species of sheep. In doing so, we have created a situation of artificial selection that is opposite to what would occur naturally in sexual selection. In 2016, researchers reported the change in horn size in Stone's sheep (*Ovis salli stonei*) based on differences in harvesting that occurred over a 37-year period across locations in British Columbia. Given that horn size is heritable in Stone's sheep, greater harvesting represents stronger selection, which should then cause a strong evolutionary response, as we discussed in **Chapter 7**. In the region of high hunting pressure, there were 2.7 times more hunters per unit area and 2.6 times more sheep harvested than the region of low-hunting pressure. Over the 37 years, the rate of horn growth declined by 12 percent in the region of high-hunting pressure, but showed no change in the region of low-hunting pressure. In short, increased hunting caused artificial selection for smaller horn size.

THE EVOLUTION OF FEMALE CHOICE

A female's preference for particular male traits should be tied to those features that improve her fitness. In terms of broad categories, we can consider two types of female preferences: *material benefits* and *nonmaterial benefits*.

Material benefits are those physical items that a male can provide a female, including a site for raising offspring, a high-quality territory, or abundant food. In these cases, the benefit to the female is straightforward—a site for raising offspring and resources for producing eggs and feeding the offspring should improve a female's fitness.

It has been more of a challenge for scientists to understand female choices when the female does not receive any material benefits from males. If the female choice is an adaptation, then there must be some benefit. One of the first demonstrations of female choice that provided no material benefit came from a study of tail length in male long-tailed widowbirds (*Euplectes progne*), a small, polygynous species that inhabits the open grasslands of central Africa. The females are mottled brown, short-tailed, and drab. In contrast, during the breeding season, males are jet black with a red shoulder patch, and they sport a half-meter-long tail that is conspicuously displayed in courtship flights (**Figure 9.19a**). The most successful males may attract up to a half dozen females to nest in their territories, but the males provide no care for their offspring. To determine what the females are choosing, researchers cut the tail feathers of some males to shorten them and glued the clipped feather ends onto the feathers of the tails of other males to lengthen them. As you can see in Figure 9.19b, males with experimentally elongated tails attracted significantly more females than those with shortened or unaltered tails.

Why would females choose a male based on his traits? There are two possible hypotheses. According to the **good genes hypothesis**, an individual chooses a mate that possesses a superior genotype. In the gray treefrog, for example, females

(a)

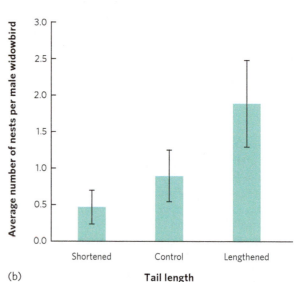

(b)

Figure 9.19 Long-tailed widowbird. (a) During the breeding season, males have exceptionally long tails. **(b)** When researchers made male tails shorter, the same length, or longer than those of typical males, females preferred males with longer tails. Error bars are standard errors. Photo by FLPA/Dickie Duckett/age fotostock. Data from M. Andersson, Female choice selects for extreme tail length in a widowbird, *Nature* 299 (1982): 818–820. Photo by FLPA/Dickie Duckett/AGE Fotostock.

prefer males that can produce the longest mating calls. Long calls can only be produced by the largest and healthiest male frogs. If male size and health have a large genetic component, choosing this trait might benefit the female's offspring. Indeed, when researchers forced females to breed with both long- and short-calling males, the offspring of long-calling males grew faster than the offspring of short-calling males. Subsequent research, illustrated in **Figure 9.20**, found that the offspring of

Good genes hypothesis The hypothesis that an individual chooses a mate that possesses a superior genotype.

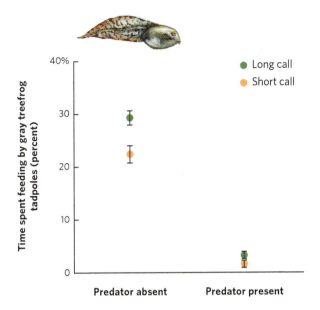

Figure 9.20 Good genes. In the gray treefrog, females who chose long-calling males produced offspring that spent more of their time eating than did the offspring of short-calling males, regardless of whether predators were present or absent. Error bars are standard errors. Data from G. V. Doty and A. M. Welch, Advertisement call duration indicates good genes for offspring feeding rate in gray treefrogs, *Behavioral Ecology and Sociobiology* 49 (2001): 150–156.

long-calling males grew faster because they spend more time feeding compared to the offspring of short-calling males.

According to a second hypothesis, known as the **good health hypothesis**, individuals choose the healthiest mates. Good health could be the outcome of either superior genetics or a superior upbringing with abundant resources. As a result, the good genes hypothesis and the good health hypothesis are not mutually exclusive. Females might prefer healthy males because such males may be both genetically superior and pose a lower risk of passing on a number of different parasites and diseases.

RUNAWAY SEXUAL SELECTION

Once female preference for a male trait has evolved, the trait may continue to evolve over time. For example, if females prefer longer tails in their mates, and there is genetic variation to select from, longer tails will continue to evolve in males. When selection for preference of a sexual trait and selection for that trait continue to reinforce each other, the result can be **runaway sexual selection**. Runaway sexual selection is thought to have favored the evolution of such extreme traits as the half-meter-long tails of the male widowbird, the giant tail fans of the peacock, and other large male ornaments such as horns, tusks, and antlers. Runaway selection continues until males either run out of genetic variation for the trait or until the fitness costs of possessing extreme traits begin to outweigh the reproductive benefits.

THE HANDICAP PRINCIPLE

If sexually selected traits indicate intrinsic attributes of male quality—at least initially, before runaway sexual selection occurs—we are then faced with a paradox. Presumably, extreme traits burden males by requiring energy and resources to maintain them, and they make males more conspicuous to predators. If this is the case, it is hard to imagine how extreme traits indicate a higher-quality mate.

One intriguing possibility is that elaborate male secondary sexual characteristics act as handicaps. If a male can survive with sexual traits that require extra energy to build or that increase the risk of predation, these traits might signal a superior genotype. This idea, known as the **handicap principle**, argues that the greater the handicap an individual carries, the greater its ability to offset that handicap with other superior qualities.

One factor that might attract females to certain males is a genetically based high resistance to parasites and pathogens. As you know, parasites evolve rapidly and thereby continually apply selection for genetic resistance in the host. Because parasites and pathogens can impair the construction of secondary sex characteristics, a male bird that has an elaborate and showy plumage might communicate to females that because he has the energy to build elaborate feathers, he can resist parasites and pathogens. If this resistance can be inherited by the offspring of the males, then the secondary sex characteristics are *honest signals* of the genetic superiority of the males and indicate that only individuals with superior genes could resist parasites and maintain a bright and showy plumage.

Numerous studies have found that parasites and pathogens affect male attractiveness. In rock doves (*Columba livia*)—also known as pigeons—hatchling birds can be infected by mites (*Dermanyssus gallinae*) that live in the nest. To determine the effect of the mites on the young doves, researchers fumigated one set of infected nests to kill the mites but did not fumigate a second set of infected nests. The nestlings living in the nests with mites experienced lower survival and slower growth as a nestling (**Figure 9.21**). Other researchers who have examined the affects of lice on rock doves have found that females prefer louse-free males to louse-infested males by a ratio of three to one. In ring-necked pheasants (*Phasianus colchicus*), females prefer males with longer spurs on their lower legs. Longer spurs are linked genetically to major histocompatibility complex (MHC) genes that influence susceptibility to disease. Males with longer spurs have MHC alleles that are linked to longer life spans. Therefore, females that choose to mate with

Good health hypothesis The hypothesis that an individual chooses the healthiest mates.

Runaway sexual selection When selection for preference of a sexual trait and selection for that trait continue to reinforce each other.

The handicap principle The principle that the greater the handicap an individual carries, the greater its ability to offset that handicap.

Figure 9.21 The effect of mites on pigeons. The pigeons on the left were raised in nests that were fumigated to reduce the mite population. The pigeons on the right were raised at the same time, but in unfumigated nests that had high mite populations. The mite infections cause lower survival, slower growth, and areas of missing feathers. Photo by Dale H. Clayton.

long-spurred males should produce offspring with a higher chance of surviving to reproduce as adults.

SEXUAL CONFLICT

Mating decisions were once thought to serve the mutual interest of both participants. More recently, scientists have come to appreciate that mating partners often behave according to their own self-interest. In lions, for example, when a new dominant male takes over the pride, he often kills newborn cubs that have been fathered by the previous dominant male so that the female loses her entire reproductive effort.

The male obtains a fitness benefit because females without newborn cubs come into breeding condition more rapidly, which allows the new dominant male to father offspring sooner. Of course, this represents a fitness cost to the mother.

One of the more dramatic examples of sexual conflict occurs in bedbugs (*Cimex lectularius*). The male bedbug has a sharp sperm-transferring appendage and fertilizes a female by stabbing her with the appendage and injecting his sperm into her circulatory system (**Figure 9.22**). These sperm make their way to the female's ovaries. As a result, the males obtain more female mates, but females fertilized by several males live shorter lives and lay fewer eggs. It is hypothesized that this aggressive behavior evolved because female bedbugs resisted copulation attempts by male bedbugs. Examples such as these demonstrate that sexual interactions can reflect different decisions that represent the self-interest of males and females.

We have seen the ecological conditions that favor sexual versus asexual reproduction as well as the factors that favor balanced versus biased sex ratios. We have also seen how the different investments into eggs versus sperm, as well as the need to care for offspring, have shaped the evolution of different mating systems and the evolution of secondary sex characteristics. The evolution of sex remains an active area of research and there is still much to be learned.

CONCEPT CHECK

1. What factors favor sexual dimorphism?
2. What traits might females prefer in a male when selecting a mate?
3. Why might exaggerated secondary sexual characteristics in males demonstrate a superior genotype to females?

Figure 9.22 Sexual conflict in bedbugs. (a) The male bedbug possesses a sharp sperm-transferring appendage that it uses to pierce the underside of females in a specific location. **(b)** Using this appendage, located on the far end of its abdomen, males can mount females and pierce the females repeatedly. This causes the females to have a shorter life and to lay fewer eggs. Photos by Andrew Syred/Science Source.

(a)

(b)

ECOLOGY TODAY   APPLYING THE CONCEPTS

MALE-HATING MICROBES

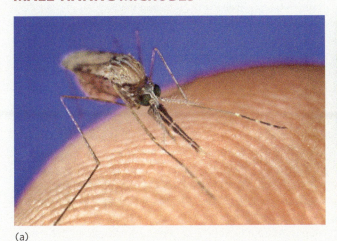

(a)

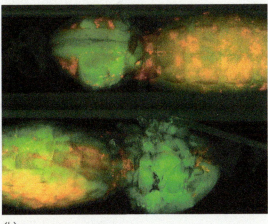

(b)

Infected with *Wolbachia.* The *Wolbachia* bacteria can infect a diversity of insect species, such as this mosquito (*Plasmodium berghei*). **(a)** The mosquito is seen under normal light. **(b)** When the mosquito and bacteria are treated to glow in different colors, the bacterial cells show up as red spots inside the mosquito's body. Photos by (a) CDC/Jim Gathany; and (b) Hughes GL, Koga R, Xue P, Fukatsu T, Rasgon JL (2011) Wolbachia infections are virulent and inhibit the human malaria parasite Plasmodium Falciparum in Anopheles Gambiae. PLoS Pathog 7(5): e1002043. doi:10.1371/journal.ppat.1002043. Image courtesy of Jason Rasgon and Ryuchi Koga.

Throughout this chapter, we have considered different mating strategies as the result of the individual's genes and its environment. However, there is a growing realization that bacteria and other microorganisms living inside an organism can control its reproductive strategies.

One of the most widespread groups of bacteria in insects belongs to the genus *Wolbachia*. Though first discovered nearly a century ago living in mosquitoes, it was not until 1990 that scientists realized the bacteria could fundamentally change an individual's reproductive strategies. *Wolbachia* infects a wide variety of invertebrates, including spiders, crustaceans, nematodes, and insects. In fact, it is currently estimated that 70 percent of all insect species experience infections by this bacteria, which can live throughout the tissues of its host. In 2007, researchers discovered that the parasite's entire genome has even been incorporated into the genome of the tropical fruit fly (*Drosophila ananassae*). An important aspect of *Wolbachia's* life history is that it is only passed down to offspring via the mother's infected egg. From the microbe's perspective, therefore, the host's female offspring are important for improving the microbe's fitness, whereas the host's male offspring are useless.

To improve its fitness, *Wolbachia* has evolved ways to exploit or eliminate these useless male hosts. In mosquitoes, *Wolbachia* alters sperm to prevent them from fertilizing the eggs of uninfected females, which ensures that subsequent generations will be dominated by offspring of infected females that carry the bacteria. In wasps, *Wolbachia* causes parthenogenesis and prevents some species from producing males. Researchers discovered this by treating some of the infected wasps with an antibiotic to kill *Wolbachia*; once the bacteria were gone, the wasps no longer produced offspring by parthenogenesis. Scientists now wonder if parthenogenesis in other species that we have discussed might also be the result of a bacterium rather than an adaptive strategy.

In woodlice (*Armadillidium vulgare*), *Wolbachia* converts males into females by suppressing male hormones. This ability for hormones to determine sex in embryos is similar to the effects of hormone manipulations by scientists investigating temperature-dependent sex determination in Jacky dragons that we saw earlier. In other species of invertebrates, the bacterium simply kills young males and prevents them from entering the adult population. In species with adults of both sexes, such as the Ugandan butterfly

(*Acraea encedon*), the increase in females relative to males has caused a reversal in the typical sex roles; females now court males and compete with other females for breeding opportunities. Note that all these sex-determining mechanisms, changes in offspring sex ratio, or mating systems are in the best interest of the bacteria, not the host.

Wolbachia are not the only bacteria with such effects. Bacteria in the genus *Rickettsia* are well-known for causing diseases such as typhus and Rocky Mountain spotted fever, but other species in this genus commonly occur in invertebrates without causing disease. In 2011, researchers found a species of *Rickettsia* living in the sweet potato whitefly (*Bemisia tabaci*). In the southwestern United States, infection rates increased from 1 percent in 2000 to nearly 100 percent by 2009. This incredibly rapid spread of the bacterium occurred because infected female flies nearly doubled their reproductive output, the survival rate of their offspring increased, and the offspring sex ratio was altered from 50 percent daughters to nearly 85 percent daughters. This is interesting from the standpoint of host–parasite interactions and the evolution of reproductive strategies. It is also important for agriculture. The sweet potato white fly is a pest that sucks the sap out of the leaves of many crops, including cotton, vegetables, and ornamental plants. Scientists worry that the massive increase in infection rates by *Rickettsia* will lead to a massive population increase of this pest.

Infections of bacteria that are passed from mother to daughter are sometimes beneficial to humans. In 2016, researchers reported that they infected mosquitoes (*Aedes aegypti*) with *Wolbachia pipientis*, a species of *Wolbachia* that does not naturally infect this mosquito. After causing the infections, the researchers then tested whether the *Wolbachia* would affect the mosquito's ability to transmit harmful human pathogens, including the yellow fever virus and the Zika virus. Amazingly, the mosquitoes infected by *Wolbachia* were less likely to be infected by and transmit the viruses. Based on this discovery, plans are being made to release large numbers of *Wolbachia*-infected mosquitoes in areas plagued by yellow fever and Zika.

Collectively, these studies suggest that there is much more to the evolution of sex than we currently know, and that understanding the evolution of sex in these organisms can have important implications for humans, including the number of pests attacking our food supply.

SOURCES:

Aliota, M. T. Aliota, S. A. Peinado, I. D. Velez, and J. E. Osorio. 2016. The wMel strain of *Wolbachia* reduces transmission of Zika virus by *Aedes aegypti*. *Scientific Reports* 6: 28792.

Himler, A. G., et al. 2011. Rapid spread of a bacterial symbiont in an invasive whitefly is driven by fitness benefits and female bias. *Science* 332: 254–256.

Knight, J. 2011. Meet the Herod bug. *Nature* 412: 12–14.

SUMMARY OF CHAPTER CONCEPTS

9.1 Reproduction can be sexual or asexual. Asexual reproduction occurs through vegetative reproduction or through parthenogenesis. Compared to asexual reproduction, sexual reproduction results in fewer copies of a parent's genes in the next generation. This can be offset by adopting a hermaphrodite sexual strategy or by providing parental care that results in raising twice as many offspring. The benefits of sexual reproduction include purging harmful mutations and creating genetic variation to help offspring deal with future environmental variation.

Key Terms: Sexual reproduction, Asexual reproduction, Vegetative reproduction, Clones, Binary fission, Parthenogenesis, Cost of meiosis, Red Queen hypothesis

9.2 Organisms can evolve as separate sexes or as hermaphrodites. If an individual possessing only male or female function can add a large amount of the other sexual function while giving up only a small amount of its current sexual function, selection will favor the evolution of hermaphrodites. If not, selection will favor the evolution of separate sexes. To avoid inbreeding depression, hermaphrodites have evolved adaptations to prevent selfing and mixed strategies for when selfing is the best option.

Key Terms: Perfect flowers, Simultaneous hermaphrodites, Sequential hermaphrodites, Monoecious, Dioecious

9.3 Sex ratios of offspring are typically balanced, but they can be modified by natural selection. Depending on the species, sex may be determined largely by genetics or by the environment. In many species, females have the ability to manipulate the sex ratio by controlling which sperm are used to fertilize the eggs or which sex chromosomes end up in the eggs, or by selective abortion of the fertilized embryos. In most organisms, the sex ratio is approximately one to one due to frequency-dependent selection.

When offspring are isolated from the rest of the population and are subjected to local mate competition, highly skewed offspring sex ratios can be adaptive.

Key Terms: Environmental sex determination, Frequency-dependent selection, Local mate competition

9.4 Mating systems describe the pattern of mating between males and females. While many species are socially monogamous, recent studies have demonstrated that many individuals participate in extra-pair copulations. As a result of this infidelity, species have evolved a variety of mate guarding behaviors to prevent a reduction in their fitness.

Key Terms: Mating system, Promiscuity, Polygamy, Polygyny, Polyandry, Monogamy, Extra-pair copulations, Mate guarding

9.5 Sexual selection favors traits that facilitate reproduction. The difference in the energetic costs of gametes and the costs of parental care typically causes female fitness to be a function of mate quality and male fitness to be a function of mate quantity. As a result, females are typically selective in choosing mates, whereas males compete strongly with each other to mate as often as possible. Male competition for female mates has favored the evolution of sexually dimorphic traits, including body size, ornaments, coloration, and courtship behaviors. Females choose particular males to obtain material benefits, such as nesting sites or food, or for nonmaterial benefits, such as good genes or good health. The best reproductive choices of males and females are often not reciprocal, which can cause conflicts between the sexes.

Key Terms: Sexual selection, Sexual dimorphism, Primary sexual characteristics, Secondary sexual characteristics, Good genes hypothesis, Good health hypothesis, Runaway sexual selection, The handicap principle

CRITICAL THINKING QUESTIONS

1. Compare and contrast the costs and benefits associated with sexual versus asexual reproduction.

2. How does the Red Queen hypothesis help us understand the fitness benefits of sexual reproduction?

3. Freshwater snails exhibit both sexual and asexual reproduction. Why might natural selection favor such a mixed breeding strategy?

4. Given that self-fertilization leads to inbreeding depression, under what conditions should a hermaphrodite use self-fertilization?

5. What strategies do monoecious plants use to avoid possible negative effects of inbreeding?

6. In addition to affecting sex ratios, what other traits might be affected by the artificial selection caused by intense fishing?

7. Turtle embryos incubated at lower temperatures typically produce males, whereas those incubated at higher temperatures produce females. In the face of global warming, how might turtles evolve to change their egg-laying behavior to maintain an even sex ratio?

8. Compare and contrast monogamy, polygyny, polyandry, and promiscuity.

9. Explain how extra-pair copulation has favored the evolution of mate guarding.

10. Why might exaggerated secondary sexual characteristics in males demonstrate a superior genotype to females?

GRAPHING THE DATA Frequency-Dependent Selection

In most species containing individuals that are either male or female, the proportion of males and females changes over time. Using a line graph, graph the proportion of male and female zebras in a population over a 10-year period. Based on these data, what happens whenever one sex becomes rare or common?

Year	Males	Females	Year	Males	Females
1	45%	55%	6	47%	53%
2	48%	52%	7	44%	56%
3	52%	48%	8	49%	51%
4	55%	45%	9	55%	45%
5	53%	47%	10	45%	55%

10 Social Behaviors

The Life of a Fungus Farmer

The leaf-cutter ant is an extraordinary farmer. Living in colonies of several million individuals, these ants leave the colony each day to harvest leaves from the surrounding forest. Using their sharp mandibles, they slice through leaves to cut off pieces that are many times larger than their bodies. They then carry the pieces back to the nest, which can rise several meters out of the ground, sink tens of meters below the ground, and expand more than 100 m wide underground. Back at the nest, the ants consume the sap in the leaves, but they do not eat the leaves. Instead, they use the leaves to grow a specialized species of fungus that they consume.

There are more than 40 species of leaf-cutter ants, which live primarily in Mexico, Central America, and South America. Like honeybees, the leaf-cutter ants form enormous societies of cooperating individuals. An ant colony normally has a single queen that can live for 10 to 20 years. Early in her life, the queen participates in a mating flight with males. The sperm she receives are then held inside her body and remain viable for the rest of her life. She uses them sparingly when fertilizing eggs to make daughters; in some species, the queen releases only one or two sperm per egg. Occasionally, she lays unfertilized eggs to make sons whose only function is to mate with other queens. The millions of individuals in the nest are daughters of the same queen and sisters to each other. They all forgo reproduction.

Daughters in this ant society are the workers and the division of labor among the workers is amazingly complex. Scientists estimate that there are nearly 30 different jobs for the leaf-cutter ants and they have determined that different workers are suited to different jobs. The worker *caste* is composed of several subcastes, known as *minims, minors, mediae,* and *majors.* Ants in each subcaste differ dramatically in size and shape. The largest workers, the majors, can be 200 times more massive than the smallest workers, the minims. The differences in subcastes are thought to be a phenotypically plastic response to different diets the ants receive as larvae. In addition to size and shape differences, the jobs of a given worker can also change during its lifetime. For example, when workers are young, they spend most of their time inside the nest, where they build tunnels, cool the nest, and raise the larvae. When large leaf pieces arrive at the nest, another group of workers cut the leaves into smaller pieces and a third group of workers mash the leaf pieces into tiny bits. The smallest workers bring a strand of fungus to the mashed bits and tend the fungus gardens, a job that includes removing other species of pathogenic fungi, which can harm the ants. In 2016, researchers discovered that in one species of leaf-cutter ant (*Mycocepurus smithii*), the workers even adjust the nutrient mixture given to the fungus. Since the ants prefer to eat the fungal threads and not the reproductive mushrooms that the fungus produces, the workers adjust the amount of carbohydrates and proteins that they give to the fungus to prevent the fungus from making mushrooms.

The oldest and largest ants act as soldiers and go out of the nest to collect leaves. The process of leaf collection is complex. Some ants climb into trees and cut large pieces of leaves that drop to the ground, while others relay leaf pieces back to the nest. When workers first begin leaf cutting, they have razor-sharp jaws, called mandibles, that are very effective at cutting tough leaves. Researchers have recently discovered that these mandibles become dull over time, causing older workers to take twice as long to cut a leaf. When an ant's cutting performance declines, it shifts

> "There are nearly 30 different jobs for the worker ants and different workers are suited to different jobs."

Leaf-cutter ants. Through an extensive division of labor, leaf-cutter ants work together to bring pieces of leaves back to their nest. In this photo, a large worker ant carries a cut piece of leaf, while a much smaller worker ant rides on top to discourage parasitoid flies from attacking the worker. Photo by Mark Bowler/Science Source.

SOURCES:

Shik, J. Z., E. B. Gomez, P. W. Kooij, J. C. Santos, W. T. Wcislo, and J. J. Boomsma. 2016. Nutrition mediates the expression of cultivar–farmer conflict in a fungus-growing ant. *Proceedings of the National Academy of Sciences* 113:10121–10126.

Hölldobler, B., and E. O. Wilson. 2011. *The Leafcutter Ants*. Norton.

Schofield, R. M. S., et al. 2011. Leaf-cutter ants with worn mandibles cut half as fast, spend twice the energy, and tend to carry instead of cut. *Behavioral Ecology and Sociobiology* 65:969–982.

jobs from leaf cutting to leaf carrying. This change in job allows older individuals to continue contributing to the ant society. The division of labor also helps the group to maintain a high foraging efficiency.

While the ants benefit from living in a very large group, the large colonies and conspicuous foraging behavior make them quite noticeable to their enemies. For example, a species of parasitoid fly that is specialized to hunt large leaf-cutter ants lays its eggs inside the necks of foraging ants. To reduce the risk of these attacks, leaf-cutter ants have evolved several tactics. Smaller individuals, which are less attractive to the fly, forage during the day. At night, when the flies do not hunt, the larger, more effective foragers go out to collect leaves. Small workers also guard larger ants by riding on leaves that the larger ants carry. When a fly approaches, these tiny guards prevent the fly from laying its eggs on the neck of the large ant that is carrying the leaf. When not repelling attacks, these hitchhiking individuals clean the leaf of undesirable microorganisms.

The leaf-cutter ants illustrate an extreme case of social behavior and living in groups. As we will see in this chapter, social groups are common in the animal world and there are both costs and benefits to social living that vary with the ecological conditions under which a species lives.

LEARNING OBJECTIVES

After reading this chapter, you should be able to:

10.1 Explain how living in groups has costs and benefits.

10.2 Illustrate the four types of social interactions.

10.3 Describe how eusocial species take social interactions to the extreme.

In the course of a lifetime, an individual typically interacts with many members of its species. Interactions with mates, offspring, other relatives, and unrelated individuals in one's species are known as **social behaviors.** Like most behaviors, social behaviors have a genetic basis and are therefore subject to natural selection. As a result, many types of social behaviors have evolved to favor the cohesiveness of family groups and populations. As we have mentioned in previous chapters, since social behaviors are traits, they are also affected by the environment in which the individuals live.

Although the study of social behavior typically focuses on animals, many other organisms interact with conspecifics in ways that might be considered social. For example, bacteria and protists can sense the presence of individuals of the same species, often through chemical secretions, and react in "friendly" or "aggressive" ways. During parts of their life cycles, free-living slime molds respond to others when they aggregate to form large fruiting bodies. Even plants communicate with one another. When one plant is damaged by herbivores, it emits volatile compounds. Other individuals detect the compounds and respond to them by producing chemical or structural defenses against future herbivore attacks.

In this chapter, we will focus on the social behaviors of animals at the individual, population, community, and ecosystem levels. We will explore some of the implications of interacting within social groups, and describe various ways that individuals manage social relationships. We will also examine how different ecological conditions affect the evolution of social behaviors.

10.1 Living in groups has costs and benefits

Animals are social for a variety of reasons. In some cases, offspring do not disperse but remain with their parents to form family groups. In other cases, individuals are attracted to each other for breeding. Individuals can also aggregate because they are independently attracted to the same habitat or resource. For instance, vultures gather around a carcass and dung flies congregate on cow pies. In this section, we will examine the costs and benefits of living in social groups and then discuss how animals use *territories* and *dominance hierarchies* in social interactions.

BENEFITS OF LIVING IN GROUPS

Animals generally form groups to increase their survival, rate of feeding, or success in finding mates.

Survival

While a single individual might not be able to fend off the attack of a predator, a group of individuals can be quite effective at doing so (**Figure 10.1**). Another survival mechanism available to social groups is a phenomenon known as the *dilution effect*. The **dilution effect** refers to the reduced, or diluted, probability of predation to a single animal when it is in a group. In an aggregation

Social behaviors Interactions with members of one's own species, including mates, offspring, other relatives, and unrelated individuals.

Dilution effect The reduced, or diluted, probability of predation to a single animal when it is in a group.

Figure 10.1 Group defense. Adult muskox (*Ovibos moschatus*), like these from Victoria Island, Canada, form an outward-facing circle and place the calves inside the circle, where they are safe from approaching predators. Photo by Eric Pierre/NHPA/Science Source.

of prey, the predator has many prey choices, so the individual living in a group has a lower probability of being caught. The dilution effect is an important benefit of large groups such as herds of mammals, flocks of birds, and schools of fish.

A lower probability of predation in groups also allows individual prey animals to spend less time watching for predators. Consider the case of the European goldfinch (*Carduelis carduelis*), a small bird that feeds on the seed heads of plants in open fields and hedgerows. If you watch closely as the birds feed, you will notice that they raise their heads and look around for predators. The total number of head raises conducted by the group increases with flock size; as you can see in **Figure 10.2a**, the larger the group, the more eyes are on the lookout for predators. As flock size increases, however, *each individual* can raise its head less frequently, as shown in Figure 10.2b. Because each individual spends less time looking for predators, it can spend more time feeding. The data in Figure 10.2c show that when a goldfinch spends less time looking for predators, it can husk a seed much faster and therefore consume seeds more quickly.

Feeding

Living in groups can also help animals locate and consume resources. Having many conspecifics all searching for food means that there are many more sets of eyes that might find food when it is rare. Furthermore, animals may find food easily but have difficulty in capturing and killing it when they are alone. In lions, for example, a lone female has a low probability of capturing and killing a zebra, but if she hunts with other lions, the probability of a capture goes up dramatically.

Mating

Socializing can also provide mating benefits since being social makes it easier to locate potential mates. An extreme

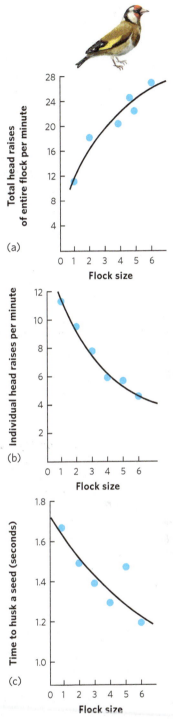

Figure 10.2 Increased vigilance when living in a group. In the European goldfinch, an increase in flock size results in **(a)** an increase in the total number of head raises performed by the flock, **(b)** a decrease in the number of head raises performed by an individual, and **(c)** a decrease in the time required to husk a seed. Data from E. Glück, Benefits and costs of social foraging and optimal flock size in goldfinches (*Carduelis carduelis*), Ethology 74 (1987): 65–79.

example of socializing for mating benefits occurs when animals aggregate in large groups to attract members of the opposite sex by making calls or displaying in ways that capture the attention of potential mates. The location of the aggregation, known as a **lek**, is used only for displaying. The site has no other value either to the displaying sex or to the attracted sex. For example, males of the ruff (*Philomachus pugnax*)—a medium-sized wading bird that lives in northern Europe and Asia—come together at a lek and participate in mating displays to attract females. On the island of Gotland in Sweden, researchers observed ruff leks to determine whether lek size affected ruff mating. As you can see in **Figure 10.3a**, males in larger leks were more successful at attracting females. In addition, as illustrated in Figure 10.3b, males in larger leks experienced a higher percentage of successful copulations with females, which confirms that forming social groups provides fitness benefits to the male birds.

COSTS OF LIVING IN GROUPS

The benefits of group living can certainly be substantial for many species, but group living can also come with costs that include predation and competition.

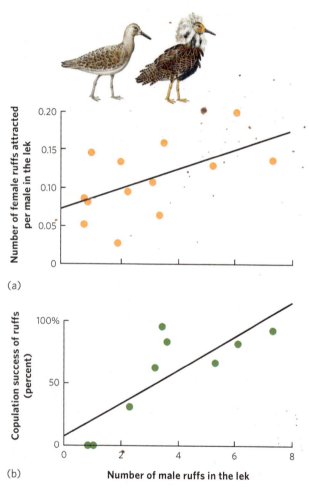

(a)

(b)

Number of male ruffs in the lek

Figure 10.3 Breeding benefits in a lek. Among displaying ruff males, those displaying to females in larger groups **(a)** are more likely to attract females and **(b)** have a higher probability of successfully copulating. Data from J. Högland, R. Montgomerie, and F. Widemo, Costs and consequences of variation in the size of ruff leks, *Behavioral Ecology and Sociobiology* 32 (1993): 31–39.

Predation

Groups of animals are much more conspicuous to predators than are individual animals. In a grassland, for example, it is easier for a predator to spot a herd of antelopes than an individual antelope. Given the propensity of antelopes to live in herds, this cost of being detected is outweighed by the benefits of the dilution effect and of more eyes to detect approaching predators.

The risk of parasites and pathogens can also increase when living with conspecifics. Many species of parasites and pathogens spread from one host to another. High population density can increase the rate at which disease spreads and can lead to epidemics. For instance, coral reefs that experience fishing pressure generally have fewer fish compared to reefs that are protected from fishing. As a result, researchers examined the fish parasites from a protected and unprotected coral reef in the central Pacific Ocean. As shown in **Figure 10.4**, they found that fish on the protected reef were infested with a higher number of parasite species than the same species living in a coral reef that was fished. In addition, fish living in the protected reef frequently carried higher numbers of each species of parasite.

The parasite and disease costs of group living are also readily observed in modern aquaculture operations, which farm aquatic species for human consumption. These operations raise oysters, salmon, catfish, shrimp, and other edible species at very high densities. Under such conditions, a single infected individual can rapidly spread parasites and pathogens to the rest of the group.

Increased transmission of parasites and diseases to groups that live in higher densities makes it undesirable for people to feed wild animals such as deer. When there is a steady, easily available source of food, deer form large aggregations around the food. This aggregating behavior makes it more likely that the animals will experience outbreaks of parasites compared to when they live in smaller family groups. Similar concerns exist for livestock operations in which animals are raised under very high densities. In this situation, diseases can jump to native wildlife populations and have dramatic effects with such viral diseases as rinderpest, avian influenza, and West Nile virus.

Competition

Another major cost of living in groups is competition for food. While larger groups are better at locating food, the food must then be shared among all the individuals in the group. Returning to the example of the European goldfinch, which experiences benefits from living in large groups, **Figure 10.5** shows one consequence of sharing the food. Larger flocks consume the seeds in an area much faster than small flocks do, so larger flocks have to spend more time flying between patches of seeds. This causes each bird to spend more time and energy looking for food.

Lek The location of an animal aggregation to put on a display to attract the opposite sex.

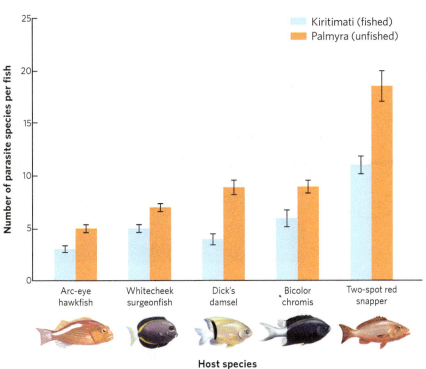

Figure 10.4 Parasite occurrence in coral reef fishes. Coral reefs protected from fishing have higher fish densities. A survey of five different species of fish found that fish living in a reef that is subject to fishing pressure contained fewer species of parasites than the same species living in a reef that is protected from fishing. Error bars are 95% confidence intervals. Data from K. D. Lafferty, J. C. Shaw, and A. M. Kuris, Reef fishes have higher parasite richness at unfished Palmyra Atoll compared to fished Kiritmati Island, EcoHealth 5 (2008): 338–345.

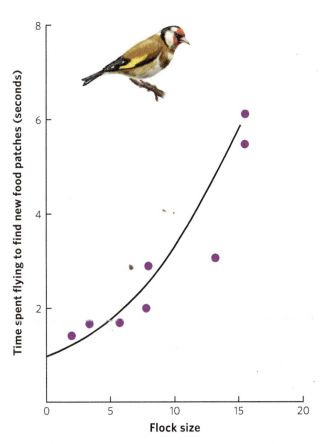

Figure 10.5 Large groups face increased competition for food. In the European goldfinch, larger flocks run out of food faster and must spend extra time and energy flying to find new patches of food. Data from E. Glück, Benefits and costs of social foraging and optimal flock size in goldfinches (Carduelis carduelis), Ethology 74 (1987): 65–79.

Each species that has evolved to live in groups faces different costs and benefits, which depend on the ecological conditions under which it lives. Assuming a genetic component for such social behavior, we expect natural selection to favor the evolution of group sizes that balance the costs and benefits for each species. A nice example of optimal group size can be found in yellow baboons (*Papio cynocephalus*). In 2015, researchers reported their observations of baboons for 11 years in East Africa (**Figure 10.6**). While baboon groups can range from 20 to 100 individuals, medium-size groups with 50-75 individuals have the lowest levels of stress, which is assessed by hormone levels found in the baboon feces. Stress is known to harm an individual's health and survival. Medium-size groups have the lowest stress levels because they travel less than small and large groups. Large groups must travel farther to find food due to the large number of mouths to feed, whereas small groups must travel farther because they get pushed out of areas by large groups and by predators.

TERRITORIES

Many species of animals have evolved to live near their conspecifics by establishing a *territory* or a *dominance hierarchy*. A **territory** is any area defended by one or more individuals against the intrusion of others. Territories can be either transient or relatively permanent, depending on the stability of the resources in the territory and how long an individual needs those resources. For example, many migratory species establish summer breeding territories and defend them for several months. Defending a high-quality

Territory Any area defended by one or more individuals against the intrusion of others.

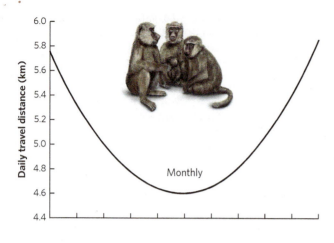

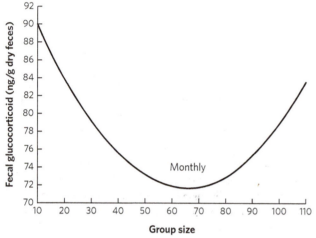

Figure 10.6 Optimal group size in yellow baboons. Groups of intermediate size have the lowest levels of stress and travel the shortest distances in search of food. Data from A. C. Markham, L. R. Gesquiere, S. C. Alberts, and J. Altmann, Optimal group size in a highly social mammal, *Proceedings of the National Academy of Sciences* 112 (2015): 14882–14887.

territory generally ensures greater resources, such as abundant food or nest sites. This typically improves the attractiveness of a territory holder as a mate, and therefore its fitness. When breeding is complete for the season, migratory species move on to their wintering grounds, where they establish new territories. Shorebirds that stop at several points along the way of their long migration defend feeding areas for a few hours or days and then continue their migratory trip. Hummingbirds and other nectar feeders defend individual flowering bushes and abandon them when those bushes cease producing flowers. As long as a resource can be defended and the benefits of defending the resource outweigh the costs, animals are likely to maintain territories.

DOMINANCE HIERARCHIES

In some situations, defending a territory is impractical. This can occur when an individual is surrounded by so many conspecifics that it becomes impractical to defend a territory against them all, when resources are available for only short periods of time, or when the benefits of living in a group override the benefits of defending a territory.

Figure 10.7 A dominance hierarchy. Bighorn sheep establish a dominance hierarchy. In this photo from Alberta Canada, the dominant male is followed by subordinate males. Photo by Mark Newman/ Getty Images

In such circumstances, individuals of many species form dominance hierarchies. A **dominance hierarchy** is a social ranking among individuals in a group, typically determined through a brief period of fighting or other contests of strength or skill (**Figure 10.7**). Once individuals order themselves into a dominance hierarchy, subsequent contests among them are resolved quickly by and in favor of higher-ranking individuals. In a linear dominance hierarchy, the first-ranked member dominates all other members of the group, the second-ranked dominates all but the first-ranked, and so on down the line to the last-ranked individual, who dominates no one else in the group. In this way, all members of the group can benefit from group living while accepting a dominance hierarchy within the group.

CONCEPT CHECK

1. How is the dilution effect a benefit of group living?
2. Why are optimal group sizes often a compromise between the costs and benefits of group size?
3. How does group size affect disease transmission?

10.2 There are many types of social interactions

Most social interactions can be considered as an action by one individual, the **donor** of the behavior, directed toward another individual, the **recipient** of the behavior.

Dominance hierarchy A social ranking among individuals in a group, typically determined through fighting or other contests of strength or skill.

Donor The individual who directs a behavior toward another individual as part of a social interaction.

Recipient The individual who receives the behavior of a donor in a social interaction.

One individual delivers food, the other receives it; one individual attacks, the other is attacked. When the attacked individual responds—by standing its ground or by fleeing—it becomes the donor of this subsequent behavior. Every interaction between two individuals has the potential to affect the fitness of both individuals, either in a positive or negative way. To understand how an interaction affects both participants, it can be useful to categorize the interactions. In this section, we will explore the four types of social interactions between donors and recipients and then examine the conditions that favor a donor helping or harming a recipient.

THE TYPES OF SOCIAL INTERACTIONS

Social behaviors can be placed into one of four categories, as illustrated in **Figure 10.8**: *cooperation, selfishness, spitefulness,* and *altruism*. When the donor and the recipient both experience increased fitness from the interaction, we call it **cooperation.** For instance, when one lion helps another kill a gazelle, and both feed from the kill, both individuals experience a fitness benefit. When the donor experiences increased fitness and the recipient experiences decreased fitness, we call it **selfishness.** Selfishness is a common interaction between two conspecifics that compete for a resource such as food, such as when a bald eagle attacks another to steal away a captured fish. The winner of the competition receives a fitness benefit, while the

loser experiences a fitness loss. For both cooperation and selfishness, the interactions benefit the donor. We would therefore expect natural selection to favor any donors that engage in either cooperation or selfishness. **Spitefulness** occurs when a social interaction reduces the fitness of both donor and recipient. Spitefulness cannot be favored by natural selection under any circumstance given that both participants experience lower fitness. Consistent with this prediction, spitefulness is not known to occur in wild animal populations. The fourth type of interaction, **altruism,** increases the fitness of the recipient but decreases the fitness of the donor. Explaining the evolution of altruism presents a unique challenge because it requires natural selection to favor individuals who improve the fitness of others while reducing their own fitness. We will explore this challenge in the next section.

ALTRUISM AND KIN SELECTION

Altruistic behavior is an interesting evolutionary behavior because it does not lead to an increase in *direct fitness*. **Direct fitness** is the fitness that an individual gains by passing on copies of its genes to its offspring. We would expect selfish individuals to prevail over altruistic individuals because selfishness directly increases the fitness of the donor, whereas altruism does not. Despite this expectation, altruism has evolved in many species. For example, some of the most extreme cases of altruism occur in colonial species, such as leaf-cutter ants and honeybees, in which workers forgo personal reproduction to rear the offspring of the dominant female.

We can explain altruistic behavior by looking beyond direct fitness. When an individual has an altruistic interaction with a relative, it increases the fitness of the relative. Because you and your relative share some genes, due to sharing a common ancestor, when you help a relative improve its fitness, you are indirectly passing on more copies of your genes, which gives you **indirect fitness.** The key to understanding the evolution of altruism is to consider an individual's **inclusive fitness,** which is the sum of its direct fitness and indirect fitness. When considering how selection operates, we say that

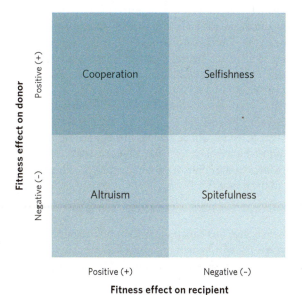

Figure 10.8 The four types of social interactions. Cooperation occurs when the donor and recipient of a behavior both obtain a positive fitness effect. Selfishness occurs when the donor obtains a positive fitness effect while the recipient obtains a negative fitness effect. Altruism occurs when the donor obtains a negative fitness effect while the recipient obtains a positive fitness effect. Spitefulness occurs when the donor and recipient both obtain a negative fitness effect.

Cooperation When the donor and the recipient of a social behavior both experience increased fitness from an interaction.

Selfishness When the donor of a social behavior experiences increased fitness and the recipient experiences decreased fitness.

Spitefulness When a social interaction reduces the fitness of both donor and recipient.

Altruism A social interaction that increases the fitness of the recipient and decreases the fitness of the donor.

Direct fitness The fitness that an individual gains by passing on copies of its genes to its offspring.

Indirect fitness The fitness that an individual gains by helping relatives pass on copies of their genes.

Inclusive fitness The sum of direct fitness and indirect fitness.

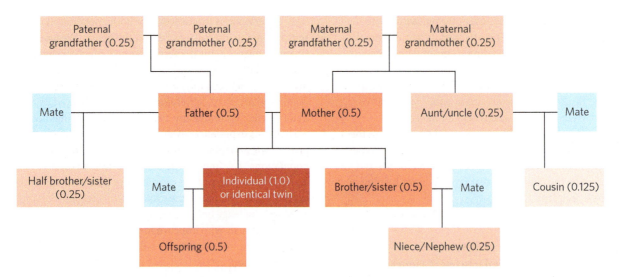

Figure 10.9 Coefficients of relatedness. The coefficient of relatedness is the probability that one individual possesses the same copy of a gene as another individual through a shared relative. In this family tree, we see that the individual in the red box has a 0.5 coefficient of relatedness with its parents, siblings, and offspring. More distant relatives have lower coefficients of relatedness, as indicated by the boxes with lighter shades of red. The coefficients of relatedness are based on the assumption that none of the mates are related to the highlighted individual.

direct fitness is favored by **direct selection.** Indirect fitness through relatives is favored by **indirect selection,** also known as **kin selection.**

As we have noted, indirect or kin selection occurs because an individual and its relatives carry copies of some of the same genes inherited from a recent common ancestor. The probability that copies of a particular gene are shared by relatives is known as the **coefficient of relatedness.** As depicted in **Figure 10.9**, its value for diploid organisms depends on the degree of relatedness between two individuals. If we focus on the individual in the red box of the family tree, we see that the coefficient between this individual and its offspring is 0.5 because a parent has two sets of genes but gives only one set to its offspring. As a result, the parent and offspring only have half of their genes in common. This also means that the coefficient of relatedness between our focal individual and its parent is also 0.5. If we next consider the focal individual and its siblings, we see that these two individuals have a 0.5 probability of receiving copies of the same gene from a parent. In the case of two cousins, the probability drops to 0.125 (one in eight) of inheriting copies of the same gene from one of their grandparents, which are their closest shared ancestors. Using these coefficients of relatedness, we can calculate the indirect fitness as the benefit given to a recipient relative (B) multiplied by the coefficient of relatedness between the donor and the recipient relative (r):

$$\text{Indirect fitness benefit} = B \times r$$

In the case of nonrelatives, there is a zero probability that an individual carries the same genes from a recent common ancestor. In examining these different coefficients of relatedness, we can see that an individual has a higher probability of leaving more copies of its genes in the next generation by promoting the fitness of its closest relatives and gains nothing by promoting the fitness of nonrelatives.

Understanding the role of kin selection and coefficients of relatedness helps resolve the puzzle of how altruistic social interactions can evolve. Whereas selfish interactions provide direct fitness benefits to the donor, altruistic interactions provide indirect fitness benefits to the donor, weighted by the coefficient of relatedness between the donor and the recipient. If the inclusive fitness of altruistic behaviors exceeds the inclusive fitness of selfish behaviors, then altruism will be favored by natural selection.

The evolution of altruistic behavior becomes clear when we examine the costs and benefits in an equation. Genes for altruistic behavior will be favored in a population when the benefit to the recipient (B) times the recipient's coefficient of relatedness to the donor (r) is greater than the direct fitness cost to the donor (C):

$$B \times r > C$$

If we rearrange this equation, we can show that for altruism to evolve, the cost-benefit ratio must be less than the coefficient of relatedness between donor and recipient:

$$C/B < r$$

Direct selection Selection that favors direct fitness.

Indirect selection Selection that favors indirect fitness. *Also known as* **Kin selection.**

Coefficient of relatedness The numerical probability of an individual and its relatives carrying copies of the same genes from a recent common ancestor.

Figure 10.10 shows this relationship graphically. Based on this equation and the figure, we can see that altruism is favored when the cost to the donor is low, the benefit to the relative is high, and the donor and its relative are closely related.

A study of wild turkeys (*Meleagris gallopavo*) in California has shown how altruistic behavior can be maintained through kin selection. Male turkeys display at leks by puffing up their feathers and strutting back and forth to attract females. Males may either display alone or with another male (**Figure 10.11**). When a pair of males displays together, only the dominant male copulates with the females that they attract. This raises the question of why the subordinate male in a pair spends its time and energy displaying when it will not be given a chance to produce any offspring with the female. Researchers obtained their first clue by using genetic data. Males in a coalition were more closely related than males drawn at random from the population. Indeed, the average coefficient of relatedness between groups of displaying males was 0.42, which suggests that paired males represent a mixture of full brothers ($r = 0.5$) and half brothers ($r = 0.25$).

The researchers then determined the average number of offspring sired by the different types of males. Dominant males in a coalition sired an average of 6.1 offspring; subordinate males in a coalition sired 0 offspring; and males that displayed alone sired an average

Figure 10.11 Wild turkey coalition. When two or more male turkeys display together to attract a female, only the dominant male sires the offspring. These individuals live in Texas. Photo by © Larry Ditto/AGE Fotostock

of 0.9 offspring. With these data, we can evaluate the fitness of the subordinate male in the coalition. By being a part of the coalition, the subordinate male forgoes the ability to breed as a solo male, which would have allowed him to sire 0.9 offspring. Therefore, his cost of being altruistic is 0.9 offspring. By helping his brother or half brother to become highly successful at attracting females, he allows his brother to sire 6.1 offspring. As we have learned, the subordinate male's average coefficient of relatedness to his brother is 0.42. Therefore, the indirect fitness benefit to the subordinate male can be calculated as

$$\text{Indirect fitness benefit} = B \times r = 6.1 \times 0.42 = 2.6$$

This means that a subordinate male obtains greater inclusive fitness by helping his brother than by going out on his own to attract females.

The concept of kin selection has given ecologists a better understanding of the evolutionary reasons underlying a wide variety of altruistic and selfish behaviors in animals. In the next section, we will explore the evolution of an extreme form of altruistic behavior in which individuals completely forgo their reproduction to help others.

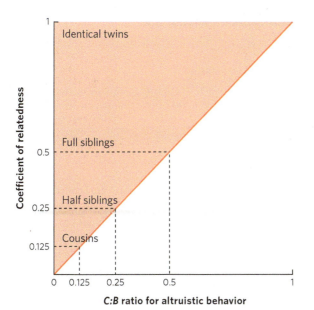

Figure 10.10 Conditions that favor the evolution of altruistic behaviors. An altruistic behavior will evolve whenever the ratio of donor costs and recipient benefits (*C/B*) is less than the coefficient of relatedness between the donor and recipient. The region in red indicates the conditions that favor the evolution of altruistic behaviors.

CONCEPT CHECK

1. Why can altruism not be explained by direct fitness alone?
2. In the kin selection explanation for the evolution of altruism, why is the benefit to the recipient weighted by its coefficient of relatedness to the donor?
3. How do direct fitness and inclusive fitness differ?

Calculating Inclusive Fitness

Eco TV

 macmillanlearning.com/ ricklefsvideo

In the pied kingfisher (*Ceryle rudis*), a fish-eating bird from Africa and Asia, adult males often forgo their own reproduction to help their parents raise offspring. Researchers have identified *primary helpers* and *secondary helpers*. Primary helpers are sons of the parents, and they work hard to protect the nest, and to bring food to the chicks. In some cases, one of the parents disappears and is replaced by an unrelated mate, so the son ends up helping a parent and a stepparent. Secondary helpers—which are unmated males from other families—are not related to the parents and do not work as hard to feed and protect the offspring. After helping for one year, both types of helpers set up their own nest the following year. A third group of males, known as *delayers*, do not help but simply delay reproduction until their second year.

Researchers followed several nests of kingfishers and determined how much each helper improved the fitness of the parents in the first year (B_1), the probability of surviving and finding a mate in the subsequent year (Psm), and the fitness of the helper when he bred independently in the second year (B_2). They also quantified the coefficients of relatedness between the helpers and the parents being helped in the first year (r_1) and between the helpers and their own offspring produced in the second year (r_2). The coefficient of relatedness for the primary helpers was 0.32 in year 1. This was the result of some nests retaining both parents of the helper ($r = 0.5$) and other nests having one parent of the helper plus a stepparent ($r = 0.25$).

Based on these data, the researchers calculated the inclusive fitness of the primary helpers, secondary helpers, and delayers.

As you can see, the primary helper had inclusive fitness that was a bit higher after 2 years. The primary helpers obtain about half of their inclusive fitness by helping their parents raise their siblings in year 1 and the other half by having their own offspring in year 2. In contrast, the secondary helpers did not gain any indirect fitness in year 1, but had a higher probability of surviving and finding a mate in year 2, which improved their direct fitness. The delayers obtained no indirect fitness in year 1 and had a poor ability to attract mates in year 2, leading to a low inclusive fitness.

YOUR TURN Use the table below to calculate the change in inclusive fitness if the primary helpers only improved the fitness of their parents by 1.0 rather than 1.8. Under this scenario, which strategy of helping would be most favored by natural selection?

	Year 1			Year 2				
MALE ROLE	B_1	r_1	INDIRECT FITNESS	B_2	r_2	Psm	DIRECT FITNESS	INCLUSIVE FITNESS
PRIMARY HELPER	1.8 × 0.32 = 0.58			2.5 × 0.5 × 0.32 = 0.41				0.58 + 0.41 = 0.99
SECONDARY HELPER	1.3 × 0.00 = 0.00			2.5 × 0.5 × 0.67 = 0.84				0.00 + 0.84 = 0.84
DELAYER	0.0 × 0.00 = 0.00			2.5 × 0.5 × 0.23 = 0.29				0.00 + 0.29 = 0.29

Source: Data from H.-U. Reyer, Investment and relatedness: A cost/benefit approach of breeding and helping in the pied kingfisher, *Animal Behaviour* 32 (1984): 1163–1178.

10.3 Eusocial species take social interactions to the extreme

We have seen that many animals are social and interact with conspecifics in a variety of ways. Some animals are so extremely social that we call them *eusocial*. **Eusocial** (i.e., "truly" social) species are distinguished by their display of all four of the following characteristics:

1. Several adults living together in a group
2. Overlapping generations of parents and offspring living together in the same group
3. Cooperation in nest building and brood care
4. Reproductive dominance by one or a few individuals, and the presence of sterile individuals

Among the insects, eusocial species are limited to the order Hymenoptera, which includes bees, ants, and wasps, and the order Isoptera, which includes termites (**Figure 10.12**). Not only are these social insects of evolutionary interest, they are also major players in ecosystem processes. Eusocial insects pollinate plants, consume plant and animal material on a large scale, and recycle wood and organic detritus. Their dominance in the world is due in large part to the immense success of eusociality. Other than insects, the only animals known to be eusocial are two species of mammals that live

Eusocial A type of animal society in which individuals live in large groups with overlapping generations, cooperation in nest building and brood care, and reproductive dominance by one or a few individuals.

Figure 10.12 Eusocial species. These Texas leaf-cutter ants (*Atta texana*) are one of many highly social species. Pictured here is a queen and her young daughters that are tending a fungus garden deep inside their underground nest. Photo by Alexander Wild

in underground tunnels in Africa: the naked mole rat (*Heterocephalus glaber*) and the Damaraland mole rat (*Fukomys damarensis*).

Eusocial species are fascinating to ecologists because most individuals in a eusocial group forgo sexual maturation and reproduction. Instead, they specialize at tasks that include defending the group, foraging for the group, or taking care of the subsequent offspring of their parents. Because

nonreproductive individuals are specialized for these tasks, they are known as sterile *castes*. A **caste** consists of individuals within a social group who share a specialized form of behavior. For example, we saw in **Chapter 9** that worker honeybees are a caste that works for the hive but typically does not reproduce. Similarly, the workers of the leaf-cutter ants represent a variety of castes that do different jobs but they do not reproduce.

How can natural selection produce individuals that lack any reproductive output and therefore lack any direct fitness? To help answer this question, we will start by examining the unique breeding habits of the hymenopteran insects.

EUSOCIALITY IN ANTS, BEES, AND WASPS

In **Chapter 9**, we talked about the unique reproductive behavior of the hymenopteran insects. You may recall that this group of insects produces sons by laying unfertilized eggs and produces daughters by laying fertilized eggs. This method of reproduction is detailed in **Figure 10.13**. Because one sex is haploid and the other sex is diploid, the hymenopterans are said to have a **haplodiploid** sex-determination system. As we will see, this sex-determination system helps to favor the evolution of eusociality.

Caste Individuals within a social group sharing a specialized form of behavior.

Haplodiploid A sex-determination system in which one sex is haploid and the other sex is diploid.

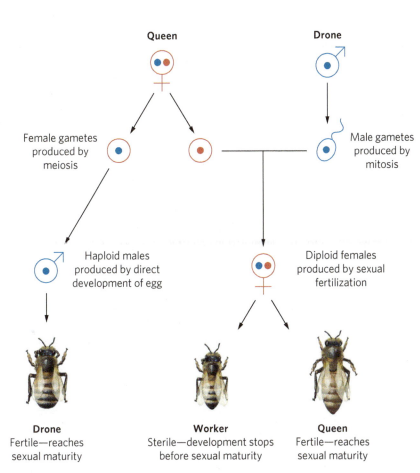

Drone Fertile—reaches sexual maturity

Worker Sterile—development stops before sexual maturity

Queen Fertile—reaches sexual maturity

Figure 10.13 The haplodiploid mating system of hymenopteran insects. In ants, bees, and wasps, sons are produced when a queen's haploid gamete remains unfertilized and it develops into a haploid son. Daughters are produced when a queen's haploid gamete is fertilized by a drone's haploid gamete and it develops into a diploid daughter.

Social Organization

Eusocial insect societies are dominated by one egg-laying female or a few egg-laying females, referred to as **queens.** The queens in colonies of ants, bees, and wasps mate only once during their lives and store enough sperm to produce all their offspring—up to a million or more offspring over a period of 10 to 15 years in some ant species. The nonreproductive progeny of a queen gather food and care for developing brothers and sisters, some of which become sexually mature, leaving the colony to mate and to establish new colonies.

As we discussed in **Chapter 9**, honeybee societies have a simple organization. The offspring of a queen are divided among a sterile worker caste that is female and a reproductive caste that consists of drones and future queens. Whether an individual becomes a sterile worker female or a fertile female is determined by the length of time the larva is fed royal jelly. As a result, the worker caste represents an arrested stage in the development of reproductive females; they stop developing before sexual maturity. Eusocial behavior appears in a large number of hymenopteran species.

Coefficients of Relatedness

Haplodiploidy is important to the evolution of eusocial animals because it creates strong asymmetries in coefficients of relatedness. The queen has the same genetic relationship to sons as she has to daughters ($r = 0.50$), so she can be relatively indifferent to the sex of her reproductive offspring—the drones and new queens that leave the nest—especially when the sex ratio among reproductive individuals in other local hives is nearly equal. However, the relatedness among siblings is unique in hymenopterans. If a queen has mated with a single drone, then all females have the same set of genes from their haploid father and they have a 50 percent probability of sharing the same genes from their mother. As a result, the coefficient of relatedness between a given female and her sisters is 0.75 in the haplodiploid mating system. This causes females to be more closely related to each other than is typical in a diploid mating system, where it is only 0.50. In contrast, her brothers are haploid, so the brothers receive no genes from the drone, but they have a 50 percent probability of sharing the same genes from their mother. As a result, the average coefficient of relatedness between a given female and her brother is 0.25. This skewed relatedness favors the evolution of eusocial groups.

To understand how being haplodiploid favors eusociality, we can compare the different options for obtaining fitness in these organisms. For example, a female worker that raises a fertile sister receives the fitness benefit of that individual multiplied by the coefficient of relatedness ($r = 0.75$). When a female raises its own offspring, it receives the same fitness benefit multiplied by the coefficient of relatedness between a mother and daughter ($r = 0.50$). This means that the indirect fitness obtained from caring for a sister exceeds the direct fitness that could be obtained from caring for a daughter. Thus, cooperation is likely to be greater among all-female castes than among male castes or, especially, among mixed castes. This may explain why workers in hymenopteran societies are all female.

The fact that workers are three times more closely related to their sisters than to their brothers may also explain why broods of reproductive individuals usually favor females, by about three to one on a weight basis, in spite of the queen's indifference to the sex ratio. When a female worker can help to rear more female than male reproductive individuals, her own inclusive fitness may actually be higher than it would be if she raised a brood of her own consisting of an equal number of males and females. Under these circumstances, it is not surprising that sterile castes have evolved.

EUSOCIALITY IN OTHER SPECIES

Although the hymenopterans are among the best known of the eusocial insects, there are a few other species that are also eusocial, including termites and a group of mammals known as naked mole rats. Unlike the hymenopterans, these latter two groups are both diploid, which present the challenge of understanding the conditions that favor eusociality in these species.

Termites

Termite colonies can be massive structures that are dominated by a mated pair called the king and queen (**Figure 10.14a**). Because the queen's role is almost entirely limited to egg production, her abdomen is grossly distended to allow her to produce thousands of eggs (Figure 10.14b). The king and queen produce sons and daughters by sexual reproduction. With a few exceptions, these sons and daughters act as workers in the termite society but can become sexually mature if either the king or the queen dies. Many species of termites also have a second caste of nonreproductive individuals known as *soldiers*. As their name implies, soldiers help to defend the nest against intruders such as ants. Soldiers typically have a very large head that can be used to block the openings of a termite nest to prevent ants from entering.

Mole Rats

There are dozens of species of mole rats, but only two are known to be eusocial: the naked mole rat and the Damaraland mole rat. These rodents live in underground tunnels in African grasslands in colonies of up to 200 individuals. In naked mole rats, a single queen and several kings are responsible for all the reproduction in the colony, and all individuals are diploid

---

Queen The dominant, egg-laying female in eusocial insect societies.

Figure 10.14 Termite colonies.
(a) A massive termite colony located in the Northern Territory, Australia.
(b) A queen termite (*Macrotermes gilvus*) living deep inside a colony surrounded by workers. Because her primary role is egg production, the queen's abdomen is much larger than that of workers and soldiers. Photos by (a) Thomas P. Widmann/Newscom; and (b) Anthony Bannister/NHPA/Science Source

(a)   (b)

(**Figure 10.15**). Although the workers in the group are capable of reproducing, they forgo reproduction in favor of caring for their younger siblings and taking care of the colony. Current research suggests that the subordinate males and females do not do this willingly. Instead, the dominant female harasses them, which causes them to become stressed. Stress reduces the levels of sex hormones in the subordinates and makes them less motivated to breed.

THE ORIGINS OF EUSOCIALITY

From its distribution across a variety of distantly related species, ranging from hymenopterans to mammals, it is clear that eusociality has evolved independently many times. Even within the hymenopterans, eusociality

Figure 10.15 Naked mole rats. The naked mole rat is one of two species of mammals that are eusocial. In this photo, a queen is laying on top of the workers. Photo by © Raymond Mendez/Animals Animals/ Earth Scenes.

appears to have evolved several times. It is tempting to conclude that eusociality is caused by a haplodiploid sex-determination system, but many haplodiploid species are not eusocial, and some eusocial species—such as termites and naked mole rats—are not haplodiploid. Based on these observations, we are left to conclude that while being haplodiploid appears to favor the evolution of eusocial behavior by providing large indirect fitness effects when workers forgo breeding to help their sisters, it is not required for the evolution of eusocial behavior.

Throughout our discussion, we have focused on the importance of the coefficient of relatedness with the idea that the coefficient needs to be relatively high to favor eusociality. However, this equation could also be satisfied if the cost of forgoing personal reproduction is very low. If individuals that chose to leave a colony had a very low likelihood of surviving and setting up a new colony, then direct fitness would be very low. In naked mole rats, for example, some individuals leave the home colony to form new colonies, but most of these small new colonies do not persist beyond one year. As a result, the cost of forgoing reproduction—in terms of lost direct fitness—would be very small. When this cost is small, a large coefficient of relatedness is no longer required to favor eusocial behavior.

The origins of eusocial behavior are still actively debated. Many researchers have argued that it has been favored by the inclusive fitness of the altruistic sterile offspring. Others have argued that it has evolved to enhance the direct fitness of parents that force the offspring to forgo reproduction. Like many biological questions, continued research will likely provide new insights that will help us to understand eusocial societies.

Behavioral relationships among the social insects represent one extreme along a continuum of social organization, from animals that live alone (except to breed) to those that aggregate in large groups organized by complex behavior. Regardless of their complexity, all social behaviors balance costs and benefits to the individual, and the magnitude of these costs and benefits is often determined by the ecological conditions in which these behaviors exist. Like morphology and physiology, behavior is strongly influenced by genetic factors and therefore is subject to evolutionary modification by natural selection. The evolution of behavior becomes complicated when individuals interact within a social setting, and the interests of individuals within a population may either coincide or conflict. Understanding the evolutionary resolution of social conflict in animal societies continues to be one of the most challenging and important concerns of biology.

CONCEPT CHECK

1. What are the four characteristics of a eusocial species?
2. How does a haplodiploid sex-determination system favor the evolution of eusociality?
3. How does a low probability of successfully starting a new colony favor the evolution of eusocial behavior in naked mole rats?

ECOLOGY TODAY   APPLYING THE CONCEPTS

Hen-Pecked Chickens

Domesticated chickens. Chicken survival and egg production depend on not only the traits of each chicken but also the way each chicken interacts with other chickens in its cage. Photo by Phillip Hayson/Science Source.

The social behavior of animals in the wild is inherently interesting, but understanding social behavior also has practical applications for domesticated animals. For example, there are costs and benefits of raising domesticated animals in large social groups. Raising animals such as pigs and chickens in high densities is more cost effective because it requires less space. However, as we have seen, living at high densities has downsides that include increased risks of disease transmission, increased competition, and increased fighting. For farmers who raise domesticated animals, raising animals in high densities translates into fewer healthy animals and smaller profits. However, during the past two decades, researchers with insights into group living have helped breed more productive domesticated animals. Some of the leading research has been conducted with chickens (*Gallus gallus*).

Chickens are well known for fighting when raised under crowded conditions. For example, if one chicken receives a small injury that causes a spot of blood on its feathers, other chickens will begin to peck at the spot, causing greater injury and often death through cannibalism. In the process, some chickens will become dotted with blood spots and other chickens, in turn, will also peck them. This incessant behavior is the origin of the phrase "hen-pecked." Fighting among chickens can reduce overall health and egg production, and can even cause death of the chickens. In fact, fighting can cause so much lost profit to poultry farms that there is a long-standing practice of cutting off the ends of chicken beaks to reduce injuries. Although beak trimming is effective at reducing fighting and cannibalism, it is a controversial practice and has been recently banned in several countries. This has motivated animal breeders to find alternative ways to reduce fighting among crowded domesticated animals.

The traditional method of breeding animals has been to select for those individuals with the best survival, growth, or other traits such as increased egg production. After decades of selection on traits, genetic variation declines and becomes quite low, which makes it difficult to conduct further selection. In the mid-1990s, however, researchers discovered that they could cause large responses to selection if they based their selection on the best-performing social groups of chickens living together rather than just the best-performing individual chickens. Groups of hens were caged together and groups with the highest survival were used to generate offspring for the next generation. At the start of the experiment, annual hen mortality was 68 percent. After six generations of selection for the best surviving groups, annual hen mortality dropped to 9 percent, which was similar to the mortality rate of hens that were raised alone with no opportunities to fight. Moreover, lifetime egg production increased dramatically, from 91 eggs per hen in the first generation to 237 eggs per hen by the sixth generation. The increase in egg production was caused by the hens living longer and being in better physical condition. These data suggest that selection for the best social groups could reduce fighting to a level that would eliminate the need to trim chicken beaks.

During the past 5 years, researchers have begun to understand why selection based on social groups of animals could cause such dramatic increases in animal production. Biologists have traditionally focused only on the heritability of traits for an isolated individual. However, the performance of an individual depends on not only the individual but also the way in which it interacts with other individuals. For example, if an individual is particularly aggressive and lives in a group of other aggressive individuals, the group will spend more time fighting and more deaths will occur. In contrast, a social group with individuals who fight less with each other will have lower stress, better health, and a reduced mortality rate. Because social animals interact with conspecifics, we need to consider the heritability of traits not just when an individual lives alone but when an individual lives in a social group. In chickens, for example, the heritability of survival for an individual can be quite low, but the heritability of survival when social interactions are included can be two to three times higher because social interactions affect survival. Similar results have been observed in other farmed animals, including domesticated pigs (*Sus scrofa*) and mussels (*Mytilus galloprovincialis*). All this suggests that when we understand the social interactions of animals, we can dramatically improve agricultural production of domesticated animals.

SOURCES:

Muir, W. M. 1996. Group selection for adaptation to multiple-hen cages: Selection program and direct responses. *Poultry Science* 75: 447–458.

Wade, M. J., et al. 2010. Group selection and social evolution in domesticated animals. *Evolutionary Applications* 3: 453–465.

SUMMARY OF LEARNING OBJECTIVES

10.1 Living in groups has costs and benefits. The benefits of social living include the dilution effect in which large groups of prey have a reduced likelihood of being killed by a predator, reduced need for personal vigilance, and increased ability to find food and mates. The costs include increased visibility to predators, increased risk of parasite and pathogen transmission, and increased competition for food. In response to social living, many species have evolved the ability to establish territories and dominance hierarchies to manage individual interactions.

Key Terms: Social behaviors, Dilution effect, Lek, Territory, Dominance hierarchy

10.2 There are many types of social interactions. When we envision interactions in terms of donors and recipients, we can devise four types of social interactions: cooperation, selfishness, spitefulness, and altruism. Cooperation and selfishness of donors should be favored by natural selection, whereas spitefulness should not. Altruism can be favored when the recipient of an altruistic act is closely related to the donor, as measured by the coefficient of relatedness. As a result, altruism evolves because individuals experience an increase in inclusive fitness, which is the sum of direct and indirect fitness.

Key Terms: Donor, Recipient, Cooperation, Selfishness, Spitefulness, Altruism, Direct fitness, Indirect fitness, Inclusive fitness, Direct selection, Indirect selection, Coefficient of relatedness

10.3 Eusocial species take social interactions to the extreme. Eusocial animals consist of many individuals living together with dominant individuals reproducing and subordinate individuals forgoing reproduction. Eusocial species are common among the haplodiploid species of bees, ants, and wasps, but also exist in diploid species of termites and at least two species of mammals. A high coefficient of relatedness favors the evolution of eusocial behavior, but it is not required. Equally important may be the presence of a low cost of lost fitness in species that have a low probability of leaving the group and reproducing on their own.

Key Terms: Eusocial, Caste, Haplodiploid, Queen

CRITICAL THINKING QUESTIONS

1. If living in large groups has costs and benefits, under what conditions would natural selection favor group living?

2. Why might individuals give up defending territories if the density of their population increases?

3. Explain the costs and benefits that might influence the optimal flock size in birds.

4. Compare and contrast the conditions under which natural selection will favor cooperative versus altruistic behavior.

5. How could helping raise the offspring of another couple improve the helper's fitness if the helper is not related?

6. Why are selfish behaviors less favored when the donor and recipient are related to each other?

7. Compare the coefficient of relatedness between brothers and sisters in diploid organisms with that of haplodiploid organisms.

8. Given that termites are diploid, what would you predict about the costs and benefits of the workers forgoing reproduction, which would favor the evolution of eusociality?

9. What evidence is there that a haplodiploid sex-determination system is not required for the evolution of eusociality?

10. Why might selection based on groups of domesticated goats result in greater meat production than selection on individual goats?

GRAPHING THE DATA How Living In Groups Affects Predation Risk

As we have discussed, living in groups has a number of potential costs and benefits. To determine if living in schools provided an antipredator benefit to minnows, researchers placed different numbers of minnows into aquaria and determined how often schools of different size were approached by a larger species of predatory fish. Using the data in the table and your knowledge about calculating sample standard deviations from Chapter 2,

calculate the means and the standard deviations for the number of approaches by a predator per minute as a function of different school sizes. Then graph the means and sample standard deviations using a line graph.

Based on these data, what can you conclude about the effect of school size on the likelihood of predation by the larger species of fish?

	Minnow school size				
Trial	3	5	10	15	20
1	0.9	0.7	0.4	0.4	0.1
2	0.8	0.8	0.5	0.5	0.1
3	0.7	0.9	0.6	0.3	0.2
4	1.1	0.6	0.8	0.2	0.3
5	1.0	1.0	0.7	0.6	0.3
Mean Number of Approaches/ Minute					
Standard Deviation					

11 Population Distributions

Bringing Back the Mountain Boomer

When Alan Templeton was a Boy Scout in 1960, he encountered his first collared lizards (*Crotaphytus collaris*), also known as mountain boomers, in the Ozark Mountains of Missouri. He was struck by the brightly colored males that ran around in the forest openings that dotted the mountains. Two decades later, as a biology professor at Washington University in St. Louis, he returned to the Ozark Mountains and was shocked to find that most of the lizards were gone. He began a course of research to identify the causes of the decline and to determine the steps that could restore the lizard population.

The collared lizard is a fascinating animal that feeds on insects such as grasshoppers. It prefers to live in dry, open areas and has a geographic range that spans from Kansas to Mexico; Missouri lies on the eastern edge of this range. Although much of the Ozark Mountains are forested, there are openings in the forest, known as glades, that contain exposed bedrock with the hot and dry conditions that provided suitable habitat for the lizards. These small, patchy habitats were once surrounded by savannas and forests with open understories. Over time, these habitats began to change.

In the Ozark Mountains, forest fires were historically common. These fires removed the small understory trees and accumulating leaves scattered across the forest floor. Forest fires were also important for maintaining the glades as open, sunny areas. However, beginning in the 1940s, a national campaign to suppress forest fires caused the glade habitats to be invaded by eastern red cedar trees (*Juniperus virginiana*). The cedar trees shaded the glades and made them cooler, which is not conducive to the growth of an ectotherm such as the collared lizard. The shade also reduced the number of insect prey, including grasshoppers. With the passage of time, only a few glades continued to support lizard populations and the geographic range shrank. Moreover, there appeared to be no movement of the lizards from these suitable glades to other suitable glades that were unoccupied.

> "The great success in restoring the collared lizard required researchers to understand its habitat needs and to recognize that the regional population was actually composed of many small populations interconnected over a large spatial area."

Templeton and his colleagues began their research in 1979. They discovered that many glades once occupied by lizards no longer had any lizard populations. They took several steps to restore those populations. First, they cut down the cedar trees in several glades to make the habitat more suitable. Next, they reintroduced 28 lizards to three sites. They marked the animals to estimate the abundance and density of each population and to determine whether the lizards they introduced would disperse to colonize the neighboring glades. Dispersal events were rare and no neighboring glades were colonized, despite the fact that some were only 50 m away. Although several glades had been cleared of cedar trees, conditions were still not suitable; the thick forest understory and accumulating leaf litter shaded much of the ground, which made the forest cooler and caused lower abundances of insect prey. At this point, the researchers realized that they had two tasks. They needed to improve both the habitat of the glades, where the lizards spent most of their time, and the forested habitat between the glades through which the lizards traveled when dispersing from one glade to another.

In cooperation with the Missouri Department of Conservation, the team began using controlled fires in 1994 to burn away the cedars in many glades as well as the small understory trees and leaves in the forest habitat between the glades. The lizards responded so well that 2 months after the burn, they began to colonize new glades. Two decades later, Templeton and colleagues reported that from the original three glades where lizards were reintroduced, there were more than 500 lizards spread across more than 140 glades. Burning the forested habitat allowed more sunlight to come in and dramatically increased the number of grasshoppers for the lizards to eat when they traveled between glades.

A male collared lizard. Males of this species have a bright orange throat that intimidates other males and attracts females. Photo by Francois Gohier/Science Source

A glade in the Ozarks.
Photo by Alan Templeton/
Department of Biology,
Washington University.

SOURCES
Neuwald, J. L., and A. R. Templeton.
2013. Genetic restoration in the
eastern collared lizard under
prescribed woodland burning.
Molecular Ecology 22:3666–3479.
Templeton, A. R. 2011. The transition
from isolated patches to a metapopulation in the eastern collared
lizard in response to prescribed fires.
Ecology 92:1736–1747.
Restoration as science: The case of the
collared lizard. 2011. *Science Daily*,
August 22. http://www.sciencedaily.com/
releases/2011/08/110822091918.htm.

The population increase and enlarged geographic range also improved the genetic diversity of the lizards. Prior to the controlled burning of the glades, the lizard populations remained small and contained, and their genetic diversity remained low due to the founder effect and genetic drift, which we discussed in Chapter 7. Today, the future of the collared lizard looks bright in Missouri as the use of controlled fire has brought back the required habitat, the genetic diversity is higher, and dispersal among glades allows any local populations that might go extinct to be recolonized by new dispersers.

The great success in restoring the collared lizard required researchers to understand its habitat needs and to recognize that the regional population was actually composed of many small populations interconnected over a large spatial area. In this chapter, we will explore how spatial distribution and the movement of individuals among habitats influence the long-term persistence of species.

LEARNING OBJECTIVES

After reading this chapter, you should be able to:

11.1 Explain why the distribution of populations is limited to ecologically suitable habitats.

11.2 Give the five important characteristics of population distributions.

11.3 Describe how distribution properties of populations can be estimated.

11.4 Recognize that population abundance and density are related to geographic range and adult body size.

11.5 Explain why dispersal is essential to colonizing new areas.

11.6 Describe how populations commonly live in distinct patches of habitat.

The story of the collared lizard demonstrates that studying ecology at the population level is fascinating and can have real-world applications, just as the prior chapters on ecology at the individual level showed. In this chapter, we will focus on the **spatial structure** of populations, which is defined as the pattern of density and spacing of individuals. We will begin by examining how the distribution of suitable habitat affects the distribution of populations. We will then discuss the many properties of population distributions and examine how we estimate these properties. Next, we will investigate the importance of individuals moving between patches of habitats. Finally, we will learn how providing strips of favorable habitat can facilitate

Spatial structure The pattern of density and spacing of individuals in a population.

the movement of individuals between habitats to help ensure the persistence of populations over time.

11.1 The distribution of populations is limited to ecologically suitable habitats

At the beginning of this chapter, we saw that collared lizards inhabited habitat patches, known as glades, interspersed within a forest. As these glades became invaded by cedar trees, however, they became less suitable for the collared lizards. In this section, we will explore how ecologists determine the suitability of habitats. We will see that understanding habitat suitability is critical for understanding where a species is capable of living and the extent to which a species can expand its geographic range.

DETERMINING SUITABLE HABITATS

In Chapter 1, we mentioned that a species niche includes the range of abiotic and biotic conditions it can tolerate. We are now ready to elaborate on this point. To explore the concept of niche, it is useful to draw a distinction between the *fundamental niche* and the *realized niche* of a species.

The **fundamental niche** of a species is the range of abiotic conditions under which species can persist. This includes the range of temperature, humidity, and salinity conditions that allow a population to survive, grow, and reproduce.

Although a species can potentially live under the conditions of its fundamental niche, many favorable locations can remain unoccupied because of other species in those locations. For example, the presence of competitors, predators, and pathogens can often prevent a population from persisting in an area, despite the existence of favorable abiotic conditions. The range of abiotic and biotic conditions under which a species persists is its **realized niche.** The realized niche determines the *geographic range* of a species or of various populations that compose a species. The **geographic range** is a measure of the total area covered by a population. For example, the American chestnut tree (*Castanea dentata*) was once a very common tree in the eastern forest of the United States because it grew and reproduced well under the abiotic and biotic conditions that existed in this region for thousands of years. Around 1900, however, a fungus (*Cryphonectria parasitica*) introduced from Asia caused a deadly disease known as chestnut blight. The fungus spread rapidly throughout the eastern forests and killed billions of trees. As a result of this biotic interaction between the chestnut trees and the fungus, few places remain where adult trees can persist. In short, the fungus

Figure 11.1 The geographic range of the sugar maple. The range is limited in the north by low winter temperatures, to the south by high summer temperatures, and to the west by droughts.

has caused a major reduction in the realized niche of the American chestnut.

When we think about the geographic range of a species or population, we need to realize that individuals often do not occupy every location within the range. This is because climate, topography, soils, vegetation structure, and other factors influence the abundance of individuals. Consider the case of the sugar maple tree. As shown in **Figure 11.1**, its geographic range includes the midwestern United States, the northeastern United States, and southeastern Canada. Its distribution is limited by cold winter temperatures at the northern extent of its range, by summer droughts at the western extent, and by hot summer temperatures at the southern extent. Throughout this entire range, however, the sugar maple does not live everywhere. For instance, it cannot live in marshes, newly formed sand dunes, recently burned areas, and a variety of other habitats that lie outside its fundamental niche. As a result, the geographic range of the sugar maple is actually composed of a patchwork of smaller occupied and unoccupied areas.

The distribution of a shrub known as Fremont's leather flower (*Clematis fremontii*) provides an excellent example

Fundamental niche The range of abiotic conditions under which species can persist.

Realized niche The range of abiotic and biotic conditions under which a species persists.

Geographic range A measure of the total area covered by a population.

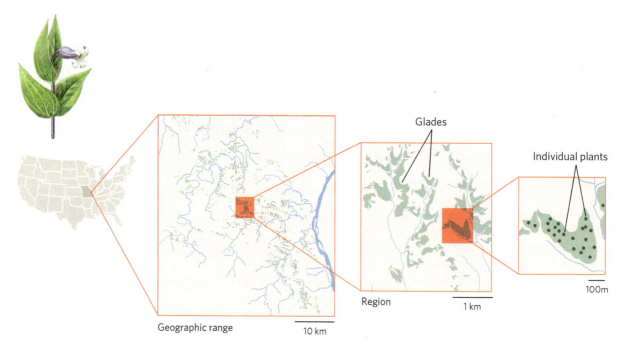

Figure 11.2 **The geographic range of Fremont's leather flower.** A survey found that this shrub species only existed in three counties in the state of Missouri. Within this range, individuals were only found living in the outcroppings of limestone, known as glades. Data from R. O. Erickson, The *Clematis fremontii* var. *Reihlii* population in the Ozarks, *Annals of the Missouri Botanical Garden* 32 (1945): 413–460.

of how small-scale variation in the environment can create geographic ranges that are composed of small patches of suitable habitat. As illustrated in **Figure 11.2**, the plant's geographic range is just three counties in the state of Missouri. This small range is thought to be the result of climatic conditions and competitive interactions with ecologically similar plants. Within its geographic range, the plant is restricted to dry, rocky soils on outcroppings of limestone, known as *limestone glades,* which are similar to the glades frequented by the collared lizard discussed at the beginning of this chapter. Small variations in elevation and soil quality further confine these plants within each limestone glade to sites with suitable soil structure, moisture, and nutrients. Local aggregations occurring on each of these sites consist of individuals that are fairly evenly distributed in space. In other words, while Fremont's leather flower has a geographic range that includes three counties, its distribution is spotty in this region due to its narrow habitat requirements.

Although patterns of distribution would suggest that only certain habitats are suitable, we can test whether this is the case. Consider the case of two species of wildflowers that live at distinct elevations in the Sierra Nevada of California. One species, known as Lewis' monkeyflower (*Mimulus lewisii*), lives at higher elevations. The other species, known as the scarlet monkeyflower (*Mimulus cardinalis*), lives at lower elevations. Both species occur at mid-elevations. To determine if

environmental conditions cause these different species distributions, researchers planted the two species at locations within and outside of the elevations where they grow in nature. The results of this experiment are shown in **Figure 11.3**. If we examine plant survival, we see that Lewis' monkeyflower survives well at high elevations but survives poorly at low elevations. The opposite is true for the scarlet monkeyflower. If we examine plant growth, a similar pattern emerges. Lewis' monkeyflower grows better as elevation increases, although growth declines under the extreme conditions of the highest elevation. The scarlet monkeyflower grows the best at low elevations and its growth declines with each increase in elevation. For each species, the survival and growth of the transplanted populations were the highest when grown within its normal elevation range. When grown outside this range, both species experienced lower survival and slower growth.

ECOLOGICAL NICHE MODELING

As a general rule, the more suitable the habitat, the larger a population can grow within that habitat. This fundamental relationship allows ecologists to predict the actual or potential distributions of species, which has a number of important applications. For example, to bring species back from the brink of extinction, we need to know the habitat conditions that the species requires. This knowledge is used to determine the locations that would provide the

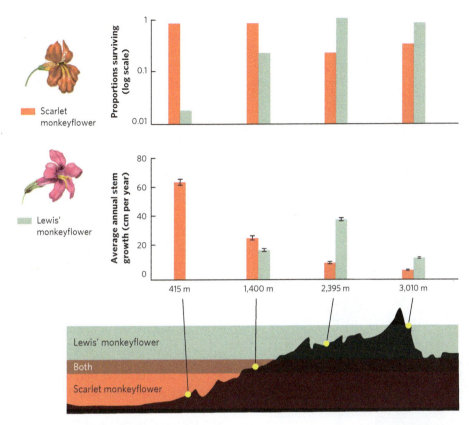

Figure 11.3 The distribution of two monkeyflower species. The scarlet monkeyflower is found at low elevations, whereas Lewis' monkeyflower grows at high elevations, as shown in the cross-section of the mountain. When transplanted to gardens at different elevations, the scarlet monkeyflower survives best at low elevations, while Lewis' monkeyflower survives best at high elevations. Similar patterns occur for plant growth. Error bars are standard errors. Data from A. L. Angert and D. W. Schemske, The evolution of species' distributions: Reciprocal transplants across the elevation ranges of *Mimulus cardinalis* and *M. lewisii, Evolution* 59 (2005): 1671–1684.

highest probability of successful reintroductions. Similarly, if a new pest species is accidentally introduced to a continent, its suitable habitats must be assessed to predict the area over which it might spread and the extent of the damage it might cause.

Predicting the potential geographic range of a single population or of all populations of a given species is a major challenge when there are few individuals living in the wild. One way to overcome this challenge is by using historic data on the distributions of populations. Such data are often available from collections of preserved organisms stored in museums and herbariums. In addition, when species have been introduced from other continents, researchers can try to determine the suitable habitat conditions found on the originating continent.

The process of determining the suitable habitat conditions for a species is known as **ecological niche modeling.** Because temperature and precipitation have a dominant influence on the distribution of biomes, modelers often begin by mapping the known locations of a species and then by quantifying the ecological conditions at the locations where the species has been recorded. The modeler can potentially include many

additional variables such as different soil types and the presence of potential predators, competitors, and pathogens that might limit the population's distribution. The range of ecological conditions that are predicted to be suitable for a species is the **ecological envelope** of the species. The concept of the ecological envelope is similar to the realized niche, but the realized niche includes the conditions under which a species currently lives, whereas the ecological envelope is a prediction of where a species could potentially live.

Modeling the Spread of Invasive Species

Ecological niche modeling can be a useful way to predict the expansion of pest species introduced to a continent where they have not previously lived. One example is the Chinese bushclover (*Lespedeza cuneata*), which is native to eastern Asia. In the late 1800s, this species of

Ecological niche modeling The process of determining the suitable habitat conditions for a species.

Ecological envelope The range of ecological conditions that are predicted to be suitable for a species.

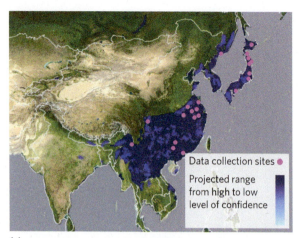

(a)

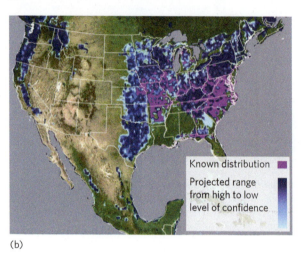

(b)

Figure 11.4 Ecological modeling of an invasive species. (a) Researchers collected data on the environmental conditions where the Chinese bushclover lives in Asia and used these data to predict the entire native range of the plant in Asia. **(b)** They then used the data to predict the future geographic range of Chinese bushclover in North America, where the plant was introduced 100 years ago and has continued to spread slowly. Data from A. T. Peterson et al., *Predicting the potential invasive distributions of four alien plant species in North America, Weed Science* 51 (2003): 863–868.

bushclover was brought to North Carolina to help control erosion, reclaim land that had been mined, and provide food for cattle. Over time, though, the plant quickly spread into the Great Smoky Mountains National Park and throughout the grasslands of the United States, where it displaced native plants. To determine the likely extent of its future spread, ecologists gathered data on the environmental conditions under which the bushclover lived in 28 locations in Asia. As illustrated in **Figure 11.4a**, they then quantified the ecological envelope for the clover to

predict the entire geographic range in Asia. They used these data to predict the potential geographic range in North America. As you can see in Figure 11.4b, they successfully predicted all locations where the clover had already spread. The model also predicted that the clover has the potential to live in many other locations, with widespread negative effects on the plant communities. The lack of Chinese bushclover in these areas suggests either that the model doesn't include additional ecological factors important to the clover or that the clover simply has not had enough time to disperse to these more distant locations.

HABITAT SUITABILITY AND GLOBAL WARMING

Knowledge of the ~~environmental conditions~~ that make a habitat suitable can also be used to understand the shift

in the geographic ranges of species as the environmental conditions of the world change. During the past century, for example, the average temperature of Earth has increased by 0.9°C. Some regions of the world—such as Alaska and northern Canada—have warmed as much as 4°C. In the relatively shallow North Sea, situated between the United Kingdom and Norway, temperatures in the bottom waters have increased more than 2°C since the 1970s, as shown in **Figure 11.5a**. Given that most species of fish have optimal temperature ranges, we might predict that warming ocean waters could cause southern fish species, which live in warm waters, to move north.

During the same period that temperatures in the North Sea were being monitored, the International Council for the Exploration of the Sea (ICES) compiled data on fish distributions by pulling large nets (called "trawls") along the ocean floor. From 1985 to 2006, scientists from six countries worked together to fish the bottom of the ocean at 300 locations distributed throughout the North Sea. Based on 7,000 trawl samples, the researchers reported that fish species richness in the North Sea had increased steadily over 22 years. As you can see in Figure 11.5b, there were about 60 species in the mid-1980s, but this grew to nearly 90 species 2 decades later. This list of species included dozens of more southerly species that had expanded their ranges northward.

The increase in species richness was positively correlated with the increase in bottom-water temperatures in the North Sea. This correlation—shown in Figure 11.5c—suggests that warmer temperatures are more hospitable to a greater variety of species, and that the warming of the North Sea has allowed more southerly species to expand the northern edges of their ranges into the area. Hence, this is a case of global warming increasing the diversity of species in a region.

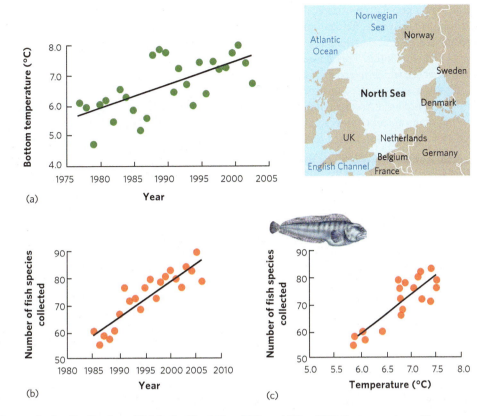

Figure 11.5 Changes in the distribution of fish in the North Sea. (a) From 1977 to 2003, the temperature of the water at the bottom of the North Sea increased by 2°C. **(b)** Surveys of fish trawls from 1985 to 2006 found that the total number of fish species collected per year in the North Sea increased from 60 to approximately 90. **(c)** The total number of fish species collected per year was positively correlated with the average sea temperatures during the previous 5 years.
Data from J. G. Hiddink and R. ter Hofstede, Climate induced increases in species richness of marine fishes, *Global Change Biology* 14 (2008): 453–460.

The increase in the diversity of species in the North Sea not only is dramatically changing the fish community but also may affect the important commercial fisheries that depend on this community. For example, the three species whose ranges have contracted as the North Sea became warmer—the wolffish (*Anarhichas lupus*), the spiny dogfish (*Squalus acanthias*), and the ling (*Molva molva*)—are all commercially important, whereas more than half of the southerly species with expanded ranges have little or no commercial value. As a result, while these shifts in species distributions with warming temperatures may increase fish diversity, it could actually decrease the value of North Sea commercial fisheries.

CONCEPT CHECK

1. What is a species' realized niche?
2. Why are not all locations in a species' geographic range occupied by the species?
3. What two environmental factors are commonly used when modeling the niche of a terrestrial species?

11.2 Population distributions have five important characteristics

To study the distribution of a population like the collared lizard in the Ozark Mountains, several characteristics of the population must be considered, including *geographic range, abundance, density, dispersion,* and *dispersal*. As we will see, each of these properties tells us something important about how individuals are distributed.

GEOGRAPHIC RANGE

We saw an example of geographic range in our discussion of the sugar maple tree (see Figure 11.1). The geographic range of a species includes all the areas its members occupy during their life. For example, the geographic range of sockeye salmon includes not only the rivers of western North America and eastern Asia, which are their spawning grounds, but also vast areas of the North Pacific Ocean, where individuals grow to maturity before making the long migration back to their birthplace. The geographic range is an important measure because it tells us how large an area a population occupies. If a population is restricted to a small area, for example, it may be very susceptible to a natural disaster such as a hurricane or a fire that can wipe it out. This is a serious challenge for **endemic** species, which live in a single, often isolated, location, such as the Galápagos finches that live on islands off the coast of South America. Populations with a larger geographic range are less vulnerable to such events because much of the population would remain unaffected. Species with very large geographic ranges that can span several continents are known as **cosmopolitan** species.

ABUNDANCE

The **abundance** of a population is the total number of individuals that exist within a defined area. For example, we might count the total number of lizards on a mountain, the number of sunfish in a lake, or the number of coconut trees on an island. The total abundance of a population is important because it provides a measure of whether a population is thriving or on the brink of extinction.

DENSITY

The **density** of a population is the number of individuals in a quantified area or volume. If we know the abundance of a population in a given area and we know the size of the area, then we can calculate density by dividing the abundance by the area. Examples of density include the number of bears per square kilometer in Alaska, the number of cattails per square meter in a pond, or the number of bacteria per milliliter of water. Density is a valuable measure because it tells ecologists how many individuals are packed into a particular area. If a habitat can support a higher density than currently exists, the population can continue to grow in the area. If the population density is greater than what the habitat can support, either some individuals will have to leave the area or the population will experience lower growth and survival.

Although individuals live only in suitable habitats, not all habitats are equal in quality because the environment is inherently variable. Some habitat patches have abundant resources that support a large number of individuals, while others have scarce resources and can support only a few individuals. Across a large geographic area, the highest concentrations of individuals are typically near the center of a population's geographic range. As one moves closer to the periphery of the geographic range, biotic and abiotic conditions become less ideal and support fewer individuals. Consider, for example, the geographic range of the dickcissel (*Spiza americana*), a small songbird related to the cardinal and found in North American prairies and grasslands. As illustrated in **Figure 11.6**, this bird has its highest densities in the center of its geographic range and its lowest densities near the periphery. However, because environmental conditions do not vary

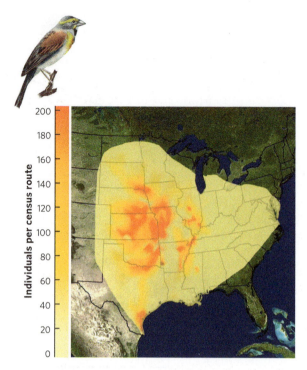

Figure 11.6 Densities across a geographic range. In the dickcissel bird, a relative of the cardinal, the highest densities are near the center of its geographic range and the lowest densities are near the periphery of its geographic range. Data from B. McGill and C. Collins, A unified theory for macroecology based on spatial patterns of abundance, *Evolutionary Ecology Research* 5 (2003): 469–492.

smoothly, the pattern of the dickcissel's preferred habitat is highly irregular.

DISPERSION

Dispersion of a population describes the spacing of individuals with respect to one another within the geographic range of a population. As shown in **Figure 11.7**, dispersion patterns can either be *clustered, evenly spaced*, or *random*.

Clustered Dispersion

In **clustered dispersion,** individuals are aggregated in discrete groups. Some clustered dispersions result from individuals living in social groups, as we discussed in

---

Endemic species Species that live in a single, often isolated, location.

Cosmopolitan Species with very large geographic ranges that can span several continents.

Abundance The total number of individuals in a population that exist within a defined area.

Density In a population, the number of individuals in a quantified area or volume.

Dispersion The spacing of individuals with respect to one another within the geographic range of a population.

Clustered dispersion A pattern of population dispersion in which individuals are aggregated in discrete groups.

Clustered

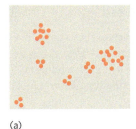

(a)

Evenly spaced

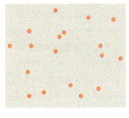

(b)

Random

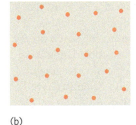

(c)

Figure 11.7 Three types of dispersion patterns. (a) Clustered dispersion is characterized by individuals that are aggregated. For example, sulphur tuft fungi (*Hypholoma fasciculare*) grow in clusters anywhere there is a rotting stump, such as this site in the United Kingdom. **(b)** Evenly spaced dispersion is characterized by each individual maintaining a minimum distance between neighbors, as is the case for these blue-eyed cormorants (*Phalacrocorax atriceps*) nesting in Patagonia, Argentina. **(c)** Random dispersion is characterized by each individual's position being independent of the location of other individuals, such as these dandelions growing in a pasture in Bulgaria. Photos by (a) Gary K. Smith/naturepl.com; (b) Juan Carlos Munoz/naturepl.com; (c) FLPA/Bob Gibbons/AGE Fotostock.

Chapter 10. For instance, some birds live in flocks and some fish live in schools. Other clustered dispersions occur because individuals stay near clustered resources. For example, salamanders and sow bugs aggregate under logs because individuals of both species are attracted to dark, moist places. Another cause of clustered dispersions is offspring that remain close to their parents. Some species of trees, such as the quaking aspen, form clusters of stems because a parent tree gives rise to offspring by sending up new stems from its roots, a form of vegetative reproduction. As a result, we commonly see clusters of aspen trees that are composed of a parent tree surrounded by its offspring. Within such a cluster, however, the stems tend to be *evenly spaced*. As a result, a population can exhibit one pattern of dispersion at a large scale but a different pattern of dispersion at a smaller scale.

Evenly Spaced Dispersion

In **evenly spaced dispersion,** each individual maintains a uniform distance between itself and its neighbors. In agricultural settings, we can observe even spacing in crops such as corn or apple trees because farmers want each plant to have sufficient resources to maximize

Evenly spaced dispersion A pattern of dispersion of a population in which each individual maintains a uniform distance between itself and its neighbors.

crop production. In natural settings, even spacing most commonly arises from direct interactions between individuals. For example, plants positioned too close to larger neighbors often suffer from shading and root competition. In addition, some plants can emit chemicals from their leaves and roots that inhibit the growth of other plants around them. As these crowded individuals die, the remaining individuals become more evenly spaced. We can also observe evenly spaced distributions in animals that defend territories, such as birds and lizards. Because territory size commonly depends on the amount of resources available, neighboring territory holders commonly defend areas that are similar in size, which causes the territory holders to be evenly spaced.

Random Dispersion

In **random dispersion,** the position of each individual is independent of the position of other individuals in the population. Random dispersions are not common in nature, largely because abiotic conditions, resources, and interactions with other species are not randomly distributed. However, if we wished to know whether the dispersion pattern of a population at a particular spatial scale is clustered or evenly spaced, we would have to demonstrate statistically that its distribution was significantly different from random.

DISPERSAL

Dispersal is the movement of individuals from one area to another. Dispersal is distinct from migration, which is the seasonal movement of individuals back and forth between habitats, such as birds flying north and south with the changing seasons. In contrast, dispersal involves individuals leaving their habitat of origin—where a seed was made or where a squirrel was born—and typically not returning. As we saw at the beginning of this chapter, dispersal is of great interest to ecologists because it is the mechanism by which individuals can move between suitable habitats and, in some cases, colonize suitable habitats that are not already inhabited by the species. Dispersal can also be a way to avoid areas of high competition or high predation risk. For example, when a fish arrives in a section of a stream, many of the aquatic insects in the stream disperse by floating downstream at night to avoid being eaten.

11.3 The distribution properties of populations can be estimated

Thus far, we have focused on the conceptual basis of the five properties—abundance, density, geographic range, dispersion, and dispersal. For a more complete understanding of how populations are distributed, quantitative measures must also be considered.

QUANTIFYING THE LOCATION AND NUMBER OF INDIVIDUALS

One way to determine the number of individuals in an area is to conduct a **census,** which means counting every individual in a population. Every 10 years, for example, the United States government conducts a census with the goal of counting every person living in the country. For most species, however, it is not feasible to count every individual in the population. As a result, scientists must conduct a **survey,** in which they count a subset of the population. Using these samples, they estimate the abundance, density, geographic range, and distribution of the population. Scientists have developed a variety of ways to make these estimates, including *area-* and *volume-based surveys, line-transect surveys,* and *mark-recapture surveys* (**Figure 11.8**).

Area- and Volume-Based Surveys

Area- and volume-based surveys define the boundaries of an area or volume and then count all the individuals within that space. The size of the defined space is typically related to the abundance and density of the population. For example, researchers who wish to know the number of bacteria in the soil might collect samples of soil that are only a few cubic centimeters in volume. In contrast, researchers who wish to know the number of individual corals on a coral reef might sample areas that are 1 m². At the most extreme, researchers who want to estimate the abundance, density, and distribution of large mammals might count the number of individuals in aerial photos that cover hundreds of square meters. By taking multiple samples, scientists can determine how many individuals are present in an average sample area of land or volume of soil or water.

Line-Transect Surveys

Line-transect surveys count the number of individuals observed as one moves along a line. There are many variations on this technique. For example, researchers might

---

Random dispersion A pattern of dispersion of a population in which the position of each individual is independent of the position of other individuals in the population.

Dispersal The movement of individuals from one area to another.

Census A count of every individual in a population.

Survey Counting a subset of the population.

Area- and volume-based surveys Surveys that define the boundaries of an area or volume and then count all the individuals in the space.

Line-transect surveys Surveys that count the number of individuals observed as one moves along a line.

CONCEPT CHECK

1. What mechanisms could cause individuals in a population distribution to be either evenly spaced or clustered?
2. Why are endemic species at a high risk of extinction?
3. What is the difference between dispersal and dispersion when describing population distributions?

(a) (b) (c)

Figure 11.8 Estimating population abundance, density, and distribution. (a) Area-based studies count the number of individuals within a fixed area, such as this researcher working on the Dendles National Natural Reserve in England. **(b)** Linear-transect studies, including this one at the Great Barrier Reef in Australia, count the number of individuals that are observed along a predefined line. **(c)** Mark-recapture studies collect a sample of the population, mark them, and then return them. For example, horseshoe crabs (*Limulus polyphemus*) in Delaware are marked with small circular tags. A short time later, a second sample is collected to determine the proportion of marked animals in the population, which can be used to estimate the size of the total population. Photos by (a) Paul Glendell/Alamy; (b) Suzanne Long/Alamy; (c) Patrick W. Grace/Science Source.

survey small plants in a field or forest by tying a long string between two fixed points and counting the number of individuals the string crosses. From this, we can determine the abundance in a given linear distance. Alternatively, researchers might count all individuals that are observed within a fixed distance of a line, such as the number of trees on a savanna located within 100 m of a line, which would provide us with an estimate of abundance in a given area. A similar approach has been used in surveys of amphibians. In this case, observers count the number of frogs that can be heard along a predetermined path. If we know how far, on average, a person can hear a frog call, we can estimate the number of frogs calling in an area that includes both sides of the path. Such line-transect data can be converted into area estimates.

One of the most famous line-transect studies is the annual Christmas bird count. The bird count began in 1900 when 27 volunteers from the Audubon Society positioned themselves at different locations in North America and counted all the birds they saw in one day. Today, tens of thousands of volunteers go outside during their winter holidays and follow a predetermined path that covers a 24-km circle. Throughout the day, the volunteers count the number of individuals of every bird species they can see or hear within this circle. This long-term survey of birds in

North America has provided incredibly valuable data that has helped scientists determine which species of birds have populations that are increasing, stable, or declining.

Mark-Recapture Surveys

Area- and volume-based and line-transect studies are very useful for organisms that do not move—such as plants and corals—or animals that are not easily disturbed—such as snails—and therefore less likely to leave the area during the survey. Some animals, however, are very sensitive to the presence of researchers and will leave the area, while other species are well camouflaged and difficult to find. Both situations can cause us to underestimate the number of individuals in a population. For these situations, we need a different type of sampling technique. One effective method is the use of **mark-recapture surveys.** As the name implies, mark-recapture surveys collect a number of individuals from a population and mark them. These individuals are then returned to the population. Once enough time has passed for the marked individuals to mingle

Mark-recapture survey A method of population estimation in which researchers capture and mark a subset of a population from an area, return it to the area, and then capture a second sample of the population after some time has passed.

ANALYZING ECOLOGY

Mark-Recapture Surveys

To estimate the number of individuals in a population using mark-recapture surveys, we need to know how many individuals were initially sampled and marked. For example, let's imagine that crayfish researchers collected 20 crayfish from a 300-m² stretch of stream and marked them with a dot of red fingernail polish. Once the fingernail polish was dry, the crayfish were returned to the stream. After waiting one day for the marked crayfish to move around the stream, the researchers collected another sample of crayfish. In this second sample, they captured 30 crayfish, 12 of which were marked. Based on these data, how many crayfish were in the 300 m² of stream? To find out, we can use an equation that considers how many individuals are marked and—after the marked individuals are released back into the population—the ratio of total individuals to marked individuals in the entire population.

Let's look at how we estimate the size of the population. First, note that that we capture and mark a number of individuals (M) from an entire population whose size is defined as N. Therefore, the fraction of marked individuals in the entire population is $\frac{M}{N}$.

When we go back and capture individuals the second time, we record the number of individuals in the second capture (C) and the number of marked individuals that are recaptured (R). The fraction of marked individuals in the recaptured population is $\frac{R}{C}$.

The fraction $\frac{R}{C}$ and our first fraction, $\frac{M}{N}$, actually represent the same fraction, which is the proportion of marked individuals in a sample. Given that these two fractions should represent the same number, we can set the two ratios equal to each other and solve for the unknown variable (N), which is the total size of the population:

$$\frac{M}{N} = \frac{R}{C}$$

$$N = M \times C \div R$$

Applying this equation to our crayfish data, the estimated number of crayfish in the stream is

$$N = 20 \times 30 \div 12$$

$$N = 50$$

YOUR TURN What is the estimated size of the crayfish population if the second sample collected 48 crayfish and the number recaptured in that sample was 24?

Based on your estimate of crayfish abundance and the data provided on stream area, what is your estimate of crayfish density?

throughout the population, a second sample of the population is made. Based on the number originally marked, the total number collected the second time, and the number of marked animals collected the second time, we can estimate the size of the population. Mark-recapture studies are commonly conducted on birds, fish, mammals, and highly mobile invertebrates. The actual calculations for arriving at this estimate are discussed in "Analyzing Ecology: Mark-Recapture Surveys."

QUANTIFYING THE DISPERSAL OF INDIVIDUALS

Quantifying the dispersal of individuals from a population requires identifying the source of individuals. As we will see, this can be done by ensuring that there is only one possible source of individuals and then determining how far individuals disperse from this single location. In other cases, individuals are marked and then observed or recaptured at some later time to determine how far they moved from the location where they were marked (**Figure 11.9**). In animal studies, possible marks include ear tags, radio transmitters, or leg bands. In plant studies, researchers

Figure 11.9 Measuring dispersal. This California condor (*Gymnogyps californianus*), which was captured in the Grand Canyon National Park in Arizona, has been marked using a wing band that has a unique number to identify where it hatched and how far it dispersed. Photo by Thomas & Pat Leeson/Science Source.

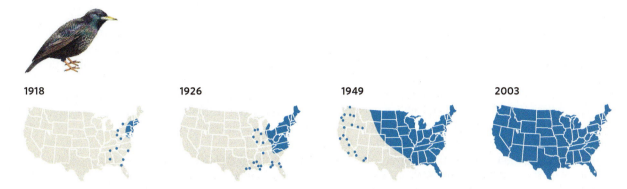

1918   1926   1949   2003

Figure 11.10 The rapid dispersal of European starlings across the United States. After being introduced near New York City in 1890 and 1891, the bird rapidly spread across the United States over a 60-year period. The blue-shaded area represents the range of breeding populations. The blue dots indicate detections of starlings that exhibited unusually long dispersal, thereby facilitating the rapid spread of the species. Today, the starling population covers more than 7 million km^2 of North America from coast to coast. Data from B. Kessel, Distribution and migration of the European starling in North America, *Condor* 55 (1953): 49–67; and the U.S. Geological Survey, 2003, http://www.mbrpwrc.usgs.gov/bbs/htm03/ra2003_red/ra04930.htm

can mark pollen with fluorescent powders and then examine surrounding flowers to determine how far the pollen grains have been moved either by the wind or by pollinators.

A common measure of dispersal is the **lifetime dispersal distance,** which is the average distance that an individual moves from where it begins its life to where it reproduces. By knowing the lifetime dispersal distance, we can also estimate how rapidly a growing population can increase its geographic range. For example, when researchers marked eight species of songbirds with leg bands, they found that lifetime dispersal distances averaged between 344 and 1,681 m. A lifetime dispersal distance of about 1 km per generation is therefore not unusual for populations of songbirds. At this rate, the descendants of an average individual might traverse an entire continent in a thousand generations or so.

These calculations suggest that it will take an average species of songbird more than 1,000 generations of dispersal to travel across a continent. However, it can actually happen much faster because a few individuals in a population can disperse much farther than the average bird in the population. An excellent example of this occurred following the introduction of the European starling to the United States. In 1890 and 1891, 160 European starlings were released in the vicinity of New York City. Within 60 years, the population had spread 4,000 km from New York to California, at an average rate of about 67 km per year. You can see this rapid spread of starlings in **Figure 11.10.** The expansion occurred rapidly because a few individuals dispersed much longer distances than the average and established new populations beyond the range boundary of the species. Although the few individuals that might move over such long distances

are rare, they can have large effects on population distributions. Today, the starling lives throughout most of the United States.

CONCEPT CHECK

1. Why are surveys, rather than censuses, used to quantify the abundance of many animals?
2. How can line-transect surveys be used to estimate the number of animals per unit area?
3. Why is the average dispersal distance a misleading estimate of how rapidly a population can move across very long distances over time?

11.4 Population abundance and density are related to geographic range and adult body size

Given the five properties of population distributions, all of which can span a wide range of values, ecologists often search for relationships that might explain the underlying causes of variation in these values. Common patterns include the relationship between population abundance and geographic range and the relationship between population density and adult body size.

POPULATION ABUNDANCE AND GEOGRAPHIC RANGE

Ecologists have consistently found that populations with high abundance also have a large geographic range. This pattern has been observed in plants, mammals, birds, and protists. Consider the data on the birds of North America

---

Lifetime dispersal distance The average distance an individual moves from where it was hatched or born to where it reproduces.

high abundance & high resc

in **Figure 11.11**. We can see a positive relationship; species with the highest abundances also are the species with the largest range sizes. This same pattern was observed in collared lizards; as their abundance declined, fewer glades were occupied and the geographic range of the population shrank. In the bird data, however, there is a great deal of variation around the regression line. If we were to quantify the coefficient of determination, we would find that abundance of the populations only explained 13 percent of the variation in the geographic ranges of the populations. To review the concept of coefficient of determination, see "Analyzing Ecology: Coefficients of Determination" on page 182 in Chapter 8.

The causes of the range–abundance relationship are still widely debated. There is general consensus that resource availability plays an important role. For example, if a species relies on resources that are only available in a small geographic area, the species will only inhabit a small geographic range. If resources are abundant over a large geographic area, we might expect that species to cover a large geographic range and to be abundant. In short, the distribution of resources should cause a positive relationship between abundance and geographic range.

Although we do see this positive relationship, we are left to wonder about all the unexplained variation in the relationship shown in Figure 11.11. In some cases, we might be observing fluctuations within a geographic range. For example, individuals might disperse to more marginal habitats during years that favor especially high reproduction and survival. The range would then contract during years that have low reproduction and survival. Such year-to-year fluctuations can cause the data on the abundance and geographic range of a population to vary a great deal.

The relationship between population abundance and distribution suggests that reducing the geographic range of a population—for example, by converting habitat to agricultural purposes or to housing—will also reduce the size of the population. Similarly, factors that reduce the overall size of the population will simultaneously reduce the geographic range of the population because marginal habitats will no longer receive as many dispersing individuals.

POPULATION DENSITY AND ADULT BODY SIZE

Another common pattern across species is the relationship between population density and body size. Generally, the density of a population is negatively correlated to the body size of the species. **Figure 11.12** shows that in herbivorous animals, the smallest-bodied species—such as mice—live at the highest densities, and the largest-bodied species—such as elephants—live at the lowest densities. Part of this relationship is a matter of body size relative to space. A square meter of soil harbors hundreds of thousands of small arthropods, whereas a single elephant simply would not fit into this space. Even if it could be squeezed in, a square meter of soil would not produce enough food to sustain the elephant. A large individual requires more food and other resources than a small one. As a result, we expect a given plot of habitat to support fewer large individuals than it does small individuals.

> ### CONCEPT CHECK
>
> 1. What is the relationship between population abundance and geographic range across many species?
> 2. What is the relationship between population density and adult body size across many species?

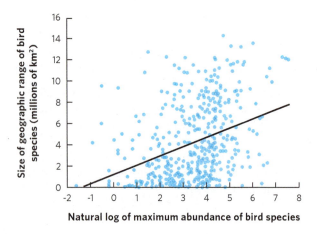

Figure 11.11 Abundance and range size of birds. Data from 457 species of North American birds show that bird species with higher abundances are generally more widely distributed. However, this relationship contains a great deal of variation around the line of best fit due, in part, to annual variations in abundance and geographic ranges. Data from B. McGill and C. Collins, A unified theory for macroecology based on spatial patterns of abundance, *Evolutionary Ecology Research* 5 (2003): 469–492.

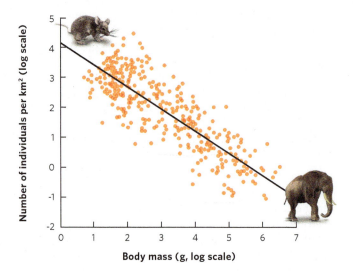

Figure 11.12 Body mass and population density. Across 200 species of herbivorous mammals, population density declines as a function of adult body mass. Data from J. Damuth, Interspecific allometry of population density in mammals and other animals: The independence of body mass and population energy-use, *Biological Journal of the Linnean Society* 31 (1987): 193–246.

11.5 Dispersal is essential to colonizing new areas

As we mentioned earlier in this chapter, dispersal plays a key role in allowing individuals to move between patches of habitat and to colonize suitable habitats that are not inhabited. In this section, we will discuss how suitable habitats may remain unoccupied due to *dispersal limitation*; we will also examine how *habitat corridors* can play an important role in facilitating dispersal.

DISPERSAL LIMITATION

In some cases, as we saw with the Chinese bushclover, not enough time has passed since the introduction of a species for individuals to spread to every suitable habitat. In other cases, such as the collared lizards, there can be substantial barriers that prevent dispersal between suitable habitats. A common barrier to dispersal is the presence of large expanses of inhospitable habitat that an organism cannot cross, such as an ocean that a plant seed cannot cross or a large desert that an amphibian cannot cross. Recall that the sugar maple has a preferred range of temperatures and precipitation that favors its presence in the northern temperate forests of North America. Although there are no other suitable areas for sugar maples to grow in North America, there is an abundance of suitable habitat in Europe and Asia. Indeed, several other species of the maple genus (*Acer*) live in habitats that should be suitable for the sugar maple. Sugar maple does not exist in Europe and Asia because its seeds are not capable of dispersing from North America across the oceans to these distant regions. Sometimes the inhospitable habitat is not particularly expansive but is still an effective barrier to dispersal, as was the case with fire-suppressed forests that surrounded glades and prevented collared lizard dispersal. The absence of a population from suitable habitat because of barriers to dispersal is called **dispersal limitation.**

Occasionally, individuals cross formidable barriers and disperse long distances without assistance from humans. We know this because many species of plants and animals have populated remote islands, such as the Hawaiian Islands, before humans arrived. However, humans have affected the dispersal of many species. For example, human activities such as road building and forest clearing have created barriers to dispersal for some species. Humans have also assisted in the dispersal of plants and animals for thousands of years. For instance, Aboriginal peoples brought dogs to Australia and Polynesians distributed pigs and rats throughout the small islands of the Pacific. In more recent times, foresters have transplanted fast-growing eucalyptus trees from Australia and pines from California to points all over the world for timber and fuel wood. Other species have been intentionally moved to assist them in dispersing over barriers created by humans, such as roads. Still other species have been accidentally moved over large distances by riding along in the ballast water of cargo ships or by attaching themselves to the outside of ship hulls. In most cases, individuals are introduced into a new area but never establish a viable population. In some cases, however, the introduced individuals are able to develop into a population that can grow and expand its geographic range over time. We'll have much more to say about such invasive organisms later in this book, but their success in many places outside their native ranges emphasizes the role of barriers to dispersal in limiting species' distributions.

HABITAT CORRIDORS

In some landscapes, dispersal is helped when strips of favorable habitat known as **habitat corridors** are located between large patches of habitat. For example, two forests might be separated by an open field with a stream that has a narrow band of trees along its banks. This band of trees along the stream can serve as a corridor of suitable habitat between the forests and allow individuals to easily disperse between them.

In recent years, large manipulative experiments have tested the importance of habitat corridors. For example, in the pine forests of South Carolina, storms and fires often create cleared areas that are suitable habitat for many species of understory plants and animals. To test the importance of habitat corridors in facilitating dispersal between these cleared areas, researchers cleared large (1.375 ha) patches of forest. One of the five cleared areas, called the central patch, was the source of dispersers. Of the remaining four patches, one was connected to the central patch with a cleared path that could serve as a corridor. The other three outer patches included a simple rectangular-shaped patch and two patches with "wings," meaning that they were patches with corridors but did not connect to other patches (**Figure 11.13**). The researchers replicated this experimental design eight times.

Dispersal limitation The absence of a population from suitable habitat because of barriers to dispersal.

Habitat corridor A strip of favorable habitat located between two large patches of habitat that facilitates dispersal.

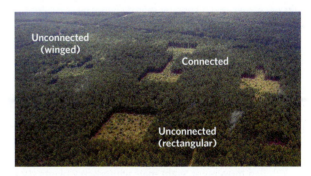

Figure 11.13 Manipulating habitat corridors. Researchers cleared patches in the pine forests of South Carolina. Using groups of five patches, the central patch served as a source of dispersers. The outer patches included one patch that was connected to the central patch by a cleared habitat corridor, while the other three were not connected. One unconnected patch was a simple rectangle. The other two patches had cleared corridor paths, termed "wings," that were not connected to another patch. Photo by Ellen Damschen.

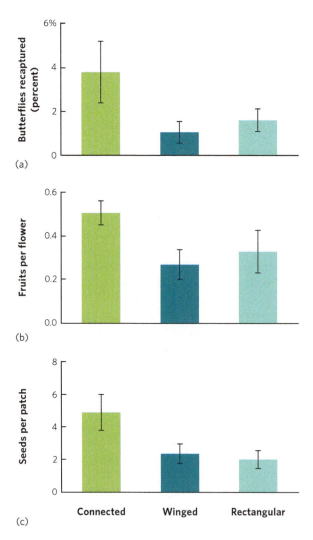

(a)

(b)

(c)

| | Connected | Winged | Rectangular |

Figure 11.14 The effects of manipulated corridors on dispersal. Compared to unconnected patches that were either winged or rectangular in shape, patches connected by a habitat corridor experienced **(a)** a higher number of recaptured butterflies, **(b)** greater pollination of plants, leading to more fruits per flower, and **(c)** greater seed dispersal by birds. Error bars are standard errors. Data from J. J. Tewksbury et al., Corridors affect plants, animals, and their interactions in fragmented landscapes, *Proceedings of the National Academy of Sciences USA* 99 (2002): 12923–12926.

A number of methods were used to measure dispersal. The movements of the common buckeye butterfly (*Junonia coenia*) were followed by marking individuals in the central patch and recapturing them in the four outer patches. To track pollen movement, eight male winterberry plants (*Ilex verticillata*) were planted in each central patch and three mature female plants were planted in each of the four outer patches. Fertilized flowers producing fruits in the outer patches would indicate that pollen had moved from the central patch. To track seed and fruit movements, a local holly (*Ilex vomitoria*) and wax myrtle (*Myrica cerifera*) were planted in the central patch. In some cases, the fruits in the central patch were dusted with colored fluorescent powder.

The researchers then collected samples of bird droppings in traps placed under artificial perches in each outer patch. Fecal matter that fluoresced under ultraviolet light indicated that birds had consumed fruits in the central patch and dispersed them to the outer patches.

The results of these experiments are shown in **Figure 11.14**. The outer patch connected by a habitat corridor had much more dispersal from the central patch, including more dispersal by butterflies, more traffic by insect pollinators—which caused more fruits to be produced—and greater movement of fruits and seeds found in bird droppings. Indeed, the movement of fruits and seeds through the corridors was so frequent that the number of species of herbs and shrubs increased faster in the connected patches. Experiments such as these confirm the importance of habitat corridors. Indeed, conservation efforts have increasingly considered the preservation of corridor habitats. Along the Rio Grande in Texas, for example, state and federal biologists, in collaboration with conservation organizations, have pushed to protect river-side habitats that would allow species to move easily among large patches of protected land. As you can see in **Figure 11.15**, in the United States, this land includes Big Bend National Park, Big Bend Ranch State Park, and the Black Gap Wildlife Management Area. Across the river in Mexico the protected lands include the Cañon de Santa Elena Flora and Fauna Protected Area, the Ocampo Flora and Fauna Protected Area, and the Maderas del Carmen Flora and Fauna Protected Area.

CONCEPT CHECK

1. Why are many species absent from continents that contain suitable habitat?
2. How do habitat corridors facilitate dispersal?

11.6 Many populations live in distinct patches of habitat

In our discussion of the collared lizard, we saw that individuals in the region live in many small populations, with each group restricted to a particular glade. The lizards can also move between habitats, but such movement depends on the quality of the forested habitat connecting the glades. When we want to understand the movement of individuals among patches of habitats and how this affects the abundance of animals in each patch of habitat, we need to consider the quality of the habitat. We also need to consider the difficulty in reaching the habitat, which is determined by the distance to the habitat and the difficulty in crossing the area between habitat patches. In this section, we will explore how habitat quality affects the distribution of individuals among habitats and how ecologists have conceptualized the distribution and movement of individuals among suitable patches of habitat.

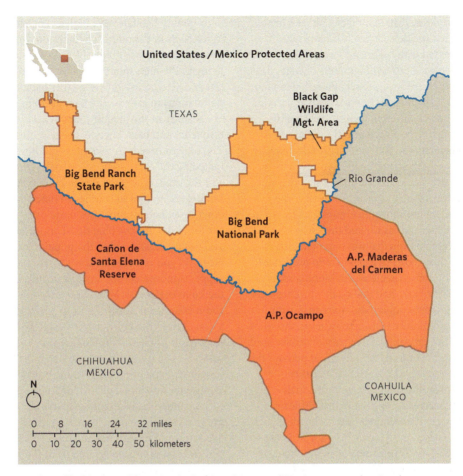

Figure 11.15 Conservation of habitat corridors. On the border of Texas and Mexico, state and national governments have protected a massive corridor of more than 1.3 million ha of land along the Rio Grande.

THE IDEAL FREE DISTRIBUTION AMONG HABITATS

When habitats differ in quality and individuals can move easily among habitat patches, natural selection should favor those individuals that can choose the habitat that provides them with the most energy; this would improve their fitness. If all individuals are able to distinguish between high- and low-quality habitats, and assess the benefit of living in a particular habitat, then all individuals should move to the high-quality habitats, which are represented by a green line in **Figure 11.16**. However, as more and more individuals choose the high-quality habitat, the resources available must be divided among many individuals. This reduces the resources available to each individual, also known as the *per capita benefit*. At some point, the per capita benefit in the high-quality habitat falls so low that an individual would be better off if it moved over to the low-quality habitat, which is indicated by the orange line in Figure 11.16. Continued increases in the population size would continue to add individuals to both the high- and low-quality habitats such that individuals in both habitats would experience the same per capita benefit. When all individuals have perfect knowledge of habitat variation and they distribute themselves in a way that allows them to have the same per capita benefit, we call it the **ideal free distribution.**

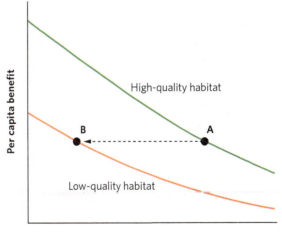

Figure 11.16 The ideal free distribution. Given the existence of high- and low-quality habitats, the first individual to arrive should select the high-quality habitat. As more individuals choose the high-quality habitat, however, the per capita benefit of the habitat declines. At some point (point A), it is equally beneficial for an individual to move to the low-quality habitat (point B) because the per capita benefit equals that experienced in the high-quality habitat. Although the low-quality habitat will have fewer individuals than the high-quality habitat, all individuals will experience the same per capita benefits.

Ideal free distribution When individuals distribute themselves among different habitats in a way that allows them to have the same per capita benefit.

An example of ideal free distribution can be seen in an experiment in which stickleback fish (*Gasterosteus aculeatus*) were presented with high- and low-quality habitats. Researchers placed different numbers of water fleas (*Daphnia magna*), which are a type of plankton that is a food source for stickleback fish, at opposite ends of an aquarium, thereby creating habitats of different quality at each of its ends. When the first fish were placed into the aquarium with no water fleas present, the fish were distributed equally between the two halves of the aquarium. Then researchers added 30 water fleas per minute to one end of an aquarium and 6 per minute to the other, a ratio of five to one. The results of this experiment are shown in **Figure 11.17**. Within 5 minutes, the fish had distributed themselves between the two halves in a ratio that was approximately four to one, which is close to the five to one ratio predicted by an ideal free distribution. When the provisioning ratio was changed—that is, the high- and low-quality patches were switched between ends of the aquarium—the fish quickly adjusted their distribution ratio. How they achieved this ideal free distribution was not determined, but they may have used the rate at which they encountered food items, or perhaps the number of other fish close by, as cues to patch quality.

The ideal free distribution tells us how individuals should distribute themselves among habitats of differing quality, but individuals in nature rarely match the ideal expectations. In some cases, individuals may not be aware that other habitats exist. Also, the fitness of an individual is not determined solely by maximizing its resources. Other factors that influence an individual's use of a particular habitat include the presence of predators or of a territory owner that precludes moving into a high-quality habitat.

When reproductive success has been measured in the field, ecologists commonly find that individuals living in high-quality habitats have higher reproductive success, while individuals in the poorest habitats do not produce enough offspring to replace themselves. As a result, the populations living in the high-quality habitats are a source of dispersing offspring that move to low-quality habitats, which allows populations in low-quality habitats to persist.

A study of the blue tit (*Parus caeruleus*), a small songbird from southern Europe, helps to illustrate this point. The blue tit breeds in two kinds of forest habitat, one dominated by the deciduous downy oak (*Quercus pubescens*) and the other by the evergreen holm oak (*Quercus ilex*). Long-term studies in southern France have revealed that the downy oak habitat produces more caterpillars, an important food item for the blue tits. This difference in caterpillar availability is reflected in the population densities of the birds. As you can see in **Figure 11.18**, the downy oak forest supports more than six times as many adult breeding pairs as the holm oak forest. Moreover, each breeding pair produces about 60 percent more offspring per year in the downy oak forest than in the holm oak forest. If we assume that juvenile survival is 20 percent in the first year and that annual adult survival is 50 percent—values that are typical of temperate songbirds—the population in the downy oak habitat would experience an annual growth of 9 percent per year if the surplus individuals did not disperse out of the habitat. At the same time, the population in the low-quality, holm oak habitat would experience an annual decline of 13 percent per year if no birds emigrated from the high-quality habitat. This would cause the population in the low-quality habitat to rapidly go extinct. In reality, the populations persist over time because the low-quality habitat receives surplus individuals that disperse in from the high-quality habitat.

CONCEPTUAL MODELS OF SPATIAL STRUCTURE

The ideal free distribution describes how individuals should distribute themselves when they are aware of the quality of different habitats and can freely move among them. However, as mentioned, it is not always easy to move from one patch of habitat to another. In such scenarios, most individuals in the patch stay put, only occasionally dispersing to other habitat patches. As a result, the larger population is broken up into smaller groups of conspecifics that live in isolated patches, called **subpopulations.**

When individuals frequently disperse among subpopulations, the whole population functions as a single structure and they all increase and decrease in abundance synchronously. When dispersal is infrequent, however, the abundance of individuals in each subpopulation can fluctuate independently of one another. In considering the spatial structure of subpopulations, ecologists have devised three models: the *basic metapopulation model*, the *source–sink metapopulation model*, and the *landscape metapopulation model*. We will explore these models in more detail in Chapter 13.

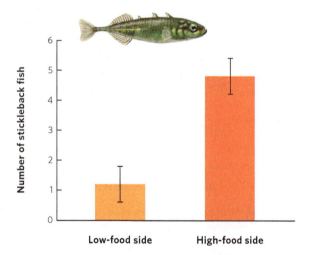

Figure 11.17 The ideal free distribution in stickleback fish. When six fish were placed in an aquarium with the two ends receiving a five-to-one difference in food rations, the fish distributed themselves in a four-to-one ratio. As a result, each fish received a similar per capita amount of food. Error bars are standard deviations. Data from M. Milinski, An evolutionarily stable feeding strategy in sticklebacks, *Zeitschrift für Tierpsychologie* 51 (1979): 36–40.

Subpopulations When a larger population is broken up into smaller groups that live in isolated patches.

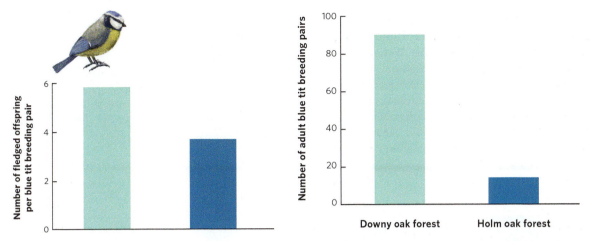

Figure 11.18 The effects of habitat quality. In blue tit songbirds, deciduous downy oak forests support many more adult breeding pairs than evergreen holm oak forests. In addition, birds living in the downy oak forest produce about 60 percent more fledglings per nest. Data from J. Blondel, Habitat heterogeneity and life-history variation of Mediterranean blue tits (*Parus caeruleus*), *The Auk* 110 (1993): 511–520.

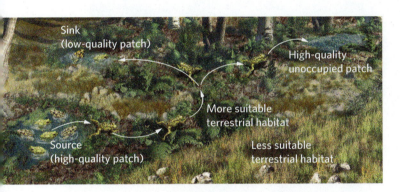

Figure 11.19 The landscape metapopulation model. The landscape model contains the most realistic conditions of habitat patches that differ in quality and a surrounding matrix that is variable in its suitability for dispersal and the presence of dispersal barriers.

The Basic Metapopulation Model

The **basic metapopulation model** describes a scenario in which there are patches of suitable habitat embedded within a matrix of unsuitable habitat. All suitable patches are assumed to be of equal quality. Some suitable patches are occupied while others are not, although the unoccupied patches can be colonized by dispersers from occupied patches. The basic metapopulation models emphasize how colonization and extinction events can affect the proportion of total suitable habitats that are occupied.

The Source–Sink Metapopulation Model

The **source–sink metapopulation model** builds on the basic metapopulation model and adds the reality that different patches of suitable habitat are not of equal quality. As we saw in the study of blue tits, it is common that occupants of high-quality habitats are a source of dispersers. Such populations are referred to as **source subpopulations.** At the same time, there can be low-quality habitats

that rarely produce enough offspring to produce any dispersers. These habitats depend on outside dispersers to maintain the subpopulation. These subpopulations are known as **sink subpopulations.**

The Landscape Metapopulation Model

The **landscape metapopulation model** is even more realistic than the source–sink model because it incorporates differences in the quality of the suitable patches and the quality of the surrounding matrix. As we discussed earlier in this chapter, the habitat in the surrounding matrix can vary in quality for dispersing organisms. For example, for a regional population of collared lizards to persist and grow requires both high-quality glade habitats and a high-quality matrix of open forest that allow for dispersal. Similarly, consider the challenge that metamorphosing frogs encounter when they move from their natal pond. They face risks of both predation and desiccation. Moving through a grassy field poses a much higher risk of predation and drying out than moving through a humid forest. Although neither field nor forest is a suitable habitat for the frog to reproduce, each is a dispersal barrier. The landscape model, which is illustrated in **Figure 11.19**, represents the most realistic, yet also the most complex, spatial structure of populations.

Basic metapopulation model A model that describes a scenario in which there are patches of suitable habitat embedded within a matrix of unsuitable habitat.

Source–sink metapopulation model A population model that builds on the basic metapopulation model and accounts for the fact that not all patches of suitable habitat are of equal quality.

Source subpopulations In high-quality habitats, subpopulations that serve as a source of dispersers within a metapopulation.

Sink subpopulations In low-quality habitats, subpopulations that rely on outside dispersers to maintain the subpopulation within a metapopulation.

Landscape metapopulation model A population model that considers both differences in the quality of the suitable patches and the quality of the surrounding matrix.

Throughout this chapter, we have examined the spatial structure of populations by considering the characteristics of abundance, density, and geographic range. We have also focused on the importance of suitable habitats and dispersal and how ecological niche modeling can help us predict the distributions of populations in the future. In doing so, we have largely ignored the fact that populations increase and decrease in abundance, but in the next chapter, we will take an in-depth look at how populations grow and how this growth is regulated.

CONCEPT CHECK

1. Why do many species fail to exhibit an ideal free distribution?
2. What reality does the source–sink metapopulation model involve that the basic metapopulation does not?
3. What reality does the landscape metapopulation model include that the source-sink metapopulation does not?

ECOLOGY TODAY   APPLYING THE CONCEPTS

The Invasion of the Emerald Ash Borer

The emerald ash borer. After being accidentally introduced to the United States in the 1990s, the insect has decimated ash tree populations. Photo by David Cappaert, Bugwood.org.

In 2002, a beautiful green insect, the emerald ash borer (*Agrilus planipennis*), was observed for the first time in North America. A native of eastern Asia, it had never lived in North America and was unfamiliar to most people. Over the next decade, its population grew and the emerald ash borer became one of the most numerous invasive pests on the continent.

The expanse of inhospitable oceans had previously prevented the emerald ash borer from dispersing to North America. Researchers suspect that the insect arrived in the 1990s from Asia by accidentally hitching a ride inside wooden crates that were shipped from Asia to Detroit, Michigan. Once the insects arrived in Michigan, they found an abundant supply of their favorite food: ash trees. The adult insect does little harm to ash trees, but it lays eggs in the cracks of the tree's bark. When the eggs hatch, the larvae live in clustered distributions and they consume the underlying cambium and phloem of the trees. Because the cambium is essential to tree growth and the phloem is essential for transporting nutrients, the consumption of this tissue by the larvae causes an ash tree to die within 2 to 3 years. Mature larvae metamorphose into adult beetles and the cycle starts all over again.

Midwestern forests have an abundance of ash trees, and the beetle population grew rapidly in just a few years. However, the beetle has a fairly short dispersal distance, rarely moving more than 100 m from where it first emerges as an adult, so one would expect the geographic range of this pest to expand slowly due to its dispersal limitation. But the beetle's range expanded rapidly through unintentional human assistance. As ash trees were dying, many of them were cut down for firewood. Much of this firewood was carried long distances for events such as camping trips, which allowed the beetle to colonize new subpopulations. In response, many states are now requiring that firewood not be transported long distances.

The impact of this assisted dispersal has been dramatic. In 2012, only 10 years after its initial discovery, the emerald ash borer spread to a geographic range from Ontario and Quebec to Missouri and Tennessee and from Wisconsin and Iowa to Virginia and Pennsylvania. By 2017, the beetle had spread from Quebec and Ontario all the way down to Florida and dispersed to the west as far away as Texas, Colorado, and Minnesota. In all these states and provinces, biologists have set up traps containing beetle sex attractants to estimate the abundance and density of beetles in the area.

American ash tree. The larvae of the emerald ash borer consume the underlying cambium and phloem under the bark of the trees, causing the trees to die. Photo by Art Wagner, Washington State Department of Agriculture, Bugwood.org.

The emerald ash borer has already killed hundreds of millions of ash trees; the financial impact is in the hundreds of millions of dollars. As a result, biologists from around North America are working rapidly to determine how far the insect may potentially spread and how it might be controlled. The estimates of its spread are being made using ecological niche modeling based on the environments that the insect inhabits in Asia. Unfortunately, current models, based on the fundamental niche of the pest, predict that the beetle will be able to live across a very large geographic range that includes most of North America. However, there are several efforts to reduce this range by introducing natural enemies of the beetle from Asia, including a parasitoid wasp and a pathogenic fungus. The hope is that these enemies of the beetle will be successful in reducing its range to a much smaller realized niche. The invasion of the emerald ash borer will likely remain a major killer of ash trees in North America for many years to come, but an understanding of its population structure has helped biologists predict its spread and develop strategies to reduce its future impact on the forest.

SOURCES:

Kovacs, K. F., et al. 2011. The influence of satellite populations of emerald ash borer on projected economic costs in U.S. communities, 2010–2020. *Journal of Environmental Management* 92: 2170–2181.

Emerald Ash Borer Information Network: http://www.emeraldashborer.info/index.cfm.

SUMMARY OF LEARNING OBJECTIVES

11.1 The distribution of populations is limited to ecologically suitable habitats. The range of suitable abiotic conditions where individuals can persist represents the fundamental niche of a species. The subset of conditions where a species actually persists due to biotic interactions is known as the realized niche. With an understanding of the realized niche, ecologists can use ecological niche modeling to predict the areas in which a species could persist if it were to be introduced.

Key Terms: Spatial structure, Fundamental niche, Realized niche, Geographic range, Ecological niche modeling, Ecological envelope

11.2 Population distributions have five important characteristics. The geographic range of a population is a measure of the total area it covers. The abundance of a population is the total number of individuals that exist within a defined area. The density of a population is the number of individuals per unit area or volume. Dispersion of a population describes the spacing of individuals with respect to one another. Dispersal is the movement of individuals from one area to another.

Key Terms: Endemic, Cosmopolitan, Abundance, Density, Dispersion, Clustered dispersion, Evenly spaced dispersion, Random dispersion, Dispersal

11.3 The distribution properties of populations can be estimated. These properties are typically measured by using a variety of survey techniques, including area- and volume-based studies, line-transect studies, and mark-recapture studies.

Key Terms: Census, Survey, Area- and volume-based surveys, Line-transect surveys, Mark-recapture survey, Lifetime dispersal distance

11.4 Population abundance and density are related to geographic range and adult body size. There is generally a positive relationship between the abundance of a population and the size of its geographic range, although many other biotic and abiotic factors play a role in determining geographic range size. There is commonly a negative relationship between adult body size and the density of a population because larger individuals require more energy.

11.5 Dispersal is essential to colonizing new areas. Many populations do not inhabit suitable habitats because they are dispersal limited. One of the key ways to facilitate dispersal is through the creation of habitat corridors.

Key Terms: Dispersal limitation, Habitat corridor

11.6 Many populations live in distinct patches of habitat. The ideal free distribution makes predictions about how individuals should distribute themselves if they were to equalize the per capita benefits, although this is rarely observed in nature due to the importance of other factors, such as predators and territoriality. Ecologists have used three types of population structure models: the metapopulation model, the source–sink model, and the landscape model.

Key Terms: Ideal free distribution, Subpopulations, Basic metapopulation model, Source–sink metapopulation model, Source subpopulations, Sink subpopulations, Landscape metapopulation model

CRITICAL THINKING QUESTIONS

1. Why is the realized niche considered to be a subset of the fundamental niche?

2. How might we use an ecological niche model to predict the future spread of the emerald ash borer?

3. The American bullfrog is native to eastern North America, but it has been moved by humans and now thrives in western North America. What does this suggest about the cause of the bullfrog's historical geographic range?

4. With continued global warming, what dispersal barriers might arise for reptiles that try to move to higher latitudes?

5. In a forest, why might you observe young maple trees exhibit a clustered pattern of dispersion?

6. Suppose that 100 cattle were allowed to graze in either one of two pastures. If the grass was three times more productive in pasture A than in pasture B, how many cattle would be in each pasture if they followed an ideal free distribution? What might prevent this distribution of cattle from happening?

7. How does knowledge of the landscape metapopulation model provide guidance in how we need to preserve habitat for species?

8. You gather a sample of 20 horseshoe crabs, and mark release them. If you return the following week and collect 30 horseshoe crabs and 6 possess marks, how large would you estimate the crab population to be?

9. Why do birds have their highest densities near the center of their geographic range?

10. Why might you expect to find a negative correlation between adult body size and population density in fish?

GRAPHING THE DATA An Ideal Free Distribution

Math Tutorial

www.pretend_url_graphic

Using the data for high- and low-quality habitats, create a graph that represents how per capita benefit changes as a function of the number of individuals in the patch.

Based on these data, how many individuals must move to the high-quality patch before an individual moves to the low-quality patch? If there were 12 individuals and they followed an ideal free distribution, approximately how many individuals would be in each habitat?

Number of Individuals	High-Quality Patch Per Capita Benefit	Low-Quality Patch Per Capita Benefit
1	10.0	5.0
2	7.9	4.3
3	6.5	3.7
4	5.4	3.2
5	4.5	2.9
6	3.9	2.6
7	3.4	2.4
8	3.0	2.2
9	2.7	2.0
10	2.4	1.9

12 Population Growth and Regulation

Putting Nature on Birth Control

We often hear about the decline of species around the world due to human activities, but some species are doing extraordinarily well. In particular, the populations of many large herbivores have increased to unprecedented numbers. Hundreds of years ago, white-tailed deer lived at relatively low densities throughout the eastern United States when most of the habitat was forested and top predators, such as wolves and mountain lions, roamed throughout the region. Over the past two centuries, much of the forested habitat gave way to farms, and many of these farms have been subsequently converted into housing developments with a variety of vegetation that deer like to eat. At the same time, there was a push to exterminate the top predators throughout most of the eastern United States. With more food and less predation, deer today live at densities that are approximately 20 times higher than before European colonists arrived. Such high densities cause an over-grazing of plants, increased consumption of agricultural crops, and 1.5 million vehicle collisions and 150 human deaths each year.

The story of the white-tailed deer is repeated around the world in many other species. For example, sika deer (*Cervus nippon*) in Nara, Japan, live in a region that is protected as a World War II monument. Tourists feed the deer, the native wolf is extinct, and the deer have experienced a population explosion to the point where they are stripping the bark off spruce trees and destroying the forest. In Australia, eastern gray kangaroos (*Macropus giganteus*) have also experienced a dramatic population increase with the extinction of the Tasmanian wolf (*Thylacinus cynocephalus*) and they are now grazing on agricultural crops and golf courses. In some cases, the growing populations of large herbivores are not native, such as the wild horses (*Equus ferus*) that were introduced to the American West, and the European and Russian hogs (*Sus scrufa*) that were introduced to the southern United States as domesticated hogs. In the case of the hogs, they have been spreading rapidly into 39 states over the past 30 years, and their rooting of soil and consumption of crops are causing an estimated $1.5 billion in damage. In all these cases, the current challenge is how to control populations that are experiencing unnatural population explosions and negatively impacting ecosystems and human activities.

> "The current challenge is how to control populations that are experiencing unnatural population explosions and negatively impacting ecosystems and human activities."

Several different strategies have been proposed for controlling overabundant wildlife populations. In localized areas, it can help to eliminate the feeding of animals. In large regions, some of the overabundant species are hunted in an effort to replace the historic role of top predators. However, in many regions of the world, the number of hunters has declined over the decades. In addition, some members of the public do not support hunting as a population control method. As a result, it would be more palatable to the public if we developed methods of animal contraception.

Research on wildlife birth control has been of interest for decades, but only recently has it become a more viable option. In the case of cats and dogs, there is a history of surgically sterilizing the animals, but this is rarely an option for wild animals in nature, so researchers have focused on nonsurgical contraception. Current approaches

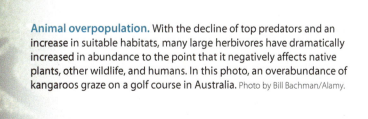

Animal overpopulation. With the decline of top predators and an increase in suitable habitats, many large herbivores have dramatically increased in abundance to the point that it negatively affects native plants, other wildlife, and humans. In this photo, an overabundance of kangaroos graze on a golf course in Australia. Photo by Bill Bachman/Alamy.

SOURCES

Klein, A. 2016. Cont-roo-ception: Hormone implants bring kangaroos under control. *New Scientist*, June 20. https://www.newscientist.com/article/2094401-cont-roo-ception-hormone-implants-bring-kangaroos-under-control/.

Nordstrum, A. 2014. Can wild pigs ravaging the U.S. be stopped? *Scientific American*, October 21. https://www.scientificamerican.com/article/can-wild-pigs-ravaging-the-u-s-be-stopped/.

Gammon, K. 2011. Approved for use: The first birth control for wildlife. *Popular Science*, September 22. http://www.popsci.com/science/article/2011-08/birth-control-wildlife.

for controlling kangaroos are focusing on treating animals with hormones to trick the females into not ovulating. For deer, horses, and elephants, researchers have focused on injecting females with a protein from pig ovaries (porcine zona pellucida, or PZP) that cause the females to make antibodies that stick to an egg, thereby preventing sperm from fertilizing the egg. The researchers used dart guns to inject PZP into the animals and found that a high percentage of the females did not conceive in the subsequent year.

At first glance, using birth control to control wild animal populations seems to be an excellent option given the lack of public support for culling animal populations through hunting. However, some of the contraceptives remain effective for only one year, finding and injecting individuals can cost $200 to $1,000 per animal, and the number of animals that must be treated to reduce a regional population can be thousands of individuals. Currently, researchers are working to develop longer-lasting contraceptives and new ways to deliver them, including the use of feeders connected to image-recognition systems to avoid giving drugs to the wrong animal. These research efforts are increasingly providing ways to reduce the problem of animal overpopulation.

LEARNING OBJECTIVES

12.1 Explain why populations can grow rapidly under ideal conditions.

12.2 Explain why populations have growth limits.

12.3 Describe how the factors of age, size, and life-history influence population growth.

In the previous chapter, we examined the spatial distribution of populations. In this chapter, we will examine changes in population size, known as population dynamics. We will look at how populations grow, and we will explore the factors that regulate their growth.

Historically, studies of population growth focused on human populations. Although estimates are understandably crude, experts believe that a million years ago, the ancestors of modern humans numbered a million individuals. With the advent of agriculture 10,000 years ago, greater food production allowed further population growth so that, by the year 1700, the human population had grown to about 600 million people. During the Industrial Revolution, increased wealth brought better nutrition, sanitation, and improvements in medical treatments. As a result, the mortality rate of children fell and the life span of adults increased. An exponential increase in population growth followed these developments. According to the United Nations, the world population hit 1 billion in 1804. Although it took more than 1 million years to reach this mark, it only took 123 years to reach the 2 billion mark in 1927. In recent decades, a billion people have been added to the planet every 12 to 13 years, and the population has doubled in just 40 years. By 2017, our planet held 7.5 billion people.

The rapid growth of the human population in the past has led many scientists to predict population sizes in the future. In the eighteenth century, the British economist Thomas Malthus examined data on human population growth and concluded that the rapid rate of growth would

cause humans to outstrip the capacity of the food supply. This insight influenced nineteenth-century scientists, including Charles Darwin, who realized that this reasoning could apply to every organism on Earth.

Understanding population growth and regulation is important because it allows us to predict future population growth. This predictive ability helps us manage the population sizes of species that are harvested by humans, species that are declining and need to be saved from extinction, and pest species that invade new regions and need to be controlled. The study of populations is known as **demography.** In this chapter, we will explore key demographic models of how populations grow.

12.1 Populations can grow rapidly under ideal conditions

Populations of any species can grow at an incredibly rapid rate, given the right conditions. In this section, we will explore two mathematical models of population growth.

THE EXPONENTIAL GROWTH MODEL

To understand how populations grow, we must first understand the concept of population *growth rate*. The **growth rate** of a population is the number of new individuals

---

Demography The study of populations.

Growth rate In a population, the number of new individuals that are produced in a given amount of time minus the number of individuals that die.

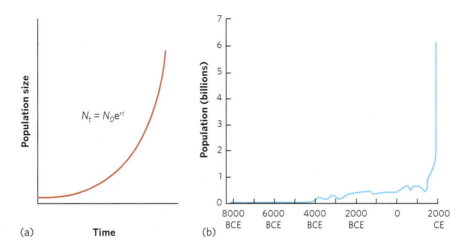

Figure 12.1 The exponential growth model. (a) Over time, a population living under ideal conditions can experience a rapid increase in population size. This produces a J-shaped curve. **(b)** Human population growth is an excellent example of a J-shaped curve. Data from http://www2.uvawise. edu/pww8y/Supplement/ConceptsSup/PopulationSup/ChartWldPopGro.html.

that are produced in a given amount of time minus the number of individuals that die. We typically consider the growth rate on an individual, or per capita, basis. Under ideal conditions, such as abundant resources, available mates, and favorable abiotic conditions, individuals can have their maximum reproductive rates and minimum death rates. When this happens, a population achieves its highest possible per capita growth rate, which is called the **intrinsic growth rate,** denoted as *r.* For example, abundant resources allow a white-tailed deer to have twins, a western mosquito (*Aedes sierrensis*) to lay up to 200 eggs in a single clutch, and a white oak tree (*Quercus alba*) to produce more than 20,000 acorns in one reproductive season. Ideal conditions also lower death rates because stresses, such as hunger and disease, are decreased.

Once we know the intrinsic growth rate for a population, we can estimate how a population will grow over time under ideal conditions. To do this, we use the **exponential growth model,** so named because it assumes that growth is exponential under ideal conditions:

$$N_t = N_0 e^{rt}$$

where e is the base of the natural log (e = 2.72). This equation tells us that when conditions are ideal, the size of the population in the future (N_t) depends on the current size of the population (N_0), the population's intrinsic growth rate (*r*), and the amount of time over which the population grows (*t*). Populations with higher intrinsic growth rates or a larger number of reproductive individuals will experience a greater rate of increase in population size.

We can see how the exponential model behaves with an example using mice. If we begin with 100 mice and their intrinsic rate of growth is 0.4, we can forecast that the mouse population will increase to 739 mice in 5 years:

$$N_t = N_0 e^{rt}$$

$$N_5 = 100 e^{0.4 \times 5}$$

$$N_5 = 100 e^2$$

$$N_5 = 100 \times 7.39$$

$$N_5 = 739$$

The exponential growth model produces a **J-shaped curve,** as shown in **Figure 12.1a**. As an example, Figure 12.1b illustrates that the human population has been growing exponentially during the past 300 years.

Using the exponential model, we can also determine the rate of growth at any point in time by taking the derivative of the exponential growth equation, as shown in the following equation:

$$\frac{dN}{dt} = rN$$

where $\frac{dN}{dt}$ represents the change in population size per unit of time. In words, this equation tells us that the rate of change in population size at any particular point in time depends on the population's intrinsic growth rate and the population's size at that point in time. Another way to think about this equation is that it tells us the slope of the line relating population size to time at any given point. Looking at the graphs in Figure 12.1, for example, the slope of the line is very shallow early in time and becomes very steep later in time.

Intrinsic growth rate (*r*) The highest possible per capita growth rate for a population.

Exponential growth model A model of population growth in which the population increases continuously at an exponential rate.

J-shaped curve The shape of exponential growth when graphed.

This confirms that the population initially grows more slowly because there are a small number of reproductive individuals, but then grows much faster as the number of reproductive individuals increases.

The exponential growth model for a population resembles the way money grows when it is earning interest in a bank account. Imagine that you have $1,000 in your account at 5 percent annual interest, that you don't add or withdraw money from the account, and that any interest earned is deposited directly into the account at the end of the year. After 1 year, the balance would be $1,050, which represents an annual growth of $50. The balance after the second year would be $1,102.50—an annual growth of $52.50. In the tenth year, the annual growth would be $77.57, and in the twentieth year the annual growth would be $126.35. As you can see, a constant interest rate results in an ever-increasing balance. Similarly, a constant intrinsic growth rate results in ever-increasing numbers of individuals in a population and the population experiences exponential growth.

THE GEOMETRIC GROWTH MODEL

The exponential growth model applies to species, such as humans, that reproduce throughout the year. However, most species of plants and animals have limited breeding seasons. For example, most birds and mammals reproduce in the spring and summer when there are abundant resources available for their offspring. As an example, let's look at the California quail (*Callipepla californica*), a bird from western North America that lays one or two clutches of eggs in the spring. The quail population experiences a large boost in its population size in the spring but then the population slowly declines over the summer, fall, and winter due to deaths. You can see the changes in population sizes in **Figure 12.2**, where each color in the graph represents the new generation of young quail that are produced each spring. When we examine the pattern of population abundance over several years, we can see that the exponential growth model, which assumes continuous births and deaths throughout the year, does not describe animals such as quail that have a distinct breeding period. For these species, ecologists use the **geometric growth model** because it compares population sizes at regular time intervals. In the case of the California quail, for instance, the geometric growth model allows us to compare population sizes at yearly intervals.

The geometric growth model is expressed as a ratio of a population's size in 1 year to its size in the preceding year (or some other time interval). This ratio is assigned the symbol λ, which is the lowercase Greek letter lambda. A λ value greater than 1 means the population size has increased from 1 year to the next because there have been more births than deaths. When λ is less than 1, the population size has decreased from 1 year to the next because there have been fewer births than deaths. Because there cannot be a negative number of individuals, the value of λ is always positive.

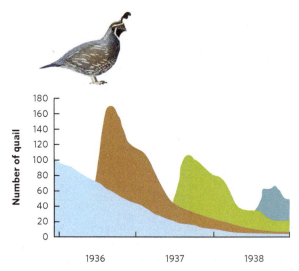

Figure 12.2 The discrete breeding events of the California quail. Each generation of new quail produced every spring is represented by a different color. Because births only happen in the spring, the growth of the California quail is better modeled by a geometric growth model than by an exponential growth model. Data from J. T. Emlen Jr., Sex and age ratios in survival of California quail, *Journal of Wildlife Management* 4 (1940): 92–99.

If N_0 is the size of a population at time 0, then its size one time interval later would be

$$N_1 = N_0\lambda$$

Next we can examine how we predict population size after more than one time interval. For example, if we wanted to estimate population size after two time intervals:

$$N_2 = (N_0\lambda)\lambda$$

which can be rearranged to

$$N_2 = N_0\lambda^2$$

In more general terms then, we can express the geometric growth model as

$$Nt = N_0\lambda^t$$

where t equals time.

Imagine that we start with a population of 100 quail and we have an annual growth rate of $\lambda = 1.5$. After 5 years, the size of the population is

$$N_5 = N_0 \times \lambda^5$$

$$N_5 = 100 \times 1.5^5$$

$$N_5 = 759$$

Geometric growth model A model of population growth that compares population sizes at regular time intervals.

COMPARING THE EXPONENTIAL AND GEOMETRIC GROWTH MODELS

Notice that the equation for exponential growth is identical to the equation for geometric growth, except that e^r takes the place of λ. Thus, geometric and exponential growth are related by

$$\lambda = e^r$$

which can be rearranged to

$$\log_e \lambda = r$$

From this relationship, we can see that λ and r are directly related to each other. Indeed, if we were to graph the growth of populations over time using both models and setting $\lambda = e^r$, we would find identical growth curves. We see this comparison in **Figure 12.3**. Although the exponential model has continuous data points whereas the geometric model has discrete data points, the two models show the same pattern in population growth. **Figure 12.4** compares the relationships between λ and r by examining the values of λ and r when populations are decreasing, constant, or increasing. When a population is decreasing, $\lambda < 1$ and $r < 0$. When a population is constant, $\lambda = 1$ and $r = 0$. When a population is increasing, $\lambda > 1$ and $r > 0$.

POPULATION DOUBLING TIME

We can appreciate the capacity of a population for growth by observing the rapid increase of organisms introduced into a new region with a suitable environment with abundant resources. In 1937, for example, two male and six female ring-necked pheasants were released on Protection Island, Washington. Within 5 years, the population increased to 1,325 adult birds, which means it experienced an annual growth rate of $\lambda = 2.78$. In other words, the population almost tripled, on average, each year. Populations maintained under optimal conditions in laboratories can have very high growth rates. Under ideal conditions, the value of λ can be as high as 24 for field voles (*Microtus agrestis*), a small

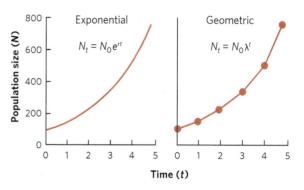

Figure 12.3 Comparing the growth of populations using the exponential and geometric growth models. The exponential model uses continuous data, whereas the geometric model uses discrete data that are calculated at each time point. However, the two models reveal the same increase in population size over time.

mouse-like mammal, 10 billion (10^{10}) for flour beetles, and 10^{30} for water fleas.

One way of appreciating the potential growth rate of populations is to estimate the time required for a population to double in abundance, known as the **doubling time**. To understand how we determine doubling time, we can start by rearranging the exponential growth equation:

$$N_t = N_0 e^{rt}$$

$$e^{rt} = N_t \div N_0$$

When a population doubles, its size is twice its original size at time 0. As a result, we can replace $(N_t \div N_0)$ with the value 2 and determine the time required (t_2) for a population to double in size:

$$e^{rt} = 2$$

$$rt = \log_e 2$$

$$t = \log_e 2 \div r$$

Doubling time The time required for a population to double in size.

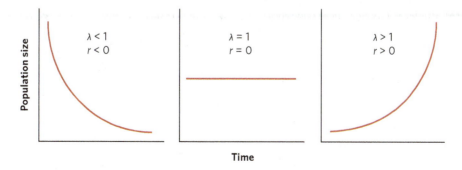

Figure 12.4 A comparison of λ and r values when populations are decreasing, constant, or increasing. When the population is decreasing, $\lambda < 1$ and $r < 0$. When the population is constant, $\lambda = 1$ and $r = 0$. When the population is increasing, $\lambda > 1$ and $r > 0$.

For the geometric model, the equation is nearly the same, except r is replaced by $\log_e \lambda$:

$$t_2 = \log_e 2 \div \log_e \lambda$$

Given that the value of $\log_e 2$ is 0.69, the doubling time for the ring-necked pheasant, which has an annual growth rate of $\lambda = 2.78$, can be calculated as

$$t_2 = 0.69 \div \log_e 2.78$$
$$t_2 = 0.67 \text{ years}$$
$$t_2 = 246 \text{ days}$$

The same calculation gives doubling times of 79 days for the field vole, 11 days for the flour beetle, and just 3.6 days for the water flea. Using this calculation, demographers have determined that the current doubling time for the human population is 40 years.

The exponential and geometric growth models are excellent starting points for understanding how populations grow. There is abundant evidence that real populations do initially grow rapidly, just as these models suggest. As we will see in the next section, however, no population can sustain exponential growth indefinitely. As populations become more abundant, they are limited by other factors such as competition, predation, and pathogens.

CONCEPT CHECK

1. Why does the graph of population increase produce a J-shaped curve even though the intrinsic growth rate is a constant?
2. In what situations should we use the geometric population growth model?
3. Using the exponential growth model, what equation can we use to estimate the population doubling time?

12.2 Populations have growth limits

In nature, we commonly observe that there are limits on how large a population can grow. Ecologists categorize these limits as being either *density independent* or *density dependent*. In this section, we will review these two types of limits on population growth and discuss how we can incorporate them into population growth models.

DENSITY-INDEPENDENT FACTORS

As the name suggests, **density-independent** factors limit population size regardless of the population's density. Common density-independent factors include natural disasters such as tornadoes, hurricanes, floods, and fires. However, other less dramatic changes in the environment, including extreme temperatures and droughts, can also limit populations. In all these cases, the impact on the population is *not* related to the number of individuals in the population.

Consider, for example, what happens when a hurricane hits a coastal forest. The number of trees that survive the hurricane is independent of the number of trees that were present before the hurricane arrived.

A classic study on density-independent factors was conducted by James Davidson and Herbert Andrewartha in Australia. The apple thrip (*Thrips imaginis*) was a common insect pest in Australia that would occasionally increase to very large numbers and devastate apple trees and rose bushes throughout large regions of the country. From 1932 to 1946, the researchers surveyed thrip populations on rose bush flowers in an attempt to understand the causes of the large variations in population size. They suspected that these population changes were the result of density-independent factors, such as seasonal variation in temperature and rainfall, so they included these factors in their population growth model. **Figure 12.5** shows the population sizes the researchers predicted and the population sizes they observed. As you can see, the predicted changes in thrip abundance closely matched the actual number of thrips they counted in their surveys.

Understanding the role of density-independent factors remains an important goal of ecologists today, particularly because of the impact of global change. For example, in 2011, researchers from the U.S. Forest Service and the Canadian Forest Service reported on the impacts of bark beetles in western North America. Bark beetles are native insects that consume the phloem tissues of trees, causing them to die. During the past

Density independent Factors that limit population size regardless of the population's density.

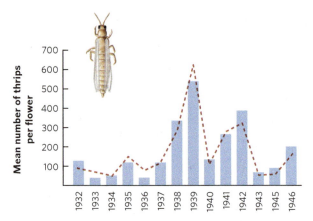

Figure 12.5 Predicting density-independent growth of the apple thrip. Scientists surveyed thrip populations on the flowers of rose bushes from 1932 to 1946. The dashed line represents the predicted number of thrips each year based on a model that included temperature and rainfall variables that cause density-independent effects on thrips. The bars represent the number of thrips actually observed each year. Data from J. Davidson and H. G. Andrewartha, The influence of rainfall, evaporation and atmospheric temperature on fluctuations in the size of a natural population of *Thrips imaginis* (Thysanoptera), *Journal of Animal Ecology* 17 (1948): 200–222.

several decades, bark beetles have killed billions of coniferous trees covering millions of hectares from Mexico to Alaska (**Figure 12.6**). Under warmer temperatures, bark beetles develop faster and have more generations per year, but they can die from unusually cold temperatures. As a result, temperature is a major determinant of population size and its effects on the beetle population are density independent. Concerned about the possible effect of climate change on the population of bark beetles, researchers applied expectations of future climate warming and their knowledge of bark beetle ecology to project future growth. They concluded that bark beetle populations are likely to increase dramatically and cause more damage to conifer trees in the future.

DENSITY-DEPENDENT FACTORS

Density-dependent factors affect population size in relation to the population's density. It is useful to break density-dependent factors down into two categories. **Negative density dependence** occurs when the rate of population growth decreases as population density increases. **Positive density dependence** (also known as **inverse density dependence**) occurs when the rate of population growth increases as population density increases. Because positive density dependence was first proposed by the ecologist Warder Allee in 1931, it is also known as the **Allee effect.**

Some of the most common negative density-dependent factors include a limited supply of resources, such as food, nesting sites, and physical space. When a population is small, there is an abundance of resources for all individuals, but as the population increases, such resources are divided among more individuals. As a result, the per capita amount of resources declines and at some point it reaches a level at which individuals find it difficult to grow, reproduce, and survive. This was the concern Malthus expressed in the eighteenth century about the rapid growth of the human population. Crowded populations also can have higher levels of stress, transmit diseases at a greater rate, and attract the attention of predators. All these factors contribute to slowing, and finally halting, population growth.

Negative Density Dependence in Animals

In a classic investigation of negative density dependence, Raymond Pearl raised breeding pairs of fruit flies (*Drosophila melanogaster*) at different densities in laboratory bottles containing identical amounts of food. With an increased number of adults, competition for food was more intense and the daily number of progeny produced per pair of adult flies decreased, as shown in **Figure 12.7**. In addition, the life span of the adults also sharply declined.

Although negative density dependence is easily demonstrated in the laboratory, a number of field experiments confirm that negative density dependence occurs in nature. For example, the common tern

---

Density dependent Factors that affect population size in relation to the population's density.

Negative density dependence When the rate of population growth decreases as population density increases.

Positive density dependence When the rate of population growth increases as population density increases. *Also known as* **Inverse density dependence** *or the* **Allee effect**.

Figure 12.6 Bark beetles. Many of these lodgepole pines in central British Columbia, Canada, have been severely damaged by bark beetles, which consume the phloem tissues of the pine trees, causing them to turn brown and die. The mountain pine beetle (*Dendroctonus ponderosae*) is one of several species of bark beetles that is predicted to become more abundant with warmer global temperatures in the future. Photos by Tom Nevesely/AGE Fotostock ; and (inset) Keith Douglas/AGE Fotostock.

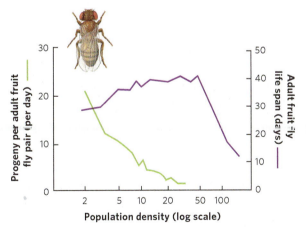

Figure 12.7 Limits on population growth. Larger populations cause increased competition. This graph shows that as competition among fruit flies increases, they experience a decline in the number of progeny produced per adult breeding pair per day. Increased competition also leads to a decline in the life span of adult flies. Data from R. Pearl, The growth of populations, *Quarterly Review of Biology* 2 (1927): 532–548.

(*Sterna hirundo*) is a bird that nests on beaches. In the 1970s, the tern population on the eastern coast of North America began to expand into an area known as Buzzards Bay in Massachusetts. It first colonized Bird Island, where its numbers quickly grew from about 200 to 1,800 individuals by 1990, as shown in **Figure 12.8**. For these terns, the factor limiting population growth appears to be the availability of suitable nesting sites. By 1991, most suitable breeding sites on Bird Island were occupied and the population leveled off. The following year, birds began to colonize Ram Island, where the population increased to just over 2,000 birds and then leveled off by the year 2000. Around this time, the terns began to colonize Penikese Island. These data demonstrate that although the terns have a high intrinsic growth rate, a limited number of nesting sites ultimately limits the population size.

Negative Density Dependence in Plants

The survival, growth, and reproduction of plants are also limited at high population densities. When plants are grown at high densities, each plant has access to fewer resources, such as sunlight, water, and soil nutrients. We can see this outcome in a study of flax plants (*Linum usitatissimum*) that were grown at a wide range of densities and then dried to determine their mass. The data from this experiment are shown in **Figure 12.9**. When seeds were sown at a density of 60 seeds per square meter, the average dry weight of individuals was between 0.5 and 1 g. When seeds were sown at densities of 1,440 and 3,600 per square meter, most of the individuals weighed less than 0.5 g. Plants with reduced growth typically experience reduced fecundity. As a result, the competition caused by high plant densities will cause a population of flax plants to grow more slowly.

Under very high densities, competition among conspecifics can cause plants to die. In an experiment using horseweed (*Erigeron canadensis*), seeds were sown

at a single extremely high density of 100,000 per square meter. Over time, competition among the tiny seedlings became intense, as you can see in **Figure 12.10a**, and over an 8-month period, most of them died. The two *y* axes in the figure reveal that there was a hundredfold decrease in population density over time, but a thousandfold increase in the average mass of the surviving individuals. As a result, the total mass of all surviving plants increased tenfold over time.

Figure 12.10b, shows the same data, but now with the changing density of surviving plants over time versus the change in average dry mass per plant over time, using a log scale. When we graph these data, they fall on a line with a negative slope. This is called a **self-thinning curve**, which is a graphical relationship that shows how decreases in population density over time lead to increases in the mass of each individual in the population. The self-thinning phenomenon has been observed across a wide variety of species. The insights that emerge from the self-thinning curve have a number of practical uses. For example, the curve can be used to predict survival and growth of crop plants that might be sown at different densities in agricultural fields or the growth and survival of tree seedlings that might be planted at different densities in forest plantations.

In plants, animals, and other taxonomic groups, negative density dependence tends to bring populations under control and to maintain their abundances at a level close to the maximum number that can be supported by the environment. In many cases, populations are affected by density-independent and density-dependent factors. Next we consider the fascinating process of positive density dependence.

Self-thinning curve A graphical relationship that shows how decreases in population density over time lead to increases in the mass of each individual in the population.

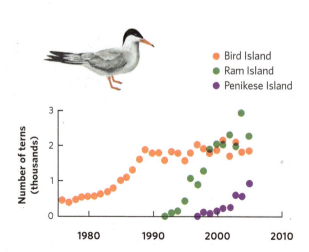

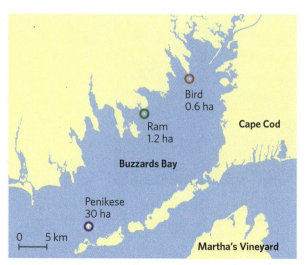

Figure 12.8 Negative density dependence in terns. As the tern population expanded in Buzzards Bay, Massachusetts, it colonized Bird Island. Rapid growth of the population filled most of the available nesting sites and the birds then colonized Ram Island. Once again, the population grew and occupied most of the available nesting sites. The terns then colonized Penikese Island. Data courtesy of Ian C. T. Nisbet.

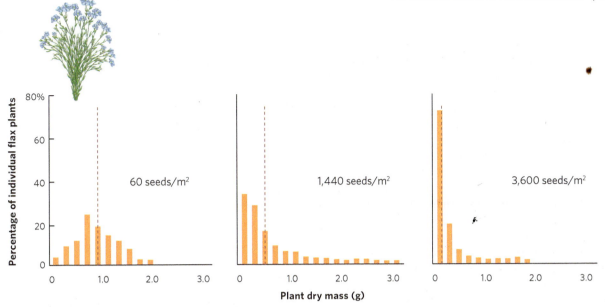

Figure 12.9 Negative density dependence in flax. When flax seeds are sown at higher densities, the average plant is smaller. Smaller plants are less fecund, so high densities cause plant populations to increase at a slower rate. The dashed red line in each panel represents the mean dry mass. Data from J. Harper, A Darwinian approach to plant ecology, *Journal of Ecology* 55 (1967): 247–270.

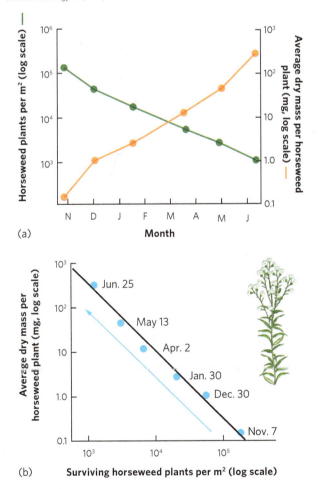

(a)

(b)

Figure 12.10 Self-thinning curve. Horseweed seeds were sown at a density of 100,000 per square meter. **(a)** Over time, the number of survivors declined by 100, while the average mass of the surviving plants increased by 1,000. **(b)** When we plot plant density against average mass per plant on log scales, we see it follows a line with a negative slope. Data from J. Harper, A Darwinian approach to plant ecology, *Journal of Ecology* 55 (1967): 247–270.

POSITIVE DENSITY DEPENDENCE

Whereas negative density dependence causes population growth to decrease as population density increases, positive density dependence causes population growth to increase as population density increases. Positive density dependence typically occurs when population density is very low and it can be caused by several different factors. For example, very low densities make it hard for individuals to find mates or, in the case of flowering plants, to obtain pollen, which can reduce reproductive success and slow population growth. As a result, a small increase in population density can have a positive effect on the population's growth. Very low densities can also lead to the harmful effects of inbreeding, as discussed in Chapter 4. There can also be problems associated with small population sizes that, by chance, have uneven sex ratios. Small populations with a low proportion of females can suffer low population growth rates. Finally, individuals living in smaller populations can face a higher predation risk than those living in large populations, as we discussed in Chapter 10. In short, while negative density dependence causes slow population growth due to overcrowding, positive density dependence causes slow population growth due to undercrowding.

Positive density dependence has been demonstrated in a wide variety of species. Many plant species, for example, avoid the costs of inbreeding through a number of mechanisms that prevent self-fertilization. They depend on receiving pollen from other plants, but this can be difficult when individuals live at low densities and are widely spread over an area. In a study of cowslip (*Primula veris*), a plant of European grasslands, researchers were interested in determining why many of the smaller populations were declining. When they looked at reproduction in populations of different sizes, they found that populations of fewer than 100 individuals produced fewer seeds per plant, as shown in **Figure 12.11**. This probably occurred because the small populations were not as good at attracting pollinators. As a result, plants in low-density populations received less pollen and therefore had fewer fertilized flowers and produced fewer seeds.

Positive density dependence also occurs in parasites. In 2012, researchers in British Columbia reported on reproduction in salmon lice (*Lepeophtheirus salmonis*), which feed on the skin of salmon. When breeding, male and female lice form a mating pair. The research team counted the number of mating pairs on each fish and found that the probability of an individual forming a mating pair is positively correlated with the number of lice of the opposite sex. Individuals living at low densities have a harder time finding a mate than individuals living at higher densities; therefore, the lice experienced positive density dependence.

Although we often think of positive and negative density dependence as isolated phenomena, populations can be regulated by both processes, as illustrated in **Figure 12.12**. As we move from low densities to intermediate densities, the effects of positive dependence can play an important role. Increased densities provide more individuals for breeding and the growth rate of the population can improve. Above some intermediate density, resources start to become limiting and negative density dependence begins to play a role. As densities continue to increase, negative density dependence continues to slow population growth and, eventually, the population growth rate falls to zero.

An example of the dynamic between positive and negative effects of density on population growth can be found in a population of herring (*Clupea harengus*) that spawns near Iceland. As you can see in **Figure 12.13**, the herring experience positive density dependence when the population is low in abundance, but negative density dependence when the populations is high in abundance

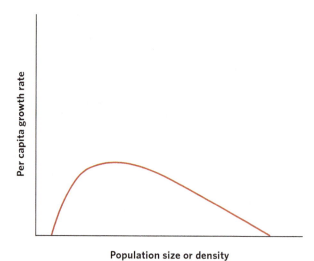

Figure 12.12 Positive and negative density dependence. When populations are very small, an increase in density can cause an increase in the population's per capita growth rate. However, when populations are large and resources start to become limiting, further increases in density can cause a decrease in the population's per capita growth rate.

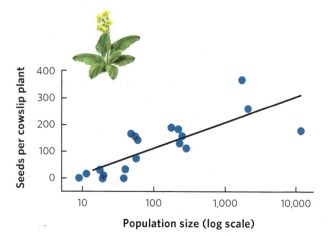

Figure 12.11 Positive density dependence in cowslip plants. At low densities, plants appear to be pollen-limited, which causes them to produce fewer seeds. At higher densities, each plant produces many more seeds. Data from M. Kéry et al., Reduced fecundity and offspring performance in small populations of the declining grassland plants *Primula veris* and *Gentiana lutea, Journal of Ecology* 88 (2000): 17–30.

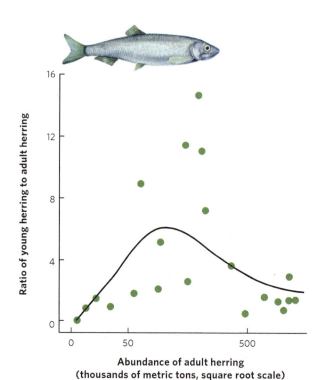

Figure 12.13 Positive and negative density dependence in herring. The highest population growth is measured as a ratio between the number of young fish and the number of adult fish. The highest growth rate of the population occurs at intermediate densities. There is positive density dependence at low herring abundance but negative density dependence at high herring abundance. Data from R. A. Myers et al., Population dynamics of exploited fish stocks at low population levels, *Science* 269 (1995): 1106–1108.

The existence of positive density dependence in the herring population concerns fisheries managers because it indicates that if the herring were driven to low densities by fishing, it would be difficult for the population to bounce back. In fact, the populations could experience negative growth rates and become extinct. Fortunately, most fish populations do not appear to experience positive density dependence.

The herring study highlights why those who manage populations in the wild need to understand positive density dependence. For example, when a species has declined to very low numbers and we wish to improve its numbers, management strategies need to consider how to avoid inbreeding and how to ensure that each female can encounter a sufficient number of mates to fertilize all her eggs. The concept of positive density dependence also offers an opportunity for ecologists to help control undesirable pest species. For example, several pest insect populations have been controlled by releasing sterile males into the population. This skews the sex ratio and causes the females to breed with sterile males, leading to a lower growth rate in the pest population. In other cases, such as in pest control using pesticides, researchers simply reduce the size of an undesirable population to the point that individuals have a hard time finding mates.

THE LOGISTIC GROWTH MODEL

Although positive and negative density dependence both occur in nature, population modelers have focused more on the effects of negative density dependence. As a result, they have developed growth models that mimic the behavior of many natural populations: rapid initial growth followed by slower growth as populations grow toward their maximum size. The maximum population size that can be supported by the environment is called the **carrying capacity** of the population, denoted as **K**.

To model the slowing growth of populations at high densities, we use the **logistic growth model.** The logistic growth model builds on the exponential growth model, but adds a term, (1 − N/K), that accounts for a decline in growth rate as the population approaches its carrying capacity:

$$\frac{dN}{dt} = rN\left(1 - \frac{N}{K}\right)$$

When the number of individuals in the population is small relative to the carrying capacity, the fraction $\frac{N}{K}$ is close to 0 and the term inside the parentheses approaches 1. When this happens, the equation becomes nearly identical to the exponential growth model. However, when the number of individuals in the population approaches the carrying capacity, the term inside the parentheses approaches 0. As a result, the population's rate of growth approaches 0. When we plot the growth of a population over time using the logistic growth curve, we obtain a *sigmoidal,* or **S-shaped curve,** which is shown in **Figure 12.14**. The middle point on the curve, where the population experiences its highest

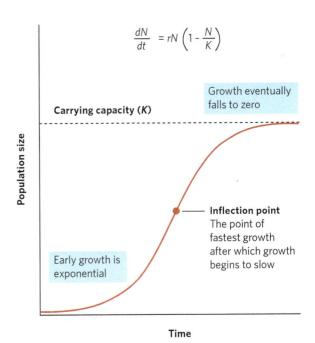

Figure 12.14 The logistic growth model. When the population is small, growth is exponential. When the population is large, growth slows. The inflection point indicates the point at which the population experiences its fastest rate of growth. Above the inflection point, growth begins to slow. When the population reaches its carrying capacity, *K*, growth is zero. As a result of this growth pattern, the logistic growth model is represented by an S-shaped growth curve.

growth rate, is known as the **inflection point.** Above the inflection point, the population growth begins to slow.

We can further understand how the logistic growth model behaves by examining the effect of population size on the rate of increase of the population ($\frac{dN}{dt}$). As shown in **Figure 12.15a**, as the population increases from a very small size, the rate of increase grows because the number of reproductive individuals increases. After reaching one-half of the carrying capacity, which corresponds to the inflection point of the S-shaped curve, the rate of increase begins to slow because the reproductive individuals are each obtaining fewer resources.

We can also examine how population size affects the rate of increase on a per capita basis, which is

$$\left(\frac{1}{N}\right)\left(\frac{dN}{dt}\right)$$

Carrying capacity (K) The maximum population size that can be supported by the environment.

Logistic growth model A growth model that describes slowing growth of populations at high densities.

S-shaped curve The shape of the curve when a population is graphed over time using the logistic growth model.

Inflection point The point on a sigmoidal growth curve at which the population achieves its highest growth rate.

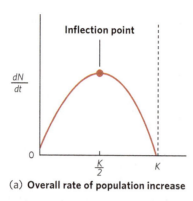

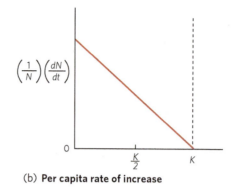

(a) **Overall rate of population increase**

(b) **Per capita rate of increase**

Figure 12.15 The effect of population size on the rate of increase and the per capita rate of increase.
(a) As populations grow from a small number to their carrying capacity, the rate of population growth increases until it reaches the inflection point, after which it decreases. (b) On a per capita basis, the rate of increase continually declines.

As you can see in Figure 12.15b, individuals in the population continually decline in their ability to contribute to the growth of the population. Therefore, the logistic growth curve shows an initial rapid increase in growth due to the increasing number of individuals in the population, followed by a slowing rate of growth as per capita resources become limited.

The logistic equation describes the growth of many populations. One of the classic studies was conducted by Russian biologist Georgyi Gause in 1934. Gause raised two species of protists, *Paramecium aurelia* and *Paramecium caudatum*, in test tubes and added a fixed amount of food each day. Both populations initially grew exponentially, and the growth then slowed until they reached carrying capacity. Because the two species differ in many ways, their populations stabilized at different carrying capacities, which are illustrated as blue lines in **Figure 12.16**. Gause suspected that the cause of this maximum population size was the amount of available food. To test this hypothesis, he did the experiment again, but this time doubled the amount of food. As you can see from the orange line in each graph, the two species once again experienced rapid initial growth followed by slower growth that stabilized. With twice as much food available, however, the two species stabilized at population sizes that were twice as large as in the first experiment. This early experiment confirmed that logistic growth can happen in real organisms and that the increased availability of a limiting resource can increase the carrying capacity of a population.

PREDICTING HUMAN POPULATION GROWTH WITH THE LOGISTIC EQUATION

The logistic growth model was first developed in 1838 by the Belgian mathematician Pierre François Verhulst. Verhulst had read Thomas Malthus's 1798 essay and sought to formulate a natural law governing the growth of populations. Nearly a century later, in 1920, Raymond Pearl and Lowell Reed independently confirmed the logistic growth model as it applies to human population growth.

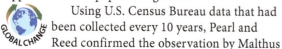 Using U.S. Census Bureau data that had been collected every 10 years, Pearl and Reed confirmed the observation by Malthus

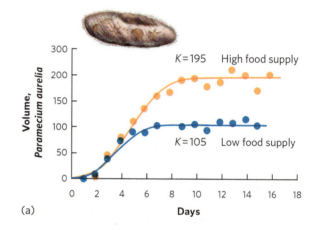

(a)

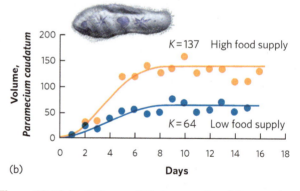

(b)

Figure 12.16 Logistic growth in two species of *Paramecium*.
(a) When *Paramecium aurelia* was raised under low or high food supply, the population size—measured in terms of the total volume of protist cells—grew rapidly at first and then slowed as they reached their carrying capacity. When twice as much food was provided, the species achieved a population size that was twice as large. (b) A similar pattern was found for another species of protist, *Paramecium caudatum*. Data from G. Gause, *The Struggle for Existence* (Zoological Institute of the University of Moscow, 1934).

that the population in the United States initially grew exponentially from 1790 to 1910. They noticed, however, that by 1910 this growth began to slow, as illustrated in

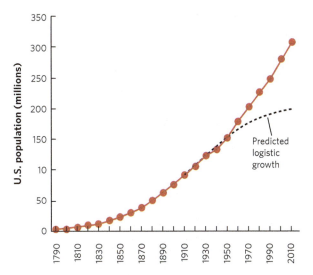

Figure 12.17 The growing human population in the United States. A census conducted every 10 years from 1790 to 1910 indicated that the U.S. population was continuing to increase. However, the rate of increase started to decline by 1910. Based on these data, Pearl and Reed predicted that the population would reach a maximum of 197 million people. In reality, the U.S. population has continued to grow. Data from R. Pearl and L. J. Reed, On the rate of growth of the population of the United States since 1790 and its mathematical representation, *Proceedings of the National Academy of Sciences* 6 (1920): 275–288; and U.S. Census Bureau, http://2010.census.gov/2010census/data/apportionment-pop-text.php.

Figure 12.17, and that the rate of growth appeared to be declining over time. Pearl and Reed applied their logistic growth equation to the census data and predicted that the population of the United States, which was 91 million in 1910, had a carrying capacity of 197 million. However, they were careful to note that this prediction depended on the population supporting itself using only the land of the United States.

As you can see in the figure, the U.S. population has greatly exceeded this prediction; by 2017, the population had grown to more than 325 million people. There are several reasons for this. One important reason is that technological advances have allowed farmers in the United States to produce food much more efficiently. Moreover, a substantial amount of the food consumed by the U.S. population comes from other regions of the world. In addition to the increases in available food, we have had major improvements in public health and medical treatment that have increased survival rates substantially, particularly for infants and children. Finally, the logistic equation did

CONCEPT CHECK

1. How do density independent effects alter population sizes?
2. What is the connection between negative density dependence and the self-thinning rule in plants?
3. Under what range of densities do we typically observe positive density dependence?

not incorporate the millions of immigrants who came to the United States after 1910.

12.3 Population growth rate is influenced by the proportions of individuals in different age, size, and life history classes

In our discussion of population models so far, we have assumed that all individuals in the population have an identical intrinsic growth rate—that is, they have the same birth rates and death rates. While this assumption is helpful to generate simple models of population growth, we know that survival and fecundity vary with an individual's age, size, and life history stage. In terms of age, individuals cannot reproduce until they have achieved reproductive maturity. In terms of individual size, individuals with greater mass typically have higher fecundity (see Chapter 8). In terms of life history stages, we know that different organisms experience different fecundity rates during each life stage. For example, tadpoles experience low survival and no fecundity, whereas later in life, as frogs, they experience high survival and high fecundity. Many perennial plants alternate between years in which they are reproductive and years in which they are not. In this section, we will examine how changes in survival and fecundity among different classes in a population affect population growth. In doing so, we will focus on age classes while recognizing that the same principles apply to size classes and life history classes.

AGE STRUCTURE

The **age structure** of a population is the proportion of individuals that occurs in different age classes. A population's age structure can tell us a great deal about its past growth and its potential for future growth. Consider the different patterns of age structure among human populations in 2017 that are displayed in **Figure 12.18**. In these figures, known as *age structure pyramids,* we see that the nations of India, the United States, and Germany all have declining numbers of people after 50 years of age, due to senescence. But each has a very different pyramid shape below 50 years of age. Age structure pyramids with broad bases reflect growing populations, pyramids with straight sides reflect stable populations, and pyramids with narrow bases reflect declining populations.

In the case of India, for example, people in the younger age classes far outnumber those in the middle age classes. If the young age classes experience good nutrition and health care, they will survive well and the number of reproductive individuals 2 decades from now will be much greater than it is today, which will cause the population to grow.

We can contrast this with the age structure pyramid of the United States. The number of people in the younger age classes is very similar to those in the middle age classes. In this case, we would predict that 2 decades from now, we would have a similar number of reproductive individuals, so the population should remain relatively stable.

Age structure In a population, the proportion of individuals that occurs in different age classes.

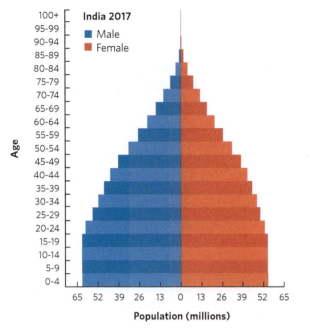

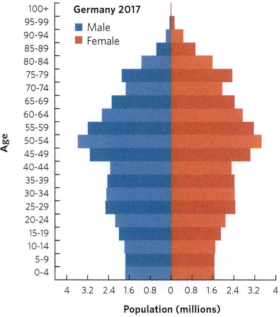

Figure 12.18 Age structure for human populations in India, the United States, and Germany in 2017. The age structure of India has a high proportion of young people, which indicates a growing population. The age structure of the United States has very similar numbers of people from 0 to 50 years of age, which indicates a stable population. Germany's age structure has fewer young people than middle-aged people, an indication that its population is declining. Data from http://www.census.gov/population/international/data/idb/informationGateway.php.

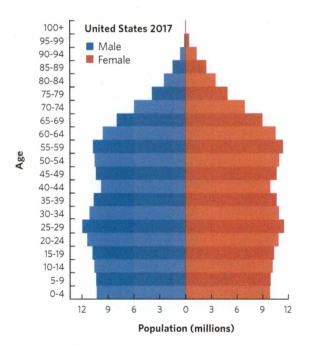

curves as type I, type II, or type III. A type I survivorship curve depicts a population with low mortality early in life and high mortality late in life. Examples include humans, elephants, and whales. A type II survivorship curve occurs when a population experiences relatively constant mortality throughout its life span. Organisms with type II curves include squirrels, corals, and some species of songbirds. A type III survivorship curve depicts a population with high mortality early in life and high survival later in life. This is a common pattern in many species of insects and in plants such as dandelions and oak trees that produce hundreds or thousands of seeds. In reality, most populations have survivorship curves that combine features of type I and type III curves, with high mortality early and late in life.

LIFE TABLES

Because both birth and death rates can differ among age, size, and life history classes, we need a way to incorporate this information into our estimates of population growth rates. In this way, we can better predict how a population will change in abundance over time, which is important for managing populations that humans consume, populations of species that we wish to conserve, and populations of pests that we wish to control. We can achieve this goal if we know the proportions of individuals in different classes and the birth and death rates of each age class. Collectively, such data

The other extreme is represented by Germany. In Germany, the current reproductive age classes have experienced reduced reproduction such that the number of people in the younger age classes is less than the number in the middle age classes. If we project ahead 2 decades, we would predict that there will be fewer reproductive individuals in the future and, as a result, the population in Germany should decline.

SURVIVORSHIP CURVES

To understand the differences in survival among different age classes, we can graph survival over time, as illustrated in **Figure 12.19**. We can then categorize survivorship

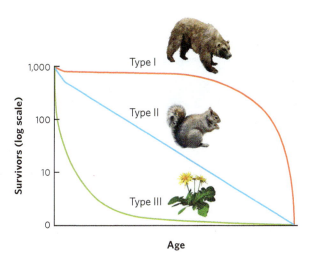

Figure 12.19 Survivorship curves. Individuals that have a type I curve experience high survival until later in life. Individuals with a type II curve experience a steady decline in survival throughout life. Individuals that have a type III curve experience low survival when young and high survival until later in life.

can tell us whether a population is expected to increase, decrease, or remain unchanged.

To determine how age, size, or life history classes affect the growth of a population, we use **life tables,** which compile class-specific survival and fecundity data. Because it can be hard to ascertain paternity in many species, life tables are typically based on females and fecundity is defined as the number of female offspring per reproductive female. For some populations with highly skewed sex ratios or unusual mating systems (see Chapter 9), using only females can pose problems, but in most cases a female-based life table provides a useful model of population growth.

To demonstrate the use of life tables, let's consider a hypothetical population composed of 100 individuals distributed into different age classes, as illustrated in **Table 12.1**. In this table, the age classes are denoted by x and the number of individuals in each age class is

denoted by n_x. The value of n_x represents the number of individuals present immediately after the population has produced offspring. In the table, survival rate from one age class to the next age class is denoted as s_x. (You can think of "s" for survival.) In our example, the column labeled as s_x indicates that the new offspring have a 50 percent survival rate, 1-year-olds have an 80 percent survival rate, 2-year-olds have a 50 percent survival rate, and no 3-year-olds survive to become 4-year-olds. The fecundity of each age class is denoted by b_x. (You can think of "b" for birth.) In the table, the column labeled as b_x indicates that the new offspring cannot reproduce, but 1-year-olds can each produce one offspring, 2-year-olds can each produce three offspring, and 3-year-olds can each produce two offspring.

The data in this life table allow us to calculate the expected size of the population after one year. **Table 12.2** shows us the calculations for the population in Table 12.1. To calculate the number of individuals that survive to the next age class, we multiply the number of individuals in an age class by the annual survival rate of that age class, as shown in

Life tables Tables that contain class-specific survival and fecundity data.

TABLE 12.1

A Hypothetical Population of 100 Individuals with Age-Specific Rates of Survival and Fecundity

Age (x)	Number of individuals (n_x)	Survival rate (s_x)	Fecundity (b_x)
0	20	0.5	0
1	10	0.8	1
2	40	0.5	3
3	30	0.0	2
4	0	—	—
Total	100		

TABLE 12.2

The Number of Survivors and New Offspring After 1 Year for the Hypothetical Population in Table 12.1

Age (x)	Number of individuals (n_x)	Survival rate (s_x)	Number surviving to next age class (n_x) × (s_x)	Fecundity (b_x)	Number of new offspring produced (n_x) × (s_x) × (b_x)	Census of population 1 year later
0	20 ×	0.5 =		0	0	74
1	10	0.8	10 ×	1 =	10	10
2	40	0.5	8	3	24	8
3	30	0.0	20	2	40	20
4	0		0			0
Total	$N_0 = 100$		38	+	74	=	$N_1 = 112$

the red-shaded region of the table. For example, if we start with 20-year-olds and their survival rate is 50 percent, then we can calculate the number of 1-year-olds that we will have in the following year:

$$\text{Number surviving to next age class} = (n_x) \times (s_x)$$
$$\text{Number surviving to next age class} = 20 \times 0.5$$
$$\text{Number surviving to next age class} = 10$$

To determine the number of new offspring that will be produced next year, we multiply the number of individuals that survive to the next age class—as shown in the orange-shaded region—by the fecundity of that age class. For example, if 10 individuals will survive to be 1-year-olds and each 1-year-old can produce one offspring, then we can calculate the number of new offspring that this age class will produce, as shown in the yellow-shaded region of the table:

$$\text{Number of new offspring produced} = (n_x) \times (s_x) \times (b_x)$$
$$\text{Number of new offspring produced} = 20 \times 0.5 \times 1$$
$$\text{Number of new offspring produced} = 10$$

If we do these calculations for every age class, we find that next year we will have a total of 74 new offspring. In addition, we will have 10 individuals that are now 1 year old, 24 individuals that are 2 years old, and 40 individuals that are now 3 years old. You can see a summary of this in **Table 12.3**.

When we add up all the age classes, we find that these survival and fecundity rates have caused the population to grow from a total size of 100 to a new total size of 112 individuals 1 year later. As a result, the geometric growth rate of the population is

$$\lambda = \frac{N_1}{N_0} = \frac{112}{100} = 1.12$$

We can continue these calculations to estimate the size of the population for many years into the future. **Table 12.4**, for example, shows the results of these same calculations for 8 years into the future. We can also continue to calculate the geometric rate of growth of the population for each year. If we examine these values of λ, we can see that they initially fluctuate widely between 1.05 and 1.69. As the years pass, however, the values of λ settle down to 1.49. Provided that the survival and fecundity

TABLE 12.3

Population Size Initially and 1 Year Later Based on the Hypothetical Population in Table 12.1

Age (x)	Number of individuals (n_x)	Number of individuals 1 year later ($n_x + 1$)
0	20	74
1	10	10
2	40	8
3	30	20
4	0	0
Total	$N_0 = 100$	$N_1 = 112$

of each class stay constant over time, λ will stabilize at a single value and the proportion of individuals in each age class also will stabilize. When the proportion of individuals in each age class does not change over time, we say it has a **stable age distribution.** However, if the survival and fecundity rates were to change, the value of λ and the proportion of individuals in each age class would be altered as well.

Calculating Survivorship
We have examined life tables using the survival rate (s_x) from one age class to the next age class. Another useful measure of survival is the probability of surviving from birth to any later age class, which we call survivorship and denote as l_x. (You can think of "l" for living.) Survivorship in the first age class is always set at 1 because all individuals in the population are initially alive. Survivorship at any given age class is the product of the prior year's survivorship and the prior year's survival rate. For example, survivorship to the second year is calculated as

$$l_2 = l_1 s_1$$

and survivorship to the third year is calculated as

$$l_3 = l_2 s_2$$

The calculated survivorships in our hypothetical population are shown in **Table 12.5**.

Stable age distribution When the age structure of a population does not change over time.

TABLE 12.4

Projecting Population Growth Over 8 Years Based on the Hypothetical Population in Table 12.1

	Year 0	Year 1	Year 2	Year 3	Year 4	Year 5	Year 6	Year 7	Year 8
n_0	20	74	69	132	175	274	399	599	889
n_1	10	10	37	34	61	87	137	199	299
n_2	40	8	8	30	28	53	70	110	160
n_3	30	20	4	4	15	14	26	35	55
N	100	112	118	200	279	428	632	943	1403
λ	1.12	1.05	1.69	1.40	1.53	1.48	1.49	1.49	

TABLE 12.5

Calculating the Survivorship (l_x) of the Hypothetical Population in Table 12.1

Age (x)	Number of individuals (n_x)	Survival rate (s_x)	Survivorship (l_x)
0	20	0.5	1.0
1	10	0.8	0.5
2	40	0.5	0.4
3	30	0.0	0.2
4	0	—	0.0
Total	100		

TABLE 12.6

Calculating the Net Reproductive Rate Based on the Hypothetical Population in Table 12.1

Age (x)	Survival rate (s_x)	Survivorship (l_x)	Fecundity (b_x)	(l_x) × (b_x)
0	0.5	1.0	0	
1	0.8	0.5	1	0.5
2	0.5	0.4	3	1.2
3	0.0	0.2	2	0.4
4	—	0.0	—	0.0
		NET REPRODUCTIVE RATE (R_0) = $\Sigma l_x b_x$ = 2.1		

Calculating the Net Reproductive Rate

The **net reproductive rate** is the total number of female offspring that we expect an average female to produce over the course of her life. If this value is greater than 1, then each female more than replaces herself in the population and the population will grow. The net reproductive rate, denoted as R_0, is calculated in two steps. First, we multiply the probability of living to each age class (l_x) by the fecundity of the respective age classes (b_x). Second, we sum these products, as shown in the following equation:

$$R_0 = \Sigma l_x b_x$$

An example is shown in **Table 12.6** for our hypothetical population. In this case, the net reproductive rate is 2.1, which means that each female is producing 2.1 daughters, on average, so the population will grow.

Calculating the Generation Time

We can also use life tables to calculate the **generation time** (**T**) of a population, which is the average time between the birth of an individual and the birth of its offspring. To calculate generation time, we must make three sets of calculations, as shown in **Table 12.7**. First, within each age class, we multiply the age (x), survivorship (l_x), and fecundity (b_x). Second, we take the sum of these products. This provides us with the expected number of births for a female, weighted by the ages at which she produced the offspring.

Using these calculations, we can now calculate generation time as

$$T = \frac{\Sigma x l_x b_x}{\Sigma l_x b_x}$$

$$T = \frac{4.1}{2.1}$$

$$T = 1.95 \text{ years}$$

This tells us that in our hypothetical population, the average time between the birth of an individual and the birth of its offspring is about 2 years.

Calculating the Intrinsic Rate of Increase

We can make connections between life tables and our earlier population growth models by using life table data to estimate the population's intrinsic rate of increase (λ or r). When an intrinsic rate of increase is estimated from a life table, we assume that the life table has a stable age distribution. However, stable age distributions rarely occur in nature because the environment varies from year to year in ways that can affect survival and fecundity. As a result, any approximation of λ or r is necessarily restricted to the set of environmental conditions that the population experiences.

Net reproductive rate The total number of female offspring that we expect an average female to produce over the course of her life.

Generation time (T) The average time between the birth of an individual and the birth of its offspring.

TABLE 12.7

Calculating the Generation Time, *T*, from the Hypothetical Population in Table 12.1

Age (x)	Survival rate (s_x)	Survivorship (l_x)	Fecundity (b_x)	(l_x) × (b_x)	(x) × (l_x) × (b_x)
0	0.5	1.0	0		0.0
1	0.8	0.5	1	0.5	0.5
2	0.5	0.4	3	1.2	2.4
3	0.0	0.2	2	0.4	1.2
4	—	0.0	—	0.0	0.0
				(R_0) = $\Sigma l_x b_x$ = 2.1	$\Sigma x l_x b_x$ = 4.1

There are complicated equations to estimate λ and r. However, we can provide close approximations, denoted as λ_a and r_a (where the letter "a" indicates an approximation), based on our estimates of net reproductive rate (R_0) and generation time (T):

$$\lambda_a = R_0^{\frac{1}{T}}$$

which given our earlier calculations simplifies to

$$\lambda_a = 2.1^{\frac{1}{1.95}}$$
$$\lambda_a = 1.46$$

Note that this value of 1.46 is close to our observed λ of about 1.49 after the population achieved a stable age distribution (see Table 12.3). If we wanted to calculate r_a, we could do so using the following equation:

$$r_a = \frac{\log_e R_0}{T}$$
$$r_a = \frac{\log_e 2.1}{1.95}$$
$$r_a = 0.38$$

You can see that a population's intrinsic rate of increase depends on both the net reproductive rate (R_0) and the

generation time (T). The greater the net reproductive rate and the shorter the generation time, the higher will be the intrinsic rate of population increase. A population grows when R_0 exceeds 1, which is the replacement level of reproduction for a population. In contrast, a population declines when $R_0 < 1$. The rate at which a population can change increases with shorter generation times.

COLLECTING DATA FOR LIFE TABLES

To determine the age structure of a population and predict future population growth, we need to collect data or organisms from different ages. This objective can be met by constructing either a *cohort life table* or a *static life table*. A **cohort life table** follows a group of individuals born at the same time from birth to the death of the last individual. In contrast, a **static life table** quantifies the survival and fecundity of all individuals in a population during a single time interval. As we will see, the two types of life tables are used in different situations and have different advantages and disadvantages.

Cohort life table A life table that follows a group of individuals born at the same time from birth to the death of the last individual.

Static life table A life table that quantifies the survival and fecundity of all individuals in a population during a single time interval.

ANALYZING ECOLOGY

Calculating Life Table Values

Eco TV

macmillanlearning.com/ricklefsvideo

Life table data are collected for many different species. For example, a group of researchers spent a decade following a population of gray squirrels (*Sciurus carolinensis*) in North Carolina to quantify the survival and fecundity of

the squirrels. Below is a table of the age-specific survival and fecundity.

YOUR TURN Using the data in the table, calculate net reproductive rate (R_0), generation time (T), and the intrinsic rate of increase (λ).

GRAY SQUIRREL LIFE TABLE					
(x)	(n_x)	(l_x)	(b_x)	$(l_x) \times (b_x)$	$(x) \times (l_x) \times (b_x)$
0	530	1.000	0.05		
1	134	0.253	1.28		
2	56	0.116	2.28		
3	39	0.089	2.28		
4	23	0.058	2.28		
5	12	0.039	2.28		
6	5	0.025	2.28		
7	2	0.022	2.28		

SOURCE: Data from F. S. Barkalow, The vital statistics of an unexploited gray squirrel population, *Journal of Wildlife Management* 34 (1970): 489–500.

Cohort Life Tables

Cohort life tables are readily applied to populations of plants and sessile animals in which marked individuals can be continually tracked over the course of their entire life. The cohort life table does not work well for species that are highly mobile or for species with very long life spans, such as trees. One of the problems in using a cohort life table is that a change in the environment during one year can affect survival and fecundity of the cohort that year. This makes it difficult to disentangle the effects of age from the effects of changing environmental conditions. For example, imagine that you mark a population of plants that lives for several years. However, in the middle of this life span, the population experiences a severe drought that causes sharp reductions in survival and fecundity. The following year is wet, which causes high survival and fecundity. When you construct the life table, you cannot determine whether the age-specific changes in survival and fecundity are due to a change in age or a change in the environment.

An excellent example of a cohort life table comes from Peter and Rosemary Grant, who have studied several species of ground finches in the Galápagos Islands off the coast of Ecuador. On the small island of Daphne Major, the Grants were able to capture all the birds on the island and mark them with uniquely colored plastic leg bands. Because this island is so isolated, few birds left the island or arrived from elsewhere. Using 210 cactus finches (*Geospiza scandens*) that fledged in 1978, the researchers constructed a cohort life table for the population. In **Figure 12.20,** you can see the annual survival rates for these birds. As is the case for many species, survival rate was low in the population's first year and then remained high for several subsequent years. However, survival was quite variable throughout the life of the birds. This variation in survival reflects variation in the environment due to climatic changes during El Niño years (see Chapter 5). Because El Niño years are wet years, the vegetation produces abundant food for the finches and this causes high survival. Following El Niño years are periods of several dry years. During this time, food becomes scarce for the finches and this causes low survival. These data emphasize a disadvantage of cohort life tables; it is difficult to determine whether low survival at a particular age is due to the age of the individuals or due to the environmental conditions that occur during that year.

Static Life Tables

Using a static life table avoids many of the problems of the cohort life table. By considering the survival and fecundity of individuals of all ages during a single time interval, differences among the age classes are quantified under the same environmental conditions, so age is not confounded with time. In addition, static life tables allow us to look at the survival and fecundity of all individuals during a snapshot in time, which means we can examine species that are highly mobile as well as species that have long life spans.

To construct a static life table, you must be able to assign ages to all individuals since they are not marked at birth. A number of techniques are available to do this. For example, counting annual growth rings in the trunks of trees, in the teeth of mammals, and in the ear bones of fish. The one major concern when using static life tables is that the age-specific data on survival and fecundity only apply to the environmental conditions that existed at the time the data were collected. Because the life table may not be representative of years in which the environmental conditions are quite different, it is helpful to construct static life tables for multiple years to assess how much environmental variation affects the predicted population growth.

A classic study using a static life table was conducted by Olaus Murie, who examined the survival of Dall mountain sheep (*Ovis dalli*) at a site in Alaska during the 1930s. Murie knew that the horns of the sheep contained annual growth rings that could be used to age the sheep. He also knew that following a cohort of highly mobile sheep for 15 years in Alaska was not feasible, so he took a static approach by conducting a search of all sheep skeletons in the area. He found a total of 608 skeletons from sheep that had died in recent years and assigned an age to each individual based on its horns. Using these estimates, he determined how many of the original 608 sheep died in each age class. His results, depicted in **Figure 12.21,** show a low survival rate during the first year, followed by high survival for the next 7 years. After 7 years of age, the annual survival rate began to drop, with the exception of those at 12 years of age. However, only four animals made it to 12 years, which gives us a poor estimate of the typical survival rate for 12-year-old sheep. Because these data were taken from the skeletons of sheep that presumably died over a relatively short period, we do not see the large variation in survival rate that we saw in the cohort life table of the cactus finch.

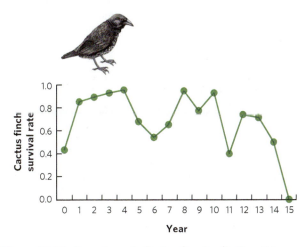

Figure 12.20 Annual survival rate of cactus finches. Using a cohort of cactus finches, researchers were able to calculate the annual survival over a 15-year period. Data from P. R. Grant and B. R. Grant, Demography and the genetically effective sizes of two populations of Darwin's finches, *Ecology* 73 (1992): 766–784.

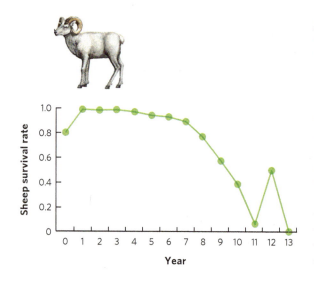

x	n_x	s_x	l_x
0	608	0.801	1.00
1	487	0.986	0.801
2	480	0.983	0.789
3	472	0.985	0.776
4	465	0.961	0.765
5	447	0.937	0.735
6	419	0.931	0.689
7	390	0.892	0.641
8	348	0.770	0.572
9	268	0.575	0.441
10	154	0.383	0.253
11	59	0.068	0.097
12	4	0.500	0.007
13	2	0.000	0.003
14	0		0.000

Figure 12.21 A static life table and survival rate curve for Dall sheep. Using the ages of Dall sheep skeletons found in Alaska, researchers calculated the survival and survivorship of different age classes. When graphed, these data show that the survival rate of the sheep remains quite high for the first 7 years and then rapidly declines. There were only four animals in the 12-year age class, and this resulted in an unreliable high estimate of survival rate. Based on data from O. Murie, *The Wolves of Mt. McKinley* (U.S. Department of the Interior, National Park Service, Fauna Series No. 5, Washington, D.C., 1944), as quoted by E. S. Deevey Jr., Life tables of natural populations of animals, *Quarterly Review of Biology* 22 (1947): 223–314.

Today, researchers continue to examine life tables for species, and some data come from unexpected sources. For example, for decades, paleontologists have been excavating fossil bones of dinosaurs in Montana, including the tibiae, or shin bones, of a single species known as the "good mother lizard" (*Maiasaura peeblesorum*). Finding the fossilized bones of a single individual of any dinosaur species is an unusual event, but these researchers found an amazing 50 fossils of a single species. They also recognized that they could age the fossil bones—based on the size and structures found in the tibiae—and this would provide data for a static life table. In 2015, they published their data as a survival curve, which is shown in **Figure 12.22**. Resembling a type I survivorship curve, the youngest age class individuals had high death rates, the middle-aged individuals had high survival rates, and the oldest age classes had high death rates.

Throughout this chapter, we have examined the mathematical models that mimic natural patterns of population growth. Although populations commonly grow rapidly when at low densities, they become limited as the populations grow larger. The simplest population models are helpful starting points, but they are not sufficient for most species that have survival and fecundity rates that vary with age, size, or life history stage. Using life tables helps to incorporate this complexity. As we will see below in "Ecology Today: Applying the Concepts," the analysis of life tables can be very helpful in setting priorities for management strategies to save species from extinction.

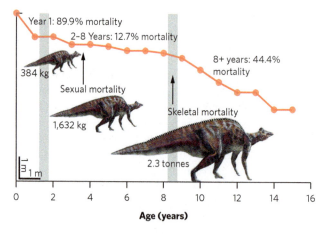

Figure 12.22 A survival rate curve for the "good mother dinosaur." Based on the tibia sizes and characteristics of 50 fossils, research estimated how many individuals were in each age class. From these static life table data, they produced a survivorship curve for the dinosaur. Based on data from H. N. Woodward, et al., *Maiasaura*, a model organism for extinct vertebrate population biology: A large sample statistical assessment of growth dynamics and survivorship, *Paleobiology* (2015), doi:10.1017/pab.2015.19.

CONCEPT CHECK

1. Why might a model that contains age structure better reflect the reality of population growth than a model that lacks age structure?
2. What are the patterns of survival with age in the three types of survivorship curves?
3. How would you describe the generation time of a population?

ECOLOGY TODAY   APPLYING THE CONCEPTS

Saving the Sea Turtles

Under the cover of darkness, the swimmers come to shore. After spending a year out in the ocean, female sea turtles crawl up the shore and dig a nest to lay their eggs. After about 2 months, these eggs will hatch and the hatchlings will scramble toward the ocean, where they will face a gauntlet of predators. If they survive, they will need 2 or 3 decades to become reproductively mature. They may live for 80 years.

Not long ago, many more females came to shore. Of the six species of sea turtles, all have declined over the past several centuries. For example, scientists estimate that there were once 91 million green turtles (*Chelonia mydas*) in the Caribbean; today, there are about 300,000, which is a decline of more than 99 percent. Sea turtles, which live in temperate and tropical waters, are declining worldwide. Historically, people hunted them for food and used turtle shells for decorations; more recently, land development has caused the loss of beach habitat for egg laying. Commercial fishing operations accidentally catch and kill turtles. Across all six species, more than 300,000 were caught along the U.S. coast in the 1980s, and more than 70,000 of these died before being returned to water. About 98 percent of these accidental catches were from shrimp trawlers pulling large nets through the water.

Attempts to reverse the decline of the sea turtles have been ongoing for many years. One attempt was to protect nesting areas and to incubate large numbers of eggs artificially. Then hatchlings were released into the oceans to improve their probability of survival. But after spending a great deal of time and money on this effort, population modelers had an interesting insight. When they began constructing life tables for turtles—a real challenge,

Sea turtles. Population models have identified the stages in life that need the most protection to allow turtle populations to recover. These models are relevant to all species of sea turtles, including this green sea turtle (*Chelonia mydas*) off the coast of Hawaii. Photo by Masa Ushioda/Image Quest Marine.

given that the turtles spend nearly all their life at sea—they discovered that improving the survival of the hatchlings would have little positive impact on the growth of the population. Because so few hatchlings survive to adulthood naturally, adding even thousands more would have only a small positive impact on the turtle population. Rather, the life tables indicated that improving the survivorship of the adult turtles was the key to growing the population.

This insight completely changed the emphasis of sea turtle conservation. The focus shifted to the commercial fishing operations that were accidentally killing adult sea turtles. Pressure mounted for commercial shrimp fishermen to install turtle excluder devices on their trawling nets. These devices work like a sieve; small shrimp pass through the sieve and get captured in the net, while sea turtles, which are too big to pass through the sieve, are shunted out of the net. The installation of these devices has been a tremendous success. In 2011, researchers reported that the number of turtles accidentally netted had been reduced by 60 percent and the number of turtles killed had declined by 94 percent.

Turtle researchers continue to explore ways to reduce accidental captures in fishing nets. In 2016, researchers reported an experiment in which they hung inexpensive green lights on 500-m fishing nets to see if it would make the nets more visible to turtles so they would veer away. They deployed pairs of nets. Within each pair, one net had green lights attached, while the other did not. After several trials, the researchers found that nets with the lights caught the same number of fish, but half as many sea turtles. Given this encouraging result, their follow-up research will examine different colors of lights to see when colors are the most effective at deterring the turtles.

Scientists continue to work on obtaining age-specific survival rates on the six species of sea turtles, which is a challenge since the animals can migrate thousands of miles and males do not return to nesting beaches. Indeed, the most recent population models indicate that improving the survival of the hatchlings does have a small positive effect on the population, although improving the survival of older animals continues to have a much greater effect. The story of the sea turtle makes it clear that being able to model population growth can be a critical step in bringing back species from the brink of extinction.

SOURCES:

Ortiz, N., et al. 2016. Reducing green turtle by-catch in small-scale fisheries using illuminated gillnets: The cost of saving a sea turtle. *Marine Ecology Progress Series* 545: 251-259.

Finkbeiner, E. M., et al. 2011. Cumulative estimates of sea turtle bycatch and mortality in USA fisheries between 1990 and 2007. *Biological Conservation* 144: 2719–2727.

Marzaris, A. D., et al. 2006. An individual based model of a sea turtle population to analyze effects of age dependent mortality. *Ecological Modelling* 198: 174–182.

SUMMARY OF CHAPTER CONCEPTS

12.1 Populations can grow rapidly under ideal conditions. When populations grow at their intrinsic growth rate, they can initially increase at exponential rates, which can be modeled using either the exponential growth model or the geometric growth model.

Key Terms: Demography, Growth rate, Intrinsic growth rate (*r*), Exponential growth model, J-shaped curve, Geometric growth model, Doubling time

12.2 Populations have growth limits. The limits can be due to density-independent factors, which regulate population sizes regardless of the population's density. The limits can also be due to density-dependent factors, which affect population growth in a way that is related to the population's density. Negative density dependence causes populations to grow more slowly as they become larger, whereas positive density dependence causes populations to grow faster as they become larger. Ecologists use the logistic growth model to demonstrate negative density dependence. The logistic model mimics rapid population growth when populations are small and slow population growth when populations approach their carrying capacity. The logistic growth model has been used to predict human population growth, but human populations have exceeded these predictions due to improvements in food production, international trade, and public health.

Key Terms: Density independent, Density dependent, Negative density dependence, Positive density dependence, Self-thinning curve, Carrying capacity (*K*), Logistic growth model, S-shaped curve, Inflection point

12.3 Population growth rate is influenced by the proportions of individuals in different age, size, and life history classes. Most organisms have rates of survival and fecundity that change over their lifetime, as illustrated by survivorship curves. Life tables were developed to incorporate age-, size-, or life history-specific rates of survival and fecundity. Using life tables, we can determine survivorship (l_x), net reproductive rates (R_0), generation times (T), and approximations of the intrinsic growth rates (r_a and λ_a). The data needed for life tables can be collected by following a cohort and building a cohort life table or by examining all individuals during a snapshot in time and developing static life tables.

Key Terms: Age structure, Life tables, Stable age distribution, Net reproductive rate, Generation time (T), Cohort life table, Static life table

CRITICAL THINKING QUESTIONS

1. Compare and contrast the approaches of the geometric growth model and the exponential growth model.

2. Given the relationship between λ and *r* in the geometric and exponential growth equations, can you demonstrate mathematically why λ must be 1 when *r* is 0?

3. Given that white-tailed deer give birth to fawns each spring, which population growth model would be the most appropriate and why?

4. Contrast the concepts of a stable population versus a stable age distribution.

5. If a life table projects a population size of 100 females and the sex ratio of the population is one-to-one, how large is the entire population?

6. In a life table, what is the fundamental difference between survival rate (s_x) and survivorship (l_x) in words and in terms of calculations?

7. Compare and contrast a cohort life table and a static life table.

8. What is the relationship between generation time and the rate of population growth?

9. When using the logistic population growth model, what are the different causes of slow population growth at low population sizes versus high population sizes?

10. What evidence would you need to determine whether a population experiences negative density dependence or positive density dependence?

GRAPHING THE DATA SURVIVORSHIP CURVES

Survivorship data are collected for a wide variety of organisms. At a U.S. research site in the state of Georgia, researchers examined the survivorship of two plants: elf orpine (*Sedum smallii*) and oneflower stitchwort (*Minuartia uniflora*). For the following survivorship data, plot the survivorship curves using a log-scale *y* axis.

SURVIVORSHIP DATA IN TWO SPECIES OF PLANTS		
Month	Elf orpine	Oneflower stitchwort
1	1.00	1.00
5	0.81	0.37
7	0.09	0.05
8	0.03	0.02
10	0.02	0.01
12	0.01	0.01
13	0.00	0.00

Source: Data estimated from R. R. Sharitz and J. F. McCormick, Population dynamics of two competing annual plant species, *Ecology* 54 (1973): 723–740.

13 Population Dynamics over Space and Time

Monitoring Moose in Michigan

Isle Royale, which is part of the state of Michigan and located off the northern shore of Lake Superior, has long served as a natural laboratory for ecologists. In the early 1900s, the island was colonized by moose from the mainland. This unlikely feat probably occurred during winter when the lake was frozen and the moose could travel the 40 km from the shores of the province of Ontario in Canada. The moose found an abundance of food on the island, and during subsequent decades the population grew rapidly. For more than 100 years, ecologists have watched the number of moose fluctuate widely due to changes in food, predators, and pathogens.

The initial changes in the population size of moose were driven by changes in the abundance of food. The population grew to approximately 3,000 animals by the early 1930s, but in 1934, when the population finally exceeded the island's carrying capacity, the moose suffered starvation and many died. In 1936, the island experienced extensive forest fires that stimulated new plant growth and, as a result, the moose population began to increase. By the late 1940s, the moose once again exceeded their carrying capacity and experienced another decline due to starvation.

The population would probably have continued this cycle of growth and decline, but sometime in the late 1940s, gray wolves (*Canis lupus*) arrived on Isle Royale by crossing the ice during the winter. Beginning in 1958, ecologist David Mech and his colleagues started estimating the number of moose and the number of wolves on the island. This was a unique research opportunity. Because both immigration and emigration of the wolves and moose were extremely rare, the researchers could watch the dynamics of the two populations unfold over time. They have estimated the wolf and moose populations every year since 1958 and have continued to make new discoveries.

During the 1960s and early 1970s, there were about 24 wolves on the island while the moose population increased from about 600 to 1,500. In the 1970s, there was a series of winters with deep snow that made it difficult for the moose to move around and locate food. At this time, the population of wolves grew to 50 animals. These two factors caused a sharp drop in the moose population.

> **"There has been tremendous public debate regarding what to do about the rapidly declining wolf population on Isle Royale."**

In 1981, a lethal virus (canine parvovirus) arrived on the island, probably carried to the island by a visitor's domestic dog. The virus caused the wolf population to experience a major decline, dropping from approximately 50 to 14 animals. With fewer wolves on the island, the moose population rebounded and it continued to grow. In 1996, the moose once again exceeded their carrying capacity and experienced widespread starvation, causing the population to drop from approximately 2,400 to 500 individuals. From 1997 to 2011, the number of moose fluctuated between 500 and 1,000, while the number of wolves fluctuated between 15 and 30. By 2016, only two wolves remained.

During the past few years, there has been tremendous public debate regarding what to do about the rapidly declining wolf population on Isle Royale. Some people have argued that we should let nature take its course. Their position is that wolves were

The wolves and moose of Isle Royale in Lake Superior. For more than 50 years, researchers have tracked large swings in moose and wolf population sizes. Photo by Steven J. Kazlowski/Alamy

SOURCES:

Daley, J. 2016, Park Service may boost wolf pack on Isle Royale. *Smithsonian*, December 23. http://www.smithsonianmag.com/smart-news/park-service-may-boost-wolf-pack-isle-royale-180961562/.

Peterson, R. O. 1999. Wolf-moose interaction on Isle Royale: The end of natural regulation? *Ecological Applications* 9: 10–16.

Vucetich, J. A., and R. O. Peterson. 2004. The influence of top-down, bottom-up and abiotic factors on the moose (*Alces alces*) population of Isle Royale. *Proceedings of the Royal Society of London*-Series B 271: 183–189.

Vucetich, J. A., and R. O. Peterson. 2011. *Ecological Studies of Wolves on Isle Royale: Annual Report 2010–2011*, School of Forest Resources and Environmental Science, Michigan Technological University.

not originally on the island, so it is acceptable if wolves become extinct on the island. Other people argue that global warming has made it less common for ice to form between the mainland and the island, so the wolves are less likely to naturally recolonize. This group believes that because the problem was caused by humans, humans should solve it by introducing more wolves to the island. Still others argue that our attention needs to be on the whole ecosystem; bringing in more wolves will help keep the moose population in check and prevent the moose from decimating the plants on the island. After considering all these positions, the National Park Service, which manages the island, announced in 2016 that they would introduce 20 to 30 wolves to the island over a period of 3 years.

The story of the moose and wolf populations on Isle Royale during the past 100 years highlights the fact that populations in nature often experience large fluctuations in size. These fluctuations can have many different causes that operate either separately or together. In this chapter, we will examine how populations fluctuate over space and time. Understanding the causes of natural fluctuations of organisms in nature enables us to predict the future abundances of populations and the species with which they interact.

LEARNING OBJECTIVES

After reading this chapter, you should be able to:

13.1 Recognize that population size fluctuates naturally over time.

13.2 Explain how density dependence with time delays can cause population size to be inherently cyclic.

13.3 Describe how chance events can cause small populations to go extinct.

13.4 Illustrate how metapopulations are composed of subpopulations that can experience independent population dynamics across space.

In the previous chapter, we saw that populations can be regulated by density-dependent and density-independent factors. In this chapter, we look at the causes of variation in population size, an issue that has become increasingly relevant as many species dwindle toward extinction and human activities fragment habitats into ever smaller, more isolated patches. Because these fluctuations include both random and cyclical changes over time, we will incorporate them into population models. We will also investigate how population fluctuations can cause small populations to be more prone to extinction. With this understanding of population variation over time, we will move on to examine population fluctuations over space by expanding our discussion of metapopulations from Chapter 12 to include metapopulation models and current research in metapopulations.

13.1 Populations fluctuate naturally over time

All populations experience fluctuations in size over time. There are many reasons for these fluctuations, such as changes in the availability of food and nesting sites, predation, competition, disease, parasites, weather, and climate. Despite these fluctuations, some populations tend to remain relatively stable over long periods. For example, fluctuations in a population of red deer—a close relative of the North American elk—on the Isle of Rum in Scotland have been followed for over 30 years. During that time, the population has

been relatively stable, fluctuating between approximately 200 and 400 individuals, as shown in **Figure 13.1**.

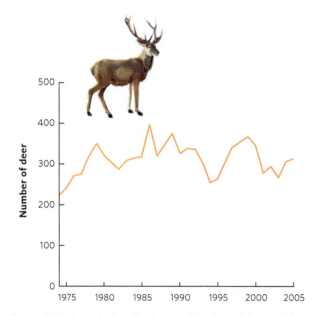

Figure 13.1 Population size fluctuations in red deer on the Isle of Rum. From 1974 to 2005, the population of the deer herd was relatively stable, ranging from approximately 200 to 400 individuals. Data from F. Pelletier et al., Decomposing variation in population growth into contributions from environment and phenotypes in an age-structured population, *Proceedings of the Royal Society* Series B 279 (2012): 94–401.

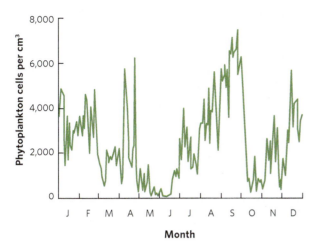

Figure 13.2 Population size fluctuations in the algae of Lake Erie. The number of algal cells in the water has fluctuated widely over time. The populations of green algae and diatoms that compose the phytoplankton varied by several orders of magnitude on a scale of days, weeks, and months. Data from C. C. Davis, Evidence for the eutrophication of Lake Erie from phytoplankton records, *Limnology and Oceanography* 9 (1964): 275–283.

Other populations exhibit much wider fluctuations. In Lake Erie, for example, researchers monitored how the numbers of waterborne algal cells, known as phytoplankton, fluctuate throughout the year. As **Figure 13.2** shows, the algae population fluctuated widely from nearly 0 cells per cubic centimeter in June to more than 7,000 cells per cubic centimeter in September. These wide fluctuations occurred across scales of days, weeks, and months.

Red deer populations are inherently more stable than algae. We can explain this by looking at differences in body size, population response time, and sensitivity to environmental change. Small organisms such as algae can reproduce in a matter of hours, which means that their populations respond very quickly to both favorable and unfavorable environmental conditions. Their small bodies and the associated high surface-area-to-volume ratio cause them to be much more affected by unfavorable environmental changes, including abiotic conditions such as temperature. As a result, their rates of survival and reproduction can decline quickly. With larger animals that live for several years and have longer generation times, a population at a given time includes individuals born over a long period, which tends to even out the effects of short-term fluctuations in birth rate. In addition, organisms with large body sizes can maintain homeostasis in the face of unfavorable environmental changes and therefore have higher survival rates.

FLUCTUATIONS IN AGE STRUCTURE

We have observed population fluctuations over time by looking at surveys of population size. In some cases, however, we can detect fluctuations over time by examining a population's age structure. When a certain age group contains an unusually high or low number of individuals, it suggests that the population experienced unusually high birth or death rates in the past.

A classic example of age structure as an indication of population fluctuations comes from data of the commercial harvest of whitefish (*Coregonus clupeaformis*) in Lake Erie from 1945 to 1951. Biologists determined the age of each fish by examining the scales of the harvested fish. As **Figure 13.3** shows, in 1947, there was a particularly large number of 3-year-old whitefish. This suggests that in 1944 the whitefish population experienced a very high rate of reproduction. This cohort continued to dominate the population's age structure in subsequent years, with high numbers of 4-year-old fish in 1948 and 5-year-old fish in 1949. Even in 1950, the 6-year-old age class contained more individuals than had been observed in that age class in previous years. From these data, we can see that a year of high birth or low death continues to show up as an abundant age class for many years.

The age structure of a forest is also readily analyzed. As you may know, the age of a tree can be determined by counting the number of rings in its trunk; under most circumstances, one ring is added each year. To examine how fluctuations affect the age structure of forest populations,

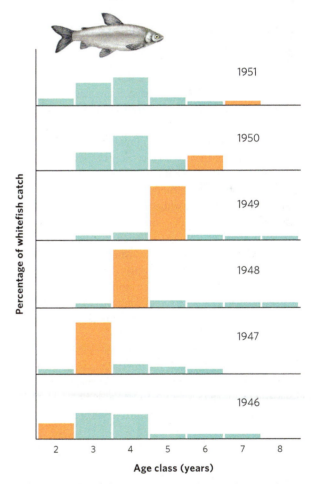

Figure 13.3 Whitefish age structure. The whitefish population in Lake Erie experienced an unusually high amount of reproduction in 1944. However, young fish are not captured by fishing nets until they are 2 years old, so the large 1944 cohort was not detected until 1946. This large reproduction event led to age structures that were dominated by that cohort of fish in subsequent years. Data from G. H. Lawler, Fluctuations in the success of year-classes of whitefish populations with special reference to Lake Erie, *Journal of the Fisheries Research Board Canada* 22 (1965): 1197–1227.

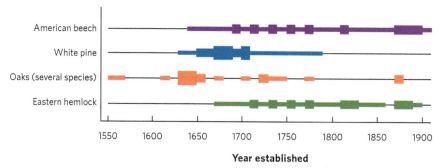

Figure 13.4 Age structure in an old-growth forest. By dating the ages of existing trees in an old-growth forest in Pennsylvania, researchers could determine the years in which populations added individuals. For simplicity, not all species of trees are shown. The thickness of the bars indicates the relative abundance of each species. Data from A. F. Hough and R. D. Forbes, The ecology and silvics of forests in the plateaus of Pennsylvania, *Ecological Monographs* 13 (1943): 299–320.

researchers work at sites that have not been logged, because logging alters the age structure of the current tree population. In one such study, a small area of old-growth forest in Pennsylvania, named Hearts Content, has been protected from logging for more than 400 years. Researchers drilled into the trunks of trees to remove a sample of wood that contained tree rings. Using these samples, they determined the age of each tree and, therefore, the time when each tree started its life.

The data from this survey of trees are illustrated in **Figure 13.4**. The data revealed that in the 1500s the forest was largely composed of several species of oak trees. A fire and drought in the mid-1600s created openings in much of the forest. Following the fire, oak seedlings reestablished themselves in the forest, although there was also an increase in the proportion of new white pines (*Pinus strobis*), which grow well under low-shade conditions. As the white pines grew large, however, they cast so much shade that new white pine seedlings had a hard time surviving, leading to a decline in the recruitment of white pine during the late 1700s. As a result, most white pine trees in the modern forest dated from the late 1600s and early 1700s. In contrast, American beech trees (*Fagus grandifolia*) and eastern hemlock trees (*Tsuga canadensis*) are very tolerant of high-shade environments; new individuals of these species began to grow as the white pines came to dominate and shade the forest. These two species continued to recruit new individuals over time, which caused beech and hemlock to have a population age structure that was more evenly distributed than that of white pine.

OVERSHOOTS AND DIE-OFFS

In Chapter 12, we saw that populations experiencing density dependence grow quite rapidly at first, but then the growth rate slows as the population reaches its carrying capacity. However, populations in nature rarely follow a smooth approach to their carrying capacity. In many cases, they grow beyond their carrying capacity, a phenomenon referred to as an **overshoot**. An overshoot can occur when the carrying capacity of a habitat decreases from one year to the next. For example, if one year has abundant rainfall and the next year

has a drought, the habitat will produce less plant mass in the second year. As a result, the carrying capacity for herbivores that rely on the plants for food will be reduced. A population that overshoots its carrying capacity is living at a density that cannot be supported by the habitat. Such populations experience a **die-off,** which is a substantial decline in population density that typically goes well below the carrying capacity. **Figure 13.5** illustrates an overshoot and die-off in a population of a hypothetical organism. The population begins small but grows at such a rapid rate that it exceeds the carrying capacity.

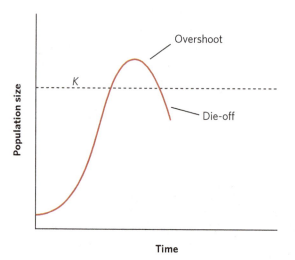

Figure 13.5 Population overshoots and die-offs. Some populations can overshoot their carrying capacity either because the carrying capacity has decreased or because the population can increase by large amounts in a single breeding season. Populations that overshoot their carrying capacity subsequently experience a die-off that causes a rapid decline in the population.

Overshoot When a population grows beyond its carrying capacity.

Die-off A substantial decline in density that typically goes well below the carrying capacity.

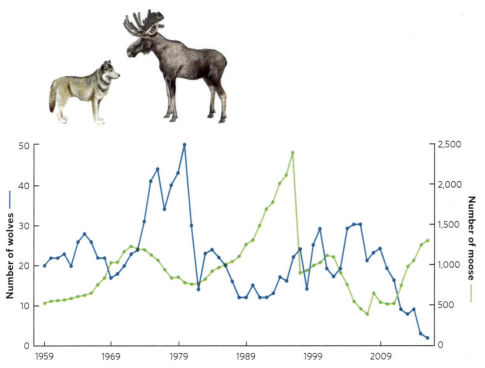

Figure 13.6 The population dynamics of wolves and moose on Isle Royale. Population estimates from 1959 to 2016 indicated that moose and wolf populations both experience wide fluctuations. Data from R. O. Peterson and J. A. Vucetich, *Ecological Studies of Wolves on Isle Royale: Annual Report 2015–2016,* School of Forest Resources and Environmental Science, Michigan Technological University.

Following the overshoot, the population experiences a rapid die-off to a point that is below the carrying capacity. We saw an example of this in the story at the beginning of this chapter that described wolves and moose on Isle Royale. Over time, the moose exceeded their carrying capacity and the population underwent massive die-offs from starvation. You can see these data in **Figure 13.6**.

An experiment with reindeer in Alaska is another good example of overshoots and die-offs. In 1911, the U.S. government introduced 25 reindeer to St. Paul Island, Alaska, to provide a source of meat for the local population. The island contained no predators of reindeer, and the reindeer population quickly began to reproduce. The reindeer fed on a variety of items during the spring, summer, and fall, but they relied on lichens to get them through the winter. By 1938, the reindeer population had swelled to more than 2,000 individuals. As you can see in **Figure 13.7**, this rate of growth followed a J-shaped growth curve that indicates exponential growth. As the population grew in the early years, the lichens they consumed in the winter remained abundant. However, with continued growth of the reindeer population, the lichens became rare, which suggests that the reindeer had far exceeded the carrying capacity of the island. Following a peak in 1938, the reindeer population began to experience a massive die-off, probably from a combination of scarce winter food and unusually cold winters. In 1940 and 1941, the government culled several hundred reindeer in an attempt to reduce the size of the herd to get it closer to

the island's now much-reduced carrying capacity. Despite this effort, the population continued to decline and, by 1950, only 8 individuals remained. In 1951, 31 new reindeer were brought to the island to supplement the population. Since

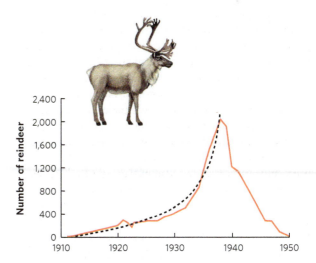

Figure 13.7 The overshoot and die-off of a reindeer population. A herd of 25 reindeer was introduced to St. Paul Island in Alaska in 1911. The population experienced a rapid increase in size that approximates an exponential J-shaped growth curve, shown by the dashed black line. After growing to nearly 2,000 animals in 1938, the population crashed, probably because the animals exhausted the food supply. Data from V. B. Scheffer, The rise and fall of a reindeer herd, *Scientific Monthly* 73 (1951): 356–362.

1980, the population has been managed by the St. Paul Tribal Government, which maintains the population at just a few hundred animals. In 2017, the population was estimated to be 400 reindeer. The reindeer, maintained at a sustainable population size, can now continue to provide valuable and affordable meat to the local residents.

CONCEPT CHECK

1. What is the relationship between species' life spans and the degree to which their population size fluctuates over time?
2. What insights can we gain about population dynamics when we examine fluctuations in age structure?
3. What are the causes of population overshoots and die-offs?

13.2 Density dependence with time delays can cause population size to be inherently cyclic

As we have just seen, populations can experience large fluctuations over time, including overshoots and die-offs. But regular patterns known as **population cycles** can be seen in some populations studied over many decades.

An interesting example of a population cycle comes from long-term records on the gyrfalcon. Falconry was a popular pastime among European nobility during the seventeenth and eighteenth centuries, and the gyrfalcon was especially prized because it is one of the largest and most beautiful falcons. During this time, Danish royalty imported gyrfalcons from Danish territories in Iceland, where the falcons were trapped and then transported to Copenhagen. The Danish royalty would then present the falcons as diplomatic gifts to the royal courts of Europe. The governor of Iceland wrote the export permits, which provide a detailed historical record of gyrfalcon exports over several decades. As shown in **Figure 13.8**, the number of gyrfalcons exported from Iceland between 1731 and 1793 reveals 10-year cycles of abundance, which reflects natural fluctuations in the abundance of the gyrfalcons in nature. After 1770, falconry became less popular, so exports declined to very low levels regardless of the bird's abundance in nature.

In some cases, cyclic population fluctuations occur among species and across large geographic areas. In Finland, for example, biologists conducted annual surveys in 11 provinces to determine the abundance of three species of grouse: capercaillie (*Tetrao urogallus*), black grouse (*Tetrao tetrix*), and hazel grouse (*Bonasa bonasia*). After monitoring the birds for 20 years and graphing the data—as shown in **Figure 13.9**—they found that all three species experienced natural population cycles every 6 to 7 years. Moreover, these three species appear to exhibit high and low populations at the same time and across all the provinces. This suggests that the drivers of natural population cycles can happen over large areas.

These examples demonstrate that some populations can exhibit remarkably regular fluctuations. The cause of such cycles, and of their synchrony, has been an interesting and

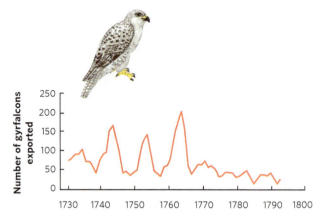

Figure 13.8 Cyclic fluctuations in gyrfalcon's population size. From 1730 to 1770, falconry was popular in Denmark and the number of falcons captured and exported from Iceland showed regular cycles that occurred approximately every 10 years. After 1770, the popularity of falconry declined and the low demand for falcons could be met even in low-population years. Data from Ó. K. Nielsen and G. Pétursson, Population fluctuations of gyrfalcon and rock ptarmigan: Analysis of export figures from Iceland, *Wildlife Biology* 1 (1995): 65–71.

persistent question in ecology. In this section, we will examine the inherent cycling behavior of populations that can be caused by fluctuating resources. In doing so, we will apply a modified version of the logistic growth equation. In subsequent chapters, we will look at how interactions with other species also influence population cycles.

THE CYCLING OF POPULATIONS AROUND THEIR CARRYING CAPACITIES

As we have just seen, populations have an inherent periodicity and tend to fluctuate up and down, although the time required to complete a cycle differs among species. To help us understand such behavior, we can think of a population as being analogous to a swinging pendulum. We know that a pendulum is stable when hanging straight up and down. The pull of gravity will cause any movement of the pendulum to the right or left to move back toward the center. However, since this movement back to the center has momentum, the pendulum overshoots the stable, center position and swings to the other side. Gravity then pulls it back toward the center, where momentum once again causes it to overshoot the stable, center position.

Populations behave like the pendulum; the momentum of increases and decreases in a population causes it to oscillate. Populations are stable at their carrying capacity. Whenever the carrying capacity increases or the size of the population decreases due to predation, disease, or a density-independent event, the population responds by growing. If the growth is sufficiently rapid, the population can grow beyond its carrying capacity. We see this phenomenon when there is a delay between the initiation of breeding and the time that offspring are added to the population. Populations

Population cycles Regular oscillation of population size over a long period of time.

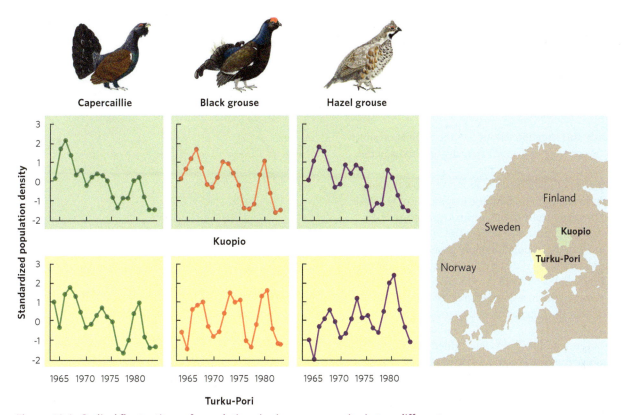

Figure 13.9 Cyclical fluctuations of population size in grouse species in two different provinces of Finland. The three species of grouse experience cycles every 6 to 7 years and appear to fluctuate in synchrony with each other and across the provinces. Data from J. Lindström et al., The clockwork of Finnish tetraonid population dynamic, *Oikos* 74 (1995): 185–194.

that overshoot their carrying capacity subsequently experience a die-off that causes the population to swing back toward its carrying capacity. Because death rates are high and birth rates are low, the population can experience a large reduction and undershoot its carrying capacity.

DELAYED DENSITY DEPENDENCE

We can model population cycles by starting with the logistic growth model introduced in Chapter 12. You may recall that this model incorporates density dependence, which causes population growth rates to slow as the population increases in size. When density dependence occurs based on a population density at some time in the past, it is called **delayed density dependence.** The key to modeling delayed density-dependent populations cycles is to incorporate a delay between a change in carrying capacity and the time the population reproduces.

Delayed density dependence can be caused by any number of factors. For example, large herbivores such as moose often breed in the fall but do not give birth until the following spring. If food is abundant in the fall, the carrying capacity is high, but by the time the offspring are born in the spring, the carrying capacity of the habitat could be much lower. The offspring will still be born, but the population will now exceed the carrying capacity of the habitat because the amount of reproduction was based on the earlier environmental conditions. We can also think about time delays

for predators. When predators experience an increase of prey, their carrying capacity increases. However, it may take weeks or months for the predators to convert abundant prey into higher reproductive rates. By this time, the prey may no longer be abundant. The lack of prey will cause the carrying capacity of the predator to decline just as the predator population is increasing. In both scenarios, the population experiences a time delay in density dependence.

Modeling Delayed Density Dependence

Delayed density dependence can be modeled with a modified form of the logistic growth model equation. As you may recall, the logistic model uses the following equation:

$$\frac{dN}{dt} = rN\left(1 - \frac{N_t}{K}\right)$$

where $\frac{dN}{dt}$ is the rate of change in the population size, r is the intrinsic growth rate, N_t is the current size of the population at time t, and K is the carrying capacity.

To incorporate a time delay, we begin by defining the amount of time delay as τ, which is the Greek letter tau.

Delayed density dependence When density dependence occurs based on a population density at some time in the past.

Now we can rewrite the logistic growth equation by making the density-dependent part of the equation (i.e., the portion of the equation in parentheses) based on the population's size ($N_{t-\tau}$) at τ time units in the past:

$$\frac{dN}{dt} = rN\left(1 - \frac{N_{t-\tau}}{K}\right)$$

In words, this equation tells us that the population slows its growth when the population's size, at τ time units in the past, approaches the carrying capacity.

Whether a population cycles above or below the carrying capacity depends on both the magnitude of the time delay and the magnitude of the intrinsic growth rate. As the time delays increase, density dependence is further delayed, making the population more prone to overshooting and undershooting the carrying capacity. In addition, having a high intrinsic rate of growth allows a population to grow more rapidly in a given amount of time, making it more likely that the population will overshoot the carrying capacity.

Population modelers have determined that the amount of cycling in a population experiencing delayed density dependence depends on the product of r and τ, as illustrated in **Figure 13.10**. As you can see in Figure 13.10a, when this product is a low value ($r\tau < 0.37$), the population approaches the carrying capacity without any oscillations. If this product is an intermediate value ($0.37 < r\tau < 1.57$), as shown in Figure 13.10b, the population initially oscillates, but the magnitude of the oscillations declines over time, a pattern known as **damped oscillations.** When the product is a high value ($r\tau > 1.57$), as shown in Figure 13.10c, the population continues to exhibit large oscillations over time, a pattern known as a **stable limit cycle.**

POPULATION SIZES CYCLE IN LABORATORY POPULATIONS

Population models that incorporate delayed density dependence help us understand how time delays cause populations to oscillate in regular cycles. Although the models do not identify specific mechanisms by which time delays occur, ecologists have investigated real populations using laboratory experiments, where it is easier to observe the cycles and identify the underlying mechanisms that drive the cycles

In some cases, delayed density dependence occurs because the organism can store energy and nutrient reserves. The water flea *Daphnia galeata*, for example, is a tiny zooplankton species that lives in lakes throughout the Northern Hemisphere. When the population is low and there is an abundance of food in a laboratory experiment, individuals can store surplus energy in the form of lipid droplets. As the population grows over time to the carrying capacity and food becomes scarce, adults with stored energy can continue to reproduce. *Daphnia*

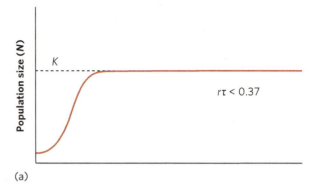

(a)

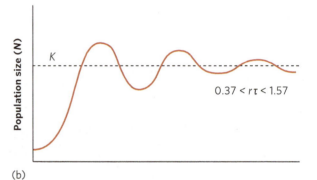

(b)

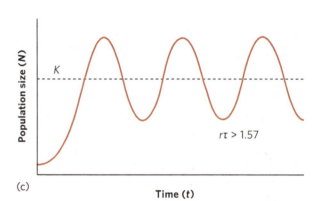

(c)

Figure 13.10 Population size cycles in models containing delayed density dependence. (a) In population models where the product of $r\tau$ is a low value ($r\tau < 0.37$), the population approaches the carrying capacity without any oscillations. **(b)** When the product of $r\tau$ is an intermediate value ($0.37 < r\tau < 1.57$), the population will exhibit damped oscillations. **(c)** When the product is a high value ($r\tau > 1.57$), the population will oscillate over time as a stable limit cycle.

mothers can also transfer some of these lipid droplets to their eggs, which allows their offspring to grow well even if the carrying capacity of the lake has been exceeded. Eventually, the stored energy is used up and

Damped oscillations A pattern of population growth in which the population size initially oscillates, but the magnitude of the oscillations declines over time.

Stable limit cycle A pattern of population growth in which the population size continues to exhibit large oscillations over time.

ANALYZING ECOLOGY

Delayed Density Dependence in the Flixweed

Eco TV

macmillanlearning.com/
ricklefsvideo

Flixweed (*Descurainia sophia*) is a weed that is native of Europe but has been introduced into North America. Studies of this weed show that the number of plants per square meter of soil fluctuates in a cyclical manner over time. The population grows according to a delayed density dependence model, where $K = 100$, $r = 1.1$, and $\tau = 1$.

The flixweed, a common plant in Europe and North America. Photo by Nigel Cattlin /Alamy.

From plant surveys, we know that there were 10 plants per square meter in year 1 and 20 plants per square meter in year 2. Based on these data, we can calculate the expected change in population size in year 3:

$$\frac{dN}{dt} = rN_2\left(1 - \frac{N_1}{K}\right)$$

$$\frac{dN}{dt} = (1.1)(20)\left(1 - \frac{10}{100}\right)$$

$$\frac{dN}{dt} = 20$$

Rounded off to the nearest whole number, the flixweed will add 20 individuals to the population in year 3. Given that the population in year 2 is 20 individuals, adding 20 more individuals in year 3 will produce a total population size of 40 individuals.

We can continue the calculations to determine the population size in year 4:

$$\frac{dN}{dt} = rN_3\left(1 - \frac{N_2}{K}\right)$$

$$\frac{dN}{dt} = (1.1)(40)\left(1 - \frac{20}{100}\right)$$

$$\frac{dN}{dt} = 35$$

As we can see, the flixweed population will increase by another 35 individuals in year 4, producing a total population size of 75 individuals.

YOUR TURN Using the data provided, calculate the population sizes of the plant from year 5 through year 15. Based on the product of $r\tau$, what type of oscillating behavior do you expect to see in this population even before doing the calculations? Graph the results to confirm your prediction.

the *Daphnia* population crashes to low numbers. When the population is low, the food can once again become abundant and the cycle begins once more. You can see these oscillations in **Figure 13.11a**.

We can contrast the story of the *Daphnia* water flea with the story of another species of water flea, *Bosmina longirostris. Bosmina* do not store as many lipid droplets as *Daphnia*, so when their population approaches the lake's carrying capacity, they have little energy to buffer the reduction in food. As a result, *Bosmina* populations do not exhibit large oscillations. Instead, as illustrated in Figure 13.11b, they grow to their carrying capacity and remain there.

Delayed density dependence can also occur when there is a time delay in development from one life stage to another. In a classic study of developmental delays, A. J. Nicholson examined the effect of time delays

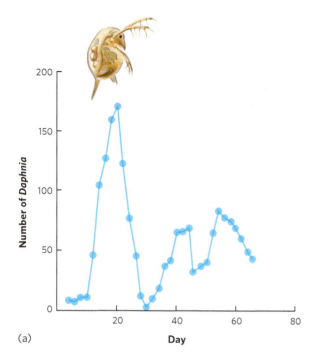

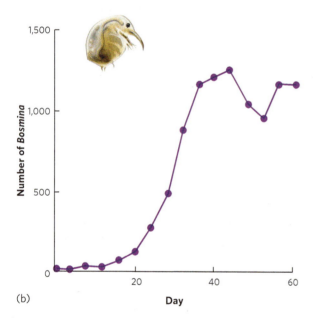

Figure 13.11 The importance of energy reserves in causing population size cycling. (a) *Daphnia galeata* water fleas can store high amounts of energy, which allows them to survive and reproduce even after reaching carrying capacity. When energy reserves run out, the population crashes to very low numbers and then rebounds and continues to oscillate. **(b)** *Bosmina longirostris* water fleas can only store a low amount of energy, so as the population nears carrying capacity, they experience reduced survival and reproduction. As a result, the population remains near its carrying capacity and oscillates much less. Data from C. E. Goulden and L. L. Hornig, Population oscillations and energy reserves in planktonic cladocera and their consequences to competition, *Proceedings of the National Academy of Sciences* 77 (1980): 1716–1720.

between the larval and adult stages of the sheep blowfly (*Lucilia cuprina*), an insect that feeds on the flesh of domestic sheep.

In his first laboratory experiment, Nicholson fed the larvae a fixed amount of food—thereby setting a carrying capacity for the larvae—but he fed the adults an unlimited amount of food. At the start of the experiment, the larvae began to metamorphose into the first set of adults, as shown in the orange line in **Figure 13.12a**. These adults then laid eggs that hatched into more larvae that eventually became adults. The adult fly population rapidly increased to more than 4,000 individuals.

As the adult population increased, the unlimited food supply allowed them to continue laying eggs. The large number of eggs hatched into a large number of larvae, but because the larvae had a limited food supply, they did not grow well enough to metamorphose into adults. The larvae died and no new adults were produced. Eventually, the adult population crashed. However, before the last few adults died, they laid a small number of eggs. When these eggs hatched, the fixed food supply provided an abundance of food for the low number of larvae. As a result, the larvae had a high rate of survival and most of them metamorphosed into adults. These new adults then laid a large number of eggs and the cycle began again. In short, there was a delay between the time that the adults produced a large number of eggs and the time these eggs hatched into larvae, died from high larval competition, and failed to produce new adults. This time delay appeared to cause the adult population cycles.

Nicholson then reasoned that if limited food for offspring caused the time delay, limiting the food for the adults should eliminate the time delay and reduce the extreme fluctuations in the adult population. To test this hypothesis, he ran the experiment again by starting with unlimited adult food, but halfway through the experiment, he limited the adult food. Under these conditions, the adults experienced density dependence without delay. As you can see in Figure 13.12b, although the abundance of the adult population still fluctuated, it no longer exhibited regular population cycles. These laboratory studies confirmed that time delays between life stages caused population cycles.

CONCEPT CHECK

1. Why does delayed density dependence cause population size to cycle?
2. How do we adjust the population growth model to account for delayed density dependence?
3. Why does the ability to store large amounts of energy reserves lead to delayed density dependence?

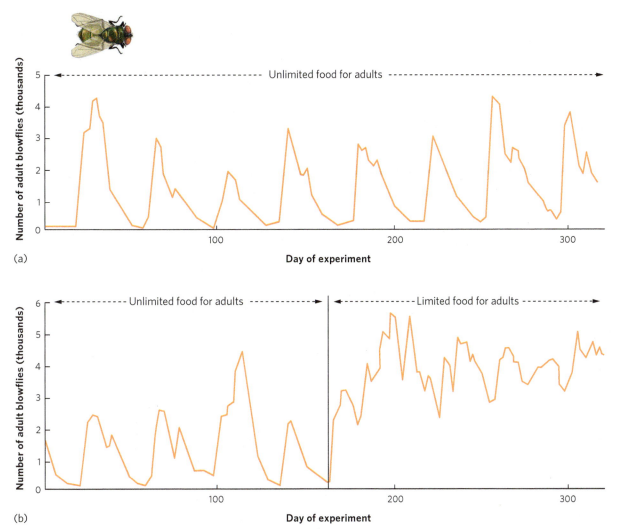

(a)

(b)

Figure 13.12 Population size cycling in sheep blowflies. (a) When researchers limited food for larvae but not adults, they observed a delay between the time that the adults produced a large number of eggs and the time these eggs hatched into larvae, died from high larval competition, and failed to produce new adults. As a result, the adult population experienced regular cycles. **(b)** When adults were initially raised with unlimited food, but then given limited food halfway through the experiment, they began to experience density dependence without a time delay. As a result, the adult population still fluctuated but no longer experienced regular cycles. Data from A. J. Nicholson, The self-adjustment of populations to change, *Cold Spring Harbor Symposia on Quantitative Biology* 22 (1958): 153–173.

13.3 Chance events can cause small populations to go extinct

In our models of population growth, we have seen that when populations are large, density-dependent factors cause slower growth, and when populations are small, density-dependent factors cause faster growth. Given these outcomes, it is hard to see how populations could go extinct, yet we know that extinction does occur in real populations in nature. In this section, we will explore the relationship between population size and the probability of extinction. We will then examine the underlying causes of this relationship.

EXTINCTION IN SMALL POPULATIONS

In nature, we find that smaller populations are more vulnerable to extinction than larger populations. To study this phenomenon, biologists conducted surveys of birds on the Channel Islands, which are located off the coast of California and range in size from 2.6 to 249 km². At different times over a period of approximately 80 years, researchers studied the number of breeding pairs of different species and the extinction rates of populations on particular islands. They counted the number of breeding pairs and then determined the extinction probability for the populations. As shown in **Figure 13.13**, the smallest populations experienced the highest probability of extinction and the

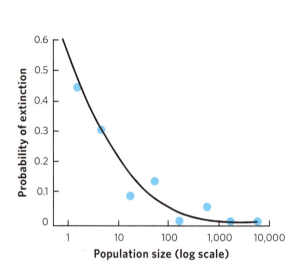

Figure 13.13 Smaller populations and the likelihood of extinction. Bird populations of the Channel Islands off the coast of California were measured in terms of the number of breeding pairs. Islands with larger population sizes had a reduced probability of going extinct over a period of **80 years.** Data from H. L. Jones and J. M. Diamond, Short-time-based studies of turnover in breeding bird populations on the California Channel Islands, *Condor* 78 (1976): 526–549.

largest populations experienced the lowest probability of extinction. Similar patterns have been observed for many other groups of animals, including mammals, reptiles, and amphibians.

The increased extinction rate of small populations is also seen in plants. For example, researchers in Germany examined the persistence of 359 populations of plants that spanned eight different species. In 1986, they counted the number of individuals in each population. They returned 10 years later and found that 27 percent of the populations had gone extinct. As you can see in **Figure 13.14**, when the researchers placed the populations into one of six population size categories and averaged the extinction rates across the eight species, they found that the probability of extinction was high for smaller population sizes and low for larger populations.

EXTINCTION DUE TO VARIATION IN POPULATION GROWTH RATES

You have seen evidence that small populations are more likely to go extinct, yet the density-dependent population models we have examined show that small populations have more rapid growth rates than large populations. This suggests that a small population should quickly rebound and grow into a larger population. By this reasoning, small populations should be resistant to extinction. How do we resolve the fundamental difference between these models and our observations of actual populations?

The models that we have studied to this point have assumed a single birth rate and death rate for each individual in the population. When a model is designed to predict a result without accounting for random variation in population growth rate, we say it is a **deterministic model.**

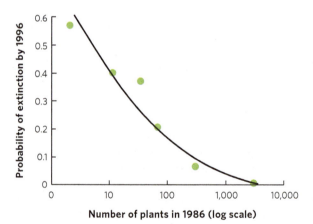

Figure 13.14 The probability of plant populations going extinct. Researchers surveyed 359 plant populations in Germany in 1986 and then again in 1996. They subsequently placed the populations into six size categories and averaged extinction rates across the eight species. They found that populations with the lowest number of individuals experienced the highest probability of extinction over a 10-year period. Data from D. Matthies et al., Population size and the risk of local extinction: Empirical evidence from rare plants, *Oikos* 105 (2004): 481–488.

Although deterministic models are simpler to work with, not every individual in the real world has the same birth rate or the same probability of dying. Rather, because we see random variation among individuals in a population, the growth rate is not constant; it can vary over time. Models that incorporate random variation in population growth rate are known as **stochastic models.** When random variation in birth rates and death rates is due to differences among individuals and not due to changes in the environment, we call it **demographic stochasticity.** In contrast, when random variation in birth rates and death rates is due to changes in the environmental conditions, we call it **environmental stochasticity.** Examples of environmental stochasticity include changes in the weather, which can cause small changes in the growth rate of a population, and natural disasters, which can cause large changes in the growth rate of a population.

When we use stochastic models, there is an average growth rate with some variation around that average. The actual growth rate experienced by the modeled population can take any of the values within the range of this variation. As a result, a population's growth rate may be above or below the average growth rate. If a population experiences a string of years with above-average growth rates, it will have faster growth. If the population happens to have a string of years with below-average growth rates, it will have slower growth. Which outcome actually occurs is determined strictly by chance.

We can model stochastic population growth by modifying the exponential growth model discussed in Chapter 12. For example, we can set the birth rate and death rate to 0.5, which are reasonable values for adult mortality and recruitment in a population of terrestrial vertebrates. We can let the values for average birth and death rates vary at random. When a population randomly experiences years of low birth rates or high death rates, it is more likely to go extinct. A string of bad years can drive the population extinct faster in small populations than in large populations. Moreover, at a given population size, as time passes, there is an increased chance that a population will have a string of bad growth years and go extinct. You can see the effect of population size and time on the average probability of extinction in **Figure 13.15**. For example, in a population with 10 individuals, the average probability of extinction within 10 years is 0.16, the average probability of extinction within 100 years is 0.82, and extinction becomes virtually certain (0.98) within 1,000 years. In contrast, a population with an initial size of 1,000 has an average probability of extinction of only 0.18 within 1,000 years. By adding the reality of stochasticity to our population models, we see that although populations of any size eventually have some chance of going extinct, the smallest populations face the greatest risk of extinction.

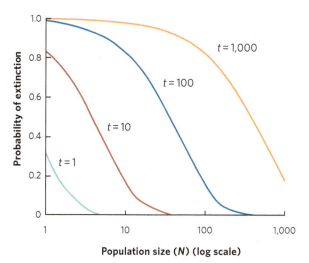

Figure 13.15 **Population models predict the probability of extinction.** Using an exponential growth model and stochastic rates of birth and death, the average probability of extinction increases with decreasing population sizes (*N*) and longer times into the future (*t*).

CONCEPT CHECK

1. What is the relationship between population size and probability of extinction?
2. What is the difference in approach between deterministic and stochastic population growth models?
3. What is the difference between demographic stochasticity and environmental stochasticity?

13.4 Metapopulations are composed of subpopulations that can experience independent population dynamics across space

We have considered population dynamics over time, but population dynamics also occur over space. Whether on land or in water, species that have a particular geographic range are commonly subdivided into smaller subpopulations. This happens because preferred habitat is not continuous but occurs as patches of suitable habitat that are surrounded by a matrix of unsuitable habitat. When individuals are able to move between patches, we consider the collection of subpopulations to be a metapopulation, a concept first

---

Deterministic model A model that is designed to predict a result without accounting for random variation in population growth rate.

Stochastic model A model that incorporates random variation in population growth rate.

Demographic stochasticity Variation in birth rates and death rates due to random differences among individuals.

Environmental stochasticity Variation in birth rates and death rates due to random changes in environmental conditions.

introduced in Chapter 11. As you may recall, the example of the collared lizards in Chapter 11 involved a metapopulation that did not live in the entire forest, but rather in glades, where temperatures were more favorable and the food supply was more abundant. In this section, we will examine why populations are broken up into metapopulations and review the three types of metapopulation models. We will then examine studies of real metapopulations in nature.

THE FRAGMENTED NATURE OF HABITATS

Metapopulations occur when a habitat is fragmented. A good example of a fragmented habitat is the wetlands that are dotted across the landscape of many parts of North America (**Figure 13.16**). Each wetland can contain populations of many different species of amphibians, crustaceans, snails, and aquatic plants. The terrestrial habitat located between these wetlands is generally inhospitable to these organisms, yet individuals of many species are able to disperse across these intervening regions to arrive at another wetland. As you might guess, the amount of movement among wetlands depends on the distance between neighboring wetlands and how far and fast an individual can move.

Metapopulations also occur as a result of human activities such as clearing forests, draining large wetlands, and constructing roads, housing developments, and commercial properties—all of which contribute to breaking up large habitats into a number of smaller habitats (**Figure 13.17**). Because the small habitats represent only fragments of the original habitat, this process is called **habitat fragmentation.** Small habitats

Figure 13.16 Wetlands within a terrestrial landscape. In many regions of the world, such as this site in North Dakota, wetlands provide patches of habitat that are hospitable to aquatic organisms such as amphibians, crustaceans, snails, and aquatic plants. However, the terrestrial habitat between the wetlands is often inhospitable to these organisms. Photo by Thomas & Pat Leeson/ Science Source.

typically support small populations, which, as we have seen, are more prone to extinction. However, a group of small populations that is interconnected by occasional dispersal have a unique dynamic in that dispersers can form new subpopulations. This balance of extinctions and colonizations allows the metapopulation to persist over time.

Habitat fragmentation The process of breaking up large habitats into a number of smaller habitats.

Figure 13.17 Habitat fragmentation due to human activities. Activities such as construction, harvesting of lumber, and agriculture have created fragmented habitats throughout the world. For example, the originally forested landscape on this site in Exmoor, England, is currently composed of cleared fields with fragments of forests between the fields. Photo by Adam Burton/ AGE Fotostock.

In Chapter 11, we examined three different types of metapopulation models. The basic metapopulation model assumes that all habitat patches are equal in quality and that the habitat matrix between patches is inhospitable. The source–sink metapopulation model builds on the basic model but incorporates the fact that habitat patches differ in quality. High-quality patches are known as *sources* because they produce high numbers of individuals that can disperse to other patches. In contrast, low-quality patches are known as *sinks* because they produce few individuals and rely on dispersers coming in from other patches to keep the subpopulation from going extinct. The landscape metapopulation model is even more realistic because it acknowledges that both the patches and the habitat matrix can vary in quality.

The dynamics of a metapopulation have a range of possible outcomes. If the subpopulations rarely exchange individuals, the fluctuations in abundance will be independent among subpopulations. Some subpopulations will increase over a period of time, while others will decrease or stay relatively constant. At the other extreme, if subpopulations are highly connected by individuals frequently moving among habitat patches, the subpopulations will act as one large population, with all experiencing the same fluctuations. Between these two extremes is the scenario in which individuals occasionally move between habitat patches, such as when juvenile animals disperse away from their family to find a mate. In this case, fluctuations in abundance in one subpopulation can influence the abundance of other subpopulations.

THE BASIC MODEL OF METAPOPULATION DYNAMICS

Having reviewed the spatial structure of metapopulations, we can now examine the population dynamics that occur in metapopulations. We will start with the basic metapopulation model that contains a number of simplifying assumptions. Although these assumptions are not realistic, they can help us understand the basic dynamics of a metapopulation.

In this model, we begin with a population that is divided into several subpopulations, each occupying a distinct patch of habitat. We assume that these habitat patches are of equal quality, each occupied patch has the same subpopulation size, and each subpopulation supplies the same number of dispersers to other habitat patches.

For the entire collection of habitat patches that exist, we will assume that some fraction of them is occupied, which we denote as *p*. We will also assume that there is a fixed probability of each patch becoming unoccupied—that is, that subpopulation going extinct—in a given amount of time, which we shall refer to as *e*. Finally, we will assume that there is a fixed probability that each unoccupied patch could be colonized, which we will denote as *c*. Using these variables, the proportion of occupied patches when at equilibrium, as indicated by $\hat{p}$, is given by the following equation:

$$\hat{p} = 1 - \frac{e}{c}$$

This basic model indicates how the number of occupied habitat patches could increase, which would increase the total number of individuals in the metapopulation (assuming that each occupied patch has the same number of individuals). One way would be to provide corridors between neighboring populations, thereby increasing the rate of colonization, as we saw in the case of the collared lizards. A second way would be to decrease the rates of extinction by reducing the major causes of population decline in subpopulations.

The concept of the metapopulation is important for the conservation of species around the world as humans continue to fragment terrestrial and aquatic habitats. We know that a key to preserving populations is to maintain large fragments of habitat whenever possible because populations in large habitats are less likely to go extinct. When we can only preserve small fragments, we must ensure that individuals can disperse to and from them so they can be colonized, which will help prevent small, declining subpopulations from going extinct.

OBSERVING METAPOPULATION DYNAMICS IN NATURE

The basic metapopulation model indicates that a metapopulation persists because of a balance between extinction of the subpopulation in some habitat patches and the colonization of others. Although these are the predictions of the model, we need to see if these processes occur in nature. One of the most extensive studies of metapopulations has been conducted on the Glanville fritillary butterfly (*Melitaea cinxia*) by ecologist Illka Hanski and colleagues. On the Åland Islands of Finland, this butterfly lives in isolated patches of dry meadows (**Figure 13.18**). The researchers found 1,600 suitable meadows on the islands, but determined that only 12 to 39 percent of these patches were

Figure 13.18 Metapopulation dynamics of a butterfly. The Glanville fritillary butterfly lives in isolated meadows on the Åland Islands of Finland. Here, the butterflies exist as a metapopulation. Photo by Robert Thompson/naturepl.com.

occupied in any given year. Over a 9-year period, they observed that more than 100 of the occupied patches experienced extinction each year and more than 100 of the unoccupied patches experienced colonization. Based on their observations, it became clear that no single patch was safe from extinction, but the metapopulation persisted because patch extinctions were offset by continued patch colonizations.

THE IMPORTANCE OF PATCH SIZE AND PATCH ISOLATION

We have noted that the basic model of metapopulation dynamics does not consider variations found in nature. For example, habitat patches are rarely equal in quality. Some patches are larger or contain a higher density of needed resources. In Southern California, for example, the California spotted owl (*Strix occidentalis occidentalis*) lives along the coast in small habitat fragments. As shown in **Figure 13.19**, these fragments are of very different sizes, and the estimates of how many owls could be supported in each fragment

range from 6 to 266. Moreover, not all patches are equally distant from all other patches.

When each patch in a metapopulation supports a different number of individuals, we expect the small patches to experience higher rates of extinction. As a result, small patches are less likely to be occupied than large patches. At the same time, we might also predict that more distant patches will have a lower probability of being occupied than closer patches. We make this prediction because successful dispersal is a function of the distance an individual has to travel. Therefore, unoccupied patches that are close to occupied patches have a better chance of being colonized. Moreover, subpopulations on the brink of extinction can be supplemented by the arrival of dispersers from other subpopulations. The phenomenon of dispersers supplementing a declining subpopulation that is headed toward extinction is called the **rescue effect.** These two mechanisms of colonization and the rescue effect should result in a higher probability that less isolated patches will be occupied.

Figure 13.19 A metapopulation of the California spotted owl. Along the coast of Southern California, the owl lives in small fragments of forested habitat, as indicated by the black regions. The lines connecting the habitat patches indicate potential dispersal paths of owls between the patches of forest. The numbers next to each patch indicate researcher estimates of how many owls could live in each of the patches. Data from W. S. Lahaye et al., Spotted owl metapopulation dynamics in Southern California, *Journal of Animal Ecology* 63 (1994): 775–785; and W. D. Shuford and T. Gardali (Eds.), *California bird species of special concern: A ranked assessment of species, subspecies, and distinct populations of birds of immediate conservation concern in California, Studies of Western Birds: Vol. 1, Western field ornithologists,* Camarillo, California, and California Department of Fish and Game, Sacramento, 2008.

The effect of patch size and patch isolation on patch colonization has been tested by studying populations of the common shrew (*Sorex araneus*) on a series of islands in two lakes in Finland. The islands vary in size from about 0.1 to 1,000 ha, and they vary in isolation from less than 0.1 to more than 2 km from other islands or from the shore of the lake. As shown in **Figure 13.20**, the researchers found that shrews were much less likely to occupy the smaller islands and those that were more isolated.

Similar patterns have been found in other species and other areas of the world. In Britain, for example, skipper butterflies (*Hesperia comma*) prefer to live in calcareous grasslands that are heavily grazed by rabbits. The grassland patches varied from about 0.01 to 10 ha in area and the distances between patches ranged from 0.02 to 100 km. As shown in **Figure 13.21**, the researchers found that the largest and least isolated patches were occupied, whereas the smallest and most isolated patches were not occupied. As we will see in Chapter 22, patch size and isolation are not only important to the metapopulation of one species but also to the total number of species that live on habitat islands.

Throughout this chapter, we have examined how populations vary over time and space. Fluctuations over time are quite common in nature, and some of these fluctuations exhibit cyclic patterns that are due to delayed density dependence. All populations have some chance of going extinct given enough time, but the smallest populations face a much higher risk of extinction due to demographic

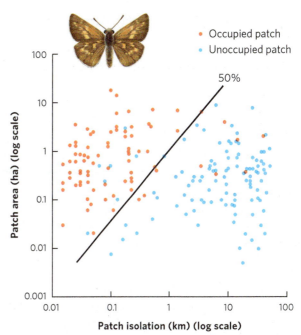

Figure 13.21 Butterfly occupation of habitat patches that differ in size and isolation. The skipper butterfly in Britain occupies the largest patches of grasslands and those that are the least isolated. The line indicates the combinations of patch area and patch isolation that correspond to 50 percent probability of patch occupancy. Data from C. D. Thomas and T. M. Jones, Partial recovery of a skipper butterfly (*Hesperia comma*) from population refuges: Lessons for conservation in a fragmented landscape, *Journal of Animal Ecology* 62 (1993): 472–481.

and environmental stochasticity. Population dynamics over time can be offset by population dynamics over space, as observed in metapopulations in which habitat fragments can experience extinctions and colonizations that allow the entire metapopulation to persist over time. As we will see in "Ecology Today: The Recovery of the Black-Footed Ferret," understanding this interplay of population dynamics over time and space can be critical in bringing species back from the brink of extinction.

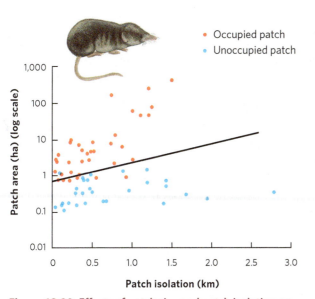

Figure 13.20 Effects of patch size and patch isolation on patch occupancy by the common shrew. The shrew was much less likely to occupy patches that were small or more isolated. Data from A. Peltonen and I. Hanski, Patterns of island occupancy explained by colonization and extinction rates in shrews, *Ecology* 72 (1991): 1698–1708.

CONCEPT CHECK

1. How are human activities causing large populations to be broken up into metapopulations?
2. Why is the proportion of occupied patches in a metapopulation at equilibrium determined by the rates of patch colonization and extinction?
3. How does distance between habitable patches affect the rate of patch colonization?

Rescue effect The phenomenon of dispersers supplementing a declining subpopulation that is headed toward extinction.

ECOLOGY TODAY   APPLYING THE CONCEPTS

The Recovery of the Black-Footed Ferret

Black-footed ferrets. An understanding of how populations fluctuate over time and space has allowed scientists to facilitate the recovery of the black-footed ferret. Photo by Shattil & Rozinski/naturepl.com.

The black-footed ferret (*Mustela nigripes*) is an elusive animal of the American West. As a nocturnal predator and a member of the weasel family, it feeds almost entirely on prairie dogs. In fact, an average ferret can consume 125 to 150 prairie dogs per year. The ferret's life is so tightly connected to prairie dogs that it lives in old prairie dog burrows within aggregations of prairie dogs known as "prairie dog towns." Historically, the ferret is thought to have had a geographic range that spanned from Texas to Arizona and north all the way up to the Canadian border. No data exist on historic population sizes, but their numbers were probably in the tens of thousands and likely varied in relation to the fluctuating abundance of prairie dogs. By the 1980s, however, the ferret was thought to have gone extinct in the wild.

The decline of the black-footed ferret occurred over the course of more than a century. With settlement of the West in the 1800s, farmers converted a large portion of the land occupied by prairie dogs to cropland. This loss of habitat substantially reduced the carrying capacity of rangelands for prairie dogs, which in turn caused a large decline in the carrying capacity of ferrets. By the 1920s, a campaign of poisoning prairie dog towns was initiated to reduce the number of prairie dogs, which were thought to compete with cattle and sheep for rangeland plants. The poisons killed many of the ferrets, but they also died from canine distemper and sylvatic plague. Canine distemper

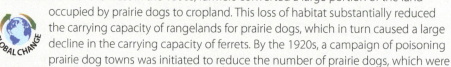

is a viral disease that is native to North America and deadly to members of the weasel family. Sylvatic plague is a disease introduced to North America from Asia in 1900 and is highly lethal to ferrets. A combination of habitat loss, poisoning, and disease caused prairie dog and ferret populations to decline rapidly.

The situation of the ferret started to receive attention in the 1960s when a single, small population was discovered in South Dakota. In a scramble to save the species from extinction, it was classified as an endangered species in 1967, which meant that efforts would be taken to promote the recovery of the species. Nine animals were brought into captivity for a breeding program. This initial attempt at captive breeding was not very successful; a few offspring were born but none survived beyond a few days. At the same time, the habitat at the South Dakota site continued to become more fragmented. As the ferret populations became smaller, the ferret faced a higher risk of extinction due to stochastic factors. Indeed, the small ferret population in nature went extinct in 1974 and the last ferret in captivity died in 1979. Everyone thought the ferret was extinct.

Two years later, a report that a rancher's dog killed a ferret near the town of Meeteetse, Wyoming, led to the discovery of a population of more than 120 ferrets living in a nearby prairie dog town. However, in 1985, this population was infected by canine distemper and sylvatic plague and began a rapid decline. Because small populations are prone to extinction from stochastic factors, biologists captured the remaining 24 ferrets and brought them into a new captive breeding program in Wyoming. Although only 18 of the 24 ferrets survived, this time the captive breeding program was so successful that additional programs were started in other locations. These locations included several zoos to ensure against a stochastic catastrophe at the Wyoming facility.

Within 6 years of bringing ferrets into captivity, biologists began to re-introduce animals into the wild where prairie dogs were abundant. By this time, because the concept of meta-population was much better understood, they chose to conduct several re-introductions throughout western North America. This strategy accounted for the possibility that should some subpopulations go extinct due to habitat loss, disease, or other stochastic factors, other subpopulations could persist and serve as future sources of new re-introductions. The recovery program has been an incredible success. As of 2016, ferrets have been introduced to 28 sites throughout the United States, Canada, and Mexico. In fact, the 28th site of re-introduction was in Meeteetse, Wyoming, where the last remaining population was discovered. Across their native range, their numbers have grown from the original 18 individuals that survived in captivity to hundreds of individuals today. They still face challenges of drought years causing a decline in food and wet years causing increases in fleas that carry the lethal sylvatic plague. However, the success of the recovery program reflects a tremendous effort by scientists working to identify and improve the factors that determine ferret population abundance over time and space.

SOURCES:

Robbins, J. 2008. Efforts on 2 fronts to save a population of ferrets. *New York Times*, July 15.

Black-footed ferret recovery program. http://blackfootedferret.org/.

Roelle, J. E., et al. (Eds.). 2006. Recovery of the black-footed ferret—progress and continuing challenges. U.S.G.S. Scientific Investigations Report 2005-5293.

Rogers, N. 2016. Black-footed ferret recovery comes full circle. http://wildlife.org/black-footed-ferret-recovery-comes-full-circle/.

SUMMARY OF CHAPTER CONCEPTS

13.1 Population size fluctuates naturally over time. This phenomenon occurs because density-dependent and density-independent factors can change from year to year and from place to place. The magnitudes of the fluctuations are often related to the ability of a species to resist changes in the environment and to differences in their life histories, including reproductive rates and life span. In some species, the population can overshoot its carrying capacity and then experience a rapid die-off. In populations that have an age structure, fluctuations in population size over time can be detected by disproportionate numbers of individuals in particular age classes. Many species experience cyclic fluctuations in population size.

Key Terms: Overshoot, Die-off, Population cycles

13.2 Density dependence with time delays can cause population size to be inherently cyclic. Delayed density dependence allows populations to fluctuate above and below their carrying capacity. Delayed density dependence can be incorporated into our population models by having the population's growth rate depend on the population density that occurred at some time in the past. Using these models, we find that the magnitude of the fluctuations depends on the product of the intrinsic growth rate (r) and the time delay (τ). Increasing values of this product causes the population to shift from experiencing no oscillations to damped oscillations to a stable limit cycle. Experiments have confirmed that time delays due to

energy reserves or development times between life stages can cause cyclic fluctuations.

Key Terms: Delayed density dependence, Damped oscillations, Stable limit cycle

13.3 Chance events can cause small populations to go extinct. Smaller populations are more likely to go extinct than large populations. This occurs due to demographic and environmental stochasticity.

Key Terms: Deterministic model, Stochastic model, Demographic stochasticity, Environmental stochasticity

13.4 Metapopulations are composed of subpopulations that can experience independent population dynamics across space. Metapopulations exist when a habitat exists in small fragments, either naturally or from human activities. The basic model of metapopulation dynamics informs us that metapopulations persist due to a balance between extinctions in some habitat patches and colonizations that occur in other patches. Although the basic metapopulation model assumes that all patches are equal, in reality, larger patches generally contain larger subpopulations and patches that are less isolated and more likely to be occupied as a result of both the rescue effect and higher rates of recolonization.

Key Terms: Habitat fragmentation, Rescue effect

CRITICAL THINKING QUESTIONS

1. If you examined a large number of similar species that differed in how much each species stored energy, what would be the likely relationship between the amount of stored energy and the likelihood of the species' population growth to exhibit delayed density dependence?

2. How does your knowledge of small island populations and the importance of the rescue effect help explain the likelihood of extinction of the wolves on Isle Royale?

3. In models of delayed density dependence, why do r and τ work together to determine the magnitude of population oscillations?

4. When predator and prey populations cycle, what are the likely causes of the cycling for the prey versus the predator?

5. How should the probability of extinction due to stochastic processes vary with population size in the California spotted owl?

6. What are the differences between demographic stochasticity and environmental stochasticity?

7. Given what you now know about the decline in the wolf population on Isle Royale, formulate arguments for and against the re-introduction of wolves.

8. In a metapopulation of the collared lizards discussed in Chapter 11, how would decreasing the distance between habitat patches affect the synchrony of fluctuations among subpopulations?

9. If you were trying to save an endangered species that lives in a metapopulation, how might you attempt to increase the proportion of occupied patches?

10. In the basic model of metapopulation dynamics, how might the rescue effect alter both the probability of colonization and the probability of extinction?

GRAPHING THE DATA Exploring the Equilibrium of the Basic Metapopulation Model

We have seen that the basic metapopulation model allows us to calculate the proportion of occupied patches based on the probability of extinction (e) and on the probability of colonization (c).

1. Based on the probabilities of extinction and colonization provided in the table below, calculate the rates of extinction and the rates of colonization.

2. Use a line graph to plot the relationship between the proportion of occupied patches and the rate of extinction.

3. On the same graph, plot the relationship between the proportion of occupied patches and the rate of colonization.

4. Based on the two line graphs, where do the extinction rates and colonization rates come into equilibrium?

PARAMETER VALUES FROM THE BASIC METAPOPULATION MODEL				
PROPORTION OF OCCUPIED PATCHES p	PROBABILITY OF EXTINCTION e	RATE OF EXTINCTION $e \times p$	PROBABILITY OF COLONIZATION c	RATE OF COLONIZATION $(c \times p) \times (1 - p)$
0.1	0.25		0.50	
0.2	0.25		0.50	
0.3	0.25		0.50	
0.4	0.25		0.50	
0.5	0.25		0.50	
0.6	0.25		0.50	
0.7	0.25		0.50	
0.8	0.25		0.50	
0.9	0.25		0.50	
1.0	0.25		0.50	

14 Predation and Herbivory

A Century-Long Mystery of the Lynx and the Hare

For centuries, naturalists, hunters, and trappers have noticed that populations of many species experience large fluctuations, and that some species fluctuate at regular intervals. In 1924, the ecologist Charles Elton drew attention to regular population fluctuations in many species of high-latitude animals in Canada, Scandinavia, and Siberia. In particular, he focused on snowshoe hares (*Lepus americanus*) and Canada lynx (*Lynx canadensis*). Elton examined data that had been compiled from the Hudson's Bay Company, a Canadian firm that had been purchasing pelts from trappers for more than 70 years. He assumed that the number of purchased pelts over time reflected the abundance of the two species. Elton and his fellow ecologists were fascinated by regular cycles of high and low density among lynx and hare populations that occurred approximately every 10 years. The 10-year cycles in the abundance of lynx and hares were clear, but the mechanisms causing these cycles have been debated for nearly a century.

There have been numerous hypotheses for these 10-year cycles. When Elton wrote his classic paper, some biologists hypothesized that animals possessed a "physiological rhythm" that caused both the lynx and the hare to reproduce abundantly in some years and sparingly in others. Elton rejected this hypothesis because it was very unlikely that such a rhythm would be synchronized for all individuals of different ages and for all individuals across large regions. Instead, he favored an explanation related to the 9- to 13-year cycle of sunspots, which are periods of increased solar activity. If the cycle of sunspots could substantially affect the climate and, therefore, the growing conditions of the plants that hares consumed, it could explain the cycles of the hares. The lynx cycle, which occurred about 2 years later than the hare cycle, was thought to reflect the fact that lynx primarily consume hares. When hares are abundant, lynx have more food and therefore reproduce more in subsequent years, but when hares are rare, the lynx reproduce poorly and many of them starve, which causes the lynx population to decline.

Since Elton's original work, ecologists have determined that although the sunspot cycle is similar in length to the hare cycle, it has never closely matched the timing of the hare cycle. Nor have they been able to find a climate-driven mechanism that connects the sunspot cycle to the hare cycle. With these hypotheses eliminated, researchers turned their attention to competition and predation. For many decades, a lively debate focused on the possibility that the hare cycles were caused by the hares exceeding their carrying capacity, which could explain the observation that hare reproduction declines as the hare population grows. Another hypothesis suggested that lynx predation caused the cycles. When lynx were rare, the hares survived better. As lynx became more numerous, they began consuming hares faster than the hares could reproduce, thereby causing the hare population to decline.

It seemed impossible to determine the answer without conducting some experiments. From 1976 to 1985, a large experiment manipulated the presence or absence of supplemental food for the hares. Although the supplemental food increased the carrying capacity for hare populations, they still cycled in synchrony with populations that were not fed. This suggested that the hare population does not decline due to a lack of food. However, if the hares do not experience a lack of food, why does their rate of reproduction decline at higher densities?

In a subsequent experiment, researchers built large fences so they could manipulate the presence or absence of supplemental food and the presence or absence of lynx predation. Both excluding lynx and adding food increased the peak population size of the hares, but the hare populations

> "The 10-year cycles in the abundance of lynx and hares have been clear, but the mechanisms causing these cycles have been debated for nearly a century."

Canadian lynx and snowshoe hare. For nearly 100 years, ecologists have been examining the regular fluctuations in the populations of these species to determine the causes. Photo by Tom & Pat Leeson/AGE Fotostock.

still cycled. However, the fence that excluded the lynx did not exclude other predators, including owls and hawks, which could fly into the fenced areas and continue to kill hares. Among the hares that died, more than 90 percent died from predation, whereas few died from starvation, which further confirmed that predation rather than food availability contributed to the cyclical decline of hares.

A new insight came in 2009 when researchers discovered that the declining rate of hare reproduction under high hare densities is caused by high densities of predators, which induce high levels of stress in the hares. Eventually, the stress of the predation threat becomes so high that the hares experience reduced reproduction. Moreover, the faster that the hare population declines due to predators, the greater stress that the surviving hares experience. Once the hare population declines, there are many fewer predators so the hare stress level is much reduced and hare reproduction returns to a high level. However, if the hares experienced especially high predation rates, the associated higher stress can affect them for multiple generations and cause the hare population to remain low in abundance for a longer amount of time. In short, while the abundance of food can affect the number of hares in the population, it appears that the lynx–hare population cycles can be attributed to a combination of direct predation and the indirect effects of predator stress that cause reduced hare reproduction.

The century-long investigation of the lynx–hare cycles illustrates that consumers and the resources that they consume can interact in complex and interesting ways. In this chapter, we will examine how predators and herbivores can affect the populations of the species they consume, including how the abundance of predators and prey populations can cycle over time, how consumers catch their prey, and how prey defend themselves.

SOURCES:

Elton, C. S. 1924. Periodic fluctuations in the numbers of animals: Their causes and effects. *British Journal of Experimental Biology* 2: 119–163.

Sheriff, M. J. 2009. The sensitive hare: Sublethal effects of predator stress on reproduction in snowshoe hares. *Journal of Animal Ecology* 78: 1249–1258.

Krebs, C. J. 2011. Of lemmings and showshoe hares: The ecology of northern Canada. *Proceedings of the Royal Society B.* 278: 481–489.

Sheriff, M. J., et al. 2015. Predator-induced maternal stress and population demography in showshoe hares: The more severe the risk, the longer the generational effect. *Journal of Zoology* 296: 305–310.

LEARNING OBJECTIVES

After reading this chapter, you should be able to:

14.1 Demonstrate how predators and herbivores can limit the abundance of populations.

14.2 Illustrate how populations of consumers and consumed populations fluctuate in regular cycles.

14.3 Explain how predation and herbivory favor the evolution of defenses.

Most species consume resources and are also a resource for other species to consume. For example, plants and algae consume nutrients, water, and light; these resources allow plants and algae to photosynthesize and grow. While plants and algae are alive, they are consumed by herbivores, parasites, and pathogens. After they die, these producers are consumed by detritivores and decomposers. Similarly, animals consume a wide variety of other organisms at the same time that they are subject to consumption by carnivores, parasites, and pathogens. After they die, animals are consumed by scavengers, detritivores, and decomposers. As you can see, a tremendous number of interactions occur among species in nature. These interactions, critical to the composition of species in different communities, are the subjects of the next four chapters.

In this chapter, we will focus on interactions between predators and their prey and between herbivores and producers. We will examine the conditions under which predators and herbivores can limit the population sizes of the species they consume. We will also look at models of predators and herbivores to help us understand how populations of these consumers fluctuate in relation to the populations of the species they consume. We will conclude the chapter by exploring how predators and herbivores have favored the evolution of defenses in prey and plants.

14.1 Predators and herbivores can limit the abundance of populations

Are populations limited primarily by what they eat or by what eats them? Studies of predation and herbivory attempt to answer this question by looking at whether or not predators and herbivores reduce the size of prey and producer populations below the carrying capacities set by resources. Understanding these relationships is of great practical concern to those interested in the management of crop pests, game populations, and endangered species. It also has far-reaching implications about the interactions among species that share resources—knowledge that helps us to understand the structure of ecological communities.

PREDATORS

As we noted with lynx and hares at the beginning of this chapter, ecologists have long been fascinated by predator–prey interactions, in both nature and in manipulative experiments. For example, a survey of

93 Caribbean islands revealed that the smaller islands contained spiders, whereas the larger islands contained spiders and lizards, which are predators of the spiders. When the researchers compared the density of spiders on islands with and without the predatory lizards, they found that spiders were about 10 times more abundant on islands without the lizards.

While these observations suggest that predator lizards play an important role in controlling spider populations, an experimental test would provide more definitive evidence that the predators were truly the cause. To that end, researchers conducted a manipulative experiment on islands in the Bahamas that ranged in size from about 200 to 4,000 m². They selected five islands with brown anole lizards (*Anolis sagrei*) and five islands without lizards. On these 10 islands, they introduced a species of orb-weaving spider (*Metepeira datona*) as a prey that was not originally present on the islands. They introduced the spiders on two occasions, for a total of 20 spiders per island. Then they surveyed the spider populations 4 days after each introduction and detected lower spider densities on the islands containing lizards. This suggested that predation from lizards had already started to affect the spider populations. Over the next

5 years, the spider populations that lived on islands without lizards became 10 times more abundant, as you can see in **Figure 14.1**. In contrast, the spider populations living with lizards became either rare or extinct, which confirmed that predatory lizards reduce the densities of spiders.

Sometimes the introduction of a predator is the result of an accident. When a species is introduced to a region of the world where it has not historically existed, we say that it is an **introduced species,** which is also known as an **exotic species** or **non-native species.** If the introduced species spreads rapidly and has negative effects on other species, human recreation, or human economies, we say that it is an **invasive species.** On the island of Guam, for example, many species of birds, bats, and lizards lived without predatory snakes for thousands of years. Sometime after World War II, however, the brown tree snake

---

Introduced species A species that is introduced to a region of the world where it has not historically existed. *Also known as* **Exotic species** or **Non-native species.**

Invasive species An introduced species that spreads rapidly and has negative effects on other species, human recreation, or human economies.

Figure 14.1 Lizard predation on spiders. (a) In the Bahamas, researchers introduced a total of 20 spiders to each of five small islands with predatory lizards and to five small islands without lizards. **(b)** Spiders were introduced to each island on two dates, indicated by the red arrows. The spiders remained rare or absent on islands with lizards but increased 10-fold on islands without lizards. Error bars are standard errors. Data from T. W. Schoener and D. A. Spiller, Effect of predators and area on invasion: An experiment with island spiders, *Science* 267 (1995): 1811–1813. Photo by Jason J. Kolbe.

(a)

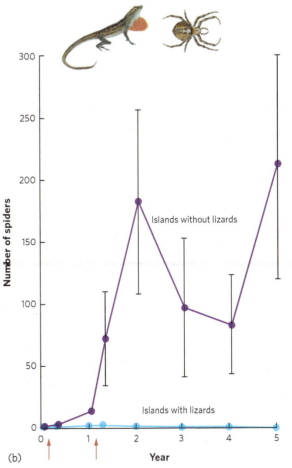

(b)

(a)

(b)

Figure 14.2 The introduced brown tree snake. (a) The introduction of this snake to the island of Guam led to the decline and local extinction of birds, bats, and lizards. **(b)** An example of the streamers that were attached to dead mice containing acetaminophen to poison the invasive snakes.
Photos by (a) John Mitchell/Science Source; and (b) USDA Photo.

(*Boiga irregularis*) was accidentally introduced to Guam by ships carrying supplies (**Figure 14.2**). While the snake was native to the South Pacific, the animals on Guam had no evolutionary history of living with snakes and many had no defenses against them. Over the next 20 years, as the population of the brown tree snake grew exponentially, it had a significant and devastating effect on the island fauna. Guam has experienced sharp declines or extinctions in nine species of forest-dwelling birds, all three species of bats, and several species of lizards.

A variety of efforts have been made to reduce the snake population, including the use of snake traps and snake-detecting dogs. In a recent and very creative experiment, thousands of dead mice were injected with acetaminophen (the active ingredient in Tylenol, which is toxic to the snakes, but not other animals) and attached to streamers, which act as parachutes. These parachuted mice were then dropped from helicopters around the island. The hope is that the snakes will find the mice dangling from trees, consume the mice, and then die from the drug.

As we mentioned in Chapter 1, parasitoids are a unique type of predator that lives within and consumes the tissues of a living host, eventually killing it. Like many other predators, parasitoids also can limit the abundance of their prey. This can be seen in an example of wasps and scale insects. The California red scale insect (*Aonidiella aurantii*) is a worldwide pest in citrus orchards. It feeds on the leaves and fruits of the trees, which causes a great deal of damage and makes the fruit unmarketable. Fortunately, a small species of parasitoid wasp (*Aphytis melinus*) is able to control the abundance of the red scale insect by laying eggs inside the scale insects and ultimately killing them.

To demonstrate the magnitude of this control, researchers in California mimicked an insect outbreak by adding large numbers of scale insects to four trees. Ten other trees with typical low numbers of scale insects were the controls. You can see the results of this experiment in **Figure 14.3**. Shortly after adding large numbers of scale insects, there was a significant increase in the number of juvenile and adult wasps. As the wasps became more abundant, the population of scale insects declined rapidly. Within a few months, the large population of scale insects was reduced to the same level found on trees that had never received an experimental addition of scale insects. Although the wasps were not able to completely eliminate the scale insects, they held the population of scale insects to a level that minimally harmed the citrus crops.

MESOPREDATORS

Two levels of predators often exist in ecological communities: *mesopredators* and *top predators*. **Mesopredators** such as coyotes, weasels, and feral cats are relatively small carnivores that consume herbivores. In contrast, **top predators** such as wolves, mountain lions, and sharks typically consume both herbivores and mesopredators. Throughout history, top predators have interfered with human activities such as ranching, farming, hunting, and fishing. To protect our livelihood, we have reduced or eliminated the world's top predators, but this has caused unintended consequences. It is estimated that the decline of

Mesopredators Relatively small carnivores that consume herbivores.

Top predators Predators that typically consume both herbivores and mesopredators.

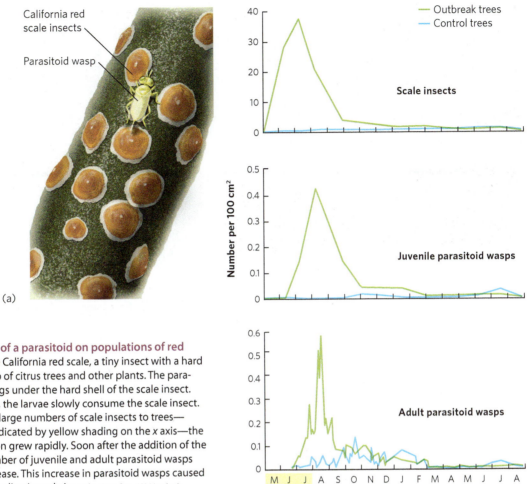

California red
scale insects

Parasitoid wasp

(a)

Figure 14.3 Effects of a parasitoid on populations of red scale insects. (a) The California red scale, a tiny insect with a hard shell, feeds on the sap of citrus trees and other plants. The parasitoid wasp lays its eggs under the hard shell of the scale insect. When the eggs hatch, the larvae slowly consume the scale insect. **(b)** After introducing large numbers of scale insects to trees—during the months indicated by yellow shading on the *x* axis—the scale insect population grew rapidly. Soon after the addition of the scale insects, the number of juvenile and adult parasitoid wasps began to rapidly increase. This increase in parasitoid wasps caused a subsequent rapid decline in scale insects. Data from W. Murdoch et al., Host suppression and stability in a parasitoid–host system: Experimental demonstration, *Science* 309 (2005): 610–613.

the top predators in North America had allowed 60 percent of all mesopredators to expand their geographic ranges.

The expansion in the range and abundance of mesopredators has had dramatic effects on the prey they consume. For example, the reduction of shark populations in the Atlantic Ocean from overharvesting has led to an increase in the cownose ray (*Rhinoptera bonasus*), a major mesopredator that consumes bay scallops (*Argopecten irridians*). This has caused a large reduction in bay scallops. In some cases, researchers have found that the benefits to humans of removing a top predator are much smaller than the damage inflicted by a mesopredator that became more abundant after the top predator declined. In Australia, for example, there has long been a campaign to remove dingoes and feral dogs because they kill sheep. However, removing these top predators has caused an increase in red foxes (*Vulpes vulpes*), which also eat sheep. This increase in red foxes has led to more than triple the loss in sheep that occurred when dingoes and feral dogs helped to control the red fox population.

HERBIVORES

Like predators, herbivores can have substantial effects on the species they consume. One of the classic examples is the control of the prickly pear cactus in Australia. The prickly pear cactus is a group made up of several different species that are native to North and South America. In the nineteenth century, many of these species were brought to Australia for a variety of reasons, including use as ornamental plants and as "living fences" for pastures. The cacti rapidly spread across the continent to the point that they dominated thousands of hectares of pasture and rangeland. To combat the spread of the cactus, biologists in the 1920s collected cactus moths (*Cactoblastis cactorum*) in South America and introduced the moths to Australia. These moths are natural herbivores of the cactus; the caterpillar stage of the moth consumes a portion of the cactus and the injuries it causes allow pathogens to infect the plants. As you can see in the

(a) (b) (c)

Figure 14.4 The prickly pear cactus in Australia. Following the import of the cactus from South America to Australia, its numbers increased dramatically. **(a)** To reduce the abundance of the cactus, the cactus moth was introduced to Australia from South America. **(b)** A site in Queensland, Australia, in 1926 prior to the introduction of the moth. **(c)** The same site 3 years after the introduction of the moth. Photos courtesy of (a) USDA/ARS, photographer Peggy Greb; (b and c) Department of Fisheries and Forestry; Queensland.

before and after photographs of **Figure 14.4**, the moths quickly reduced the cactus to very low abundances. While the moth has not completely eradicated the cactus because the cactus is able to disperse to moth-free areas, today, the prickly pear cactus exists only in small pockets around Australia.

Herbivorous insects have been used in similar ways in North America. In California, a plant known as Klamath weed

(*Hypericum perforatum*), which is toxic to livestock, was accidentally introduced from Europe in the early 1900s. By 1944, the weed had spread over nearly a million hectares of rangeland in 30 counties. In the 1950s, biologists decided to introduce a leaf-feeding beetle (*Chrysolina quadrigemina*) that consumed Klamath weed in Europe. After introduction of the beetle, the weed rapidly declined in abundance, as you can see in **Figure 14.5**. Biologists estimate that the beetle has now eliminated 99 percent of the Klamath weed population in North America.

The effects of herbivores are easily observed. For example, when domesticated animals such as cattle or sheep are raised at very high densities, little plant life remains. Similarly, deer and geese can consume large amounts of plants. Their impact can be demonstrated by fencing off areas to exclude animals to prevent grazing; the fenced areas have a greater total mass, or *biomass*, of plants and a higher composition of plants that are preferred by the herbivores (**Figure 14.6**). The plant species that remain in the areas where grazing occurs are those that the herbivores prefer not to eat.

Herbivore effects also occur in aquatic habitats. In rocky shore communities, for example, sea urchins control populations of algae. When sea urchins are removed from an area, the

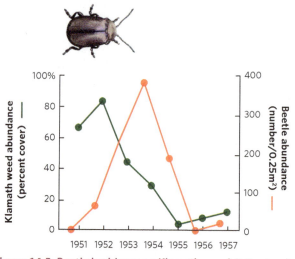

Figure 14.5 Beetle herbivory on Klamath weed. Following the introduction of leaf-feeding beetles, the beetle population initially increased and the plant began a rapid decline. Once the plant population declined, the beetle population also declined. Data from C. B. Huffaker and C. E. Kennett, A ten-year study of vegetational changes associated with biological control of Klamath weed, *Journal of Range Management* 12 (1959): 69–82.

biomass of algae increases rapidly, while the species composition of the algae also changes. In the presence of sea urchins, the remaining algae consist mostly of those species the sea urchins do not like and those that can withstand intense grazing. When sea urchins are removed, however, more palatable species such as the large brown algae become more abundant and can shade out the other species of algae. Such studies demonstrate that the influence of herbivores on the abundance of producers affects the species composition of the entire community.

Figure 14.6 Fencing out deer. In this deer enclosure in Gwaii Haanas National Park Reserve and Heritage Site, Haida Gwaii, BC, Canada, the long-term fencing of an area to prevent deer herbivory has allowed a much greater abundance of plants to grow. Photo by Jean-Louis Martin RGIS/CEFE-CNRS.

CONCEPT CHECK

1. What evidence is there that predators can control the abundance of prey?
2. How has the reduction of top predators had unintended consequences on the abundance of prey?
3. What evidence is there that herbivores can control the abundance of plants?

14.2 Populations of predators and prey fluctuate in regular cycles

In the previous chapter, we discussed how populations can fluctuate over time and space, and we saw that some populations cycle. At the beginning of this chapter, we introduced a study of population cycles for lynx and snowshoe hares. You can see the data from this study in **Figure 14.7**. The hare and lynx populations both experienced cycles of 9 to 10 years, with the lynx cycles lagging about 2 years behind the hare cycles.

It turns out that other large herbivores in the boreal and tundra regions of Canada—such as muskrat, ruffed grouse, and ptarmigan—also have population cycles of 9 to 10 years. Smaller herbivores, such as voles, mice, and lemmings, tend to have 4-year cycles. Studies of predators in these regions have revealed that some predators—including red foxes, marten, mink, goshawks, and horned owls—feed on larger herbivores and have long population cycles. In contrast, other predators—including Arctic foxes (*Vulpes lagopus*), rough-legged hawks (*Buteo lagopus*), and snowy owls (*Bubo scandiacus*)—feed on small herbivores and have short population cycles. The close synchrony of population cycles between predators and the prey they consume suggests that these oscillations are the result of interactions between them.

To understand the mechanisms underlying the predator–prey cycles, it is useful to examine them in the context of population models.

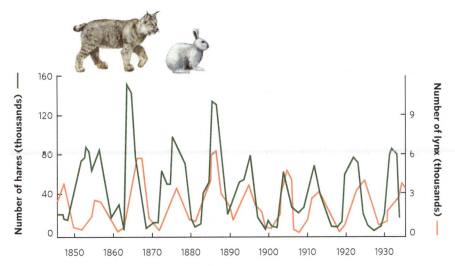

Figure 14.7 Cyclic fluctuations in the populations of snowshoe hares and lynx. Records from the Hudson's Bay Company of pelts purchased from trappers demonstrate that the populations of hares and lynx went through cycles approximately every 10 years. Data from D. A. MacLulich, *Fluctuations in the number of the varying hare (Lepus americanus),* University of Toronto Studies, Biological Series No. 43 (1937).

CREATING PREDATOR–PREY CYCLES IN THE LABORATORY

During the early twentieth century, biologists became interested in using predators and pathogens to control populations of crop and forest pests. One of the leading researchers in this effort was Carl Huffaker, a biologist at the University of California at Berkeley who pioneered the biological control of crop pests. Huffaker sought to understand the conditions that cause predator and prey populations to fluctuate. He chose two species of mites that lived on citrus trees: the six-spotted mite (*Eotetranychus sexmaculatus*) was the prey and the western predatory mite (*Typhlodromus occidentalis*) was the predator. In a series of experiments, he established populations on large trays that contained oranges—which were both habitat and food for the prey—with rubber balls interspersed among the oranges, as illustrated in **Figure 14.8**. On each tray, he varied the number and distribution of the oranges.

In most experiments, Huffaker introduced 20 female prey per tray and then introduced two female predators 11 days later. Because both species reproduce parthenogenetically, no males were added. When the prey were introduced to trays without predators, the prey population leveled off at between 5,500 and 8,000 mites. When predators were added, the predator population increased rapidly and soon wiped out the prey population. Without prey to feed on, though, predators quickly became extinct. However, it took longer for predators and prey to go extinct if the oranges were spread far apart from each other because it took longer for the predators to locate the prey. Under these experimental conditions, predators and prey could not coexist over time.

Huffaker reasoned that if predator dispersal could be further impeded, the two species might be able to coexist. To accomplish this, he introduced barriers to predator dispersal. The predatory mites disperse by walking, but the prey mites use a silk line that they spin to float on wind currents. Based on these differences, Huffaker modified his trays to give the prey a dispersal advantage by placing a mazelike pattern of Vaseline barriers among the oranges to slow the dispersal of the walking predators. He also placed vertical wooden pegs throughout the trays for the six-spotted mites to use as jumping-off points. This arrangement produced a series of three population cycles during the 8-month experiment, as depicted in **Figure 14.9**. The distribution of predators and prey throughout the trays continually shifted as the prey, which were on their way to extermination on one orange, recolonized another orange and stayed one step ahead of their predators. In short, Huffaker created a metapopulation in the laboratory.

Huffaker's experiment demonstrates that predators and prey cannot coexist in the absence of suitable refuges for the prey. However, the predator and prey populations can coexist through time in a spatial mosaic of suitable

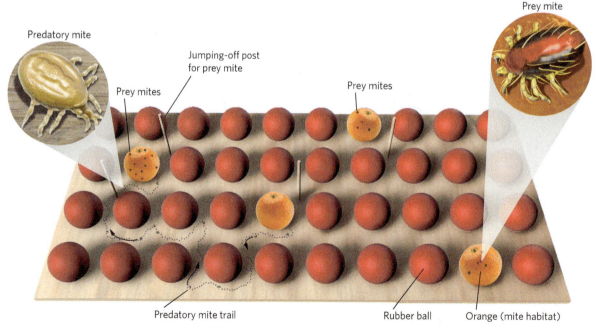

Figure 14.8 Huffaker's predator–prey lab experiment. To determine the factors that cause predator–prey cycles, Huffaker manipulated the number and distribution of oranges in a tray filled with oranges and rubber balls. He also added posts that only the prey species were able to climb. The prey drifted from the posts to colonize new oranges. Because the predators had to walk from one orange to another and avoid Vaseline barriers, the prey stayed one step ahead of the predators.

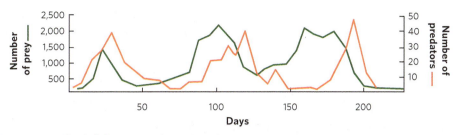

Figure 14.9 Predator–prey cycles in laboratory mites. When the prey species was given a dispersal advantage, it could colonize predator-free oranges and avoid going extinct. The cycling of the predator population lagged behind the prey population because it required more time to disperse to oranges containing prey and it required time to reproduce after finding prey. Data from C. B. Huffaker, Experimental studies on predation: Dispersion factors and predator-prey oscillations, *Hilgardia* 27 (1958): 343–383.

habitats that provided a dispersal advantage to the prey. Two kinds of time delays caused the populations to cycle: one was the result of predators dispersing more slowly between food patches than their prey, and the other was the result of time needed for predator numbers to increase through reproduction. From this, we can conclude that stable population cycles can be achieved when the environment is complex enough that predators cannot easily find scarce prey.

MATHEMATICAL MODELS OF PREDATOR–PREY CYCLES

Even before Huffaker's laboratory experiments on predator–prey cycles, mathematicians Alfred Lotka and Vito Volterra were developing models of predator–prey interactions. The **Lotka-Volterra model** incorporates oscillations in the abundances of predator and prey populations and shows predator numbers lagging behind those of their prey. It does this by calculating both the rate of change in the prey population and the rate of change in the predator population.

Let's begin with the prey population. Following the convention of our population models in Chapter 12, we will designate the number of prey as N and the number of predators as P. In words, the growth rate of the prey population depends on the rate of individuals being added to the prey population minus the rate of individuals being killed by predators:

$$\frac{dN}{dt} = rN - cNP$$

The first term in this equation (rN) represents the exponential growth of a prey population based on the intrinsic growth rate (r), as we saw in Chapter 12. For simplicity, this term does not include density dependence. The second term (cNP) represents the loss of individuals due to predation. The model assumes that the predation rate is determined by the probability of a random encounter between predators and prey (NP) and the probability of such an encounter leading to the prey's capture (c) (think of c as "capture efficiency").

Next we turn to the predator population. The equation for the predator population is similar to that for the prey population in that it has two terms, one term that represents

the birth rate of the predator population and a second term that represents the death rate of the predator population:

$$\frac{dP}{dt} = acNP - mP$$

The first term in the equation ($acNP$) represents the birth rate of the predator population. It is determined by the number of prey consumed by the predator population (cNP), which we saw in the prey equation above, multiplied by the efficiency of converting consumed prey into predator offspring (a). The second term (mP) represents the death rate of the predator population and is determined by the per capita mortality rate of predators (m) multiplied by the number of predators (P).

Changes in the Prey Population

We can use the Lotka-Volterra equations to determine the conditions that allow the prey population to be stable. By definition, a population is stable when its rate of change is zero. We can write this as

$$0 = rN - cNP$$

We can rearrange this as

$$rN = cNP$$

This indicates that the prey population becomes stable when the addition of prey (rN) is equal to the consumption of prey (cNP). We can further simplify this equation as

$$P = r \div c$$

In other words, the prey population will be stable when the number of predators equals the ratio of the prey's growth rate and the predator's capture efficiency.

Now that we know the conditions that make the prey population stable, we can also explore the conditions that cause the prey population to increase or decrease. The prey

Lotka–Volterra model A model of predator–prey interactions that incorporates oscillations in the abundances of predator and prey populations and shows predator numbers lagging behind those of their prey.

population will increase when the addition of prey (rN) exceeds the consumption of prey (cNP), which we can write as

$$rN > cNP$$

We can rearrange this inequality as

$$P < r \div c$$

This inequality represents the number of predators that the population of prey can support and still increase. This number is higher when the growth potential of the prey population (r) is higher or when the predators are less efficient at capturing prey (c). Using the same logic, the prey population will decrease whenever

$$P > r + c$$

Changes in the Predator Population

We can also use these equations to understand the conditions that make the predator population remain stable, increase, or decrease. Once again, the population is stable when the rate of change is zero:

$$0 = acNP - mP$$

We can rearrange this as

$$acNP = mP$$

This indicates that the predator population becomes stable when the production of new predators ($acNP$) is equal to the mortality of existing predators (mP). We can simplify this equation as

$$N = m \div ac$$

Given that this is the condition needed for a stable predator population, we can predict that the predator population will increase when the predator's production of new predators

exceeds the mortality of existing predators, which we can write as

$$acNP > mP$$

We can rearrange this as

$$N > m \div ac$$

This inequality represents the number of prey required to support the growth of the predator population. This number is higher when the death rate of predators (m) is higher, and it is lower when predators are more efficient at capturing prey (c) and converting them into offspring (a). Using the same logic, the predator population will decrease whenever

$$N < m \div ac$$

Trajectories of Predator and Prey Populations

Knowing the conditions under which predator and prey populations increase, decrease, or remain stable helps us understand why predator and prey populations sometimes cycle. **Figure 14.10a** plots the abundance of both populations. We can draw a horizontal line at the point where $P = r \div c$, which is the number of predators associated with a stable prey population. This line is called the **equilibrium isocline,** or **zero growth isocline,** for the prey because it indicates the points at which a population is stable. At any combination of predator and prey numbers in the region below the equilibrium isocline, there are relatively few predators and the prey population increases. In the

---

Equilibrium isocline The population size of one species that causes the population of another species to be stable. *Also known as* **Zero growth isocline.**

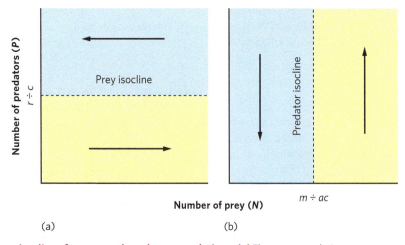

Figure 14.10 Equilibrium isoclines for prey and predator populations. (a) The prey population is stable when the number of predators is equal to $r \div c$. A higher number of predators causes the prey population to decrease, whereas a lower number of predators causes the prey population to increase. **(b)** The predator population is stable when the number of prey is equal to $m \div ac$. A higher number of prey causes the predator population to increase, whereas a lower number of prey causes the predator population to decrease. In both figures, population decreases occur in the yellow-shaded areas, whereas population increases occur in the blue-shaded areas.

region above this line, the prey population decreases because predators remove them faster than they can reproduce.

We can also plot the equilibrium isocline for the predator population, as shown in Figure 14.10b. In this graph, we draw a vertical line at the point where $N = m \div ac$, which is the number of prey that causes the predator population to be stable. Any combination of predator and prey numbers that lies in the region to the right of this line allows the predator population to increase because there is an increased abundance of prey to consume. In the region to the left, the predator population decreases because there is not enough prey available.

We can now combine our understanding of how predator and prey populations change in abundance by considering the trajectory of both populations simultaneously, which is called a **joint population trajectory.**

Figure 14.11a traces the path of the joint population trajectory. Beginning with the lower right region, predators and prey both increase, and their joint population trajectory moves up and to the right. In the upper right region, prey are still abundant enough that predators can increase, but the increasing number of predators depresses the prey population. Accordingly, the joint population trajectory moves up and to the left. In the upper left region, the continued decline in prey causes the predator population to decline, so the trajectory moves down and to the left. In the lower left region, the continued decline in predators allows the prey population to start increasing, which causes the trajectory to move down and to the right and completes the cycle. Together, the trajectories in the four regions define a counterclockwise cycling of predator and prey populations.

At the center of Figure 14.11a, you can see the **joint equilibrium point,** which is the point at which the equilibrium isoclines for predator and prey populations cross. The joint equilibrium point represents the combination of predator and prey population sizes that falls exactly at this point and

---

Joint equilibrium point The point at which the equilibrium isoclines for predator and prey populations cross.

Joint population trajectory The simultaneous trajectory of predator and prey populations.

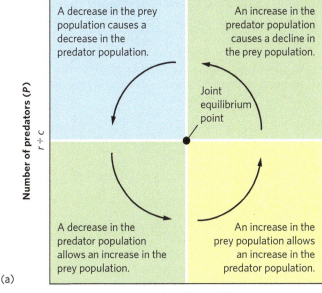

(a)

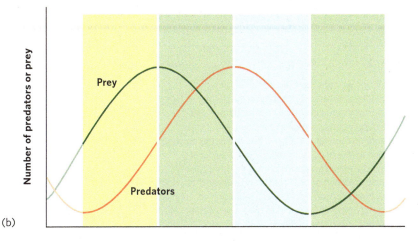

(b)

Figure 14.11 Predator–prey oscillations. The Lotka–Volterra model illustrates how the abundance of predator and prey populations cycle. (a) With a decline in predators, prey can increase in abundance. The increase in prey then provides more food for predators; this allows predators to increase their population size through reproduction. As predators grow more numerous, they start to kill prey at such a high rate that the prey population begins to decline. A decline in available prey reduces the survival and reproduction of the predator population, which starts to decline. (b) Over time, both populations cycle up and down in abundance, with predator abundance lagging behind prey abundance.

will not change over time. If either of the populations strays from the joint equilibrium point, the populations will oscillate around the joint equilibrium point rather than returning to it.

Based on the joint population trajectory of the Lotka-Volterra model, we can plot the changes in sizes of the two populations over time, as shown in Figure 14.11b. This visualizes how the two populations cycle and reveals that the prey population remains one-fourth of a phase ahead of the predator population. Although some populations cycle due to time delays, as we discussed in Chapter 13, the predator–prey equations do not cycle due to time delays; the predator–prey equations are continuous equations in which population changes are immediate. The cycling in this case is the result of each population responding to changes in the size of the other population. This pattern is reminiscent of the lynx population, discussed at the beginning of this chapter, which follows the fluctuations in the hare population.

FUNCTIONAL AND NUMERICAL RESPONSES

The Lotka-Volterra model provides an explanation for population cycles that relies on a very simplified version of nature. As we have already discussed, it does not include time delays or density dependence, and it does not incorporate the real foraging behavior of most predators. To get a more realistic picture of predator and prey relationships, we have to consider the *functional response* and the *numerical response* of predators. As we will see, both of these responses help to stabilize the cycling of predator and prey populations.

The Functional Response

The relationship between the density of the prey population and an individual predator's rate of food consumption is known as the **functional response** of the predator. There are three potential categories of functional responses, as illustrated in **Figure 14.12**. For each of these categories, we can examine the *number of prey* consumed by each predator, shown in Figure 14.12a, or the *proportion of prey* consumed by each predator, shown in Figure 14.12b. A key point to remember is that whenever prey population density increases and a predator can consume a higher proportion of those prey, the predator has the ability to regulate the growth of the prey population.

A **type I functional response,** indicated by the purple line, occurs when a predator's rate of prey consumption increases linearly with an increase in prey density until the predator is satiated. As shown in Figure 14.12a, the increase in prey density results in an ever-increasing number of prey consumed by a predator until the predator becomes satiated and can consume no additional prey. Some species of predators, such as web-building spiders that catch an

---

Functional response The relationship between the density of prey and an individual predator's rate of food consumption.

Type I functional response A functional response in which a predator's rate of prey consumption increases in a linear fashion with an increase in prey density until satiation occurs.

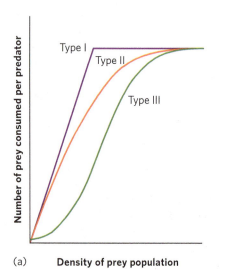

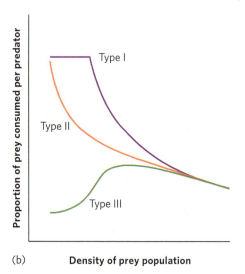

(a)   **Density of prey population**   (b)   **Density of prey population**

Figure 14.12 Functional responses of predators. (a) When we consider the number of prey consumed per predator, we see that a type I response exhibits a linear relationship between the density of the prey population and the number of prey a predator consumes until eventually the predator is satiated. In contrast, a type II response exhibits a slowing rate of prey consumption as prey density increases; the slowing rate is due to an increase in the time spent handling the additional prey. Eventually, the predator achieves satiation. A type III response also shows a slowing rate of consumption when prey density is high. In addition, it exhibits the effect of a predator learning a search image for prey as the prey density increases across the lower range. **(b)** When we consider the proportion of prey consumed, we see that the type I response results in a constant proportion of prey being consumed—prior to satiation—as prey density increases. A type II response results in a decreasing proportion of prey being consumed. Finally, a type III response initially causes an increased proportion of prey consumed, followed by a decrease.

increasing number of prey as the density of the prey population increases, exhibit a type I functional response. As you can see in Figure 14.12b, this means that as prey population density increases, each predator continues to consume a constant proportion of the prey up until the point of predator satiation. Once the prey population becomes so dense that predator satiation occurs, the predators consume a continuously decreasing proportion of the prey. This is the functional response used by the Lotka–Volterra model that we discussed.

The **type II functional response,** shown by the orange line, occurs when the number of prey consumed slows as prey population density increases and then plateaus when predator satiation occurs. The number of prey consumed slows because as predators consume more prey, they must spend more time handling the prey. For example, when a pelican catches a fish, it must take the time to manipulate the fish in its mouth and position the fish so that it will slide down the pelican's throat. The more fish that the pelican catches, the more time it must spend handling fish, which leaves less time available to hunt for fish. Eventually, because the predator is spending so much of its time handling large numbers of fish and has little time remaining to catch additional fish, its predation rate levels off. In Figure 14.12b, we can see that in a type II functional response, the slowing rate of prey consumption causes a decline in the proportion of prey consumed by each predator.

In a **type III functional response,** illustrated by the green line in Figure 14.12a, prey consumption is low when prey population density is low, consumption is rapid when prey population density is moderate, and prey consumption slows when prey population density is high. Figure 14.12b shows how this type of functional response affects the proportion of prey consumed. As prey population density increases, there is an initial increase in the proportion of prey consumed. However, as the predators spend more time handling prey and become satiated, this proportion subsequently declines—just as we saw in the type II response.

Low consumption at low prey population density can be the result of three factors. First, at very low prey density, the prey can hide in refuges where they are safe from predators. A predator can consume prey only after the prey become so numerous that some individuals are unable to find a refuge.

Second, at low prey population density, predators have less practice locating and catching the prey and therefore are relatively poor at doing it. As prey population density increases, however, the predators learn to locate and identify a particular species of prey, a phenomenon known as a *search image.* A **search image** is a learned mental image that helps the predator locate and capture food, just as a person might locate a can of cola in a grocery store by searching for a small, red cylinder.

A third factor that can cause relatively low consumption of prey at low population density is the phenomenon of prey switching. This occurs when one prey species is rare and a predator changes its preference to another prey species that is more abundant. If the population density of the first species then increases, the predator can switch back to it.

Many laboratory and field studies have demonstrated type III functional responses. For example, researchers examined the feeding preferences of a predatory insect known as a backswimmer (*Notonecta glauca*). **Figure 14.13** shows what happened when the predator was offered two types of prey—isopods (*Asellus aquaticus*) and larval mayflies (*Cloeon dipterum*)—and manipulated the proportions of each prey species. If the predator could not develop a search image, it would consume the two prey species in proportion to their availability. If the predator could develop a search image, it would eat fewer than expected of the rare prey and more than expected of abundant prey. As predicted, when the mayflies were rare, the backswimmer consumed fewer mayflies than expected, given their abundance. When

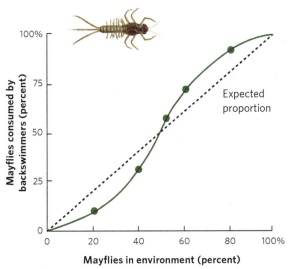

Figure 14.13 A type III functional response. Using predatory insects known as backswimmers, researchers manipulated the proportion of available prey: larval mayflies (shown above the graph) and isopods. The dashed black line indicates the expected proportion of mayflies consumed if the predator's preference was not affected by the proportional abundance of mayflies. When mayflies were rare, the predator consumed fewer mayflies than expected at random. When mayflies were common, the predator consumed more mayflies than expected at random. Data from M. Begon and M. Mortimer, *Population Ecology*, 2nd ed. (Blackwell, 1981), after J. H. Lawton, J. R. Beddington, and R. Bonser in M. B. Usher and M. H. Williamson (Eds.), *Ecological Stability* (Chapman & Hall, 1974), pp. 141–158.

Type II functional response A functional response in which a predator's rate of prey consumption begins to slow down as prey density increases and then plateaus when satiation occurs.

Type III functional response A functional response in which a predator exhibits low prey consumption under low prey densities, rapid consumption under moderate prey densities, and slowing prey consumption under high prey densities.

Search image A learned mental image that helps the predator locate and capture food.

mayflies were abundant, however, the backswimmers consumed more mayflies than expected, based on their abundance.

The reason that the predator switched to the more abundant prey was because its success rate was higher with the prey that were abundant. For example, when larval mayflies were rare, predator attacks were successful less than 10 percent of the time. When the mayflies were common, predator attacks were successful nearly 30 percent of the time. Researchers attributed the improved capture rate to practice; the backswimmer became more proficient at catching the mayflies when it had more opportunities to catch them. The backswimmer showed no innate preference for either prey species, only a preference for the one that was more abundant.

The Numerical Response

As we have seen, the functional response considers changes in the number of prey consumed by each predator. A predator's functional response tells us how many prey can be consumed by predators and therefore the conditions under which predators can regulate prey populations. The **numerical response** is a change in the number of predators through population growth or population movement due to immigration or emigration. Populations of predators usually grow slowly relative to populations of their prey, although the movement of mobile predators from surrounding areas can occur quite rapidly when the population density of prey changes.

An example of numerical response occurs in local populations of the bay-breasted warbler (*Dendroica castanea*), a small insectivorous bird of eastern North America. During outbreaks of the spruce budworm, populations of the warbler increase dramatically. In most years, the warbler populations live at densities of about 25 breeding pairs per square kilometer. However, during outbreak years, the warblers congregate in areas where the spruce budworm is abundant and warbler population densities can reach 300 breeding pairs per square kilometer. As a result of this rapid numerical response, the predator has the potential to reduce prey densities quickly and to regulate the prey's abundance.

14.3 Predation and herbivory favor the evolution of defenses

Given the large effects that predators can have on their prey and herbivores can have on producers, it is not surprising that many species have evolved strategies to defend themselves. In this section, we will review the types of defenses that prey have evolved and how some predators have evolved counterdefenses.

DEFENSES AGAINST PREDATORS

To understand the defenses prey use against predators, we first need to understand predator hunting strategies. Predator hunting strategies can be categorized as either *active hunting* or *ambush hunting,* also known as *sit-and-wait hunting.* A predator that uses active hunting spends most of its time moving around looking for potential prey. For example, American robins actively hunt when they move around a lawn searching for earthworms. In contrast, a predator that uses ambush hunting lies in wait for a prey to pass by. Chameleons can sit very still as they wait for an insect to pass. When the insect is close enough, the chameleon shoots out its long tongue and the sticky, prehensile tip grabs the unsuspecting prey.

Regardless of hunting mode, we can think of hunting by predators as a series of events: detecting the prey, pursuing the prey, catching the prey, handling the prey, and consuming the prey. As we will see, prey have evolved defenses to thwart the predator at different points in this series of events.

Behavioral Defenses

Some of the most common behavioral defenses against predators include alarm calling, spatial avoidance, and reduced activity. Alarm calling is used by many species of birds and mammals to warn their relatives that predators are approaching. Prey that use spatial avoidance move away from the predator. Prey that follow a strategy of reduced activity reduce their activity when predators are detected, making them less likely to come into contact with a predator. In a study of six species of tadpoles, the animals were placed in tubs of water with one of two treatments. The first treatment contained a caged predator that could not kill the tadpoles but could emit chemical cues that tadpoles can detect. The second treatment was a control, which consisted of an empty cage. Once the experiment was set up, the tadpoles were watched to determine their activity level, which was defined as the percent of the time that the animals spent moving. As shown in **Figure 14.14**, each species exhibited a different level of activity when predators were absent and all species reduced their level of activity in the presence of the predator.

Crypsis

Another way to reduce the probability of being detected by a predator is through camouflage that either matches the environment or breaks up the outline of an individual

---

CONCEPT CHECK

1. How does a prey population's ability to disperse allow the prey to persist in the presence of predators?
2. Based on the predator–prey population equations, why is the prey population stable when $rN = cNP$?
3. What causes the difference between a predator exhibiting type II versus type III functional responses?

Numerical response A change in the number of predators through population growth or population movement due to immigration or emigration.

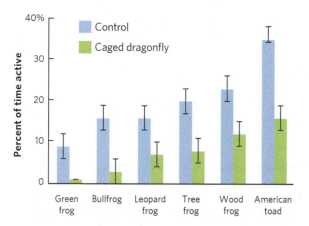

Figure 14.14 Behavioral defenses. Tadpoles commonly avoid predation by becoming less active, which is measured as the portion of the day that they spend moving. Being less active reduces the probability of being detected by a predator. Error bars are standard errors. Data from R. A. Relyea, The relationship between predation risk and antipredator responses in larval anurans, *Ecology* 82 (2001): 541–544.

to blend in better with the background environment, a phenomenon known as **crypsis.** Various animals resemble sticks, leaves, flower parts, or even bird droppings. These organisms are not so much concealed but mistaken for inedible objects and passed over by predators. Common species that use crypsis include stick insects, katydids, and horned lizards (**Figure 14.15**). Some species have a fixed color pattern that aids in crypsis, while other species, such as the octopus, are able to rapidly change color in ways that make themselves match their background.

Structural Defenses

Although some prey use behavior and crypsis to avoid being detected, other species employ mechanical defenses that reduce the predator's ability to capture, attack, or handle the prey. One of the best-known examples of a mechanical defense is the barbed quills of the porcupine; more than 30,000 quills cover the porcupine's body and can penetrate the flesh of an attacking predator. In other species, the structural defenses are phenotypically plastic and therefore induced only when the prey detects a predator in the environment. For example, water fleas—tiny freshwater crustaceans—that detect the chemical cues of predators early in their life can develop spines along different parts of their body to deter predators from consuming them.

Other mechanical defenses involve changes in the overall shape of the body. For example, the crucian carp (*Carassius carassius*), a species of fish that lives in Europe and Asia, develops a deep, hump-shaped body over a period of many weeks when the carp smells a predatory fish in the water. As you can see in **Figure 14.16**, carp with a hump-shaped body have greater muscle mass and can accelerate more quickly away from a predator.

Chemical Defenses

Prey can also use chemical defenses to deter a predator. Skunks, well known for using this strategy, spray potential threats with foul-smelling chemicals from posterior glands. Many insects also use chemical defenses. For example, when monarch butterfly caterpillars feed on milkweed, they store some of the milkweed toxins in their

---

Crypsis Camouflage that either allows an individual to match its environment or breaks up the outline of an individual to blend in better with the background environment.

(a)

(b)

Figure 14.15 Cryptic prey. Some prey avoid detection by blending into the environment. Crypsis is a strategy used by a number of animals including those shown here: **(a)** a katydid and **(b)** a horned lizard (*Phrynosoma platyrhinos*). Photos by (a) GEORGE GRALL/National Geographic Creative; and (b) Jason Bazzano/Alamy.

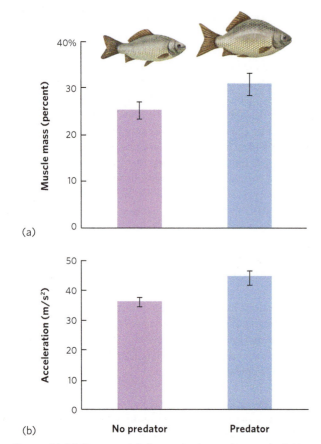

(a)

(b)

No predator **Predator**

Figure 14.16 Structural defenses. In the crucian carp, individuals living with predatory fish develop a deep, hump-shaped body that allows the carp to escape predation. These predator-induced fish have **(a)** greater muscle mass, which allows **(b)** greater acceleration while swimming away from predators. Error bars are standard errors. Data from P. Domenici et al., Predator-induced morphology enhances escape locomotion in crucian carp, *Proceedings of the Royal Society B* 275 (2008): 195–201.

body, which make the butterfly very distasteful to predatory birds. Bombardier beetles take a different approach. Their abdomens contain two glands, each of which makes a distinct chemical. When agitated, the beetle mixes these two chemicals, causing a reaction that makes the liquid approach 100°C. They shoot the boiling hot liquid out of the abdomen, causing pain or death to small predators, as illustrated in **Figure 14.17**.

While chemical defenses are often effective in deterring predation, these defenses are even more effective if the prey can communicate their distastefulness to the predator before an attack occurs. In many species, distastefulness has evolved in association with very conspicuous colors and patterns, a strategy known as **warning coloration,** or **aposematism.** Predators quickly learn to avoid markings such as the black and orange stripes of monarch butterflies; the insect tastes so bitter that a single experience is well remembered. Conspicuous combinations of black, red, and yellow adorn such diverse animals as bombardier beetles, yellow-jacket wasps, and coral snakes. These color combinations consistently

Figure 14.17 Chemical defenses. The bombardier beetle (*Stenaptinus insignis*) mixes two chemicals inside its abdomen that react to release a boiling hot fluid to deter predators.

advertise distastefulness or a threat of harm; some predators have evolved innate aversions to such prey and, therefore, do not need to learn to avoid them.

Mimicry of Chemical Defenses

When predators avoid aposematic species, any individuals of palatable species that resemble the distasteful, aposematic species would also be favored by selection. Over generations, a palatable species can evolve to more closely resemble an aposematic species, a phenomenon called **Batesian mimicry,** named for Henry Bates, the nineteenth-century English naturalist who first described it. In his journeys to the Amazon region of South America, Bates found numerous cases of palatable insects that did not retain the cryptic patterns of their close relatives but instead had evolved to resemble brightly colored, unpalatable species (**Figure 14.18**).

Studies have demonstrated that mimicry does indeed confer an advantage to the mimics. For example, toads that consumed live bees and received a sting on the tongue subsequently avoided palatable drone flies, which mimic the appearance of bees. In contrast, when naive toads consumed dead bees with stingers removed, they consumed the bees and the drone flies. This result indicates that toads learned to associate the conspicuous and distinctive color patterns of live bees with an unpleasant experience.

Another type of mimicry, called **Müllerian mimicry,** occurs when several unpalatable species evolve a similar pattern of warning coloration. Müllerian mimicry is named

---

Warning coloration A strategy in which distastefulness evolves in association with very conspicuous colors and patterns. Also known as aposematism.

Batesian mimicry When palatable species evolve warning coloration that resembles unpalatable species.

Müllerian mimicry When several unpalatable species evolve a similar pattern of warning coloration.

(a)

(b)

(c)

Figure 14.18 Batesian mimicry. (a) The common wasp (*Vespula vulgaris*) possesses aposematic coloration as a warning to predators that an attack could result in being stung and injected with painful chemicals. Other harmless species that do not possess any stinging ability have evolved to resemble the color patterns of the wasp. These include **(b)** a species of hover fly (*Helophilus pendulus*) and **(c)** the hornet clearwing (*Sesia apiformis*), which is a species of moth. Their resemblance to the wasp reduces their risk of predation. Photos by (a) Nick Upton/naturepl.com; (b) Geoff Dore/naturepl.com; and (c) FLPA/Gianpiero Ferrar/AGE Fotostock.

after its discoverer, the nineteenth-century German zoologist Fritz Müller. When multiple species of prey have conspicuous color patterns and all are unpalatable, a predator that learns to avoid one prey species will later avoid all prey species with a similar appearance. For example, most of the bumblebees and wasps that co-occur in mountain meadows share a pattern of black and yellow stripes and they all have the ability to sting a predator. Similarly, in Peru, several

species of poison dart frogs, all in the genus *Ranitomeya*, closely resemble each other. In four regions of that country, researchers have found three species that vary in coloration according to location, including one species (*R. variabilis*) that looks very different across two locations. A fourth species, *R. imitator*, is also unpalatable and has populations in each of these four locations that closely resemble the other species that is present (**Figure 14.19**).

Sauce, San Martin, Peru

R. summersi   *R. imitator*

Cainarachi Valley, San Martin, Peru

R. variabilis   *R. imitator*

Varadero, Loreto, Peru

R. fantastica   *R. imitator*

Pongo de Cainarachi, San Martin, Peru

R. variabilis   *R. imitator*

Figure 14.19 Müllerian mimicry. Müllerian mimics are a collection of unpalatable species that share a pattern of warning coloration. At four locations in Peru, researchers have found pairs of species of poison dart frogs that are unpalatable and closely resemble each other. For each location, the frog on the left is a particular species of unpalatable frog. In all four locations, the frog on the right is the unpalatable *R. imitator,* so named because in each location it has evolved to imitate the coloration of the other frog. Data from M. Chouteau et al., Advergence in Müllerian mimicry: The case of the poison dart frogs of Northern Peru revisited, *Biology Letters* 7 (2011): 796–900. Photos by Evan Twomey, East Carolina University.

Costs of Defenses Against Predators

Many types of defenses against predators can be costly, as we discussed in our coverage of predator- and herbivore-induced defenses in Chapter 4. For example, behavioral defenses such as spatial avoidance can result in reduced feeding or increased crowding as prey move to locations away from predators. In such cases, behavioral defenses often come at the cost of reduced growth and development. Similarly, most mechanical defenses, such as the 30,000 quills of a porcupine, are energetically expensive to produce. When the costs of defense are so high that they come at the cost of growth and reproduction, the presence of predators can cause smaller prey population sizes even when they do not consume the prey.

Less is known about the costs of chemical defenses in prey, but studies suggest that these defenses are also energetically costly to produce. In ladybugs, which are technically known as ladybird beetles, many species are red with black spots. These warning colors communicate to predators that the ladybugs taste bad because of chemicals in their bodies known as alkaloids. However, there is a large amount of variation in the concentration of alkaloids that each ladybug can produce. In 2012, researchers reported that only beetles that consumed large amounts of food had sufficient energy to produce high concentrations of carotenoids, which give the ladybugs a more intense red color, as shown in **Figure 14.20a**. Moreover, as illustrated in Figure 14.20b, beetles that produced more carotenoids also produced higher concentrations of alkaloids. As a result, ladybugs with the highest energy diet can better advertise their level of toxicity to predators and thereby reduce their chances of being attacked.

Counter Adaptations of Predators

If predation can select for prey to evolve a wide range of defenses, then prey defenses should favor the selection for counter-adaptation in predators. In this way, predators and prey experience an evolutionary arms race between prey defenses and predator offenses. When two or more species affect each other's evolution, we call it **coevolution.** In the case of the porcupine, for example, the spines deter most predators. However, bobcats (*Lynx rufus*) and wolverines (*Gulo gulo*) have an effective solution. When these predators find a porcupine, they flip the porcupine on its back and attack the belly, which is not defended by spines. Other common predator adaptations include high-speed locomotion to catch their prey and camouflage that allows them to ambush their prey.

Some predators can also evolve to handle the toxic chemicals produced by prey. For example, the cane toad (*Bufo marinus*) is a species that was introduced into Australia in 1935. Like other species of toads, the cane toad contains toxins in its skin that can cause predators to become sick or die. As a result, predators in the native range of cane toads do not attack cane toads even though these

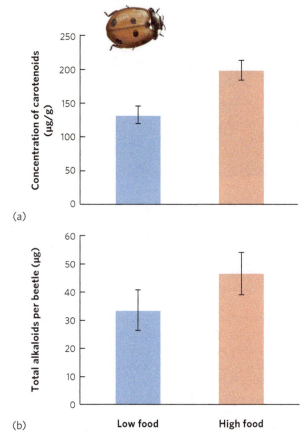

(a)

(b)

| Low food | High food |

Figure 14.20 Costly chemical defenses in ladybugs. In male ladybugs, individuals given a high amount of food produced (a) a higher concentration of carotenoids that make the ladybug's body more intensely red, and (b) a greater amount of defensive alkaloid chemicals. Error bars are standard errors. Data from J. D. Blount et al., How the ladybird got its spots: Effects of resource limitation on the honesty of aposematic signals, *Functional Ecology* 26 (2012): 334–342.

same predators regularly consume other species of amphibians. When the toads were introduced to Australia, predators of amphibians, such as black snakes (*Pseudechis porphyriacus*), had no evolutionary experience with cane toads. When they ate the toads, most snakes had no resistance to the toxins and they died. However, some populations of the black snakes were consuming cane toads, which suggests that while most black snakes that consumed cane toads died, some snakes must have been resistant to the toad's toxins and survived. Over time, the selection for snake resistance to the toad toxin must have resulted in the evolution of resistant populations. To test this hypothesis, the researchers fed samples of toad skin to snakes from different populations around Australia. They then examined how much the toad skin reduced the swimming speed of each snake, which indicates the susceptibility of each snake to the toxin.

Coevolution When two or more species affect each other's evolution.

Understanding Statistical Significance

Eco TV
macmillanlearning.com/
ricklefsvideo

In examining adaptations of prey, counter-adaptations of predators, or any other ecological measurements, we often consider experiments in which researchers find differences in the outcomes of experimental manipulations. Until now, we have not explored how ecologists assess when such differences are meaningful versus when the differences are due to chance.

For any group of measurements, such as the concentration of toxins in ladybugs fed high and low amounts of food, there will be variation among the individuals of each group. If we were to sample ladybugs at random from the high- and low-food treatment groups, we would find that the mean toxin concentration is higher in the high-food treatment group. However, the measurements taken on some of the individuals in the high-food treatment group might overlap with the measurements taken on some of the individuals in the low-food treatment group. When the means are similar and the distribution of the data from two groups is almost entirely overlapping, we would have to conclude that the two groups are nearly identical in whatever we are measuring, as shown in the figure below.

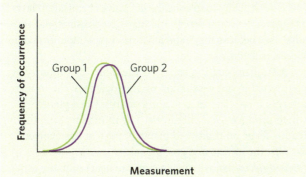

In contrast, when the means are very far apart and the distribution of the data from two groups shows no overlap, as in the case of the next graph, we would feel confident that the two groups are completely different in whatever we are measuring.

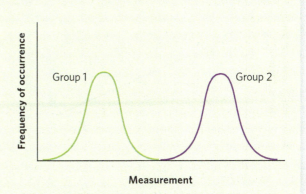

Although it is rare for two groups to have completely overlapping or completely nonoverlapping distributions, we need to know if the degree of overlap between the two data sets is acceptable in order to conclude that the groups are different from each other, with regard to the variable being measured.

Scientists agree that two distributions can be considered "significantly different" if we can sample the two distributions many times and find that the means of those distributions overlap less than 5 percent of the time. This somewhat arbitrary, but widely accepted, cutoff value is known as alpha (α). Thus, we say that our cutoff for statistical significance is $\alpha < 0.05$. Determining that something has statistical significance is not the same as stating that a difference between two means is large, substantial, or important. In other words, the everyday use of "significant" is not synonymous with the scientific use of "significantly different."

YOUR TURN In Chapter 2, "Analyzing Ecology: Standard Deviation and Standard Error," we mentioned that when data have a normal distribution, about 68 percent of the data fall within 1 standard deviation of the mean, 95 percent of the data fall within 2 standard deviations of the mean, and 99.7 percent of the data fall within 3 standard deviations of the mean. Based on this information, if you had two groups of data with identically shaped distributions of data, approximately how many standard deviations apart would they have to be in order for the two groups to be considered significantly different?

As you can see in **Figure 14.21**, the snake populations that coexisted with cane toads for the longest amount of time evolved the lowest susceptibility to the cane toad toxin. This evolution would have occurred in less than 70 years, which is a remarkably short time.

DEFENSES AGAINST HERBIVORES

Just as selective pressure from predators has caused the evolution of prey defenses, selective pressure from herbivores has caused the evolution of defenses against herbivory. In some cases, these defenses are induced by an herbivore

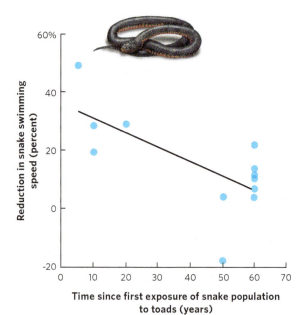

Figure 14.21 Counter-adaptations of predators to prey chemical defenses. Researchers measured the swimming speed of black snakes before and after being fed the skin of cane toads, which contains toxins. They found that snake populations that had coexisted for the longest times with the toad had the lowest susceptibility to the toad toxin. Data from B. L. Phillips and R. Shine, An invasive species induces rapid change in a native predator: Cane toads and black snakes in Australia, *Proceedings of the Royal Society B* 273 (2006): 1545–1550.

attack and are therefore phenotypically plastic, as we saw in Chapter 4. In other cases, plant defenses are fixed and therefore expressed whether or not an herbivore has attacked the plant. In both cases, some herbivore species have evolved counter-adaptations. In fact, some species of herbivores are so specialized in countering the defenses of a particular plant species that they do not consume any other plants.

Structural Defenses

When it comes to structural defenses, plants have evolved a variety of traits to deter herbivores from consuming their leaves, stems, flowers, or fruit. Some plants, such as cacti, roses, and blackberries, have sharp spines and prickles that inflict pain on the mouths of herbivores. Other plants grow a wooly layer of hair over their leaf surfaces to make it difficult for herbivorous insects to penetrate them.

Chemical Defenses

A wide variety of chemical defenses have evolved in plants. Plant chemicals include sticky resins and latex compounds that are hard to consume. Some plants also produce alkaloids—including caffeine, nicotine, and morphine—that have a wide range of toxic effects on herbivores. Other chemicals in plants, such as tannins, are difficult for herbivores to digest. It is generally thought that the chemicals produced by plants are the by-products of a plant's physiology.

Figure 14.22 Chemical plant defenses and counter-adaptations of specialized herbivores. The Tahitian noni, also known as the vomit flower, is a shrub that produces such a foul smell that most species of fruit flies avoid it. However, one species of fruit fly has evolved that tolerates the chemicals and therefore lays its eggs on the plant. Photo by U.S. National Park Service/Bryan Harry.

For many of these chemical defenses, one or more species of herbivore has evolved tolerance. For example, a shrub in Polynesia called the Tahitian noni (*Morinda citrifolia*) produces toxic chemicals with such a foul smell that the plant has the nickname "vomit fruit" (**Figure 14.22**). Most species of fruit flies avoid this plant because if they were to land on it, they would die. However, one species of fruit fly (*Drosophila sechellia*) has evolved the ability to tolerate the defensive chemicals. It lays its eggs on the vomit flower, and they have a distinct fitness advantage because they experience no competition from other species of fruit flies.

A similar situation exists for the monarch butterfly. As mentioned in our discussion of prey defenses, the monarch caterpillar specializes in feeding on milkweed plants, which produce toxic chemicals. The monarch caterpillar readily feeds on milkweed plants because it has evolved to resist the effects of a group of chemicals known as *cardiac glycosides,* which can stop the heart of many other herbivores. It also sequesters some of the chemicals to use as a defense against its own predators. In 2012, researchers made a striking discovery about the evolution of insects' ability to consume the plants containing cardiac glycosides. They found that a diversity of insects from different orders—including flies, beetles, true bugs, and butterflies—independently evolved the same changes in a gene that offers resistance to the effects of the toxin. In short, the different groups of insects exhibited convergent evolution.

Tolerance to Being Eaten

Some plants that have not evolved extensive defenses against herbivores take an alternative strategy of tolerating herbivory. Plants taking this strategy are able to grow

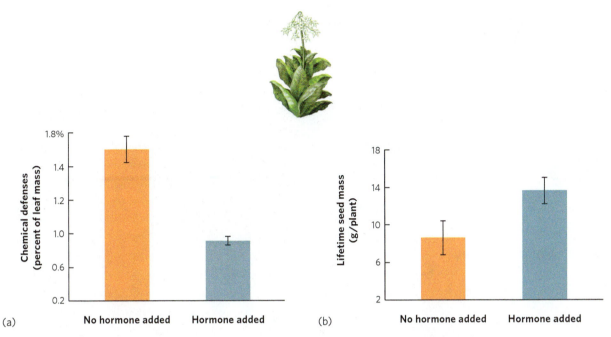

Figure 14.23 Costs of defenses against herbivores in tobacco plants. Researchers damaged tobacco plants to simulate the act of herbivory. They then prevented a chemical response in some individuals by applying plant hormones to the damaged area of the leaves. **(a)** Damaged plants treated with the plant hormone produce a lower amount of chemical defense than damaged plants not treated with the hormone. **(b)** Damaged plants treated with the plant hormone also experienced higher lifetime fitness, as measured by the mass of seeds produced. Error bars are standard errors. Data from I. T. Baldwin et al., The reproductive consequences associated with inducible alkaloid responses in wild tobacco, *Ecology* 71 (1990): 252–262.

new tissues rapidly to replace those that are consumed. For example, herbivores often consume the top meristem of a plant, which is the region of the plant where most growth occurs. When this meristem is removed, the meristems of lower stems begin to experience increased rates of growth, which still allows the plant to experience relatively high fitness despite being partially consumed by an herbivore.

Costs of Herbivore Defenses

For decades, researchers have investigated whether plant defenses come at the cost of reduced fitness. When defense traits are phenotypically plastic, researchers can compare the fitness of individuals with induced defenses against the fitness of noninduced individuals. For example, tobacco plants (*Nicotiana sylvestris*) respond to herbivory by producing chemical defenses including nicotine. Researchers damaged one group of tobacco plants to induce an increase in chemical defenses. As a control, they also damaged a second group of plants and then treated the damaged areas with a plant hormone that prevented the chemical defenses from responding. When they later counted the number of seeds produced by the two groups, they found that the group with increased chemical defenses produced fewer seeds, as shown in **Figure 14.23**.

A second approach to quantifying the costs of defenses against herbivores is to make the genes responsible for defenses nonfunctional. For instance, researchers have examined the growth of different genotypes of mouse-ear cress, a tiny plant native to Europe and Asia. They reported that individuals with intact defense genes commonly grew more slowly than individuals with nonfunctional defense genes. This confirmed that plants pay a cost for defenses against herbivores.

CONCEPT CHECK

1. What are four ways in which prey have evolved to reduce their risk of being killed by predators?
2. Why does natural selection favor counter-adaptations to prey defenses?
3. What are three ways in which plants have evolved to reduce their risk of being killed by herbivores?

ECOLOGY TODAY   APPLYING THE CONCEPTS

The Trouble with Cats and Rabbits

Introduced species on Macquarie Island. Introduced rabbits have caused devastating effects on the plants of the island, which do not have an evolutionary history of living with rabbits. Attempts to reduce the rabbit populations have caused feral cats to switch from eating rabbits to eating the island's native birds. Photo by James Doust, Courtesy Australian Antarctic Division.

Islands throughout the world often contain a variety of endemic species that have evolved together for millions of years. As we have seen in this chapter, new species that are introduced to islands can have devastating effects on the native plants and animals. Because the native species do not share an evolutionary history with the introduced species, they have not evolved defenses against introduced predators and herbivores. These introductions happen commonly on islands. In response, we often attempt to remove the intruders to reverse the harmful effects, although the actual outcomes of these efforts can lead to unintended consequences.

An example of this occurred on Macquarie Island, a small island located halfway between Australia and Antarctica. Island plants and animals existed together for eons in a tundra biome that contained a large number of seabirds and land birds, as well as a diversity of plants that included tall grasses. Humans began to visit the island in the 1800s, and they brought several species of animals with them.

In the early 1800s, the island was used by seal hunters as a place to rest and to resupply their ships. These visitors introduced house cats (*Felis catus*), which soon became feral. While there are no data on the impact of the feral cats, it is generally assumed that they were mesopredators and preyed on the abundant island birds. In 1878, the seal hunters introduced European rabbits (*Oryctolagus cuniculus*) to the island to serve as a food source whenever the sailors returned to the island. Despite the fact that the cats fed on the rabbits, over time the rabbit population grew to very large numbers. Data collected from the 1950s through the 1970s suggest that the rabbit population experienced large fluctuations approximately every 10 years, similar to those seen in hares and other animals that live in high northern latitudes. Because the plants on the island had no evolutionary history with the rabbits, periods of high rabbit numbers caused devastating effects on the abundance of palatable plant species; when the rabbit population experienced large declines, the vegetation rebounded.

(a)

(b)

Removal of introduced species on Macquarie Island. (a) By 2011, severe degradation of vegetation existed on Macquarie Island, resulting from introduced rabbit populations. (b) The problem was corrected with the removal of every last non-native mammal on the island. In 2014, this effort was declared a success with every rabbit and mouse removed, and a rapid improvement in the vegetation, native birds, and several native invertebrate animals that depend on the vegetation. Photos by Ivor Harris/Australian Antarctic Division.

Because periods of high rabbit populations caused such adverse effects on the island's vegetation, scientists introduced the European rabbit flea (*Spilopsyllus cuniculi*) in 1968. The fleas can carry a *Myxoma* virus that causes the disease myxomatosis, which is fatal to rabbits. The scientists quickly learned that the virus did not persist well on the island, so beginning in 1978, they reintroduced it every year. The virus had its desired effect and caused the rabbit populations to plunge from a high of 130,000 individuals in 1978 to fewer than 20,000 by the 1980s. As the rabbit population declined, the vegetation of the island began to recover. However, the changes to the island community did not end there.

With the rapid decline of the rabbits, the cats of the island exhibited prey-switching behavior—described in our discussion of predator functional responses—and added more birds to their diet. They consumed an estimated 60,000 seabirds per year and caused the extinction of two species of endemic land birds. To help save the island's undefended birds, government officials decided to eradicate all the cats from the island.

Starting in 1985, scientists removed 100 to 200 cats from the island each year. An analysis of the cat stomachs in 1997 indicated that while the cat population was killing birds, the cat population was still consuming approximately 4,000 rabbits per year, so the cats—in combination with the *Myxoma virus*—were probably still regulating the rabbit population.

When the last cat was removed in 2000, the island community responded rapidly. For example, five species of seabirds started breeding on the island, some of which had not bred there for more than 80 years. The rabbits also responded. From 1985 to 2000, the rabbit population increased from approximately 10,000 to more than 100,000 individuals. This increase was probably caused by a combination of fewer predatory cats, a rebounding abundance of plants, and a decline in the amount of virus distributed each year. By 2006, there were 130,000 rabbits, and these herbivores had consumed so much vegetation that parts of the island resembled a closely cropped lawn.

These changes on the island underscored the importance of understanding all the interactions that introduced species can have on a community, including interactions among the introduced species themselves. In the case of Macquarie Island, it has become clear that removing the introduced cats was not enough to return the island community to its former condition. The introduced rabbits had to be removed as well. As a result, the Australian government agreed to spend $24 million to remove the rabbits from the island as well as several other introduced mammals, including mice and rats. In 2011, grain-containing poison was distributed around the island by helicopter, and this was very successful in killing the vast majority of the introduced mammal species. From 2012 to 2013, hunters with specially trained dogs were brought to the island to eliminate the few remaining rabbits, mice, and rats. It is estimated that the hunters walked a total of 92,000 km in their search for every last non-native mammal on the island. In 2014, this effort was declared a success with every rabbit and mouse removed and a rapid improvement in the vegetation, native birds, and several native invertebrate animals that depend on the vegetation.

SOURCES:

Success on Macquarie Island. 2014. *Macquarie Island Pest Eradication Project Newsletter,* Issue 14, July. http://www.parks.tas.gov.au/file.aspx?id=36472.

Bergstrom, D. M., et al. 2009. Indirect effects of invasive species removal devastate World Heritage Island. *Journal of Applied Ecology* 46: 83–81.

Dowding, J. E., et al. 2009. Cats, rabbits, *Mxyoma* virus, and vegetation on Macquarie Island: A comment on Bergstrom et al. (2009). *Journal of Applied Ecology* 46: 1129–1132.

Bergstrom, D. M., et al. 2009. Management implications of the Macquarie Island trophic cascade revisited: A reply to Dowding et al. (2009). *Journal of Applied Ecology* 46: 1133–1136. http://www.abc.net.au/news/2012-04-19/rabbit-hunters-head-to-macquarie-island/3961270.

SUMMARY OF LEARNING OBJECTIVES

14.1 Predators and herbivores can limit the abundance of populations. Using observations in nature and manipulative experiments, ecologists have found that predators commonly limit the abundance of prey and herbivores commonly limit the abundance of producers.

Key Terms: Introduced species, Exotic species, Non-native species, Invasive species, Mesopredators, Top predators

14.2 Populations of consumers and consumed populations fluctuate in regular cycles. Cycling populations have been observed frequently in nature and recreated in laboratory experiments. Lags in the response times of predator movement and reproduction linked to changes in the abundance of prey cause these cycles. Mathematical models have been developed to mimic the cycling behavior of predator and prey populations.

Key Terms: Lotka–Volterra model, Equilibrium isocline, Joint population trajectory, Joint equilibrium point, Functional response, Type I functional response, Type II functional response, Type III functional response, Search image, Numerical response

14.3 Predation and herbivory favor the evolution of defenses. Prey have evolved a wide variety of defenses, including behavioral defenses, mechanical defenses, chemical defenses, crypsis, and mimicry. Producers have evolved defenses against herbivores, including mechanical defenses, chemical defenses, and tolerance. Evolved defenses are commonly costly and can sometimes be countered by subsequent adaptations in predators.

Key Terms: Crypsis, Warning coloration, Aposematism, Batesian mimicry, Müllerian mimicry, Coevolution

CRITICAL THINKING QUESTIONS

1. How could you experimentally test whether herds of African antelope affect the abundance of plants on which they graze?

2. Explain why an herbivore that consumes many different species of plants might be less successful at regulating the abundance of a well-defended plant species compared to an herbivore that specializes on eating a single species of plant.

3. How do the results of the classic experiments of C. F. Huffaker, using mites and oranges in the lab, inform us about how predator and prey populations are able to persist in nature?

4. In evolutionary terms, explain why introduced species can often have harmful effects on native species but can also be controlled by an enemy that comes from the introduced species' native region.

5. Using the lynx–hare interaction, explain in words the equations of the Lotka–Volterra model for the change in the population sizes of prey and predators.

6. According to the Lotka–Volterra model of predator–prey interactions, why do the populations of foxes and rodents cycle?

7. Compare and contrast a predator's numerical response and functional response.

8. How might a type II functional response prevent a predator from controlling a large prey population?

9. What are the causes of an evolutionary arms race between consumers and the species that they consume?

10. For each of the five stages in a predation event, how might prey evolve a defense against predators?

GRAPHING THE DATA The Functional Response of Wolves

In Gates of the Arctic National Park in Alaska, researchers monitored the densities of wolves and their major prey, including caribou (*Rangifer tarandus*). To understand whether the wolves could potentially regulate the growth of the caribou population, they wanted to know the shape of the wolves' functional response. They could do this by determining the number of caribou killed by wolves in different areas and at different times of the year.

Using the data from this study, plot the relationship between caribou density and the number of caribou killed per wolf. Then plot the relationship between caribou density and the proportion of caribou killed per wolf. Based on your graphs, what type of functional response do the wolves have?

WOLF AND CARIBOU DATA		
CARIBOU DENSITY (NUMBER/km²)	NUMBER OF CARIBOU KILLED PER WOLF (PER DAY)	PROPORTION OF CARIBOU KILLED PER WOLF (PER DAY)
0.1	0.50	1.80
0.2	0.70	0.90
0.3	0.90	0.50
0.4	0.95	0.30
0.5	0.98	0.22
1.0	1.00	0.15
1.5	1.01	0.10
2.0	1.02	0.07
2.5	1.03	0.05
3.0	1.03	0.04

15 Parasitism and Infectious Diseases

The Life of Zombies

Zombie films scare us because they depict zombies as the walking dead in search of victims they will turn into more zombies. But something similar to this happens commonly in nature; some species of parasites infect a host and take control of its life for their own benefit.

Consider the case of the amber snail (*Succinea putris*) from Europe. This animal lives along the edges of streams and ponds. It normally spends its time in the shade of terrestrial vegetation, where it chews on leaves and remains hidden from the eyes of predatory birds. An amber snail will occasionally consume bird feces, which sometimes contain the eggs of a parasitic flatworm (*Leucochloridium paradoxum*). These eggs hatch inside the snail and grow, but to reproduce, the parasites must spend the next stage of their lives inside a bird. To achieve this goal, the parasitic larvae slowly make their way into the snail's eyestalks. These eyestalks are normally pale and slender, but the parasitic infection causes the eyestalks to become enlarged and banded with colors that pulsate in a way that resembles a moving caterpillar. The parasites also take control of the snail's brain and force the snail to move up a plant's stem, which is not something an amber snail normally does. Snails that move up plant stems are more easily noticed by predatory birds and, because the eyestalks look like caterpillars, birds consume the infected snails. The parasite completes the second stage of its life inside the bird and the cycle continues. The flatworm reproduces, its eggs leave the bird through feces, and snails consume them.

Parasites can control the behavior of many different animals. In Thailand, for example, carpenter ants (*Camponotus leonardi*) normally spend their time living in nests in the hot, dry rainforest canopy. Occasionally, the carpenter ants travel from the canopy to the ground, where they can be exposed to the spores of the fungus *Ophiocordyceps unilateralis*, the so-called zombie-ant fungus, before heading back up to their nests. An ant that is infected subsequently moves down to the

> "The ability of parasites to act as puppet masters over the behavior of their victims is just one way that parasites have evolved to improve their fitness."

An infected carpenter ant. In Thailand, carpenter ants that become infected by a fungus crawl down from the canopy and attach themselves to the underside of a leaf by biting the leaf vein and then dying. After death, a spore-producing stalk grows out of the ant's head and releases its spores into the environment. Photo by David Hughes/Penn State University.

A parasitized amber snail. The snail on the right has one normal eye stalk that is pale and slender and another that is infected by a parasitic flatworm, which causes the eye stalk to become enlarged and colorful. It also pulsates in a way that is attractive to predatory birds. Photo by Alex Teo Khek Teck via Flickr.

SOURCES:

Andersen, S. B., et al. 2009. The life of a dead ant: The expression of an adaptive extended phenotype. *American Naturalist* 174: 424–433.

Hughes, D. P., et al. 2011. Behavioral mechanisms and morphological symptoms of zombie ants dying from fungal infection. *BMC Ecology* 11: 13.

Lefevre, T., and F. Thomas. 2008. Behind the scene, something else is pulling the strings: Emphasizing parasitic manipulation in vector-borne diseases. *Infection, Genetics and Evolution* 8: 504–519.

Wesolowska, W., and T. Wesolowski. 2014. Do *Leucochloridium* sporocysts manipulate the behavior of their snail hosts? *Journal of Zoology* 292: 151–155.

humid, understory vegetation. It stops at about 25 cm above the ground in a place that many previously infected ants have also gone, known to researchers as the "ant graveyard." Here, it bites the underside of a leaf and maintains a death grip on a leaf vein as it dies. After the ant dies, the fungus grows a spore-producing structure out of the ant's head and releases its spores into the environment.

This unusual ant behavior does not benefit the ant, but it greatly benefits the fungus. The higher humidity found lower in the rainforest is more suited to fungal growth than the drier conditions high in the canopy. Moreover, if the infected ant were to die in its nest in the canopy, its nest mates would remove the corpse before the fungus could grow the spore-producing structure that is critical to its reproduction.

The ability of parasites to act as puppet masters over the behavior of their victims is just one way that parasites have evolved to improve their fitness. As we will discover in this chapter, parasites come in a wide variety of forms, and their effects on hosts can range from mild to lethal. Adaptations that enable parasites to infect hosts and adaptions that help hosts to resist infections by parasites offer intriguing insights into the strategies of parasite–host interactions.

LEARNING OBJECTIVES

After reading this chapter, you should be able to:

15.1 Identify the many different types of parasites affecting the abundance of host species.

15.2 Describe how parasite and host dynamics are determined by the parasite's ability to infect the host.

15.3 Illustrate how parasite and host populations commonly fluctuate in regular cycles.

15.4 Explain the process of parasites evolving offensive strategies, while hosts evolve defensive strategies.

The struggle between parasites and hosts has produced many fascinating examples of ecological interactions and evolutionary adaptations. In Chapter 1, we defined a parasite as an organism that lives in or on another organism, called the host, and causes harmful effects as it consumes resources from the host. It is estimated that approximately half of all species on Earth are parasites. Some hosts have **infection resistance,** which is the ability of a host to prevent the parasite from causing an infection, while other hosts have **infection tolerance,** which is the ability of a host to minimize the harm once an infection has occurred. The number of parasites of a given species that an individual host can harbor is known as the host's **parasite load.** A parasite typically infects only one or a few species of hosts, whereas a given host species can contain dozens of species of parasites.

Parasites that can cause an infectious disease are called *pathogens.* Infection by a pathogen does not always result in a disease. For example, humans can be infected with human immunodeficiency virus (HIV), but they may never experience the disease symptoms known as Acquired Immune Deficiency Syndrome (AIDS). In many cases, it is not known what causes a host to transition from being infected by a pathogen to having the disease.

Infectious diseases take a large toll on people; the World Health Organization estimates that more than 25 percent of all human deaths are caused by infectious diseases. Note that only infectious diseases are caused by pathogens; there are many noninfectious diseases in which pathogens do not play a role, such as heart disease.

In this chapter, we will focus on the interaction between parasites and their hosts. In later chapters, we will discuss the larger role that parasites can play in communities and ecosystems. We will begin by looking at the many different types of parasites that exist, including those that have large effects on crops, domesticated animals, and human health. We will then examine the factors that determine whether parasites can infect hosts, spread rapidly through a population, and cause widespread harmful effects. Because mathematical models can help us understand the population dynamics of interacting species, we will also discuss parasite–host models. Finally, we will consider how parasites have evolved to increase their chances of infecting hosts, and how hosts have evolved to combat the risk of infection.

15.1 Many different types of parasites affect the abundance of host species

Parasites typically have specific habitat needs and, as a result, often live in particular places on a host organism. In humans, for instance, parasites find certain parts of the body

---

Infection resistance The ability of a host to prevent an infection from occurring.

Infection tolerance The ability of a host to minimize the harm once an infection has occurred.

Parasite load The number of parasites of a given species that an individual host can harbor.

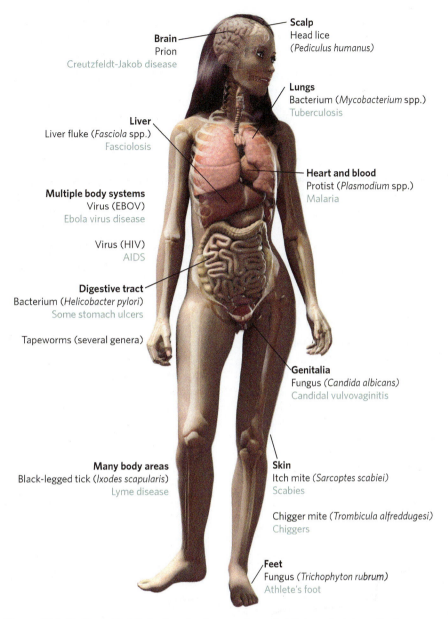

Brain
Prion
Creutzfeldt-Jakob disease

Scalp
Head lice
(*Pediculus humanus*)

Lungs
Bacterium (*Mycobacterium* spp.)
Tuberculosis

Liver
Liver fluke (*Fasciola* spp.)
Fasciolosis

Heart and blood
Protist (*Plasmodium* spp.)
Malaria

Multiple body systems
Virus (EBOV)
Ebola virus disease

Virus (HIV)
AIDS

Digestive tract
Bacterium (*Helicobacter pylori*)
Some stomach ulcers

Tapeworms (several genera)

Genitalia
Fungus (*Candida albicans*)
Candidal vulvovaginitis

Many body areas
Black-legged tick (*Ixodes scapularis*)
Lyme disease

Skin
Itch mite (*Sarcoptes scabiei*)
Scabies

Chigger mite (*Trombicula alfreddugesi*)
Chiggers

Feet
Fungus (*Trichophyton rubrum*)
Athlete's foot

Figure 15.1 Preferred habitats. Parasites have preferred habitats on their host. The human body, for example, offers a wide range of habitats for parasites.

to be highly suitable habitats. As illustrated in **Figure 15.1**, head lice insects live in the hair; liver flukes, which are flatworms, reside in the liver; the fungus that causes athlete's foot resides in the feet; and so on.

We can categorize the wide variety of parasites common to both plants and animals as either *ectoparasites* or *endoparasites*. **Ectoparasites** live on the outside of organisms, whereas **endoparasites** live inside organisms. Each lifestyle has advantages and disadvantages, as summarized in **Table 15.1**. For example, because ectoparasites live on the outside of their hosts, they do not have to combat the immune system of a host, and they can easily move on and off a host. A disadvantage is that

ectoparasites are exposed to the variable conditions of the external environment, including natural enemies, and they must find a way to pierce the flesh of their host to feed. In contrast, because endoparasites live inside their hosts, they must contend with the immune system of their host, and they may have a difficult time getting in and out of their host's body. Endoparasites have the advantage of being protected from the external environment and are therefore not exposed to most of their enemies. In addition, living inside the host gives endoparasites easy access to the host's

Ectoparasite A parasite that lives on the outside of an organism.

Endoparasite A parasite that lives inside an organism.

TABLE 15.1

Comparing the Consequences of Endoparasite and Ectoparasite Lifestyles

Factor	Ectoparasites	Endoparasites
Exposure to natural enemies	High	Low
Exposure to external environment	High	Low
Difficulty of movement to and from host, for parasite or its offspring	Low	High
Exposure to host's immune system	Low	High
Ease of feeding on host	Low	High

body fluids on which they feed. In this section, we will discuss some of the most common types of ectoparasites and endoparasites and the effects they have on their hosts.

ECTOPARASITES

A variety of organisms live as ectoparasites, as shown in **Figure 15.2**. Most ectoparasites that attack animals are arthropods, including two groups of arachnids—ticks and mites—and two groups of insects—lice and fleas. There are many other groups of ectoparasites, including leeches and some species of lamprey fish. Each of these animals attaches to a host and consumes the host's blood and other body fluids.

Plants also have ectoparasites, but they are most commonly either nematodes—also known as roundworms—or other species of plants. Nematodes that act as plant

ectoparasites live in the soil and feed on plant roots. The 1-mm-long worms attach to a plant root, inject digestive enzymes that break down root cells, and then consume the resulting slurry. This parasitic behavior can reduce the growth, reproduction, and survival of the plant.

Approximately 4,000 species of plants make their living as ectoparasites on other plants. For example, mistletoes embed rootlike organs into the branches of trees and shrubs. Because such plants have their roots inside the host and their shoots outside the host, they are sometimes categorized as *hemiparasites.* The mistletoe leaves conduct photosynthesis but obtain water and minerals from the host plants. Some species, such as dwarf mistletoes (in the genus *Arceuthobium*), cause death in conifer trees—particularly under drought conditions—by extracting too much of the host's water and nutrients. In some species of mistletoe, the plant's fruit is consumed by birds, which inadvertently disperse the seeds from the fruit when they defecate onto the branches of other trees and shrubs. Other species of mistletoe are able to project seeds laterally up to 15 m.

ENDOPARASITES

Endoparasites can be categorized as either *intracellular* or *intercellular.* As the names imply, intracellular parasites live inside the cells of a host, whereas intercellular parasites live in the spaces between cells that include the cavities of a host's body. Intracellular parasites are very small; examples include viruses, small bits of proteins known as *prions,* and some types of bacteria and protists. Intercellular parasites are much larger; examples include some types of protozoa and bacteria, fungi, and a group of worms known as *helminths,*

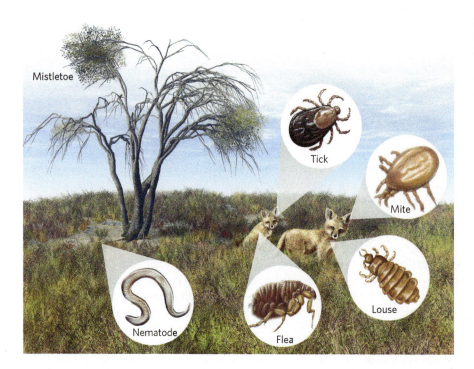

Figure 15.2 Ectoparasites.
Ectoparasites are parasites that live attached to the outside of their host. Common ectoparasites include ticks, fleas, lice, mites, mistletoes, and nematodes.

Mistletoe

Tick

Mite

Louse

Nematode

Flea

which include the nematodes. Each of these groups of endoparasites can have substantial effects on their host's survival, growth, and reproduction. Because endoparasites often cause fatal diseases, they can alter the abundance of host species and change the composition of ecological communities.

Viruses

Late in the nineteenth century, researchers struggled to find the cause of a disease striking tobacco plants. They recognized that the pathogen was too small to be a bacterium. After many years of research, they were able to identify it as a virus, which they named tobacco mosaic virus. Plants that are infected exhibit a mosaic of colors in their leaves and may also have growths that resemble blisters (**Figure 15.3**). Although this virus was first discovered in tobacco, we now know that it can infect more than 150 species of plants. Viruses that infect plants can have a devastating effect on food production, which makes them a matter of great concern to farmers and consumers.

 Animals can also be infected by many different pathogenic viruses. For example, there are different genera of the poxvirus that can infect mammals: cowpox in cattle, monkeypox in primates and rodents, and smallpox in humans. Hoofed animals such as deer, cattle, and sheep are susceptible to bluetongue virus, which can cause high mortality rates. Similarly, the West Nile virus—named for the location of the first known human case in the West Nile region of Uganda in 1937—can be highly lethal to some species of birds. The mosquitoes that carry this virus usually transmit the disease to birds, but occasionally to horses and humans, both of which can die from the infection. West Nile virus arrived in the United States in 1999 and spread rapidly. Several species of North American birds experienced high rates of death,

including blue jays (*Cyanocitta cristata*), American robins, and American crows (*Corvus brachyrhynchos*). At the same time, horses and people started to become infected. In the decade following the introduction of the virus into the United States, more than 1,000 people died. As a result, public health officials in the United States have made great efforts to reduce mosquito populations that spread the disease. These efforts, possibly combined with the evolution of the virus to be less virulent, have caused the number of infections and deaths to decline dramatically from 2003 to 2011, as illustrated in **Figure 15.4**. However, in 2012, there was a sharp increase in the number of infections and deaths in the United States. The sharp increase is thought to be due to unusually warm and humid conditions combined with abundant rain, which created a lot of standing water where mosquitoes can breed and subsequently build up a large adult population. In subsequent years, the number of infections and deaths has declined.

 Both mammals and birds are susceptible to strains of influenza virus. For example, a strain of influenza known as the *Spanish flu* is caused by the H1N1 virus and normally infects only birds, but in 1918 it also infected humans. Researchers hypothesize that the virus experienced a mutation that allowed it to survive in humans and to infect other humans directly. They believe that humans initially contracted the disease by handling domesticated ducks and chickens. However, once the virus found its way into humans, it rapidly spread around the world and killed up to 100 million people. More recently, the H1N1 virus has been found on pig farms, which is why the pathogen also has the name *swine flu*.

Figure 15.3 Tobacco mosaic virus. Plants infected with the tobacco mosaic virus, such as this tobacco plant, develop blisters and light areas on their leaves. These effects can cause substantial reductions in plant growth. Photo by Nigel Cattlin/Science Source.

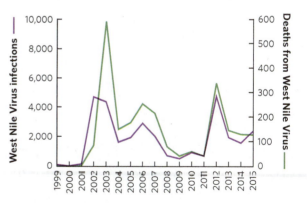

Figure 15.4 Human infections and deaths from West Nile virus in the United States. Following the introduction of the virus to the United States in 1999, the number of human infections and deaths rose rapidly through 2003. Efforts to control mosquito populations, which carry the virus, caused a sharp decline in these effects on humans through 2011. In 2012, however, unusually warm temperatures with high precipitation in some regions of the country resulted in large mosquito populations and a new surge in West Nile infections and deaths. After 2012, infections and deaths declined substantially. Note that the two *y* axes have different scales. Data from https://www.cdc.gov/westnile/resources/pdfs/data/1-WNV-Disease-Cases-by-Year_1999-2015_07072016.pdf.

In 2006, a similar strain of virus known as *bird flu*, or H5N1, infected and killed large numbers of birds and also jumped from domesticated birds to humans. Hundreds of thousands of domesticated birds were killed in an attempt to stop the spread of the virus (**Figure 15.5**). By 2017, the World Health Organization reported that 856 people had been infected with bird flu and more than half of them had died. The ability of pathogens to jump from traditional hosts to human hosts, as we will discuss in the next section, causes a great deal of concern for human health in the future.

Prions

Prions are a more recently discovered category of pathogenic parasite. All prions begin as a beneficial protein in the brain of an animal, but occasionally a protein folds into an incorrect shape and becomes pathogenic. Prions do not contain any RNA or DNA; instead, they replicate by coming into contact with normal proteins and causing the normal proteins to fold incorrectly, with the first prion serving as a template. As the number of prions increases in the body, they can kill cells and damage tissues.

One of the best-known diseases caused by prions is bovine spongiform encephalopathy, commonly known as mad cow disease. This disease, which is always fatal, gets its name from the way that infected cows (and sheep) lose control of their body, as if they have gone mad. While cows cannot transmit prions to each other, in the 1980s it was a common practice to feed cows the ground remains of other cows. Cows that consumed the mutated prions from the dead cattle became infected and ultimately passed on the infection to humans. Mad cow disease was most prevalent in the United Kingdom during the 1990s, where more than 180,000 cases of infected animals have occurred. Around the world, more than 220 people have died from mad cow disease. Today, new rules forbid feeding dead cattle and sheep to other cattle and sheep. As a result, the incidence of infected cattle and sheep is now rare, and mad cow disease in cows, sheep, and humans has declined.

Chronic wasting disease is a more common prion disease. It infects members of the deer family, including white-tailed deer, mule deer (*Odocoileus hemionus*), elk, and moose. Individuals infected with chronic wasting disease excrete prions that are later consumed inadvertently by others. Infected individuals begin to lose weight and eventually die (**Figure 15.6**). Currently, chronic wasting disease is concentrated in Colorado and Wyoming, but small numbers of infected deer have been discovered in other regions of the United States and Canada.

Figure 15.5 Controlling the spread of bird flu. In an attempt to stop the spread of bird flu, millions of domesticated birds in Asia were killed and their bodies burned. Data from http://www.cdc.gov/ncidod/dvbid/westnile/surv&control.htm#maps. Photo by SONNY TUMBELAKA/ Getty Images.

Figure 15.6 Chronic wasting disease. Deer infected with prions that cause chronic wasting disease lose weight and eventually die. Photo courtesy of Game Warden Michael D. Hopper, Kansas Dept. of Wildlife, Parks and Tourism.

Protozoans

Protozoans are a group of parasites that can cause a variety of diseases. For example, several protozoans cause diarrhea in humans and other animals. They are also the source of malaria in humans and avian malaria in birds. Malaria is transmitted by mosquitoes that acquire the protists when they feed on an infected individual and then transfer the protists to uninfected individuals they subsequently feed on. In Hawaii, as well as in some other parts of the world, there is no historic presence of avian malaria, so the birds of the Hawaiian Islands have not evolved defenses against the protists. Mosquitoes were accidentally introduced into Hawaii in the early 1800s, and by the early 1900s the protists that cause avian malaria had also arrived, which has contributed to the decline and extinction of many native bird species in these islands.

Bacteria

Bacterial parasites can cause a wide variety of plant and animal diseases. In plants, bacterial infections can cause spotted leaves, wilted stems, scabbed surfaces, and large abnormal tissue growths known as *galls* (**Figure 15.7**). Because bacteria must enter a plant through a wound in the plant's tissues, they commonly require the assistance of herbivores that pierce a plant's tissues.

Bacterial infection in animals can also be quite harmful to the host. One of the earliest discovered species of deadly bacteria was anthrax (*Bacillus anthracis*). Anthrax, originally isolated from cows and sheep, is also highly lethal to humans. After an incubation period of about 1 week, the growing bacterial population begins to release toxic compounds that cause internal bleeding, which often brings rapid death. Other common bacterial pathogens include those that cause plague, pneumonia, salmonella, leprosy, and many sexually transmitted diseases, all of which occur in a wide variety of animals.

Fungi

Fungal parasites have large ecological impacts on a wide range of plants and animals. Fungal diseases have devastated many dominant plant species, including critical food sources for humans. A number of tree species in North America have suffered tremendous declines as a result of pathogenic fungi that have been introduced from Europe and Asia. The American chestnut tree was once one of the tallest tree species in the temperate forests of the eastern United States and composed up to 50 percent of all trees in the forest. It was also one of the most prized trees for lumber as well as for the crop of edible nuts that it produced every autumn. Around 1900, however, species of Asian chestnut trees that were imported to New York inadvertently carried a fungus (*Cryphonectria parasitica*). While Asian chestnut trees had a long evolutionary history with the fungus and were resistant to its harmful effects, American chestnut trees had no such history. As a result, the American chestnut rapidly succumbed to the fungal disease, known as chestnut blight. The decline of the chestnut trees also likely had widespread effects on the many animal species that relied on the tree's large crop of nuts for food. Today, the American chestnut tree is quite rare; young seedlings can emerge from seeds, but they ultimately become infected with the fungus and die early in life. In response, researchers are currently breeding more resistant varieties that they hope can someday be planted throughout forests in the eastern United States.

Similar fungal disease problems have occurred with other trees, including the American elm tree (*Ulmus americana*). The elm was widely planted along North American streets

(a)   (b)

Figure 15.7 Bacterial disease of plants. Common plant diseases that are caused by bacteria include **(a)** shot-hole such as this *Pseudomonas syringae mors-prunorum* on a cherry tree and **(b)** Crown gall (*Rhizobium radiobacter*) shown on a birch tree in England. Photos by (a) Nigel Cattlin/Science Source; and (b) Caroline Morgan.

(a)

(b)

Figure 15.8 Declining tree populations in North America. The introduction of a fungus that causes Dutch elm disease killed many American elms, which were once common along North American streets. **(a)** A street in Detroit, Michigan, in 1974 before Dutch elm disease. **(b)** The same street in 1981 after Dutch Elm disease killed all the elm trees. Photos by Jack H. Barger, U.S. Forest Service.

because of its attractive arching branches that provided shade (**Figure 15.8a**). In the 1930s, a fungus from Asia was introduced into Europe and North America that caused a rapid decline in elm trees. Because the fungal disease was first described by researchers in Holland, it is known as Dutch elm disease. The American elm had no evolutionary history with the Asian fungus, and 95 percent of the elm trees across North America have succumbed (Figure 15.8b). The small number of remaining trees appears to be resistant to the fungal disease, and researchers are currently working to plant these resistant genotypes back into eastern forests.

Some fungal parasites also damage crop plants. One of the best-known examples is a group of fungi that causes a disease known as *rust.* Throughout human history, rust diseases have affected many of our most important crops, including wheat, corn, rice, coffee, and apples. The effect of rust can range from reduced food production to a complete loss of the crop, costing millions of dollars of damage. Widespread rust infections, such as wheat rust, have caused the starvation of millions of people and domesticated animals over the past two centuries.

Animals can also be infected by fungi. One high-profile animal fungal disease is caused by a species of chytrid fungus (*Batrachochytrium dendrobatidis*) that infects amphibians, such as frogs and salamanders. During the 1990s, scientists working in Central America began to notice massive die-offs of amphibians and subsequently determined that the dead animals were infected with the fungus. The fungus, which lives in the outer layers of the amphibian's skin, causes an imbalance of ions in the body that ultimately causes the animal's heart to stop.

For 2 decades, researchers surveyed amphibians from Costa Rica to Panama and found that the fungus was moving from the northwest to the southeast, as shown in **Figure 15.9a**. Between 2000 and July 2004, at a site in El

Copé, Panama, more than 1,500 amphibians were sampled before the fungus arrived; not one individual tested positive. By October 2004, however, 21 of 27 sampled species were infected in at least 10 percent of their populations and by December 2004, 40 species tested positive for the fungus.

The sizes of amphibian populations in El Copé were also estimated, using transect surveys of amphibians that were active either during the day or during the night. After the fungus arrived in 2004, the number of live amphibians declined sharply (Figure 15.9b). The dead amphibians included 38 different frog species. Moreover, 99 percent of the 318 dead individuals collected had moderate to severe chytrid infections. In 2010, researchers reported that of the 63 species of amphibians present at the site before the fungus arrived, 30 species were gone. As of 2016, many of these species have not been seen since chytrid arrived. Relatively little is known about this fungus, although the massive die-offs suggest that amphibians have few defenses against it, which indicates that the fungus may have been introduced to Central America. The fungus is now suspected to have caused the extinction of dozens of species of amphibians around the world.

Helminths

Helminths include several groups of roundworms and flatworms that can cause serious diseases. We have already discussed the impact of nematodes as ectoparasites on plant roots. Nematodes can also live as endoparasites within the tissues of plants, where they damage the plant's growth, reproduction, and survival. In animals, disease-causing helminths include hookworms that feed on the blood of intestines, lungworms that live in the lungs, and echinostome worms that live in the kidneys. In livestock, infection by liver flukes has been a problem for centuries. Livestock

(a)

Figure 15.9 The deadly chytrid fungus in Central America. (a) Surveys of amphibians in Central America demonstrated that the fungus was spreading from the northwest to the southeast. **(b)** Surveys of amphibians at the El Copé site detected a sharp decline in amphibians, both in species that are active during the day and species that are active at night. Data from K. Lips et al., Emerging infectious disease and the loss of biodiversity in a Neotropical amphibian community, *Proceedings of the National Academy of Sciences,* USA 103 (2006): 3165–3170.

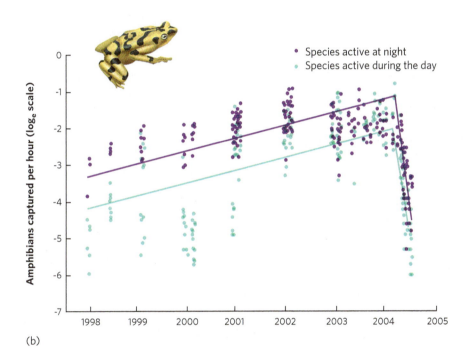

(b)

inadvertently consume these parasites when they drink water containing fluke-infected snails or when they eat grass that contains a stage of the parasite that is excreted by snails. Liver flukes are a particular problem in sheep, where they may cause liver damage, hemorrhaging, and sudden death. Left untreated, liver flukes can kill up to 10 percent of a sheep population; fortunately, drugs can cure infected animals.

EMERGING INFECTIOUS DISEASES

Many infectious diseases have been infecting hosts for thousands of years, while others have only emerged recently. When a new disease is discovered, or a formerly common disease that declined in the past suddenly becomes common again, it is called an **emerging infectious disease.**

New diseases typically emerge when a mutation allows a pathogen to jump to a new host species. We have already mentioned several emerging infectious diseases, including the chytrid fungus that has decimated amphibians around the world; the H5N1 bird flu that jumped from birds to humans; and mad cow disease that infected cattle, sheep, and humans. Since the 1970s, the world has experienced an average of one new emerging infectious disease each year.

In 2006, an emerging disease infecting bats in a cave near Albany, New York, was identified. These bats had a white-colored fungus (*Geomyces destructans*) growing on their noses, and they were dying

GLOBAL CHANGE

Emerging infectious disease A disease that is newly discovered or has been rare and then suddenly increases in occurrence.

Figure 15.10 White-nose fungus in bats. This species, known as the little brown bat, is infected by the fungus *Geomyces destructans*. One symptom of the infection is that the bat's nose turns white. When a bat colony becomes infected, a large proportion of the bats die. Photo by Photo courtesy Ryan von Linden/New York Department of Environmental Conservation.

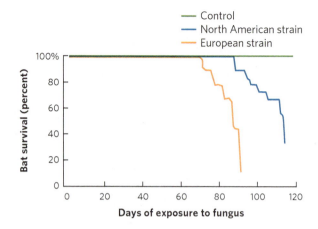

Figure 15.11 Testing North American versus European strains of white-nose fungus. Bats exposed to the control treatment did not die. In contrast, bats exposed to the North American fungus strain or the European fungus strain died in large numbers. However, bats died faster when exposed to the European strain than the American strain. Data from L. Wernicke et al., Inoculation of bats with European *Geomyces destructans* supports the novel pathogen hypothesis for the origin of white-nose syndrome, *PNAS* 109 (2012): 6999–7003.

in large numbers (**Figure 15.10**). Researchers named the disease white-nose syndrome. They hypothesized that the fungus causes bats hibernating in caves to come out of their normal torpor and to use up their fat reserves until they become emaciated and die. By 2016, the fungus had spread to 29 states and 5 Canadian provinces, and killed more than 5 million bats. In caves containing infected bats, the mortality rate can reach 100 percent. This is important because bats provide many important services, including the consumption of a large number of insects. Therefore, it is plausible that the death of so many bats will have substantial effects on many communities and ecosystems.

An interesting detail in the story of white-nose syndrome is that bats in Europe also carry the fungus but do not experience widespread death. This raises the possibility that the fungus is native to Europe and has only recently been introduced to North America. Alternatively, it could be that the fungus has always been in North America but recently mutated to become much more lethal. To test these alternative hypotheses, researchers isolated strains of the fungus from European and North American bats. They then exposed the little brown bat (*Myotis lucifugus*), a species from North America, to fungal spores from the European strain, fungal spores from the North American strain, or a control treatment containing no fungal spores. After the exposures, they monitored bat survival for 120 days. In 2012, the results were reported, which you can see in **Figure 15.11**. In the control treatment, no bats died, but bats exposed to either fungal strain experienced high rates of death, which suggests that the fungus is a recent arrival from Europe and not a mutated North American native. Moreover, the bats died more slowly when exposed to the North American strain of fungus than the European strain, which may suggest that North American bats are already beginning to evolve some level of resistance to their local strain of fungus. Unfortunately, there is currently no treatment to prevent the continued death of bats from white-nose syndrome.

15.2 Parasite and host dynamics are determined by the parasite's ability to infect the host

Parasites and their hosts have populations that fluctuate over time, much like the dynamics of predator and prey populations that we discussed in Chapter 14. Unlike predators, however, parasites generally have a higher reproductive rate than their hosts and often do not kill their hosts. If we know what causes fluctuations in parasite and host populations, we can predict when parasitism will be prevalent in a host population. This allows us to anticipate population changes in hosts and, in some cases, to intervene and reduce the harmful effects of a parasite on a species of concern. The probability that a host will become infected by a parasite depends on numerous factors that include the parasite's mechanism of transmission, its mode of entry into a host's body, its ability to jump between species, the existence of reservoir species, and the response of the host's immune system.

MECHANISMS OF PARASITE TRANSMISSION

The first factor that determines the risk of infection by a parasite is the mechanism of transmission. As illustrated in **Figure 15.12**, there are many different mechanisms. Parasites can move between hosts through *horizontal*

Figure 15.12 Mechanisms of parasite transmission. Parasites can be transmitted vertically or horizontally. When transmission is horizontal, the parasite can be transmitted through a vector such as a mosquito, transmitted directly between two conspecifics, or transmitted to other species. When transmission is vertical, a parent host transmits the parasite to its offspring, such as when a mother bird transmits lice to its hatchlings at the nest.

Labels within figure:

Vertical transmission from parent to offspring (e.g., *Chlamydia*)

Horizontal transmission through a vector (e.g., West Nile virus)

Horizontal transmission directly between conspecifics (e.g., bird flu)

Horizontal transmission through multiple hosts (e.g., some helminths)

transmission or *vertical transmission.* **Horizontal transmission** occurs when a parasite moves between individuals other than parents to their offspring. For example, horizontal transmission occurs when helminths are transmitted from snails to frogs or from frogs to birds. Horizontal transmission can also occur between conspecifics, such as the transmission of bird flu from one bird to another. Some parasites, such as the prions that cause mad cow disease, cannot be naturally transmitted from one individual to another. The risk of transmission is essentially zero in these cases. As we have seen with mad cow disease, dead infected cows were fed to other cows, which significantly raised the risk of horizontal transmission. At the other extreme of transmission are influenza viruses, which pass easily between individuals. You may have experienced the rapid transmission of the flu virus if you or someone in your family became ill with flu; most likely, other members of your family quickly became infected. For most hosts, the risk of becoming infected by a parasite generally increases with host population density because higher densities mean that individuals are likely to come into contact more often with the parasites or with infected individuals.

Some parasites require another organism, known as a **vector,** to disperse from one host to another. For example, the West Nile virus is transmitted from one host to another by a mosquito that picks up the virus from one infected bird and transmits it to an uninfected bird or other animal that it bites. In this case, the mosquito is the vector.

Still other parasites require horizontal transmission to multiple host species to complete their life cycle, as we saw in the story of the amber snail. Some species of helminths, for example, spend the first stage of their life in a snail, the second stage of their life in an amphibian, and the final stage of their life in a bird. This requirement for multiple hosts poses a substantial challenge for the parasite, which must find all its required hosts during its lifetime.

Vertical transmission occurs when a parasite is transmitted from a parent to its offspring. In this case, the parasite must evolve in such a way that it does not cause the death of its host until after the host has reproduced and passed the parasite to its offspring. Many sexually transmitted diseases can be passed by vertical transmission. For example, chlamydia is a disease caused by several different species of bacteria within two genera, *Chlamydia* and *Chlamydophila,* that infect mammals, birds, and reptiles. In humans, chlamydia causes inflammation of the urethra and cervix but is typically not lethal. While the bacterium is often horizontally transmitted between individuals, it can also be transmitted vertically from an infected mother to her fetus. When this happens in humans, the newborn baby can suffer eye infections and pneumonia.

Horizontal transmission When a parasite moves between individuals other than parents and their offspring.

Vector An organism that a parasite uses to disperse from one host to another.

Vertical transmission When a parasite is transmitted from a parent to its offspring.

Researchers continue to learn about the wide range of pathways that parasites can take to infect a host. The existence of vertical transmission and various methods of horizontal transmission can make it very challenging to predict and control the spread of infectious diseases in humans, crops, domesticated animals, and wild organisms.

MODES OF ENTERING THE HOST

The mode of entry into the host's body also affects the ability of the parasite to infect the host. As we have discussed, some species of parasites, such as leeches, are able to pierce the tissues of the host. Other parasites, such as some viruses, bacteria, and protists, rely on another organism to penetrate the host's tissues and use the damaged tissue as an entry point. The protist that causes avian malaria, for example, depends on mosquitoes to inject the protist into a bird's body after the protist completes a critical life stage inside the mosquito's body. Without the mosquito, the malaria life cycle could not be completed. Of course, another mode of entering a host occurs when the pathogen is inadvertently consumed, as we saw in the case of ducks eating infected snails in Chapter 9.

JUMPING BETWEEN SPECIES

If a parasite specializes on only one host species and is able to cause a lethal disease in only that host species, then it might eventually run out of hosts and face extinction. One solution sometimes favored by natural selection is for the parasite to be nonlethal to the host; this allows a host population to persist. We will talk more about this strategy later in the chapter.

Alternatively, the parasite might evolve the ability to infect other species. We saw an example of this with bird flu, which infected several species of birds before a mutation occurred that allowed the virus to infect humans. A similar scenario occurred with HIV. For many years, it was hypothesized that human HIV originated in chimpanzees, and in 2006 researchers identified a population of chimps in the West African nation of Cameroon that carried a genetically similar strain of the virus. The researchers suspect that the virus jumped from chimpanzees to humans when local hunters consumed the chimpanzees as food. Other examples of parasites jumping between species include the chytrid fungus, which can jump between amphibian species, and the canine parvovirus, which can jump from cats to dogs.

RESERVOIR SPECIES

One way that parasite populations persist in nature is through the use of *reservoir* species. **Reservoir species** carry a parasite but do not succumb to the disease that the parasite causes in other species. Because reservoir species do not die from the infection, they serve as a continuous source of parasites as other susceptible host species become rare. For example, some species of birds can be infected with the protist that causes avian malaria, but they are resistant to developing the disease. However, mosquitoes that feed on these resistant species of birds can pick up the protist and transfer it to other species of susceptible birds that become infected and die. In some

Figure 15.13 Reservoir species. Many species of birds such as this 'Apapane (*Himatione sanguinea*, a species of Hawaiian honeycreeper) can survive infection and subsequently serve as a reservoir of avian malaria that can be spread to other birds through the bites of mosquitoes. Photo by Jack Jeffrey Photography.

species, such as the Hawaiian honeycreepers (**Figure 15.13**), the individuals that are infected by avian malaria but survive through acquired immunity can become a reservoir for the pathogen. In this way, reservoir species and immune individuals favor the persistence of the parasite population over time.

THE HOST'S IMMUNE SYSTEM

A host's immune system can play a role in combatting an infection from endoparasites. As a result, some parasites have evolved the ability to escape the immune system by making themselves undetectable. For example, when HIV enters a human cell, it can hide from the body's immune system by living in the cytoplasm or by incorporating itself into the chromosomes of the cell. Because the body's immune system searches for infections on the outside of cells, it cannot detect the virus. Other parasites have evolved additional strategies to escape the immune system. For instance, parasitic worms known as schistosomes produce a protective outer layer around their bodies that prevents them from being detected by the host's immune system. Still others, such as the protists that cause African sleeping sickness, are able to continually change the compounds present on their outer surface so they become a moving target that stays one step ahead of the immune system as it tries to respond to the infection.

CONCEPT CHECK

1. What are the different ways in which parasites can be transmitted between hosts?
2. What is the process that allows a parasite to jump to a new species?
3. How does a reservoir species help a parasite population persist over time?

Reservoir species Species that can carry a parasite but do not succumb to the disease that the parasite causes in other species.

15.3 Parasite and host populations commonly fluctuate in regular cycles

In our discussion of predators and prey in Chapter 14, we saw that population fluctuations are common and sometimes occur in regular cycles. Because parasites and hosts represent consumers and resources, they show similar population dynamics. In this section, we will examine how parasites and hosts fluctuate over time. We will also look at a mathematical model that helps us describe and understand the behavior of infectious pathogens and hosts.

POPULATION FLUCTUATIONS IN NATURE

As we discussed in the previous chapter, the density of the host population can affect how easily parasites are transmitted from one host to another. An excellent example of this can be seen in the dynamic between the forest tent caterpillar (*Malacosoma disstria*)—a major herbivore of broadleaf trees in the United States and Canada—and a group of viruses that can infect and kill the caterpillar. The forest tent caterpillar can defoliate trees over thousands of square kilometers. In years of high caterpillar densities, they remove the majority of a tree's leaves and tree growth is reduced up to 90 percent. Canadian researchers reported that caterpillar populations had cycles that last 10 to 15 years and that these fluctuations were fairly synchronous across large geographic areas. In three different locations in the province of Ontario, the fluctuations in caterpillar populations exhibit the same pattern of growth and decline over time, as shown

in **Figure 15.14**. Although the caterpillar is susceptible to predators and parasitoids, viruses have the greatest ability to reduce its abundance. When the tent caterpillar population is high, the virus can more easily spread from one host to another, so each caterpillar is more likely to get the virus. Under these conditions, large numbers of caterpillars die. As host population density decreases, it becomes harder for the viruses to find a new host and the prevalence of the disease declines. Since fewer caterpillars become infected and die, the caterpillar population starts to increase again. These patterns are similar to the predator–prey population cycles that we discussed in the previous chapter.

Fluctuations in parasite and host populations can also be caused by changes in the proportion of the host population that has achieved immunity. When a host species is able to become immune to a parasite after an initial infection, continued infection in the population causes an increasing proportion of the population to develop immunity. When a high proportion of the population is immune to the parasite, the spread of the parasite is slowed. Measles, for example, is a highly contagious viral disease that stimulates lifelong immunity in humans. In unvaccinated populations, measles typically produces epidemics at 2-year intervals. Once most of the population is infected and develops immunity, the number of new measles cases declines sharply. As humans continue to reproduce, however, there are enough new children born without immunity to initiate another measles outbreak after 2 years. The number of measles cases in London, England,

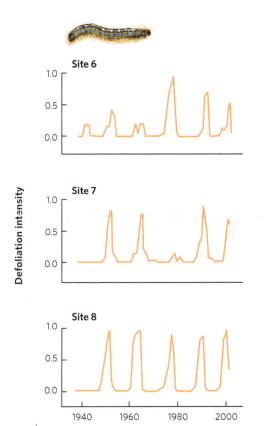

Figure 15.14 Cyclic population fluctuations of forest tent caterpillars. At three sites in the province of Ontario, researchers quantified the population sizes of tent caterpillars by measuring the intensity of leaf removal on the trees. Over a 60-year period, the caterpillars exhibited large population fluctuations every 10 to 15 years. The rapidly growing populations ultimately succumb to an outbreak of infection by a virus and the populations quickly die back. Data from B. J. Cooke et al., The dynamics of forest tent caterpillar outbreaks across east-central Canada, *Ecography* 35 (2012): 422–435.

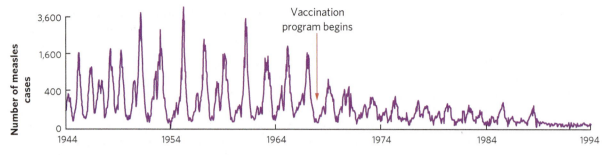

Figure 15.15 The cyclical occurrence of measles in a human population. Before vaccinations became available in 1968, the human population in London, England, experienced cycles of measles every 2 years. Once vaccination became available and the number of people vaccinated increased, the number of measles cases declined and the fluctuations no longer occurred. Data from P. Rohani et al., *Opposite patterns of synchrony in sympatric disease metapopulations, Science* 286 (1999): 968–971.

from 1944 to 1968—before vaccinations became available—shows a distinct pattern of 2-year cycles, as you can see in **Figure 15.15**. Once vaccinations became available and a higher proportion of susceptible individuals were immunized, the number of new cases dropped sharply and the fluctuations declined and eventually disappeared.

MODELING PARASITE AND HOST POPULATIONS

For pathogenic parasites, we can understand the dynamics of the parasite and the host using models that are similar to the Lotka–Volterra predator–prey model. However, the parasite–host model differs from the predator–prey model in two key ways. Parasites, unlike predators, do not always remove host individuals from a population, and hosts may develop immune responses that make some individuals resistant to the pathogen.

The simplest model of infectious disease transmission that incorporates immunity is the **Susceptible-Infected-Resistant (S-I-R) model.** In this model, all individuals begin as susceptible to the pathogen (S). Of those, some number becomes infected (I). Of the infected individuals, some number develops resistance via immunity (R). We can use this model to examine the conditions that favor an epidemic versus the conditions that cause a disease to decline. The proportions of S, I, and R individuals in a population are determined by rates of transmission of the disease and acquisition of immunity, as well as the birth of new, susceptible individuals.

In a population of hosts, the first individual to be infected by a pathogen is known as the *primary case* of the disease. Any individuals infected from this first individual are known as *secondary cases*. The rate at which the pathogen spreads through the population depends on two opposing factors. One factor is the rate of transmission between individuals (b), which includes both the rate of contact of susceptible individuals with an infectious individual and the probability of infection when there is contact. The other factor is the rate of recovery (g), which determines the period of time from when an individual is infected and can transmit the infection to when the

individual's immune system clears the infection and the individual becomes resistant to any future infections.

The rate of infection and the rate of recovery can be used to determine whether an infectious disease will spread through a population. A disease will spread whenever the number of newly infected individuals is greater than the number of recovered individuals. To determine the number of newly infected individuals, we need to know the probability that an infected individual and a susceptible individual will come into contact with each other. If we assume that individuals in the population move around randomly, the probability of contact is the product of their proportions in the population:

Probability of contact between susceptible and infected individuals = $S \times I$

Once infected and susceptible individuals come into contact, we also have to consider the rate of infection between them (b):

Rate of infection between susceptible and infected individuals = $S \times I \times b$

Next we have to determine how many individuals are recovering from the infection. We can do this by knowing the proportion of infected individuals (I) and the rate of recovery from an infection (g):

Rate of recovery of infected individuals = $I \times g$

Now we can determine whether an infection will spread through a population. To do this, we need to calculate the reproductive ratio of the infection (R_0), which is the number of secondary cases produced by a primary case during its period of infectiousness. The reproductive ratio of the infection is the ratio of new infections to recoveries:

$$R_0 = (S \times I \times b) \div (I \times g)$$
$$R_0 = S \times (b \div g)$$

Susceptible-Infected-Resistant (S-I-R) model The simplest model of infectious disease transmission that incorporates immunity.

If $R_0 > 1$, the infection will continue to spread through the population and an epidemic will occur. This happens because each infected individual infects more than one other individual before it recovers from the disease and becomes resistant. When $R_0 < 1$, the infection fails to take hold in the host population. This happens because each infected individual fails to infect another individual, on average, before it recovers and becomes resistant.

Figure 15.16 illustrates the dynamics of a typical disease. Imagine that we start with a population that is composed entirely of uninfected individuals. When an infectious disease arrives, there is initially a rapid increase in the number of infected individuals but, over time, infected individuals subsequently recover and become resistant (R). At this point, the number of susceptible individuals (S) decreases, so the value of R_0 decreases. When R_0 declines to <1, the growing epidemic can no longer sustain itself. As a result, the number of infected individuals achieves a peak in abundance and then declines.

With this understanding of how R_0 affects the spread of diseases, we can now look at the R_0 values for different diseases that are caused by parasites. HIV is transmitted through rather limited mechanisms, including direct sexual contact, blood transfusion, or perinatal transmission from mother to offspring. HIV has a relatively low range of R_0 values, from 2 to 5. Typical values for R_0 in childhood diseases of humans—measles, chicken pox, and mumps, among others—range from 5 to 18 at the time a population is initially infected. At the extreme, malaria, which is transmitted by mosquitoes, has an R_0 value greater than 100 in crowded human populations. This high value occurs because mosquitoes are excellent vectors for transmitting the parasite and infected people remain infectious for long periods of time.

Assumptions of the S-I-R Model

The basic S-I-R model has several important assumptions. For example, it assumes there are no births of new susceptible individuals and that individuals retain any resistance they develop. In such a model, the result is an epidemic that runs its course until all individuals in the population have become resistant or there are too few susceptible individuals remaining to sustain the spread of the disease. Some pathogens, such as influenza viruses, fit these assumptions. The basic model also explains why vaccinations slow or stop the spread of diseases such as influenza; by vaccinating individuals, we reduce the size of the susceptible population (S), which reduces the value of R_0. This makes it harder for an epidemic to sustain itself.

Other factors can be added to the model, including births of susceptible infants, lag times between when an individual is infected and becomes infectious to others, host mortality, host population dynamics, and transmission of disease from parent to offspring. These additional factors can have substantial effects on the predictions of the model. For example, the birth of new susceptible individuals can cause cyclic fluctuations in the model, as we observed in the case of the measles data from London (Figure 15.15).

If a pathogen can kill the host, the pathogen should increase in abundance until the hosts begin to die. As the host population declines, the pathogen population will subsequently decline, and as the pathogen population declines, the host population should subsequently recover. This is analogous to the predator–prey cycles we discussed in Chapter 14. However, some pathogens do not follow these dynamics because they do not attack a single host species. For instance, the chytrid fungus that we discussed earlier in this chapter infects dozens of species of amphibians. A host species can decline all the way to extinction, yet the fungus does not decline in abundance because it can infect other species of amphibians. Some of these newly infected species will become ill and die, while others never develop the disease and serve as reservoirs for the disease. When a pathogen is not restricted to a single host species, it has the ability to persist and spread even after it causes one of its hosts to go extinct.

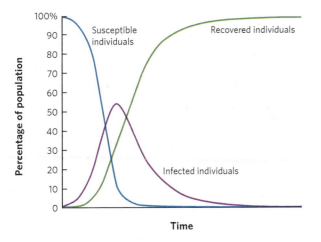

Figure 15.16 The dynamics of an infection over time. In the basic S-I-R model, all individuals in a population are initially susceptible. When the infection is introduced at the beginning of the time period, there is an initial rapid growth in the number of infected individuals. As some infected individuals recover and become resistant, there are fewer susceptible individuals left to infect, so the number of infected individuals declines.

CONCEPT CHECK

1. How do lethal parasites cause host populations to cycle?
2. What is the effect of host immunity of parasite populations?
3. Why do we want to know the reproductive ratio of a parasite?

15.4 Parasites have evolved offensive strategies, while hosts have evolved defensive strategies

Parasites gain fitness when they find a suitable host and reproduce. Hosts will experience higher fitness if they avoid being parasitized. Therefore, natural selection has favored the evolution of parasite offenses and host defenses. In this section, we will examine some of these strategies as well as look at evolutionary relationships between parasites and hosts.

PARASITE ADAPTATIONS

As we saw at the beginning of the chapter, parasites have evolved a wide array of strategies for finding and infecting hosts. Some of the most fascinating examples of parasite adaptation involve strategies that increase the probability of parasite transmission. For example, the pathogenic fungus *Entomorpha muscae* infects both houseflies (*Musca domestica*) and yellow dungflies (*Scatophaga stercoraria*). To improve its chance of transmitting its spores, it has two different strategies. When the fungus infects and kills female houseflies, the bodies become attractive to male houseflies looking for a mate. Although it is unclear what attractive cues the fungus produces, when a male housefly tries to mate with a dead female, spores of the fungus are transferred to the male's body.

When the same fungus infects the yellow dungfly, the fly climbs up nearby vegetation, moves to the upwind side of the plant, and hangs upside down on the underside of a leaf. The fly then moves its wings toward the leaf and moves its abdomen—which is swollen with fungal spores—away from the leaf. Once it reaches this position, the dungfly dies and the fungal spores erupt from its abdomen. The erupting spores can then potentially infect dungflies that pass below, as illustrated in **Figure 15.17**. When researchers compared the behavior of infected dungflies to uninfected dungflies, they found that uninfected flies perched lower on plants and never perched on the underside. You can view these data in **Figure 15.18**.

For parasites that require a series of different hosts to complete their life stages, it is challenging to find a way to get transmitted from one host to another. Some parasites—such as trematodes that first infect snails and subsequently infect tadpoles—have evolved a simple strategy of leaving the body of the first host and then searching for the second host. Other parasites have evolved ways of manipulating the behavior of the first host to ensure that the second host consumes it. We saw an example of this with the amber snail discussed at the beginning of the chapter. There are many examples of parasites taking control of the host's behavior. For example, mice can inadvertently consume a parasitic protist, *Toxoplasma gondii*, when feeding near cat feces. Once the mouse becomes infected, it no longer avoids an area marked by the smell of bobcat (*Lynx rufus*) urine but, instead, is mildly attracted to it. As a result, the mouse is more likely to be eaten by the bobcat, which is a valuable outcome for the protist since it can only reproduce in the gut of a cat.

Similarly, small crustaceans known as isopods (*Caecidotea intermedius*) normally hide in refuges from predatory fish. However, if the isopods are infected by a parasitic worm (*Acanthocephalus dirus*), they spend less time in the refuge and more time out in the open where the fish are more likely to notice them, as you can see in **Figure 15.19**. This manipulation of the isopod's behavior is beneficial to the parasite because fish serve as the parasite's second host.

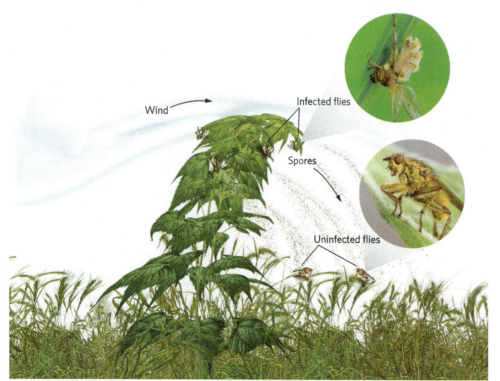

Figure 15.17 Transmission of fungal spores in the yellow dungfly. Infected flies climb up a plant, position themselves upside down on the underside of a leaf, and die as the spores erupt from their abdomen. In contrast, uninfected flies position themselves lower on plants and sit upright on the top sides of leaves. Based on D. P. Maitland, A parasitic fungus infecting yellow dungflies manipulates host perching behavior, *Proceedings of the Royal Society of London* Series B 258 (1994): 187–193. Photos by (top) Biosphoto/Julien Boisard; and (bottom) Tierfotoagentur/S. Ott/ AGE Fotostock.

Wind

Infected flies

Spores

Uninfected flies

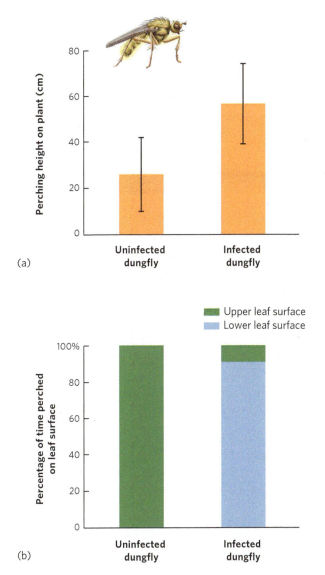

(a)

(b)

Figure 15.18 Fungal control of the perching behavior in the yellow dungfly. (a) Compared to uninfected dungflies, infected dungflies perch higher on plants. Error bars are 1 standard deviation. **(b)** Uninfected dungflies never perch on the underside of a leaf, whereas infected flies do so 91 percent of the time. Data from D. P. Maitland, A parasitic fungus infecting yellow dungflies manipulates host perching behavior, *Proceedings of the Royal Society of London* Series B 258 (1994): 187–193.

HOST ADAPTATIONS

Host species have evolved a range of mechanical and biochemical defenses to combat parasite invaders. Some species of plants and animals can produce antibacterial and antifungal chemicals that kill bacterial and fungal parasites. For example, many species of amphibians naturally release antimicrobial peptides onto their skin, which inhibit the growth of the deadly chytrid fungus.

In some cases, host species have evolved both mechanical and biochemical defenses to combat parasites. We see this with chimpanzees living in Tanzania that become infected with intestinal worms. Instead of eating their normal food diet, chimps that are ill with the intestinal worms pick a few leaves from plants in the genus *Aspilia*

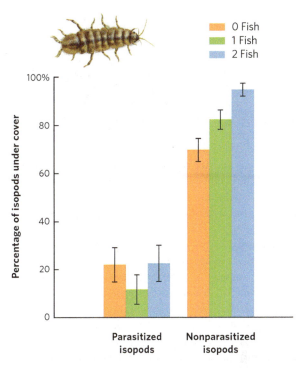

Figure 15.19 Parasite-induced changes in isopod behavior. When isopods are not infected by parasites, they generally stay under cover whether there are fish present or not. When they are infected, the isopods move out from under the cover, which makes them susceptible to predatory fish. Error bars are standard errors. Data from L. J. Hetchel et al., Modification of antipredator behavior of *Caecidotea intermedius* by its parasite *Acanthocephalus dirus*, *Ecology* 74 (1993): 710–713.

and swallow them whole (**Figure 15.20**). The leaves of these plants are covered in tiny hooks and, as the leaves pass through the chimp's digestive system, the hooks pull nematode parasites (*Oesophagostomum stephanostomum*) out of the digestive tract, so they can be expelled from the chimp's

Figure 15.20 Self-medicating chimpanzees. When chimpanzees become ill from intestinal parasites, they swallow whole leaves and chew bitter twigs of plants not normally in their diet. These leaves and twigs reduce the number of parasites in their body and help them recover. Photo by LMPphoto/Shutterstock.

ANALYZING ECOLOGY

Comparing Two Groups with a *t*-Test

Eco TV

macmillanlearning.com/ricklefsvideo

When the dungfly researchers examined the perching height of infected and uninfected flies, they concluded that the infected flies perched significantly higher on vegetation. How did they come to that conclusion? They compared the mean heights for the infected and uninfected flies and used a statistical test to determine if the two means were different.

In Chapter 14, we discussed how scientists typically consider two means to be different if we can sample their distributions many times and find that the means of those sampled distributions overlap less than 5 percent of the time. We commonly call this critical cutoff *alpha*—abbreviated as α. Therefore, we say that our critical cutoff is $\alpha = 0.05$. To determine if the means of the two groups are significantly different, we need to know three things from each group: the mean, the variance, and the sample size.

With these three parameters, we can calculate a ***t*-test,** which determines if the distributions of data from two groups are significantly different. An important assumption of the *t*-test is that the values from both groups are normally distributed.

We begin by calculating the *difference between the means* of the two groups:

$$\overline{X}_1 - \overline{X}_2$$

where $\overline{X}_1$ is the mean of group 1 and $\overline{X}_2$ is the mean of group 2.

Next we calculate the *standard error of the difference in the two means*:

$$\sqrt{\frac{S_1^2}{n_1} + \frac{S_2^2}{n_2}}$$

where S_1^2 is the sample variance of group 1, S_2^2 is the sample variance of group 2, n_1 is the sample size of group 1, and n_2 is the sample size of group 2 (you can review the concept of sample variances in Chapter 2).

The next step is to divide the first calculated value by the second calculated value:

$$t = \frac{\overline{X}_1 - \overline{X}_2}{\sqrt{\frac{S_1^2}{n_1} + \frac{S_2^2}{n_2}}}$$

As you can see from this equation, the value of t becomes larger as the difference between the means becomes larger, or the standard error of the difference in the means becomes smaller.

For example, consider two groups of data with the following parameters:

Group	Mean	Sample variance	Sample size
1	20	10	5
2	10	10	5

In this case,

$$t = \frac{20 - 10}{\sqrt{\frac{10}{5} + \frac{10}{5}}} = \frac{10}{\sqrt{2 + 2}} = 5$$

The last thing we need to calculate is known as the degrees of freedom, which is defined as the sum of the two sample sizes minus 2. In this example, the degrees of freedom is

$$5 + 5 - 2 = 8$$

With this value of t calculated, we can then determine if our value exceeds a critical level of t and therefore whether the means are significantly different. We can find the critical value of t from a statistical table. If you look at the *t*-table in the "Statistical Tables" section at the back of this book, you can find the column that uses an alpha value of 0.05 and then find the row that contains 8 degrees of freedom. The critical *t*-value is the number found in this row and column. In this case, the critical *t*-value is 2.3. Because our calculated *t*-value exceeds the critical *t*-value, we can conclude that the two means differ significantly.

YOUR TURN For the following data, calculate the *t*-value.

Group	Mean	Sample variance	Sample size
1	10	4	8
2	5	4	8

Next calculate the degrees of freedom.

Based on your calculated *t*-value and the critical *t*-value using $\alpha = 0.05$, determine if the two groups are significantly different.

***t*-test** A statistical test that determines if the distributions of data from two groups are significantly different.

body with feces. Infected chimps also chew on bitter twigs from the *Vernonia* plant, something that healthy chimps do not do. The sick chimps become well within 24 hours because the twigs contain chemicals that kill a variety of different parasites. These plants are also consumed by people in the region when they experience symptoms of parasite infections. In short, both chimps and humans use plants to medicate themselves against parasitic infections.

COEVOLUTION

When we consider the evolutionary forces that affect adaptations of parasites and hosts, we often find that as one species in the interaction adapts, the other species responds by adapting as well. As we discussed in Chapter 14, when two or more species continue to evolve in response to each other's evolution, we call it coevolution. When all species involved in an interaction coevolve together, no one species is likely to get an upper hand. For example, when a parasite evolves an advantage that makes it more effective at infecting hosts, the hosts will then be under stronger selection to evolve defenses against that parasite.

 We can see an example of coevolution between parasites and hosts in the dynamics of the rabbit population in Australia. In the 1800s, European rabbits were introduced to many regions of the world where they had not lived historically. In Australia, for instance, rabbits were first introduced from England in 1859. Within a few years, local ranchers were erecting fences to keep them out and organizing shooting parties to help control the size of the population. Eventually, hundreds of millions of rabbits ranged throughout most of the continent, destroying sheep pasturelands and thereby threatening wool production (**Figure 15.21**). The Australian government tried poisons, predators, and other control measures, all without success.

In 1950, the government tried a new approach. A virus known as *Myxoma* had been discovered in South American rabbits. The virus—which was carried by mosquitoes—causes a disease known as myxomatosis. The disease had

minor effects on South American rabbits because they had coevolved with the virus for a long time, but it turned out to be highly lethal to European rabbits; an infected rabbit dies within 48 hours. The Australian government introduced the virus and initially it had a devastating effect on the rabbits. Among infected rabbits, 99.8 percent died and the rabbit population declined to very low numbers. However, 0.2 percent of the infected rabbits proved to be resistant and passed on their resistant genes. Before the introduction of the *Myxoma* virus, a few rabbits possessed resistant genes, but they had never been favored. In a second outbreak of the disease, only 90 percent of infected rabbits died. By the third outbreak, only 40 to 60 percent of the rabbits died and the rabbit population of Australia started to increase. You can see the decline in rabbit mortality over time in **Figure 15.22**.

The decline in rabbit mortality was due partly to the evolution of increased resistance in the rabbit population, but because the rabbit population was initially killed in large numbers, the rapidly dwindling host population also favored any virus strains that infected the rabbits without killing them. Dead rabbits represent a dead end for the virus, especially if the rabbit dies before it is bitten by a mosquito, which is the only way the virus can be transmitted.

Over time, the interaction between the rabbits and the *Myxoma* virus resulted in a more resistant host population and a less lethal pathogen. This allowed the rabbit population to rebound and the virus to persist. This dynamic is probably what researchers observed in South America, where the rabbit population and *Myxoma* virus persisted

Figure 15.21 A growing rabbit population in Australia. After being introduced to Australia in 1859, the rabbit population exploded and consumed nearly all the vegetation that was needed by ranchers for raising sheep. Photo John Carnemolla/Getty Images.

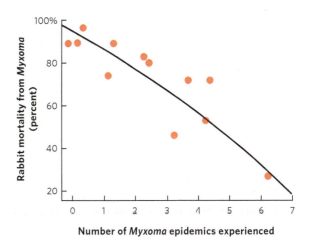

Figure 15.22 Coevolution of Australian rabbits and the *Myxoma* virus. In rabbit populations that had never been exposed to the virus, nearly all infected rabbits died. However, a small percentage of resistant rabbits survived and passed on their resistant genes to the next generation. At the same time, selection favored less lethal virus strains, which could reproduce and get transmitted by mosquitoes that only feed on live rabbits. Since rabbit populations evolved higher resistance and viruses evolved to be nonlethal, rabbit populations that experienced more epidemics of the disease experienced less mortality. Data from F. Fenner and F. N. Ratcliffe, *Myxomatosis* (Cambridge University Press, 1965).

together for a long time. To meet the Australian government's goal of controlling the rabbit population in Australia, scientists continue to introduce new strains of the virus from South America that are highly lethal to Australian rabbits because the new strains have no history of coevolution with the Australian rabbits. In this way, they maintain the effectiveness of the *Myxoma* virus as a pest control agent.

CONCEPT CHECK

1. What is an example of a parasite adaptation that helps promote its transmission from one host to another?
2. What is an example of a host adaptation that helps the host combat parasites?
3. What is the process of coevolution of parasite and host adaptations?

ECOLOGY TODAY CONNECTING THE CONCEPTS

Of Mice And Men . . . And Lyme Disease

The black-legged tick. The tick, also known as a deer tick, is the vector of the bacteria that cause Lyme disease in humans. Photo by Scott Bauer/USDA-ARS.

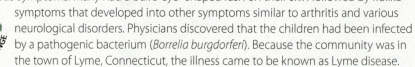

Back in the 1970s, a number of children in a Connecticut community experienced a suite of mysterious symptoms. Many had a bull's-eye–shaped rash on their skin followed by flulike symptoms that developed into other symptoms similar to arthritis and various neurological disorders. Physicians discovered that the children had been infected by a pathogenic bacterium (*Borrelia burgdorferi*). Because the community was in the town of Lyme, Connecticut, the illness came to be known as Lyme disease.

The parasite that causes Lyme disease is carried by ticks that feed on wild animals. Occasionally, these ticks latch onto people and transmit the parasite. In North America, the primary vector is the black-legged tick (*Ixodes scapularis*), also known as the deer tick. The ticks and the parasite probably existed in North America for thousands of years before the disease was identified. According to the Centers for Disease Control and Prevention, currently 20,000 to 30,000 cases of Lyme disease occur every year in the United States. Most cases occur in the Northeast, although there is a growing number of cases in the Midwest.

Over 4 decades, ecologists and medical researchers have determined that to understand Lyme disease requires a knowledge of the ecological community in which the parasite lives. First, they had to determine how ticks obtain the parasite. They found that 99 percent of newly hatched ticks do not carry the parasite, so vertical transmission is not likely. Instead,

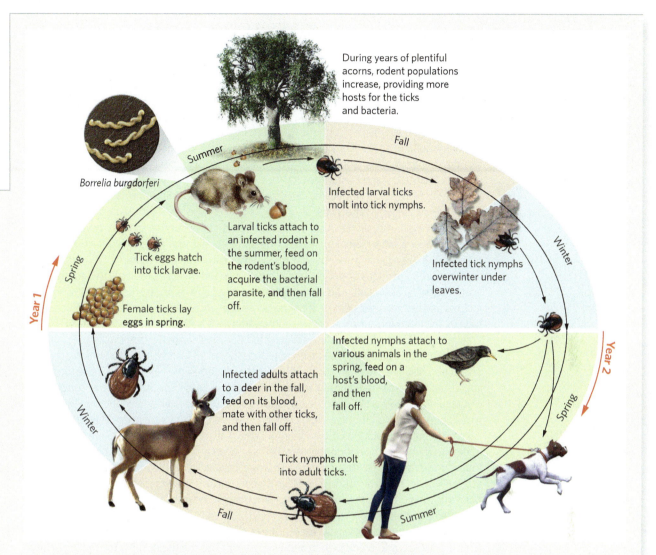

The cycle of Lyme disease. A diverse array of species affects the prevalence of the bacteria that causes Lyme disease. The cycle begins when ticks attach themselves to infected rodents and become infected with the bacteria. After they drop from the rodents, they molt into nymphs and spend the winter under leaves. In the spring, the nymphs attach themselves to hosts again and then drop and molt into adults. As adults, most ticks attach themselves to deer, search for mates, and then lay eggs. During years of high acorn production by oak trees, there is an abundance of food for rodents that produces large rodent populations. Large rodent populations lead to large tick populations that can infect a variety of animals, including deer and humans.

the tick obtains the parasite from infected hosts. But because the black-legged tick can live on a wide variety of hosts, the next step was to determine which hosts affected the abundance of infected ticks.

By examining patterns of abundance in nature and conducting manipulative experiments, the researchers were able to work out the life cycle of the black-legged ticks. The ticks are predominantly creatures of the forest. Larval ticks hatch from eggs each summer, but since they are unable to climb very high on vegetation, they can only infect small animals living close to the ground, such as birds and rodents. The most commonly infected rodents are chipmunks (*Tamias striatus*) and white-footed mice (*Peromyscus leucopis*). These two species can be infected without exhibiting any signs of disease, so they act as reservoir species that maintain the bacterial population. After feeding on the rodents for a few days, the ticks drop off their host and molt into nymphs.

As nymphs, the ticks spend the winter living under the leaves on the forest floor. The following spring, they climb a short distance up in the vegetation and wait for another host to pass by. They most commonly attach to birds, small mammals, and any humans who might be taking a walk in the woods or in a garden. After attaching themselves to a host and becoming engorged with blood, the ticks drop off again and molt into adults by autumn. Adult ticks can climb higher up on vegetation, which allows them to jump onto larger mammals such as white-tailed deer. The deer's body not only offers a blood meal but is also a place for adult male and female ticks to find each other and mate. From there, they drop off the deer and lay their eggs on the forest floor each autumn. All of this movement of ticks between multiple hosts results in many opportunities for the bacteria to be transmitted. In fact, researchers found that frequency of infection with the bacteria is 25 to 35 percent for nymphs and 50 to 70 percent for adults.

Based on this research, it became clear that deer, mice, and chipmunks were keys to the tick's life cycle. But what affects the abundances of these species? The researchers discovered one more key player: oak trees. Oak trees produce a massive number of acorns every few years, and acorns are a major food item for deer, chipmunks, and mice. In high-acorn years, tick-carrying deer gather under the oak trees, which causes an aggregation of reproducing ticks that drop their eggs under the trees. Chipmunks and mice are also attracted to the acorns since the abundant food permits higher survival and increased reproduction. More rodents and an increased density of tick eggs cause an increase in the number of infected rodents the following summer. In low-acorn years, when both deer and rodents spend more time in maple forests, the prevalence of infected ticks shifts from the oak forests.

To examine what steps can be taken to reduce the risk of Lyme disease to humans, the researchers also used mathematical models that included all the key species. These models suggested that, while reductions in deer densities would have little effect on the population of infected ticks unless the deer population was completely eradicated, reductions in the rodent population could cause a large reduction in the population of infected ticks.

With the tremendous increase in cases of Lyme disease, researchers have looked back in time to determine how long the bacteria have been around. In 2012, the bacteria were found in the tissues of a frozen mummy that was discovered in the Alps between Austria and Italy. Given that the mummy lived 5,300 years ago, it suggests that Lyme disease has been infecting humans for thousands of years, despite the fact that physicians only recently identified the disease.

The story of Lyme disease is an excellent example of how understanding the ecology of parasites allows us to predict when and where they pose the greatest risk to wild organisms and to humans.

SOURCES:

Keller, A., et al. 2012. New insights into the Tyrolean Iceman's origin and phenotype as inferred by whole-genome sequencing. *Nature Communications* 3: 698

Ostfeld, R. 1997. The ecology of Lyme-disease risk. *American Scientist* 85: 338–346.

Ostfeld, R., et al. 2006. Climate, deer, rodents, and acorns as determinants of variation in Lyme-disease risk. *PLOS Biology* 4: 1058–1068.

SUMMARY OF LEARNING OBJECTIVES

15.1 Many different types of parasites affect the abundance of host species. Ectoparasites live on host organisms, whereas endoparasites live in host organisms. As a group, parasites include a wide range of species that include plants, fungi, protozoa, helminths, bacteria, viruses, and prions. Among parasites that cause diseases—known as pathogens—those that have recently become abundant are called emerging infectious diseases.

Key Terms: Infection resistance, Infection tolerance, Parasite load, Ectoparasites, Endoparasites, Emerging infectious diseases

15.2 Parasite and host dynamics are determined by the parasite's ability to infect the host. The transmission of parasites can be horizontal—either through direct transmission or transmission by a vector—or vertical from parent to offspring. The ability to infect a host also depends on the parasite's mode of entering the host, its ability to infect reservoir species, its ability to jump to new host species, and its ability to avoid the host's immune system.

Key Terms: Horizontal transmission, Vector, Vertical transmission, Reservoir species

15.3 Parasite and host populations commonly fluctuate in regular cycles. These fluctuations occur because transmission increases with host density but decreases as an increased proportion of the host population develops immunity. These fluctuations can be modeled using the S-I-R model.

Key Term: Susceptible-Infected-Resistant (S-I-R) model

15.4 Parasites have evolved offensive strategies, while hosts have evolved defensive strategies. Natural selection has favored parasites that can improve their probability of transmission, including manipulations of host behavior. Hosts have evolved both specific and general immune responses to combat host infection. Hosts also can employ mechanical and biochemical defenses against parasites. Coevolution occurs when the parasite and host continually evolve in response to each other.

Key Term: *t*-test

CRITICAL THINKING QUESTIONS

1. Compare and contrast the advantages and disadvantages of life as an ectoparasite versus an endoparasite.

2. Why might parasites that are not very harmful to hosts in their native range be useful in controlling non-native hosts in an introduced range?

3. If a parasite has a reservoir host species, how effectively will the parasite population be controlled by immunizing a susceptible host species?

4. Given that there is currently no cure for mad cow disease, what is likely to be the most effective action to reduce its transmission?

5. Why might we continue to discover new emerging infectious diseases?

6. Compare and contrast horizontal versus vertical transmission of a parasite.

7. Using the basic S-I-R model of parasite and host dynamics, explain why the proportion of infected individuals in the population declines over time.

8. In the S-I-R model of parasite and host dynamics, how does the outcome change if we allow new susceptible individuals to be born into the population?

9. When using a *t*-test, what factors make it more likely that you will find a significan difference?

10. Explain why Dutch elm disease might become less lethal to its host over time.

GRAPHING THE DATA TIME SERIES DATA

As we saw in "Ecology Today: Applying the Concepts" at the end of this chapter, the number of black-legged ticks infected by the bacterium that causes Lyme disease is determined by the number of rodents, which in turn is affected by the production of acorns. In making these connections, researchers knew the abundance of the different species over time. Using the data provided in the table, plot the density of acorns, chipmunks, and ticks over time using a line graph.

Based on your graph, what patterns do you see when changes in acorn density affect chipmunk density and when changes in chipmunk density affect tick density?

Densities of acorns, chipmunks, and ticks over time			
Year	Acorns/m^2	Chipmunks/2.25-ha grid	Ticks/100 m^2
2002	5	1	1
2003	40	3	4
2004	3	70	5
2005	4	10	38
2006	30	5	4
2007	20	50	7
2008	1	15	35
2009	35	8	5
2010	7	62	4
2011	5	18	33
2012	3	10	5

16 Competition

Trying to Catch Up to Garlic Mustard

Garlic mustard is a powerful forest foe. It was introduced to the United States from Europe 150 years ago and has since spread throughout eastern and midwestern forests. It not only spreads, it dominates as a competitor against native plant species. Where you find garlic mustard, you often find few other forest herbs. During the past decade, researchers have determined that the plant is a dominant competitor because it possesses a novel weapon.

As mentioned in Chapter 1, many species of native plants depend on mutualistic relationships with soil fungi to obtain minerals. Without the fungi, these plants cannot grow well. The novel weapon of garlic mustard is its production of sinigrin, a chemical that it releases into the soil from its roots and that is toxic to soil fungi. Because garlic mustard does not depend on soil fungi for its own growth, it has an immediate advantage over many native plant species.

While sinigrin production helps garlic mustard survive and grow, researchers hypothesized that sinigrin production is costly. They hypothesized further that in forests where garlic mustard has been living for decades and has outcompeted most other plants, natural selection should favor reduced sinigrin production because at that point the cost comes with no benefit.

To test this hypothesis, researchers collected seeds from locations where garlic mustard had recently invaded and locations where it has been present for up to 50 years. In total, they raised seeds from 44 different locations and discovered that populations with a longer history of being present at a given location produced less of the toxic chemical. Moreover, they saved the soil from each of these 44 populations and then raised three species of native trees in the saved soils. All three tree species grew better in the soils that had contained older populations of garlic mustard than soils with younger populations of garlic mustard. This confirmed that the garlic mustard was evolving to be less toxic to the soil fungi and, in turn, a less dominant competitor.

The researchers also noticed that native plants were starting to become more abundant in forests with the oldest invasions of garlic mustard, which is consistent with the evolution of reduced sinigrin production. However, it is also possible that the native plants were simultaneously evolving a higher tolerance to competition from garlic mustard. To test this hypothesis, they collected a native forest herb from six forests that differed in their soil sinigrin concentrations. They raised the herb populations in the soil from which each had come, and then determined how well each herb population was able to compete against garlic mustard. They discovered that herb populations that derived from high-sinigrin soils were better able to compete against garlic mustard, which demonstrates that the native plants were evolving to tolerate the invader.

> "Garlic mustard is a highly effective competitor because it possesses a novel weapon."

After a decade of experiments, it has become clear that garlic mustard is a highly effective competitor because it possesses a novel weapon. However, its competitive advantage continues to evolve as it experiences changes in the intensity of competition from native plants, while the native plants continue to coevolve to combat the invader.

SOURCES:

Evans, J. A., et al. 2016. Soil-mediated eco-evolutionary feedbacks in the invasive plant *Alliaria petiolata. Functional Ecology* 30: 1053–1061.

Lankau, R. A. 2012. Coevolution between invasive and native plants driven by chemical competition and soil biota. *PNAS* 109: 11240–11245.

Lankau, R. A., et al. 2009. Evolutionary limits ameliorate the negative impacts of an invasive plant. *PNAS* 106: 15362–15367.

Garlic mustard is an invasive species that has spread throughout the eastern and midwestern United States. The invasive plant has a competitive advantage because it produces a toxin that harms native plants.

After reading this chapter, you should be able to:

16.1 Illustrate that competition occurs when individuals experience limited resources.

16.2 Explain the theory of competition as an extension of logistic growth models.

16.3 Describe how the outcome of competition can be altered by abiotic conditions, disturbances, and interactions with other species.

16.4 Distinguish between exploitation competition, interference competition, and apparent competition.

In Chapter 1, we defined competition as a negative interaction between two species that depend on the same limiting resource to survive, grow, and reproduce. Competition can help determine where a species can live in nature and how abundant a population can become. It can occur among many groups of organisms, including predators, herbivores, and parasites.

In this chapter, we will explore different types of competition and the resources for which species compete. We will then examine models of competition, which extend the Lotka-Volterra models introduced in Chapter 14. Competition models allow us to predict the conditions that determine when a species will win a competitive interaction. Knowing how species compete for a resource is important, but we must also consider how other interactions might alter or even reverse the expected outcomes of competition, including abiotic effects, disturbances, and interactions with other species. At the end of the chapter, we will examine a variety of cases that appear to describe competition but actually reflect other processes, including predation and herbivory.

16.1 Competition occurs when individuals experience limited resources

When we study competition, we differentiate between **intraspecific competition,** which is competition among individuals of the same species, and **interspecific competition,** which is competition among individuals of different species. We considered intraspecific competition in our discussion of negative density dependence in Chapter 12, and saw that an increase in a population's density causes a decline in the growth rate of the population. We have discussed interspecific competition much less. Interspecific competition can cause the population of either species to decline and eventually die out. Both intraspecific and interspecific competition play substantial roles in determining the distribution and abundance of species on Earth.

In this section, we will examine how competition for a limited resource can cause one species to outcompete another. We will explore the wide variety of resources that are available to organisms, including resources that are fixed in abundance and those that are renewable. With our understanding of resources, we will then investigate the importance of the most limiting resource in a population and examine patterns of competition between closely and distantly related species.

THE ROLE OF RESOURCES

A **resource** is anything an organism consumes or uses that causes an increase in the growth rate of a population when it becomes more available. For plants, resources generally include sunlight, water, and soil nutrients such as nitrogen and phosphorus. Each is used by most plants and plays a role in the growth of plant populations. Resources for animals generally include food, water, and space. For example, animals such as mussels and barnacles spend most of their lives attached to rocks in the intertidal zone along the ocean shores (**Figure 16.1**). Open space is a critical resource for them because as the rocks become more crowded, there is less space to grow. With less space, the growth and fecundity of adult mussels and barnacles decline, and there are few places on the rocks for offspring to settle. Similarly, many birds may compete for a limited number of nest sites or cavities, and many prey species compete for a limited number of holes and crevices in which they can hide from predators. In each of these cases, when more space becomes available, the populations can increase.

Ecological factors that cannot be consumed or used are not considered resources. Temperature, for instance, plays a major role in the growth and reproduction of organisms in nature, but temperature is not a resource because it is not consumed or used. The same is true for other abiotic factors, including pH and salinity.

Renewable Versus Nonrenewable Resources

We can categorize resources as either *renewable* or *nonrenewable*. **Renewable resources** are constantly regenerated. For example, rodents and ants often compete for seeds, and every year new plants grow and renew the seed supply. Similarly, sunlight is continually generated by the Sun.

Intraspecific competition Competition among individuals of the same species.

Interspecific competition Competition among individuals of different species.

Resource Anything an organism consumes or uses that causes an increase in the growth rate of a population when it becomes more available.

Renewable resources Resources that are constantly regenerated.

Figure 16.1 Competition for space. Sessile organisms that live in the rocky intertidal biome compete for open space on the rocks as a point of attachment. This photo from Olympic National Park in Washington shows a number of sessile organisms at low tide, including mussels, giant green sea anemones (*Anthopleura xanthogrammica*), goose barnacles (*Lepas anserifera*), and ochre sea stars (*Pisaster ochraceus*). Photo by Gary Luhm/Danita Delimont Stock Photography.

In contrast, **nonrenewable resources** are not regenerated. For instance, space is a resource that typically has a fixed availability. If we think of the rocky intertidal habitat, there is a fixed number of rocks to which algae and animals can attach; space becomes available only when a competitor leaves or dies.

Renewable resources can originate from either inside or outside the ecosystem in which the competitors live. For example, dead leaves that fall into streams from the surrounding forest serve as food for stream insects. Because these resources come from outside the system, however, competition can reduce resource abundance but cannot affect the rate of the resource supply. Moreover, resources originating from outside the system do not respond to the rate of resource consumption. Sunlight continually strikes the surface of Earth regardless of the rate at which plants and algae consume it, and the amount of local precipitation is largely independent of the rate at which plants use water.

Competitors can affect both the supply of and demand for resources that originate within the ecosystem. When competing herbivores or competing predators consume another species in the system, they reduce the supply of the species that they consumed. However, by reducing the abundance of a consumed species, the predators or herbivores also affect the future supply of that resource. For example, when rodents continue to consume seeds of abundant plants, the abundance of these plants declines over several years.

In some cases, the supply rate of a renewable resource generated within an ecosystem is only indirectly affected by competitors. For example, plants take up nitrate from the soil and use it to grow. Competing herbivores that consume the plants return large amounts of nitrogen compounds back to the soil when they defecate and after death when their bodies decompose. These nitrogen compounds are further broken down by microorganisms, which release the nitrogen as nitrate, a form plants can use. The herbivores can affect the supply of nitrate, but the chain of events has many links and takes such a long time that the herbivores do not have an immediate effect on the future plant population size through this indirect pathway.

Leibig's Law of the Minimum

Although consumers can reduce the abundance of both renewable and nonrenewable resources, not all resources limit consumer populations. For example, all terrestrial animals require oxygen, but an increase in the population of an animal species does not depress the concentration of oxygen in the atmosphere to the point that population growth is limited. Well before the concentration of oxygen can become limiting, some other resource, such as food supply, will decline in abundance to the point that it limits the population's growth.

In 1840, Justus von Liebig, a German chemist, proposed that populations were limited by the single resource that was most scarce relative to demand. This idea is now known as **Liebig's law of the minimum,** which states that a population increases until the supply of the most limiting resource prevents it from increasing further.

The amount of a resource that limits a population's growth depends on the resource. For example, the microscopic, glass-shelled algae known as diatoms require both silicate (SiO_2) and phosphate to grow and reproduce. When one diatom species, *Cyclotella meneghiniana*, is grown under different concentrations of each element, the population growth ceases whenever the concentration of silicate is reduced to 0.6 micromolar (μM) or whenever

---

Nonrenewable resources Resources that are not regenerated.

Liebig's law of the minimum Law stating that a population increases until the supply of the most limiting resource prevents it from increasing further.

the concentration of phosphate is reduced below 0.2 μM. According to Liebig's law of the minimum, whichever resource reaches its limiting value first will be the resource that regulates the growth of the diatom population.

If we know the minimum amount of a resource that is required for populations to grow, we should be able to predict which species is the best competitor for the resource. An experiment with two species of diatoms, shown in **Figure 16.2**, demonstrates how competition for silicate reduces the availability of silicate and affects the outcome of competition. When the two species are raised separately, *Asterionella formosa* and *Synedra ulna* both experience rapid population growth, followed by a plateau as they reach their carrying capacity, shown in Figures 16.2a and 16.2b. When *Asterionella formosa* reaches its carrying capacity, it drives down the abundance of silicate to 1 μM. However, when *Synedra ulna* reaches its carrying capacity, it drives down the abundance of silicate to 0.4 μM, which is not enough silicate to support the other diatom's population. As a result, *Synedra ulna* should outcompete *Asterionella formosa*. When both species were put together, that is exactly what happened. As illustrated in Figure 16.2c, the two species drove the abundance of silicate down to a level that allowed *Synedra ulna* to persist but caused *Asterionella formosa* to decline to extinction. The outcome between these two diatoms is one that is commonly observed in nature: when two species compete for a single limiting resource, the species that persists is the one that can drive down the abundance to the lowest level.

Interactions Among Resources

Leibig's law of the minimum assumes that each resource has an independent effect on the growth of a population. In other words, it assumes that if a given resource limits the growth of individuals and populations, increasing the availability of other resources will not improve such growth. However, this is not always the case, as seen in research on small balsam (*Impatiens parviflora*), a plant that is common in English woodlands. Researchers wanted to know if increasing the abundance of one resource would cause the plant to become soon limited by a second resource. They sowed seeds of the plants in pots filled with soil. Control pots received only water, while fertilized pots received a solution containing both water and fertilizer. These two groups of pots were then grown under one of four different light intensities for 5 weeks. As you can see in **Figure 16.3**, plants grown in low-nutrient soil experienced only a small increase in growth when the amount of light was increased. Similarly, plants grown at a low light intensity experienced a small growth increase when soil nutrients were increased. However, plants grown under high soil fertility and high light intensity experienced an increase in growth that was much larger than the sum of the separate effects of fertility and light intensity. This means that when the plants were given more nutrients, they soon became limited by light, whereas plants that were given more light soon became limited by nutrients.

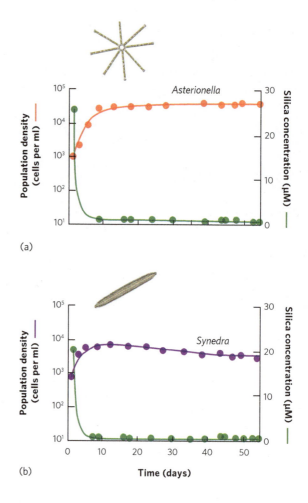

(a)

(b)

(c)

Figure 16.2 Competition for the most limiting resource. When raised separately, **(a)** *Asterionella* and **(b)** *Synedra* diatoms grow rapidly and then persist with population sizes near their carrying capacity. At their carrying capacity, both species dramatically reduce the abundance of silicate, although *Synedra* causes the silicate to be driven down to a lower abundance than *Asterionella*. **(c)** When combined, the two species again drive down the abundance of silicate to the point that there is still enough to meet the minimum requirements of the *Synedra* population but not enough to meet the minimum requirements of the *Asterionella* population. As a result, the *Asterionella* population declines. Data from D. Tilman et al., Competition and nutrient kinetics along a temperature gradient: An experimental test of a mechanistic approach to niche theory, *Limnology and Oceanography* 26 (1981): 1020–1033.

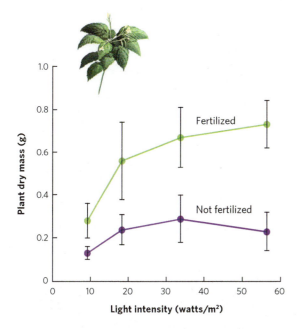

Figure 16.3 Interactions among resources. The small balsam plant experienced little growth when light intensity was increased under low soil fertility. However, when light intensity and fertility were both increased, the large amount of growth that followed was much higher than the sum of the two separate effects. Error bars are 95 percent confidence intervals. Data from W. J. H. Peace, and P. J. Grubb, Interaction of light and mineral nutrient supply in the growth of *Impatiens parviflora*, New Phytologist 90 (1982): 127–150.

THE COMPETITIVE EXCLUSION PRINCIPLE

A classic study of competition between species was conducted by Russian biologist Georgyi Gause in the 1930s. As you may recall from Chapter 12, Gause conducted laboratory experiments that explored how populations of protists in the genus *Paramecium* grew when living alone or together with limited resources. He began by growing two species—*P. aurelia* and *P. caudatum*—separately in test tubes. The test tubes contained a fixed amount of food, which was a species of bacteria (*Bacillus pyocyaneus*). When grown separately, the population of each paramecium species initially experienced rapid growth and then began to plateau as it reached its carrying capacity. You can see these data in the first two graphs of **Figure 16.4**. However, when the two species were grown together in the test tube, Gause observed a different outcome. As you can see in Figure 16.4c, the population of *P. aurelia* persisted in the test tube, but the population of *P. caudatum* declined to very low levels by the time the experiment ended. In short, *P. aurelia* was the superior competitor, presumably by driving down the abundance of the bacteria to such a low point that *P. caudatum* could not persist.

Similar competition experiments have been conducted hundreds of times using a wide variety of organisms. These experiments frequently produce the same result: one species

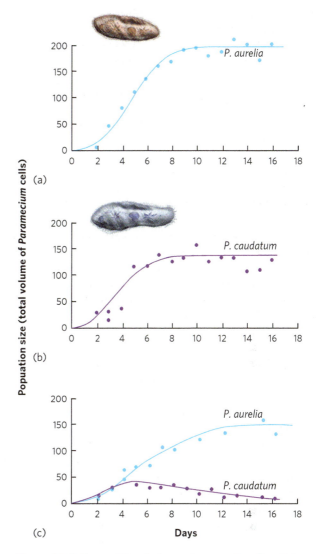

Figure 16.4 Competition between two species of protists. Gause raised different species of protists and measured the populations in terms of the total volume of cells in the container. When he raised each species separately on a fixed daily ration of bacteria, the population of **(a)** *P. aurelia* and **(b)** *P. caudatum* grew rapidly until they reached their carrying capacity. **(c)** When the two species were grown together, *P. aurelia* continued to grow well but *P. caudatum* was reduced to a very low population density. Because *P. caudatum* was still present in low numbers, *P. aurelia* did not attain the same population size as when it was raised alone. Data from G. F. Gause, The Struggle for Existence (Williams & Wilkins, 1934).

persists and the other dies out. This common pattern led to the development of the **competitive exclusion principle,** which states that two species cannot coexist indefinitely when they are both limited by the same resource. Researchers commonly find that when two species are limited by the same resource, one species either is better at obtaining the resource or is better able to survive when the resource is scarce.

Competitive exclusion principle The principle that two species cannot coexist indefinitely when they are both limited by the same resource.

Competition Among Closely Related Species

Although competition can occur between closely and distantly related, organisms, Charles Darwin believed that competition is most intense between closely related species. He argued that closely related species possess very similar traits and likely consume similar resources. As a result, they have the potential to compete strongly.

In consideration of Darwin's prediction, naturalists observed that closely related species living in the same region often grow in different habitats within that region. They hypothesized that because these species would compete strongly for the same resources, natural selection would favor differences in habitat use. These differences would allow each species to have a competitive advantage within its preferred habitat and to have a competitive disadvantage in habitats preferred by closely related species.

The first experimental test of this hypothesis was conducted in 1917 by the British botanist Arthur Tansley, who worked with two species of small perennial plants known as bedstraw. Heath bedstraw (*Galium saxatile*) typically lives on acidic soils, whereas white bedstraw (*G. sylvestre*) typically lives on alkaline soils. To determine whether each species competed best under the conditions where it naturally occurred, Tansley and his colleagues planted them separately and together in deep boxes containing either acidic or alkaline soils, as illustrated in **Figure 16.5**. Because the boxes were located at a single site— the Botanic Garden in Cambridge, England—any differences in growth would be due to the types of soils in which the bedstraw was planted.

When planted separately, each species germinated and grew in both types of soil. However, each species presented more vigorous germination and growth in the soil type characteristic of its natural habitat. When grown together on alkaline soils, the white bedstraw overgrew and shaded the heath bedstraw. When grown together on acidic soils, heath bedstraw outcompeted the white bedstraw. Tansley concluded that although both species are capable of living in both soil types when they are grown separately, intense competition from a closely related species restricts their distribution in nature to the soil type that gives them a competitive advantage. Hundreds of studies have supported Tansley's finding that many closely related species that live within the same region are intense competitors and are therefore distributed among different habitats in ways that reduce overlap, and thus competition, with each other.

Competition Among Distantly Related Species

While competition can be quite intense among closely related species, it can also be intense among distantly related species that consume a common resource. As we saw in Figure 16.1, open space on rocks in the intertidal biome is a resource used by numerous distantly related species, such as

Heath bedstraw grown alone

White bedstraw grown alone

Both species grown together

Alkaline soil   **Acidic soil**

Figure 16.5 Tansley's bedstraw experiment. White bedstraw naturally lives in alkaline soils and heath bedstraw naturally lives in acidic soils. Each species was planted either in its preferred soil or in the preferred soil of the other species; each species grew and survived best in its own preferred soil. This experiment showed that although closely related species can be intense competitors, the competition is reduced when competitors evolve to perform best in different habitats.

barnacles, mussels, algae, sponges, and others, all of which compete intensely for this limited space. As we will see later in this chapter, the outcome of this competition among so many competitors depends on their abilities to compete as well as their abilities to tolerate different abiotic conditions and predation.

Another example of competition among different species occurs among animals that eat krill (*Euphausia superba*). Krill are shrimp-like crustaceans that live in the oceans surrounding Antarctica and are consumed by virtually every type of large marine animal, including fish, squid, penguins, seals, and whales. Commercial exploitation of whales in the Southern Hemisphere has caused a decline in whale populations, while penguin and seal populations have increased. This suggests that reducing the number of whales has eased competition for krill and allowed penguin and seal populations to grow.

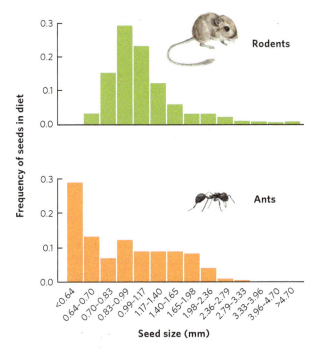

Figure 16.6 Seed size competition in rodents and ants.
The two groups of herbivores in the deserts of Arizona consume seeds of similar size, although rodents prefer somewhat larger seeds and ants prefer somewhat smaller seeds. This overlap in seed consumption means that these two distantly related species can compete with each other. Data from J. H. Brown and D. W. Davidson, Competition between seed-eating rodents and ants in desert ecosystems, *Science* 196 (1977): 880–882.

Intense competition can also occur among distantly related species in terrestrial ecosystems. On the forest floor, spiders, ground beetles, salamanders, and birds all consume the invertebrates that live in the leaf litter. In desert ecosystems, birds and lizards eat many of the same insect species. Moreover, desert ants and rodents both compete for seeds from a variety of plants. When we examine the patterns of seed sizes consumed by ants and rodents, as shown in **Figure 16.6**, we can see that the two groups both consume a wide range of seed sizes, but the ants tend to consume more of the small seeds, whereas the rodents tend to consume more of the large seeds.

CONCEPT CHECK

1. What factors affect population growth but are not consumable resources?
2. Why might we expect closely related species to experience stronger competition with each other than would distantly related species?
3. What does the competitive exclusion principle predict about the outcome of two species competing for the same resource or resources?

16.2 The theory of competition is an extension of the logistic growth model

You may recall from Chapter 12 that population growth and regulation can be modeled using the logistic growth equation. Extensions of this equation can help us understand the population dynamics of two species that are competing with each other. We will begin by examining Lotka-Volterra competition models when there is a single limiting resource and then expand our perspective and consider models that include multiple limiting resources. Although these models represent conditions that are a simplified version of what occurs in nature, they provide a useful foundation for exploring competitive interactions and the conditions under which species can coexist.

COMPETITION FOR A SINGLE RESOURCE

To understand how the populations of two species compete for a single resource, let's begin with the logistic growth equation introduced in Chapter 12:

$$\frac{dN}{dt} = rN\left(1 - \frac{N}{K}\right)$$

As you may recall, the part of the equation in parentheses represents intraspecific competition for resources. As the size of the population, N, approaches the carrying capacity of the environment, K, the term in parentheses approaches zero. At this point, the population's growth rate is zero, which means that the population has achieved a stable equilibrium.

If we wish to use this equation to model competition between two species that compete for a single resource, we need to consider the carrying capacity of the environment relative to the number of individuals from both species that are being supported. For example, imagine an environment that has a carrying capacity of 100 rabbits. Let's also assume that the food needed to support 100 rabbits could instead support 200 squirrels. The environment could also support many different combinations of rabbits and squirrels, such as 90 rabbits and 20 squirrels or 80 rabbits and 40 squirrels.

To include a second species in our equation, we need to add two pieces of information: the number of individuals of the second species and how much each individual of the second species affects the carrying capacity of the first species. For example, if we wanted to know the growth rate of a rabbit population, we would need to know the number of rabbits and squirrels in the area, and—in terms of resource consumption—how many squirrels equal one rabbit. Given that we want to model the changes in population for two species, we will use one equation for each. In these equations, we denote species 1 and 2 using the subscripts 1 and 2, respectively:

$$\frac{dN_1}{dt} = r_1 N_1 \left(1 - \frac{N_1 + \alpha N_2}{K_1}\right)$$

$$\frac{dN_2}{dt} = r_2 N_2 \left(1 - \frac{N_2 + \beta N_1}{K_2}\right)$$

The first equation tells us that the rate of change in the population of species 1 depends on its intrinsic rate of growth (r), the number of individuals of species 1 that are present (N_1), and—in parentheses—the combined effects of species 1 and species 2 in consuming the resource relative to the carrying capacity. The second equation tells us the same thing for species 2.

In these equations, we use two variables, the terms α and β, which are called *competition coefficients*. **Competition coefficients** are variables that convert between the number of individuals of one species and the number of individuals of the other species. In the preceding equations, α converts individuals of species 2 into the equivalent number of individuals of species 1. Similarly, β converts individuals of species 1 into the equivalent number of individuals of species 2. In our example of rabbits and squirrels competing for a common food resource, we can define rabbits as species 1 and squirrels as species 2. In that case, the value of α would be 0.5 because the resource consumption of one squirrel is equivalent to 0.5 rabbits. Similarly, the value of β would be 2.0 because the resource consumption of one rabbit is equivalent to two squirrels.

Using these two equations, we can create graphs that help us understand when each population will reach the point at which it no longer increases or decreases. This point is the population's equilibrium point. We can start by determining the conditions under which species 1 will have a population at equilibrium. This will happen whenever the change in population size per unit time is zero:

$$\frac{dN_1}{dt} = r_1 N_1 \left(1 - \frac{N_1 + \alpha N_2}{K_1}\right) = 0$$

$$r_1 N_1 \left(1 - \frac{N_1 + \alpha N_2}{K_1}\right) = 0$$

The equation indicates that there are two conditions that would put a population's growth at equilibrium. One condition occurs when $N_1 = 0$. Not surprisingly, if we have no individuals of species 1, there will be no

Competition coefficients Variables that convert between the number of individuals of one species and the number of individuals of the other species.

growth of species 1. For the second condition, when N_1 is not zero, we can find equilibrium by rearranging the equation:

$$r_1 N_1 \left(1 - \frac{N_1 + \alpha N_2}{K_1}\right) = 0$$

$$\left(1 - \frac{N_1 + \alpha N_2}{K_1}\right) = 1$$

$$\frac{N_1 + \alpha N_2}{K_1} = 1$$

$$N_1 + \alpha N_2 = K_1$$

$$N_1 = K_1 - \alpha N_2$$

In words, this equation tells us that the number of N_1 individuals that can be present at equilibrium depends on the total carrying capacity for species 1 minus the amount of resources consumed by some number of N_2 individuals. Based on this equation, we can draw a line that represents all combinations of N_1 and N_2 that would exist when the N_1 population is at equilibrium. We can plot this line on a graph that has N_1 on the x axis and N_2 on the y axis, as shown in **Figure 16.7a**. To draw this line, we need to identify the x and y intercepts on the graph.

The x intercept can be found by setting N_2 equal to zero. If we do this, the equation simplifies to

$$N_1 = K_1 - \alpha(0)$$

$$N_1 = K_1$$

As you can see from this equation, when there are no N_2 individuals, the N_1 population achieves equilibrium when it reaches its carrying capacity, K_1.

The y intercept can be found by setting N_1 equal to zero. If we do this, the equation simplifies to

$$0 = K_1 - \alpha N_2$$

$$\alpha N_2 = K_1$$

$$N_2 = K_1 \div \alpha$$

As you can see from this equation, when there are no N_1 individuals, equilibrium occurs when the N_2 population reaches $K_1 \div \alpha$ individuals. In our example of squirrels and rabbits, this would mean that in the absence of any rabbits, the habitat could support 200 squirrels.

Now that we have identified the x and y intercepts, we can construct the line on the graph that represents all combinations of N_1 and N_2 that cause the population of N_1 to be at equilibrium, as illustrated by the green line in Figure 16.7a. As you can see in this figure, as we increase N_2, the equilibrium population size of N_1 becomes smaller because species 2 is now consuming some of the resource that species 1 needs.

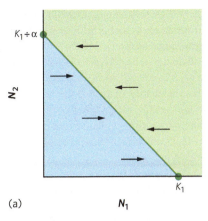

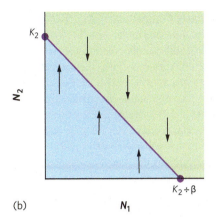

Figure 16.7 Equilibrium population sizes of two species. (a) The equilibrium isocline for species 1, shown as a green line, represents all combinations of species 1's population sizes (N_1) and species 2's population sizes (N_2) that cause species 1 to experience zero growth. When species 2 has an abundance of zero, species 1 will have an abundance of K_1, which is the carrying capacity of species 1. **(b)** The equilibrium isocline for species 2, shown as a purple line, represents all combinations of species 1's population sizes (N_1) and species 2's population sizes (N_2) that cause species 2 to experience zero growth. When species 1 has an abundance of zero, species 2 will have an abundance of K_2, which is the carrying capacity of species 2. The arrows indicate how each population either increases or decreases as it moves away from the equilibrium line.

Because this equation of a straight line represents all population sizes at which a population experiences zero growth, it is called a **zero population growth isocline.**

The zero population growth isocline for species 1 represents the values of N_1 that are in equilibrium across a range of N_2 values. As a result, when the N_1 population is to the left of the isocline, as indicated by the blue-shaded region, it will experience population growth and move to the right until it reaches the isocline. In contrast, when the N_1 population is to the right of the isocline, as indicated by the green-shaded region, it will experience a population decline, thereby moving to the left until it reaches the isocline.

Now that we have graphically represented the zero population growth isocline for species 1, we need to do the same for species 2. To do this, we set the second equation equal to zero and then rearrange the equation:

$$\frac{dN_2}{dt} = r_2 N_2 \left(1 - \frac{N_2 + \beta N_1}{K_2}\right) = 0$$

$$r_2 N_2 \left(1 - \frac{N_2 + \beta N_1}{K_2}\right) = 0$$

$$1 - \frac{N_2 + \beta N_1}{K_2} = 0$$

$$\frac{N_2 + \beta N_1}{K_2} = 0$$

$$N_2 + \beta N_1 = K_2$$

$$N_2 = K_2 - \beta N_1$$

You might notice that this equation looks very similar to the equation we derived for N_1. To determine the x and y intercepts for the equilibrium line of species 2, we can go through the same mathematical steps that we used for species 1. For example, when we set $N_1 = 0$, equilibrium will occur for species 2 when it reaches its carrying capacity:

$$N_2 = K_2 - \beta(0)$$
$$N_2 = K_2$$

If we then set $N_2 = 0$, equilibrium will occur when

$$0 = K_2 - \beta N_1$$
$$\beta N_1 = K_2$$
$$N_1 = K_2 \div \beta$$

Using these two intercepts, we can plot a zero population growth isocline for species 2, as illustrated by the purple line in Figure 16.7b. As you can see in this figure, as N_1 increases, the equilibrium population size of N_2 decreases because species 1 is consuming the resource that species 2 needs. Because this is a zero population growth isocline for species 2, when the N_2 population is above the isocline, as indicated by the green-shaded region, it will experience a population decline until it reaches the isocline. In contrast, when the population is below the isocline, as indicated by the blue-shaded region, it will experience population growth until it hits the isocline.

In drawing both of these population isoclines, you might notice that the growth rates of the two populations (r_1 and r_2) have no effect on the position of the isocline. While the growth rates determine how quickly a population can reach equilibrium, they do not affect the location of the equilibrium.

Zero population growth isocline Population sizes at which a population experiences zero growth.

Predicting the Outcome of Competition

Now that we know the conditions under which each of two species reaches an equilibrium population size, we can determine if one of the two species will win the competitive interaction or if the two species will coexist. We do this by overlapping the two zero population growth isoclines for species 1 and species 2. As illustrated in **Figure 16.8**, overlaying the two lines can produce four possible outcomes. In all four cases, when both species start with small population sizes, as indicated by the yellow-shaded regions in all four graphs, both species will experience population growth. In contrast, when both species start with large population sizes, as indicated by the green-shaded regions in all four graphs, both species will experience a population decline. It is the region that lies between these two extremes where we can determine the outcome of competition.

In Figure 16.8a, the isocline for species 1 lies farther out than the isocline for species 2. As a result, any combination of the two species numbers that falls within the blue-shaded region means that species 1 is below its isocline and its population will grow, whereas species 2 is above its isocline and its population will decline. The net effect of species 1 moving to the right and species 2 moving down—both indicated by thin arrows—is that the combination of the two species moves down and to the right, as indicated by the middle, thick arrow. This combination will continue to move until it reaches equilibrium, which is indicated by the open circle. At this point, N_1 reaches its carrying capacity and N_2 goes extinct.

In Figure 16.8b, we see the opposite situation; the isocline for species 2 lies farther out than the isocline for species 1. In this case, any combination of the two species that falls within the blue-shaded region means that species 1 is above its isocline and its population will decline, whereas species 2 is below its isocline and its population will grow. With species 1 moving to the left and species 2 moving up, the net effect is that the combination of the two species moves up and to the left.

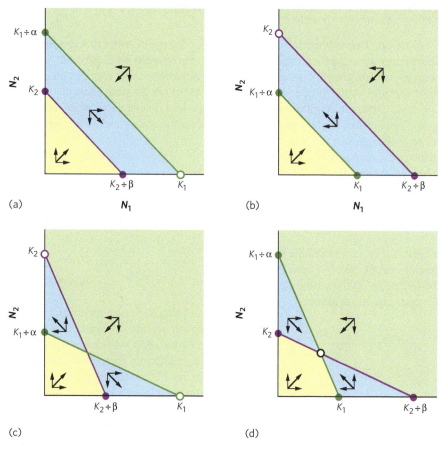

Figure 16.8 Predicting the outcome of competition for a single resource. Based on the Lotka–Volterra competition equations, there are four possible outcomes of competition. **(a)** If the zero population growth isocline for species 1 is farther out than species 2, species 1 will increase to its carrying capacity and species 2 will go extinct. **(b)** If the isocline for species 2 is farther out than species 1, species 2 will increase to its carrying capacity and species 1 will go extinct. **(c)** If the isoclines cross and the two carrying capacities are the outermost points on the two axes, the winner depends on the initial number of each species. **(d)** If the isoclines cross and the two carrying capacities are the innermost points on the two axes, then the two species will coexist. As in Figure 16.7, the green line represents the population isocline for species 1 and the purple line represents the population isocline for species 2. The thin black arrows represent increases and decreases in population sizes for each species, whereas the thick black arrows represent the net effects of population changes in both species. Open circles represent stable equilibria, which is the outcome of competition.

As this process continues, the combination will reach equilibrium, which is again indicated by the open circle; at this point, N_2 reaches its carrying capacity and N_1 goes extinct.

In Figure 16.8c, the two isoclines cross, with K_1 and K_2 as the two most extreme points on the axes. In this scenario, there are two possible equilibria, as indicated by the two open circles. If the combination of the two species falls within the blue-shaded region, the outcome of competition depends on the combination of N_1 and N_2 that are present. If the combination of N_1 and N_2 falls within the upper left blue region, the net movement is toward K_2 and species 2 wins the competitive interaction. If the combination of N_1 and N_2 falls within the lower right blue region, the net movement is toward K_1 and species 1 wins.

In Figure 16.8d, the two isoclines also cross, but this time with K_1 and K_2 as the two innermost points on the axes. In this scenario, the movement of each population relative to its respective isocline causes a net movement toward a single equilibrium point where the two lines cross. This means that the two competing species will coexist; one species will not outcompete the other.

In general, these equations tell us that coexistence of two competing species is most likely when interspecific competition is weaker than intraspecific competition—that is, when the competition coefficients α and β are less than 1. In other words, coexistence of two competing species occurs when individuals of each species compete more strongly with themselves (conspecifics) than with individuals of other species (heterospecifics).

COMPETITION FOR MULTIPLE RESOURCES

Although it is simpler to think about two species competing for one resource, in reality, species in nature compete for multiple resources. For example, grassland plants might simultaneously compete for water, nitrogen, and sunlight. Let's consider two species competing for two different resources. If species 1 is better able to sustain itself at low levels of both resources than species 2, then species 1 will win the competitive interaction. However, the outcome changes when species 1 is better at persisting at a low level of one resource but species 2 is better at persisting at a low level of another resource. In this case, because neither species can drive the other to extinction, the two species should coexist.

This principle was first demonstrated by David Tilman in a series of experiments conducted with two species of diatoms: *Asterionella formosa* and *Cyclotella meneghiniana*. We encountered these species when we discussed Liebig's law of the minimum. Both species require silicate (SiO_2) to produce their glassy outer shells and both species require phosphorus (P) for their metabolic activities. However, each species differs in how much of each resource it needs. *Cyclotella* uses silicate more efficiently, so its population can persist when there is a low abundance of silicate. In contrast, *Asterionella* uses phosphorus more efficiently, so its population can persist when there is a low abundance of phosphorus. Based on these differences, Tilman predicted that if the diatoms were raised under a low ratio of SiO_2/P, *Cyclotella* would dominate the competitive interaction because silicate would be low in abundance. If they were raised under a high ratio of SiO_2/P, *Asterionella* would dominate the competitive interaction because phosphorus would be low in abundance. He also predicted that at intermediate ratios of SiO_2/P, the two species would coexist because each is limited by a different resource.

To test this hypothesis, Tilman raised the two species separately and together under a range of different SiO_2/P ratios in laboratory containers, as shown in **Figure 16.9**. When raised separately, both species were able to survive

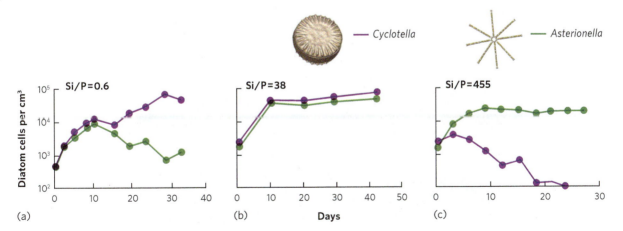

Figure 16.9 Species competing for multiple resources. In the experiment, two species of diatoms compete for silicon and phosphorus. *Cyclotella* is more limited by a low abundance of phosphorus (P), whereas *Asterionella* is more limited by a low abundance of silicate. **(a)** When raised under a low ratio of SiO_2/P, *Cyclotella* dominates the competitive interaction. **(b)** When raised under an intermediate ratio of SiO_2/P, neither species dominates. Instead, the two species coexist. **(c)** When raised under a high ratio of SiO_2/P, *Asterionella* dominates the competitive inter-action. Data from D. Tilman, Resource competition between planktonic algae: An experimental and theoretical approach, *Ecology* 58 (1977): 338–348.

under all the SiO_2/P ratios. However, *Cyclotella* dominated the containers when they were raised together at a low SiO_2/P ratio of 0.6, as shown in Figure 16.9a. In contrast, *Asterionella* dominated the containers when they were raised at a high SiO_2/P ratio of 455, as shown in Figure 16.9c. When they were raised at an intermediate SiO_2/P ratio of 38, the two species coexisted and neither became dominant, as you can see in Figure 16.9b. These were the first results to demonstrate that competing species can coexist when there are multiple resources and each species is limited by a different resource.

CONCEPT CHECK

1. What mathematical terms need to be added to the logistical growth equation to create the competition equation of population growth?
2. In the competition equation of population growth, why do we need to include competition coefficients?
3. Under what conditions can we expect the stable coexistence of two species competing for two resources?

16.3 The outcome of competition can be altered by abiotic conditions, disturbances, and interactions with other species

We have considered what happens when two species compete for shared resources. However, in nature, competition occurs in the context of many different environmental conditions and with other types of species interactions, including predation and herbivory. In this section, we will examine how competitive outcomes can be altered by abiotic conditions, disturbances, and interactions with other species.

ABIOTIC CONDITIONS

The ability to persist when resources are scarce is important to winning in a competitive situation, but it can be overwhelmed by the ability to persist under harsh abiotic conditions. We can see an example of this in a classic study of barnacles, conducted by Joseph Connell in Great Britain. Barnacles begin their lives as larvae that float on ocean currents and then settle on open, rocky spaces of coastal intertidal zones. They spend the rest of their lives as sessile adults, commonly living in dense groups. They feed on plankton that they filter from the water that washes over them. Because the barnacles cannot substantially reduce the vast amount of plankton that exists in coastal waters, plankton is not a limiting resource. Instead, barnacles are often limited by the amount of open space on rocks in the intertidal habitat.

Joseph Connell observed that the larvae of two barnacle species on the coast of Scotland are distributed broadly along the upper and lower intertidal zones. However, adults are distributed in two separate zones; Poli's stellate barnacles (*Chthamalus stellatus*) live in the upper intertidal zone, whereas rock barnacles (*Semibalanus balanoides,* formerly *Balanus balanoides*) live in the lower intertidal zone. The two species have only a small overlap in their distributions, as shown in **Figure 16.10**.

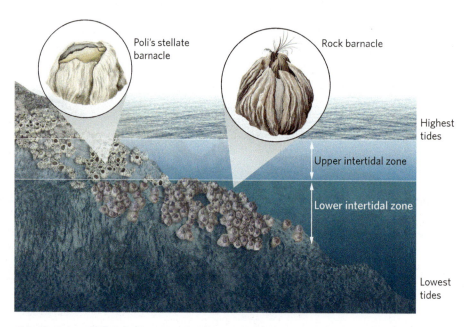

Figure 16.10 Competition between barnacles. In Great Britain, Poli's stellate barnacle lives in the upper intertidal zone because it is more resistant to desiccation and it cannot compete successfully in the lower intertidal zone. In contrast, the rock barnacle lives in the lower intertidal zone because it is the superior competitor but is not resistant to the desiccating environment of the upper intertidal zone.

To understand the reason for this distribution, Connell conducted a series of experiments in which he removed one or the other species. When he manually removed the rock barnacles from the lower part of the intertidal zone, the stellate barnacles quickly moved down and thrived. However, when he removed the stellate barnacle from rocks in the upper intertidal zone, the rock barnacles did not move up because they could not survive the desiccation that occurs in the upper intertidal zone. The stellate barnacles thrive in the upper intertidal zone because they are resistant to desiccation during periods of low tide. Though they could live in the lower intertidal zone, competition from the rock barnacle prevented it. The rock barnacle has a rapid growth rate and heavier shell than the stellate barnacle, which allows the rock barnacle to expand in size, pry the stellate barnacle off the rock, and take over the newly opened space. In short, the distribution of the two species is the result of a trade-off that many species experience—the trade-off between competitive ability and the ability to tolerate another challenging abiotic factor.

DISTURBANCES

Competitive interactions among species can also be altered by disturbances. For example, in the southeastern United States, more than 36 million ha of forest were once dominated by longleaf pine trees (*Pinus palustris*) with a very open, grassy understory. Many of the plants that live in this forest depend on frequent, low-intensity fires to persist (**Figure 16.11**). In fact, some species such as wiregrass (*Aristida beyrichiana*) can only reproduce after a fire has swept through the forest. Throughout the twentieth century, the natural fires were suppressed in much of the longleaf pine forest, which caused other species of woody plants to increase and outcompete the plants adapted to fire. Fire suppression altered the outcome of plant competition in the forest because it allowed the superior competitors previously removed by fire to persist.

PREDATION AND HERBIVORY

The outcome of competition can also be altered by the presence of predators and herbivores. We frequently find a trade-off between competitive ability and resistance to predators or herbivores; the most competitive plants are often the most susceptible to herbivores and the most competitive animals are typically the most susceptible to predators. For example, while many good competitors have high feeding rates because they move a lot in search of food, this activity makes them more noticeable to predators and thus more likely to be killed by predators.

Studies of aquatic communities have shown that predators can reverse the outcome of competition. For example, in one experiment, large outdoor tanks

Figure 16.11 Fire in a longleaf pine forest. These low-intensity fires, such as this one in Georgia, remove many species of woody plants and favor the growth of fire-adapted trees and grasses.
Photo by National Geographic Creative/Alamy.

were used to simulate ponds. Each pond was supplied with 200 tadpoles of the spadefoot toad (*Scaphiopus holbrookii*), 300 tadpoles of the spring peeper (*Pseudacris crucifer*), and 300 tadpoles of the southern toad (*Anaxyrus terrestris*). After setting up a series of identical tanks, each one was randomly assigned a treatment of 0, 2, 4, or 8 predatory newts (*Notophthalmus viridescens*).

In the absence of newts, the spadefoot toad's superior competitive ability was clear. As depicted in **Figure 16.12**, survival was high for spadefoot toad tadpoles, moderate for southern toad tadpoles, and low for spring peeper tadpoles. Spadefoot toad tadpoles had high survival because they are very active and rapidly consume whatever algae are available. In contrast, southern toad tadpoles are somewhat less active and spring peeper tadpoles are very inactive and therefore grow slowly. When the predatory newts were added, the outcome changed dramatically. As the number of newts increased, the survival of the spadefoot toads dropped precipitously and the survival of the southern

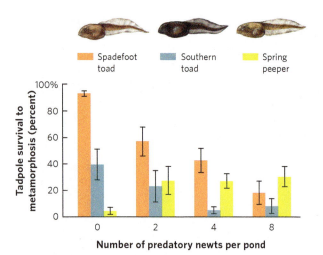

Figure 16.12 **Altering competition with predators.** In the absence of predatory newts, the spadefoot toads and southern toads survive well because they are superior competitors. With the addition of predatory newts, the survival of the two toad species decreases, while the survival of the spring peepers increases. Error bars are standard errors. Data from P. J. Morin, Predatory salamanders reverse the outcome of competition among three species of anuran tadpoles, Science 212 (1981): 1284–1286.

Figure 16.13 **Altering competition with herbivores.** Over a 10-year period, the plot on the right was sprayed with an insecticide to prevent beetles from consuming the goldenrods—the tall plants with yellow flowers—while the plot on the left was left unsprayed. The unsprayed plot experienced an outbreak of beetles, and this greatly reduced the density and height of the goldenrod plants. As a result, many species of plants that are inferior competitors became more numerous. Photo by W. Carson. W. P. Carson and R. B. Root, Herbivory and plant species coexistence: Community regulation by an outbreaking phytophagous insect, Ecological Monographs 70 (2000): 73–99.

toads dropped modestly, which is consistent with the southern toad's higher activity levels. Compared to the control treatment, the addition of eight newts reduced the survival of spadefoot toads by 75 percent and caused the survival of southern toad tadpoles to decrease by 32 percent. However, the survival of spring peepers increased by 26 percent. The reason for the improved survival of the spring peepers is that they are a very inactive tadpole species. This low activity helps them avoid being noticed by predators, but it comes at the cost of poor competitive ability. When newts are abundant and consume the superior competitors, the poorly competing spring peepers do much better and the outcome of competition among the three species of tadpoles is reversed.

Herbivores can have a similar effect on competition. For example, several species of goldenrod dominate fallow fields throughout the northeastern United States. These plants can grow more than a meter high and cast shade upon their shorter competitors. A species of beetle (*Microrhopala vittata*) specializes in consuming the goldenrods. Every 5 to 15 years, the beetle population achieves very high densities, known as an insect outbreak, during which the beetles consume large amounts of the goldenrod plants.

The impact of the beetles on the plant community was determined in a decade-long experiment in which some plots of plants were sprayed with an insecticide to kill the goldenrod-consuming beetles and prevent the insect outbreak, while other plots of plants were left unsprayed.

The results were dramatic (**Figure 16.13**). In the sprayed plots that contained few herbivorous beetles, shown on the right side of the photo, the goldenrods grew tall, cast shade on the other species of plants, and dominated the community with their superior competitive ability. In the unsprayed plots, shown on the left side of the photo, the beetles experienced a very large population increase midway through the experiment, and they consumed most of the goldenrod plants. The goldenrods that remained were substantially shorter, so they cast less shade on the inferior competitors. With fewer large goldenrods present, the inferior competitors became much more abundant. Based on these results, the researchers confirmed that the process of the herbivorous beetle eating the superior competitor caused a reversal in the outcome of competition among plants.

CONCEPT CHECK

1. How can abiotic conditions alter the outcome of competition?
2. How do disturbances alter the outcome of competition?
3. What is the underlying trade-off that allows predators and herbivores to reverse the outcome of competition?

16.4 Competition can occur through exploitation, direct interference, or apparent competition

Competition can be categorized in several ways. Thus far, we have seen numerous examples of **exploitative competition,** in which individuals consume and drive down the abundance of a resource to the point that other individuals cannot persist. For example, the different species of diatoms compete for silicate and phosphorus by consuming these resources and driving down their abundance. Sometimes competitors do not immediately consume resources but defend them, a type of competition known as **interference competition.** Whereas interference competition involves a direct interaction between two individuals, exploitative competition is considered an indirect interaction because it operates through a shared resource. Finally, two species can appear to compete for a shared resource but in fact cause a negative effect on each other due to mechanisms that are not competition, so we say that they experience *apparent competition.* Given our previous emphasis on examples of exploitative competition, in this section, we will explore the concepts of interference competition and apparent competition.

INTERFERENCE COMPETITION: AGGRESSIVE INTERACTIONS

Aggressive interactions are an effective form of interference competition that occurs between species of animals. For example, in the deserts of New Mexico, two species of ants—the long-legged ants (*Novomessor cockerelli*) and the red harvester ants (*Pogonomyrmex barbatus*)—compete for seeds and for any insects they can catch and subdue. However, the long-legged ants have a unique method of competing through interference. Early in the morning, they emerge from their nests, find the nests of the red harvester ants, and plug the nest entrances with stones and soil. As you can see in **Figure 16.14**, when researchers examined the effects of this aggression, they found that harvester ants with plugged entrances required several hours to unplug their nests before they could begin foraging for seeds. As a result, the long-legged ants can forage for several hours in the morning with minimal competition from the harvester ants.

INTERFERENCE COMPETITION: ALLELOPATHY

Another type of interference, known as **allelopathy,** occurs when organisms use chemicals to interfere with their competitors. We saw an example of this in the earlier

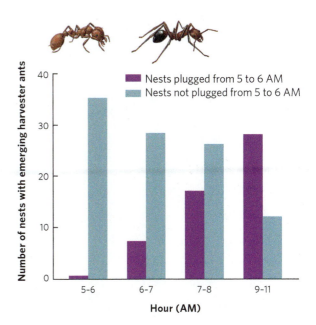

Figure 16.14 Interference competition in ants. Red harvester ants and long-legged ants compete for seeds and insects. When the harvester ant nest is not plugged by the long-legged ant, a large number of colonies emerge by 6 AM. However, if the nest entrances are plugged by the long-legged ant, it can take several hours for the harvester ants to clear the openings and start foraging. Data from D. M. Gordon, Nest-plugging: Interference competition in desert ants (*Novomessor cockerelli* and *Pogonomyrmex barbatus*), Oecologia 75 (1988): 114–118.

story of garlic mustard. As a second example, black walnut trees (*Juglans nigra*) produce *juglone,* an aromatic organic compound that inhibits certain enzymes in other plants. The compound—found in the leaves, bark, roots, and seed husks of the black walnut tree—leaches out of the plant and into the soil, making it difficult for most other plant species to germinate and grow under a walnut tree.

Allelopathy can also be an effective strategy for invasive plants, which invade and dominate a community. The wetland plant known as the common reed (*Phragmites australis*) is found throughout the world (**Figure 16.15**). In North America, some genetic strains are native and other strains have been introduced from Europe. In the Great Lakes region of North America, the introduced strain is spreading rapidly and has displaced many other species of native wetland plants. For many years, researchers hypothesized that the common reed was successful at spreading due to allelopathy, but they lacked evidence. Recently, it was discovered that the root of the common reed produced a chemical known as *gallic acid,* which is highly toxic to the roots of many other plants. By harming the roots of other species, the common reed impairs the growth of its interspecific competitors. The researchers also discovered that the root chemicals produced by the introduced strains of the reed were much more lethal to some plants than the

---

Exploitative competition Competition in which individuals consume and drive down the abundance of a resource to the point that other individuals cannot persist.

Interference competition When competitors do not immediately consume resources but defend them.

Allelopathy A type of interference that occurs when organisms use chemicals to harm their competitors.

Figure 16.15 The common reed. The common reed, a wetland plant, can be found in many locations around the world, including this site in Cape Cod, Massachusetts. In North America, some genetic strains are native, whereas others have been introduced from Europe. Photo by Ken Wiedemann/Getty Images.

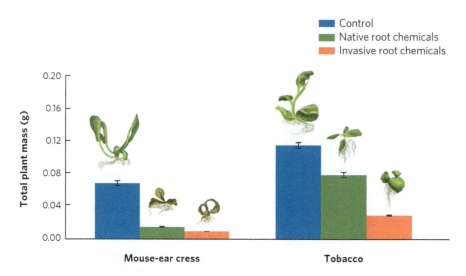

Figure 16.16 Allelopathy in the common reed. Root chemicals from native or invasive strains of the common reed were extracted and then added to pots of mouse-ear cress or tobacco plants. The addition of the root chemicals caused both plant species to grow poorly compared to the control. In the case of the tobacco plant, chemicals from the invasive strain of the reed inhibited growth more than the native strain. Error bars are standard deviations. Data from T. Rudrappa et al., Root-secreted allelochemical in the noxious weed *Phragmites australis* deploys a reactive oxygen species response and microtubule assembly disruption to execute *rhizotoxicity, Journal of Chemical Ecology* 33 (2007): 1898–1918.

root chemicals produced by the native strain, as shown in **Figure 16.16**. This result supports the observation that the introduced strains of the reed are able to invade a wetland and spread much more rapidly than the native strains.

Allelopathy does not always take the form of toxic effects on interspecific competitors. In Australia, for example, flammable oils in the leaves of some eucalyptus could be an adaptation that promotes frequent fires in the leaf litter of the forest floor, which kill seedlings of competitors. Regardless of whether organisms use direct aggressive interactions or allelopathic chemicals, it is clear that interference competition can be an important mechanism in determining the abundance and distribution of species.

APPARENT COMPETITION

Throughout this chapter, we have defined competition as a negative interaction between two individuals for a limited resource. Sometimes, however, two species can share a resource and have negative effects on each other without competing for a resource. When two species have a negative effect on each other through an enemy—including a predator, parasite, or herbivore—we call it **apparent competition.** As its name suggests, apparent competition causes an outcome that looks like competition, but the underlying mechanism is different.

Apparent competition can be observed in a variety of communities. For example, the ring-necked pheasant and the gray partridge (*Perdix perdix*) live in many of the same habitats in the United Kingdom. For 50 years, a large decline in the number of gray partridges had been attributed to increased agriculture, although researchers suspected that a parasitic nematode (*Heterakis gallinarum*) might also play a role. Infected pheasants experience few harmful effects from the parasite, but infected partridges can suffer weight loss, reduced fecundity, and death.

To determine the susceptibility of each species to infection by parasites in the environment, researchers allowed both bird species to feed either in pens containing parasite eggs scattered across the soil or in pens without the eggs. After 50 days of living in the pens without added parasite eggs, they found that both species contained low parasite infections, as shown in **Figure 16.17**. However, in the pens that contained added parasite eggs, the pheasants became infected at rates that were more than 20 times higher than those for the partridges. Moreover, when both birds were fed the same number of nematode eggs over a 100-day period, the pheasants excreted over 80 times more nematode eggs than the partridges. In short, pheasants can carry a large number of parasites without being harmed, and they excrete a large number of parasites that subsequently infect and harm partridges. The presence of pheasants is associated with partridges that gain less weight and have a low rate of survival. The relationship between the birds has the appearance of competitors for a shared resource, but the real cause of the declining partridges is a shared parasite.

There are many examples of apparent competition involving pathogens. For instance, when Europeans colonized other continents, particularly North and South America, they brought with them diseases such as smallpox. The Europeans could tolerate many of these diseases because they had a long evolutionary history of living with them, but Native Americans lacked this evolutionary history and the introduced diseases devastated their populations. While the Europeans appeared to cause competition, the reduction of the Native American population occurred primarily because the Europeans brought new pathogens to the Western Hemisphere.

Apparent competition When two species have a negative effect on each other through an enemy—including a predator, parasite, or herbivore.

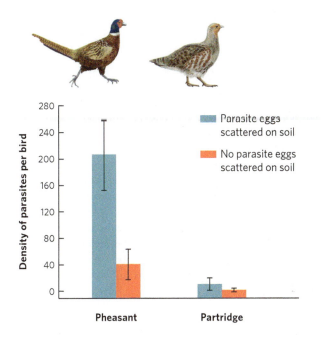

Figure 16.17 Apparent competition in pheasants and partridges. Pheasants (left) and partridges (right) were held in pens with no parasite eggs, or with parasitic eggs of nematodes scattered across the soil. In the absence of parasite eggs, some birds exhibited a low level of background infection. In the presence of parasite eggs, pheasants had a much higher number of parasites in their bodies after 50 days. Pheasants are not harmed by the parasites, but they serve as a major reservoir of new parasite eggs, which they excrete onto the soil. Partridges, which experience harmful effects from the parasite, had a low number of parasites and therefore excreted few parasite eggs into the environment. The presence of pheasants causes apparent competition because they release large numbers of parasite eggs, which leads to a decline in the partridges. Error bars are standard errors. Data from D. M. Tompkins et al., The role of shared parasites in the exclusion of wildlife hosts: Heterakis *gallinarum* in the ring-necked pheasant and the grey partridge, *Journal of Animal* Ecology 69 (2000): 829–840.

(a)

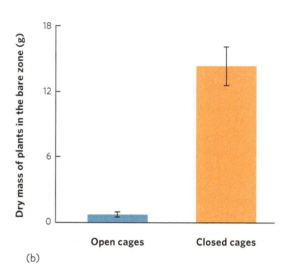

(b)

Figure 16.18 **Apparent competition between sage and grasses in California. (a)** Researchers originally hypothesized that the bare zone surrounding the purple sage shrubs (*Salvia leucophylla*) occurred because of allelopathic chemicals the shrubs produced. However, subsequent research discovered that herbivores used the shrubs as a refuge and only fed on grasses that were close to the safety of the shrub. To demonstrate that the effect was due to mice eating the plants surrounding the sage shrubs, researchers placed open-sided and closed cages in the bare zone, as shown here in a photo of the researchers in the Santa Ynez Valley, California, in 1968. **(b)** One year later they found nearly 20 times more vegetation growing in the closed cages, which confirmed that herbivores were causing the bare zone of vegetation around the shrubs. Error bars are standard errors. Courtesy Richard W. Halsey. Data from B. Bartholomew, Bare zone between California shrub and grassland communities: The role of animals, *Science* 170 (1970): 1210–1212.

In some cases, apparent competition occurs not because of a shared enemy, but because one species facilitates the enemy of another species. For example, on the coast of California, several species of shrubs commonly have a zone of bare soil immediately around them followed by a more distant zone made up of grasses (**Figure 16.18a**). Researchers initially hypothesized that the bare soil zone was caused by allelopathic chemicals produced by the shrubs, which made sense because some of the shrub species were known to produce chemicals that inhibited the growth of other plants when the chemicals were at high concentrations in the soil. However, since not all the shrub species produced these chemicals, the allelopathy hypothesis was not well supported.

Further research of the shrub growth showed that what appeared to be interference competition was not really competition at all. In one experiment, when seeds were placed on the bare zone and in the grass, 86 percent of the seeds were consumed in the bare zone but only 12 percent were consumed in the grass. When small mammal traps were set out in both areas, 28 mice were captured in the bare zone but only 1 mouse in the grass. Finally, two types of cages were placed in the bare zone: open-sided cages that allowed mice to move in and out of the cages and closed cages that excluded rodents. One year later, there was still little vegetation in the open cages but nearly 20 times more vegetation in the closed cages, as shown in Figure 16.18b. Collectively, these results demonstrate that the bare zone, which appeared to be the outcome of competition by allelopathy, actually occurred because the shrubs were a safe haven for mice from predators. The mice forage for seeds only a short distance from the shrubs and then run back to the shrub whenever predators appear. As a result, there are very few seeds surviving in close proximity to the shrubs, causing a bare zone that is created by herbivory rather than allelopathy.

ANALYZING ECOLOGY

Chi-Square Tests

Eco TV

 macmillanlearning.com/ricklefsvideo

When the researchers in California tried to determine the causes of the bare zone around coastal shrubs, they captured 28 mice in the bare zone near the shrubs and 1 mouse in the grass. Although this is a substantial difference, how do we know if this difference is statistically significant? In Chapter 14, we discussed the importance of testing hypotheses using a criterion of α < 0.05 for statistical significance. In Chapter 15, we discussed testing hypotheses using *t*-tests when we have estimates of mean values and standard deviations. Sometimes, however, our data do not consist of means and standard deviations, but rather of individuals that are counted, such as the number of mice in the bare zone versus those in the grass.

When our data consist of counts, we can test whether the data are significantly different from an expected distribution using a *chi-square test*. A **chi-square test** is a statistical test that determines whether the number of observed events in different categories differs from an expected number of events, which is based on a particular hypothesis. In the case of the mice, the researchers sampled both habitats equally and they captured a total of 29 mice from both areas. If the 29 mice exhibited no habitat preference, we would have expected to find 14.5 mice in the bare zone and 14.5 mice in the grass. As a result, we want to statistically compare the observed distribution of mice—28 mice in the bare zone and 1 mouse in the grass—to an even distribution between the two habitats. Now that we know the observed distribution and the expected distribution for mice that exhibited no preference, we can create a small table and conduct the chi-square test.

	Observed	Expected
Bare Zone	28	14.5
Grass	1	14.5

To calculate the chi-square value, we use an equation that includes our observed (*O*) and expected (*E*) values to calculate the value of chi-square (χ^2):

$$\chi^2 = \sum_{i=1}^{n} \frac{(O_i - E_i)^2}{(E_i)}$$

$$\chi^2 = \frac{(28 - 14.5)^2 + (1 - 14.5)^2}{14.5}$$

$$\chi^2 = 25.1$$

As was the case when we did a *t*-test in Chapter 15, we have to compare our calculated value to a table of values to determine if our observed distribution is significantly different from the expected distribution. The first step in that process is determining the degrees of freedom:

Degrees of freedom = (number of observed categories − 1)
= (2 − 1) = 1

Now we can examine the chi-square table provided in the statistical tables in the appendix. Using these tables, we can compare our calculated value (25.1) to the critical chi-square value when we have 1 degree of freedom and we set α = 0.05. In this case, the critical chi-square value is 3.841. Since our calculated value is greater than the critical value, we can conclude that the observed distribution of mice differed significantly from an equal distribution of mice between the two habitats.

YOUR TURN Using a chi-square test, determine whether an observed distribution of 12 mice in the bare zone and 8 mice in the grass differs significantly from an expected even distribution of mice.

Chi-square test A statistical test that determines whether the number of observed events in different categories differs from an expected number of events, which is based on a particular hypothesis.

Throughout this chapter, we have seen that competition is a major force affecting the distribution and abundance of organisms. This competition can occur by exploitation or interference and it can occur among species both closely and distantly related. As we will see in "Ecology Today: Finding the Forest in the Ferns," understanding the role of competition and how it interacts with disturbance, herbivory, and apparent competition is important for understanding how the natural world works, including biomes that provide humans with valuable natural resources.

CONCEPT CHECK

1. How is interference a form of competition?
2. Why is allelopathy considered a form of interference competition?
3. For what types of data do we use chi-square tests?

ECOLOGY TODAY   APPLYING THE CONCEPTS

Finding the Forest in the Ferns

A mature temperate forest carpeted with hay-scented fern. The dense fern layer is associated with a low number of tree seedlings, as evidenced by the lack of small trees growing in the understory of this forest in Shenandoah National Park, Virginia. Photo by GREG DALE/ National Geographic Creative.

In forests around the world, adult trees produce offspring that will become the next generation of trees. For a long time, it was assumed that the composition of trees in the next generation would be determined by the composition of seedlings produced by adult trees, the amount of shade, and the ability of the seedlings of each species to tolerate shade. For example, because seedlings of quaking aspen trees (*Populus tremuloides*) cannot tolerate shade from adult trees, they regenerate only when there are large openings in the forest that allow high amounts of sunlight to reach the ground. In contrast, seedlings of American beech trees are highly tolerant of shade and can grow in forests even under a dense canopy of leaves. While light gaps in the forest and variation in the shade tolerance of different tree species are important determinants of the composition of the next generation of forest, recent research has found that the effects of light gaps and shade tolerance can be overwhelmed by herbivory.

An excellent example of this phenomenon has occurred in the eastern deciduous forests of the United States. Historically, these forests had a variety of mature tree species in the upper canopy and younger trees in the understory. The forests also had low numbers of large white-tailed deer and occasional natural fires. Over more than a century, the forests were logged, fires were suppressed, and top predators such as mountain lions and wolves were removed. The removal of top predators caused a dramatic increase in the white-tailed deer population, which consumed large numbers of tree seedlings. Although the logging increased the amount of sunlight reaching the ground, and this should promote the germination and growth of new tree seedlings, the deer became so abundant that they greatly reduced the survival of many seedling species. However, we now know that the impact of the deer eating tree seedlings is only part of the story.

When deer eat plants, they prefer the most palatable species and leave the others unharmed. In the forests of Pennsylvania, for example, deer do not eat the unpalatable hay-scented fern (*Dennstaedtia punctilobula*). Although the fern historically comprised less than 3 percent of the understory vegetation, today it is the dominant understory plant in more than one-third of Pennsylvania forests. This is an important change in the forest because few tree seedlings can emerge out of the dense fern layer. As a result, there are now many large trees in the forest canopy but few young trees to become the next generation.

For many years, foresters were certain that few seedlings were emerging from the layer of ferns because the ferns were casting high amounts of shade over the seedlings. Competitive exclusion between these distantly related species made a great deal of sense. Recently, however, an alternative explanation has been proposed; perhaps the fern layer is not competing but is instead providing a refuge for rodents from predatory hawks and owls and that this refuge allows rodents to consume more tree seeds and tree seedlings.

Researchers reported the results of an experiment to determine which of these two hypotheses was correct: whether the negative effect of the ferns on tree seedlings was caused through competition for resources or through apparent competition by helping the rodents. They began by selecting plots of ferns within a forest, each with an area of 4 m². For each plot, they either fenced out rodents or let rodents roam throughout the plot, and they either removed the ferns or left them in place. After replicating the four different manipulations several times, they examined how many rodents were captured in each plot. The most common rodent species were deer mice, chipmunks, and red-backed voles (*Clethrionomys gapperi*). Rodents rarely got inside the fenced plots, regardless of fern coverage. However, rodents were frequently captured in the unfenced plots, with more than twice as many rodents captured when the ferns remained in place.

When beech seeds were added to the plots, only a small percentage of seeds were removed in the fenced plots—regardless of fern coverage—as shown in part (a) of the graph. However, a large percentage of seeds were removed in the unfenced plots, and even more seeds were removed when the ferns remained in place. Similar results were seen for many other species of seeds, including black cherry (*Prunus serotina*) and sugar maple (*Acer saccharum*).

When the researchers set out seedlings of sugar maple, they observed low mortality in the fenced plots—regardless of whether ferns were removed—as shown in part (b) of the graph. In the unfenced plots, however, seedling mortality was higher, but only when ferns were present. These results confirmed that the primary effect of the ferns on these tree seedlings was not as a competitor for light, but rather that they served as a refuge for rodents that eat the seeds and seedlings of beech, black cherry, and sugar maple. The results confirmed that the fern effects on American beech, black cherry, and sugar maple were due to apparent competition mediated through the fern's protection of rodents.

These experiments make clear that the regeneration of forests is complex, with multiple processes happening simultaneously. The success of new tree seedlings depends

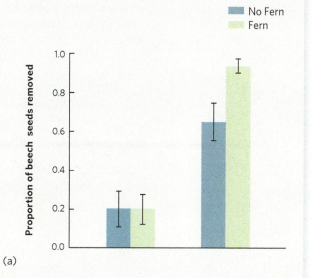

(a)

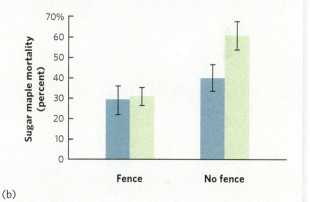

(b)

Apparent competition from ferns in a forest. To determine whether the effect of ferns was due to competition or apparent competition, the researchers used fencing to exclude rodents and manipulated the presence of ferns. **(a).** Few seeds of beech trees were removed from fenced plots, but many seeds were removed from unfenced plots, especially when ferns were present. **(b)** Similarly, sugar maple seedlings experienced low rates of mortality in fenced plots but high rates of mortality in unfenced plots in which ferns remained in place. Error bars are standard errors. Data from A. Royo and W. C. Carson, Direct and indirect effects of a dense understory on tree seedling recruitment in temperate forests: Habitat-mediated predation versus competition, *Canadian Journal of Forest Research* 38 (2008): 1634–1645.

on competition for light, but also directly by large mammal herbivory and indirectly by the altered species composition of the understory that is caused by the rodent herbivory. In turn, these altered understories can appear to determine competitive outcomes through competition, but in fact they cause apparent competition by providing refuges for rodents.

SOURCES:

Nuttle, T., et al. 2013. Historic disturbance regimes promote tree diversity only under low browsing regimes in the eastern deciduous forest. *Ecological Monographs* 83: 3–17.

Royo, A., and W. C. Carson. 2006. On the formation of dense understory layers in forests worldwide: Consequences and implications for forest dynamics, biodiversity, and succession. *Canadian Journal of Forest Research* 36: 1345–1362.

Royo, A., and W. C. Carson. 2008. Direct and indirect effects of a dense understory on tree seedling recruitment in temperate forests: Habitat-mediated predation versus competition. *Canadian Journal of Forest Research* 38: 1634–1645.

SUMMARY OF LEARNING OBJECTIVES

16.1 Competition occurs when individuals experience limited resources. Competition can either be intraspecific or interspecific and occurs when there is a limited resource. Resources can either be renewable or nonrenewable, and they can be generated either from within an ecosystem or from outside it. Leibig's law of the minimum states that a population will increase until the most limiting resource prevents further growth, although we now appreciate that different resources can have interactive effects on population growth. The competitive exclusion principle states that two species cannot coexist indefinitely when they are both limited by the same resource.

Key Terms: Intraspecific competition, Interspecific competition, Resource, Renewable resources, Nonrenewable resources, Liebig's law of the minimum, Competitive exclusion principle

16.2 The theory of competition is an extension of logistic growth models. The simplest models consider competition for a single resource and consider the zero population growth isoclines of two competing species. Using these models, we can make predictions regarding the conditions under which two species can win a competitive outcome or coexist. Under the more realistic situation of multiple limiting resources, we can have coexistence of multiple species of competitors when each species is limited by a different resource.

Key Terms: Competitive coefficients, Zero population growth isocline

16.3 The outcome of competition can be altered by abiotic conditions, disturbances, and interactions with other species. If a species is competitively superior but not tolerant of extreme abiotic conditions, it will not be able to dominate areas that experience such conditions. Similarly, competitively superior species that cannot persist with frequent disturbances, such as fire, cannot come to dominate inferior competitors. In the same way, superior competitors that are more vulnerable to herbivores or predators cannot outcompete inferior competitors because they are preferentially harmed or killed.

16.4 Competition can occur through exploitation or direct interference, or it may be apparent competition. Exploitative competition occurs when one species consumes enough of a resource that another species can no longer persist. In contrast, interference competition occurs when a species defends a resource and prevents other individuals from consuming it. Interference competition includes aggressive interactions among species and allelopathy. Sometimes species appear to be competing because the presence of one species has a negative effect on the population of the other. In cases of apparent competition, the underlying mechanism is not competition but another type of interaction, such as predation, herbivory, or parasitism.

Key Terms: Exploitative competition, Interference competition, Allelopathy, Apparent competition, Chi-square test

CRITICAL THINKING QUESTIONS

1. Compare and contrast renewable and nonrenewable resources.

2. How does Liebig's law of the minimum explain how silicate controls the growth of diatom populations?

3. If two species require the same limiting resource, what situation would favor their coexistence?

4. Why are population growth rates at zero when two species are at equilibrium?

5. Under what conditions would distantly related species compete?

6. If natural fires and herbivores can both reduce the abundance of competitively superior plants, how should this affect the number of other plant species that can persist in the community?

7. Why it is critical to include competition coefficients in the Lotka–Volterra competition equations?

8. Why is allelopathy not a form of exploitative competition?

9. Under what conditions can two species competing for two resources not coexist?

10. Explain the difference between competition and apparent competition.

GRAPHING THE DATA Competition for a shared resource

As we have seen in this chapter, the outcome of competition between two species for a common resource can be predicted if we know the carrying capacity and competition coefficients of each species. Using the following data, graph the zero population growth isoclines for each species.

Based on your graph, what is the predicted outcome of competition between these two species?

Species 1: $K_1 = 100$, $\alpha = 0.4$
Species 2: $K_2 = 50$, $\beta = 0.3$

Bathrooms with Benefits

Pitcher plants are famous for trapping insects, which are attracted by the smell and the nectar produced by the plant. Unfortunately for the insects, the cup-shaped pitcher has a slippery rim, making it nearly impossible for insects to escape once they enter. As a result, the insects die and are subsequently digested by the plant. Through this predator–prey relationship, pitcher plants obtain the nitrogen that they need to grow and reproduce, which is critical given that they often live in habitats where soil nitrogen is quite limiting. Knowing this, imagine the surprise of researchers when they discovered that a pitcher plant in the tropical forests of Asia was being used as a toilet by the local tree shrews.

The pitcher plant (*Nepenthes lowii*) is endemic to the island of Borneo in Southeast Asia. Immature plants live close to the ground and trap ants and other insects, while the mature plants live higher up in the trees. In these mature plants, researchers noted that they rarely observed any trapped insects, but they did notice the presence of feces from a small mammal. To determine the identity of the mammal, they set up video cameras around mature pitcher plants during the daylight hours. When they later checked the cameras, they discovered that mountain tree shrews (*Tupaia montana*) were visiting the pitcher plants and licking the abundant nectar that the plant produces on the overhanging pitcher lid. To reach the nectar, the shrew has to position itself with its posterior directly over the pitcher, and as it feeds, it defecates into the pitcher. The plant can then digest the shrew feces to obtain nitrogen. In fact, the plant can obtain all its nitrogen from shrew feces. This is highly beneficial to the pitcher plants given that it lives in the mountains of Borneo where insects are not particularly abundant. To encourage shrew visits, the plan has evolved to produce very large amounts of nectar, which is important to the shrew since other sources of nectar are scarce. In addition, rather than growing a slippery cup to trap insects, the rim of the mature pitchers can be easily gripped and the cups are structurally reinforced to hold the weight of the shrew. In short, the shrews and the pitcher plants are involved in a highly unusual mutualism.

The initial discovery of shrews feeding on and defecating in a species of pitcher plant in 2009 inspired much more research. In 2010, researchers discovered that two other species of large pitcher plants (*N. rajah* and *N. macrophylla*) were also visited by mountain tree shrews, and these other pitcher plants had many of the same characteristics as the first species. The following year, other researchers discovered that the pitchers, which are visited by tree shrews during the day, are also visited by summit rats (*Rattus baluensis*) during the night and the rats also leave their feces behind. Another group of researchers working in Borneo discovered that a tiny species of bat (*Kerivoula hardwickii*) was using a related species of pitcher plant (*N. hemsleyana*) as a roosting site. Bats were not known to roost in pitcher plants, but the structure of this species of pitcher plant allowed one or two bats to hide inside the cup during the day, while still staying above the level of the fluid inside the cup. In exchange for providing the bats with a safe place to hide, the plants obtained nearly a third of their nitrogen requirement from the feces that the bats left behind.

Research on the pitcher plants in Borneo has made it clear that we have a great deal to learn about how species have co-evolved mutualisms in nature. This work also underscores the fact that mutualisms are probably much more pervasive than we once thought and that many species depend on each other to ensure their growth and reproduction.

> "To reach the nectar, the shrew must position itself with its posterior directly over the pitcher, and as it feeds, it defecates into the pitcher."

SOURCES:

Clarke, C. M., et al. 2009. Tree shrew lavatories: A novel nitrogen sequestration strategy in a tropical pitcher plant. *Biology Letters* 5: 632–635.

Chin, L., et al. 2010. Trap geometry in three giant montane pitcher plant species from Borneo is a function of tree shrew body size. *New Phytologist* 186: 461–470.

Wells, K., et al. 2011. Pitchers of *Nepenthes rajah* collect faecal droppings from both diurnal and nocturnal small mammals and emit fruity odour. *Journal of Tropical Ecology* 27: 347–353.

Grafe, T. U. 2011. A novel resource-service mutualism between bats and pitcher plants. *Biology Letters* 7: 436–439.

A tree shrew feeding on a pitcher plant in Borneo. The tree shrew consumes nectar from the plant and then defecates into the pitcher, which provides nitrogen-rich nutrients to the plant. Photo by © Christian Loader/Scubazoo.com.

After reading this chapter, you should be able to:

17.1 Describe how mutualisms can provide water, nutrients, and places to live.

17.2 Explain how mutualisms can aid in defense against enemies.

17.3 Illustrate the role that mutualisms play in facilitating pollination and seed dispersal.

17.4 Describe how mutualisms can change when conditions change.

17.5 Explain how mutualisms can affect species distributions, communities, and ecosystems.

In Chapter 1, we defined mutualism as a positive interaction between two species in which each species receives benefits that only the other species can provide. Mutualisms are common in nature: Corals live with their symbiotic algae, shrews and pitcher plants exchange nutrients, and, as we saw in Chapter 10, leaf-cutter ants farm a fungus. In this chapter, we will see how species have evolved to participate in mutualistic interactions for a wide variety of benefits. Some species benefit by obtaining resources, while others benefit by obtaining a place to live, aid in defense, and assist in pollination or seed dispersal. When we consider mutualistic relationships, it is tempting to think that each species is trying to help the other. However, as you might recall from our review of natural selection in Chapter 1, selection favors any strategy that increases the fitness of the individual. Regardless of the specific benefit, there are requirements for the evolution of mutualistic interactions and conditions under which a positive, mutualistic relationship can change to a neutral or negative interaction. Finally, we are interested in mutualisms not only because they are a common interaction in nature, but also because they can affect the abundance of populations, the distribution of species, the diversity of communities, and the functioning of ecosystems.

17.1 Mutualisms for resource acquisition

We can categorize mutualisms in several ways. For example, some mutualists are **generalists,** which means that one species interacts with many other species. Other mutualists are **specialists,** which means that one species interacts with either one other species or a small number of closely related species. When two species provide fitness benefits to each other and require each other to persist, we call them **obligate mutualists.** We saw an example of this in Chapter 1 when we discussed the tubeworms and chemosynthetic bacteria that live together near deep-sea hydrothermal vents. Tubeworms provide a place for bacteria to live and bacteria provide food for the tubeworms; neither species can survive without the other. In contrast, **facultative mutualists** provide fitness benefits to each other, but the interaction is not critical to the persistence of either species. For example, a group of tiny insects known as aphids suck the sap from plants and produce a droplet rich in carbohydrates that is

consumed by several species of ants. The ants gain a source of food and in exchange they protect the aphids from predators. Although both groups benefit, it is a facultative mutualism because each can persist without the other. A mutualism between two species can be composed of two obligate mutualists, one obligate and one facultative mutualist, or two facultative mutualists.

One of the most common functions of mutualisms is to help species acquire resources they need, such as water, nutrients, and a place to live. In previous chapters, we have discussed a few examples of such mutualisms. In Chapter 1, we saw that lichens are composed of a fungus living with either green algal cells or cyanobacteria (see Figure 1.13). This fungus provides the algae with water, CO_2 from fungal respiration, and nutrients, and, in exchange, the algae provide the fungus with carbohydrates from photosynthesis. Similarly, in Chapter 2, we discussed how corals provide a home for photosynthetic algae known as zooxanthellae. As you can see in **Figure 17.1**, the coral catches bits of food with its tentacles and during digestion the coral emits CO_2, which the algae use during photosynthesis. The algae then produce sugars and O_2, some of which can be consumed by the coral. Other animals also incorporate symbiotic algae into their bodies. Similarly, in Chapter 2, we discussed how the eggs of spotted salamanders incorporate algae in the tissues of the embryo. Similarly, when the leaf sheep sea slug (*Costasiella kuroshimae*) consumes algae, it stores the chloroplasts from the algae inside its tissues and thus gains most of its energy through photosynthesis (**Figure 17.2**). In this section, we will review how several species of plants, animals, fungi, and bacteria interact in mutualisms to gain water, nutrients, and places to live.

RESOURCE ACQUISITION IN PLANTS
Although plants obtain water and soil minerals through their root systems, many plants also rely on mutualisms with fungi and bacteria to help them obtain nutrients.

Generalist A species that interacts with many other species.

Specialist A species that interacts with one other species or a few closely related species.

Obligate mutualists Two species that provide fitness benefits to each other and require each other to persist.

Facultative mutualists Two species that provide fitness benefits to each other, but whose interaction is not critical to the persistence of either species.

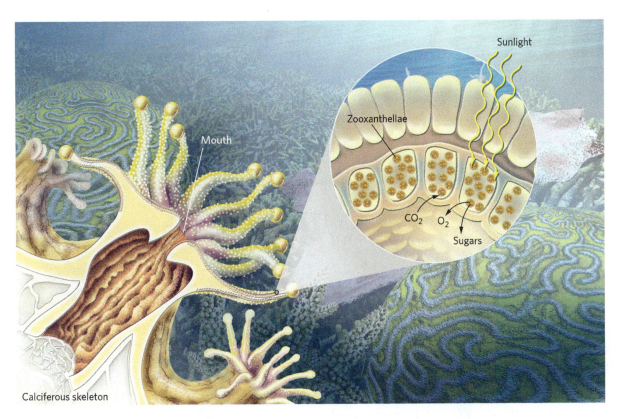

Figure 17.1 A mutualism between coral and zooxanthellae. A coral grabs food with its tentacles, which contain stinging cells, and pulls the food into its mouth. Zooxanthellae algae live along the surface of a coral's tentacles where they obtain sunlight for photosynthesis. Corals provide a place for algae to live and emit CO_2 that algae use in photosynthesis. As algae photosynthesize, they provide sugars and O_2 that corals consume.

Plants and Fungi

Fungi that surround plant roots and help plants obtain water and minerals are known as **mycorrhizal fungi.** The network of fungal hyphae provides plants with minerals, such as nitrogen and phosphorus, and water from the surrounding soil. Plants provide the fungi with the sugars they produce by photosynthesis. Because fungi can increase the amount of minerals obtained by the plants, they are able to increase the plants' tolerance to both drought and salt stress. They can also help plants combat infections from pathogens.

Mycorrhizal fungi can be divided into either *endomycorrhizal fungi* or *ectomycorrhizal fungi.* **Endomycorrhizal fungi** are characterized by hyphal threads that extend far out into the soil and penetrate root cells between the cell wall and the cell membrane (**Figure 17.3a**). One fungal species can commonly infect multiple plant species. There are several types of endomycorrhizal fungi.

Figure 17.2 The leaf sheep sea slug. When the slug first hatches from an egg in the ocean, it is brown. However, as it begins to eat algae, it stores the chloroplasts from the algae in its own tissues. As the slug accumulates a large number of chloroplasts, its body turns green and it is able to acquire most of its energy through photosynthesis rather than through herbivory. Photo by Lynn Wu/Jim & Lynn Photography.

Mycorrhizal fungi Fungi that surround plant roots and help plants obtain water and minerals.

Endomycorrhizal fungi Fungi characterized by hyphal threads that extend far out into the soil and penetrate root cells between the cell wall and the cell membrane.

Fungal hyphae grow into root cells between the cell wall and cell membrane.

Fungal hyphae grow between root cells.

Root epidermis

Fungal sheath

Fungal hyphae extend into soil.

(a) Endomycorrhizal fungus

(b) Ectomycorrhizal fungus

Figure 17.3 Mycorrhizal fungi. Mycorrhizal fungi can be categorized as endomycorrhizal fungi or ectomycorrhizal fungi. **(a)** Endomycorrhizal fungi have hyphae that penetrate the root cells of plants and reside between the cell wall and cell membrane. **(b)** Ectomycorrhizal fungi have hyphae that do not penetrate the root cells but instead grow between the root cells of plants.

The most common type is **arbuscular mycorrhizal fungi,** which infect a tremendous number of plants, including grasses and apple, peach, and coffee trees. Arbuscules are branching hyphal structures found within plant cells that help the fungus provide nutrients to the plant.

Ectomycorrhizal fungi, illustrated in Figure 17.3b, are characterized by hyphae that surround the roots of plants and enter between root cell walls, but they rarely penetrate between the cell wall and the cell membrane. These fungi are currently known to live only in mutualistic relationships with trees and shrubs. Species of ectomycorrhizal fungi also tend to form

mutualistic relationships with fewer plant species than do endomycorrhizal fungi.

The mutualistic relationship between plants and mycorrhizal fungi goes back more than 450 mya to the time when plants first evolved to live on land. This ancient interaction between the ancestral plants and fungi

Arbuscular mycorrhizal fungi A type of endomycorrhizal fungi that infects a tremendous number of plants, including apple trees, peach trees, coffee trees, and grasses.

Ectomycorrhizal fungi Fungi characterized by hyphae that surround the roots of plants and enter between root cells but rarely enter the cells.

probably explains why so many modern species of plants and fungi continue to interact as mutualists. It is estimated that mutualism between plants and fungi involves more than 6,000 species of mycorrhizal fungi and 200,000 species of plants, which is about two-thirds of all plant species.

Plants and Bacteria

In some cases, mutualistic interactions between plants and bacteria convert unusable forms of minerals into forms that plants can use. One of the best-known examples is the group of bacteria in the genus *Rhizobium* that live in a mutualistic relationship with numerous species of legumes, including important crops such as beans, peas (*Pisum sativum*), and alfalfa (*Medicago sativa*). When legumes detect the presence of *Rhizobium* bacteria in the soil or when bacteria enter the plant through an opening in the root, the plant develops small nodules that surround the bacteria on the roots and provide them with a place to live (**Figure 17.4**). Plants also provide the bacteria with the products of photosynthesis. In exchange, the bacteria do something the plant cannot do; they convert atmospheric nitrogen—a form of nitrogen plants cannot use—into ammonia, a form of nitrogen plants can readily use. This mutualism can be quite valuable to plants, especially when they are living in areas of low soil fertility. We will discuss this phenomenon in more detail in Chapter 21.

RESOURCE ACQUISITION IN ANIMALS

Animals also use a variety of organisms to help them obtain food, water, and habitat. These interactions range from protozoans living in animals to mutualisms between animals.

Animals and Protozoans

Termites are a group of insects that consume wood, which is difficult to digest since it is composed largely of lignin and cellulose. To assist in this effort, species of protozoa that are able to consume lignin and cellulose live in termite guts. In the gut, the protozoa receive a constant source of food from wood that the termite consumes and, in exchange, the termite receives nutrients from the waste products of protozoan digestion. Many other animals also contain microbes in their digestive system. Humans, for example, host hundreds of species of microbes—bacteria, fungi, and protozoa—that largely seem to be beneficial. In fact, a person's digestive system contains 10 times more bacterial cells—from more than 500 bacterial species—than the total number of human cells in that person's body.

Animals and Other Animals

Mutualisms for acquiring resources can also occur between two species of animals. A fascinating example occurs between humans and a bird known as the greater honeyguide (*Indicator indicator*). For centuries, people in Africa have consumed the honey produced by bees,

Figure 17.4 **Root nodules that contain *Rhizobium* bacteria.** Legumes such as this bean plant can enter into a mutualistic relationship with *Rhizobium* bacteria. The plant provides the bacteria with sugars from photosynthesis and a root nodule in which the bacteria can grow. In return, the bacteria convert atmospheric nitrogen to ammonia, a form of nitrogen that the plant can use. Photo by Nigel Cattlin/Science Source.

but locating the beehives is a challenge. While the greater honeyguide likes to consume bee larvae and bee's wax, it has a hard time getting into beehives. The local people and the honeyguide both obtain resources by working together. Over time, local people have learned to use calls to attract the attention of the bird and then follow the bird to the beehive. Along the way, the bird stops to perch in nearby trees. As the bird gets closer to the beehive, it flies shorter distances from one perch to the next and perches lower in trees, as shown in **Figure 17.5**. Local people have learned how to interpret the bird's behavior and they follow it to the hive. When they find the hive, they scoop out the honey and leave pieces of the honeycomb with beeswax and bee larvae on the ground for the honeyguide to consume. It is thought that the honeyguide may have originally evolved this behavior as a mutualism with other honey-consuming mammals, such as the honey badger (*Mellivora capensis*).

Some animals provide a habitat for other animals in exchange for reciprocal benefits, as we saw in the case of the pitcher plants and the roosting bats. For example, alpheid shrimp live in the ocean and have very poor vision. They burrow into the sand and allow a group of fish known as gobies to share their burrows. In contrast to shrimp, gobies have excellent vision and are able to see shrimp predators. In exchange for receiving a burrow, a goby allows a shrimp to stay in close contact by permitting the shrimp to place an antenna on it once the goby

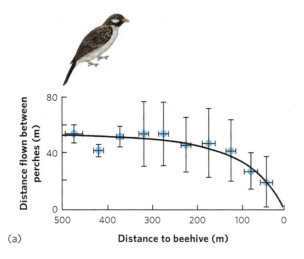

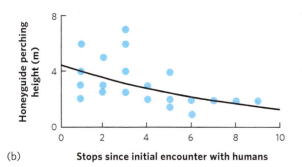

Figure 17.5 The greater honeyguide. When the greater honeyguide leads humans or other honey-consuming mammals to a beehive, the bird benefits by consuming the discarded honeycombs that contain beeswax and bee larvae. Humans have learned that as the greater honeyguide gets close to a beehive, **(a)** the bird flies shorter distances between stops and **(b)** at each stop it perches lower on trees. Error bars are standard deviations. Data from H. A. Isack and H.-U. Reyer, Honeyguides and honey gatherers: Interspecific communication in a symbiotic relationship, *Science* 243 (1989): 1343–1246.

leaves the burrow. If the goby sees a predator, it warns the shrimp by twitching. The shrimp detects the twitching via its antenna and heads back into the burrow for protection (**Figure 17.6**).

CONCEPT CHECK

1. If facultative mutualists do not require another species to help them, why do they engage in mutualism?
2. What benefits do mycorrhizal fungi provide for plants?
3. Why do termites require a mutualism with gut protists?

17.2 Mutualisms can aid in defense against enemies

As we saw with the alpheid shrimp, mutualisms can help a species defend itself against enemies. To obtain a defense benefit from a mutualistic partner, an organism must provide some type of benefit in return. In this section, we will examine a number of ways that mutualisms have evolved to benefit both the species being defended and the species providing the defense.

PLANT DEFENSE

Plants are involved in a number of mutualisms that help defend them from enemies. Two well-known examples are the mutualisms between ants and acacia trees and the mutualisms between fungi and plants.

Acacia trees are found in tropical forests throughout the world; they face a variety of herbivores and numerous competitors, including vines that try to wrap around the acacia's branches. In Central America, acacia trees are commonly inhabited by ants in the genus *Pseudomyrmex* that constantly patrol the tree branches. The acacia trees provide large thorns with pulpy centers that ants hollow out and convert into nests.

Figure 17.6 Habitat mutualist. The alpheid shrimp (*Alpheus randalli*) digs a burrow that it shares with a fish known as the pinkbar goby (*Amblyeleotris aurora*) in the waters off Indonesia. In exchange for this habitat, the goby allows the shrimp to place an antenna on its body so that the shrimp can feel the goby twitching when it sees an approaching predator. Photo by Jim Greenfield/imagequestmarine.com.

(a)

(b)

Figure 17.7 A mutualism between ants and acacia trees. (a) Acacia trees, such as this swollen thorn acacia in Panama, have large brown thorns that the ants can hollow out for nests where they can raise their larvae. The plants also have green nectaries that provide nectar the ants consume. In exchange for these benefits, the ants attack herbivores that try to eat the acacia tree and they attack vines and other plants that grow near the acacia tree. **(b)** Because the ants attack encroaching plants, acacia trees with ants are typically surrounded by bare ground.
Photos by Alex Wild Photography.

The trees also contain nectaries, which produce nectar that the ants consume (**Figure 17.7a**). In exchange, these ants bite and sting any herbivores—from small insects to large mammals—that attempt to consume the leaves. The ants also eliminate plants that attempt to grow near their home tree by chewing on them until they die (Figure 17.7b).

In a classic study, Dan Janzen compared acacia trees that had ants living on them and acacia trees from which he removed the ants. As you can see in **Figure 17.8a**, the trees with ants had a lower percentage of herbivorous insects than trees with no ants. In addition, trees with ants grew to be 14 times heavier than trees without ants. When Janzen monitored the trees over a 10-month period, he found that trees with ants had much higher survival rates than trees without ants, as shown in Figure 17.8b. In short, Janzen demonstrated that the ants are critical to the survival and growth of the acacia trees.

More recently, researchers have discovered that the ant–acacia mutualism has additional benefits. In 2010, it was reported that the chemicals in the nectaries contain numerous proteins with antibacterial properties. To determine if ants helped to distribute these chemicals on the leaves, the researchers used two sets of trees: one group had the ants present and the other group had the ants removed for 2 weeks. Researchers also examined the effect of one ant species that acted as a mutualist (*P. ferrugineus*) and another ant species that did not (*P. gracilis*). When *P. ferrugineus* was present, the plant leaves were nearly free of bacteria, but when *P. gracilis* was present, there was no significant reduction in bacteria. The researchers do not yet know how the *P. ferrugineus* causes a reduction in leaf bacteria, but it seems likely that *P. ferrugineus* is distributing some of the antibacterial nectar to the leaves, whereas *P. gracilis* does not. These results demonstrated that ants in a mutualistic relationship with trees not only defend the tree against herbivores and

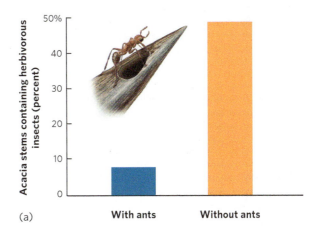

(a)

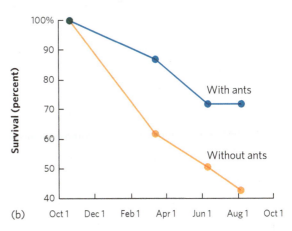

(b)

Figure 17.8 The effect of ants on acacia trees. (a) When ants were removed from acacia trees, the trees experienced a large increase in the number of herbivorous insects feeding on them. **(b)** As a result, the removal of ants caused much lower survival of the acacia trees. Data from D. H. Janzen, Coevolution of mutualism between ants and acacias in Central America, *Evolution* 20 (1966): 249–275.

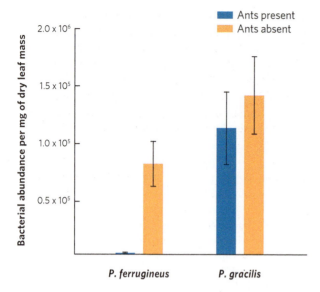

Figure 17.9 Reduction in bacteria by mutualistic ants. The presence of the mutualistic ant (*P. ferrugineus*) causes a large reduction in the abundance of bacteria on the leaves of the acacia tree. In contrast, the presence of a nonmutualistic ant (*P. gracilis*) has no effect on bacterial abundance. Error bars are standard errors.
Data from M. González and M. Heil, *Pseudomyrmex* ants and *Acacia* host plants join efforts to protect their mutualism from microbial threats, *Plant Signaling and Behavior* 5 (2010): 890–892.

competitors but appear to also defend the tree against bacteria that act as plant pathogens. You can view the results of the experiment in **Figure 17.9**.

Some plants can defend themselves from herbivores through mutualisms with fungi, known as **endophytic fungi,** which live within the plant's tissues. These fungi produce chemicals that can repel insect herbivores and also provide drought resistance by increasing the concentration of minerals in plant tissues, which increases the plant's ability to absorb and retain water from the soil. In exchange, the plant supplies the fungi with the products of photosynthesis. While endophytic fungi can be beneficial to plants, some of the chemicals they produce can be quite harmful to herbivores. For example, when the grass known as tall fescue (*Festuca arundinacea*) contains endophytic fungi, the fungi produce chemicals that are highly toxic to cattle, sheep, goats, and horses. Research is currently focused on identifying alternative strains of fungi that can offer defense against insect herbivores and drought resistance without being toxic to livestock.

ANIMAL DEFENSE

Animals also participate in mutualisms that aid their defense. An excellent example occurs in a group of fish known as cleaner wrasse. These tiny fish spend their life consuming ectoparasites that are attached to other, much larger fish (**Figure 17.10**). As the cleaner wrasse approaches, the larger fish opens its mouth and flares its gills to permit access to the many parasites that are attached to its body.

The number of parasites removed can be substantial; a single cleaner wrasse can consume more than 1,200 parasites per day. The cleaner wrasse benefits from having a large source of food and the larger fish benefit by having fewer parasites.

A similar situation exists for large terrestrial animals in Africa. Two species of birds, the red-billed oxpecker (*Buphagus erythrorhynchus*) (**Figure 17.11**). and the

Figure 17.10 Cleaner wrasse. Off the coast of Hawaii, the Hawaiian cleaner wrasse (Labroides phthirophagus) removes parasites from a yellowfin goatfish (Mulloidichthys vanicolensis). Photo by Seapics.com.

Figure 17.11 Oxpeckers. On the African savanna, oxpeckers, such as this red-billed oxpecker in Kruger National Park in South Africa, remove ticks from a variety of grazing mammals, such as this impala (*Aepyceros melampus*). Photo by robertharding/Superstock.

Endophytic fungi Fungi that live inside a plant's tissues.

yellow-billed oxpecker (*B. africanus*), perch on the backs of grazing animals such as rhinos and antelopes The birds consume ticks that are attached to the backs of mammals. However, because the birds also peck at the wounds caused by the ticks, scientists wondered whether the birds were mutualists or parasites. If oxpeckers act primarily as mutualists, their preferences for certain species of grazing mammals should be related to the number of ticks carried by each species. Alternatively, if the birds act primarily as parasites that seek to peck at mammal flesh, they should prefer mammals with thinner hides, which their beaks can penetrate more easily. A recent study examined these relationships in both species of oxpeckers using up to 15 species of grazing mammals in Africa. The researchers

quantified the oxpecker preferences for different species of mammals by observing grazing mammals of several species and dividing the number of oxpeckers on a given species by the total number of species grazing. Then they quantified the abundance of ticks on an individual of each mammal species. Although they found no relationship between the species preferences of oxpeckers and the thickness of the animals' hides, they found positive correlations between the species preferences of oxpeckers and tick abundance, as you can see in **Figure 17.12**. These results suggest that the birds are acting primarily as mutualists whose preferences are geared toward feeding on ticks. Thus, the benefits of tick removal for the grazing mammals likely outweigh the cost of having their flesh pecked at by the birds.

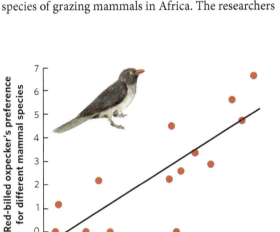

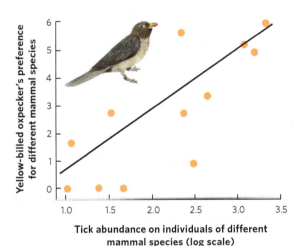

Tick abundance on individuals of different mammal species (log scale)

Figure 17.12 Oxpecker preferences for different species of mammals. The preferences of red-billed oxpeckers and yellow-billed oxpeckers are positively correlated to the abundance of ticks on the different mammals. Neither bird showed a preference for mammals with thin hides. These data suggest that oxpeckers seek out particular species of mammals primarily to consume ticks as a mutualist rather than to consume bits of mammal flesh as a parasite. Data from C. L. Nunn et al., Mutualism or parasitism? Using a phylogenetic approach to characterize the oxpecker-ungulate relationship, *Evolution* 65 (2011): 1297–1304.

> **CONCEPT CHECK**
>
> 1. What defense benefits do ants provide to acacia trees?
> 2. How do endophytic fungi help defend plants against herbivores?
> 3. What is the evidence that oxpeckers act as mutualists, not as parasites, with large grazing animals?

17.3 Mutualisms can aid with pollination and seed dispersal

In addition to providing resources and defense, mutualisms can also provide the valuable services of plant pollination and the dispersal of plant seeds. Indeed, without these services many species could not reproduce or colonize areas throughout their geographic range. In this section, we will examine the various mutualisms that serve these functions.

POLLINATION

To make seeds, flowering plants need pollen to fertilize their ovules. Some plants, such as grasses, commonly rely on the wind to blow pollen from one plant to another. Other plants rely on animals to carry the pollen. Over evolutionary time, plants have evolved a number of reward mechanisms to entice pollinators to visit their flowers. For example, the common honeybee visits flowers that offer both nectar and pollen. The bees consume both items, but the bees also inadvertently transfer some of the pollen that is attached to their bodies as they move from flower to flower. Many plants appear to have evolved flowers that are specialized in attracting one particular type of pollinator. For example, flowers pollinated by hummingbirds tend to be red—a color hummingbirds favor. Plants pollinated by hummingbirds also tend to have long tubular flowers inaccessible to most insects. However, because hummingbirds have long tongues, they can easily reach the nectar inside. Similarly, many bat-pollinated flowers are large, are often open only at night, and contain high volumes of nectar to attract the large pollinators. In some plant species, the volume of nectar increases during the night when bats are most likely to be foraging (**Figure 17.13**).

Although flowers in many species of plants can be pollinated by several different pollinator species, some plants, such as the group known as yuccas, have evolved very specific mutualisms with their pollinators. In fact, most species of yucca plants rely on a single species of yucca moth for pollination. Unlike most pollinators, however, a visit from the yucca moth is not a brief affair. A female yucca moth arrives at a yucca flower and lays eggs in the ovaries of the flower (**Figure 17.14**). To ensure that the flower makes seeds for her offspring to eat, the female moth climbs to the top of the flower and

Figure 17.13 Pollination by bats. Flowers that are pollinated by bats are typically large and contain large nectar rewards to entice the bats to visit. Pictured here is a long-nosed bat (*Leptonycteris curasoae*) pollinating a flower of the saguaro cactus (*Carnegiea gigantea*). Photo by Dr. Merlin D. Tuttle/Science Source.

Figure 17.14 Pollination by a yucca moth. A female yucca moth transfers pollen to the style of a yucca flower after she lays her eggs in the ovary of the flower. Photo by Robert and Linda Mitchell.

adds pollen grains to the flower's stigma. These pollen grains produce long tubes that grow down through the style and into the ovary, where they deposit male gametes, which then fuse with the female gametes to produce seeds. The moth eggs hatch into caterpillars that feed on the seeds that the flower forms. As a result, the plant gains a very effective pollinator at the cost of a few seeds.

Despite the consumption of seeds by the caterpillars, enough seeds remain for the yucca to reproduce. For many years researchers wondered what prevented the female moths from laying so many eggs that all the seeds would be consumed, leaving the flower with no fitness. They discovered that the plant is tolerant of up to six moth eggs per flower. Under these conditions, the plant develops 62 percent of its flowers to maturity. However, if there are more than six eggs per flower, some species of yuccas selectively abort most of their flowers and retain only 17 percent of their flowers to maturity, as shown in **Figure 17.15**. When a flower is aborted, all moth eggs and larvae within the aborted flower die. By aborting flowers whenever the moths lay more than six eggs, the plant favors those moths that do not lay too many eggs in each flower.

SEED DISPERSAL

The seeds and fruits produced by plants span a wide range of sizes, from tiny dandelion seeds that float on the wind to massive fruits such as coconuts. Many of the smallest seeds

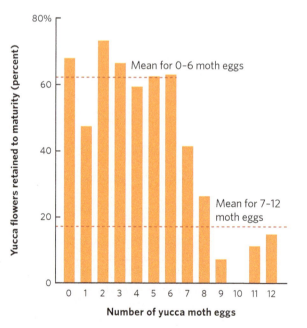

Figure 17.15 Controlling consumption of yucca seeds by yucca moth caterpillars. When female yucca moths lay 0 to 6 eggs in a yucca flower, the average plant develops 62 percent of its flowers to maturity. If a moth lays 9 to 12 eggs, the average plant only develops 17 percent of its flowers to maturity; the rest are selectively aborted. This response favors yucca moths that lay fewer eggs. Data from O. Pellmyr and C. J. Huth, Evolutionary stability of mutualism between yuccas and yucca moths, *Nature* 372 (1994): 257–260.

are easily dispersed by wind, but larger seeds are typically dispersed by animal mutualists. As we have seen, animals contribute to a mutualistic relationship if they receive a benefit in return. The most common benefit is food in the form of seeds or nutritious fruit that surrounds the seeds. When animals eat the fruits of plants, the seeds commonly pass through the digestive system unharmed and are viable after being excreted. However, when animals eat the seeds of plants, the seeds are digested and are no longer viable.

It may seem like a contradiction that plants would depend on seed-eating animals to help disperse their seeds. After all, if the seeds are eaten, there are no seeds left to disperse. However, in most cases, plants produce large numbers of seeds and not all seeds that are carried away get eaten. Many of them are stored in the ground, which means that the seeds are planted and able to germinate if they are not consumed. For example, squirrels store acorns in numerous locations and some of these acorns are never retrieved. Similarly, a single Clark's nutcracker (*Nucifraga columbiana*)—a bird that specializes on the seeds of the whitebark pine (*Pinus albicaulis*)—collects approximately 32,000 pine seeds in a season and stores them in thousands of different locations. Because this number of seeds is three to five times what the bird requires to meet its energy needs, many of the stored seeds are never eaten but rather germinate into new pine trees in locations far from the parent tree.

Some species of plants entice animals to disperse their seeds by surrounding them with a fruit that contains substantial nutrients. For example, many forest herbs, including the painted trillium (*Trillium undulatum*), produce seeds that have a lipid- and protein-rich package—known as an *elaiosome*—attached to the seed (**Figure 17.16**). Ants collect the seeds and take them back to their nests. After consuming the elaiosomes, they discard the seeds from the nest.

This process disperses the seeds away from the parent plant. The ants receive nutrition and the plants receive dispersal of their seeds.

When plants surround their seeds with a large fruit, the fruit can be a substantial reward to animal dispersers. If the seeds inside the consumed fruit have a hard coat that resists digestion, they pass through the animal's digestive system and are still able to germinate. One of the most striking examples is the African tree known as *Omphalocarpum procerum*. The tree produces fruits as large as a person's head and only the African elephant can break it open. Moreover, the seeds cannot germinate in the soil unless they have passed through an elephant's digestive system. The tree is highly dependent on elephants to consume and disperse its seeds, but as elephant populations decline, the tree is losing its only seed disperser.

A key element to this strategy for seed dispersal is for the seeds to remain inedible or hidden until they are fully developed. As a result, many fruits are green and relatively camouflaged while the seeds are developing. At this stage, the fruit is typically quite unpalatable. However, once the seeds have fully developed, the fruit becomes ripe, the tissues of the fruit become palatable, and it commonly changes color to be highly visible to animal dispersers.

CONCEPT CHECK

1. Why have some plant species evolved to have flowers of particular sizes and shapes?
2. How might an animal's storing of a large number of seeds create a mutualism between the animal and the plant?
3. How does the consumption of fruit by an animal result in improved seed dispersal for a plant?

Figure 17.16 Ants eating elaiosomes. Some forest herbs produce seeds with a packet of lipid- and protein-rich tissue attached that is known as an elaiosome. Shown here are seeds of the bloodroot (*Sanguinaria canadensis*). The ants carry these seeds to their nest, consume the elaiosome, and discard the seeds outside of the nest where they can subsequently germinate.
Photo by Alex Wild Photography.

17.4 Mutualisms can change when conditions change

Although individuals of two or more species might interact in a way that allows both to receive fitness benefits, we need to recognize that each individual participates in the mutualism to improve its own fitness and not the fitness of its partner. Therefore, when changing conditions alter the costs and benefits for each species, the interaction can change to something that is no longer a mutualism.

SHIFTING FROM MUTUALISM TO NEGATIVE INTERACTIONS

When one species in a relationship provides a benefit to another species at some cost but no longer receives a benefit in return, the interaction can shift from a positive, mutualistic interaction to a negative interaction such as herbivory, predation, or parasitism. For example, we discussed the role that mycorrhizal fungi play in helping plants obtain nutrients. Such a mutualism should be important to plants when nutrients are rare, but not when nutrients are abundant. In an experiment with citrus trees in highly fertile soils, researchers examined the effect of eliminating the mycorrhizal fungi by treating the soil with a fungicide. They found that eliminating the fungus made the trees grow up to 17 percent faster. Because the soil was fertile, the trees could grow well by collecting nutrients on their own, but when the fungi still existed in the soil, the trees still provided the fungi with the products of photosynthesis. As a result, the normally mutualistic relationship had changed into a parasitic interaction.

A similar situation exists for cleaner wrasse fish. You may recall that this fish removes ectoparasites from larger fish. Both species benefit and the interaction is a mutualism. However, it turns out that cleaner wrasse also like to consume the mucus and scales of larger fish, which is harmful to the larger fish because mucus and scales are costly to produce and offer protection against infection. Researchers working on coral reefs in the Caribbean examined whether the feeding decisions of a cleaner wrasse, the Caribbean cleaning goby (*Elacatinus evelynae*), changed when there were differences in the number of ectoparasites carried by the longfin damselfish (*Stegastes diencaeus*). They sampled the number of ectoparasites on the damselfish off the coasts of six different islands and then observed the cleaner wrasse to determine the percentage of mucus and scales in their diet. You can see their data in **Figure 17.17**. When damselfish populations had a high number of parasites, the cleaner wrasse ingested a small percentage of mucus and scales. However, when the damselfish had a low number of parasites, the cleaner wrasse ingested a much higher percentage of mucus and scales. Therefore, when the parasites are rare on the damselfish, the cleaner wrasse are forced to switch from being mutualistic to predatory and consume a higher percentage of mucus and scales.

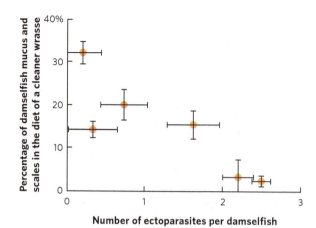

Figure 17.17 Switching from mutualism to predation. In populations of damselfish where ectoparasites are abundant, the cleaner fish primarily consume parasites of the damselfish and very little of the damselfish's mucus and scales. This is a mutualistic interaction. In populations of damselfish where the ectoparasites are not abundant, the diet of the cleaner fish includes a substantial percentage of the mucus and scales from the damselfish. This becomes more of a predatory interaction. Error bars are standard errors. Data from K. L. Cheney and I. M. Côté, Mutualism or parasitism? The variable outcome of cleaning symbioses, *Biology Letters* 1 (2005): 162–165.

DEALING WITH CHEATERS IN MUTUALISMS

When a mutualistic relationship changes into a relationship in which one species receives a benefit but does not provide one in return, natural selection should favor mechanisms that enable organisms to defend themselves. We saw an example of this with yuccas and yucca moths. When a moth lays so many eggs that the hatching larvae will eat all the developing yucca seeds, some species of yucca can respond by aborting the flower, thereby killing the moth larvae. In this way, the yucca punishes any yucca moths that act as cheaters in a mutualism.

A similar situation exists in relationships between plants and mycorrhizal fungi. When they operate as mutualists, both species provide a benefit at some cost. If a fungus reduces the benefit it provides to a plant, the plant should respond by providing a smaller benefit to the fungus. In 2009, researchers conducted a study to see if a plant could discriminate between different fungi and send products of photosynthesis to the most beneficial fungal mutualist. To test this question, individual wild onion (*Allium vineale*) plants were planted in two pots; half of the plant's roots went into a pot of soil containing a fungal species that helps the plant obtain food and the other half of the roots went into a second pot containing a fungal species that provides no benefit to the plant. This experiment is illustrated in **Figure 17.18a**. After the onion roots grew for 9 weeks, researchers measured how much photosynthetic product the plant was sending to each fungus. They did this by placing

(a)

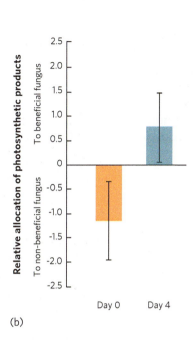
(b)

Figure 17.18 Favoring the most beneficial partner. (a) In an experiment, the wild onion plant was grown with its roots into two separate pots: one contained a beneficial mycorrhizal fungus and the other contained a non-beneficial mycorrhizal fungus. After 9 weeks of growth, researchers were able to determine whether the plant allocates more of its photosynthetic products to the beneficial or non-beneficial fungus. **(b)** Because the plant can distinguish between the two fungal species, it is able to allocate more of its photosynthetic products to the more beneficial fungus. Error bars are standard errors. Data from J. D. Bever et al., Preferential allocation to beneficial symbiont with spatial structure maintains mycorrhizal mutualism, *Ecology Letters* 12 (2009): 13–21.

a bag around each plant for 4 days and pumping in a special form of CO_2 that contained a rare isotope of carbon known as ^{14}C. Using this radioactive form of carbon, they could track the movement of carbon during the 4 days from the plant to each fungal species. They found that onion plants sent more products of photosynthesis to the beneficial fungi, as you can see in Figure 17.18b. This means that plants can distinguish among different fungi and preferentially provide greater benefits to the most beneficial fungi.

CONCEPT CHECK

1. How might an increase in soil nutrients alter a mutualism between plants and mycorrhizal fungi?
2. Based on the mutualism between cleaner wrasse and the large species of fish that they clean, what would you predict about the preference of the cleaner wrasse for parasites versus a diet of fish scales and mucus?
3. What is the evidence that plants can detect non-beneficial species of mycorrhizal fungi and respond to them appropriately?

17.5 Mutualisms can affect species distributions, communities, and ecosystems

When we think about mutualisms, we often focus on how the relationship helps each of the interacting species; it often improves the fitness and abundance of each participant. For example, corals cannot survive without zooxanthellae and many plants cannot produce offspring without pollinators. In addition to affecting abundance, mutualisms can also affect other ecological levels. In this section, we will examine how mutualisms, and the interruption of mutualisms, can alter the distributions of species, the diversity of communities, and the functioning of ecosystems.

EFFECTS ON SPECIES DISTRIBUTIONS

Given the benefits of mutualisms, we might expect natural selection to favor mutualistic relationships among species and to expand the distribution of species engaged in mutualisms. In contrast, we would expect the disruption of a mutualism to cause a decline of the species involved and a reduction in their distribution. We can see an example of this in the plant known as garlic mustard, a member of the mustard family with leaves that smell like garlic when crushed. As we discussed in Chapter 16, garlic mustard was introduced to North America more than a century ago from Native to Europe and Asia. When garlic mustard grows in North American forests, it causes young trees to grow poorly and makes them less likely to reach adult size. For a number of years, researchers sought to understand the mechanism underlying garlic mustard's harmful effects.

In 2006, researchers discovered that garlic mustard was interfering with the mutualism between the forest trees and arbuscular mycorrhizal fungi in the soil. To demonstrate this link, they examined how well three species of trees grew when raised in soil where garlic mustard had invaded and in soil where it had not invaded.

ANALYZING ECOLOGY

Comparing Two Groups of Data That Do Not Have Normal Distributions

Eco TV

 macmillanlearning.com/ricklefsvideo

As we saw in the case of the cleaner wrasse, researchers who study mutualisms often need to test whether the mutualistic interaction really provides a benefit to each species in the interaction. To do so, statistical tests are used to compare how each species performs both with and without the presence of the other species. In Chapter 15, we discussed the use of *t*-tests to compare the means of two groups. As you may recall, *t*-tests require that the data collected follow a normal distribution, which are the bell-shaped curves that we discussed in Chapter 2. In some cases, however, the data from two groups do not have normal distributions, so we cannot use the *t*-test. In these cases, we need to use the Mann-Whitney rank sum test, which is named for the statisticians who developed it.

The Mann-Whitney rank sum test begins by ranking the data—from low to high—and then taking the sum of those ranks. For example, consider the following set of data for the number of species occurring in eight coral reefs where researchers have removed the cleaner fish and eight coral reefs where researchers have allowed the cleaner fish to remain. If we were to take all 16 numbers and place them in order from the lowest value to the highest value, we could give each value a rank from 1 to 16. Whenever we have more than one occurrence of a value, we assign both values the average rank.

Number of Species Observed Without Cleaner Wrasse	Ranks	Number of Species Observed with Cleaner Wrasse	Ranks
(Group 1)	(Group 1)	(Group 2)	(Group 2)
3	1	7	6.5
4	2	8	8.5
5	3	9	10.5
6	4.5	10	12.5
6	4.5	10	12.5
7	6.5	11	14
8	8.5	12	15
9	10.5	14	16

The next step is to sum the ranks for group 1 (denoted as R_1) and group 2 (denoted as R_2):

$$R_1 = 1 + 2 + 3 + 4.5 + 4.5 + 6.5 + 8.5 + 10.5 = 40.5$$

$$R_2 = 6.5 + 8.5 + 10.5 + 12.5 + 12.5 + 14 + 15 + 16 = 95.5$$

We can then use either value of R to calculate a test statistic known as U. We will use R_2:

$$U = R_2 - (n_2 \times (n_2 + 1) \div 2)$$

where n_1 is the number of observations in group 1, and n_2 is the number of observations in group 2. Using this formula, we get

$$U = 95.5 - (8 \times (8 + 1) \div 2) = 59.5$$

Now that we know the value of U, we need to calculate the mean and standard deviation of U for the entire collection of data from both groups. The mean value of U, known as m_U, is

$$m_U = (n_1 n_2) \div 2$$

$$m_U = (8 \times 8) \div 2 = 32$$

The standard deviation of U is

$$\sigma_U = \sqrt{\frac{n_1 n_2 \times (n_1 + n_2 + 1)}{12}}$$

$$\sigma_U = \sqrt{\frac{8 \times 8 \times (8 + 8 + 1)}{12}} = 9.52$$

Using our calculated values of U, m_U, and σ_U, we can calculate z, which is defined as

$$z = \frac{U - m_U}{\sigma_U}$$

$$z = \frac{59.5 - 32}{9.52} = 2.89$$

We can look up this value in a table of z values in the Statistical Tables appendix, where we find that the probability value is 0.002. Since this probability is less than 0.05, we can conclude that the groups are significantly different from each other. This means that removing the cleaner fish causes a significant decrease in the number of fish species that live on a coral reef.

YOUR TURN When we calculate the value of z, we actually use the absolute value of z. To convince yourself that it does not matter whether we use R_1 or R_2, redo the above calculations based on the formula for U when using R_1:

$$U = R_1 - (n_1 \times (n_1 + 1) \div 2)$$

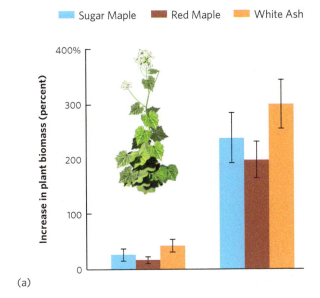

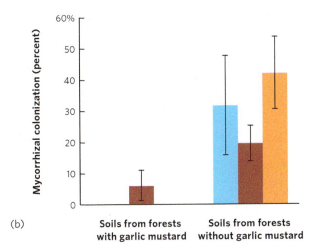

(a)

(b)

Soils from forests with garlic mustard · Soils from forests without garlic mustard

Figure 17.19 Disrupting a mutualism with garlic mustard. When researchers grew three species of trees in soils from forests with and without garlic mustard, they found that soils from forests with garlic mustard caused **(a)** much smaller increases in biomass and **(b)** little or no colonization by mycorrhizal fungi. Error bars are standard errors. Data from K. A. Stinson et al., Invasive plant suppresses the growth of native tree seedlings by disrupting belowground mutualisms, *PLOS Biology* 4 (2006): 727–731.

As you can see in **Figure 17.19a**, the increases in biomass of sugar maple, red maple, and white ash trees were many times greater when grown in soil that had not been invaded by garlic mustard. Moreover, when the researchers examined the percentage of tree roots colonized by mycorrhizal fungi, they found that soil collected from forests with garlic mustard showed little or no colonization by mycorrhizae, as illustrated in Figure 17.19b. Given that tree species depend on fungal mutualism to different degrees, garlic mustard has the largest negative effect on those tree species that have the greatest dependence on the fungi. Because garlic mustard disrupts vital mutualisms, it has the potential to alter the

distribution of a large number of other species as it spreads across North America.

MUTUALISM'S EFFECTS ON COMMUNITIES

Mutualisms that alter the abundance and distribution of one or more species can have widespread effects on the rest of the community. A community can be affected in several ways; the mutualism might cause a change in species diversity or it might alter the abundance of individuals within species in the community.

Altering Species Diversity

A generalist mutualist species interacts with many other species and offers widespread benefits. In the case of the cleaner wrasse fish, for example, a particular species of cleaner wrasse may remove parasites from many different species of larger fish. If this behavior helps species to persist on a coral reef, then the removal of cleaner wrasse should cause a decline in the total number of large fish and the number of species. Researchers working on coral reefs

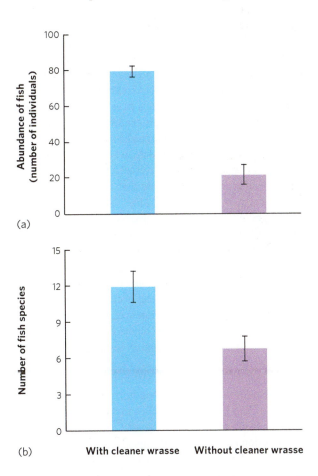

(a)

(b)

With cleaner wrasse · Without cleaner wrasse

Figure 17.20 Effects of mutualists on the abundance and diversity of a coral reef fish community. After 18 months of removing cleaner wrasse, **(a)** the abundance of individual fish declined by about three-quarters and **(b)** the number of species declined by about half. Error bars are standard errors. Data from A. S. Grutter et al., Cleaner fish drives local fish diversity on coral reefs, *Current Biology* 13 (2003): 64–67.

in Australia tested this question by removing a species of cleaner wrasse (*Labroides dimidiatus*) from nine small reefs. They designated another nine reefs as controls. After 18 months, they counted both the number of individual fish and the number of fish species on each reef. As you can see in **Figure 17.20**, removing the cleaner wrasse caused the number of other reef fish to decline by about three-quarters and the number of species to decline by half, which suggests that cleaner wrasse play a critical role in maintaining populations of reef fish.

Initiating a Chain Reaction of Species Interactions

In some cases, the community does not lose species when a mutualism is disturbed, but the abundance of many species changes through a chain of interactions. Earlier in this chapter, we discussed the role of ants in defending acacia trees against herbivores. We saw that the ants reduce herbivory on acacia trees, which improves tree survival. In return, ants benefit from having a place to make nests and a source of food in the nectaries. However, what would happen if much of the herbivory were removed, causing the ants to no longer provide a benefit to the trees?

Researchers examined this question in the savanna region of Kenya. They set aside 12 plots of 4 ha each; half of the plots were fenced to exclude all large herbivores, whereas the other half were left as unfenced controls. The plots contained 40- to 70-year-old acacia trees (*Acacia drepanolobum*). After 10 years, researchers examined how fencing the plots affected the ant–acacia mutualism and the rest of the community. You can view their data in **Figure 17.21**. In the fenced plots that lacked large herbivores, trees produced fewer swollen thorns and nectaries than trees in the control plots, as shown in 17.21a and 17.21b.

Changes in the acacia trees caused subsequent changes in the abundances of mutualistic ants. One species of ant, *Crematogaster mimosa*, relies heavily on the swollen thorns for making its nests and raising its offspring. The reduction of swollen thorns in the fenced plots caused a 30 percent reduction in the proportion of trees occupied by this ant, as illustrated in Figure 17.21c. Moreover, of those trees that were occupied, the average colony of *C. mimosa* was 47 percent smaller than in the control plots.

Another ant species, *C. sjostedti*, does not use the swollen thorns for nests but instead nests in tree cavities that are excavated in the acacia tree by long-horned beetles. Because this species does not require swollen thorns, the decline in *C. mimosa* in the fenced plots allowed *C. sjostedti* to double the proportion of trees that it occupied. This increase in *C. sjostedti* had further effects on the community. In contrast to *C. mimosa*, which works to eliminate insect herbivores, *C. sjostedti* allows long-horned beetles to live on the trees and bore holes into the trunk as it slowly consumes the trees. The beetle receives a food benefit and *C. sjostedti* receives a nest benefit, so these two species represent another

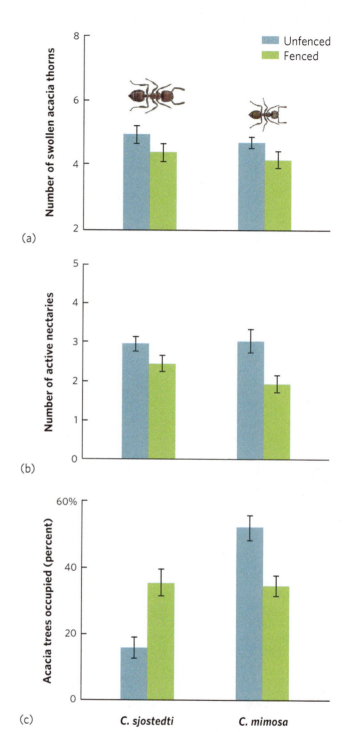

Figure 17.21 Community effects of a mutualism. The community effects of the mutualism were discovered when researchers fenced off areas of acacia trees from large mammalian herbivores. Fenced trees that were tended by either *C. sjostedti* or *C. mimosa* began to produce **(a)** fewer swollen thorns and **(b)** fewer nectaries. **(c)** In response to the changes in the trees, *C. sjostedti* began occupying many more acacia trees, while *C. mimosa* occupied fewer trees. Error bars are standard errors.

Data from T. M. Palmer et al., Breakdown of an ant-plant mutualism follows the loss of large herbivores from an African savanna, *Science* 319 (2008): 192–195.

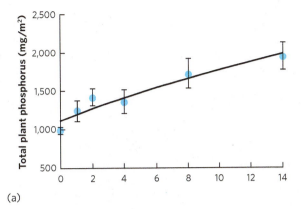

(a)

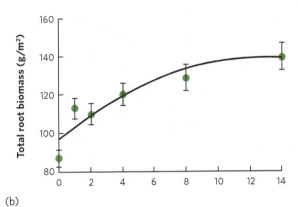

(b)

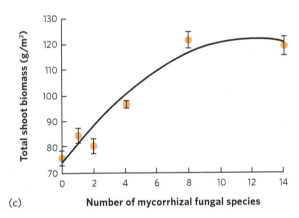

(c)

Figure 17.22 Ecosystem effects of fungal diversity. When researchers manipulated increases in the species diversity of mycorrhizal fungi, they observed an increase in **(a)** the total amount of phosphorus in the plants, **(b)** the total biomass of roots, and **(c)** the total biomass of shoots. Error bars are standard errors.

Data from A. G. A. van der Heijden et al., Mycorrhizal fungal diversity determines plant biodiversity, ecosystem variability, and productivity, *Nature* 396 (1998): 69–72.

mutualism in the community. However, beetle excavation causes the acacia tree to grow more slowly and die at about twice the rate as trees that are tended by *C. mimosa*. As a result, although larger mammal herbivores can graze the acacia trees and have negative effects on tree growth, the exclusion of large herbivores turns out to have a much larger negative effect on the trees because the trees are no longer defended from competitors or long-horned beetles. As this example demonstrates, mutualisms affect more than the species involved in the interactions; they can have far-reaching effects throughout the entire community.

MUTUALISM'S EFFECTS ON ECOSYSTEM FUNCTION

Mutualisms can also have ecological effects at the level of the ecosystem. As we discussed in Chapter 1, researchers working on ecosystems examine the movement of energy and matter among many sources, including both the biotic and abiotic realms. For example, researchers working with arbuscular mycorrhizal fungi investigated how a grassland ecosystem of 15 species would respond to different numbers of fungal species in the soil. When they examined the total amount of phosphorus that the plants took up from the soil, they found that plants living in soils that contained more fungal species took up a greater amount of phosphorus, as shown in **Figure 17.22a**. They also quantified the total biomass of roots in the soil and shoots above the soil. They found that a higher number of fungal species in the soil caused substantial increases in the biomass of the roots and shoots in the ecosystem. You can see these data in Figures 17.22b and 17.22c. As illustrated in this example, mutualisms not only affect individual species, but they can also have large effects on the functioning of ecosystems.

Throughout this chapter, we have seen how species can interact in a variety of mutualisms. These interactions fulfill a wide range of needs, including resource acquisition, places to live, defense, pollination, and dispersal. These interactions can have important effects on communities and ecosystems and they can even have major conservation implications, as you will see in the "Ecology Today" feature that follows.

CONCEPT CHECK

1. How is the invasion of garlic mustard altering the distribution of tree species?
2. Why do acacia trees that are no longer experiencing herbivory from large mammals provide smaller rewards for mutualistic ants?
3. Why does increasing the diversity of mycorrhizal fungi lead to greater plant productivity?

ECOLOGY TODAY   APPLYING THE CONCEPTS

Dealing with the Death of Dispersers

A substitute mutualist. Aldabra tortoises were introduced onto a small island in the Republic of Mauritius to act as a seed disperser for a population of critically endangered ebony trees. Photo by Olivier Naud.

The Republic of Mauritius provides an example of how important mutualisms can be to the persistence of species in nature. Mauritius, a group of islands located in the Indian Ocean southeast of the African continent, is the only place on Earth where the dodo bird (*Raphus cucullatus*) once lived. A large, flightless bird, the dodo was an easy source of food for the sailors who visited the islands, and was also harmed by the introduction of many non-native species. By the late 1600s, about 200 years after humans first set foot on Mauritius, the dodo was extinct. Numerous other species unique to Mauritius have also been driven extinct, including two species of giant tortoises: the high-backed tortoise (*Cylindraspis triserrata*) and the domed tortoise (*C. inepta*). Not only are the islands missing these species, they are also missing the services that these species provided.

Over subsequent centuries, as humans settled the islands, they cut down large swaths of tropical forest for lumber, firewood, and space for sugarcane fields. Today, only about 2 percent of the original forest remains, and some of the tree species continue to decline. For example, *Syzygium mamillatum* is a critically endangered tree species that consists of just over a hundred individuals. Scientists studying this problem realized that many declining tree species had probably once relied on the dodos and tortoises for seed dispersal. These species consumed the fruit, and the seeds were released after they passed through the animals' digestive systems. When the animals defecated, typically far away from the trees that produced the fruits, the seeds were ready to germinate. With these mutualists extinct, fewer fruits were consumed and any released seeds remained close to the parent tree. Seeds that geminate near the parent tree compete with the parent, and there are high rates of disease among conspecific trees concentrated in one place. Without dispersers, the seeds of the rare trees cannot disperse and recolonize open areas that have been logged. Researchers wondered if it was possible to bring back the process of seed dispersal even without the animals that originally did the dispersing.

The researchers came up with a radical idea hypothesis: perhaps non-native species could be introduced to fill the role of the lost mutualists and help save the endangered trees. To play the role of the dodo, they brought in domestic turkeys (*Meleagris gallopavo*), which—similar to dodos—have a large gizzard that grinds the fruits they swallow. The thought was that when the dodo consumed the fruits, the grinding of the gizzard broke down the fruits and released the seeds, but maintained the viability of the seeds as they passed through the bird's digestive system. To play the role of the two species of extinct tortoises, they brought in giant Aldabra tortoises (*Aldabrachelys gigantea*) from the nearby Seychelles Islands.

To test the potential for these species to act as substitute seed dispersers, fruits of the *S. mamillatum* tree were fed to the turkeys and Aldabra tortoises. The turkeys' large gizzards not only broke apart the fruit but also ground the seeds into small pieces, which meant that turkeys were not suitable replacements for the dodos. However, when fruits of the tree were fed to the Aldabra tortoises, the researchers found that 16 percent of the seeds inside the fruits remained whole in tortoise feces. Although these seeds exhibited lower germination success than seeds that did not pass through the tortoise's gut, they produced seedlings that grew taller and had more leaves. This meant that the Aldabra tortoises might be effective substitute dispersers.

Seed-dispersing tortoises. By consuming the large ebony fruits and dispersing the seeds through their feces, the tortoises spread the seeds throughout the island. They may represent the best hope for bringing the ebony tree back from the brink of extinction. Photo by Dennis Hansen.

A follow-up study tested whether the tortoises could be new mutualists for critically endangered trees if they were released into the wild. To test this idea, Aldabra tortoises were introduced to a small, 25-ha island that is part of the Republic of Mauritius. This island contains another rare tree species, the ebony tree (*Diospyros egrettarum*), which was once abundant and produces large, 16-g fruits. Logging for firewood had made this tree rare on the island, and large areas contained no new ebony tree seedlings. Without dispersers, the fruits landed near the parent tree and stayed there. After introducing the tortoises into pens in 2000, the researchers confirmed that the tortoises had no negative effects on the plant community. In 2005, 11 tortoises were allowed to roam freely throughout the island. In 2011, the investigators reported some remarkable results. Few fruits were now found near the parent trees because the tortoises were consuming most of them and, because the seeds passed through the tortoises when they defecated, the seeds were relocated to many places around the island. New seedlings showed up everywhere, including open areas where the tree had not existed for many decades. Moreover, ebony seeds that passed through the tortoise gut germinated better than unconsumed seeds.

In 2013, the researchers reported the results of releasing tortoises on Round Island, which is another small island off the coast of Mauritius. Introduced rabbits and goats had lived on this island until they were eliminated in the 1970s. The island also had several species of introduced plants that were outcompeting the native plants. The manual labor required to remove the introduced plants was costly, so the researchers proposed using tortoises as an ecological replacement for the extinct tortoises that would have grazed on many of the plants. The researchers initially introduced Aldabra tortoises and Madagascan radiated (*Astrochelys radiata*) tortoises to enclosures for 1 year to determine their grazing impacts on the plants. After that, they allowed the tortoises to freely roam around the island and graze on plants. In both situations, the tortoises' diets were made up of 81 to 93 percent non-native plants. Moreover, the cost of introducing the tortoises over the long term was less than the cost of paying people to control the non-native plants.

After more than a decade of research, these results suggest that, although we cannot bring back extinct species, we may be able to substitute some species to resurrect mutualisms so that the original extinctions do not cause subsequent extinctions of their mutualistic partners.

SOURCES:

Griffiths, C. J., et al. 2011. Resurrecting extinct interactions with extant substitutes. *Current Biology* 21: 1–4.

Griffiths, C. J., et al. 2013. Assessing the potential to restore historic grazing ecosystems with tortoise ecological replacements. *Conservation Biology* 27: 690–700.

Hansen, D. M., et al. 2008. Seed dispersal and establishment of endangered plants on oceanic islands: The Janzen–Connell model, and the use of ecological analogues. *PLOS One* 3: 1–13.

Seddon, P. J., et al. 2014. Reversing defaunation: Restoring species in a changing world. *Science* 345: 406–411.

SUMMARY OF LEARNING OBJECTIVES

17.1 Mutualisms can improve the acquisition of water, nutrients, and places to live. Mutualisms can be categorized as generalists, which interact with many species, or specialists, which interact with few other species. When both species require each other to persist, they are obligate mutualists. When the interaction is beneficial but not critical to the persistence of either species, they are facultative mutualists. Mutualisms for resources include the algae and fungi that compose lichens and the corals and zooxanthellae that build coral reefs. Plants also participate in this type of mutualism by interacting with endomycorrhizal fungi, ectomycorrhizal fungi, and *Rhizobium* bacteria. In most animals, protists can play an important role in digesting food. Other animals construct habitats that they share with other species in exchange for other benefits.

Key Terms: Generalists, Specialists, Obligate mutualists, Facultative mutualists, Mycorrhizal fungi, Endomycorrhizal fungi, Arbuscular fungi, Ectomycorrhizal fungi

17.2 Mutualisms can aid in defense against enemies. Plants make use of defensive mutualisms in a number of ways, including mutualisms with aggressive insects such as ants, and with endophytic fungi that produce chemicals harmful to herbivores. Animals that interact as mutualists to defend against enemies include cleaner fish that remove parasites from large fish and oxpecker birds that remove ticks from mammals.

Key Term: Endophytic fungi

17.3 Mutualisms can facilitate pollination and seed dispersal. Pollinators allow many species of plants to be fertilized, and some plants have evolved traits that favor a particular type of pollinator. When this happens, the plants and the pollinators can co-evolve. Numerous plants also depend on mutualisms to disperse their seeds. In some cases, seeds are dispersed as the result of animals storing them far from the parent plant. In other cases, animals consume the fruit of plants and the seeds are dispersed after passing through their digestive systems.

17.4 Mutualisms can change when conditions change. Although mutualisms benefit all species in the interaction, a positive mutualism can switch to a neutral or negative interaction when conditions change. In some cases, species can respond to cheaters in a mutualism by only rewarding individuals that provide benefits in return.

17.5 Mutualisms can affect communities. Mutualisms can increase or decrease the abundance of participating species. An absent mutualist can cause another species to be completely eliminated, thereby affecting the distribution of a species. Mutualists can also affect communities either by directly altering the number of species or by initiating a chain of interactions through a community. At the ecosystem level, mutualists can have additional effects, such as moving nutrients into producers and increasing the total biomass of producers.

CRITICAL THINKING QUESTIONS

1. Compare and contrast obligate mutualists and facultative mutualists.

2. If garlic mustard were to reduce arbuscular mycorrhizal fungi in an apple orchard, what effect would this have on apple crops?

3. If the mountain tree shrew and summit rat of Borneo were to go extinct, what would likely happen to the pitcher plant that currently lives in a mutualism with these two mammals?

4. If one species provides a habitat as part of a mutualistic relationship, what is the probable effect on the abundance and distribution of the other species?

5. How would you respond to someone who states that a mutualism is favored by natural selection because each species is trying to increase the other species' fitness?

6. If cleaner wrasse fish consume both parasites and scales from larger fish, what would determine whether the interaction is best categorized as mutualism or parasitism?

7. For a tree that uses a facultative mutualist animal to disperse its seeds, what would be the impact of the animal going extinct?

8. Why should wild strawberries, which have animal-dispersed seeds, not make their fruit very colorful and sweet until the seeds are fully developed?

9. What might prevent an endophytic fungus from reaping the benefits from a grass without providing a benefit in return?

10. When we introduce a non-native herbivore to replace an extinct herbivore that once served as a mutualist, what factors are likely to be important in determining the success at the community level?

GRAPHING THE DATA Ecosystem Function of Fungi

Earlier in this chapter, we saw that manipulating the number of mycorrhizal fungal species has several effects on the ecosystem, including an increase in the total phosphorus that accumulates in plants. (See Figure 17.22.) Using the data in the table, calculate the means and standard errors for the amount of phosphorus that

the researchers found remaining in the soil for each of the five fungus treatments. Using these calculated data, plot the relationship between the number of fungal species and the amount of soil phosphorus.

Number of Fungal Species	Phosphorus Remaining in the Soil (mg P/kg Soil)
0	16
0	15
0	17
2	11
2	10
2	12
4	9
4	10
4	8
8	4
8	6
8	5
14	4
14	3
14	5

18 Community Structure

Pollinating the "Food of the Gods"

Chocolate is one of the most popular food ingredients in modern cuisines. It comes from the beans of the cacao tree (*Theobroma cacao*) that originated in South and Central America. Given that ancient people used it in religious ceremonies as gifts to their gods, the genus of the plants, *Theobroma*, translates to "the food of the gods." Today, the cacao tree is grown in tropical regions around the world, including the Americas, Africa, and Asia. However, there is a widespread problem; only about 10% of cacao flowers get pollinated on modern farms. This results in fewer cocoa beans produced on each plant and less cocoa production for farmers and consumers. As a result, farmers have to hand-pollinate the flowers to improve their crop yield.

Farmers and ecologists have wondered why cacao plants experience such low pollination success. Modern plantations keep the ground cleared around cacao trees so that there are no decomposing old fruit husks. These decomposing husks can attract a fungus that can be harmful to the cacao tree, so removing the old pods reduces the fungus. However, the decomposing plant matter is also a habitat for the larval stage of the main pollinator of cacao trees, which are small flies known as midges. By cleaning up the ground around cacao trees, farmers have reduced the problem of the harmful fungus, but they may have also reduced the abundance of the pollinators.

To test this hypothesis, researchers in Australia conducted an experiment in which they manipulated the presence or absence of old cacao fruit husks around groups of eight cacao trees. Within each group of eight trees, they also manipulated the presence or absence of hand pollination. Over a 7-month period, the researchers documented the number of flowers produced, the number of fruits produced, and the mass of fruits produced on each tree. In addition, they documented the presence of any predatory spiders or skinks (a type of lizard) that might be attracted to the insects living in the decomposing husks.

In 2017, the researchers reported that the addition of old husks resulted in cacao trees producing more flowers, likely as a result of the decomposing husks providing more nutrients or water to the trees. After adjusting for this difference in total flower number, they discovered that adding old husks dramatically increased the number of flowers that were pollinated by the midges. When they examined the fruits at maturity, they found that the increase in the number of flowers pollinated by adding old husks was nearly as much as the increase observed when they used hand pollination. The total mass of the fruit on each tree followed a similar pattern. In addition, they found that adding the husks attracted more of the predatory spiders and skinks, but these predators did not appear to harm the midges.

> "By cleaning up the ground around cacao trees, farmers have reduced the problem of the harmful fungus but they may have also reduced the abundance of the pollinators."

Collectively, these results demonstrate that decomposing fruit husks of the cacao tree provide essential habitat for the eggs and larvae of the midge and other insects, which in turn improves the pollination of cacao flowers and provides insect prey for multiple species of predators. Because these decomposing husks can also serve as a habitat for fungal pathogens that are harmful to cacao trees, the researchers suggest that other sources of decomposing plant matter, such as old banana stems, might be a better way of providing midge habitat while reducing the prevalence of the fungal pathogen. This research illuminates how ecological communities are made up of interconnected species and that these connections have a major impact on natural and agricultural communities.

SOURCE
Forbes, J. F., and T. D. Northfield. 2017. Increased pollinator habitat enhances cacao fruit set and predator conservation. *Ecological Applications* 27: 887–899.

Cacao trees produce fruits that contain the cocoa beans. When old fruit husks are placed around the trees, they provide a habitat for the eggs and larvae of midges, which later turn into adult flies that pollinate the flowers of the cacao tree. Photo by urf/Getty Images.

LEARNING OBJECTIVES

After reading this chapter, you should be able to:

18.1 Illustrate how communities can have distinct or gradual boundaries.

18.2 Explain why the diversity of a community incorporates both the number and relative abundance of species.

18.3 Describe the ways in which species diversity is affected by resource availability, habitat diversity, keystone species, and disturbances.

18.4 Explain how communities are organized into food webs.

18.5 Describe how communities respond to disturbances with resistance, resilience, or switching among alternative stable states.

In Chapter 1, we defined a community as an assemblage of species living together in a particular area. We also discussed how ecologists working at the community level focus on a multitude of species interactions, and how these interactions affect the number of species and the relative population size of each species. In Chapters 14 through 17, we discussed the major types of species interactions. We are now ready to expand our understanding of communities in which these interactions take place. In this first of two chapters on ecological communities, we will examine community structure, which includes the composition of species in a community, the relative abundance of each species, and the relationships among the species. We will start by looking at a classic debate over whether or not communities are distinct entities. We will then examine how ecologists quantify patterns of species diversity in communities based on the number of species and the relative abundance of each species in the community. All these species function as producers and consumers in a *food web,* which helps to determine the number of species, the relative abundance of each species, and the stability of the community.

18.1 Communities can have distinct or gradual boundaries

Our first step is to consider how communities change across the landscape and how they are categorized. As we will see, because many species move between communities, drawing boundaries around a community can be difficult. For example, many birds migrate every spring and fall, and amphibians spend their larval life in aquatic communities and their adult life in terrestrial communities. In this section, we will discuss community boundaries and investigate how communities are affected when community boundaries are either distinct or gradual.

COMMUNITY ZONATION

One of the most noticeable features of communities is that the species composition of the community changes as one moves across the landscape. With the change in environmental conditions, some species become better able to survive and compete. For example, if you were to walk from the base to the top of a mountain in the Santa Catalina Mountains in Arizona, you would observe striking changes in the vegetation, as illustrated in **Figure 18.1**. At the base

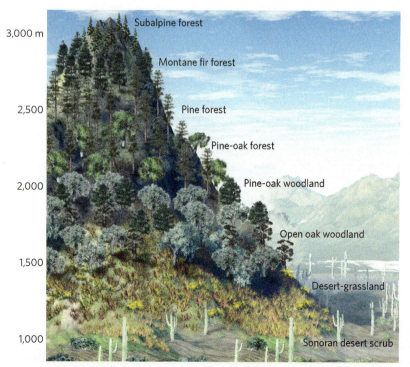

3,000 m — Subalpine forest
— Montane fir forest
2,500 — Pine forest
— Pine-oak forest
2,000 — Pine-oak woodland
— Open oak woodland
1,500 — Desert-grassland
1,000 — Sonoran desert scrub

Figure 18.1 Zones in terrestrial communities. As one climbs the Santa Catalina Mountains in Arizona, the vegetation changes from desert shrubs to large trees.

of the mountains, you would first encounter desert-adapted plants, including rabbit bush (*Franseria deltoidea*) and creosote bush. As you ascend the mountain, grasses, some shrubs such as narrowleaf goldenbush (*Haplopappus laricifolius*), and a few scattered oaks predominate. Approaching the top of the mountain, you reach a forest of ponderosa pines and southwestern white pine (*Pinus strobiformis*), followed by a forest containing Engelmann spruce (*Picea engelmanni*), white fir (*Abies concolor*), and subalpine fir. The zones in which each species flourishes reflect different tolerance ranges for temperature and moisture availability, as well as different abilities to compete with other species for resources. There are similar changes in the animal species that live at different elevations on the mountains. The changes in plants and animals at different elevations create continuous changes in community composition from the base of the mountains to the highest peaks.

Zonation also occurs in aquatic communities, as depicted in **Figure 18.2**. We saw an example of this in Chapter 16 when we discussed competition between species of barnacles on rocky coasts of Great Britain. Due to both competition for space and the need to resist desiccation, stellate barnacles live in the upper intertidal zone, whereas the rock barnacles live in the lower intertidal zone. As another example, we find kelp forests below the intertidal zone, and a variety of seaweeds, mussels, anemones, rock barnacles, and hermit crabs occupy the lower and middle intertidal zones. In the upper intertidal zone, we find limpets, which are a type of gastropod, and stellate barnacles. Higher up on the shore, an area known as the splash zone is home to more limpets and a type of snail known as the periwinkle (*Littorina littorea*). The distribution of species in the different zones of the coastline reflects a combination of tolerance to changing abiotic conditions and the outcome of species interactions that include competition, predation, and herbivory.

CATEGORIZING COMMUNITIES

We characterize communities as we characterize biomes—either by their dominant organisms or by the physical conditions that affect the distribution of species. In North America, for example, we might examine a beech-maple forest community in Ontario, a pine savanna community in Virginia, or a sagebrush community in Wyoming. Each of these communities is named for the dominant plants that are present. In aquatic systems, we focus on physical characteristics—such as a stream community, lake community, or wetland community—or on the dominant group of organisms in that system, such as a coral reef community.

Because it is not realistic to study every species in the community, which could number in the hundreds or thousands, we typically focus on a subset of species that live in an area. For example, a study of a wetland community might

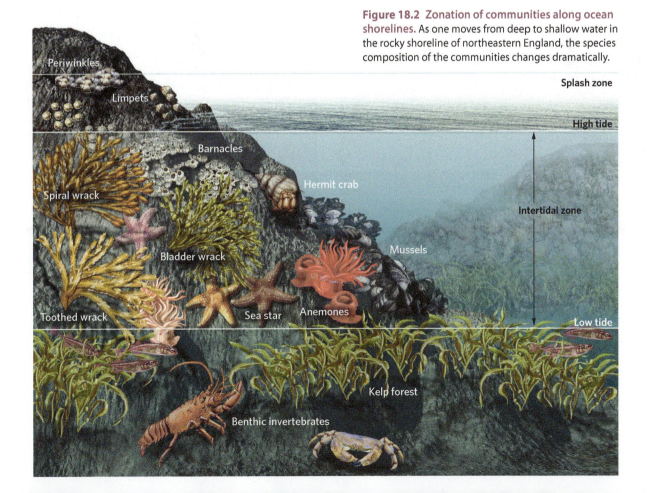

Figure 18.2 Zonation of communities along ocean shorelines. As one moves from deep to shallow water in the rocky shoreline of northeastern England, the species composition of the communities changes dramatically.

Periwinkles

Limpets

Splash zone

High tide

Barnacles

Spiral wrack

Hermit crab

Intertidal zone

Bladder wrack

Mussels

Toothed wrack

Sea star

Anemones

Low tide

Kelp forest

Benthic invertebrates

focus only on bacteria, algae, snails, crustaceans, insects, amphibians, or fish. Several of these groups might be studied simultaneously.

ECOTONES

When we consider the zonation of communities, it is clear that some have distinct boundaries that are either natural or constructed by humans. For example, the natural boundary between a lake community and a forest community can be clearly drawn at the water's edge. This boundary exists because there is a sharp change in the environmental conditions as we leave one community and enter the adjacent community. Similar sharp changes in environmental conditions occur, for example, when there is an abrupt change in the type of soil due to the underlying geology of an area or when an individual moves from north-facing to south-facing slopes of mountains that have large differences in temperatures and moisture. Humans can also create distinct community boundaries. For example, an area of forest that has been cleared for agriculture creates a clear boundary between the field community and the forest community. Sharp changes in environmental conditions over a relatively short distance, accompanied by a major change in the composition of species, creates a boundary known as an **ecotone** (**Figure 18.3**). Although some species move between the adjacent communities that come together to form the ecotone, most species live in one of the communities and spread into the ecotone. As a result, ecotones typically support a large number of species, including those from each of the adjoining habitats and species adapted to the ecotone's special conditions.

One way to document the existence of an ecotone is to use a line-transect survey (see Chapter 11) to determine the abundances of different species along an environmental gradient. When an ecotone is present, we expect to observe sharp changes in the distributions of most species as we leave one community and enter the adjacent community. An example can be found in plant communities on serpentine soils. These soils are derived from underlying rock that contains heavy metals, including nickel, chromium, and magnesium. These heavy metals are toxic to many plants, and the soils are also typically low in nutrients such as nitrogen and phosphorus. Serpentine soils exist in small patches across the landscape in many different parts of the world. Because of the harsh conditions, most plant species cannot live in serpentine soils. Yet some species of plants have evolved an ability to tolerate them and a small portion of the species from adjacent communities also inhabit the ecotone (**Figure 18.4**).

Ecotone A boundary created by sharp changes in environmental conditions over a relatively short distance, accompanied by a major change in the composition of species.

Figure 18.4 Serpentine soils. In areas where serpentine rocks emerge to the soil surface, the soils contain low concentrations of nutrients and high concentrations of metals such as nickel, chromium, and magnesium. At this site in Northern California, the forest abruptly ends where the serpentine soils begin because the trees are not able to survive in the serpentine soils. Photo by Julie Kierstead Nelson.

Figure 18.3 Ecotones. Ecotones are regions where two communities come together with a relatively sharp boundary, indicated by a rapid replacement of species. This ecotone, in Greater Sudbury, Ontario, Canada, occurs where a forest and a field come together. Photo by Don Johnston/AGE Fotostock.

In one study in southern Oregon, researchers conducted a line-transect survey that stretched from nonserpentine soils to serpentine soils. They measured the concentrations of several metals and quantified the presence of numerous plant species. Concentrations of chromium, nickel, and magnesium all increased in the shift from nonserpentine soils to serpentine soils, as you can see in the bottom half of **Figure 18.5**. The top of Figure 18.5 shows that species such as black oak (*Quercus kelloggii*) and poison oak

(*Rhus diversiloba*) did not grow on the serpentine soils, whereas canyon live oak (*Quercus chrysolepis*) and ragwort (*Senecio integerrimus*) were found almost entirely in the ecotone where the two soils come together, and species such as fireweed (*Epilobium minutum*) and knotweed (*Polygonum douglasii*) were found only in serpentine soils. These species show distinct boundaries as one moves across the gradient of nonserpentine to serpentine soils. As you can also see in the transect data, a few species such as hawkweed (*Hieracium albiflorum*) and fescue (*Festuca californica*) were found across the entire gradient. From these data, we can conclude that while some species are largely restricted to either serpentine or nonserpentine soils, the highest number of species occurs within the ecotone.

COMMUNITIES WITH INTERDEPENDENT VERSUS INDEPENDENT SPECIES DISTRIBUTIONS

As we have noted, a community is often described by the dominant species that live within it. Although we know that species interact with each other, for many years ecologists questioned whether species in a community are found together because they depended on each other or simply because they have similar habitat needs.

Interdependent communities are those in which species depend on each other to exist. In the early twentieth century, plant ecologist Frederic Clements proposed that most communities function as interdependent communities and that they act as *superorganisms*. He compared individual species to the different parts of an organism's body, all of which require each other for survival.

Independent communities are those in which the species do not depend on each other to exist. Independent communities are composed of species that live in the same place because they have similar adaptations and habitat requirements.

Although each species has a somewhat different range of conditions under which it can live, both interdependent and dependent communities reflect the overlapping ranges of the species that exist within it. In other words, similar habitat requirements happen to put various species in the same place at the same time. Plant ecologist Henry Gleason rejected Clements's superorganism metaphor and proposed that most communities consist of species with independent distributions.

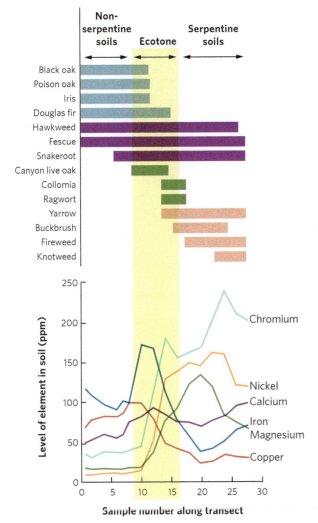

Figure 18.5 A line transect of serpentine and nonserpentine soils. As one moves along the transect, the concentration of metals in the soil exhibits a sharp increase midway along the transect. The presence of plant species along the gradient, as indicated by the bars, demonstrates that some species primarily live in nonserpentine soils (shown in blue), a few species primarily live in the ecotone (shown in green), and several species primarily live in the serpentine soils (shown in pink). Three of the species are able to live across the entire transect (shown in purple). Data from C. D. White, "Vegetation–Soil Chemistry Correlations in Serpentine Ecosystems" (PhD dissertation, University of Oregon, 1971).

Differentiating Between Interdependent and Independent Species Distributions

How can we determine whether a community is made up of an interdependent or an independent group of species?

Interdependent communities Communities in which species depend on each other to exist.

Independent communities Communities in which species do not depend on each other to exist.

One approach has been the use of line-transect studies. If species are interdependent, we should observe suites of species appearing together and disappearing together as we move along a line transect. If species distributions are independent, we should be able to observe gradual changes in species composition as we move along a line transect. Each species will appear and disappear at different points along the line because of the unique habitat requirements of each species.

A classic study using this approach was conducted during the 1950s by Robert Whittaker who surveyed the distributions of plant species in the Great Smoky Mountains on the border between Tennessee and North Carolina (**Figure 18.6a**). At different elevations that varied in temperature and moisture, he calculated the number of stems present for each plant species. As you can see in Figure 18.6b, different species appeared and disappeared at different points along the moisture gradient, and each

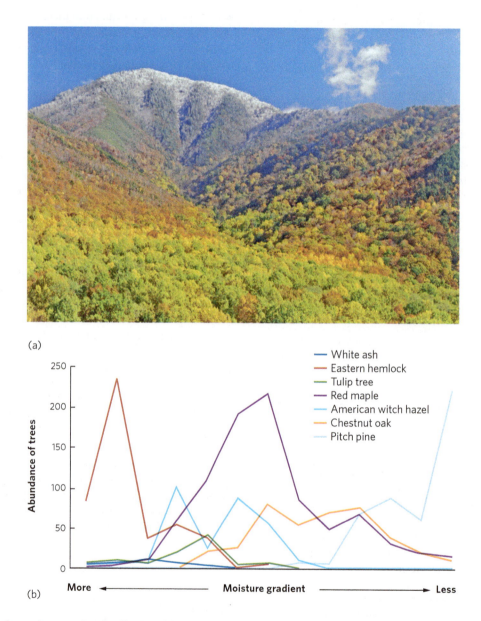

(a)

(b)

Figure 18.6 Independent species distribution. (a) The Great Smoky Mountains exhibit change in plant species composition with changes in elevation and soil moisture. **(b)** When a tree survey was conducted across a moisture gradient within an elevation of 1,067 to 1,372 m, researchers found that each species has its greatest abundance at different points along the moisture gradient. Moreover, each species appears and disappears at different points along the moisture gradient. This led researchers to reject the hypothesis that plant communities are distinct entities in which the existence of one species depends on the presence of one or more other species. Data from R. H. Whittaker, Vegetation of the Great Smoky Mountains, *Ecological Monographs* 26 (1956): 1–80. Photo by Chuck Summers/Contemplative Images.

species achieved its peak abundance at different points. This study and other studies by Whittaker provided strong evidence that when environmental conditions change gradually along a gradient, species in the community commonly show gradual changes in abundance that are independent of each other.

Observing species distributions along environmental gradients is one way to test the independence of species distributions. However, even when we see abrupt changes in species distributions, we cannot be sure whether species in a community are independent or interdependent. An abrupt transition may reflect an abrupt change in abiotic conditions whereby many species simply cannot exist under the changed conditions. For example, fish and other aquatic organisms are not restricted to live in a lake because they are necessarily interdependent; they are restricted to the lake because they lack the adaptations to live on land. Interdependency can be determined by monitoring the results of removing one or more species from the community. If species rely on each other to persist, then removing a species from a community should cause other species in the community to decline. If they don't need each other to persist, then removing a species should leave other species in the community unaffected or even improve their fitness if they happen to compete with the species that is removed.

Interdependent Species Under Harsh Environmental Conditions

Recent studies have found that some species do rely on each other to persist, particularly in communities that experience harsh environmental conditions, such as extreme high or low temperatures or low moisture. For example, a large experiment conducted in 11 locations around the world examined how 115 plant species growing at either low or high elevation in alpine tundra biomes responded when a neighboring species was removed. The researchers measured the percentage of plants that survived as well as the percentage of plants that produced flowers. Removing neighboring plants at low-elevation sites caused an increase in the survival of the remaining plants, as shown in **Figure 18.7a**. This also caused an increase in the percentage of plants that produced flowers or fruit, which you can see in Figure 18.7b. In contrast, removing neighboring plants at high-elevation sites caused a decrease both in the survival of the remaining plants and in the percentage of plants that produced flowers. Plants living under the harsher conditions of high elevations were helped by neighboring species because those species reduced the harsh winds, provided shade, or offered protection from herbivores. Such experiments teach us that while most communities appear to be composed of species with independent distributions, species living under harsh environmental conditions frequently depend on each other.

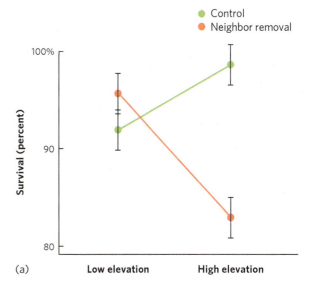

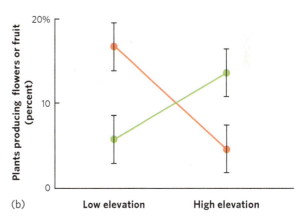

Figure 18.7 Interdependence of species under extreme environmental conditions. In alpine tundra biomes around the world, researchers measured how plants at low and high elevations responded to having neighboring plants removed in terms of **(a)** survival and **(b)** the percentage of plants that produced flowers. At low elevations, the removal of neighbors caused plants to survive better and produce more flowers or fruit, which suggests that the plants are competing. At high elevations, removal of neighbors caused the plants to experience lower survival and produce fewer flowers, which suggests the plants in the community facilitate each other. Error bars are standard errors.

Data from R. M. Callaway et al., Positive interactions among alpine plants increase with stress, *Nature* 417 (2002): 844–848.

CONCEPT CHECK

1. Why do we often observe more species living at an ecotone, than within either of the adjacent communities?
2. What are the underlying causes of community zonation of species distributions?
3. How do we determine whether the composition of species in a community is interdependent or independent?

18.2 Community diversity incorporates both the number and relative abundance of species

To understand the processes that influence the structure and functioning of communities, we need to quantify how communities differ from place to place. In this section, we will examine patterns in **species richness,** which refers to the number of species in a community. We will also examine patterns in the abundance of individuals for each species and quantifying the diversity of species in a community in terms of richness and abundance. This is an important issue because ecologists often want to compare the diversity of species among communities or assess the effects of human activities on individual abundance and the species diversity of a community.

PATTERNS OF ABUNDANCE AMONG SPECIES

Abundance can be examined in both absolute and relative terms. **Relative abundance** is the proportion of individuals in a community represented by each species. When ecologists count the number of individuals of each species in a community, they frequently find that only a few species have low or high abundance, whereas most species have intermediate abundance. In a series of classic papers, Frank Preston developed a way to visualize this pattern of variation in abundance of different species. Preston examined the abundance of different bird species near Westerville, Ohio, and plotted their abundances, as shown in **Figure 18.8a**. He found that only a few species of birds in the community had fewer than 2 individuals or more than 100 individuals; most had 4 to 64 individuals.

To plot the data, Preston used the y-axis to represent the number of species and the x-axis to represent the number of individuals that comprise each species. The key to visualizing the patterns of abundance in communities was to use categories of abundance, for example, <2, 2 to <4, 4 to <8, 8 to <16, and so on. When these categories are plotted on a $\log_2$ scale, the first value in each category translates into 0, 1, 2, 3, and so on. Data plotted in this way produce a normal, or bell-shaped, distribution such that a few species have high abundance, many species have moderate abundance, and a few species have low abundance. A normal, or bell-shaped, distribution that uses a logarithmic scale of the x-axis is a **log-normal distribution**.

Log-normal distributions of species abundances can be found across a wide variety of communities and taxonomic groups. For instance, Robert Whittaker surveyed desert plants and measured the abundance of each species by quantifying the percent of vegetation cover that each species provided. When he plotted percent cover on a $\log_2$ scale, which you can view in Figure 18.8b, he also found a log-normal distribution of abundance. Many studies since Preston's classical work have also observed log-normal distributions of species abundances.

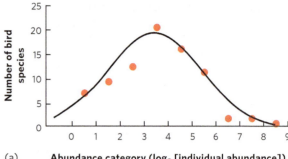

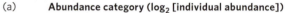

(a)

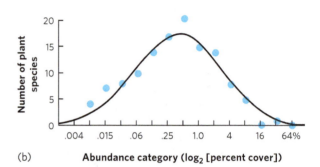

(b)

Figure 18.8 Log-normal distributions of species abundance. Log-normal distributions are found by plotting the abundance categories of each species on a $\log_2$ scale and then plotting the number of species that contain each abundance category. **(a)** A 10-year survey of birds in Westerville, Ohio, found that few species had extremely low or high abundances; most bird species had intermediate abundances. **(b)** A survey of desert plants in Arizona measured abundance using the percent cover of each species, rather than the number of individuals of each species, and found a similar log-normal distribution. Data from (a) F. W. Preston, The canonical distribution of commonness and rarity: Part I, Ecology 43 (1962): 185–215; (b) R. H. Whittaker, Dominance and diversity in land plant communities, Science 147 (1965): 250–260.

RANK-ABUNDANCE CURVES

Another way to visualize the relationship between the number of species and the relative abundance of each species is to use *rank-abundance curves.* **Rank-abundance curves** plot the relative abundance of each species in a community in rank order from the most abundant to the least abundant. Rank-abundance curves are particularly good for illustrating how communities differ in species richness and *species evenness.* **Species evenness** is a comparison of the relative abundance of each species in a community. The greatest

---

Species richness The number of species in a community.

Relative abundance The proportion of individuals in a community represented by each species.

Log-normal distribution A normal, or bell-shaped, distribution that uses a logarithmic scale on the x-axis.

Rank-abundance curve A curve that plots the relative abundance of each species in a community in rank order from the most abundant species to the least abundant species.

Species evenness A comparison of the relative abundance of each species in a community.

evenness occurs when all species in a community have equal abundances, and the lowest evenness occurs when one species is abundant and the remaining species are rare. To plot a rank-abundance curve, we rank each species in terms of its abundance; the most abundant species receives a rank of 1, the next most abundant species receives a rank of 2, and so on.

Consider two hypothetical communities. Community A has five species with relative abundances of 0.5, 0.3, 0.1, 0.06, and 0.04, and Community B has five species with relative abundances of 0.2, 0.2, 0.2, 0.2, and 0.2. From these numbers, we can see that both communities have the same species richness, but Community A has lower species evenness than Community B. The rank-abundance curves of these data are shown in **Figure 18.9a**. Both curves extend equally far to the right, which confirms that they have the same species richness. However, the slope of Community A is considerably steeper than the slope of Community B because the five species in Community A range from a very high relative abundance to a very low relative abundance. In contrast, all five species in Community B have equal relative abundances, which, when plotted, produce a flat line. Rank-abundance curves allow us to determine quickly which communities have greater species richness and evenness.

Rank-abundance curves are good indicators of how communities differ in richness and evenness. For example,

researchers in Brazil recently examined rank-abundance curves for lizards across different forested habitats. They determined the abundance of every lizard species they could find during daytime searches. They ranked each lizard species based on abundance and plotted the ranks against the relative abundance of each species, as shown in Figure 18.9b. As you can see in this figure, the rank-abundance curve for the primary forest extends farther to the right than the curve for lizards in the eucalyptus plantation, which indicates that the primary forest contains many more species of lizards than does the eucalyptus plantation. Moreover, if we consider the slopes of the curves across the ranks that the two communities have in common—from ranks 1 to 5—we see that the primary forest has a steeper slope than the curve for the eucalyptus plantation. This indicates that the lizard community in the primary forest has a lower evenness than the community in the eucalyptus plantation.

CONCEPT CHECK

1. What does the log-normal distribution of species distributions tell us about the abundance of the full suite of species in a community?
2. What does the slope of a rank-abundance curve tell us about the relative abundance of species in a community?
3. What two measures of species diversity are incorporated when calculating indices of species diversity?

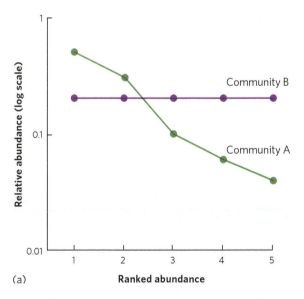

(a)

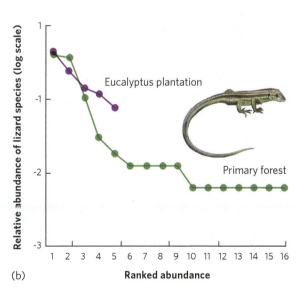

(b)

Figure 18.9 Rank-abundance curves. (a) These rank-abundance curves represent two hypothetical communities with identical species richness of five species. However, the two communities differ in species evenness. Community A has one species with a high relative abundance and other species that have moderate to low abundance, which means this community has low species evenness. In contrast, Community B contains five species that are all equally abundant so there is high evenness. **(b)** These rank-abundance curves for lizards living in Brazil show that the primary forest has a higher species richness and a lower species evenness than the eucalyptus plantation. Data from M. A. Ribiero-Júnior et al., Evaluating the effectiveness of herptofaunal sampling techniques across a gradient of habitat change in a tropical forest landscape, *Journal of Herpetology* 42 (2008): 733–749.

18.3 Species diversity is affected by resources, habitat diversity, keystone species, and disturbance

We have seen that communities differ in the number of species they contain, but we have not yet addressed the more fundamental question of why they differ. On a global scale, there is great variation in the number of species found in different locations. For example, naturalists have known for centuries that more species live in tropical regions than in temperate or boreal zones. We will discuss global patterns of species richness in later chapters and consider the effects of long-term processes such as continental drift, the time that has passed since glaciation, the emergence of new islands, the dispersal of species, and the evolution of new species. In this section, we will focus on how species richness of communities within any given region of the world is affected by the resources available, the diversity of the habitat, the presence of *keystone species,* and the frequency and magnitude of disturbances.

RESOURCES

The number of species in a community can be affected by the amount of available resources. Researchers have examined the effects of resources on species diversity by determining correlations between productivity and species richness in nature. They have also looked at how species richness changes when the productivity of a community is experimentally manipulated.

Natural Patterns of Productivity and Species Richness

For decades, ecologists have examined how the amount of biological productivity correlates with the number of species in communities. To do this, they commonly measure productivity in terms of the biomass of producers or consumers that is generated over time. Among hundreds

---

ANALYZING ECOLOGY

Calculating Species Diversity

Eco TV

macmillanlearning.com/ricklefsvideo

Ecologists agree that communities with more species and greater evenness have higher species diversity. Although rank-abundance curves are a helpful way to visualize differences in richness and evenness among communities, they do not provide a specific value for the species diversity within a community. Over the years, numerous indices of species diversity have been created. We will consider the two that are most common and equally valid: *Simpson's index* and *Shannon's*

index. Both indices incorporate species richness, which we abbreviate as *S,* and evenness. However, they do so in different ways.

To see how we calculate species diversity with either index, we can begin with data from three communities for which we have the absolute abundance for each of the species. From these data, we can then calculate relative abundance for each of the five species in the community, which is denoted as p_i. With these relative abundance data, we can calculate both Simpson's index and Shannon's index.

	The Abundance of Different Mammal Species in Three Communities					
	Community A		**Community B**		**Community C**	
Species	**Absolute Abundance**	**Relative Abundance (P_i)**	**Absolute Abundance**	**Relative Abundance (P_i)**	**Absolute Abundance**	**Relative Abundance (P_i)**
MOUSE	24	0.24	20	0.20	24	0.25
CHIPMUNK	16	0.16	20	0.20	24	0.25
SQUIRREL	8	0.08	20	0.20	24	0.25
SHREW	34	0.34	20	0.20	24	0.25
VOLE	18	0.18	20	0.20	0	0.00
TOTAL ABUNDANCE	**100**	**1.0**	**100**	**1.0**	**100**	**1.0**
S	5	5	5	5	4	4

of studies that have been conducted on animals and plants in both aquatic and terrestrial environments, researchers have found a wide range of patterns, as illustrated in **Figure 18.10a**. In rare cases, researchers have observed a U-shaped curve in which increased productivity is associated with an initial decrease in species richness that is followed by an increase in species richness. In some studies, conducted in different biomes and in different parts of the world, diversity decreases with increasing productivity. In other cases, the correlation is positive; increases in productivity are associated with increases in species richness. In still other studies, there is no relationship between productivity and species richness. Finally, in some studies, the relationship is best described by a hump-shaped curve; initial increases in productivity are associated with an increase in species richness, but further increases in productivity are associated with a decrease in species richness.

To gain a better sense of the relationship between productivity and species richness, a group of researchers compiled data from the hundreds of studies that have been conducted to determine the most common type of correlations. You can view their results in Figure 18.10b. Among studies of vertebrates, a hump-shaped curve was

the most commonly observed relationship in aquatic systems, although hump-shaped and positive relationships were observed with similar frequency in terrestrial systems. For aquatic and terrestrial invertebrates and plants, the most commonly observed relationship was a hump-shaped curve. This reflects the fact that the most productive communities are often dominated by a small number of dominant competitors. In short, although a variety of relationships have been observed in individual studies, the most common relationship across all studies is a hump-shaped curve. This means that a site with medium productivity has a higher species richness than sites with either low or high productivity.

Manipulations of Productivity and Species Richness

Since most communities exhibit a hump-shaped relationship between productivity and species richness, we ought to be able to predict how species richness will be affected if communities experience increased productivity. Numerous experiments have been conducted in communities of plants in which productivity has been manipulated by adding soil nutrients such as nitrogen and phosphorus. As predicted by the hump-shaped curve, the most common observation

---

Simpson's index, a measurement of species diversity, is given by the following formula:

$$\frac{1}{\sum_{i=1}^{s}(p_i)^2}$$

In words, this formula means that we square each of the relative abundance values, sum these squared values, and then take the inverse of this sum. For example, for Community A, Simpson's index of species diversity is

$$\frac{1}{(0.24)^2 + (0.16)^2 + (0.08)^2 + (0.34)^2 + (0.18)^2}$$

$$= \frac{1}{.24} = 4.21$$

Simpson's index can range from a minimum value of 1, which occurs when a community only contains one species, to a maximum value equal to the number of species in the community. This maximum value only occurs when all the species in the community have equal abundances.

Shannon's index (*H'*), also known as the **Shannon–Wiener index,** is another measurement of species diversity given by the following formula:

$$H' = -\sum_{i=1}^{s}(p_i)(\ln p_i)$$

In words, this formula means that we multiply each of the relative abundance values by the natural log of the relative abundance values, sum these products, and then

take the negative of this sum. For example, in Community A, Shannon's index is

$$-[(0.24)(\ln 0.24) + (0.16)(\ln 0.16) + (0.08)(\ln 0.08)$$
$$+ (0.34)(\ln 0.34) + (0.18)(\ln 0.18)]$$

$$-[(-0.34) + (-0.29) + (-0.20) + (-0.37) + (-0.31)]$$
$$= 1.51$$

Shannon's index can range from a minimum value of 0, which represents a community that contains only one species, to a maximum value that is the natural log of the number of species in the community. As we saw with Simpson's index, the maximum species diversity value occurs when all species in the community have the same relative abundances.

YOUR TURN Calculate Simpson's index and Shannon's index for Community B and Community C. Based on your calculations, how does species evenness and species richness affect the values of each index?

Simpson's index A measurement of species diversity, given by the following formula:

$$\frac{1}{\sum_{i=1}^{s}(p_i)^2}$$

Shannon's index (*H'*) A measurement of species diversity, given by the following formula:

$$H' = -\sum_{i=1}^{s}(p_i)(\ln p_i)$$

Also known as the **Shannon–Wiener index.**

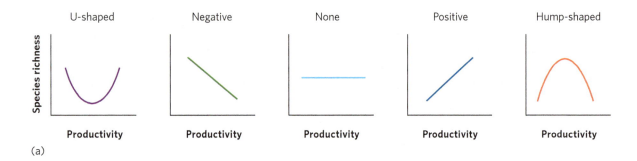

(a)

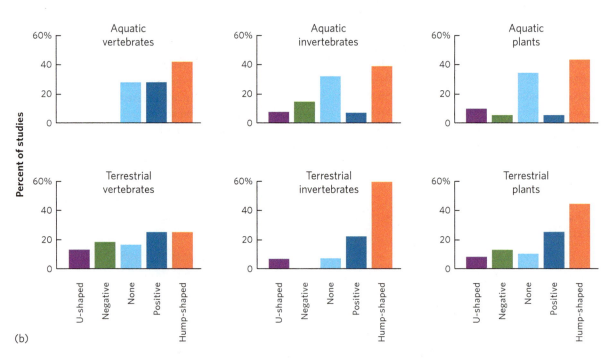

(b)

Figure 18.10 Natural patterns of productivity and species richness. (a) The relationship between productivity and species richness can be described by one of five types of curves. **(b)** Among six categories of organisms, a hump-shaped curve is the most commonly observed relationship between productivity and species richness for aquatic vertebrates, aquatic and terrestrial invertebrates, and aquatic and terrestrial plants.

Data from G. G. Mittelbach et al., What is the observed relationship between species richness and productivity?, *Ecology* 82 (2001): 2381–2396.

in such experiments is that species diversity declines over time at high levels of productivity. In a classic study that took place in England, known as the Park Grass Experiment, researchers in 1856 began adding different types of fertilizers to several plots of grass each year while leaving other plots as unfertilized controls (**Figure 18.11a**). Within 2 years of starting the experiment, species richness in the fertilized plots began to decline. This decline continued for several decades and researchers determined that the amount of decline was related to the number of different nutrients that were added, as shown in Figure 18.11b. The annual addition of nutrients continues today, over 150 years later, which makes the Park Grass Experiment one of the longest-running experiments in the world.

Although the original experiment contained treatments that were not replicated, subsequent experiments with replicated treatments have supported the conclusions of the Park Grass Experiment.

The Park Grass Experiment and others like it have demonstrated that added fertility commonly causes a decline in the species richness of producers such as plants and algae. 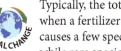 Typically, the total biomass of producers increases when a fertilizer is added, but this added fertilizer causes a few species to dominate the community, while rare species—which are often competitively inferior—begin to decline until they eventually disappear from the community. This observation has important global ramifications because humans continue to add nutrients

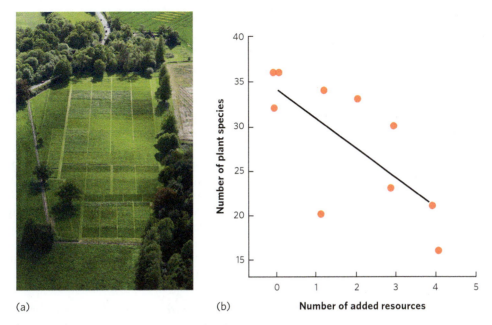

(a)　　　　　　　　　　　　　　(b)

Figure 18.11 Fertility effects on species richness. (a) Researchers of the Park Grass experiment have manipulated the addition of different types of nutrients in grassland plots over time and observed changes in species richness. **(b)** Plots that received more types of resources—such as nitrogen, phosphorus, and micronutrients—experienced a greater decline in species richness. Data from W. S. Harpole and D. Tilman, Grassland species loss resulting from reduced niche dimension, *Nature* 446 (2007): 791–793. Photo by Centre for Bioimaging/Rothamsted Research/Science Source.

inadvertently to aquatic and terrestrial communities through fertilizer runoff and air pollution.

Although we know that species richness declines with increased habitat fertility, the reasons have been unclear. In the case of plant communities, ecologists have hypothesized that the increase in nutrients caused competitively dominant plants to cast more shade on the competitively inferior plants. In one experiment, four manipulations were used, as illustrated in **Figure 18.12**: a control community, a community that received added nutrients, a community that received added light in the plant understory, and a community that received added nutrients and added light in the plant understory. As Figure 18.12 shows, while adding nutrients alone caused a decline in species richness, as expected from past experiments, adding light alone had no effect. However, adding both nutrients to the soils and light to the understory reversed the decline in species richness that was seen with the addition of nutrients alone. These results confirm that soil fertility drives down species richness because it promotes the growth of taller, competitively superior plants that cast shade over less-competitive plants.

HABITAT DIVERSITY

The number of species in a community can also be affected by the diversity of the habitat. Because different habitats provide places to feed and breed for different species, it seems reasonable that communities with a higher diversity of habitats—which should offer more potential niches—will also have a higher diversity of species.

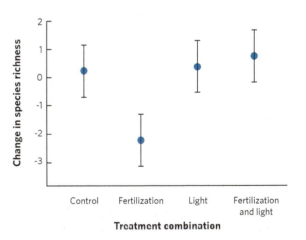

Figure 18.12 Reversing the effects of added nutrients on species richness. When fertilizer alone is added to a grassland community, species richness declines. When light is added under the canopy of the grasses, species richness is not affected. However, when fertilizer is added in combination with light under the canopy, there is no decline in richness. These results suggest that adding fertilizer causes richness to decline because increased fertility favors the tall and competitively dominant plants, which in turn shade the competitively inferior plants. Error bars are 95 percent confidence intervals. Data from Y. Hautier et al., Competition for light causes plant biodiversity loss after eutrophication, *Science* 324 (2009): 636–638.

In a classic study of the role of habitat diversity, Robert and John MacArthur investigated whether there was a relationship between the diversity of habitats among different regions of the United States and Panama and the diversity of birds in each location. To accomplish this, they measured the density of foliage at different heights above the ground in these different regions, from 0.2 to 18.3 m, and calculated the diversity of foliage heights using the Shannon index. Then they surveyed the number of breeding bird species in each area and calculated the diversity of birds, again using the Shannon index. When they graphed the relationship between the two sets of data, as shown in **Figure 18.13**, they found that habitats with greater foliage height diversity supported higher diversity of bird species.

KEYSTONE SPECIES

All species play a role in a community, but some species, known as **keystone species,** substantially affect the structure of communities despite the fact that individuals of that species may not be particularly numerous. The concept of the keystone species is a metaphor that comes from the field of architecture. In arches that are built of stone, the center stone is known as the keystone (**Figure 18.14**). The keystone does not carry much of the arch's mass, but without the keystone

the arch would collapse. Similarly, removing a keystone species can cause a community to collapse.

Keystone species affect communities in a wide variety of ways, some in their roles as predators, parasites, herbivores, or competitors. We saw an example of herbivores acting as a keystone species in Chapter 16 when we discussed an experiment in which researchers sprayed several field plots with an insecticide and left other plots unsprayed (see Figure 16.13). In the unsprayed plots, beetles consumed the competitively superior goldenrod and the competitively inferior plants survived and grew better. In the sprayed plots, the beetles remained rare, and this allowed the goldenrod plants to dominate the community. In this community, the beetle was a keystone species because its presence completely altered the structure of the field community.

A similar scenario exists in the intertidal communities along the coast of Washington State. In a classic experiment,

Keystone species A species that substantially affects the structure of communities despite the fact that individuals of the species might not be particularly numerous.

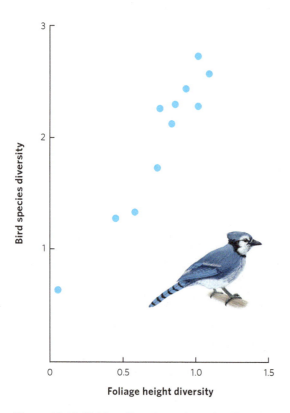

Figure 18.13 Habitat diversity and species diversity. Using the Shannon index, researchers quantified the diversity of foliage height and the diversity of bird species in different locations around the United States and Panama. They found that more diverse habitats contained a greater diversity of birds. Data from R. H. MacArthur and J. W. MacArthur, On bird species diversity, *Ecology* 42 (1961): 594–598.

Figure 18.14 Keystone. A keystone in an arch is the final stone that is inserted to prevent the arch from collapsing. Similarly, a keystone species is one whose removal causes a dramatic change in the community, despite not being particularly abundant. Photo by Nick Hawkes/Alamy.

Robert Paine built cages in the intertidal zone to exclude predatory sea stars (*Pisaster ochraceus*) from feeding on a variety of herbivores, including mussels, barnacles, limpets, and snails. Adjacent plots of similar size were left as uncaged controls. As you can see in **Figure 18.15a**, the control plots experienced little change in species composition from 1963 to 1973. In the caged plots, however, nearly 20 species declined to the point that they were eventually eliminated from the area. In their place, a single species of mussel (*Mytilus californianus*) came to dominate the rock surface. Because the mussels are the superior competitors for space on the rocks, mussels dominate the community when sea stars are absent (Figure 18.15b). When the sea stars are present, they consume large numbers of mussels, and this creates open areas of rock that many species of inferior competitors can colonize (Figure 18.15c). In short, the sea star is a keystone species in the intertidal community.

Keystone species can also affect communities by influencing the structure of a habitat. In such cases, keystone species are sometimes called *ecosystem engineers*. One of the best-known ecosystem engineers is the beaver. Beavers build dams in streams that block the flow of water and cause large ponds to develop (**Figure 18.16**). Because the flowing stream is converted into a nonflowing pond, a different community of plants and animals colonize and persist in beaver ponds than in streams. Similarly, alligators create numerous large

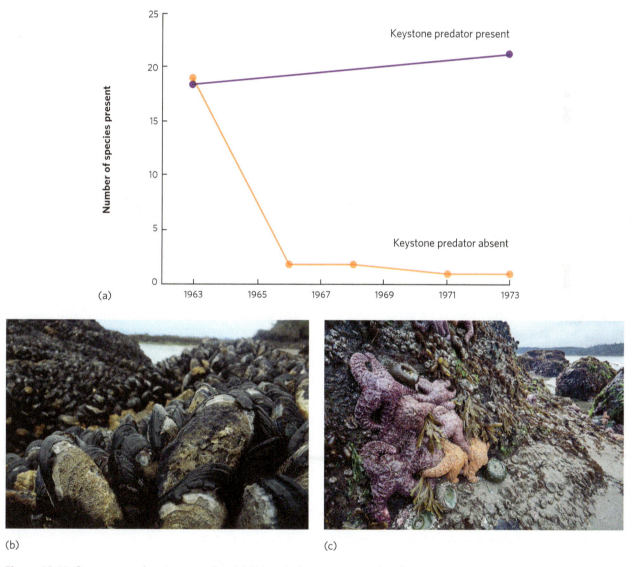

Figure 18.15 Sea stars as a keystone species. (a) Although they are not very abundant, sea stars act as keystone species in intertidal communities along the coast of Washington State because they remove competitively dominant mussels. **(b)** In areas where sea stars were removed, the competitively superior mussels came to dominate the community and caused a decline in the diversity of the competitively inferior species. **(c)** In areas where the sea stars remained, the community retained a high diversity of intertidal species. Data from R. T. Paine, Intertidal community structure: Experimental studies on the relationship between a dominant competitor and its principal predator, *Oecologia* 15 (1974): 93–120. Photos by (b) Jonathan Hucke; (c) Gary Luhm/DanitaDelimont/Newscom/

Figure 18.16 Ecosystem engineers. While beavers are not very abundant, the dams they build, like this one in Grand Teton National Park, Wyoming, have a major effect on community structure. Beaver dams flood terrestrial communities, which kills most terrestrial plants. In addition, the aquatic community changes from species that live in flowing streams to those that live in nonflowing ponds. Photos by (left) Copyright © Braud, Dominique/Animals Animals—All rights reserved; (right) Malcolm Schuyl/FLPA/Newscom.

depressions in wetlands, known as *gator holes,* that many other species use, including fish, insects, crustaceans, birds, and mammals. Although alligators and beavers are not particularly numerous, each of these species has a major effect on its community, which makes it a keystone species.

DISTURBANCES

As we discussed in previous chapters, some species are well adapted to environments that experience frequent and large disturbances, such as hurricanes, fires, or intense herbivory. When environments are rarely disturbed or the disturbances have a low intensity, populations can continue to grow, resources become less abundant, and the ability to compete becomes more important for the persistence of a species. Conversely, frequently disturbed habitats typically support species that are adapted to disturbances. However, when habitats experience disturbances at some intermediate frequency, both types of species can persist and the total number of species can be higher than it would be at either extreme. The **intermediate disturbance hypothesis** tells us that more species are present in a community that experiences occasional disturbances than in a community that experiences frequent or rare disturbances. As depicted in **Figure 18.17a**, when disturbances in a community are of low frequency or intensity, species richness is relatively low. However, when disturbances are moderate in frequency or intensity, species richness is relatively high. When disturbances are high in frequency or intensity, it declines.

Intermediate disturbance hypothesis The hypothesis that more species are present in a community that occasionally experiences disturbances than in a community that experiences frequent or rare disturbances.

The intermediate disturbance hypothesis has been observed in many different communities. One classic study was conducted by Jane Lubchenco, who manipulated the density of herbivorous periwinkle snails in tidal pools to determine how an increase in the magnitude of an herbivory disturbance would affect the species richness of algae in the pools. Her data, shown in Figure 18.17b, reveal a low richness of algal species at low snail densities, a high richness of algal species at moderate snail densities, and a low richness of algal species at high snail densities. Many similar types of studies have been conducted over several decades, and a 2012 analysis of these studies found strong support for the intermediate disturbance hypothesis with respect to species richness.

CONCEPT CHECK

1. Why does an increase in nutrients commonly lead to a decline in plant species diversity?
2. What is the relationship between habitat diversity and species diversity?
3. Why are sea stars considered a keystone species in intertidal communities?

18.4 Communities are organized into food webs

Even the simplest communities are composed of a large number of species. To understand the relationships among the species, it is helpful to categorize species in a community into

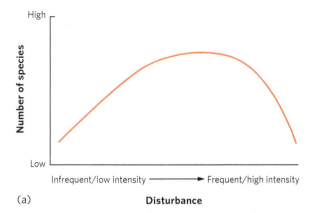

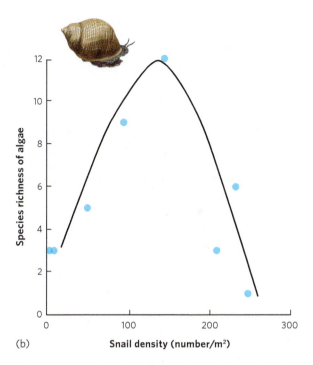

Figure 18.17 The intermediate disturbance hypothesis.
(a) When disturbances are frequent or of high intensity, only those species that are adapted to these conditions can persist. When disturbances are rare or of low intensity, competition becomes more intense and only those species that are well adapted to compete can persist. At intermediate levels of disturbance, species from both extremes can persist, which results in a higher species richness. **(b)** In a study that manipulated snail density to create different intensities of disturbance from herbivory, there was strong support for the intermediate disturbance hypothesis.
Data from J. Lubchenco. Plant species diversity in a marine intertidal community: Importance of herbivore food preference and algal competitive abilities, *American Naturalist* 112 (1978): 23–39.

food chains and food webs. **Food chains** are linear representations of how species in a community consume each other and therefore how they transfer energy and nutrients from one group to another in an ecosystem. Food chains greatly simplify species interactions in a community. In contrast, **food webs** are complex and realistic representations of how species feed on each other in a community and include links among many species of producers, consumers, detritivores, scavengers, and decomposers. We saw an example of this in our opening story of decomposing cacao husks, pollinators, predators, and cacao trees. In constructing food webs, arrows are used to indicate consumption patterns—and therefore the movement of energy and nutrients—between groups. You can see an example of a food web in **Figure 18.18**.

Food webs describe the feeding relationships of ecological communities, which is important because feeding relationships help determine whether a species can exist in a community and whether it will be rare or abundant. The species richness in a food web can often be quite high, which makes it challenging to comprehend how the abundance of one species is affected by other species in the community. To simplify this challenge, ecologists categorize species into *trophic levels*.

TROPHIC LEVELS
You may recall the concepts of *producer* and *consumer* from the overview of basic ecological concepts in Chapter 1.

These represent broad categories of **trophic levels,** which are the levels in a food chain or food web of an ecosystem. All organisms at a particular trophic level obtain their energy in a similar way. Producers are the autotrophs, including algae such as phytoplankton, and plants that convert light energy and CO_2 into carbohydrates through photosynthesis. Producers form the first trophic level of a food web. Consumers can be subdivided into *primary consumers, secondary consumers,* and *tertiary consumers.* **Primary consumers** are those species that eat producers. In the aquatic food web portrayed in Figure 18.18, the primary consumers include zooplankton that eat algae and snails that eat plants. **Secondary consumers** are those species that eat primary consumers. In the lake food web, these secondary consumers include small fish that eat the zooplankton and ducks that eat the snails. Some communities support **tertiary consumers,** which eat secondary consumers; for example, large fish in a lake are tertiary consumers because they consume small fish. In addition to these producers and consumers, food webs also include consumers of dead organic matter, such as scavengers, detritivores, and decomposers. One challenge to placing species into specific trophic levels is that many species are **omnivores,** meaning that they can feed at several trophic levels. For example, crayfish consume algae, which would make them primary consumers, but they also feed on insects and detritus, which would make them secondary consumers and detritivores.

Food chain A linear representation of how different species in a community feed on each other.

Food web A complex and realistic representation of how species feed on each other in a community.

Trophic level A level in a food chain or food web of an ecosystem.

Primary consumer A species that eats producers.

Secondary consumer A species that eats primary consumers.

Tertiary consumer A species that eats secondary consumers.

Omnivore A species that feeds at several trophic levels.

Figure 18.18 Food webs. Food webs capture the full diversity of species and species interactions in the community. In this North American lake, the major producers are the plants along the shoreline and the phytoplankton in the water. These producers are eaten by a variety of primary consumers, as indicated by the arrows that illustrate the flow of energy from an organism to the species consuming it. This food web also includes secondary consumers, such as bluegill sunfish; tertiary consumers, such as the great blue heron; and decomposers, such as bacteria.

Within a given trophic level, we can group species that feed on similar items into **guilds.** For example, a grassland might contain a wide variety of primary consumers that feed on plants in different ways; these consumers can be categorized into guilds of leaf eaters, stem borers, root chewers, nectar sippers, or bud nippers. Members of a guild feed on similar items, but they need not be closely related. In Chapter 16, for example, we discussed how ants and rodents in the deserts of the southwestern United States compete for seeds of the same size. Though ants and rodents are not closely related, their competition for the same seeds would place them in the same guild.

DIRECT VERSUS INDIRECT EFFECTS

As we have seen, a change in the abundance of any one species can affect the abundance of the other species. The simplest way that one species can affect the abundance of another is through a **direct effect,** which occurs when two species interact without involving other species. We have discussed a variety of direct effects in our chapters on predation, parasitism, competition, and mutualism.

Because so many species are interconnected in a food web, the direct effects of one species on another often set off a chain of events that affects still other species in the community. When two species interact in a way that involves one or more intermediate species, it is called an **indirect effect.** When indirect effects are initiated by a predator, it is called a **trophic cascade.** Indirect effects are widespread in ecological communities. For example, in Chapter 14, we discussed how the *Myxoma* virus was introduced onto Macquarie Island to help reduce a population of European rabbits that had been introduced by sailors in the 1800s. As the virus killed the rabbits, there were fewer rabbits remaining to consume the vegetation, which resulted in an increase in the

Guild Within a given trophic level, a group of species that feeds on similar items.

Direct effect An interaction between two species that does not involve other species.

Indirect effect An interaction between two species that involves one or more intermediate species.

Trophic cascade Indirect effects in a community that are initiated by a predator.

island's vegetation. In short, the virus caused a positive indirect effect on the vegetation.

Exploitative competition between two animals might at first glance seem like a direct effect because each species has a negative effect on the other. However, it is actually an indirect effect because the two competitors are interacting with each other by feeding on a common resource. In the case of ants and rodents in the southwestern United States, for instance, the presence of ants causes a reduction in available seeds, and this reduction causes the rodent population to decline. In short, the negative effect of one competitor on another is mediated through the abundance of a third species, which is a shared resource. As a result, exploitative competition is a negative indirect effect.

Indirect effects are common within communities, but sometimes they can occur between adjacent communities. For example, fish prey on aquatic insects, including the aquatic larvae of dragonflies. As a result, ponds containing fish typically have fewer dragonfly larvae. Dragonfly larvae eventually metamorphose into dragonfly adults that consume bees and flies, which are common plant pollinators. You can see this food web in **Figure 18.19a**. When researchers examined the dragonfly populations of ponds with fish and ponds without fish, they confirmed that the ponds without fish had more larval and adult dragonflies. Because of this higher abundance of dragonflies, the ponds without fish also had fewer bees and flies visiting flowering plants along the shoreline, as shown in Figure 18.19b. The scarcity of bees and flies reduced potential seed production in at least one plant species. Thus, a negative direct effect between fish and dragonflies caused a cascade of negative and positive indirect effects on the surrounding terrestrial community.

Ecologists have traditionally assumed that indirect effects occurred only when changes in one species altered the density of another species, such as a trophic cascade caused by predators. However, more recent research has shown that trophic cascades can also be initiated when one species causes changes in the traits of another species. We now look at a variety of indirect effects in more detail.

Density-Mediated Indirect Effects

Indirect effects that are caused by changes in the density of an intermediate species are called **density-mediated indirect effects**. For example, increased densities of sea stars in intertidal communities cause a decline in mussels, which allows other species such as snails to occupy the limited open space on the rocks. The positive indirect effect of sea stars on snails occurs because the sea stars reduce the density of mussels. Similarly, the introduction of *Myxoma* virus reduced the density of rabbits on Macquarie Island, which allowed an increase in the growth of the plants that rabbits eat.

Trait-Mediated Indirect Effects

Communities can also experience **trait-mediated indirect effects,** which are indirect effects that are caused by changes in the traits of an intermediate species. This commonly

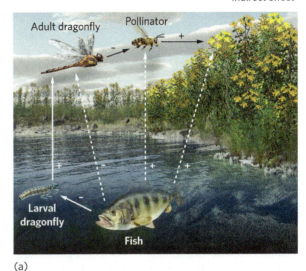

—— Direct effect
- - - - Indirect effect

(a)

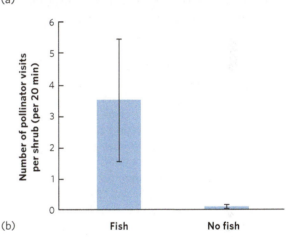

(b)

Figure 18.19 Indirect effects between communities. (a) A food web of species from the aquatic and terrestrial communities includes fish that consume larval dragonflies, adult dragonflies that consume pollinators, and pollinators that pollinate plants. Solid lines represent direct effects. Dashed lines with a minus sign represent a negative indirect effect, and dashed lines with a plus sign represent a positive indirect effect. **(b)** Researchers found that ponds without fish had higher dragonfly populations, which reduced the numbers of pollinators available to visit flowering plants on the surrounding shoreline. Error bars are standard errors.
Data from T. M. Knight et al., Trophic cascades across ecosystems, *Nature* 237 (2005): 880 883.

happens when a predator causes its prey to change its feeding behavior, which in turn alters the amount of food consumed by the prey. For example, the reintroduction of wolves into Yellowstone National Park has caused elk to feed in more protected areas. This behavior alters the amount of plant consumption that occurs in these areas, not because

Density-mediated indirect effect An indirect effect caused by changes in the density of an intermediate species.

Trait-mediated indirect effect An indirect effect caused by changes in the traits of an intermediate species.

the density of elk has changed but because the behavior of the elk has changed.

A similar example can be found in a community of spiders, grasshoppers, and grasses. Grasshoppers are insects that readily consume grasses and serve as prey for some species of spiders. There is more grass present when spiders are present in the community. The increase in grass could occur because spiders eat the grasshoppers and therefore lower the grasshopper's density, or because the spiders scare the grasshoppers so that they hide more and spend less time feeding. To determine which process is operating, researchers conducted an experiment in which they placed cages around small areas of grass in a field and then conducted one of four manipulations by adding: (1) no animals (the control), (2) grasshoppers, (3) grasshoppers and lethal spiders, and (4) grasshoppers and nonlethal spiders. Although the nonlethal spiders could not kill the grasshoppers because their mouthparts had been glued shut, the grasshoppers would still recognize them as potential predators, which should cause grasshoppers to reduce their feeding time.

When the lethal spiders were added, there was less herbivory, as you can see in **Figure 18.20a**. At first glance, this appears to be a density-mediated indirect effect because spiders kill grasshoppers. However, the density of grasshoppers was not affected by the presence of lethal spiders. This is probably because grasshopper reproduction offset spider predation. When the researchers added nonlethal spiders with glued mouthparts, the number of grasshoppers was identical to when lethal spiders were added and there was still reduced herbivory. In short, nonlethal spiders, which can alter the behavioral traits of the grasshoppers but not grasshopper density, had the same positive indirect effect on the grass as lethal spiders, which alter both the density and the traits of grasshoppers. These results illustrate that the mere presence of spiders can alter the behavioral traits of grasshoppers and initiate a trait-mediated indirect effect.

Trait-mediated effects can be far-reaching. For example, grasshoppers that are exposed to the threat of spider predation not only feed less, but their bodies also contain less nitrogen. Because nitrogen is typically a limited resource in soils, as we discussed in Chapter 3, researchers suspected that the amount of nitrogen in grasshopper bodies would affect the rate of soil decomposition when grasshoppers die and become part of the detritus. To test this hypothesis, researchers again raised grasshoppers with no spiders and spiders with glued mouthparts. After being exposed to the two treatments, the grasshoppers were killed and their carcasses were mixed with dead grass. When researchers measured the rate of grass decomposition, they found that it was approximately three times faster with grasshoppers that had been reared without spiders than grasshoppers that had been reared with spiders, as shown in Figure 18.20b. These results demonstrate that

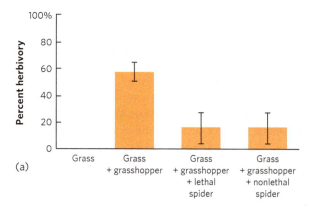

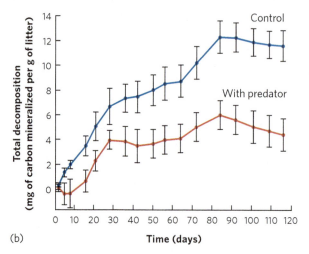

Figure 18.20 Trait-mediated indirect effects. (a) When researchers created simplified communities with only grass and grasshoppers, the amount of grass damage was relatively high. When lethal spiders were added, the grasshoppers caused less damage to the grass, which indicates that spiders had an indirect positive effect on the grass. When spiders with glued mouthparts were added, there was a similar reduction in herbivory because the grasshoppers spent less time feeding. These data confirm that the indirect effect is not mediated by reductions in grasshopper density, but by changes in grasshopper traits. **(b)** After the grasshoppers died, the bodies of grasshoppers raised with predators were lower in nitrogen, which caused the decomposition of the surrounding dead grass to occur at a slower rate. Error bars are standard errors.

Data from A. P. Beckerman, M. Uriarte, and O. J. Schmitz, Experimental evidence for a behavior-mediated trophic cascade in a terrestrial food chain, *Proceedings of National Academy of Science* 94 (1997): 10735–10738; D. Hawlena et al., Fear of predation slows plant-litter decomposition, *Science* 336 (2012): 1434–1438.

trait-mediated effects can have far-reaching consequences throughout communities.

TOP-DOWN AND BOTTOM-UP EFFECTS

The availability of resources as well as the amount of predation and parasitism a species experiences can affect the abundance of species. The same can be said for entire trophic groups. When the abundances of trophic groups in a community are determined by the amount of energy available from the producers in that community, it is referred to as

bottom-up control. When the abundance of trophic groups is determined by the existence of predators at the top of the food web, it is called **top-down control.** You can see both of these concepts illustrated in **Figure 18.21**. If we return to our lake example, we can start with four trophic groups that consist of large fish, small fish, zooplankton, and phytoplankton. If an increase in phytoplankton causes an increase in the zooplankton, small fish, and large fish, the abundance of the trophic groups experiences bottom-up control. If an increase in the abundance of large fish causes a decrease in the small fish, an increase in the zooplankton upon which the small fish feed, and a decrease in phytoplankton, the abundance of the trophic groups experiences top-down control.

For many years, ecologists debated whether communities are more commonly under top-down or bottom-up control. If we think of food webs as having three broad trophic levels, top-down control by predators would reduce the abundance of herbivores and this would result in an abundance of vegetation. In a classic paper published in 1960, Nelson Hairston, Frederick Smith, and Lawrence Slobodkin suggested that because most communities contain an abundance of vegetation, trophic groups must be controlled from the top of the food web.

Their hypothesis caused a great deal of debate among ecologists and inspired many studies of top-down and bottom-up effects. For example, researchers surveyed the abundance of zooplankton and phytoplankton and found that ponds with more phytoplankton also had more zooplankton, which suggested that zooplankton abundance is controlled from the bottom up. However, when they conducted manipulative experiments in which small fish that eat zooplankton were added to the pond, the abundance of zooplankton decreased, which caused a trophic cascade that allowed the phytoplankton to increase. This suggested that the abundance of species in the community was controlled from the top down. Over the past 2 decades, it has become clear that many communities are simultaneously controlled both from the top down by predators and from the bottom up by resources. For example, the number of zooplankton in a lake can be influenced by both the amount of phytoplankton available for consumption and by the number of small fish available to consume the zooplankton.

Bottom-up control When the abundances of trophic groups in nature are determined by the amount of energy available from the producers in a community.

Top-down control When the abundance of trophic groups is determined by the existence of predators at the top of the food web.

CONCEPT CHECK

1. What are the trophic levels of communities?
2. Why is exploitative competition considered to be an indirect effect between two competing species?
3. How can predators cause trait-mediated indirect effects?

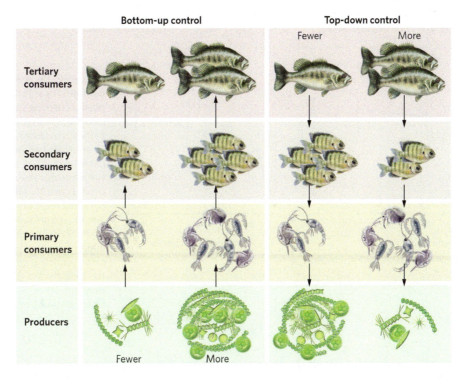

Figure 18.21 Top-down and bottom-up control of communities. When communities are controlled from the bottom up, an increase in the abundance of producers results in an increase in the abundance of higher trophic levels. When communities are controlled from the top down, a trophic cascade occurs from the top down: The trophic level just below the top predator decreases in abundance and the next lower trophic level increases in abundance.

18.5 Communities respond to disturbances with resistance, with resilience, or by switching among alternative stable states

Communities commonly experience disturbances from both natural and human causes. Whatever the reason behind the disturbance, the key question is whether the community will be affected and, if it is, whether it will bounce back or become a substantially different community. To address this question, we need to consider the concepts of *community stability* and *alternative stable states.*

COMMUNITY STABILITY

As we have seen throughout this chapter, communities can experience large changes when the abundances of particular species change. Given enough time, most altered communities can rebound and resemble their original structure of species richness, composition, and relative abundance. The ability of a community to maintain a particular structure is known as **community stability.** There are two aspects of community stability, *community resistance* and *community resilience.* **Community resistance** is a measure of how much a community changes when acted upon by some disturbance, such as the addition or removal of a species. For example, we might measure community resistance after a predator is removed by looking at how much the abundance of herbivores increases. When the impact only causes a small change in the community, we say that the community is resistant to the disturbance. **Community resilience** is the time it takes after a disturbance for a community to return to its original state.

A question of practical importance is whether having a greater variety of species helps communities bounce back faster from disturbances. A study to address this question was conducted in grasslands where researchers manipulated the diversity of plant species in a Minnesota prairie ecosystem by setting up plots containing 1, 2, 4, 8, or 16 species. For 11 years, the abundances of more than 700 species of invertebrate herbivores and invertebrate predators and parasitoids were monitored. During this time, the environment varied considerably, allowing researchers to ask whether increased plant diversity provided greater community stability. In this study, community stability was defined as the amount of year-to-year variation in the abundance and species richness of the herbivores and predators.

The researchers discovered that increasing the number of plant species increased the stability of the communities, as shown in **Figure 18.22**. When they examined the herbivores, shown in Figure 18.22a, they found that an increase in the number of plant species caused an increase in the stability of herbivore richness and herbivore abundance. When they examined the predators and parasitoids, shown in Figure 18.22b, they found that an increase in the number of

---

Community stability The ability of a community to maintain a particular structure.

Community resistance The amount that a community changes when acted upon by some disturbance, such as the addition or removal of a species.

Community resilience The time it takes after a disturbance for a community to return to its original state.

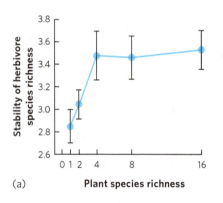

(a)

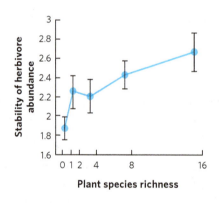

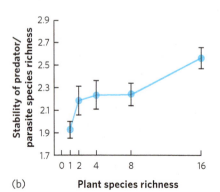

(b)

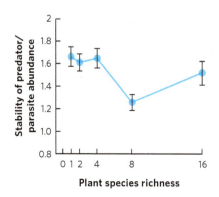

Figure 18.22 Species richness and community stability. Researchers manipulated the species richness of plants and then monitored the stability of the abundance and species richness of invertebrate herbivores, predators, and parasitoids. **(a)** An increase in the diversity of plants caused greater stability of richness and abundance of herbivores. **(b)** An increase in the diversity of plants also caused greater stability of richness of predators and parasitoids but had no effect on the stability of abundance. In all cases, stability is defined as the inverse of the coefficient of variation, which is the standard deviation divided by the mean. **Error bars are standard errors.** Data from N. M. Haddad, Plant diversity and the stability of foodwebs, *Ecology Letters* 14 (2011): 42–46.

plant species caused an increase in the stability of predator and parasitoid richness, but there was no effect on the stability of predator and parasitoid abundance. The underlying reason that plant diversity led to increased stability of higher trophic levels was because communities with high plant diversity provided more consistently available food and habitat for the herbivores, predators, and parasitoids.

ALTERNATIVE STABLE STATES

Sometimes when a stable community is strongly perturbed, it does not bounce back to its original state. Rather, it is disturbed so much that the species composition and relative abundance of populations in the community change. The new community structure is called the **alternative stable state.** For a community to move from one stable state to an alternative stable state typically requires a large disturbance, such as the removal of a keystone species or a dramatic change in the environment.

Moving between alternative stable states happens commonly where the prairie and forest biomes come together in the midwestern United States. During years of abundant rain, fire is suppressed and trees spread out into the prairie. Once trees have become established, they shade out many of the grasses and the grass species have little chance of returning. Moreover, the shade of the trees keeps the ground moist, which further reduces the likelihood of a future fire. During years of drought, however, fires are more common and large fires can kill trees, which favors the spread of grasses where trees once lived. Once grasses establish, it is difficult for trees to move into the prairie because grasses are well adapted to resprouting after fires, whereas trees are not.

Alternative stable states also occur in aquatic environments. For example, intertidal communities on the coast of Maine are typically dominated by the brown rockweed algae (*Ascophyllum nodosum*). During the winter, however, the scouring action of ice can scrape areas free of the rockweed algae, and these areas become dominated by another brown algae known as the bladder wrack (*Fucus vesiculosus*). In this case, the largest disturbances occur when the largest areas are scraped free of organisms by the ice scouring. To determine whether scraping larger areas leads to communities moving to an alternative stable state, researchers simulated ice scouring on the coast by clearing areas of 1, 2, 4, and 8 m². As shown in **Figure 18.23**, small scoured areas did not cause the communities to be dominated by bladder wrack, but large scoured areas did become dominated by bladder wrack. Moreover, the bladder wrack continued to dominate sites for at least 20 years. The switch to an alternative state occurred

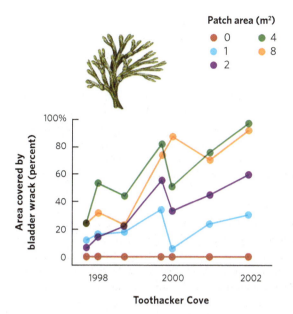

Patch area (m²)

- 0
- 1
- 2
- 4
- 8

Toothacker Cove

Figure 18.23 Alternative stable states. Intertidal communities of Maine are commonly dominated by brown rockweed algae. When researchers scoured areas of different sizes in 1996 and then monitored the colonization of the scoured areas, they found that the largest scoured areas switched to an alternative stable state in which the community came to be dominated by a different species of brown algae known as bladder wrack. Bladder wrack is a superior disperser to new scoured sites, but it is an inferior competitor once the brown rockweed algae eventually arrives. Data from P. S. Petraitis and S. R. Dudgeon, Divergent succession and implications for alternative states on rocky intertidal shores, *Journal of Experimental Marine Biology and Ecology* 326 (2005): 14–26.

because the brown rockweed algae is a poor disperser, so it is slow to reach newly scoured sites compared to bladder wrack. However, brown rockweed algae is a superior competitor for space, so over time it will dominate the disturbed community.

In this chapter, we have seen that communities are composed of large numbers of interacting species that operate within food webs. The interconnected nature of food webs means that changes in the abundance of one species can potentially affect many other species in the community. In "Ecology Today: Connecting the Concepts," we will see that understanding the connectedness of species was a key to discovering the effects of pollution on amphibians.

Alternative stable state When a community is disturbed so much that the species composition and relative abundance of populations in the community change, and the new community structure is resistant to further change.

CONCEPT CHECK

1. Why does higher plant species diversity stabilize the number of herbivore species over time?
2. How is community resilience different from community resistance?
3. How do fires occurring at the transition of grasslands and forests promote alternative stable states?

ECOLOGY TODAY   CONNECTING THE CONCEPTS

Lethal Effects of Pesticides at Nonlethal Concentrations

Leopard frog. Low concentrations of an insecticide that cannot kill the tadpoles of the leopard frog can kill other species in the community. This sets off a chain of events that alters the food web and indirectly prevents the leopard frogs from achieving metamorphosis. Photo by Lee Wilcox.

Appreciating the connectedness of species in a food web can help us understand some of the effects our activities have on ecological communities. Pesticides, for example, provide important benefits in protecting crops and improving human health, and their widespread use has made them common in ecological communities. Maintaining pesticide concentrations below levels lethal to nonpest organisms is the key to ensuring that pesticides harm only pests and not other species. Pesticides are tested on a range of species in the laboratory to determine concentrations that cause death. However, laboratory tests do not consider the fact that species are part of a food web. Therefore, the question is if a pesticide that is not directly lethal to a species can nonetheless have indirect impacts that cause mortality.

Some of the best-studied pesticides are those used to control terrestrial insects that damage crops, as well as insects such as mosquitoes that carry infectious diseases. Of those insecticides, the most commonly applied is malathion, which impairs the nervous system of animals and is highly lethal to insects and other invertebrates. Based on single-species laboratory tests, scientists thought that concentrations found in nature were not lethal to vertebrate animals.

Researchers examined the effects of malathion in large outdoor tanks that contained many of the components of natural wetlands, including algae, zooplankton, and tadpoles (see Figure 1.18b). It had been long known that insecticides such as malathion were highly toxic to zooplankton, which are tiny crustaceans, but little was known about how large reductions in zooplankton might affect other species in the food web.

In one such experiment, researchers added tadpoles of wood frogs and leopard frogs (*R. pipiens*) to tanks and then manipulated malathion in one of four ways: no malathion to serve as a control, 50 parts per billion (ppb) at the beginning of the experiment, 250 ppb at the beginning of the experiment, or 10 ppb once per week. These manipulations allowed the researchers to ask whether single, large applications and multiple, small applications could alter the food web in ways that could affect the amphibians.

Soon after malathion was applied, the zooplankton populations declined to very low numbers in all the tanks treated with the insecticide. This decline was not surprising since zooplankton were known to be quite sensitive to insecticides, but it set off a chain of events that ultimately affected the amphibians. Zooplankton feed on phytoplankton. When malathion was added and the zooplankton were killed, the phytoplankton populations rapidly grew over the next few weeks. This rapid growth, known as an algal bloom, caused the water to turn pea green.

When the water turned green with phytoplankton, sunlight could no longer reach the bottom of the water column, where another group of algae, known as *periphyton*, grows attached to objects such as soil and leaves. The periphyton could not grow very well with less available sunlight, so within a few weeks after the start of phytoplankton bloom, the periphyton declined in abundance, especially in those communities that received 10 ppb of malathion every week. In this treatment, the decimated zooplankton populations could never exhibit resilience because the weekly additions of malathion continually reduced the rebounding populations.

Periphyton is the primary source of food for amphibians; without it, they cannot grow and metamorphose. Fortunately, the wood frogs develop rapidly. In the experiment, wood frogs metamorphosed after only 5 weeks, which meant that they were able to metamorphose before the bloom of phytoplankton caused a decline in their periphyton food source. As a result, wood frog growth and survival were not affected by the insecticide.

The leopard frogs, however, were not so fortunate. Because leopard frogs need 7 to 10 weeks of growth as tadpoles before they can metamorphose, they experienced the full effect of the declining periphyton when the insecticide was present. In fact, the leopard frogs had such stunted growth in the tanks with the weekly addition of 10 ppb of malathion that almost half of them failed to metamorphose before their aquatic environment dried in late summer.

This research demonstrated for the first time that a low concentration of an insecticide that is not directly lethal to amphibians can be indirectly lethal. The insecticide initiated a chain of events that moved through the food web and caused nearly half of the leopard frogs to die. Moreover, the smallest application of the insecticide, which was applied each week, had a much larger effect on the leopard frogs than single applications containing 25 times more insecticide. Quantifying the insecticide's direct effects on the amphibians suggested that the animals would not be harmed, but tests that incorporated a food web approach with indirect effects found that the insecticide could kill nearly half of the animals.

Subsequent experiments have demonstrated that these outcomes are common across a wide range of wetland communities containing zooplankton, algae, and amphibians. However, a new insight arrived when researchers also incorporated aquatic plants into the experiments. Aquatic plants are present in many wetlands, although not in forested wetlands that receive little sunlight due to shading by overhead trees. When researchers examined the effect of adding aquatic plants, they discovered that plants can make malathion-contaminated water safe for zooplankton. The plants accomplish this by elevating the pH of the water (via photosynthesis), which causes the pesticide to break down much more rapidly (i.e., within hours rather than days). As a result, the zooplankton do not die and the lethal chain of events through the food webs no longer occurs.

These results underscore the importance of understanding ecology at the community level. By understanding direct and indirect effects in food webs, we can better understand and predict the impacts of human activities on ecological communities.

(a)

(b)

Effects of an insecticide on an aquatic community. **(a)** In the absence of an insecticide, zooplankton are abundant and they consume phytoplankton. This consumption keeps the water relatively clear and allows sunlight to reach the bottom of a wetland where periphyton grow and serve as food for tadpoles. **(b)** When an insecticide is added, the zooplankton populations are dramatically reduced, which allows an increase in phytoplankton. Abundant phytoplankton cause the water to turn green and prevent sunlight from reaching the periphyton at the bottom of the wetland. With less light available, the periphyton grow poorly, which limits tadpole growth. The number of individuals pictured in each group indicates the relative change in biomass.

SOURCES:

Brogan III, W. R., and R. A. Relyea. 2017. Multiple mitigation mechanisms: Effects of submerged plants on the toxicity of nine insecticides to aquatic animals macrophytes. *Environmental Pollution* 220: 688–695.

Hua, J., and R. A. Relyea. 2012. East Coast versus West Coast: Effects of an insecticide in communities containing different amphibian assemblages. *Freshwater Science* 21: 787–799.

Relyea, R. A., and N. Diecks. 2008. An unexpected chain of events: Lethal effects of pesticides on frogs at sublethal concentrations. *Ecological Applications* 18: 1728–1742.

SUMMARY OF LEARNING OBJECTIVES

18.1 Communities can have distinct or gradual boundaries.
Communities frequently exist in zones that reflect changes in biotic and abiotic conditions. Based on the aggregations of species, ecologists categorize communities in terms of the dominant biological or physical conditions. Although some communities have distinct boundaries as a result of sharp changes in environmental conditions, most communities contain species with geographic ranges that are independent of each other, which produces gradual changes in the composition of communities as one moves across the landscape.

Key Terms: Ecotone, Interdependent communities, Independent communities

18.2 The diversity of a community incorporates both the number and relative abundance of species.
Species richness is the number of species in a community, whereas species evenness is the similarity of relative abundance among species in a community. Rank-abundance curves are graphical representations of richness and evenness. When we plot categories of individual abundance against the number of species that fall within each category, communities commonly exhibit a log-normal distribution, which means that few species have either few or many individuals and that most species have an intermediate number of individuals.

Key Terms: Species richness, Relative abundance, Log-normal distribution, Rank-abundance curve, Species evenness

18.3 Species diversity is affected by resources, habitat diversity, keystone species, and disturbance.
Increases in the abundance of resources and in the number of different resources can alter species diversity. Increases in habitat diversity offer a greater diversity of niches, which favors higher species diversity. Keystone species can alter the composition of species in a community because they have large effects, even though they are not particularly numerous. Disturbances also affect the species diversity of communities, with the greatest diversity occurring in communities experiencing disturbances that are intermediate in frequency or intensity.

Key Terms: Simpson's index, Shannon's index, Keystone species, Intermediate disturbance hypothesis

18.4 Communities are organized into food webs.
Food chains are linear representations of how different species in a community feed on each other, whereas food webs are more complex and realistic representations. Species in a food web can be categorized into trophic levels and guilds according to how they obtain their energy as producers, primary consumers, secondary consumers, and tertiary consumers. Because species exist within food webs, the abundance of each species is affected not only by direct effects but also by density- and trait-mediated indirect effects. Because of these indirect effects, communities can experience trophic cascades from the top of the food chain down, and from the bottom up.

Key Terms: Food chain, Food web, Trophic level, Primary consumer, Secondary consumer, Tertiary consumer, Omnivore, Guild, Direct effect, Indirect effect, Trophic cascade, Density-mediated indirect effect, Trait-mediated indirect effect, Bottom-up control, Top-down control

18.5 Communities respond to disturbances with resistance, resilience, or switching among alternative stable states.
Resistant communities show little or no response to a disturbance. Resilient communities can be affected by a disturbance, but they bounce back to their original states relatively quickly. Some communities respond to large disturbances by moving to alternative stable states in which the community persists with a different composition of species for a relatively long time.

Key Terms: Community stability, Community resistance, Community resilience, Alternative stable state

CRITICAL THINKING QUESTIONS

1. In a desert biome, how could you determine if the distribution of plant species indicates whether the community is independent or interdependent?

2. Given that the distributions of many animals are determined by the species composition of the plant community, what might you predict about the diversity of animals in an ecotone, compared to each adjacent habitat?

3. If you observe zones of different plant species as you move uphill from the edge of a pond, what ecological processes would you hypothesize might underlie these plant distributions?

4. Why do ecologists consider both species richness and species evenness when they quantify species diversity?

5. Compare and contrast log-normal abundance distributions and rank-abundance curves.

6. Why might an increase in total resources or number of added resources lead to declines in species diversity in a lake?

7. In areas experiencing intermediate disturbance, how might species richness and species evenness compare to areas experiencing no disturbances?

8. Rabbits in Australia were once overabundant and decimated the vegetation. Why are the rabbits in Australia not an example of a keystone species?

9. Compare and contrast density- and trait-mediated indirect effects.

10. What evidence would convince you that a grassland was experiencing top-down control versus bottom-up control?

GRAPHING THE DATA Log-Normal Distributions and Rank-Abundance Curves

You have been asked to assess the diversity of species in a grassland plot and you find 16 species that vary in relative abundance.

Using the data in the table, create a rank-abundance curve and a log-normal distribution.

Species Rank	Absolute Abundance	Relative Abundance	Log-Normal Category
1	308	0.385	≥128
2	121	0.151	64 to <128
3	65	0.081	64 to <128
4	62	0.078	32 to <64
5	51	0.064	32 to <64
6	41	0.051	32 to <64
7	31	0.039	16 to <32
8	30	0.038	16 to <32
9	27	0.034	16 to <32
10	18	0.023	16 to <32
11	15	0.019	4 to <8
12	13	0.016	4 to <8
13	8	0.010	4 to <8
14	5	0.006	2 to <4
15	4	0.005	2 to <4
16	1	0.001	<2
TOTAL =	800		

19 Community Succession

Retreating Glaciers in Alaska

In the panhandle of Alaska near Juneau, a stretch of land has undergone incredible changes during the past 200 years. In 1794, Captain George Vancouver found an inlet that headed toward modern-day Juneau. To the north of the inlet was a massive glacier that measured more than 1 km thick and 32 km wide. When the naturalist John Muir visited the site 85 years later, he was shocked to find that the glacier was not located where Captain Vancouver had indicated on his map. Instead, it had receded nearly 80 km and left behind a large bay. Moreover, while most bays in Alaska were heavily forested, the shores of this new bay were relatively barren. As Muir explored, he found several locations with large stumps that were the remnants of hemlock, black cottonwood (*Populus trichocarpa*), and Sitka spruce trees that had been sheared off by the glacier as it advanced centuries ago. Muir had no way of knowing that he was looking at one of the fastest-melting glaciers in the world.

John Muir's writings about the natural world were widely read. In 1916, William Cooper, an ecologist who was particularly intrigued by Muir's stories of Alaska, made the first of several expeditions to the site that came to be known as Glacier Bay. On these expeditions, he surveyed the plants growing at various sites along the shore of the bay, including sites where the glacier existed during Vancouver's visit, sites where the glacier existed during Muir's visit, and sites where the glacier had receded only recently. He also set up quadrats around the bay to record the changes in vegetation at various locations that he planned to check when he returned on subsequent expeditions.

> "When the naturalist John Muir visited the site 85 years later, he was shocked to find that the glacier was not located where Captain Vancouver had indicated on his map."

Cooper reasoned that sites exposed for the longest duration would have the most time to grow back into the kind of forest that had existed before the glaciers had advanced centuries earlier. In contrast, he expected recently exposed sites to be bare rock and gravel, representing the first stages of a developing forest. On these recently exposed sites, Cooper found mosses, lichens, herbs, and low shrubs. Sites that had been exposed for 35 to 45 years had tall species of willow and alder shrubs and black cottonwood trees. Sites older than 100 years contained Sitka spruce trees, and sites older than 160 years had hemlock in the understory. By examining changes in the plant community at sites that had been exposed for different lengths of time, he was able to hypothesize how the forests of Alaska responded to the massive disturbance of an advancing and retreating glacier.

The observations of Vancouver, Muir, and Cooper paved the way for more than a century of subsequent ecological research at Glacier Bay. In fact, the original terrestrial quadrats set up by Cooper a century ago are still monitored today, as are local streams that have been created by the melting glacier. Researchers are also examining the soils and finding that as the plant communities shift from mosses to shrubs to trees, the soil becomes more nutrient rich. As a result, they found that forbs and grasses grow better in soils containing more nutrients. As we will see in this chapter, long-term changes in ecological communities follow predictable patterns, which are important for us to understand the ways in which terrestrial and aquatic communities change over time and the processes that underlie these changes.

Glacier Bay, Alaska. In 1794, a glacier that was more than 1,200 m thick covered all but the inlet of the bay. Since then, the glaciers have melted and retreated, leaving behind a large body of water. Photo by Accent Alaska.com/Alamy.

The rate of glacier retreat. Historic observations by Captain George Vancouver, John Muir, William Cooper, and current researchers indicate that the glacier at Glacier Bay has retreated rapidly during the past 200 years.

SOURCES

Castle, C., et al. 2016. Soil biotic and abiotic controls on plant performance during primary succession in a glacial landscape. *Journal of Ecology* 104: 1555–1565.

Cooper, W. S. 1923. The recent ecological history of Glacier Bay, Alaska II: The present vegetation cycle. *Ecology* 4: 223–246.

Fastie, C. L. 1995. Causes and ecosystem consequences of multiple pathways of primary succession at Glacier Bay, Alaska. *Ecology* 76: 1899–1916.

LEARNING OBJECTIVES

After reading this chapter, you should be able to:

19.1 Discuss how succession occurs in a community.

19.2 Describe the multiple mechanisms by which succession occurs.

19.3 Explain the ways in which succession does not always produce a single climax community.

Because most ecological communities typically experience little change over weeks or months, it is tempting to think of a community as a static collection of species. However, communities commonly experience changes in species composition and relative abundance over longer periods. Such changes in communities over time are especially evident when a community experiences a major disturbance, like the receding glacier at Glacier Bay. Other examples include a field that has been plowed, a forest that has experienced an intense fire, or a pond that has dried during a drought and then refills. In these cases, the community slowly rebuilds and, given sufficient time, it often resembles the original community that existed prior to the disturbance. In this chapter, we will explore how the composition of species in communities changes over time. We will also discuss the evidence that scientists have used to quantify these changes in both terrestrial and aquatic communities and describe how modern studies have altered original hypotheses. Understanding such changes is critical for scientists trying to predict how communities might respond to future environmental disturbances, such as hurricanes and fires, or the results of anthropogenic activity such as logging and mining. Such an understanding is also critical for scientists who try to predict the impacts of larger disturbances such as global climate change.

19.1 Succession occurs in a community when species replace each other over time

The process of **succession** in a community is the change in species composition over time. For example, when a field is plowed but not planted, grasses and wildflowers soon colonize it. In climates with sufficient precipitation, the grasses and wildflowers will eventually be replaced by shrubs and then by large trees. Each stage of community change during the process of succession is known as a **seral stage.**

Succession The process by which the species composition of a community changes over time.

Seral stage Each stage of community change during the process of succession.

The earliest species to arrive at a site are known as **pioneer species.** These species typically have the ability to disperse long distances and arrive quickly at a disturbed site. The final seral stage in this process of succession is known as the **climax community.** A climax community is generally composed of the group of organisms that dominate in a given biome. As we will see, a climax community can be achieved through a variety of different sequences over time. Also, the climax community may continue to experience change.

We begin this section by considering that succession can be observed either directly or indirectly. We will then look at how we examine successional patterns in a variety of terrestrial and aquatic environments. While most studies of succession focus on the changes in plant communities, there are also associated changes in the species of animals that depend on plants for food and habitat. Much less studied is the succession that occurs on dead organic material, such as rotting logs and animal carcasses.

OBSERVING SUCCESSION

Observing succession in a community can take different amounts of time, depending on the life histories of the species involved. For example, the succession of decomposers on a dead animal can happen in a matter of weeks or months. In contrast, the succession of a forest can take hundreds of years. When succession occurs over long time horizons, modern scientists can sometimes return to sites that have been studied by others decades or even centuries

earlier and observe how the species composition of an area has changed over time. In other cases, when there are no historical data for an area, more indirect methods are used to estimate the pattern of succession. As we will see, taking both direct and indirect approaches to examine succession provides the most complete picture of how communities change over time.

Direct Observations

The clearest way to document succession in a community is by direct observation of the changes over time. We have already described the research at Glacier Bay, a well-known example of directly observed succession. Another example comes from the small island of Krakatau, Indonesia.

In 1883, a massive volcanic eruption on Krakatau blew away nearly three-quarters of the island, as depicted in **Figure 19.1a.** The remaining part of the island was covered with a layer of volcanic ash that obliterated all life. Researchers began to visit the island after the eruption to document how and when various species would return and whether the developing community would eventually be similar to the community that had existed on the island before the eruption. By 1886, 24 species of plants had colonized Krakatau, as you can see in Figure 19.1b. Ten of these species were sea-dispersed plants that are common on tropical shores throughout the

---

Pioneer species The earliest species to arrive at a site.

Climax community The final seral stage in the process of succession.

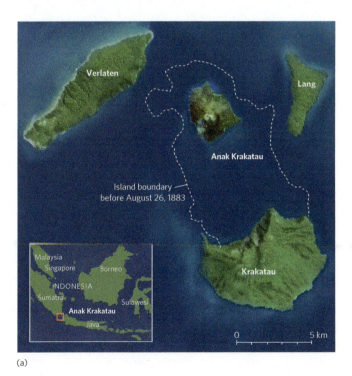

(a)

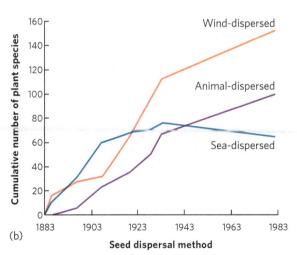

(b)

Figure 19.1 Direct observations of succession. (a) The island of Krakatau in Indonesia experienced a massive volcano in 1883 that blew away much of the original island and obliterated all life. **(b)** Since then researchers have documented the colonization of new plant species on the island over time. Initially, most plants colonizing the island as seeds were sea- and wind-dispersed. As the forests developed, birds and bats that came to the island carried the seeds of many other species of plants. Data from R. J. Whittaker et al., Plant recolonization and vegetation succession on the Krakatau Islands, Indonesia, *Ecological Monographs* 59 (1989): 59–123.

region. Many of the other plants were wind-dispersed grasses and ferns whose seeds and spores had been blown to the island from surrounding islands. With more time, the seeds of trees arrived. By 1920, most of the island had developed into a forest community. The presence of forest habitats made the island more hospitable to many species of birds and bats. Bird and bat species that were fruit eaters brought a variety of seeds with them in their digestive systems. Many of the seeds that were excreted while the birds and bats were on the island subsequently germinated. Today, direct observations continue to monitor how succession on the island is affected by additional disturbances, such as volcanic eruptions, erosion of ash deposits by the ocean waves, and strong storms that pass through the region.

Succession is much easier to observe when it takes place over short timescales, such as the succession of species that occurs on decomposing organisms. For example, when a mammal dies, the dead tissues are rapidly colonized by bacteria and fungi. Within a few minutes, insects also begin to locate the carcass, often attracted by odors that are emitted by the dead animal. One of the first groups of insects to show up are the blowflies; they lay their eggs on the carcass and these hatch into larvae known as maggots (**Figure 19.2**). Given that insects are ectotherms, the rate at which the maggots grow, develop, and consume the carcass depends on the environmental temperature. Over time, more species of flies arrive as well as several species of beetles and predators. Once the tissues are consumed and only bones and hair remain, the number of insects drops sharply and the rate of decomposition slows. Scientists have directly observed such successions for centuries, and today this knowledge is also used to estimate the time of death for human carcasses based on the insects found on a body and the environmental temperature. These observations have formed the basis for the field known as forensic entomology.

Indirect Observations

Because it is a challenge to directly observe succession in many types of communities, researchers have sought ways to determine the patterns of succession indirectly. The two most common methods attempt to look back in time from the present day. One approach examines regional communities that began succession at different times. For example, classic research by Henry Cowles in the late 1800s examined the succession of sand dunes in Indiana along the southern shore of Lake Michigan. Cowles knew that the water level in Lake Michigan had fallen since the last glaciation and that new sand dunes had formed along the edges of the receding shoreline. At the time Cowles visited Lake Michigan, the lake was surrounded by multiple ridges of sand dunes. The oldest dunes were far away from the current shoreline, while the youngest dunes were close to the shoreline. On these younger dunes, he found scattered plants such as beachgrass (*Ammophila breviligulata*) and bluestem grasses that represented the earliest stages of succession, much like the wind-dispersed grasses that appeared on Krakatau during the earliest years of succession. Farther away from the water, older dunes contained larger and more abundant plants that included herbs and several species of shrubs. Beyond these plants, still older dunes had pine trees, while the oldest dunes had beech, oak, maple, and hemlock trees (**Figure 19.3**).

Figure 19.3 Chronosequences. The dunes of Lake Superior increase in age the farther one moves away from the water. Based on this relationship, researchers have examined the plant life on dunes of different ages as a way of estimating how succession proceeds over time. Photo by BambiG/Getty Images.

Figure 19.2. Blowfly. The green bottle fly (*Lucilia sericata*) is a species of blowfly that is among the first groups of insects to arrive when an animal dies. Photo by Howard Marsh /Alamy.

These observations led Cowles to propose a technique called **chronosequencing,** which is used to create a model of the sequence of communities that exist over time at a given location. The model, called a chronosequence, helps ecologists understand how succession has progressed over time in an area. Many other ecologists subsequently used chronosequencing, including William Cooper, a graduate student of Cowles who examined chronosequences in Glacier Bay.

It is also possible to look back in time by examining pollen and other plant parts preserved in distinct layers of lake and pond sediments. Flowering plants produce pollen grains with distinct sizes and shapes. When these pollen grains travel through the air and land on the surface of a lake, they sink and, over time, become preserved in layers of sediment at the bottom of the lake. Researchers can determine the age of each of these layers, by taking a sample that penetrates through many layers of mud on the lake bottom and then using a technique known as *carbon dating* that identifies the age of the pollen in each layer. Dating the pollen helps determine changes in the species of plants around the lake over hundreds or even thousands of years.

SUCCESSION IN TERRESTRIAL ENVIRONMENTS
Researchers investigating terrestrial environments have focused primarily on the succession of plant communities. In terrestrial environments, we can categorize succession into two types that are based on their starting conditions: *primary succession* and *secondary succession*. In both cases, we will begin by discussing a simplified version of terrestrial succession that is represented by an ordered progression over time. Later in this section, we will see that succession in terrestrial communities can be much more complex.

Primary Succession
Primary succession is the development of communities in habitats that are initially devoid of plants and organic soil, such as sand dunes, lava flows, and bare rock. These inhospitable environments are colonized by species, such as lichens and mosses, that require no soil, and that can live on the surfaces of rocks, and drought-tolerant grasses that are able to colonize dry sand dunes (**Figure 19.4**). The species that first colonize these places produce bits of organic matter that combine with the processes of rock weathering and microbial activity to create soils that make the site more hospitable for other species.

Secondary Succession
Secondary succession is the development of communities in habitats that contain no plants but do have organic soil. For example, secondary succession occurs in fields that have been plowed or forests that have been uprooted by a hurricane. Such habitats typically contain well-developed soils. These soils may also include plant roots and seeds, both of which contribute to rapid growth of new plants after the disturbance.

Secondary succession can be observed in a chronosequence of abandoned agricultural fields. An excellent

(a)

(b)

Figure 19.4 Primary succession. Primary succession occurs when organisms colonize sites that are initially devoid of life. **(a)** Bare rocks, such as these rocks in Wisconsin, are initially colonized by lichens and mosses. **(b)** Sand dunes, like these dunes along the shores of Lake Michigan, are initially colonized by grasses able to tolerate dry, sandy soils with little organic matter. Photos by Lee Wilcox.

example of this is Duke Forest, an outdoor laboratory in the Piedmont region of North Carolina composed of 1,900 ha of abandoned agricultural fields and forests that Duke University purchased in 1931. To understand the process of succession in this area, ecologist Henry Oosting visited multiple farm sites, each abandoned for different lengths of time. He found a clear pattern of succession that starts with annual plants and ends with large deciduous trees, as illustrated in **Figure 19.5**. Even before abandonment, crabgrass (*Digitaria sanguinalis*) is common in the fields. In the first summer after abandonment, crabgrass and horseweed dominate the fields. By the second summer, ragweed (*Ambrosia artemisiifolia*) and heath asters (*Symphyotrichum ericoides*) dominate

Chronosequence A sequence of communities that exist over time at a given location.

Primary succession The development of communities in habitats that are initially devoid of plants and organic soil, such as sand dunes, lava flows, and bare rock.

Secondary succession The development of communities in habitats that have been disturbed and include no plants but still contain an organic soil.

| Field:
Crabgrass | Year 1
Crabgrass,
horseweed | Year 2
Ragweed,
heath aster | Years 3-25
Broomsedge,
perennial flowers,
shrubs, pines | Years 25-100
Pine forest,
hardwood
understory | Years 100-200
Remnant pines
with young oak
and hickory trees | Years 200+
Oak-hickory
climax forest |

Figure 19.5 Secondary succession. Secondary succession begins with a well-developed soil. In a temperate seasonal forest, such as this one from the Piedmont region of North Carolina, the climax community is a forest dominated by deciduous trees.

the fields, and by the third summer, broomsedge (*Andropogon virginicus*) dominates. Next come shrubs, followed by pines. Within about 25 years, pines eventually crowd out earlier successional species. As the decades pass, deciduous tree species arrive and start displacing the pines. After about 100 years, deciduous trees dominate and constitute the last seral stage in the successional sequence.

We distinguish between primary and secondary succession by the starting point of the community. For example, in the midwestern United States, an area containing bare rock will pass through seral stages of lichens and mosses, annual weeds, perennial weeds, shrubs, several species of pioneer tree species, and then a beech-maple forest. In contrast, an area containing bare soil will start with annual weeds and then pass through many of the same seral stages on its way to form a beech-maple forest. As we will see later in the chapter, this beech-maple forest can vary somewhat in the composition of dominant tree species.

Sometimes the distinction between primary and secondary succession is unclear because the intensity of a disturbance can vary. For instance, because a tornado that levels a large area of forest usually does not harm soil nutrients and there are often seeds and living roots in the soil, succession follows quickly. In contrast, a severe fire that burns through organic layers of the forest soil requires the community to start over from scratch in a way that resembles primary succession; there is soil, but it contains few seeds or roots that can immediately sprout after the fire.

The Complexity of Terrestrial Succession

A given biome has a characteristic climax community, but the sequence of seral stages through which a single site passes on its way to this climax community can differ, depending on the initial conditions. For example, consider the succession of a field, a sand dune, and a wetland near Lake Michigan in Indiana. The abandoned field is typically colonized by asters

that include horseweed and goldenrod. The sand dune is typically colonized by beachgrass and bluestem grasses, which are perennial grasses that can stabilize the sand dunes and add organic matter to the soil. The wetland supports plants such as cattails, which produce organic matter; over many years, this organic matter can fill in the wetland and create conditions that allow terrestrial plants to colonize the site. In all three cases, the progression of seral stages begins with a different community, but ends with the same climax community of a forest dominated by beech and maple trees.

Although it is tempting to think that terrestrial succession is a simple linear process, this is often not the case. The sequence of seral stages can be quite variable. For example, the concept of chronosequences relies on the assumption that older sites pass through the same stages as the younger sites. It also assumes that sites of different ages do not differ in other aspects, such as historic abiotic conditions, soil fertility, and human or natural disturbances. When chronosequences go back hundreds of years, it can be difficult or impossible to confirm that the sites did not differ in ways that affect succession. For example, sites of similar age in a given area sometimes contain important differences in species composition because of local disturbances, such as a tornado. Modern researchers will observe the changes in species composition, but they may not know that a tornado had passed through the area.

The research at Glacier Bay provides a good example of the complexity of succession. For decades, ecologists presented a simplified scenario of how the communities in Glacier Bay have changed over time. Based on Cooper's original observations, they identified primary succession as a linear process that begins with lichens, mosses, and herbs. Next is a seral stage containing low shrubs, followed by a stage containing tall shrubs that include willows, cottonwoods, and alders. The next stage is dominated by spruce trees, and the final seral stage is dominated by hemlock trees.

More recent research has used tree rings to date the colonization of successional sequences and has reached a different conclusion regarding the path of succession. The number of growth rings tells us the age of the tree. From the tree's age, we can determine when the tree first appeared at a site relative to when the retreating glacier opened the site to succession. For example, we now know that the older sites probably never had a seral stage that included cottonwood trees, whereas the younger sites currently pass through a stage of abundant cottonwoods. Furthermore, spruce and hemlock trees rapidly colonized sites that were exposed in the early 1800s, but not sites that were exposed in the late 1800s and early 1900s. You can view the data for Sitka spruce trees in **Figure 19.6**. This suggests that the soils in the older and younger sites may not have experienced the same changes as they underwent succession.

Because the assumptions underlying chronosequences may not always be supported, the best approach is to use a combination of data and research methods, including chronosequences, pollen records, and long-term studies of single locations undergoing succession. Collectively, these studies provide the most accurate description of terrestrial succession.

Animal Succession

Ecologists have traditionally focused on changes in plant species when describing succession in terrestrial environments. However, the changes in the plant community cause substantial changes in the habitats that are available to animals, which in turn cause changes in the animal community. For example, a classic study by David Johnston and Eugene Odum examined the distribution of bird species in the Piedmont region of Georgia, along the same successional seral stages that Oosting surveyed in the Piedmont region of North Carolina. As illustrated in **Figure 19.7**, grasshopper sparrows (*Ammodramus savannarum*) and eastern meadowlarks (*Sturnella magna*) dominate the early-succession stages that contain annual plants. With the colonization of shrubs into the abandoned fields comes the arrival of many different bird species, including field sparrows (*Spizella pusilla*) and yellowthroats (*Geothlypis trichas*). As we move from the pine forest to the oak-hickory climax forest, other species appear that include the red-eyed vireo (*Vireo olivaceus*) and the wood thrush (*Hylocichla mustelina*). Although some species of birds specialize on a narrow range of the plant successional stages, many species inhabit multiple seral stages.

SUCCESSION IN AQUATIC ENVIRONMENTS

Succession also occurs in aquatic environments. In this section, we will examine how succession proceeds in three types of aquatic environments: intertidal zones, streams, and ponds.

Intertidal Communities

In contrast to most terrestrial communities, succession in intertidal communities can occur much more quickly, in part because the generation time of the dominant species

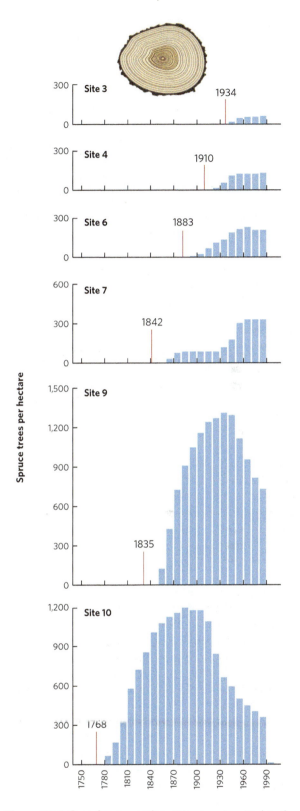

Figure 19.6 Complex succession. Using tree rings to date the Sitka spruce trees from different sites in Glacier Bay, we can see that older sites experienced rapid colonization by spruce trees after the glaciers retreated, whereas the younger sites experienced slow colonization by spruce trees after the glaciers retreated. Vertical red lines indicate the approximate year of glacial retreat at each site.

Data from C. L. Fastie, Causes and ecosystem consequences of multiple pathways of primary succession at Glacier Bay, Alaska, *Ecology* 76 (1995): 1899–1916.

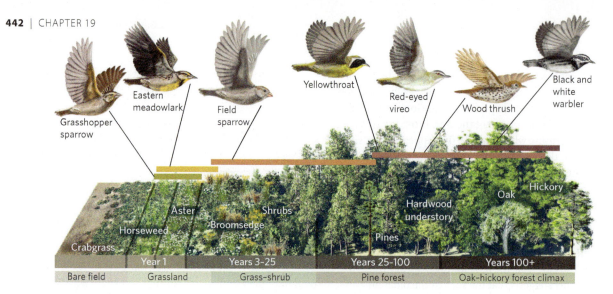

Figure 19.7 Animal succession. As the plant community experiences secondary succession, the habitat available to birds changes. In response to changes in the habitat, the species of birds that live in the community change. Data from D. W. Johnston and E. P. Odum, Breeding bird populations in relation to plant succession on the Piedmont of Georgia, *Ecology* 37 (1956): 50–62.

is much shorter. In intertidal communities, powerful waves that occur during storms commonly remove organisms that are attached to boulders. For example, ecologist Wayne Sousa examined the succession of different species of algae on boulders of the intertidal zone in a classic experiment

in Southern California. He examined some boulders that were not overturned by storms and others that had been overturned and had areas of bare rock exposed that could be colonized by algae. As shown in **Figure 19.8**, Sousa observed that the first species to arrive was a green alga known as

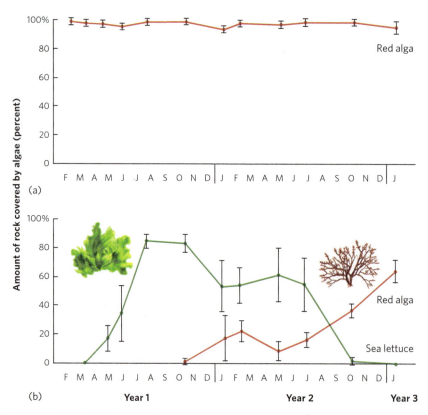

Figure 19.8 Intertidal succession. (a) On undisturbed boulders in the intertidal zone, communities are dominated by a species of red alga, shown in the drawing on the left. **(b)** When strong waves overturn boulders, bare areas of rock are exposed and quickly colonized by sea lettuce, shown in the drawing on the right. Sea lettuce initially dominates the bare rock surface, as measured by the percent of the rock covered by the algae. As time passes, crabs begin to consume the sea lettuce, which creates space that allows the red algae to colonize the rock and ultimately dominate the boulder's surface. Error bars are standard errors. Data from W. P. Sousa, Experimental investigations of disturbance and ecological succession in a rocky intertidal algal community, *Ecological Monographs* 49 (1979): 227–254.

sea lettuce (*Ulva lactuca*). Over the next year, sea lettuce came to dominate the rocky habitat and largely prevented a competing species of red alga, *Gigartina canaliculata,* from colonizing. However, as the sea lettuce became more dominant, it attracted crabs that eat it, which cleared areas on the boulders for the less edible red algae to colonize. Over time, red algae came to dominate the community.

Stream Succession

Like intertidal habitats, streams also experience rapid succession largely because organisms can move downstream from sites that are less disturbed. Streams can experience major disturbances during heavy rainfall that increases both the volume of water the streams carry and the speed at which the water moves. With the greater speed of the rushing water, sand and rocks can tumble downstream and wipe out most plants, animals, and algae. In one study, researchers looked at the effects of a flood event in Sycamore Creek in Arizona. The floodwater scoured the creek and eliminated nearly all of the algae and 98 percent of the invertebrates, leaving behind rocks and bare sand. The researchers then monitored how the community changed over the subsequent 2 months; you can view their data in **Figure 19.9**. Within just a few days after the flood, the stream became colonized with several species of algae known as diatoms. Within 5 days, the diatoms covered nearly 50 percent of the creek bottom, and after 13 days, the diatoms covered nearly 100 percent. After 3 weeks, cyanobacteria started to colonize the stream, followed by a species of filamentous green

alga (*Cladophora glomerata*) along with associated diatoms that live as epiphytes on this green alga. As the three types of algae rebounded, adult insects from the surrounding terrestrial environment began to lay eggs in the stream. This returned a diverse group of larval insect species to the stream, which remained there until they metamorphosed into terrestrial adults.

Lake Succession

For decades, ecologists have explained pond and lake succession using a paradigm of slow transformation, as depicted in **Figure 19.10a**. If we imagine a pond or shallow lake created by a receding glacier thousands of years ago or a beaver dam converting a stream into a pond, we start with a basin full of water. Over time, the erosion of soil and the growth and death of organisms in the lake form sediments that gradually fill the basin. In addition, plants living along the shoreline slowly extend themselves out into the water to form a floating mat of vegetation. Underneath this live mat of vegetation is an accumulating layer of partially decomposed vegetation known as *peat*. Eventually, the entire basin has a floating mat of vegetation that continually contributes detritus to the peat layer in the basin. Microbial decomposition of the dead vegetation is slow because the water underneath the floating mat has little oxygen. As a result, detritus accumulates on the bottom of the basin, and over hundreds or thousands of years, the pond or lake becomes a bog. In North American bogs, sphagnum moss, sedges, and shrubs—such as

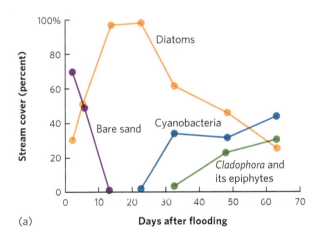

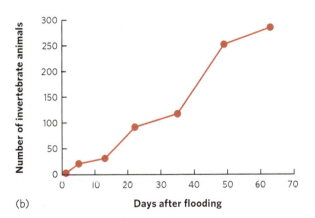

Figure 19.9 Stream succession. After a major flood event, Sycamore Creek in Arizona experienced a nearly complete elimination of algae, leaving behind only rocks and bare sand. **(a)** Within just a few days, several species of diatoms came to dominate the bare sand and rocks in the stream. Later in the summer, other groups of algae colonized the stream and became more abundant, including cyanobacteria and a species of green filamentous alga (*Cladophora*) and its epiphytes. **(b)** As the different types of algae returned, adult flying insects from the terrestrial environment began to lay their eggs in the stream. As a result, the stream experienced a rapid increase in the number of invertebrate animals. Data from S. G. Fisher et al., Temporal succession in a desert stream ecosystem following flash flooding, *Ecological Monographs* 52 (1979): 93–110.

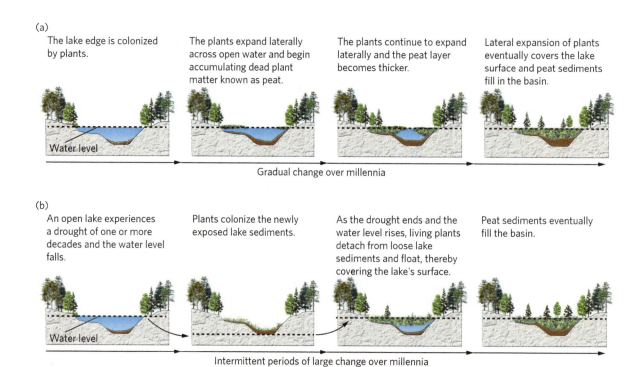

(a)

The lake edge is colonized by plants.

The plants expand laterally across open water and begin accumulating dead plant matter known as peat.

The plants continue to expand laterally and the peat layer becomes thicker.

Lateral expansion of plants eventually covers the lake surface and peat sediments fill in the basin.

Water level

Gradual change over millennia

(b)

An open lake experiences a drought of one or more decades and the water level falls.

Plants colonize the newly exposed lake sediments.

As the drought ends and the water level rises, living plants detach from loose lake sediments and float, thereby covering the lake's surface.

Peat sediments eventually fill the basin.

Water level

Intermittent periods of large change over millennia

Figure 19.10 Succession in shallow lakes and ponds. (a) The classical explanation for succession in these habitats describes a gradual and steady accumulation of organic matter that eventually fills in the basin and converts it to a terrestrial habitat. **(b)** More recent studies have demonstrated that the process can occur in occasional large bursts when multi-year droughts allow vegetation to extend along the dry part of the basin. When water becomes abundant again, the extended vegetation floats on the water's surface and grows in thickness. Multiple drought events allow the vegetation to expand; eventually, it covers the water's surface and fills in the basin. Based on A. W. Ireland, Drought as a trigger for rapid state shifts in kettle ecosystems: Implications for ecosystem responses to climate change, *Wetlands* 22 (2012): 989–1000.

leatherleaf and cranberry—become established along the edges and add to the development of a soil with progressively more terrestrial qualities. At the edges of the bog, shrubs may be followed by black spruce (*Picea mariana*) and tamarack trees (*Larix laricina*), which eventually give way to birch, maple, and fir trees, depending on the locality. In short, the classic explanation for lake succession was that it happened very slowly over long periods of time.

Recently, researchers proposed a new model of pond and lake succession. This model is based on carbon dating of core samples of a bog to determine when each plant species lived in the area. Contrary to the classic model of slow and continuous succession, the study illustrated that ponds and lakes can experience long periods of several hundred years in which little succession occurs, followed by brief episodes of rapid change. During times of prolonged drought that last for a decade or more, water levels drop and vegetation grows down onto the newly exposed shoreline. As you can see in Figure 19.10b, when the drought ends and the water level rises again, the mat of vegetation releases its hold from the lake bottom and

floats up to the water's surface. With continued growth of the vegetation, this floating mat becomes thicker and deposits dead organic matter into the water below it. For example, after studying a 16-ha bog in Pennsylvania, it was estimated that 50 percent of the former lake became covered by bog vegetation in just a few decades during a severe drought in the late sixteenth century. In short, lake succession does not always have to be slow and steady; it can happen in brief bursts during rare, but prolonged, periods of drought.

CHANGE IN SPECIES DIVERSITY

Across both terrestrial and aquatic habitats, the process of succession exhibits consistent effects on species richness. In most cases of succession, we begin with few or no species and then species richness increases rapidly at first. This is followed by a plateau and a small decline, as shown in **Figure 19.11**. Oosting's survey of the Duke Forest, for example, found a rapid increase in the species richness of woody plants during the first 25 years and then a gradual decline in the rate of increase over the next 125 years. Similar patterns of richness over time can be

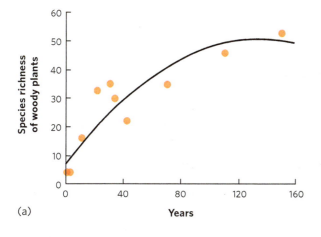

(a)

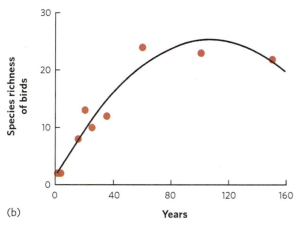

(b)

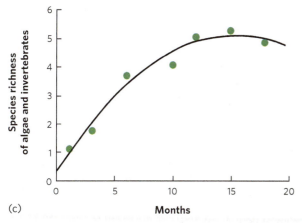

(c)

Figure 19.11 Succession effects on species richness. Across a variety of communities, succession is associated with a rapid increase in species richness that slows over time and eventually plateaus. Examples shown here include **(a)** woody plants in old fields in North Carolina, **(b)** birds in old fields in Georgia, and **(c)** algae and invertebrates from the intertidal boulders of southern California. Data from (a) H. J. Oosting, An ecological analysis of the plant communities of Piedmont, North Carolina, *American Midland Naturalist* 28 (1942): 1–126; (b) D. W. Johnston and E. P. Odum, Breeding bird populations in relation to plant succession on the Piedmont of Georgia, *Ecology* 37 (1956): 50–62; (c) W. P. Sousa, Disturbance in marine intertidal boulder fields: The nonequilibrium maintenance of species diversity, *Ecology* 60 (1979): 1225–1239.

seen in the number of bird species observed by Johnston and Odum in a forest community, and the number of algae and invertebrates observed by Sousa in an intertidal community.

CONCEPT CHECK

1. What is the fundamental difference in primary and secondary succession?
2. Why is it difficult to directly observe succession in most ecological communities?
3. How are chronosequences used to understand succession?

19.2 Succession can occur through different mechanisms

Now that we have an idea of what succession looks like across a variety of communities, we can explore how succession actually occurs. In this section, we consider the traits of early- and late-succession species, compare the different mechanisms of succession, and then examine succession studies to determine which mechanisms are common in ecological communities.

TRAITS OF EARLY- VERSUS LATE-SUCCESSION SPECIES

Early-succession and late-succession species possess different traits important to their performance. For example, pioneer species of terrestrial plants are typically better at dispersing seeds to newly created or disturbed sites. They produce many small seeds that are easily dispersed by wind or that stick to passing animals. These seeds can also persist in the soil for years and then germinate when a disturbance occurs. When they do germinate, these early-succession plants invest more resources into their shoots than their roots and grow rapidly and reproduce quickly. They are also typically quite tolerant of the harsh abiotic conditions that can exist in newly disturbed sites, including full sun and widely fluctuating temperatures and water availability. However, they are not tolerant of the high-shade conditions of late-succession plant communities.

In contrast, climax species produce a relatively small number of large seeds that disperse poorly; some simply drop to the ground, whereas others are consumed by animals. The seeds have a relatively short viability, and once they germinate, they grow slowly, but their shade tolerance as seedlings and their large size as mature plants give them a competitive edge over early-succession species.

As succession progresses, we see a shift in the balance between adaptations that promote dispersal, rapid growth, and early reproduction, and adaptations that enhance

ANALYZING ECOLOGY

Quantifying Community Similarity

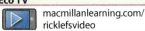
When ecologists examine the species living in different communities, such as when examining chronosequences, they often quantify species abundance and richness. Although such data tell us about the species living in each community, they do not provide a measure of comparison between communities. To address this need, we use several indices of community similarity that can range from zero to one; a value of zero indicates that two communities have no species in common, whereas a value of one indicates that two communities have an identical composition of species.

One of the most common ways to quantify similarity is Jaccard's index of similarity, developed by the Swiss botanist Paul Jaccard in 1901. Jaccard's index is calculated using the following equation:

$$J = \frac{X}{A + B + X}$$

where A represents the number of species that are *only* present in Community A, B represents the number of species that are *only* present in Community B, and X represents the number of species present in *both* communities. For example, consider the table below that lists the species of fish found in each of three stream communities that are at different stages of succession.

We can now use Jaccard's index to calculate the similarity between Community A and Community B:

$$J = \frac{X}{A + B + X}$$
$$J = \frac{3}{1 + 4 + 3}$$
$$J = 0.33$$

This value indicates that there is relatively low similarity in the species composition of Communities A and B.

YOUR TURN Use Jaccard's index to calculate the similarity between Communities A and C and between Communities B and C. Based on these calculations, which communities are the most similar to each other?

Fish Species	Stream Community A	Stream Community B	Stream Community C
Rainbow Trout	X	X	
Brook Trout		X	X
Brown Trout	X	X	
Mudminnow			X
Common Shiner	X	X	
Creek Chub	X		
White Sucker		X	X
Johnny Darter		X	X
Smallmouth Bass			X
Mottled Sculpin		X	X

competitive ability. **Table 19.1** summarizes the traits of early-succession and late-succession plants.

The traits of early-succession and late-succession plants are different because they face inherent trade-offs, similar to the life history trade-offs that we discussed in Chapter 8. To test for trade-offs between early-succession and late-succession species, researchers examined the mass of a single seed from each of nine species of trees. They raised the seeds under low-light conditions to simulate the high amount of shade in mature forests. After 12 weeks, they examined the mortality of the germinated seedlings, which you can view in **Figure 19.12**. The large-seeded species, which are common in old forests, had low rates of mortality under low-light conditions since large seeds provide their seedlings with ample nutrients to get started in the light-limited environment of the forest floor. In contrast, small-seeded species, which are common pioneer species that disperse well in large numbers, had high rates of mortality under low-light conditions. As a result, pioneer species cannot establish in mature forests.

FACILITATION, INHIBITION, AND TOLERANCE

The ability to disperse and the ability to persist under existing abiotic and biotic conditions determine which species will appear in different seral stages during succession. Organisms that disperse well and grow rapidly have an initial advantage and therefore dominate the early stages of succession. Species that disperse slowly or grow slowly once they colonize an area typically become established later in succession. Early-succession species can also modify the environment in ways that affect whether late-succession species can become

TABLE 19.1

General Characteristics of Early-Succession and Late-Succession Plants

Trait	Early Succession/ Pioneer Species	Late Succession/ Climax Species
Number of seeds	Many	Few
Seed size	Small	Large
Mode of dispersal	Wind or stuck to animals	Gravity or eaten by animals
Seed viability	Long	Short
Root:shoot ratio	Low	High
Growth rate	Fast	Slow
Size at maturity	Small	Large
Shade tolerance	Low	High

Figure 19.13 Bryozoans. Bryozoans, such as these colorful individuals from Australia, are tiny invertebrate animals that live attached to rocks in the ocean. If they colonize the rocks first, they can prevent colonization by tunicates and sponges, which compete for space on the rocks. Photo by © Gary Bell/OceanwideImages.com.

established. We must therefore consider whether a species has a positive, negative, or neutral effect on the probability of a second species becoming established. The mechanisms can be categorized as *facilitation, inhibition,* and *tolerance.*

Facilitation is a mechanism of succession in which the presence of one species increases the probability that a second species can become established. Early-succession species do this by altering the environmental conditions of the site in a manner that makes it more suitable for other species to establish and less suitable for themselves. For example, alder shrubs, which are legumes, live in a mutualistic relationship with nitrogen-fixing bacteria in their roots, as we discussed in Chapter 17. This relationship produces additional

nitrogen in the soil, which facilitates the establishment of nitrogen-limited plants such as spruce trees. Over time, the spruce trees eventually grow tall and cast a deep shade that is not a favorable environment for the alder shrubs.

Inhibition is a mechanism of succession in which one species decreases the probability that a second species will become established. Common causes of inhibition include competition, predation, and parasitism. That is, individuals of one species can inhibit those of other species by outcompeting them for resources, eating them, or attacking them with noxious chemicals or antagonistic behavior. Early in succession, inhibition can prevent movement toward a climax community, while late in succession, inhibition can prevent the pioneer species from colonizing and surviving. For example, in a mature forest in the northeastern United States, adult maple and beech trees cast a deep shade that prevents pioneer tree species from surviving.

When inhibition occurs in a seral stage, the outcome of an interaction between two species depends on which species becomes established first. The arrival of one species at a site that affects the subsequent colonization of other species is known as a **priority effect.** We can see an example of the priority effect in the subtidal habitats of South Australia with bryozoans, a group of tiny invertebrate animals that live in colonies attached to rocks and that feed by filtering the water (**Figure 19.13**). If bryozoans become established first, they can prevent the establishment of tunicates and sponges—two other groups of filter-feeding animals that attach themselves to rocks. Sometimes the priority effect occurs because the first species to arrive has grown to a competitively superior adult stage, whereas the second species to arrive is in the competitively inferior immature stage. For example, if a beech tree gets

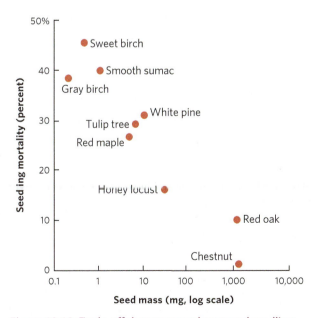

Figure 19.12 Trade-offs between seed mass and seedling mortality under high-shade conditions. Species with large seeds, which are common in late-succession forests, produce seedlings that survive well in high shade. Species with small seeds, which are common in early-succession forests, produce seedlings that have a poor survival rate in high shade. Data from J. P. Grime and D. W. Jeffrey, Seedling establishment in vertical gradients of sunlight, *Journal of Ecology* 53 (1965): 621–642.

Facilitation A mechanism of succession in which the presence of one species increases the probability that a second species can become established.

Inhibition A mechanism of succession in which one species decreases the probability that a second species will become established.

Priority effect When the arrival of one species at a site affects the subsequent colonization of other species.

established in a forest and grows to an adult stage, it shades the ground below, preventing the seedlings of most other tree species from conducting sufficient photosynthesis to survive. In short, the mechanism of inhibition can make the path of succession dependent on which species arrives at the site first.

Tolerance is a mechanism of succession in which the probability that a species can become established depends on its dispersal ability and its ability to persist under the physical conditions of the environment. For example, species that can tolerate the stressful environmental conditions of early succession—such as low moisture or more extreme temperature fluctuations—can become established quickly and dominate early stages of succession. Similarly, plants that can tolerate high-shade environments can become established in forests that cast deep shade. These species do not alter the environment in ways that either help or inhibit other species, but once a stress-tolerant species becomes established, it can be affected by interactions with other species. For example, superior competitors that arrive later will eventually replace the stress-tolerant species.

TESTS FOR THE MECHANISMS OF SUCCESSION

For many years, ecologists debated which of the three succession mechanisms were the most important determinants of the pattern of species replacement over time. Knowing the dominant mechanisms would allow scientists to predict how changes in communities proceed over time, especially since succession is often not a simple linear progression toward a climax community. To address this question, a great deal of research had to be done across a variety of biomes. Here, we discuss two such studies: one conducted in an intertidal community and one conducted in a forest community.

Succession in Intertidal Communities

Sousa's research in Southern California showed that green algae prevented the colonization of red algae (see Figure 19.8). This suggested that the succession of intertidal communities is determined by inhibition. However, the intertidal communities off the Oregon coast consist not only of large species of algae—known as macroalgae—but also of several common species of invertebrates that include the little brown barnacle (*Chthamalus dalli*), the common acorn barnacle (*Balanus glandula*), and several species of limpets, which are gastropods that consume algae. Observations of these communities showed that they were dominated by the common acorn barnacle and a brown macroalgae, *Pelvetiopsis limitata*.

To determine how these communities came to be dominated by the two species, researchers scraped areas of rock clean and watched the succession of species over the subsequent 2.5 years. As illustrated in **Figure 19.14a**, the bare rocks were first colonized by the little brown barnacle. However, as more time passed, the larger acorn barnacle became established on the sites. As it became established, it slowly crushed the smaller little brown barnacle. Over time, the little brown barnacle became rare. Thus, the colonization of the acorn barnacle fit the tolerance model, but the decline of the brown barnacle fit the inhibition model. As the acorn barnacle became abundant, numerous species of macroalgae colonized the site and also became abundant. In fact, after

3 years, communities in the cleared sites closely resembled the control sites that had not been cleared.

Which mechanisms of succession allowed the increase of macroalgae? To answer this question, researchers removed different species from the community. They knew that although the little brown barnacle would be the first to arrive due to its superior dispersal ability, the acorn barnacle would eventually arrive and be a superior competitor for space. They hypothesized that the presence of the acorn barnacles facilitated the colonization and survival of the macroalgae. The researchers began the experiment by once again scraping multiple sites clear of all organisms. Each site then received one of five manipulations: (1) no subsequent removal of any organisms, (2) removal of any little brown barnacles that colonized, (3) removal of any acorn barnacles that colonized, (4) removal of both species of barnacles that colonized, or (5) attachment of empty acorn barnacle shells. The final manipulation was used to test whether the mere physical presence of the acorn barnacle facilitated the colonization and growth of macroalgae by providing the macroalgae with protected crevices to which they could attach. The density of the macroalgae was then monitored for 2 years.

The results of this experiment are shown in Figure 19.14b. Starting near the bottom of the figure, you can see that the removal of the acorn barnacle, or both species of barnacles, caused a lower abundance of macroalgae than the control. In other words, the acorn barnacle helped the macroalgae establish itself. In contrast, the removal of the little brown barnacle had little effect on the density of the macroalgae compared to the control. This confirmed that the little brown barnacle neither helps nor inhibits the macroalgae. However, when empty acorn barnacle shells were added to an intact community that already included live acorn barnacles, there was a large increase in macroalgae. Additional experiments revealed that the acorn barnacle helps the macroalgae by providing crevices where young macroalgae can attach to the rocks without being consumed by herbivorous limpets.

Collectively, these experiments in the intertidal zone indicate that the little brown barnacle persists because it is a good disperser that can rapidly colonize the frequently disturbed intertidal rocks and tolerate the conditions present on a bare rock. This is an example of tolerance. However, once the acorn barnacle arrives, it outcompetes the little brown barnacle, which is an example of inhibition. Finally, the succession of the macroalgae depends on the acorn barnacle, which is an example of facilitation. Thus, a given community can include all three mechanisms of succession.

Succession in Forest Communities

Like intertidal communities, forest communities can also exhibit a mixture of successional mechanisms. At Glacier Bay, for example, it was long hypothesized that each seral stage facilitates the establishment of species in the subsequent stage. This made sense because the soils of later seral stages contained more organic matter, nitrogen, and moisture.

Tolerance A mechanism of succession in which the probability that a species can become established depends on its dispersal ability and its ability to persist under the physical conditions of the environment.

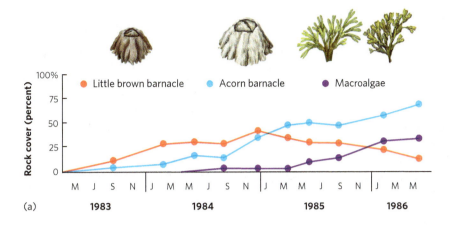

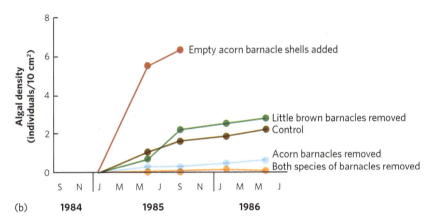

Figure 19.14 Tests of succession mechanisms in an intertidal community. (a) When rocks were scraped bare of all organisms, the first species to dominate was the little brown barnacle. Over time, the acorn barnacle became more abundant, followed by an increase in several species of macroalgae. **(b)** To determine which mechanisms were responsible for these successional changes, the presence of each barnacle species was manipulated. The acorn barnacle was critical to facilitate the macroalgae, whereas the little brown barnacle was not. Data from T. M. Farrell, Models and mechanisms of succession: An example from a rocky intertidal community, *Ecological Monographs* 61 (1991): 95–113.

If facilitation were the most important mechanism of succession, we would expect that a given species would grow and survive well in a seral stage where it dominates, but have a difficult time growing and surviving in earlier stages.

To test this hypothesis, a group of researchers planted spruce seeds and seedlings in four seral stages of Glacier Bay: an early-succession pioneer stage containing lichens, mosses, and herbs; a low-shrub stage dominated by Drummond's avens (*Dryas drummondii*); an alder stage dominated by dense thickets of tall alder shrubs (*Alnus sinuata*); and a spruce stage dominated by spruce trees. You can see the results in **Figure 19.15a**. When spruce seeds were planted in bare plots within each seral stage, germination of these seeds was high in the pioneer stage, low in the low-shrub and alder stages, and high in the spruce stage. When spruce seedlings were planted in plots within each seral stage, seedling growth was high in the pioneer and low-shrub stages, none survived in the alder stage, and growth was low in the spruce stage, as shown in Figure 19.15b. In short, the spruce seeds and seedlings germinated and grew quite well in the earliest seral stages at Glacier Bay, yet spruce was a rare plant in these early stages. Moreover, whereas spruce seeds and seedlings grew well in the low-shrub and alder stages, seedling growth was inhibited in the spruce stage.

These observations refuted the hypothesis that each seral stage facilitates the next stage, but it raised the question of why spruce trees were not common in the earlier stages. Researchers decided to measure how many spruce seeds were dispersed by wind to each site. They discovered that very few spruce seeds arrived to colonize the pioneer and low-shrub stages. There were many more spruce seeds dispersed into the alder stage and even more dispersed into the spruce stage, as shown in Figure 19.15c. The researchers concluded that the dominance of spruce in later seral stages had little to do with facilitation or inhibition, but was due to differences in how many dispersing spruce seeds arrived in each location.

CONCEPT CHECK

1. Why are early-succession species typically better at dispersing than late-succession species?
2. Why are priority effects considered part of the inhibition mechanism of ecological succession?
3. How did the spruce experiments determine that spruce trees did not depend on the facilitation by early-succession species?

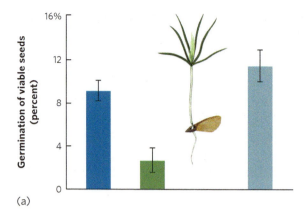

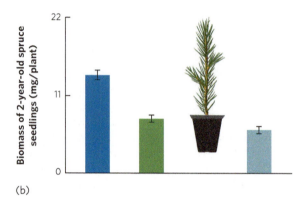

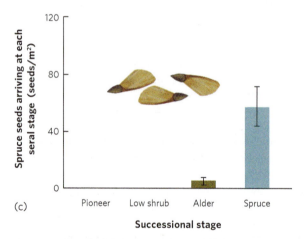

Figure 19.15 Testing the mechanism of spruce succession. Across four successional seral stages in Glacier Bay, Alaska, researchers examined the **(a)** germination success of planted spruce seeds and **(b)** biomass of 2-year-old spruce seedlings. They also measured **(c)** the number of spruce seeds arriving naturally at each seral stage. These data demonstrate that while spruce trees can germinate and grow well in earlier seral stages, they are absent from the early seral stages because very few seeds are dispersed to these sites. Error bars are standard errors. Data from F. S. Chapin III et al., *Mechanisms of primary succession following deglaciation at Glacier Bay, Alaska, Ecological Monographs* 64 (1994): 149–175.

19.3 Succession does not always produce a permanent single climax community

The traditional view of succession was that a series of stages end with a climax community, which remains constant over space and time unless a major disturbance occurs. As we saw in the case of sand dune succession around Lake Michigan, a site can follow different pathways of succession and end up at the same climax community. However, the species composition of a climax community can still exhibit variation over space and time within a given biome. Moreover, a climax community can be short-lived if a disturbance wipes it out. In this section, we will examine how climax communities can change over time and how their composition varies along environmental gradients

CHANGES IN CLIMAX COMMUNITIES OVER TIME

When succession occurs in a community, we typically observe changing environmental conditions and a progression from small life forms to large life forms. For example, primary succession on land begins with lichens and mosses and progresses to grasses and herbs. When sufficient moisture is available, as is the case in eastern North America, succession can continue to a stage that includes large trees. As succession occurs, the abiotic conditions are rapidly altered; areas with trees have less light at ground level, lower ground temperatures during hot summer days, and higher soil moisture. However, once a point is reached where the community contains the largest plants it can support, changes in environmental conditions occur more slowly. As a result, the changes in the community become less dramatic once the climax community develops.

When environmental conditions become relatively stable, the composition of plant species that dominate the community also becomes relatively stable. However, the species found in a climax community can continue to change. For example, northern deciduous forests have a climax community dominated by large trees, but the composition of large tree species can slowly change over time. Initially, these large trees are mostly oak, hickory, and tulip poplar (*Liriodendron tulipifera*). However, over time the dominant species can change to sugar maple and beech. In an old-growth forest in Pennsylvania, researchers surveyed the adult trees that lived in the canopy and the sapling trees that lived in the understory. They then calculated the *importance value* of each tree, which incorporates both the abundance and total area of the trunks for each species. As you can see in **Figure 19.16**, the canopy contained a variety of species with high importance values that included sugar maple, American beech, tulip poplar, several species of oaks, and hickories.

In contrast to this adult distribution, there were few oaks and no tulip poplars or hickories in the understory. The oaks, tulip poplars, or hickories were not in the understory because these species are not very tolerant of the deep shade that occurs in a mature forest and they are also susceptible to being eaten by deer. In comparison, maple and beech trees are very tolerant of shade and less susceptible to being eaten by deer. As a result, maple and beech trees can survive and grow in the understory of large trees, while other species cannot. Over time, the climax community of large trees will experience a gradual shift in the composition of the dominant species; as the current canopy species gradually die, there will be no younger oaks, tulip poplars, or hickories to replace them. In short, the composition of a climax community of large trees in a northern deciduous forest can continue to change over time.

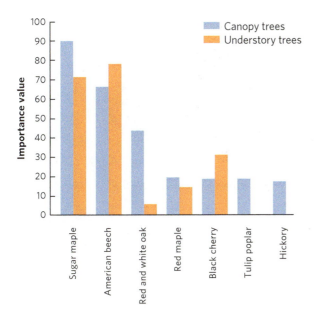

Figure 19.16 Changing species composition in a climax forest. There are many species of trees in the canopy of a forest in Pennsylvania, as indicated by the importance value of each species. However, in the understory, there are few oak trees and no tulip poplar or hickory trees because when they are young, these species are not tolerant of deep shade and they are susceptible to herbivory by deer. This suggests that the future forest canopy will experience a substantial change in the composition of dominant trees. Data from Z. T. Long et al., The impact of deer on relationships between tree growth and mortality in an old-growth beech-maple forest, *Forest Ecology and Management* 252 (2007): 230–238.

CHANGES IN CLIMAX COMMUNITIES OVER SPACE

We have seen that the composition of a climax community can vary over time. Climax communities can also vary in composition as one moves along environmental gradients. For instance, in the 1930s, plant ecologists described the climax vegetation of much of Wisconsin as a sugar maple–basswood (*Tilia americana*) forest. However, they later determined that the climax forest exhibited differences in various locations around the state. In the southern part

of the state, beech trees were more common, while in the north, birch, spruce, and hemlock trees were more common. In the drier regions bordering prairies to the west, oaks became prominent. On drier upland sites, quaking aspen, black oak, and shagbark hickory—long recognized as successional species on moist, well-drained soils—came to be accepted as climax species.

TRANSIENT CLIMAXES

Sometimes a climax community is not persistent, a phenomenon known as a **transient climax community.** A transient climax occurs when a site is frequently disturbed so that the climax community cannot continue to perpetuate itself. A common example of a transient climax occurs in small wetlands, sometimes known as *vernal pools,* which fill with water each spring and then either dry up in summer or freeze solid in winter (**Figure 19.17**). Although drying and freezing events eliminate most species that comprise a pond community, some species have resting stages and persist in the soil until the wetland refills again in the spring. For example, many species of zooplankton produce resting eggs that can persist in the dry bottom of a wetland and then hatch when the wetland fills with water again. Similarly, some species of snails can aestivate, as discussed in Chapter 4, which allows them to live under the soil surface of the dry pond with their metabolic processes largely shut down. When the water returns, plants, animals, and microbes come back to life from their resting stages. Many other species that live on land as adults, such as frogs, salamanders, aquatic beetles, and dragonflies, lay their eggs in the wetland. In this way, the community once again starts the process of succession, only to be destroyed by summer drying or winter freezing.

CREATING GAPS IN A CLIMAX COMMUNITY

Sometimes climax communities contain species that are not considered climax species. These species fill relatively large gaps created by small-scale disturbances in an

---

Transient climax community A climax community that is not persistent.

(a)

(b)

Figure 19.17 Transient climaxes. Vernal pool communities undergo rapid succession of the aquatic community, but **(a)** the climax community that is achieved throughout the spring and summer is frequently destroyed by **(b)** drying in the late summer or freezing in the winter. This vernal pool is in the French Creek Wildlife Area of Wisconsin. Photos by Lee Wilcox.

Figure 19.18 Gaps in a climax community. When a large tree dies and falls to the ground, it opens a gap in the forest that allows large amounts of light to reach the forest floor and favors the growth of species from earlier seral stages. This photo is from the Corcovado National Park in Costa Rica. Photo by Chris Gallagher/Science Source.

Figure 19.19 Grazer-maintained climax. In areas of dry grassland, such as this site in Catalina, Arizona, intense grazing eliminates many species of grasses and favors the growth of cactus and mesquite trees. Photo by Susan E. Swanberg.

area. In mature forests, for example, adult trees eventually die and leave a gap in the canopy that lets in sunlight (**Figure 19.18**). If the gap is not large, the surrounding branches of neighboring trees will likely grow in and close the gap. However, if the gap is large, the area of intense sunlight provides local conditions favoring species from earlier seral stages that have widely dispersed seeds and an ability to grow rapidly under high sunlight conditions. As a result, a mature forest that contains mostly climax species of trees can also contain a few early-succession trees. Gaps can occur in a variety of terrestrial and aquatic biomes. For example, in Chapter 18, we saw how gaps of increasing size in the intertidal zone favored the formation of alternative stable states (see Figure 18.23).

CLIMAX COMMUNITIES UNDER EXTREME ENVIRONMENTAL CONDITIONS

As we have seen throughout this chapter, the composition of a climax community is determined by the environmental conditions that develop over time, including temperature, light, nutrients, and moisture. In some areas, however, additional environmental conditions also play a role, such as in communities affected by fire or grazing.

Biomes in which fires occur at regular intervals favor the persistence of fire-tolerant species. For example, in Chapter 16, we saw that the pine forests in the southeastern United States experience periodic fires that kill oak trees and other species of broadleaf trees, but not pine trees. In fact, some species of pines do not even release seeds from their cones unless triggered by the heat of a fire. After a fire, pine seedlings grow rapidly because there is little or no competition from other understory species. As a result, forest succession reaches a climax that is dominated by pine trees. When a successional stage persists as the final seral stage due to periodic fires, we call it a **fire-maintained climax community.**

The chaparral vegetation found in California is another community that has a fire-maintained climax. As we discussed in Chapter 6, the California chaparral is an example of the woodland/shrubland biome, which has cool, wet conditions in the winter and hot, dry conditions in the summer. As a result, plants can produce thick layers of detritus that become very susceptible to fires during the dry summers.

When a successional stage persists as the final seral stage due to intense grazing, it is called a **grazer-maintained climax community.** Grazers can create a different climax community because they preferentially consume the most palatable plants and do not consume the less palatable or better-defended plants. In the dry grasslands of Arizona, for example, intense cattle grazing can kill or severely damage many species of grass and leave behind less palatable plants, such as mesquite and cactus (**Figure 19.19**).

In western North America, cattle grazing allows invasion by cheatgrass (*Bromus tectorum*), a grass native to Europe, Asia, and North Africa that was introduced to North America in the late 1800s. Cheatgrass is able to colonize when grazing removes much of the competing grasses. Once it establishes, the detritus it produces is quite susceptible to fire. In fact, areas dominated by cheatgrass burn every 3–10 years, rather than the natural fire cycle of every 30–70 years. These more frequent fires promote the long-term persistence of cheatgrass as well as making the area more susceptible to other invasive plant species. In this case, the change in climax community is maintained by a combination of grazing and fire.

Fire-maintained climax community A successional stage that persists as the final seral stage due to periodic fires.

Grazer-maintained climax community When a successional stage persists as the final seral stage due to intense grazing.

Throughout this chapter, we have discussed how communities change over time after a major disturbance. We have emphasized that communities are always changing, both when we move from early to late seral stages and even when a community appears to have attained a climax state. By observing both what happens after a disturbance and the underlying mechanisms that cause the changes, we obtain new insights into the processes that regulate the structure of communities. It is also important to remember that the climate that influences succession is also changing, which means that succession toward a climax stage is actually heading toward a perpetually moving target.

CONCEPT CHECK

1. What evidence is there that climax communities can continue to change over time?
2. How do grazers cause changes in the composition of the climax community?
3. Why is a vernal pool an example of a transient climax community?

ECOLOGY TODAY   APPLYING THE CONCEPTS

Promoting Succession on a Strip Mine

Coal has long been a major source of energy in North America and Europe. For several centuries, humans have mined coal near the surface of the ground by removing the upper layers of soil and rock and then digging out the coal underneath. This process, known as strip mining, surface mining, or mountaintop-removal mining, is an efficient way of mining coal, but these mines eventually run out of coal and leave an immense barren landscape. Many areas of the world require the coal mining industry to return the barren land to a more natural condition.

Since the ground may have little or no organic soil remaining after strip mining, it is a challenge to manipulate the primary succession of communities to reclaim the stripped land. Preventing soil erosion is critical and can be accomplished by rapidly colonizing the area with plants that will hold the soil and ultimately undergo succession in a way that promotes species richness and improves the functioning of the ecosystem. To promote succession on former mined areas, we need to understand how it works by observing the succession changes in plant communities.

In Spain, researchers examined a chronosequence of 26 strip mines in the northern region of the country that had been abandoned for 1 to 32 years. The mines all had similar soils and climate. When the researchers examined changes in species richness, they found that mines abandoned for 1 year had 8 plant species and that species richness peaked at 28 species after 10 years of being abandoned. This peak was caused by the persistence of some of the pioneer species combined with the colonization by later succession species. After 10 years, species richness exhibited a gradual decline; mines abandoned for 32 years had only 7 to 8 species. As with other communities discussed in this chapter, most of the species in the young abandoned sites were annual wind-dispersed plants. These species were able to tolerate the harsh conditions of the newly abandoned mine. After 10 years, there was an increase in the number of perennial herbs, and woody plants started to

Strip mine succession.
When strip mines are no longer used for the extraction of coal, grasses are commonly planted to rapidly hold the soil and prevent erosion, such as at this site in Muskingum County, Ohio. Over time, other plants will slowly colonize the area and succession will occur.
Photos by Michael Hiscar, Office of Surface Mining Reclamation and Enforcement, US Department of the Interior.

colonize the area. These included nitrogen-fixing plants that contributed large amounts of organic matter and nitrogen to the soil, which facilitated subsequent woody plants, including trees that began to dominate after 20 years. These changes correlated with increased nitrogen in the soil, which suggested that the plants that formed the earliest communities facilitated the colonization and growth of later species.

Similar results were found in a 2016 study of succession in coal mines of the Czech Republic. Over several decades, subsoils with little or no organic matter were placed into piles next to the mine to get to the underlying coal. Because the mines were active for decades, these soil piles had experienced succession for 12, 20, 30, or 50 years. Researchers examined how well bushgrass (*Calamagrostis epigejos*) grew on these soils in the presence and absence of arbuscular mycorrhizal (AM) fungi, which we discussed in Chapter 17. The youngest soils contained low concentrations of phosphorus, and adding AM fungi caused greater plant growth. In contrast, the older soils contained higher concentrations of phosphorus, and adding AM fungi caused no additional growth improvement. Thus, AM fungi play a key role in promoting primary succession on the newly mined soils that lack abundant phosphorus.

Understanding how succession proceeds helps scientists develop recommendations for speeding up succession. For example, researchers in Germany examined how succession would be affected under three different manipulations: let the site naturally undergo succession; sow the area with a mixture of herb and grass seeds; or cover the area with newly cut hay, which helps to reduce erosion and provides the seeds of dozens of herbs and grasses contained in the soil. After 4 years, the plots experiencing natural succession had only 35 percent plant cover, while the plots treated with either sown seeds or hay had more than 80 percent plant cover. These substantial differences in plant cover affected soil erosion; plots with sown seeds or green hay had channels that had eroded less than 5 cm deep, while plots undergoing natural succession had channels that had eroded up to 1.5 m deep. The plots experiencing natural succession also had lower species richness during the first year and a lower index of similarity, although the richness of the three plots converged by the end of the 9-year experiment. Collectively, these data confirm that while primary succession will naturally occur on abandoned strip mines, we can use our knowledge of succession to speed up the process and rapidly move a barren landscape toward a much more natural community. However, as we have learned throughout this chapter, terrestrial communities may require centuries of succession after a disturbance such as strip mining before they can approximate the communities that were originally present.

SOURCES

Alday, J. G., et al. 2011. Functional groups and dispersal strategies as guides for predicting vegetation dynamics on reclaimed mines. *Plant Ecology* 212: 1759–1775.

Baasch, A., et al. 2012. Nine years of vegetation development in a postmining site: Effects of spontaneous and assisted site recovery. *Journal of Applied Ecology* 49: 251–260.

Rydlová, J., et al. 2016. Nutrient limitation drives response of *Calamagrostis epigejos* to arbuscular mycorrhiza in primary succession. *Mycorrhiza* 26: 757–767.

SUMMARY OF LEARNING OBJECTIVES

19.1 Succession occurs in a community when species replace each other over time. The process of succession can be observed through either direct observations over time or indirect observations that use chronosequences or parts of organisms, such as pollen, that have been naturally preserved over time. Succession occurs on land, where we can distinguish between primary and secondary succession, and in the water. Succession often does not follow a simple linear path of species replacements, but there is a common pattern of rapidly increasing species richness over time that plateaus and can subsequently exhibit a small decline.

Key Terms: Succession, Seral stage, Pioneer species, Climax community, Chronosequence, Primary succession, Secondary succession

19.2 Succession can occur through different mechanisms. The mechanisms of succession can be categorized as facilitation, inhibition, and tolerance. More than one mechanism can operate in a community experiencing succession, and the traits of species help determine the mechanisms that develop and where each species occurs along the successional stages.

Key Terms: Facilitation, Inhibition, Priority effect, Tolerance

19.3 Succession does not always produce a permanent single climax community. As succession proceeds, the environment continues to change until conditions reach a point of relative stability and the dominant species appear to be persistent. However, the climax community can continue to experience slow changes over time. The climax community can also differ within a region along environmental gradients, such as temperature and moisture. Some climax communities are transient because they experience regular disturbances that reset succession, such as vernal pools that dry each summer. Extreme conditions, including fires and intense grazing, can also alter climax communities to produce a different composition of dominant organisms.

Key Terms: Transient-climax community, Fire-maintained climax community, Grazer-maintained climax community

CRITICAL THINKING QUESTIONS

1. In a pond that experiences succession, what relationship would you expect regarding changes in species richness over time?

2. If you were to use a chronosequence to document the pathway of succession in a tropical forest, what are the limitations of this method?

3. How would a trade-off between dispersal ability and competitive ability affect which types of species could colonize small versus large gaps in a community?

4. Why should we not expect a single climax community on recovering coal mines?

5. Compare and contrast the concepts of facilitation, inhibition, and tolerance in the context of ecological succession.

6. If two plant species have similar dispersal and competitive abilities, what factor might help determine which species occupies an early seral stage?

7. Compare and contrast the classic and modern explanations for the succession of ponds and lakes.

8. If two locations in the northeastern United States follow different paths of succession but end up at the same climax community, how will Jaccard's coefficient of similarity change over time?

9. Why do early- and late-succession species tend to possess different adaptations?

10. Why might the community of insects on a decomposing animal be considered a transient climax?

GRAPHING THE DATA Species Richness at Glacier Bay

macmillanlearning.com/ricklefsvideo

We typically think of succession as a series of different species dominating a site over time. Ecologist William Reiners and his colleagues visited Glacier Bay sites that had been exposed by the retreating glacier at different times. They quantified the species richness for each of five types of vegetation. Using their data, given in the table, create a stacked bar graph that shows the species richness for each of the five types of vegetation at a given seral stage.

Based on this graph, what happens to total species richness of the Glacier Bay sites as succession proceeds?

Successional Stage	Mosses, Liverworts, and Lichens	Low Shrubs and Herbs	Tall Shrubs	Trees
Pioneer	2	9	0	0
Low Shrub	5	9	3	0
Tall Shrub	3	9	6	0
Spruce Forest	18	8	3	2
Hemlock Forest	20	15	4	2

SOURCE
Data from Reiners, W., et al. 1971. Plant diversity in a chronosequence at Glacier Bay, Alaska. *Ecology* 52: 55–69.

20 Movement of Energy in Ecosystems

Worming Your Way into an Ecosystem

Earthworms play an important role in terrestrial ecosystems. As they burrow in the ground to consume detritus, they aerate compact soils, which allows water to better percolate into the ground. Earthworms seem to be everywhere in North America; you may have seen them on sidewalks and roads during rainy days or in the soil if you have ever worked in a garden. Therefore, it is surprising that many common species of earthworms are not native to North America but were introduced from Europe and Asia in the eighteenth century. The northern temperate and boreal forests of North America did not previously have any earthworms, so these forests are not evolved to the altered environmental conditions caused by earthworms. As a result, the introduced worms are now having a profound effect on the movement of energy in these northern ecosystems.

Scientists hypothesize that northern forests completely lack native species of earthworms because the glaciers that advanced over the region during past ice ages eliminated all life. Since the ice receded 10,000 years ago, many other animal species have since returned to the area. However, the native earthworms that survived in the southern United States have a very slow rate of dispersal and have not yet arrived in many habitats in the northern United States. When there are no worms present, the leaf litter in these forests is decomposed primarily by soil fungi and microbes. However, this changed when European colonists inadvertently introduced European and Asian worms to the northern regions of North America. These worms have a high rate of dispersal and they tolerate a wide range of ecological conditions.

The introduced earthworms are still in the process of spreading into northern forests, assisted by the construction of logging roads and anglers who use the worms as bait. As consumers of dead leaves, the invasive worms can consume much of the energy available in the detritus and move many of the nutrients deeper into the soil, where young plants cannot access them. The worms also leave much less energy for other organisms, such as the soil fungi that specialize on decomposing dead leaves. This is a major shift in the movement of energy. In addition to moving nutrients away from the forest floor, earthworm activity leaves a thinner layer of leaves and other organic matter, which leads to drier soil conditions. These changes in abiotic conditions dramatically alter the soil food web. For example, researchers recently reported that the recent arrival of introduced worms into the forests of New York and Pennsylvania caused a nearly 50 percent reduction in leaf litter. This, in turn, resulted in a substantial decline in the abundance of soil insects, such as springtails, ants, and beetles. The decline in the insects also meant that there was less energy available for predators of the insects, such as the red-backed salamander (*Plethodoncinereus*). In areas with high densities of introduced earthworms, the decline in leaf litter and soil insects resulted in an 80 percent reduction in the number of salamanders. While adult salamanders can consume earthworms, young salamanders consume the much smaller soil insects.

> "Scientists hypothesize that northern forests completely lack native species of earthworms because the glaciers that advanced over the region during past ice ages eliminated all life."

The red-backed salamander. The introduction of earthworms from Europe and Asia to northern forests that historically lacked earthworms has caused a decline in leaf litter and soil insects. This has led to a sharp decline in the abundance of red-backed salamanders. Photo by GEORGE GRALL/National Geographic Stock.

Introduced earthworms. Northern temperate and boreal forests have lacked native earthworms since glaciers eliminated them during past ice ages. Today, these forests are being invaded by species of earthworms that have been inadvertently introduced from Europe and Asia.

Photo by Oxford Scientific/Getty Images.

Salamanders aren't the only consumer affected by the earthworm invasion. In 2012, researchers reported that a forest-dwelling ovenbird (*Seiurusaurocapilla*) was declining in areas of Wisconsin and Minnesota that the introduced earthworms had invaded. While the exact mechanisms were not clear in this case, the researchers suspected that the decline in the ovenbirds occurred, in part, because earthworms reduced the availability of insects on the forest floor, something similar to what caused the decline of the red-backed salamander.

In addition to affecting the animals of the forest, earthworms can also affect the plants. In 2017, researchers brought together all the studies of non-native earthworm effects on North American plants, and what they found was striking. While there was no overall effect of species richness or evenness, a decrease in the abundance of native plants and an increase in the abundance of non-native plants took place. The worms also favored an increase in grasses, but not an increase in herbs, shrubs, or trees. The reason is likely because grasses are especially good at rapidly absorbing available nutrients and tolerating dry summer conditions. Some species of introduced earthworms may also consume small seeds and seedlings of the particular plants.

The story of the invasive earthworms demonstrates that species depend on energy flowing between producers, detritivores, and consumers. Changes in energy flow within and between trophic groups can have major impacts on the species that inhabit an ecosystem. In this chapter, we will explore the flow of energy through food webs and the dynamics of energy movement through the ecosystem.

SOURCES

Craven, D., et al. 2017. The unseen invaders: Introduced earthworms as drivers of change in North American forests (a meta-analysis). *Global Change Biology* 23: 1065–1074.

Loss, S. R., et al. 2012. Invasions of non-native earthworms related to population declines of ground-nesting songbirds across a regional extent in northern hardwood forests of North America. *Landscape Ecology* 27: 683–696.

Maerz, J. C., et al. 2009. Declines in woodland salamander abundance associated with non-native earthworm and plant invasions. *Conservation Biology* 23: 975–981.

LEARNING OBJECTIVES

After reading this chapter, you should be able to:

20.1 Describe how primary productivity provides energy to the ecosystem.

20.2 Compare the net primary productivity among different ecosystems.

20.3 Explain how the movement of energy depends on the efficiency of energy flow.

In Chapter 1, we noted that the ecosystem approach to ecology focuses on the transfer of energy and matter among living and nonliving components within and between ecosystems. The amount of energy that fuels ecosystems and the efficiency with which energy is transferred through trophic levels determine the number of trophic levels in communities and ecosystems. The amount of energy available and the efficiency of its transfer determine the biomass of organisms that exist at each trophic level and the amount of energy that is left behind for scavengers, detritivores, and decomposers. In this chapter, we will discuss the movement of energy in ecosystems, including the importance of primary producers and the flow of energy throughout the food web. In the next chapter, we will concentrate on how matter, in the form of key chemical elements, cycles around ecosystems.

20.1 Primary productivity provides energy to the ecosystem

The vast majority of all energy that moves through ecosystems originates as solar energy that powers the photosynthesis of producers. As we saw in Chapter 1, some communities—such as those that form around thermal vents in the deep ocean—rely on chemosynthesis as their source of energy. Regardless of origin, producers harness energy and form the basis of food webs. Producers use this energy for respiration, growth, and reproduction; the amount used for growth and reproduction represents the energy that is available to consumers. In this section, we will examine how to quantify the amount of energy that producers use for different functions, how to measure the energy of producers in different types of ecosystems, and how the amount of energy available in producers affects the growth and reproduction of consumers.

PRIMARY PRODUCTIVITY

Primary productivity is the rate at which solar or chemical energy is captured and converted into chemical bonds by photosynthesis or chemosynthesis. Primary productivity tells us how much energy is available in an ecosystem. A related concept is the **standing crop** of an ecosystem, which is the biomass of producers present in the ecosystem in a given area at a given time. For example, the standing crop of a forest is the total mass of trees, shrubs, herbs, and grasses that is present in an area of the forest on a particular day. Ecosystems with high primary productivity may or may not have a high standing mass. In lakes where algae experience high productivity, consumers often eat it nearly as quickly as it grows, and so the standing crop of algae remains low.

Ecologists identify two types of primary productivity: *gross primary productivity* and *net primary productivity*. **Gross primary productivity (GPP)** is the rate at which energy is captured and assimilated by producers in a given area. We often express productivity in units of Joules (J) or kilo Joules (kJ) per square meter per year. From this total, producers use some of the assimilated energy for their metabolism, which is measured in terms of the amount of respiration. The rest of the assimilated energy is converted into their biomass, which includes growth and reproduction. The rate at which energy is assimilated by producers and converted into producer biomass in a given area is **net primary productivity (NPP).** We can also show this in the form of an equation:

$$NPP = GPP - Respiration$$

Considering the rate at which the Sun provides energy, photosynthesis is not a very efficient process. As illustrated in **Figure 20.1**, approximately 99 percent of all solar energy that is available to producers either reflects off them or passes through their tissues without being absorbed.

Primary productivity The rate at which solar or chemical energy is captured and converted into chemical bonds by photosynthesis or chemosynthesis.

Standing crop The biomass of producers present in a given area of an ecosystem at a particular moment in time.

Gross primary productivity (GPP) The rate at which energy is captured and assimilated by producers in a given area.

Net primary productivity (NPP) The rate at which energy is assimilated by producers and converted into producer biomass in a given area.

1% of solar energy is captured by photosynthesis; this is gross primary productivity.

60% of gross primary productivity is respired.

40% of gross primary productivity is used for producer growth and reproduction; this is net primary productivity.

99% of solar energy is reflected or not absorbed by plants.

Figure 20.1 Gross primary productivity and net primary productivity. About 99 percent of solar energy is reflected by producers or passes through their tissues without being absorbed. Only about 1 percent is captured by photosynthesis for gross primary productivity (GPP). Of the total GPP, about 60 percent is used by producers for respiration. The remaining 40 percent is available to producers for growth and reproduction, which is known as net primary productivity (NPP).

Of the 1 percent of the solar energy that producers absorb and use for photosynthesis—that is, gross primary productivity—approximately 60 percent is used for respiration. This means that only 40 percent of absorbed solar energy is used for net primary productivity, which represents the growth and reproduction of producers. For example, if we measure the productivity of a forest in North America in units of kilograms of carbon per square meter per year, such a forest might have a GPP of 2.5 kg C/m²/year. Of this total, the forest will use about 1.5 kg C/m²/year for respiration, leaving 1.0 kg C/m²/year for growth and reproduction, which represents the forest's NPP.

MEASURING PRIMARY PRODUCTIVITY

Primary productivity is the foundation for the flow of energy through food webs and ecosystems. Being able to measure primary productivity allows us to track how it changes in an ecosystem over time and how it varies in different ecosystems around the globe. Since primary productivity is a rate that is achieved by the benefits of respiration minus the costs of respiration, we can measure it by quantifying the change in the biomass of producers over time, the movement of carbon dioxide over time, or the movement of oxygen over time. In terrestrial ecosystems, we commonly measure the productivity of just the plants, but in aquatic systems, we might measure the productivity of plants, large kelp, or tiny species of algae. As we will see, the choice of how to measure primary productivity depends on the particular ecosystem being studied.

When measuring primary productivity, it is critical to determine whether our measurements represent gross or net primary productivity. In some cases, we can measure net

primary productivity and respiration separately and then use these values to estimate gross primary productivity. As we will see in the next section, there is a variety of methods for measuring primary productivity in terrestrial and aquatic ecosystems.

Measuring Changes in Producer Biomass

One of the simplest ways to measure NPP is by measuring producer biomass in an area at the beginning and at the end of a growing season (**Figure 20.2**). For example, in prairie ecosystems, researchers commonly measure the amount of new plant growth that has accumulated by the end of the summer growing season, while in aquatic ecosystems, ecologists measure the biomass of large aquatic producers such as kelps. A season's growth can be harvested, dried to eliminate any water, and then weighed to determine how much growth occurred during the season. When a harvest is measured to determine biomass, the assumption has to be made that no substantial herbivory or tissue mortality has occurred during the period of producer growth. Alternatively, the amount of biomass lost to herbivory or tissue mortality can be estimated and included in the estimate of net primary productivity.

In some NPP studies, only the part of the plant that lives aboveground is harvested. However, the amount of biomass belowground for some plants can be substantial. For example, perennial grasses with extensive root systems have twice as much belowground biomass as aboveground biomass. In contrast, most trees have about five times more aboveground biomass than belowground biomass.

Measuring underground biomass is quite a challenge. Retrieving the rhizomes, tubers, and roots of plants from the

(a)

(b)

Figure 20.2 Harvesting primary productivity. Net primary productivity is commonly measured by harvesting a defined area of an ecosystem at the end of the growing season and determining the accumulated biomass of producers. This technique can be used on **(a)** small scales, such as by these students, who are measuring the aboveground biomass of plants, or on **(b)** large scales, such as the harvesting of crops in a field. Photos by (a) Jamie Shady and (b) Leonid Shcheglov/Shutterstock.

terrestrial soils or the aquatic benthos is often not feasible because they are either very deep or the root systems are composed of many fine roots that break off when harvested. In many plants, these fine roots die and are replaced with new fine roots, which makes it difficult to estimate how much biomass a group of producers has accumulated. It is also difficult to account for plants in mutualistic relationships with mycorrhizal fungi. These plants may provide carbohydrates to the fungi, but because these carbohydrates are no longer a part of the plant, it is hard to get an accurate estimate of the plant's productivity. Therefore, we have to be cautious when making conclusions about differences in productivity when we measure only changes in the aboveground biomass to estimate net primary productivity.

Measuring CO_2 Uptake and Release

Because producers take up CO_2 during photosynthesis and produce CO_2 during respiration, we can measure primary productivity in terrestrial ecosystems by quantifying the uptake and release of CO_2 by plants. One way to measure these changes in CO_2 is by placing a small plant or a leaf into a sealed container with a highly sensitive CO_2 sensor, as illustrated in **Figure 20.3**. When the container is placed in front of a light that simulates sunlight, the plant consumes CO_2 as it conducts photosynthesis. The plant simultaneously produces CO_2 as it metabolizes some of its carbohydrates through respiration. Since CO_2 uptake from photosynthesis

exceeds CO_2 release from respiration, the net uptake of CO_2 represents net primary productivity.

We can also estimate GPP using this technique. Since GPP is the sum of net primary productivity and respiration, we can determine GPP by combining our estimate of NPP with an estimate of how much respiration is occurring in a plant. We can do this by placing leaves or whole plants in chambers without sunlight. If we measure the plant's rate of respiration in the dark and add it to the plant's NPP, we can estimate the gross primary productivity using the following rearranged equation:

$$GPP = NPP + Respiration$$

Because these experiments are conducted using some chambers placed in the light and other chambers placed in the dark, they are sometimes referred to as *light-dark bottle experiments*.

Several other techniques can also be used to measure CO_2 uptake and release. In one technique, a plant or a leaf is placed in a sealed container and then CO_2 containing a rare isotope of carbon, such as ^{14}C, is added to the container. The net movement of ^{14}C from the air to the plant tissues during photosynthesis and from the plant tissues back to the air during respiration can be tracked. This net movement of ^{14}C is a measure of net primary productivity.

On a larger scale, such as a grassland or forest, CO_2 uptake and release can be measured using towers that

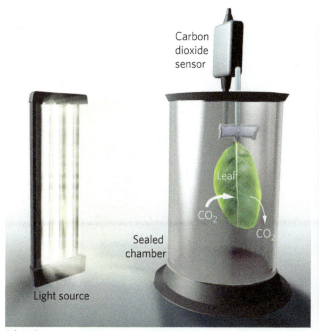

(a) Light

(b) Dark

Figure 20.3 Measuring CO_2 uptake and release. (a) In a sealed container illuminated by a light, sensors can detect a decline in CO_2 in the air as a leaf conducts photosynthesis. This decline occurs because plant photosynthesis consumes CO_2 at a faster rate than plant respiration produces CO_2. **(b)** When the light is turned off, sensors detect an increase in CO_2 in the air because the leaf only undergoes respiration.

sample the concentration of CO_2 at different heights above the ground. The concentration of CO_2 moving up from the vegetation, compared to the concentration of CO_2 in the atmosphere, gives a measure of the rates of photosynthesis and respiration that are occurring in an area.

Measuring O_2 Uptake and Release

In aquatic ecosystems, the dominant producers are typically algal cells. Because most algae are tiny and rapidly consumed, it is not always feasible to harvest and measure their biomass as a way to measure net primary productivity. Quantifying changes in CO_2 concentration is also not a viable way to measure primary productivity because CO_2 dissolved in water rapidly converts into bicarbonate ions, as we discussed in Chapter 2. However, because producers release O_2 during photosynthesis and take up O_2 during respiration, we can estimate NPP and GPP by measuring the uptake and release of O_2.

To estimate primary productivity using the O_2 concentration in the water, we can use a process that is similar to estimating primary production based on CO_2. That is, we can conduct a light-dark bottle experiment. We begin by immersing two bottles under the surface of the water to collect the algae. One bottle is clear, allowing sunlight to penetrate. In this bottle, we measure the net increase in O_2

production that occurs with the combined effects of photosynthesis and respiration by the algae. The other bottle is opaque, so sunlight cannot penetrate. In this bottle, the algae cannot photosynthesize but can only respire and drive down the concentration of O_2. Because the clear bottle measures NPP and the opaque bottle measures only respiration, we can estimate GPP by adding the values obtained from each.

Remote Sensing

The techniques we have discussed measure gross and net primary productivity on relatively small spatial scales, ranging from a single leaf to a small area of land to a small volume of water. What if we want to assess productivity on very large spatial scales, including changes in productivity across continents or oceans? One solution to this challenge is *remote sensing*. **Remote sensing** is a technique that measures conditions on Earth from a distant location, typically using satellites or airplanes that take photographs of large areas of the globe, as shown in **Figure 20.4**. These images reveal how different wavelengths of light are reflected or absorbed. As we discussed in Chapter 3, chlorophyll pigments

Remote sensing A technique that measures conditions on Earth from a distant location, typically using satellites or airplanes that take photographs of large areas of the globe.

Figure 20.4 Remote sensing of primary productivity. Using aerial images from planes or satellites, scientists can determine the wavelengths of light that are absorbed and reflected by the chlorophyll pigments of algal blooms. Here we see satellite images of an algal bloom in Lake Erie. Photo by NASA.

absorb wavelengths in the red and blue range but reflect wavelengths in the green range. Therefore, satellite images of ecosystems that show a pattern of high absorption of blue and red light and high reflectance of green light indicate ecosystems with high standing crops of producers. Changes in producer biomass over time can then be used to estimate net primary productivity.

SECONDARY PRODUCTION

Primary productivity is the foundation of the food web because it represents the source of energy for herbivores. To understand how energy moves from producers to consumers, we need to consider a number of different pathways, shown in **Figure 20.5**. We begin with producers that herbivores consume. Herbivores consume only a small fraction of the total amount of producer biomass available, and they can only digest a portion of the energy they consume. For example, many fruits contain hard seeds that herbivores cannot digest, so the seeds are excreted whole. We saw an example of this at the end of Chapter 17 when we discussed Aldabra tortoises that had been introduced to an island in the Republic of Mauritius. These tortoises consume fruits of the ebony tree and disperse the seeds when they defecate. The portion of consumed energy that is excreted or regurgitated is known as **egested energy.** In contrast, the portion of energy that a consumer digests and absorbs is known as **assimilated energy.**

Of the energy assimilated by a consumer, the portion used for respiration is known as **respired energy.** The rest can be used for growth and reproduction. For example, consumers such as fruit-eating birds need a great deal of energy for respiration to maintain a constant body temperature, whereas fruit-eating tortoises of the same mass need much less energy for respiration because they are ectotherms. As a result, birds devote more of their assimilated energy to respiration than tortoises do. Because the rate of assimilated energy of consumers is the amount of energy that is used for respiration, growth, or reproduction, it is analogous to the concept of GPP for producers.

If we consider the assimilated energy of consumers and subtract the energy used for respiration, we are left with the energy used for biomass accumulation; the rate of biomass accumulation of consumers in a given area is called **net secondary productivity.** Because net secondary

Egested energy The portion of consumed energy that is excreted or regurgitated.

Assimilated energy The portion of energy that a consumer digests and absorbs.

Respired energy The portion of assimilated energy a consumer uses for respiration.

Net secondary productivity The rate of consumer biomass accumulation in a given area.

Figure 20.5 The path from primary to secondary productivity. Herbivores gather energy by ingesting producers. Of the energy gained by ingestion, some is lost in the form of egested energy, which represents the indigestible tissues of the producers. The rest of the energy is assimilated. Of the assimilated energy, some is used for respiration. The rest is used for growth and reproduction, which is secondary productivity.

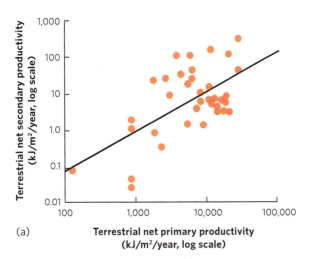

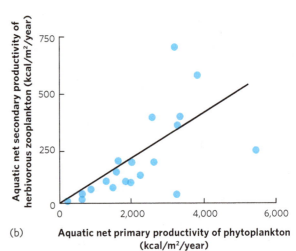

Figure 20.6 Secondary productivity. (a) In terrestrial ecosystems, an increase in primary productivity is positively correlated to an increase in secondary productivity. **(b)** In aquatic ecosystems, a similar relationship is found between the primary productivity of phytoplankton and the secondary productivity of the herbivorous zooplankton that consume the phytoplankton.

Data from S. McNaughton et al., Ecosystem-level patterns of primary productivity and herbivory in terrestrial habitats, *Nature* 341 (1989): 142–144; M. Brylinsky and K. H. Mann, An analysis of factors governing productivity in lakes and reservoirs, *Limnology and Oceanography* 18 (1973): 1–14.

productivity depends on primary productivity as its source of energy, we might expect that an increase in net primary productivity would cause an increase in net secondary productivity. Researchers have compiled net primary and net secondary productivity estimates from a wide range of terrestrial biomes. As shown in **Figure 20.6a**, increases in net primary productivity are positively correlated with increases in net secondary productivity. In aquatic ecosystems, researchers have also found that increases in the productivity of phytoplankton—a dominant producer in many aquatic biomes—are positively correlated to the productivity of the herbivorous zooplankton that eat the algae, which you can see in Figure 20.6b.

Quantifying net secondary productivity has many of the same challenges as quantifying net primary productivity, including the need to account for herbivores that are removed from the ecosystem by predators or diseases. In addition, the positive correlation between net primary and secondary productivity suggests an important role of bottom-up control of communities, a concept we discussed in Chapter 18. However, we know that the top-down effects of predators also affect some communities.

CONCEPT CHECK

1. Why do algae commonly experience a high productivity but a low standing crop?

2. Gross primary production can be used to provide energy for what two processes?

3. What are some key assumptions when measuring the net primary productivity of terrestrial plants?

20.2 Net primary productivity differs among ecosystems

Now that we understand net primary productivity and how it affects the herbivores that consume it, we can examine how net primary productivity varies among ecosystems around the world. Understanding these patterns is important because ecosystems with greater primary productivity should generally support greater secondary productivity, which means that the most productive places will likely have either a high abundance or a high diversity of consumers. In this section, we will investigate patterns of primary productivity in different ecosystems and discuss the abiotic factors that determine these patterns of net primary productivity.

PRIMARY PRODUCTIVITY AROUND THE WORLD

When we look at patterns of NPP around the world in **Figure 20.7**, we see that net primary productivity varies with latitude. The most productive terrestrial ecosystems occur in the tropics, and productivity declines as we move to temperate and polar regions. The most productive ocean ecosystems are found along coasts, while primary productivity is low in the open ocean.

We can also consider the differences in NPP among various aquatic and terrestrial ecosystems. These ecosystem categories coincide well with the biomes described in Chapter 6. **Figure 20.8** shows that tropical rainforests are the most productive terrestrial ecosystems. The least productive ecosystems include those that are very cold, such as tundra, and those that are very dry, such as deserts. Among the

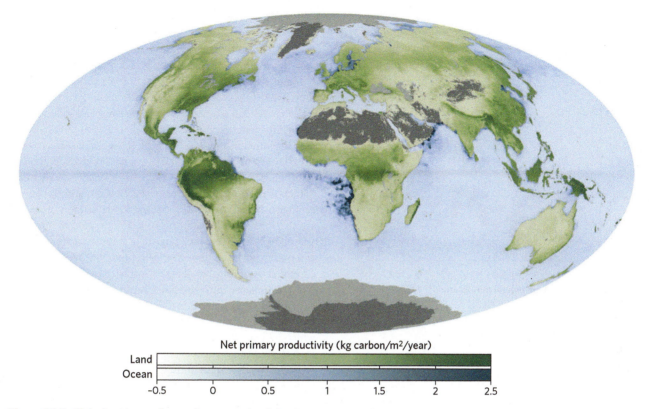

Figure 20.7 Global patterns of net primary productivity. Among terrestrial ecosystems, the highest productivity occurs in the tropics, which experience warm temperatures throughout the year and have abundant rainfall. The most productive aquatic ecosystems include the shallow water near the edges of continents and large islands. NASA Earth Observatory.

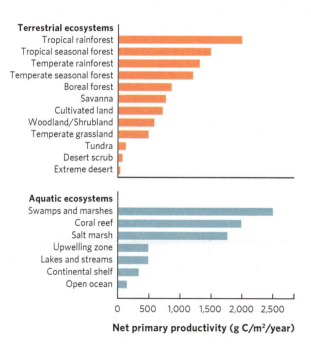

Figure 20.8 The net primary productivity of different ecosystems around the world. Among terrestrial ecosystems, tropical rainforests are the most productive, whereas deserts are the least productive. Among aquatic ecosystems, swamps, marshes, and coral reefs are the most productive, whereas the open ocean is the least productive. Data from R. H. Whittaker and G. E. Likens, Primary production: The biosphere and man, *Human Ecology* 1 (1973): 357–369.

freshwater ecosystems, lakes and streams, on average, are on the low end of NPP, although NPP can vary widely in these systems. In contrast, marine ecosystems exhibit a wide range of productivity, ranging from high productivity in coral reefs and salt marshes to low productivity in the open ocean. Many abiotic factors are responsible for these patterns of productivity.

DRIVERS OF PRODUCTIVITY IN TERRESTRIAL ECOSYSTEMS

As you may recall from our discussion of terrestrial biomes in Chapter 6, dominant plant forms are determined by patterns of annual temperature and precipitation. Similarly, temperature and precipitation are major drivers of NPP. The most productive terrestrial ecosystems occur in tropical areas because those areas have the most intense sunlight, warm temperatures throughout the year, high amounts of precipitation, and rapidly recycled nutrients that support growth. At higher latitudes, such as in temperate and polar regions, productivity is much lower due to shorter periods of sunlight and lower temperatures during the winter. In the deserts that occur at 30° N and 30° S latitude, productivity is primarily constrained by the lack of precipitation.

One way to examine the effects of temperature and precipitation on NPP is by looking at a large number of studies that have measured NPP in different parts of the world.

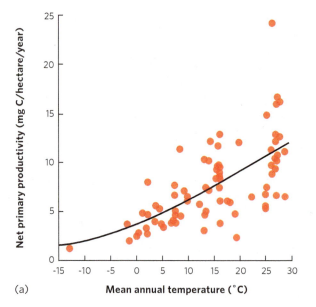

(a)

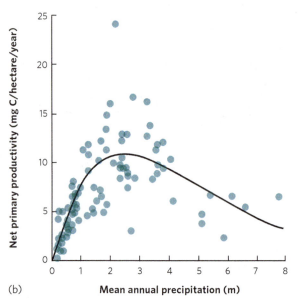

(b)

Figure 20.9 The effects of temperature and precipitation on the net primary productivity of terrestrial ecosystems. **(a)** Locations on Earth that have higher mean annual temperatures have higher NPP. **(b)** Locations that have higher mean annual precipitation have higher NPP but at locations with the highest amounts of precipitation, increased leaching and reduced decomposition of organic matter cause NPP to decline. Data from E. A. G. Schuur, Productivity and global climate revisited: The sensitivity of tropical forest growth to precipitation, *Ecology* 84 (2003): 1165–1170.

For example, **Figure 20.9** displays the results of studies from 96 locations. In Figure 20.9a, you can see that an increase in mean annual temperature is positively correlated with an increase in NPP. This reflects the fact that lower latitudes with warmer temperatures favor plant growth and have a longer growing season. In Figure 20.9b, you can see that an increase in mean annual precipitation shows a positive correlation with NPP until 3 m of annual precipitation is reached. Ecosystems

that receive 3 m or more of precipitation experience a decline in NPP because nutrients leach away from the soil, and the rates of decomposition are reduced because of waterlogged soils. When organic matter is broken down more slowly, fewer nutrients are available in the soil for primary productivity. In short, while temperature and precipitation are the primary drivers of NPP in terrestrial ecosystems, they also influence the availability of nutrients in ways that affect NPP.

Given that nutrients affect NPP in terrestrial ecosystems, which nutrients are the most important? For decades, ecologists believed that nitrogen was the most important element that constrained NPP in terrestrial ecosystems. However, ecologists began to discover that some terrestrial ecosystems are also limited by phosphorus or a combination of nitrogen and phosphorus. To obtain a more accurate overview of whether nitrogen or phosphorus limited NPP, researchers compiled data from 141 separate terrestrial experiments that had manipulated nitrogen, phosphorus, or both. For each experiment, they determined the ratio of NPP in treatments with added nutrients, compared to NPP in the control, which had no added nutrients. With a response ratio for each study, the researchers could then determine the average response for all studies in three categories of terrestrial ecosystems: grasslands, forests and shrublands, and tundra. As you can see in **Figure 20.10**, all three categories of terrestrial ecosystems experienced increases in NPP with the addition

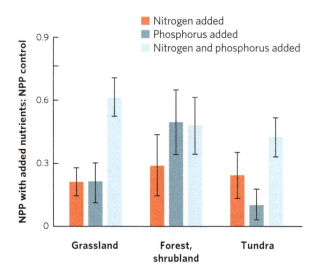

Figure 20.10 Increases in net primary productivity when nutrients are added to terrestrial ecosystems. Based on 141 experiments conducted around the world, researchers examined how NPP responds to different nutrient additions. To quantify the change in NPP, they determined the ratio of NPP in treatments with added nutrients, compared to NPP in the control, which had no added nutrients. The addition of nitrogen and phosphorus increased the NPP of grassland, forest, shrublands, and tundra ecosystems. When both nutrients were added, the response ratio of NPP was often larger than when either nutrient was added separately. Error bars are standard errors. Data from J. J. Elser et al., Global analysis of nitrogen and phosphorus limitation of primary producers in freshwater, marine, and terrestrial ecosystems, *Ecology Letters* 10 (2007): 1135–1142.

of nitrogen or phosphorus. In grasslands and tundra, adding both nitrogen and phosphorus caused a greater increase in NPP than either alone, suggesting that adding more of one nutrient causes the growth of plants to soon be limited by the other nutrient. From these data, we can see that nitrogen and phosphorus are both important nutrients that constrain NPP.

DRIVERS OF PRODUCTIVITY IN AQUATIC ECOSYSTEMS

We have seen that terrestrial ecosystems are primarily constrained by temperature, precipitation, and nutrients. In addition to these factors, aquatic ecosystems are also constrained by light since the transmission of light down through the water is required for photosynthesis. Indeed, abundant light is one reason that coral reefs, which exist in shallow tropical waters, are such productive ecosystems. However, within aquatic ecosystems that have similar temperatures and similar light levels, the most important driver of NPP is the amount of nutrients.

The limiting role of nutrients can be seen in a variety of aquatic ecosystems. For example, in the open ocean, the remains of dead animals sink to the bottom, where they decompose and release nutrients. Because this regeneration of nutrients is far below the ocean's surface, the surface of the open ocean experiences low NPP. Small streams also are typically low in nutrients and experience low NPP. Moreover,

if the small streams are in a forest, they receive little sun because of the shade from trees, and this restricts their productivity. As we discussed in Chapter 6, a large fraction of the energy and nutrients that exist in small streams enters the stream in the form of allochthonous inputs, such as dead leaves, that drop in from the surrounding terrestrial environment. In contrast, estuaries and coral reefs receive abundant nutrients in the form of runoff from rivers and the adjacent land, which allows these ecosystems to have very high primary productivity. Among all aquatic ecosystems, NPP is most commonly constrained by the availability of phosphorus and nitrogen, although silicon and iron can be limiting in some areas of the open ocean.

Limitation by Phosphorus and Nitrogen

For many years, it was thought that phosphorus is the most important nutrient that limits the NPP of aquatic ecosystems. For example, in a classic experiment, David Schindler and his colleagues selected an hourglass-shaped lake in Ontario and placed a plastic curtain at the constriction that divided the lake into two halves (**Figure 20.11a**). On one side carbon and nitrogen were added, and on the other side carbon, nitrogen, and phosphorus were added. These additions were continued from 1973 to 1980 as the researchers monitored the two sides of the lake. They also monitored a second lake that had an initial fertility similar to the divided

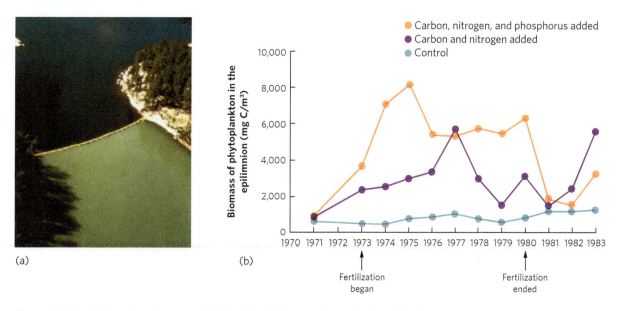

(a)
(b)

Figure 20.11 Adding phosphorus to half of a lake. (a) Researchers split a Canadian lake in half and then added carbon and nitrogen to one side and carbon, nitrogen, and phosphorus to the other side. **(b)** Compared to a reference lake that received no nutrient additions, the side that received additions of carbon and nitrogen from 1973 to 1980 experienced a modest increase in NPP, as measured by the biomass of phytoplankton in the epilimnion. The side of the lake that received additions of carbon, nitrogen, and phosphorus experienced a large increase in NPP. When the addition of phosphorus stopped in 1980, NPP in the fertilized side of the lake declined to levels that were similar to those of the control. Data from D. L. Findlay and S. E. M. Kasian, Phytoplankton community responses to nutrient addition in Lake 226, Experimental Lakes Area, Northwestern Ontario, *Canadian Journal of Fisheries and Aquatic Sciences* 44 (Suppl. 1) (1987): 35–46. Photo Courtesy of David W. Schindler. *Science*, 184:897–899.

lake to serve as a control. In the half of the lake that received carbon and nitrogen, there was a modest increase in NPP compared to the control lake, measured by the growth of cyanobacteria. However, in the half of the lake that received carbon, nitrogen, and phosphorus, there was a large increase in NPP, as shown in Figure 20.11b. After the treatments were terminated in 1980, the NPP of the side of the lake with added phosphorus rapidly declined. This experiment confirmed that human inputs of excess phosphorus—in the form of fertilizers that wash off farms and various household detergents—can have a major effect on the productivity of aquatic ecosystems.

To gain a broader perspective on how nitrogen and phosphorus affect aquatic ecosystems, let's revisit the average NPP responses from experiments conducted around the world. As part of the same study discussed earlier, researchers compiled data from 928 experiments that manipulated the addition of nitrogen, phosphorus, or both in freshwater and marine ecosystems. Among freshwater ecosystems, as shown in **Figure 20.12a**, nitrogen and phosphorus both caused an increase in NPP, although phosphorus has a much larger effect on the benthic ecosystem of lakes than does nitrogen. Among marine ecosystems, shown in Figure 20.12b, nitrogen and phosphorus additions had similar effects on NPP in ocean ecosystems with soft bottoms, such as estuaries made up of

seagrass and algae. However, the addition of nitrogen had a much larger effect on NPP than that of phosphorus in ecosystems with hard bottoms, such as coral reefs and rocky intertidal biomes, and in the open water of oceans. Across all freshwater and marine ecosystems, adding nitrogen and phosphorus together often caused an NPP response that was much larger than adding either nutrient separately. Once again, this suggests that adding more of one nutrient causes the growth of producers to soon be limited by the other nutrient. The results of this research suggest that the availability of nitrogen and phosphorus can constrain the NPP of aquatic ecosystems, as well as that of terrestrial ecosystems.

Limitation by Silicon and Iron in the Ocean

While the primary productivity of aquatic ecosystems is typically limited by the availability of nitrogen and phosphorus, primary productivity remains low in approximately 20 percent of the open ocean even though nitrogen and phosphorus are abundant there. This suggests that productivity in these areas is limited by other nutrients that are unusually scarce, such as silicon and iron.

Silicon is the raw material for the silicate shells of diatoms (see Figure 2.5), which comprise most of the phytoplankton in some regions of the ocean. Silicon is lost from the surface waters when diatoms die and their dense shells

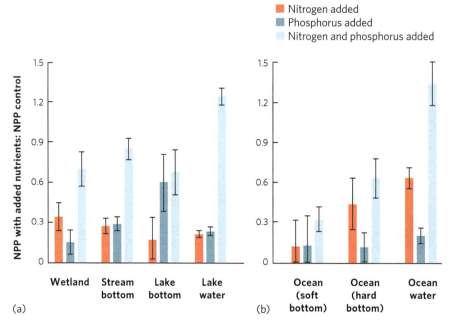

Figure 20.12 Increases in net primary productivity when nutrients are added to aquatic ecosystems. Using data from 928 experiments conducted around the world, researchers determined how NPP responds to different nutrient additions. NPP was examined as a ratio between productivity when nutrients were added and a control treatment in which nutrients were not added. In **(a)** freshwater and **(b)** marine ecosystems, adding nitrogen or phosphorus caused an increase in NPP, although nitrogen has a larger effect than phosphorus in the soft-bottom-ocean ecosystems and the ocean–water ecosystems. Error bars are standard errors. Data from J. J. Elser et al., Global analysis of nitrogen and phosphorus limitation of primary producers in freshwater, marine, and terrestrial ecosystems, Ecology Letters 10 (2007): 1135–1142.

fall to the ocean bottom. For example, the area west of southern South America, between 40° S and 50° S latitude, appears to have too little silicon, probably because silicon-containing particles sink below the photic zone more rapidly than do nitrogen and phosphorus across this long stretch of the southern Pacific Ocean.

Iron is an important component in many metabolic pathways, but it is lost from the ocean's surface when it combines with phosphorus and precipitates. Rivers are a source of iron, which explains the low concentrations of iron in regions of the ocean that are far from continents. In these areas, the only iron inputs occur in the form of windblown dust.

If iron limits the growth of phytoplankton in the ocean, then additions of iron should act as a fertilizer and cause large increases in primary productivity. In a 1993 experiment, 450 kg of iron—roughly the amount in an automobile— was added to over 64 km² of ocean off the west coast of South America. The treatment caused a 100-fold increase in the concentration of iron and, within just a few days, the concentration of phytoplankton tripled. Although this increase in phytoplankton was of a short duration, the experiment confirmed that a lack of iron limited the producers in this area of the ocean. In subsequent experiments conducted off the coast of Antarctica, similar increases in productivity were observed (**Figure 20.13**).

Fertilizing the ocean with iron has implications beyond confirming that iron is the limiting nutrient in these areas. If primary productivity in the ocean can be increased, it has the potential to draw down the amount of CO_2 in the water. Because CO_2 is exchanged between the water and the air, reducing CO_2 in the water would subsequently reduce the amount of CO_2 in the atmosphere. Since CO_2 is a greenhouse gas, some scientists have hypothesized that fertilizing the oceans with iron could be a way to counteract global warming.

Although adding iron increased primary production, in some cases the large increase in phytoplankton caused a subsequent large increase in the zooplankton populations that consume phytoplankton. When this happened, the zooplankton that was produced caused an increase in the amount of CO_2 due to increased respiration, thereby counteracting the beneficial effects of the phytoplankton. When researchers conducted experiments that monitored the amount of carbon that precipitated out of the water column during iron additions, they found that there was only a small increase in the amount of carbon that precipitated out of the water column.

Other researchers have examined historic variation in iron inputs on ocean productivity. During five major glacial periods, about 2.5 times more iron dust traveled through the atmosphere and arrived in the Pacific Ocean near the equator. In 2016, researchers reported their results of examining cores from the ocean floor in this region of the Pacific, which consists of layers of sediments deposited over centuries. They found no increase in productivity during past periods of increased iron availability due to increased iron dust. Based on the observations from experiments and historic data, the long-term effects of iron enrichment on CO_2 remain uncertain, as does the question of whether large-scale applications of iron in the ocean could have potential adverse effects on the ocean ecosystem.

CONCEPT CHECK

1. Which ecosystems have the highest NPP?
2. Why does adding a combination of nitrogen and phosphorus often result in higher producer growth than adding either nutrient alone?
3. Why have researchers hypothesized that adding iron to certain areas of the ocean might help offset elevated CO_2 concentrations in the atmosphere?

20.3 The movement of energy depends on the efficiency of energy flow

As we have discussed, the organisms in each trophic level must obtain energy, assimilate some fraction of the energy they secure, and then use some of the assimilated energy for respiration. Whatever remains goes to growth and reproduction. At each step, different organisms vary in the efficiency with which they obtain and retain energy. These efficiencies, combined with trophic interactions that affect the abundance of other trophic levels, affect the amount of energy and biomass that can be supported in a given trophic level. These factors also affect how many trophic levels can exist in an ecosystem. In this section, we will explore patterns in the amounts of energy and biomass that exist in different trophic levels. We will then examine the various efficiencies of energy transfer in trophic groups and how these differences affect the number of trophic levels found in an ecosystem.

Figure 20.13 Fertilizing the oceans. In the Southern Ocean near Antarctica, researchers lower water-sampling equipment to measure the effects of adding iron to the surface waters. McLane Research Laboratories Sediment Trap, photo courtesy Renata Giulia Lucchi and Leonardo Lagnone, ISMAR.

TROPHIC PYRAMIDS

A useful way to think about the distribution of energy or biomass among the trophic groups in an ecosystem is by drawing a **trophic pyramid,** which is a chart composed of stacked rectangles representing the relative amount of energy or biomass in each trophic group. The first person to consider the flow of energy between trophic levels was Raymond Lindeman, an ecologist who conducted his doctoral dissertation on Cedar Bog Lake in Minnesota during the 1930s. As Lindeman collected data on the energy and biomass produced by different trophic levels in the lake, he concluded that energy must be lost as it moves from one trophic level to the next. He demonstrated that this was the case when he created a trophic pyramid that displayed the percentage of the total energy existing at each trophic level, known as a **pyramid of energy.** As you can see from his data in **Figure 20.14**, each trophic level in the lake contained less energy than the trophic level below it.

Pyramids of energy are just one way of representing the distribution of organisms in an ecosystem. We can also create a trophic pyramid that represents the standing crop of organisms present in different trophic groups, which is known as a **pyramid of biomass.** In terrestrial ecosystems, the distribution of biomass among trophic levels looks quite similar to the pyramid of energy. As **Figure 20.15a** shows, the greatest amount of biomass occurs in producers, with less biomass in primary and secondary consumers.

For example, consider the biomass in a forest ecosystem. The greatest portion of the biomass occurs in producers, which include trees, shrubs, and wildflowers. There is considerably less biomass in primary consumers, which include herbivorous birds, mammals, and insects. There is even less biomass in secondary consumers, which include hawks, owls, and carnivorous mammals. A similar scenario exists in the grassland ecosystem of Africa. All the grass in Africa piled together would dwarf a mound of all the grasshoppers, gazelles, zebras, wildebeests, and other animals that eat grass. That mound of herbivores, in turn, would dwarf the relatively tiny mound of all the lions, hyenas, and other carnivores that feed on them.

In aquatic ecosystems, the pyramid of biomass has a very different shape. In these ecosystems, the major producers are the phytoplankton, the tiny algae that float or swim through the water. Unlike trees and shrubs, algae have short lives with rapid reproduction, and they are consumed in large numbers. As a result, although the productivity of algae is

---

Trophic pyramid A chart composed of stacked rectangles representing the amount of energy or biomass in each trophic group.

Pyramid of energy A trophic pyramid that displays the total energy existing at each trophic level.

Pyramid of biomass A trophic pyramid that represents the standing crop of organisms present in different trophic groups.

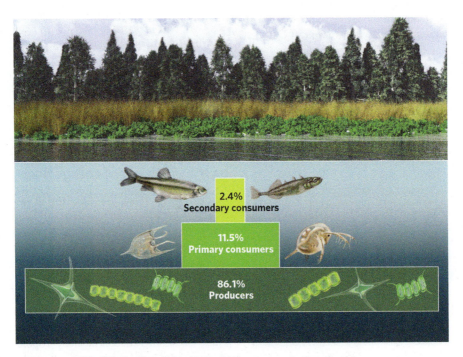

Figure 20.14 The energy pyramid of Cedar Bog Lake. Most of the energy assimilated in this lake ecosystem is found in producers such as algae. Because energy is lost as it is converted from one trophic level to the next, considerably less energy is found in primary consumers such as zooplankton and secondary consumers such as fish. Data from R. L. Lindeman, The trophic-dynamic aspect of ecology, *Ecology* 23 (1942): 399–417.

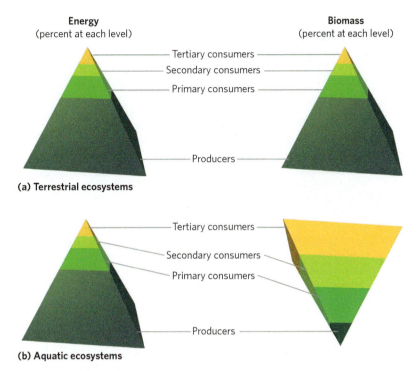

Figure 20.15 Pyramids of energy and biomass. (a) In terrestrial ecosystems, pyramids of energy and biomass have similar shapes because most of the energy and most of the standing biomass are found in the producers. **(b)** In aquatic ecosystems, most of the energy is still found in the producers, but these producers are primarily tiny algae that do not live very long because they are rapidly consumed by herbivores. This continual rapid consumption results in a large biomass of consumers in these systems.

much higher than the consumers of algae, the standing biomass of algae is often much lower than the standing biomass of the consumers of algae. This creates an inverted pyramid, as illustrated in Figure 20.15b.

THE EFFICIENCIES OF ENERGY TRANSFERS

As we discussed, the amount of energy that moves from one trophic level to another determines how much energy or biomass can exist at each trophic level. The amount of energy that is transferred between adjacent trophic levels depends on several steps that occur within each trophic level and include consumption, assimilation, and production. When we consider the transfer of energy from one trophic level to another, we can quantify it as the percentage of available energy that is transferred, which is a measure of efficiency.

Consumption Efficiency

Turning back to Figure 20.4, we see that the first step in the transfer of energy from one trophic level to another is the consumption of energy from the lower trophic level. Some of the total amount of energy available in the lower trophic level is consumed and the rest is left to become dead organic matter. The percentage of energy or biomass in a trophic level that is consumed by the next higher trophic level is

known as the **consumption efficiency.** Consumption efficiency is calculated using the following equation:

$$\text{Consumption efficiency} = \frac{\text{Consumed energy (J)}}{\substack{\text{Net production energy of the} \\ \text{next lower trophic level (J)}}}$$

For example, there might be 10 J of energy available in a field of wildflowers, but the herbivores might consume only 1 J of that energy because many of the plant species possess anti-herbivore defenses. In this case, the consumption efficiency of the herbivores would be 10 percent. In ecosystems containing producers with fewer defenses against herbivores, the consumption efficiency would be much higher and more energy would enter the consumer trophic level.

Assimilation Efficiency

As we have discussed, energy that is consumed is subsequently either assimilated or egested. In the case of plants, especially terrestrial plants, many components such as cellulose and lignin are not easily digested. Similarly, the feathers, bones, exoskeletons, and hair of animals are not easily digestible by the predators that consume them. Owls provide

---

Consumption efficiency The percentage of energy or biomass in a trophic level that is consumed by the next higher trophic level.

(a)

(b)

Figure 20.16 Undigested energy. (a) Owls such as this barn owl (*Tytoalba*) consume small rodents but they do not digest the hair and bones. **(b)** Instead, owls regurgitate a pellet that contains these indigestible materials. Consumers that eat large amounts of indigestible materials have low assimilation efficiencies. Photos by **(a)** Frederic Desmette/Biosphoto, and **(b)** Philippe Clement/naturepl.com.

an excellent example. All species of owls are predators that commonly feed on small mammals. They swallow their prey whole, digest it, and then regurgitate the hair and small bones in the form of compact pellets (**Figure 20.16**).

The percentage of consumed energy that is assimilated is known as the **assimilation efficiency** and is calculated using the following equation:

$$\text{Assimilation efficiency} = \frac{\text{Assimilated energy (J)}}{\text{Consumed energy (J)}}$$

Assimilation efficiency varies a great deal from one trophic level to another. For example, herbivores that feed on seeds, such as many species of birds and rodents, have assimilation efficiencies as high as 80 percent. In contrast, herbivores, such as horses that feed on grasses and woody vegetation, have assimilation efficiencies of only 30 to 40 percent. Secondary consumers, which are predators on the herbivores, typically have high assimilation efficiencies, ranging from 60 to 90 percent. These high assimilation efficiencies occur because consumed animal tissues are generally more digestible than consumed plant tissues.

Net Production Efficiency

Ultimately, we want to know how much assimilated energy is converted into the growth and reproduction of the organisms in a trophic level. The percentage of assimilated energy that is used for growth and reproduction is the **net production efficiency**, which is calculated using the following equation:

$$\text{Net production efficiency} = \frac{\text{Net production energy (J)}}{\text{Assimilated energy (J)}}$$

Put another way, net production efficiency is the percentage of assimilated energy that remains after respiration. For active homeothermic animals that must spend a large fraction of their energy to maintain a constant

body temperature, move about, circulate their blood, and balance their salts, net production efficiency can be as low as 1 percent. In contrast, sedentary poikilothermic animals, particularly aquatic species, channel as much as 75 percent of their assimilated energy into growth and reproduction.

Understanding net production efficiency also has practical applications. For example, livestock producers understand that if they keep their animals indoors at high densities during the winter, the animals will spend less of their assimilated energy on respiration to maintain a constant body temperature. As a result, the animals can devote more of their assimilated energy to growth, and faster-growing livestock result in more income for the livestock producer.

Ecological Efficiency

Now that we understand the three efficiencies that occur within each trophic level, we can explore the efficiency of energy transfer between adjacent trophic levels. **Ecological efficiency,** also known as **food chain efficiency,** is the percentage of net production from one trophic level, compared to the next lower trophic level:

$$\text{Ecological efficiency} = \frac{\text{Net production energy of a trophic level (J)}}{\text{Net production energy of the next lower trophic level (J)}}$$

Assimilation efficiency The percentage of consumed energy that is assimilated.

Net production efficiency The percentage of assimilated energy that is used for growth and reproduction.

Ecological efficiency The percentage of net production from one trophic level, compared to the next lower trophic level. *Also known as* **Food chain efficiency.**

When we consider all the efficiencies in the chain of events that starts with consumption and ends with net production, we can see that each plays a role in determining the ecological efficiency between trophic levels. To confirm that this is true, we can multiply the efficiencies in each step of the process to produce the equation for ecological efficiency:

$$\text{Ecological efficiency} = \text{Consumed energy (J)} \times$$
$$\text{Assimilation efficiency} \times$$
$$\text{Net production efficiency}$$

$$\text{Ecological efficiency} = \frac{\text{Consumed energy (J)}}{\text{Net production energy of the next lower trophic level (J)}} \times$$

$$\frac{\text{Assimilated energy (J)}}{\text{Consumed energy (J)}} \times$$

$$\frac{\text{Net production energy (J)}}{\text{Assimilated energy (J)}}$$

$$\text{Ecological efficiency} = \frac{\text{Net production energy (J)}}{\text{Net production energy of the next lower trophic level (J)}}$$

Because energy is lost at each of these steps, ecological efficiencies are typically quite low and range from 5 to 20 percent. Given this range, ecologists often use 10 percent as a rule of thumb. Such low ecological efficiencies help us understand why each trophic level in the pyramid of energy becomes much smaller as we move from producers to primary consumers to secondary consumers.

An ecological efficiency of 10 percent between adjacent trophic levels means that only 10 percent of the total energy present in the producer trophic level will be found in primary consumers, and only 1 percent will be found in secondary consumers. As you can see, an ecological efficiency of 10 percent makes it difficult to have long food chains in an ecosystem because there is not enough energy to support additional higher trophic levels. The only way to support additional trophic levels is by increasing the amount of energy at the producer level or by increasing the ecological efficiencies moving between adjacent trophic levels.

A 10 percent ecological efficiency also applies to humans and where we feed along a food chain. For example, when we act as a primary consumer and eat plants, we can assimilate 10 percent of the energy available from plants. However, when we act as a secondary consumer and consume herbivores, we assimilate 1 percent of the original energy that was in the plants. As a result, we would expect a diet composed of more plants and less meat to dramatically increase the amount of food available to humans.

Ecological Efficiency and the Number of Trophic Levels
Aquatic ecosystems typically have more trophic levels than terrestrial ecosystems, in part because of differences in ecological efficiencies. In terrestrial ecosystems,

producers are composed primarily of plants that range in size from wildflowers to trees. In many of these plants, a large proportion of tissue is dedicated to deterring herbivores from consuming them. Other plants contain an enormous proportion of biomass that herbivores cannot consume, such as the wood of trees. As a result, there is low consumption efficiency in terrestrial ecosystems, so a large fraction of the producer biomass ultimately becomes detritus. In contrast, aquatic ecosystems are composed primarily of unicellular algae that contain fewer defenses and are relatively easy for herbivores to digest. Therefore, algae provide higher consumption and assimilation efficiencies to their herbivores, which leads to higher ecological efficiencies. These higher ecological efficiencies mean that a higher fraction of the ecosystem's energy can move up the food chain and support additional trophic levels in aquatic ecosystems, compared to terrestrial ecosystems.

In addition to having higher ecological efficiencies, aquatic ecosystems also contain herbivores that are generally quite small, such as the tiny zooplankton that consume algae. These zooplankton are consumed by secondary consumers, such as small species of fish (**Figure 20.17**). These small fish are consumed by medium fish, which in turn are consumed by large fish. Because the producers are mostly single-cell algae, and because each successive consumer is only a bit larger than the diet it consumes, aquatic ecosystems can often contain five trophic levels.

We can contrast this with the situation that occurs in terrestrial ecosystems. In terrestrial ecosystems, the producers are relatively large, ranging in size from wildflowers to large trees. In addition, many of the major herbivores, such as deer and antelope, are quite large and can

Figure 20.17 Aquatic ecosystems begin with small herbivores, which are consumed by small fish that are consumed by larger fish. As a result, aquatic ecosystems can often have more trophic levels than terrestrial ecosystems. Photo by Picavet/Getty Images.

ANALYZING ECOLOGY

Quantifying Trophic Efficiencies

Eco TV

 macmillanlearning.com/ricklefsvideo

Trophic efficiency calculations derive from data collected on consumption efficiency, assimilation efficiency, net production efficiency, and ecological efficiency.

Energy	Terrestrial Ecosystem	Lake Ecosystem	Stream Ecosystem
Net Production Available In Lower Trophic Level (J)	1,000	1,000	400
Consumed Energy (J)	250	650	260
Assimilated Energy (J)	120	450	100
Net Production Energy (J)	50	190	40

We can calculate these efficiencies using the data from the terrestrial ecosystem.

$$\text{Consumption efficiency} = \frac{\text{Consumed energy (J)}}{\text{Net production energy of the next lower trophic level (J)}}$$

$$\text{Consumption efficiency} = \frac{250\ J}{1,000\ J} = 25\%$$

$$\text{Assimilation efficiency} = \frac{\text{Assimilated energy (J)}}{\text{Consumed energy (J)}}$$

$$\text{Assimilation efficiency} = \frac{120\ J}{250\ J} = 48\%$$

$$\text{Net production efficiency} = \frac{\text{Net production energy (J)}}{\text{Assimilated energy (J)}}$$

$$\text{Net production efficiency} = \frac{50\ J}{120\ J} = 42\%$$

$$\text{Ecological efficiency} = \frac{\text{Net production energy of a trophic level (J)}}{\text{Net production energy of next lower trophic level (J)}}$$

$$\text{Ecological efficiency} = \frac{50\ J}{1,000\ J} = 5\%$$

YOUR TURN Use the data for the lake and stream ecosystems to calculate the four different efficiencies for each ecosystem.

Based on these calculations, why do the two aquatic ecosystems have higher ecological efficiencies than the terrestrial ecosystem? Which efficiencies cause the stream ecosystem to have lower net production energy than the lake ecosystem?

---

be consumed only by very large secondary consumers, such as wolves and lions. Because terrestrial ecosystems contain many large producers and large herbivores, they are less likely to have a fourth and fifth trophic level. In short, the low ecological efficiency of terrestrial ecosystems combined with the large size of many producers and herbivores results in terrestrial ecosystems that commonly contain only three or four trophic levels.

RESIDENCE TIMES

Ecological efficiency tells us the proportion of energy that moves from producers into the higher trophic levels. We also want to examine the rate of energy movement between trophic levels, which tells us how long energy remains in a given trophic level and, therefore, how much energy can accumulate at that level. We can define the length of time that energy spends in a given trophic level as the **energy residence time.** Energy residence time is directly related to the amount of energy that exists in a trophic level; the longer the residence time, the greater the accumulation of energy in that trophic level. The average residence time of energy

at a particular trophic level equals the energy present in the tissues of organisms divided by the rate at which energy is converted into biomass, or net productivity:

$$\text{Energy residence time (years)} = \frac{\text{Energy present in a trophic level (J/m}^2)}{\text{Net productivity (J/m}^2\text{/year)}}$$

If we substitute biomass for energy in this equation, we can determine **biomass residence time,** which is the length of time that biomass spends in a given trophic level:

$$\text{Biomass residence time (years)} = \frac{\text{Biomass present in a trophic level (kg/m}^2)}{\text{Net productivity (kg/m}^2\text{/year)}}$$

Energy residence time The length of time that energy remains in a given trophic level.

Biomass residence time The length of time that biomass remains in a given trophic level.

For example, plants in humid tropical forests produce dry matter at an average rate of 1.8 kg/m²/year and have an average living biomass of 42 kg/m². Inserting these values into the preceding equation gives a biomass residence time of 23 years. Average biomass residence times for primary producers range from more than 20 years in forest ecosystems to less than 20 days in aquatic phytoplankton-based ecosystems. The much shorter biomass residence time in aquatic ecosystems is the reason these ecosystems frequently have inverted pyramids of biomass, like the one shown in Figure 20.15b; the biomass produced by algae is rapidly consumed by zooplankton.

These residence times track the movement of energy that is consumed from one trophic level to another by primary, secondary, and tertiary consumers, but these estimates do not take into account the residence time of the dead organic matter that is consumed by scavengers, detritivores, and decomposers. We calculate the residence time in dead organic matter by using a variation of the equation for energy residence time:

$$\text{Dead organic matter residence time (years)} = \frac{\text{Dead organic matter present in a trophic level (kg/m}^2\text{)}}{\text{Dead organic matter productivity (kg/m}^2\text{/year)}}$$

For example, the residence time of dead leaf litter is 3 months in humid tropical ecosystems, 1 to 2 years in dry tropical ecosystems, 4 to 16 years in temperate forest ecosystems in the southeastern United States, and more than 100 years in temperate mountains and boreal ecosystems (**Figure 20.18**). As you can see from the preceding equation, these differences in the residence times of leaf litter are a function of how much litter falls each year and how rapidly decomposition can occur. As we discussed in our tour of the biomes in Chapter 6, warm temperatures and abundant moisture allow the rapid decomposition of litter in lowland tropical regions, while colder and drier conditions cause slow decomposition and litter accumulation in temperate and boreal ecosystems.

STOICHIOMETRY

In addition to obtaining energy, organisms also must have the correct balance of nutrients to grow and reproduce. Ideally, the ratio of nutrients an organism needs must match the ratio of nutrients it consumes, but sometimes this is a challenge. The study of the balance of nutrients in ecological interactions, such as between an herbivore and a plant, is called **ecological stoichiometry.** Understanding ecological stoichiometry is useful in explaining variation in the ecological efficiencies that we have just discussed.

The balance of nutrients required by different species depends on their biology. For example, the composition of an organism's body can affect the types of nutrients it needs to obtain. Diatoms have high requirements for silicon because they produce glass shells for protection (see Figure 2.5), whereas vertebrates require large amounts of calcium and phosphorus for growing bones and scales. In the case of birds and mammals that primarily consume fruits, which are typically low in calcium and phosphorus, they often must supplement their diets with snail shells or

---

Ecological stoichiometry The study of the balance of nutrients in ecological interactions, such as between an herbivore and a plant.

(a)

(b)

Figure 20.18. Residency times for dead organic matter. (a) In tropical forests, such as this site in Thailand, warm temperatures and high precipitation cause a rapid breakdown in dead organic matter. **(b)** In temperate forests, such as this site in Ohio, the colder temperatures cause a slower breakdown of dead organic matter and therefore a longer residency time. Photos by **(a)** Pakorn Lopattanakij/Alamy; and **(b)** Steve and Dave Maslowski/Getty Images.

bits of limestone to take in sufficient amounts of calcium and phosphorus.

Growth rates and other life-history traits also can influence the nutrient composition of organisms. For instance, if we compare two types of zooplankton, we find that slowly growing marine copepods (**Figure 20.19a**) have nitrogen-to-phosphorus ratios as high as 50:1, whereas rapidly growing freshwater water fleas (Figure 20.19b) have ratios below 15:1. The faster-growing water fleas have a lower ratio because they must maintain a higher concentration of phosphorus in their tissues to synthesize the large amounts of proteins necessary for rapid growth.

To understand how stoichiometry affects the efficiencies of energy transfer between adjacent trophic groups, consider

(a)

(b)

Figure 20.19 Effects of growth rate on stoichiometry. (a) Slow-growing organisms, such as this marine copepod, have tissues with a high ratio of nitrogen to phosphorus of approximately 50:1. **(b)** Fast-growing organisms, such as this water flea (*Daphnia magna*), require much more phosphorus so that they can produce nucleic acids. As a result, their tissues have a lower ratio of nitrogen to phosphorus, approximately 15:1. Photos by **(a)** Peter Parks/Image Quest Marine, and **(b)** Laguna Design/Science Source.

again the water flea that has tissues containing a 15:1 ratio of nitrogen and phosphorus. This becomes challenging when the ratio of ingested nutrients in the algae does not match the ratio of nutrients needed by the water fleas. When this occurs, consumers must process larger amounts of food to obtain sufficient amounts of the most limiting nutrient. If the water flea consumed algae that contained a 30:1 ratio of nitrogen and phosphorus, it would have to consume twice as much algae to meet its phosphorus need. Moreover, when it consumed algae containing a 30:1 ratio, it would have to excrete the excess nitrogen. In this example, the assimilation efficiency of the water flea, in terms of nitrogen, declines and therefore the ecological efficiency of the water flea's trophic level declines. As you can see, tracking the ecological stoichiometry helps us understand why there can be a low ecological efficiency whenever a nutrient-poor producer is consumed by an herbivore that requires a nutrient-rich diet. In short, understanding ecological stoichiometry helps us explain why ecological efficiencies can vary between trophic levels.

In this chapter, we have explored how energy moves through ecosystems. We have seen that this process begins with producers capturing the energy of the Sun and converting it into producer biomass. The rate of primary production influences the rate of production for higher trophic levels. We have seen that the amount of primary productivity differs among ecosystems around the world because of variations in temperature, precipitation, light, and nutrients. The amount of energy that passes through the trophic levels of ecosystems depends on the efficiency of each step in the chain, with all the steps producing an overall ecological efficiency between adjacent trophic levels and different residence times of energy, biomass, and dead organic matter.

CONCEPT CHECK

1. Why does the energy pyramid exhibit large reductions in energy level as we move up through the trophic levels?
2. What are the two possible fates of assimilated energy in a primary consumer?
3. Why does organic matter accumulate in cold, dry biomes?

ECOLOGY TODAY   APPLYING THE CONCEPTS

Feeding an Ocean of Whales

Humpback whales. Researchers have determined the number of whales living off the West Coast of the United States, such as this humpback whale. When whale numbers were combined with data on the diet of each species and the ecological efficiencies between adjacent trophic levels, the researchers estimated that whales consume approximately 12 percent of the NPP of the ocean off the West Coast of the United States. Photo by Masa Ushioda/AGE fotostock.

The ocean contains a diversity of species feeding at different trophic levels, and whales sit at or near the top trophic level. Although whales are a top predator, it has been a long-standing challenge for us to understand how much NPP is required to provide the energy that whale populations need. Knowing the NPP requirement would tell us how much energy passes through this trophic level. It would also allow us to estimate how much NPP whales used prior to the commercial hunting that has caused many whale populations to decline during the past century. The reduction in consumed NPP could potentially be consumed by other species in the ecosystem.

Given that most whales live in the open ocean, how does one begin to estimate the percentage of NPP that whales consume? A group of researchers decided to tackle this challenge by starting with information about the whales and working down to the producers. First, they settled on a particular study area: the region of ocean within 550 km from the western coast of the United States. Next, they estimated how many whales lived in this area. Fortunately, line transects had been conducted in this region of the ocean for 15 years and these transects provided abundance estimates for 21 species of whales that included dolphins, porpoises, sperm whales (*Physetermacrocephalus*), and humpback whales (*Megapteranovaeangliea*). By knowing the typical mass of an individual whale for each species and how much energy whales of different sizes consume, assuming an assimilation efficiency of 80 percent, the researchers could estimate the total energy consumed annually by each species of whale.

The next step was to determine the diet of each whale species. Each species has a unique diet of prey made up of different proportions of primary consumers, such as krill, and secondary consumers, such as fish. Using existing data on whale diets, the researchers knew the proportion of primary and secondary consumers that each whale species eats. They then used the ecologist's rule of thumb that adjacent trophic levels passed along energy with an ecological efficiency of 10 percent. Based on this rule of thumb, they could determine how much NPP was required to produce the prey consumed by the whales.

The final step was to determine how much NPP was available in this region of the ocean. They measured NPP using satellites and the remote sensing techniques for detecting chlorophyll concentrations that we discussed earlier in the chapter. Once NPP was quantified, they determined that whales consumed about 12 percent of the ocean's NPP. For comparison, this is about half of the NPP required to support fish populations that are commercially harvested.

 The percentage of NPP consumed by this group of mammals is likely to increase dramatically in the coming decades. For example, during the 1980s and early 1990s, many dolphins were accidentally killed by commercial fishing operations. Because this source of mortality has declined in recent years, we anticipate an increase in the number of some types of whales, such as dolphins. This may cause a higher proportion of the NPP to be consumed. Moreover, because many of the larger species of whales are now protected around the world, their populations are rising. Scientists expect populations of some of the larger species to eventually triple or quadruple, which will result in an even greater consumption of the ecosystem's NPP.

A current challenge is estimating how this large increase in the biomass of top predators will affect the ocean's distribution of energy among the trophic levels. An insight on this question came in 2014 when researchers reported the results of their studies on whales transporting nutrients. For many years, the paradigm had been that whales consume a great deal of primary productivity and, as a result, may be competing with recreational anglers and commercial fishing operations. By radio-tagging whales to see where they travel, the researchers discovered that many whales are hunting for food in deep waters and then defecating when they come up to the surface. As a result, the whales are transporting nutrients from the deep waters to the surface waters, where their egested food can provide nutrients for the algae. The researchers noted that this "whale pump" of nutrients might provide a large benefit to the surface-water food web and help feed the fish that humans catch and pursue. As whale populations continue to recover, these plumes of egested nutrients should be even more abundant, and we will continue to learn about the key role that whales play in ocean ecosystems.

SOURCES:

Barlow, J., et al. 2008. Cetacean biomass, prey consumption, and primary production requirements in the California Current ecosystem. *Marine Biology Progress Series* 371: 285–295.

Roman, J., et al. 2014. Whales as ecosystem engineers. *Frontiers in Ecology and the Environment* 12: 377–385.

SUMMARY OF LEARNING OBJECTIVES

20.1 Primary productivity provides energy to the ecosystem.
Primary productivity is the process of capturing solar or chemical energy and converting it into chemical bonds by photosynthesis or chemosynthesis over a given amount of time. We can distinguish between gross primary productivity, which is the total amount of energy assimilated, and net primary productivity, which is the assimilated energy that is converted into producer biomass. Net primary productivity can be measured in numerous ways, including measuring plant biomass, measuring CO_2 uptake and release in terrestrial ecosystems, measuring O_2 uptake and release in aquatic ecosystems, and using remote sensing. Across all ecosystems, the amount of net primary productivity has a direct positive relationship with the amount of net secondary productivity.

Key Terms: Primary productivity, Standing crop, Gross primary productivity (GPP), Net primary productivity (NPP), Remote sensing, Egested energy, Assimilated energy, Respired energy, Net secondary productivity

20.2 Net primary productivity differs among ecosystems.
Net primary productivity differs a great deal among ecosystems around the world. In terrestrial ecosystems, major drivers of this productivity include temperature, precipitation, nitrogen, and phosphorus. In aquatic ecosystems, major drivers include temperature, light, nitrogen, and phosphorus.

20.3 The movement of energy depends on the efficiency of energy flow.
The energy of ecosystems exists in different trophic levels and moves between these trophic levels with different efficiencies. Pyramids of energy exhibit similar distributions among ecosystems, with producers having the most energy and each higher trophic group possessing less energy. Pyramids of biomass show a similar trend in terrestrial ecosystems, but they are often inverted in aquatic ecosystems. To understand how energy moves between trophic levels, we can calculate the consumption efficiency, assimilation efficiency, and net production efficiency, all of which can be multiplied to determine the overall ecological efficiency of transferring energy between adjacent trophic groups. The efficiencies can be affected by the stoichiometry of consumer tissues relative to the consumer's diet and can affect the number of links in an ecosystem's food chain. Efficiencies can also affect the residence time of energy and biomass in ecosystems.

Key Terms: Trophic pyramid, Pyramid of energy, Pyramid of biomass, Consumption efficiency, Assimilation efficiency, Production efficiency, Ecological efficiency, Energy residence time, Biomass residence time, Ecological stoichiometry

CRITICAL THINKING QUESTIONS

1. Why is the efficiency of energy transfer between grasses and gazelles quite low?

2. How would you distinguish between gross primary productivity and net primary productivity in a desert ecosystem?

3. Compare and contrast the measurement of primary productivity in terrestrial versus aquatic ecosystems.

4. Compare and contrast the factors that limit the net primary productivity in terrestrial versus aquatic ecosystems.

5. How might remote sensing be used to track changes in the standing crop of aquatic ecosystems in response to global warming?

6. Why might assimilation efficiencies be much higher for herbivores eating seeds than for herbivores eating leaves?

7. What is the likely shape of the pyramid of biomass in a lake?

8. Why would you expect larger changes in stoichiometry between predators and prey to alter assimilation efficiencies?

9. Why are residence times much longer in forest ecosystems than in aquatic phytoplankton-based ecosystems?

10. How might the carrying capacity for people in the United States change if we ate more plant products than animal products?

GRAPHING THE DATA NPP Versus the Total Primary Productivity of Ecosystems

Figure 20.7 shows estimates of NPP for various ecosystems around the world. However, these data tell us only the amount of primary production per square meter. We can also examine the amount of primary production in terms of the total amount produced in a year among the different ecosystems. Using the data on NPP and area in the table, calculate total production for each ecosystem by multiplying NPP and area. Then plot the total production values using a bar graph.

Discuss how the NPP and the area of the different ecosystems affect total production.

SOURCE:
Data from Whittaker, R. H., and G. E. Likens. 1973. Primary production: The biosphere and man. *Human Ecology* 1: 357–369.

Terrestrial Ecosystems	Npp (g/m^2/year)	Area (10^6 km^2)	Total Production (10^{12} kg/year)
Tropical Rainforest	2,000	17.0	
Tropical Seasonal Forest	1,500	7.5	
Temperate Rainforest	1,300	5.0	
Temperate Seasonal Forest	1,200	7.0	
Boreal Forest	800	12.0	
Savanna	700	15.0	
Cultivated Land	650	14.0	
Woodland/Shrubland	600	8.0	
Temperate Grassland	500	9.0	
Tundra	140	8.0	
Desert Shrub	70	18.0	

Aquatic Ecosystems	Npp (g/m^2/year)	Area (10^6 km^2)	Total Production (10^{12} kg/year)
Swamp and Marsh	2,500	2.0	
Coral Reef	2,000	0.6	
Salt Marsh	1,800	1.4	
Upwelling Zones	500	0.4	
Lake and Stream	500	2.5	
Continental Shelf	360	26.6	
Open Ocean	125	332.0	

21 Movement of Elements in Ecosystems

Living in a Dead Zone

Each summer, as the Mississippi River flows into the Gulf of Mexico, an area develops where animals can't survive. While fish, crawfish, and crabs remain abundant in other parts of the Gulf, the summer algal bloom in this area makes it uninhabitable. In many cases, the rapidly increasing populations of algae contain green pigments that turn the water green. When the algae contain red pigments, the algal bloom is called a red tide.

Algal blooms can have both direct and indirect effects on aquatic organisms. A direct effect occurs when the species of algae or cyanobacteria that bloom produce toxins. At high algal densities, these toxins can accumulate to concentrations that impair the survival, growth, and reproduction of other species living in the area. One prominent example happened in Lake Erie in 2014, when tremendous amounts of algae grew in the bay next to Toledo, Ohio. The water was so green that is resembled pea soup, and the algae were producing such high concentrations of toxins that the city of Toledo had to shut off its water intake valves from the lake for several days to avoid harming its population of nearly 500,000 people. In response, the Ohio National Guard had to truck drinkable water into the city and many businesses and restaurants had to close.

An indirect effect of algal blooms occurs after the algae, also known as phytoplankton, bloom and die. While live algae produce oxygen during photosynthesis, bacteria that consume the enormous biomass of dead algae use large amounts of oxygen. This leads to a dramatic reduction in oxygen in the water and causes many of the animals in the water to die from oxygen deprivation. Aquatic ecosystems that experience algal blooms and large animal die-offs are called dead zones.

What causes the large algal blooms? Researchers have discovered that many rivers carry large amounts of nutrients, such as nitrogen and phosphorus, that come from fertilizers that run off lawns and agricultural fields when it rains. This water runoff enters streams and rivers that join before emptying into the ocean. Other sources of nutrients include wastewater with a range of components, including detergents that contain phosphorus and sewage, that is released from wastewater treatment systems when they are overwhelmed by large rain events. The nutrients that rivers dump into the ocean allow rapid algal growth, particularly during the warm summer months and this causes algal blooms. Because warm temperatures play a key role, global warming will likely favor an increase in the frequency and intensity of algal blooms today and in the future.

The Mississippi River drains 41 percent of the contiguous United States, so it carries nutrients gathered from a very large area. The resulting dead zone in the Gulf of Mexico can cover more than 22,000 km² during the summer—an area the size of New Jersey. As autumn weather comes, fewer nutrients enter the Mississippi River and the temperatures in the Gulf of Mexico become cooler, conditions that make it more difficult for algal populations to increase. As a result, the dead zone disappears each winter.

Human activities cause most dead zones, although some do have natural causes. The abundance of dead zones around the world is growing rapidly. In the 1910s, only four dead zones were known to exist. This number increased to 49 dead zones in the 1960s and 87 in the 1980s. The number of dead zones rose to 305 in 1995 and increased further to 405 by 2008, the most recent date for which there is an estimate. For example, a large dead zone occurs in the Chesapeake Bay on the east coast of North America, where up to 40 percent of the bay can become *hypoxic,* meaning the water is low in oxygen. Similarly, the bottom of Lake Erie becomes hypoxic each summer. Around the world, dead zones cover a total area of more than 205,000 km².

> "The resulting dead zone in the Gulf of Mexico can cover more than 22,000 km² during the summer—an area the size of New Jersey."

The effects of a dead zone on fish populations. Because algae that bloom will eventually die, the decomposition uses up nearly all the oxygen in the water. These fish died from a dead zone that occurred in Lake Trafford, Florida. Photo by Michele and Tom Grimm/Alamy.

Drainage basin of the Mississippi River

Missouri River
Mississippi River
Ohio River

Oxygen concentration
Low   High

A dead zone at the mouth of the Mississippi River. The Mississippi River drains 41 percent of the contiguous United States and carries nutrients from fertilizers leached from people's yards and agricultural fields and from wastewater from communities located throughout the region. These nutrients facilitate the rapid growth of algae that ultimately cause a dead zone, which you can see at the mouth of the river. Dead zone after NOAA.

The existence of dead zones illustrates why we need to understand how nutrients—including water, nitrogen, and phosphorus—move within and between ecosystems and the important role that decomposition plays in recycling these nutrients. In this chapter, we will examine the movement of nutrients and how nutrients are regenerated in terrestrial and aquatic ecosystems.

SOURCES:

Diaz, R. J., and R. Rosenberg. 2008. Spreading dead zones and consequences for marine ecosystems. *Science* 321: 926–929.

Neuhaus, L. 2016. A menace afloat. *New York Times,* July 19.

Rabalais, N. N., et al. 2002. Gulf of Mexico hypoxia, a.k.a. The dead zone. *Annual Review of Ecology and Systematics* 33: 235–263.

LEARNING OBJECTIVES

After reading this chapter, you should be able to:

21.1 Describe how the hydrologic cycle moves many elements through ecosystems.

21.2 Explain why the carbon cycle is closely tied to the movement of energy.

21.3 Illustrate the ways in which nitrogen cycles through ecosystems in many different forms.

21.4 Describe how the phosphorus cycle moves between land and water.

21.5 Explain why most nutrients regenerate in the soil in terrestrial ecosystems.

21.6 Illustrate why most nutrients regenerate in the sediments in aquatic ecosystems.

Unlike energy—which moves through ecosystems—elements such as hydrogen, oxygen, carbon, nitrogen, and phosphorus cycle among the biotic and abiotic components of ecosystems. The movement of these elements through ecosystems is affected by chemical, physical, and biological processes. To understand the movement of these elements, which exist in a variety of chemical forms, it is helpful to think of different pools in which a given element resides, as well as the different processes that are responsible for moving an element from one pool to another. For example, two important pools for carbon are the CO_2 that exists in the atmosphere and the biomass of producers that use carbon to build their tissues. In this example, the process that causes carbon to move from the atmosphere to producers is photosynthesis.

Organisms contain large amounts of hydrogen, oxygen, and carbon. However, as we noted in Chapter 2, organisms also need seven major nutrients: nitrogen, phosphorus, sulfur, potassium, calcium, magnesium, and iron. Some elements are required in much smaller amounts; these include silicon, manganese, and zinc. In this chapter, we will examine the biogeochemical cycles of some of the major elements on Earth by looking at interactions among biological, geological, and chemical processes. We will also explore how human activities are currently altering these cycles in ways that have far-reaching effects on ecosystems.

21.1 The hydrologic cycle moves many elements through ecosystems

Throughout the previous chapter, we saw that water plays a critical role at every level of ecological study.

In Chapter 2, we discussed how water is a key compound involved in the many chemical transformations that happen in the living and nonliving components of ecosystems. Chapter 5 examined the processes that determine global climates, including patterns of precipitation, and in Chapter 6, we discussed the various types of aquatic biomes. We begin this chapter by tracking the movement of water through ecosystems. Once we understand how water moves through ecosystems, we will look at the role water plays as it moves elements through ecosystems.

THE HYDROLOGIC CYCLE

The movement of water through ecosystems and atmosphere, known as the **hydrologic cycle,** is driven largely by evaporation, transpiration, and precipitation. The largest pool of water, approximately 97 percent of all water on Earth, is found in the oceans. The remaining water exists in lakes, streams, rivers, wetlands, underground aquifers, and soil.

The hydrologic cycle is illustrated in **Figure 21.1**. Evaporation of water occurs from bodies of water, soil, and plants that experience evapotranspiration, which we discussed in Chapter 5. Solar energy provides the energy for the process of evaporation and evapotranspiration, which changes water from a liquid to a gas in the form of water vapor. There is a limit to the amount of water vapor that the atmosphere can contain. As additional water continues to evaporate, the water vapor in the atmosphere condenses into clouds that ultimately create precipitation in the form of rain, hail, sleet, or snow.

When precipitation drops from the atmosphere, it can take several paths. Some precipitation falls directly onto the surface of aquatic ecosystems and the rest falls onto terrestrial ecosystems. Water that falls onto terrestrial ecosystems can travel along the surface of the ground or it can infiltrate the ground, where it is either absorbed by plants or moves deeper into the ground and becomes part of the underlying groundwater. The surface runoff and some of the groundwater will eventually find their way back into water bodies, thereby completing the cycle.

Precipitation that falls on land either runs off along the surface or infiltrates the soil. In the soil, this water may evaporate, be taken up by plants, or enter groundwater. Excess water ultimately returns to the ocean.

Hydrologic cycle The movement of water through ecosystems and atmosphere.

Figure 21.1 The hydrologic cycle. The movement of water is driven by the energy of the Sun, which causes evaporation from soil and water bodies, and evapotranspiration from plants. Evaporated water condenses into clouds that eventually return the water to Earth as precipitation. Water from precipitation either runs off on the surface or infiltrates the soil. Runoff flows along the surface of the ground until it enters streams and rivers. Water in the soil is taken up by plants or enters the groundwater. Ultimately, excess water returns to the ocean.

The rate of evaporation must balance the rate of precipitation or water would continually accumulate in one part of the cycle. When we consider the hydrologic cycle on a global scale, we find that precipitation exceeds evaporation in terrestrial ecosystems, whereas evaporation exceeds precipitation in aquatic ecosystems. To help maintain an overall balance, the excess water that is evaporated from aquatic ecosystems is transported by the atmosphere and falls onto terrestrial ecosystems. At the same time, the excess water that falls on terrestrial ecosystems is transported in the form of runoff and groundwater into aquatic ecosystems.

HUMAN IMPACTS ON THE HYDROLOGIC CYCLE

Water cycles through Earth's biosphere with no net loss or gain over the long term. Therefore, any change in one part of the water cycle influences the other parts. For example, in large developed areas, construction materials such as roofing and paved parking lots are impervious to water infiltration. The amount of water that can percolate into the soil is significantly reduced, and we see an increase in surface runoff. Less water is able to infiltrate the soil for plants to use or to replenish the groundwater that many people rely on for drinking water. An increase in surface runoff also increases soil erosion. A similar effect occurs when we reduce the amount of plant biomass in a terrestrial ecosystem, as occurs during logging. Where there are fewer trees and other plants, much less precipitation is taken up by plant roots and subsequently released to the atmosphere through evapotranspiration. Consequently, the amount of surface runoff increases, which often causes severe soil erosion and flooding (**Figure 21.2**). Finally, when we pump out groundwater for irrigation or household use, we sometimes reduce the amount of groundwater at a rate that exceeds its replenishment. For example, in the Great Plains of the United States, a large supply of groundwater, known as the Ogallala aquifer, extends from South Dakota to Texas (**Figure 21.3**). This groundwater supplies about 30 percent of all water used for irrigation in the United States and provides drinking water for 82 percent of the people who live in the region. However, the extraction of this water has exceeded its rate of replenishment, and scientists are concerned that this critical supply of water for industry, households, and irrigation could run out sometime during this century.

Humans also alter the hydrologic cycle through activities that contribute to global warming. Scientists expect that as air and water temperatures rise, there will be an increase in the rate of water evaporation. An increase in the evaporation rate will cause water to move through the hydrologic cycle more quickly, potentially leading to an increased intensity of rain and snowstorms in various parts of the world.

CONCEPT CHECK

1. What are the pathways that water can take when precipitation falls on terrestrial ecosystems?
2. How does an increase in the construction of buildings and water-impervious parking lots affect the hydrological cycle?
3. How does tree removal affect the hydrological cycle?

Figure 21.2 Altering the hydrologic cycle. When forests are logged, as in this site in Haiti, fewer plant roots are available to hold soil and the soil absorbs less rainwater. These changes cause increased surface runoff, more severe floods, and large amounts of soil erosion.

Photo by REUTERS/Daniel Morel.

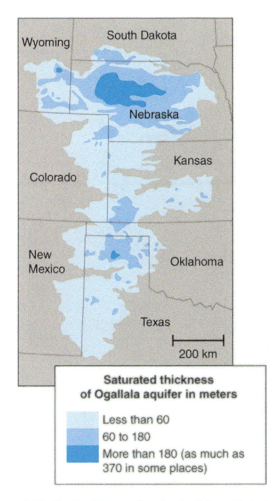

Figure 21.3 The Ogallala aquifer. The Ogallala aquifer is the largest source of ground water in the United States. Data from U.S. Geological Survey, Department of the Interior.

21.2 The carbon cycle is closely tied to the movement of energy

Because all organisms are composed of carbon, the movement of carbon in ecosystems largely follows the same paths as the movement of energy. In this section, we consider the many pools and processes that are involved in the carbon cycle. We will then examine how human activities have altered the carbon cycle.

THE CARBON CYCLE

To understand how the carbon cycle works, we need to consider six types of transformations: photosynthesis, respiration, sedimentation and burial, exchange, extraction, and combustion. The carbon cycle is illustrated in **Figure 21.4**.

We begin our examination with the processes of photosynthesis and respiration. As we have discussed in previous chapters, producers use photosynthesis in terrestrial and aquatic ecosystems to take CO_2 from the

air and water and to convert it to carbohydrates. These carbohydrates are used to make other compounds, including proteins and fats. The carbon that is locked up in producers can then be transferred to consumers, scavengers, detritivores, and decomposers. All these trophic groups experience respiration, which releases CO_2 back into the air or water.

In some habitats, such as the waterlogged sediments of swamps or marshes, oxygen is not available for respiration. Under such anaerobic conditions, some species of archaea use carbon compounds for respiration. For example, some archaea use methanol (CH_3OH) during respiration to produce CO_2 in the following reaction:

$$4\,CH_3OH \rightarrow CO_2 + 2\,H_2O + 3\,CH_4$$

As you can see from this reaction, the products are carbon dioxide, water, and methane. The methane that is released from swamps during anaerobic respiration is known as swamp gas. The production of methane through the process of anaerobic respiration is a growing concern because methane is a greenhouse gas and, on a per-molecule basis, it is 72 times more effective at absorbing and radiating infrared radiation back to Earth than CO_2.

Carbon dioxide is also exchanged between aquatic ecosystems and the atmosphere, as shown in Figure 21.4. The exchange occurs in both directions at a similar magnitude, which means there is little net transfer over time. As we discussed in Chapter 2, when CO_2 diffuses from the atmosphere into the ocean, some of it is used by plants and algae for photosynthesis and some is converted to carbonate (CO_3^{2-}) and bicarbonate ions (HCO_3^-). The carbonate ions can then combine with calcium in the water to form calcium carbonate ($CaCO_3$). Calcium carbonate has a low solubility in water, so it precipitates out of the water and becomes part of the sediments at the bottom of the ocean. Over millions of years, the calcium carbonate sediments that accumulate in the ocean bottoms, combined with the calcium carbonate skeletons from tiny marine organisms, can develop into massive sources of carbon in the form of rocks known as dolomite and limestone. Humans mine dolomite and limestone for use in making concrete and fertilizer as well as for numerous industrial processes.

Carbon can also be buried as organic matter before it fully decomposes. Over millions of years, some of this organic matter is converted to fossil fuels such as oil, gas, and coal. The rate of carbon burial is slow, and it is offset by the rate of carbon released into the atmosphere by the weathering of limestone rock and during volcanic eruptions. Because the process of sedimentation and burial can lock up carbon for millions of years, carbon moves through these pools very slowly.

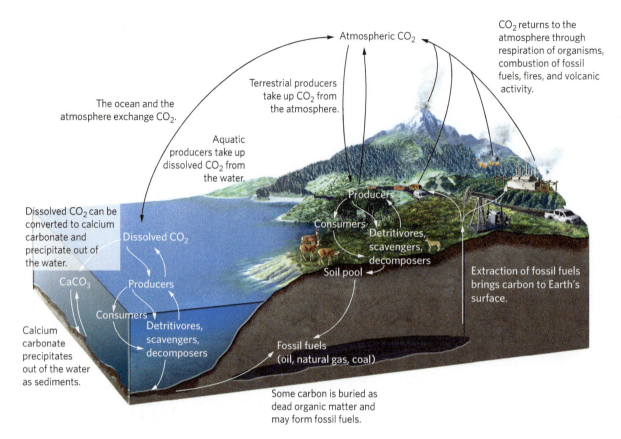

Atmospheric CO$_2$

CO_2 returns to the atmosphere through respiration of organisms, combustion of fossil fuels, fires, and volcanic activity.

Terrestrial producers take up CO$_2$ from the atmosphere.

The ocean and the atmosphere exchange CO$_2$.

Aquatic producers take up dissolved CO$_2$ from the water.

Dissolved CO$_2$ can be converted to calcium carbonate and precipitate out of the water.

Dissolved CO$_2$

Producers

Consumers

Detritivores, scavengers, decomposers

Soil pool

CaCO$_3$

Producers

Consumers

Detritivores, scavengers, decomposers

Calcium carbonate precipitates out of the water as sediments.

Fossil fuels (oil, natural gas, coal)

Extraction of fossil fuels brings carbon to Earth's surface.

Some carbon is buried as dead organic matter and may form fossil fuels.

Figure 21.4 The carbon cycle. In the carbon cycle, producers take up CO$_2$ from the atmosphere and the water. They transfer assimilated carbon to consumers, detritivores, scavengers, and decomposers. These organisms return CO$_2$ to the atmosphere and oceans through respiration. Throughout an ecosystem, CO$_2$ is exchanged between the atmosphere and the ocean and between the ocean and sediments. Carbon that has been stored underground for long periods turns into fossil fuels, which can be extracted. CO$_2$ is returned to the atmosphere through the combustion of fossil fuels, burning in terrestrial ecosystems, and volcanic activity.

HUMAN IMPACTS ON THE CARBON CYCLE

Now that we understand the pools and processes in the carbon cycle, we can examine how human activities have altered it. One major way in which humans have altered the carbon cycle is through the extraction and combustion of fossil fuels. During the past 2 centuries, extraction and combustion have been happening at an increasing rate to meet growing energy demands. Other forms of combustion by humans include burning land to prepare it for agriculture.

In Chapter 4, we discussed the recent rise in atmospheric CO$_2$ seen in measurements made atop Mauna Loa on the island of Hawaii. These measurements document an increase in CO$_2$ from 316 ppm in 1958 to more than 405 ppm in 2017—a 28 percent increase in just 59 years. Although these measurements at Mauna Loa began only in 1958, human activities have affected CO$_2$ concentrations in the atmosphere for much longer. To measure the concentrations of CO$_2$ that were in the atmosphere hundreds of thousands of years ago, researchers have traveled

to some of the coldest places on Earth. In locations such as Greenland and Antarctica, snowfall slowly compresses into ice that has tiny bubbles of air trapped inside. Because these bubbles contain tiny samples of the air from thousands of years ago, they can tell us about the climate conditions in the distant past. Each year, ice is formed by adding a new layer; the surface layers contain the youngest ice and the deepest layers contain the oldest ice. To sample the air that is trapped in the ice, researchers drill far down into the ice and extract long cylinders of ice known as ice cores (**Figure 21.5**).

Ice cores contain ice that has been formed as far back as 500,000 years ago. After determining the age of different ice core layers, each layer is melted, which allows the release of trapped air bubbles of which the concentration of CO$_2$ can be measured. **Figure 21.6** shows some of the data from these ice cores. As you can see, CO$_2$ concentrations in the atmosphere during the past 400,000 years have varied a great deal, from about 180 to 300 ppm. However, since 1800, as humans increasingly burned fossil fuels, CO$_2$

Figure 21.5 Ice cores. Researchers from Britain drill far down into ancient ice to collect ice cores that have been created in layers over the past 500,000 years. Photo by British Antarctic Survey/Science Source.

concentrations have increased exponentially to the current value of 405 ppm. This means that the current concentration of CO_2 in our atmosphere is 35 percent higher than the highest concentrations that existed during the past 400,000 years.

The rise in atmospheric CO_2 is of great importance to humans because CO_2 is a greenhouse gas that absorbs infrared radiation and radiates some of it back to Earth. Having CO_2 in the atmosphere helps keep our planet warm, but an excessive amount of CO_2 and other greenhouse gases in the atmosphere will cause our planet to become much warmer than it has been in a very long time. We know that the mean temperature on Earth is now 0.8°C warmer than it was when the first temperature measurements were taken in the 1880s. While a mean increase

of 0.8°C may not seem like much, we can find some dramatic changes in specific locations. Some regions, such as parts of Antarctica, have experienced cooler temperatures, while other regions, such as the high latitudes of Alaska, Canada, and Russia, are now 4°C warmer than they were a century ago.

These high-latitude regions contain large deposits of frozen peat, which is a mixture of dead sphagnum moss and other plants. Peat thaws and decomposes more easily with warmer temperatures. Because peat decomposes under anaerobic conditions, the decomposition produces methane, which is a greenhouse gas. This means that the rise in temperatures due to increased atmospheric CO_2 causes the release of additional greenhouse gases from the decomposing peat that exacerbate the problem.

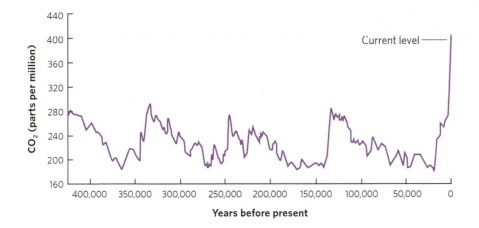

Figure 21.6 Atmospheric concentrations of CO_2 over time. Using measurements of air bubbles from ice cores and, in modern times, direct measurements, researchers have documented that for more than 400,000 years CO_2 concentrations never exceeded 300 ppm. During the past 200 years, the concentration of CO_2 in the atmosphere has increased rapidly and is now more than 400 ppm. Data from http://climate.nasa.gov/evidence.

Temperature increases can have numerous effects around the world, such as reducing the size of the polar ice sheets (see Chapter 5), altering the length of plant growing seasons, and changing the timing of plant and animal life histories. We will have much more to say about global warming in our discussion of global biodiversity conservation in Chapter 23.

CONCEPT CHECK

1. What form of carbon is produced during anaerobic respiration?
2. What form of carbon precipitates out of water?
3. How has increased combustion of carbon sources over the past two centuries contribute to global warming?

21.3 Nitrogen cycles through ecosystems in many different forms

Nitrogen is an important component of amino acids, the building blocks of proteins, and nucleic acids, which are the building blocks of DNA. Nitrogen exists in many different forms and is part of a complex set of pathways. In this section, we will examine the cycling of nitrogen and explore the ways that human activities have altered the nitrogen cycle.

THE NITROGEN CYCLE

A large pool of nitrogen gas (N_2) exists in the atmosphere, where it comprises 78 percent of all atmospheric gases. Nitrogen moves through five major transformations, shown in **Figure 21.7**: *nitrogen fixation, nitrification, assimilation, mineralization,* and *denitrification.*

Nitrogen Fixation

The process of converting atmospheric nitrogen into forms producers can use is known as **nitrogen fixation.** Nitrogen fixation converts nitrogen gas into either ammonia (NH_3), which is rapidly converted to ammonium (NH_4^+), or into nitrate (NO_3^-). The compound that is formed depends on whether nitrogen fixation occurs by organisms, lightning, or the industrial production of fertilizers.

As we discussed in previous chapters, some organisms are able to convert nitrogen gas into ammonia. Nitrogen fixation occurs in some species of cyanobacteria, in some free-living species of bacteria such as *Azotobacter,* and in mutualistic bacteria such as *Rhizobium* that live in the root nodules of some legumes and other plants (see Figure 17.4). Nitrogen fixation is an important source of required nitrogen, especially for early-succession plants colonizing habitats that have little available nitrogen. The process of nitrogen fixation requires a relatively high amount of energy, which nitrogen-fixing organisms obtain either by metabolizing organic matter from the environment or by acquiring carbohydrates from a mutualistic partner.

As you can see in Figure 21.7, nitrogen fixation can also occur through abiotic processes. For example, lightning provides a high amount of energy that can convert nitrogen gas into nitrate in the atmosphere. Similarly, combustion that occurs during wildfires or when fossil fuels are burned also produces nitrates. In both cases, nitrates, which are suspended in the air after combustion, fall to the ground with precipitation.

The industrial production of fertilizers that improve crop productivity converts nitrogen gas to either ammonia or nitrates. Like all nitrogen fixation, this process requires a great deal of energy and is powered mostly by the combustion of fossil fuels. The manufacture of nitrogen fertilizer has developed into such a large commercial endeavor that fixation conducted by humans now exceeds the nitrogen fixation that occurs through all natural processes.

Nitrification

Another process in the nitrogen cycle is **nitrification,** which converts ammonium to nitrite (NO_2^-) and then converts nitrite to nitrate (NO_3^-):

$$NH_4^+ \rightarrow NO_2^- \rightarrow NO_3^-$$

These conversions release much of the potential energy that is contained in ammonium. Each step is carried out by specialized bacteria and archaea in the presence of oxygen. In terrestrial and aquatic ecosystems, the conversion of ammonium to nitrites is carried out by *Nitrosomonas* and *Nitrosococcus* bacteria, whereas the conversion of nitrites to nitrates is carried out by *Nitrobacter* and *Nitrococcus* bacteria. Although nitrites are not an important nutrient for producers, plants can take them up and use them.

Assimilation and Mineralization

Producers can take up nitrogen from the soil or water as either ammonium or nitrates. Once producers take up nitrogen, they incorporate it into their tissues, a process known as assimilation, which was described in Chapter 20. When primary consumers ingest producers, they can either assimilate nitrogen from the producers or excrete it as waste. The same process occurs again with secondary consumers. Animal waste, as well as the biomass of dead

---

Nitrogen fixation The process of converting atmospheric nitrogen into forms producers can use.

Nitrification The final process in the nitrogen cycle, which converts ammonium (NH_4^+) or ammonia (NH_3) to nitrite (NO_2^-) and then to nitrate (NO_3^-).

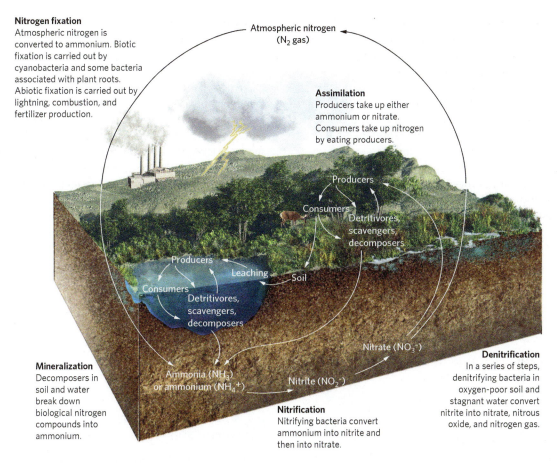

Nitrogen fixation
Atmospheric nitrogen is converted to ammonium. Biotic fixation is carried out by cyanobacteria and some bacteria associated with plant roots. Abiotic fixation is carried out by lightning, combustion, and fertilizer production.

Atmospheric nitrogen
(N_2 gas)

Assimilation
Producers take up either ammonium or nitrate. Consumers take up nitrogen by eating producers.

Producers

Consumers

Detritivores, scavengers, decomposers

Producers

Leaching

Soil

Consumers

Detritivores, scavengers, decomposers

Nitrate (NO_3^-)

Denitrification
In a series of steps, denitrifying bacteria in oxygen-poor soil and stagnant water convert nitrite into nitrate, nitrous oxide, and nitrogen gas.

Mineralization
Decomposers in soil and water break down biological nitrogen compounds into ammonium.

Ammonia (NH_3) or ammonium (NH_4^+)

Nitrite (NO_2^-)

Nitrification
Nitrifying bacteria convert ammonium into nitrite and then into nitrate.

Figure 21.7 The nitrogen cycle. The nitrogen cycle begins with nitrogen gas in the atmosphere. The process of nitrogen fixation converts it into a form that producers can use. The fixed nitrogen can then be assimilated into producers and consumers; it ultimately decomposes into ammonium through the process of mineralization. The ammonium can be converted into nitrite and then nitrate through the process of nitrification. Under anaerobic conditions, the nitrate can be converted into nitrogen gas through the process of denitrification.

producers and consumers, is broken down by scavengers, detritivores, and decomposers. The fungal and bacterial decomposers break down biological nitrogen compounds into ammonia. The process of breaking down organic compounds into inorganic compounds is known as **mineralization.**

Denitrification

Because nitrates produced by nitrification are quite soluble in water, they readily leach out of soils and into waterways, where they settle in the sediments of wetlands, rivers, lakes, and oceans. These sediments are typically anaerobic. Under anaerobic conditions, nitrates can be transformed back into nitrites, which are then transformed into nitric oxide (NO):

$$NO_3^- \rightarrow NO_2^- \rightarrow NO$$

This reaction is accomplished by bacteria such as *Pseudomonas denitrificans*. Additional chemical reactions under anaerobic conditions in soils and water subsequently convert nitric oxide to nitrous oxide (N_2O) and then to nitrogen gas, thereby completing the nitrogen cycle:

$$NO \rightarrow N_2O \rightarrow N_2$$

This process of converting nitrates into nitrogen gas is known as **denitrification.**

Denitrification is necessary for breaking down organic matter in oxygen-depleted soils and sediments. However, as you can see from the above reaction, it produces nitrogen gas (N_2). Since N_2 cannot be taken up by producers, the denitrification process causes nitrogen to leave the waterlogged soils and aquatic ecosystems in the form of a gas.

HUMAN IMPACTS ON THE NITROGEN CYCLE

 Before human activities began dramatically altering the environment, the production of usable forms of nitrogen through the process of fixation was approximately offset, on a global scale, by the loss of usable nitrogen through denitrification. However, during the last 3 centuries, and especially during the past 50 years,

---

Mineralization The process of breaking down organic compounds into inorganic compounds.

Denitrification The process of converting nitrates into nitrogen gas.

human activities have nearly doubled the amount of nitrogen put into terrestrial ecosystems. These activities include combustion of fossil fuels that add nitric oxide to the air, production of nitrogen fertilizers, and planting nitrogen-fixing crops.

As nitric oxide enters the atmosphere from combustion, it reacts with water in the air to form nitrates, which then fall to the ground during precipitation events. Because nitrogen is often a limiting nutrient, we would expect the addition of nitrates to affect a variety of ecosystems.

Over the years, there have been numerous investigations into whether adding nitrogen to terrestrial ecosystems in North America affects productivity and species richness. These study sites, which range north to south from Alaska to Arizona and west to east from California to Michigan, were recently compiled in an effort to determine whether their results showed a general pattern. When nitrogen was added in the form of nitrates and ammonium, all sites experienced an increase in primary productivity, shown in **Figure 21.8a**. This confirmed that nitrogen was a limiting resource at all

sites. However, the sites differed in the proportion of species that were eliminated over time, as you can see in Figure 21.8b. The researchers explored a wide range of potential causes for this variation in species loss among sites. In addition to differences in temperature and soil among the sites, they found that the sites with the largest increases in productivity experienced the largest reduction in species richness. Adding nitrogen to these communities commonly caused a few plant species to grow very large and to dominate the community. These large plants shaded the less-competitive smaller plants, which caused the smaller species to decline. These results demonstrate that the increases in nitrogen in the environment due to human activities can reduce the species diversity of ecosystems.

CONCEPT CHECK

1. What are three processes that cause nitrogen fixation?
2. Under what abiotic conditions does the process of denitrification occur?
3. Why does human production of nitrogen fertilizers alter plant species richness?

21.4 The phosphorus cycle moves between land and water

Phosphorus is a critical element for organisms because it is used in bones and scales, teeth, DNA, RNA, and ATP, a molecule involved in metabolism. As we discussed in Chapter 20, phosphorus is also a common limiting nutrient in aquatic and terrestrial ecosystems. For this reason, phosphorus is a component of most fertilizers manufactured to boost the growth of crops. In this section, we will examine the phosphorus cycle and explore how human activities have altered it in ways that affect ecosystems.

THE PHOSPHORUS CYCLE

The phosphorus cycle is considerably less complicated than the nitrogen cycle. As you can see in **Figure 21.9**, the atmosphere is not an important component of this cycle because phosphorus does not have a gas phase; phosphorus can only enter the atmosphere in the form of dust. Unlike nitrogen, phosphorus rarely changes its chemical form and typically moves as a phosphate ion (PO_4^{3-}). Plants take up phosphate ions from soil or water and incorporate them directly into various organic compounds. Animals eliminate excess phosphorus in their diets by excreting urine containing either phosphate ions or phosphorus compounds that are converted into phosphate ions by phosphatizing bacteria.

We begin our exploration of the phosphorus cycle by examining phosphate rocks, which are the major source of phosphate. As you can see in Figure 21.9, over time calcium phosphate ($Ca(H_2PO_4)_2$) precipitates out of ocean water and slowly forms sedimentary rock. Later,

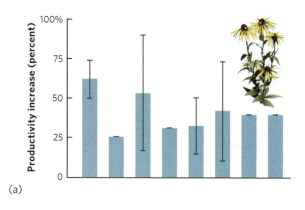

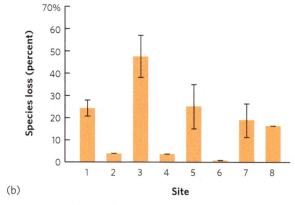

Figure 21.8 Effects of nitrogen addition on plant productivity and species richness. Twenty-three experiments at eight locations around the United States added nitrogen to plant communities. **(a)** All sites experienced increases in primary productivity. **(b)** All sites experienced a loss in species richness, but the magnitude of the loss varied greatly among sites due to differences in existing nitrogen concentrations and differences in the cation exchange capacity of the soil. Error bars are standard errors. Data from C. M. Clark et al., Environmental and plant community determinants of species loss following nitrogen enrichment, *Ecology Letters* 10 (2007): 596–607.

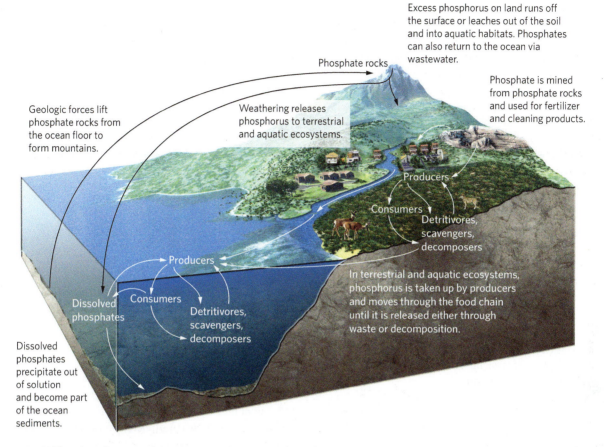

Excess phosphorus on land runs off the surface or leaches out of the soil and into aquatic habitats. Phosphates can also return to the ocean via wastewater.

Phosphate rocks

Phosphate is mined from phosphate rocks and used for fertilizer and cleaning products.

Geologic forces lift phosphate rocks from the ocean floor to form mountains.

Weathering releases phosphorus to terrestrial and aquatic ecosystems.

Producers

Consumers

Detritivores, scavengers, decomposers

Producers

Consumers

Detritivores, scavengers, decomposers

In terrestrial and aquatic ecosystems, phosphorus is taken up by producers and moves through the food chain until it is released either through waste or decomposition.

Dissolved phosphates

Dissolved phosphates precipitate out of solution and become part of the ocean sediments.

Figure 21.9 The phosphorus cycle. Phosphate rocks that are uplifted by geologic forces naturally weather over time to release phosphorus. These rocks are the source of phosphorus used in fertilizer and detergents. Phosphorus is taken up by producers and moves through the food chain until it is released either through waste or decomposition. Excess phosphorus on land runs off the surface or leaches out of the soil and into aquatic habitats. In the ocean, phosphorus combines with calcium or iron and precipitates out of the water, ultimately to form phosphate rocks again.

some of this rock is uplifted by geologic forces. Exposed rocks experience weathering, which causes them to slowly release phosphate ions. Phosphate rocks are also mined for phosphate that is used in fertilizers and in a variety of detergents.

When phosphate ions enter terrestrial ecosystems, they can be either bound strongly to the soil or taken up by plants and passed through the terrestrial food web. Animal excretions and the decomposition of all terrestrial organisms release phosphorus back to the soil. Excess phosphorus that is not bound to the soil or taken up by plants either moves across the surface of the land during a rainstorm as runoff or leaches from the soil. When soil erosion occurs, the phosphorus that is bound to the soil is carried away with the eroding soil particles. In either case, the phosphorus can be carried to a variety of aquatic ecosystems.

When phosphate ions enter aquatic ecosystems, they are taken up by producers and enter the food web in a manner that is similar to the terrestrial food web. In well-oxygenated

waters, phosphorus binds readily with calcium and iron ions and precipitates out of water to become part of the sediments. Thus, marine and freshwater sediments act as phosphorus sinks by continually removing phosphorus from the water. Under low-oxygen conditions, iron tends to combine with sulfur rather than phosphorus, so phosphorus remains more available in the water. Over time, the phosphate that precipitates down to ocean sediments is converted into calcium phosphate rocks, and the phosphorus cycle begins again.

HUMAN IMPACTS ON THE PHOSPHORUS CYCLE

In Chapter 20, we discussed how phosphorus is commonly a limiting nutrient in terrestrial and aquatic ecosystems. Yet, adding phosphorus to these ecosystems can have harmful effects. As we saw at the beginning of this chapter, phosphorus, sometimes in combination with excess nitrates, contributes to algal blooms that cause dead zones where rivers empty into oceans. This phenomenon happens in locations around the world, as shown

Figure 21.10 Dead zones. In 2008, more than 400 dead zones were identified around the world. This represents an increase of 33 percent from 1995. Data from R. J. Diaz and R. Rosenberg, Spreading dead zones and consequences for marine ecosystems, *Science* 321 (2008): 926–929.

in **Figure 21.10**. An increase in the productivity of aquatic ecosystems is called **eutrophication.** An increase in the productivity of aquatic ecosystems caused by human activities is called **cultural eutrophication.**

From the 1940s to the 1990s, household detergents contained phosphates to improve their cleaning effectiveness. These detergents became part of the wastewater that traveled through public sewage systems, ultimately emptying into rivers, lakes, and oceans. People began to realize that these detergents significantly increased phosphorus in waterways, which contributed to eutrophication and dead zones. In 1994, the United States banned phosphates in laundry detergents after several states had already done so. In 2010, 16 states banned phosphates in dishwashing detergents. In 2011, the European Union agreed to similar restrictions on phosphates in laundry and dishwasher detergents to help reduce the problems of cultural eutrophication and dead zones.

CONCEPT CHECK

1. What are the two paths that dissolved phosphates can take in aquatic ecosystems?
2. In what ways is mined phosphate used by people?
3. How has laundry detergent led to cultural eutrophication?

21.5 In terrestrial ecosystems, most nutrients regenerate in the soil

We have seen that elements cycle through the ecosystem and how they are both used and regenerated. In this section, we will examine how nutrients are regenerated in terrestrial ecosystems by the weathering of bedrock and by the breakdown of organic matter.

THE IMPORTANCE OF WEATHERING

Terrestrial ecosystems constantly lose nutrients because many are leached out of the soil and transported away in streams and rivers. To maintain a stable level of productivity, the loss of nutrients from an ecosystem must be balanced by an input of nutrients. For some nutrients, such as nitrogen, inputs come from the atmosphere. For most other nutrients, such as phosphorus, the inputs come from the weathering of bedrock beneath the soil. As we described in Chapter 5, weathering is the physical and chemical alteration of rock material near Earth's surface. Substances such as carbonic acid in rainwater and organic acids produced by the decomposition of plant litter react with minerals in the bedrock and release various elements that are essential to plant growth.

Determining the rate of weathering can be difficult because bedrock often exists far below the surface of the soil. One solution has been to measure the nutrients that enter a terrestrial ecosystem from precipitation and the nutrients that leave an ecosystem by leaching out of the soil and into a stream. You can see a diagram of nutrient inputs and outputs in **Figure 21.11**. If the system is in equilibrium—if nutrient inputs equal nutrient outputs—the difference between the nutrients entering the system by precipitation and particulate matter and the amount of nutrients leaving the system by leaching and runoff should equal the amount of nutrients made available by weathering.

Ecologists commonly determine the rate of nutrient regeneration through weathering by quantifying nutrient inputs and outputs from a *watershed*. A **watershed** is an area of land that drains into a single stream or river, as illustrated in **Figure 21.12**. In a watershed, the rate of weathering is

---

Eutrophication An increase in the productivity of aquatic ecosystems.

Cultural eutrophication An increase in the productivity of aquatic ecosystems caused by human activities.

Watershed An area of land that drains into a single stream or river.

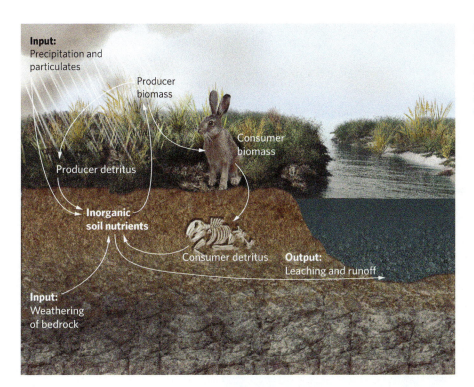

Figure 21.11 Quantifying nutrient weathering. Soil nutrients cycle through producers, consumers, and detritus. This cycle also has inputs from precipitation and weathering, and outputs of groundwater and surface water runoff.

Input: Precipitation and particulates

Producer biomass

Consumer biomass

Producer detritus

Inorganic soil nutrients

Consumer detritus

Output: Leaching and runoff

Input: Weathering of bedrock

Figure 21.12 A watershed. A watershed is an area of land that drains down into a single stream or river. The dashed black line indicates the boundaries of the watershed and the black arrows indicate the directions of water movement down the mountain.

estimated by measuring the net movement of several highly soluble nutrients, such as calcium (Ca^{2+}), potassium (K^+), sodium (Na^+), and magnesium (Mg^{2+}). These nutrients easily leach out of the soil and move into streams, where their concentrations can be measured as the stream leaves the watershed.

An example of this approach was recently reported for 21 small watersheds in the Canadian province of Quebec. Each watershed contained a small lake with a stream flowing out of the lake that drained the watershed. The watersheds were all forested and had little human activity, so the

researchers assumed that nutrient movement was at equilibrium. Measurements were made of the amount of calcium, potassium, sodium, and magnesium entering each watershed through precipitation, the amount of each element present in the soil and bedrock, and the amount of each element coming out of the watershed in the stream that drained it. By knowing the inputs and outputs of a watershed, scientists were able to determine rates of bedrock weathering. When the data were placed on a map of the study area, as shown in **Figure 21.13**, they found that weathering rates varied geographically. The rate of weathering was highest in the southwest region of

Figure 21.13 Weathering rates in 21 Canadian watersheds in Quebec. When researchers quantified the combined inputs and outputs of calcium, magnesium, potassium, and sodium, they found that the rates of weathering differed a great deal across the landscape. After D. Houle et al., Soil weathering rates in 21 catchments of the Canadian Shield, *Hydrology and Earth System Sciences* 16 (2012): 685–607.

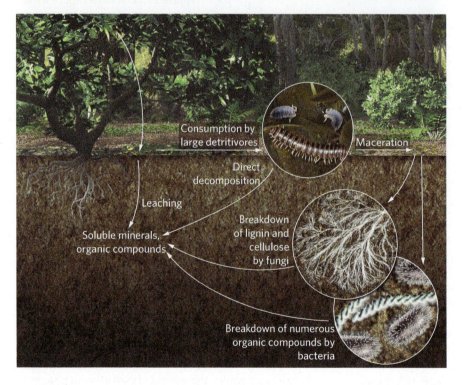

Figure 21.14 Decomposition of organic matter. Organic matter decomposes through the leaching action of water, consumption by invertebrates, mineralization by fungi, and mineralization by bacteria. Photo by Philippe Clement/naturepl.com.

the province, probably because of regional differences in temperatures, precipitation, and soil conditions.

THE BREAKDOWN OF ORGANIC MATTER

Although weathering of bedrock provides nutrients to terrestrial ecosystems, it is a very slow process. Therefore, primary production largely depends on rapid regeneration of

nutrients from decomposition, a process that breaks down organic matter into smaller and simpler chemical compounds and is conducted primarily by bacteria and fungi. As we discussed in Chapter 1, scavengers consume dead animals and detritivores break up organic matter into smaller particles.

As illustrated in **Figure 21.14**, the breakdown of plant matter in a forest occurs in four ways: soluble minerals and small organic compounds leach out of organic matter, large detritivores consume organic matter, fungi break down the woody components and other carbohydrates in leaves, and bacteria decompose almost everything. Leaching removes 10 to 30 percent of soluble substances from organic matter, which includes most salts, sugars, and amino acids. What remains behind are complex carbohydrates, such as cellulose, and other large organic compounds, such as proteins and lignin. Lignin determines the toughness of leaves and many of the structural qualities of wood. The lignin content of plants is a particularly important determinant of decomposition rate because it resists decomposition more than cellulose. However, some fungi and bacteria can break down cellulose and lignin. They secrete enzymes that break down the plant matter into simple sugars and amino acids that they then absorb. Some portion of the lignins, as well as other plant compounds that resist decomposition, may never break down in the surface soils but can form fossil fuels when buried for millions of years.

Large detritivores, including millipedes, earthworms, and wood lice, also play an important role in decomposition. These animals can consume 30 to 45 percent of the energy available in leaf litter, but they consume a much lower fraction of the energy available in wood. The importance of large detritivores is twofold; they decompose organic matter directly and they macerate organic matter into smaller pieces of detritus, which have a greater surface-area-to-volume ratio. This gives bacteria and fungi more surfaces on which to act and increases the rate of decomposition.

Bacteria and fungi play an important role in decomposition because they help convert organic matter into inorganic nutrients. Fungi play a special role because the hyphae of fungi can penetrate the tissues of leaves and wood that large detritivores and bacteria cannot penetrate on their own. If you have ever walked through a forest, you may have seen the fruiting bodies of many different fungi, including the impressive shelf fungi that emerge from the sides of dead logs (**Figure 21.15**).

In terrestrial ecosystems, 90 percent of all plant matter produced in a given year is not consumed directly by herbivores but is ultimately decomposed. Many plants resorb some of the nutrients from their leaves before the leaves are dropped. The aboveground dead plant biomass, combined with the organic matter of dead animals and animal waste, drops onto the soil surface where nutrients are leached. Here, decomposition is primarily aerobic, and plant roots and their associated mycorrhizal fungi have ready access to the nutrients that are released by the decomposers.

Because plant growth and decomposition are biochemical processes, nutrient cycling between producers and decomposers in terrestrial ecosystems is influenced by temperature, pH, and moisture. The rate of decomposition is also affected by the ratio of carbon and nitrogen

Figure 21.15 Fungi decomposing a dead log. Fungi play a key role in the decomposition of organic matter in terrestrial ecosystems by penetrating the dead tissues of plants. The fruiting bodies of this oyster bracket fungus (*Pleurotus ostreatus*), commonly referred to as mushrooms, are emerging from a dead log in Belgium. Photo by Philippe Clement/naturepl.com.

in the organic matter. As we discussed in Chapter 20, differences in the stoichiometry of an organism's food can affect the consumption of the food and the number of consumers that can be supported by the food supply. In the case of decomposition, if the decomposers require high amounts of nitrogen, then low nitrogen availability in the organic matter can cause slower rates of decomposition.

A study of leaf decomposition in Costa Rica provides insights into how several factors affect the rate of decomposition. In this study, leaves recently shed from 11 species of tropical trees were placed into coarse-mesh bags, as shown in **Figure 21.16**. The bags were positioned on the ground in the forest where invertebrates had access to the leaves, while the leaves remained in the bag. These

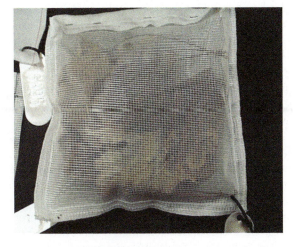

Figure 21.16 Researchers collected recently fallen leaves and placed them into mesh bags to determine the rate of daily mass loss over time in the forest. Photo by Mark Harmon, College of Forestry, Dept. of Forest Ecosystems and Society.

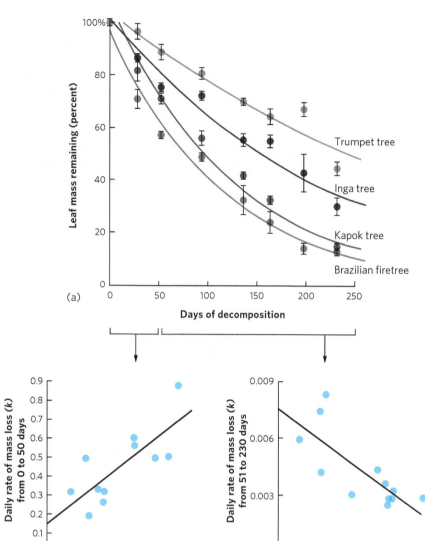

(a)

(b)

(c)

Figure 21.17 Decomposition rates of leaves in a tropical forest. **(a)** A sample of 4 of the 11 species examined reveals a range of different decomposition rates. Error bars are standard errors. **(b)** During the first 50 days of decomposition, the decomposition rates of the 11 leaf species were positively correlated with the solubility of compounds that could be leached from each species. **(c)** After 50 days, the decomposition rates of the 11 leaf species were considerably lower and negatively correlated to the amount of lignin in each leaf species relative to the amount of nitrogen. Data from W. R. Wieder et al., Controls over leaf litter decomposition in wet tropical forests, *Ecology* 90 (2009): 3333–3341.

invertebrates shredded the leaves into smaller pieces with an increased surface area, which led to more rapid decomposition. The leaf bags were weighed at different times over a 230-day period to determine the rate of leaf decomposition. You can view the data for several of these species in **Figure 21.17a**. Leaf species with curves that decline faster over time, such as the Brazilian firetree (*Schizolobium parahyba*), have the highest rate of decomposition.

Such species have relatively low amounts of lignin and cellulose and they have high leaf solubility, which allows more nutrients to be leached out of the leaf. The daily rate of mass loss, denoted as k, can be calculated for each curve. Species that decompose faster have a higher value of k. In "Analyzing Ecology: Calculating Decomposition Rates of Leaves," we discuss these calculations in more detail.

To examine how leaf traits affected decomposition rate, measured as the daily rate of mass loss (k), the researchers focused on distinct time periods: from day 0 to 50 when most decomposition would occur due to leaching, and from day 51 to day 230 when most decomposition would be due to invertebrates, fungi, and bacteria. During the first 50 days, there was a positive relationship between the decomposition rate and the fraction of soluble compounds in the different leaf species that can be leached, as shown in Figure 21.17b. During the latter portion of the experiment, there was a negative relationship between the decomposition rate and the ratio of lignin to nitrogen, as you can see in Figure 21.17c. This means that leaf species with a relatively high amount of lignin experience a slower rate of decomposition. From this study, we see that the decomposition of organic matter depends on both the environmental conditions and chemical traits of the organic matter.

Once researchers calculated values of k for each species, they could examine how both precipitation and the traits of the leaves affect the leaf decomposition rate. For example, they manipulated the amount of precipitation that some of the mesh bags received. When precipitation was reduced to 50 or 25 percent of normal levels, it caused a 10 to 20 percent decline in the rate of decomposition.

ANALYZING ECOLOGY

Calculating Decomposition Rates of Leaves

Eco TV
macmillanlearning.com/ricklefsvideo

As we discussed in the study of leaf decomposition in Costa Rica, researchers often want to examine the rate of leaf litter decomposition to determine how fast nutrients can be regenerated so they are available to producers, such as algae that form the base of an aquatic food web, plants that form the basis of a terrestrial food web, or last year's crops that decompose and provide nutrients to the current year's crop. The rate of decomposition commonly follows a negative exponential curve; there is initially a rapid loss in mass that slows over time. This negative exponential curve can be described by the following equation:

$$m_t = m_o e^{-kt}$$

where m_t is the mass of leaf litter that remains at a particular time, m_o is the original mass of leaf litter, e is the base of the natural log, k is the daily rate of mass loss, and t is time, which is measured in days. The decay constant k is the key parameter in this equation because it determines the shape of the curve; leaves that decompose at a faster rate have a larger value of k.

The value of k can be estimated using statistical software that determines the line that best fits a set of data for decomposition over time. Once the value of k is known, we can estimate the amount of leaf litter at any point in time, provided that the decomposition occurs under similar environmental conditions. For example, if we start with 100 g of leaves and $k = 0.01$, we can estimate the mass of leaves that have not decomposed after 10, 50, and 100 days:

After 10 days: $m_t = m_o e^{-kt} = 100 \ e^{-(0.01)(10)} = 90$ g
After 50 days: $m_t = m_o e^{-kt} = 100 \ e^{-(0.01)(50)} = 61$ g
After 100 days: $m_t = m_o e^{-kt} = 100 \ e^{-(0.01)(100)} = 37$ g

YOUR TURN Estimate the remaining mass of leaves that has not decomposed after 10, 50, and 100 days for two other leaf species: one with a daily decomposition rate of $k = 0.05$, and the other with a daily decomposition rate of $k = 0.10$.

DECOMPOSITION RATES AMONG TERRESTRIAL ECOSYSTEMS

Because environmental conditions are a key determinant for rates of decomposition, terrestrial ecosystems differ a great deal in their decomposition rates. Comparative studies of temperate and tropical forests show that detritus in the tropics decomposes more rapidly because of warmer temperatures and higher amounts of precipitation. For example, we can compare the amount of dead plant matter on the forest floor—including leaves, branches, and logs—versus the total biomass of vegetation and detritus in a forest. The proportion of dead plant matter is about 20 percent in temperate coniferous forests, 5 percent in temperate hardwood forests, and only 1 to 2 percent in tropical rainforests. Of the total organic carbon in terrestrial ecosystems, more than 50 percent occurs in soil and litter in northern forests, but less than 25 percent occurs in tropical rain forests, where the majority of the organic matter exists in the living biomass.

These differences in litter decomposition rates mean that tropical forests have a much larger proportion of the total organic matter in living vegetation than in detritus. This has important implications for tropical agriculture and conservation. For example, when tropical forests are cleared and burned, a large fraction of the nutrients are mineralized by burning and by subsequent high rates of decomposition. Together, these processes create an abundance of nutrients during the first 2 to 3 years of crop growth, but any surplus nutrients not taken up by the crops quickly leach away. Traditionally, tropical areas burned for agricultural fields would be farmed for 2 to 3 years and then left to undergo natural succession for 50 to 100 years to rebuild the fertility of the soil. However, many regions have human populations that are too dense to allow rotation of agriculture into different areas over several decades. Without rotation, the soils cannot replenish their nutrients and the fertility of the land rapidly degrades.

CONCEPT CHECK

1. What causes bedrock weathering?
2. How does taking a whole watershed approach aid researchers in quantifying rates of bedrock weathering?
3. What factors influence the rate at which organic matter is broken down?

21.6 In aquatic ecosystems, most nutrients regenerate in the sediments

Because most cycling of elements takes place in an aqueous medium, the chemical and biochemical processes involved are similar in terrestrial and aquatic ecosystems. However, the location of decomposition differs between terrestrial and aquatic ecosystems. In terrestrial ecosystems, nutrients regenerate close to the location, where they are taken up by producers. In aquatic ecosystems, however, most nutrients regenerate in sediments, which are often far below the surface waters that contain dominant producers, such as phytoplankton. In addition, in terrestrial ecosystems, aerobic decomposition is most common, whereas decomposition in the sediments and deep waters of aquatic ecosystems is typically anaerobic, which is considerably slower. In this section, we will investigate how decomposition operates in streams and wetlands, which receive much of their energy from leaves that blow in from the surrounding terrestrial environment. We will then examine the important role of sedimentation in rivers, lakes, and oceans and explore how stratification of water affects the movement of regenerated nutrients in lakes and oceans.

ALLOCHTHONOUS INPUTS TO STREAMS AND WETLANDS

As we have discussed in Chapter 6, streams and small, forested wetlands receive a major portion of their energy from the surrounding terrestrial environment from allochthonous inputs in the form of dead leaves that fall into the water. The decomposition process of leaves in a stream is similar to the process on land. As the leaves settle onto the bottom of the stream, the first stage is the leaching of soluble compounds, followed by the shredding of the leaves into smaller pieces by invertebrates such as amphipods, isopods, and larval caddisflies. At the same time that leaves are being shredded, fungi and bacteria are working to decompose the leaves much as they do on land.

Given the similar processes in both the terrestrial and stream ecosystems, it is perhaps not surprising that the rate of leaf decomposition depends on the temperature of the water and the chemical composition of the leaves. To determine the rate at which leaves decompose in streams,

aquatic ecologists follow a protocol that is similar to the protocol used by terrestrial ecologists. Newly fallen leaves are collected, weighed, and placed into mesh bags that are submerged in the stream. The mesh bag allows aquatic invertebrates to enter the bag without losing any of the leaves. Over time, the bags are removed from the stream, dried, and reweighed to determine the amount of leaf mass that remains.

To determine how leaf composition affects leaf decomposition rates, researchers placed coarse-mesh bags of nine leaf species into a stream located in the Black Forest of Germany and removed the bags over time. The nine leaf species differed a great deal in their nitrogen, phosphorus, and lignin content. The findings were that leaf decomposition rate is not related to the amount of nitrogen or phosphorus in the leaves but it is strongly associated with the lignin content of the leaves, as shown in **Figure 21.18a**.

The researchers also were interested in knowing how important the invertebrates were in the decomposition process. To answer this question, a second set of leaves was placed into fine-mesh leaf bags that prevented invertebrates from entering. As you can see in Figure 21.18b, leaf decomposition was about 20 percent higher in bags where invertebrates had been allowed to enter. This demonstrates that the invertebrates play an important role in breaking down the organic matter.

The small wetlands that exist in forests also receive a large proportion of their energy from leaves. Similar to leaves in streams, the decomposition of leaves in forested wetlands is largely tied to the lignin content. Moreover, the breakdown rate of the leaves has widespread effects on the entire food web and functioning of the ecosystem.

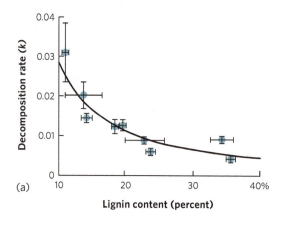

(a)

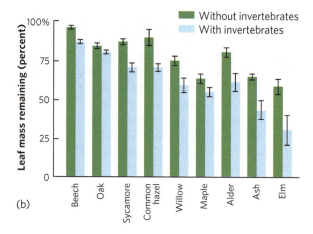

(b)

Figure 21.18 Leaf decomposition in a stream. (a) When nine species of leaves in mesh bags were added to a stream in Germany, the decomposition rate was slower in leaf species that contained more lignin in their tissues. **(b)** The researchers also assessed the contribution of aquatic invertebrates to leaf decomposition by using large-mesh and fine-mesh leaf bags. After 55 days, the leaves in the fine-mesh bags, which excluded invertebrates, experienced 20 percent less decomposition compared to the large-mesh bags, which allowed invertebrates to enter. Error bars are standard errors. Data from M. H. Schindler and M. O. Gessner, Functional leaf traits and biodiversity effects on litter decomposition in a stream, *Ecology* 90 (2009): 1641–1649.

DECOMPOSITION AND SEDIMENTATION IN RIVERS, LAKES, AND OCEANS

In most rivers, lakes, and oceans, organic matter sinks to the bottom and accumulates in deep layers of sediments. While some nutrients are recycled in the surface waters when animals excrete waste or when microbes in the surface water decompose organic matter, most organic matter sinks to the sediments. As a result, most nutrients must come from the sediments, although they will return slowly to the productive surface waters.

The process of nutrient regeneration in sediments of aquatic ecosystems helps us understand many patterns of ecosystem productivity. For example, you might recall from Chapter 20 that the least productive aquatic ecosystems are the deep oceans where the benthos is far from the surface waters (see Figure 20.6). Shallow oceans are more productive, in part, because the sediments are much closer to the surface waters and can therefore regenerate nutrients to the surface water more quickly through decomposition. The upwelling of water from the deep sediments to the surface brings nutrients from the site of regeneration to the site of algal productivity. In addition, some of the most productive regions of the ocean occur at the upwelling of water along the coasts of continents, where currents draw deep, nutrient-rich water up to the surface (see Figure 5.11).

STRATIFICATION OF LAKES AND OCEANS

The stratification of water, a phenomenon that we discussed in Chapter 6, further affects the availability of nutrients in aquatic ecosystems. The vertical mixing of water from the sediments to the surface can be hindered whenever surface waters have a different temperature and therefore a different density than that found in deep waters. The vertical mixing can affect primary production in two opposing ways: it can bring the deep, nutrient-rich water to the surface where phytoplankton can use it; but it can also carry phytoplankton down to the deep water, where they will die because of the low light conditions. When phytoplankton die, primary production may shut down in the deep water and little primary production will occur in the nutrient-rich waters at the surface.

Stratification happens in temperate and tropical lakes when the surface waters are warmed by the summer sunlight, while the deeper waters stay cold and dense. Such stratification does not happen in polar lakes because their surfaces never become warm enough. In estuaries and oceans, stratification of the water happens when an input of less dense freshwater from rivers or melting glaciers is positioned above a layer of the denser ocean saltwater.

Occasionally, stratified aquatic ecosystems experience periods of vertical mixing. For example, temperate lakes in the spring and fall experience changes in the temperature of surface waters that eventually match the temperature of the deeper waters. When the temperatures become equal, spring and fall winds that blow along the surface of the lakes cause the entire lake to mix. Mixing can also happen in oceans. In areas where the surface water is less salty than the deep water, sunlight can slowly cause the surface waters to evaporate and leave the salt behind. At some point, the surface water becomes saltier than the deeper water and the surface water sinks, thereby causing the ocean water to circulate.

In this chapter, we have learned that elements cycle within and between ecosystems and that this cycling determines the availability of elements to organisms. We have also seen that human activities commonly affect element cycles in ways that are harmful to ecosystem function. Finally, we have seen how elements regenerate through weathering and the decomposition of organic matter. In "Ecology Today: Applying the Concepts," we will apply these concepts to understand how logging and global change alter the functioning of an entire forest.

CONCEPT CHECK

1. Why is ocean upwelling an important process in regenerating nutrients in deep ocean waters?
2. What processes contribute to the breakdown of allochthonous inputs in streams?
3. Why is most decomposition typically anaerobic in the deep waters of lakes and oceans?

ECOLOGY TODAY   APPLYING THE CONCEPTS

Nutrient Cycling in New Hampshire

Monitoring nutrient flows from a watershed. At the Hubbard Brook Experimental Forest in New Hampshire, the soils lie on bedrock that prevents water from percolating down, so all nutrients leached from the soils find their way into the streams that drain the watershed. At the bottom of the watershed, researchers constructed a catchment device to monitor water volumes and nutrient flow from the watershed. Photo by U.S. Forest Service.

As we have seen throughout this chapter, producers assimilate nutrients from the environment and hold the nutrients in their tissues to be passed on to consumers and decomposers. However, assessing the degree to which plants affect nutrient cycling in an ecosystem is a daunting task, in part because ecosystems are large and complex. Fifty years ago, researchers at the Hubbard Brook Experimental Forest in New Hampshire took on this challenge by selecting a number of forested watersheds and logging some of them. Then they monitored the changes in how nutrients moved in each watershed.

The Hubbard Brook Experimental Forest was an ideal location for this grand experiment. A layer of impenetrable rock under the soil prevents water from percolating into the deep groundwater. Because of this, all the precipitation falling on the watershed is either taken up by plants or passes over and through the soil, ultimately ending up in the stream that leaves the watershed. This allows the researchers to monitor the stream to measure the amount of water and nutrients leaving the ecosystem.

In 1962, researchers removed all the trees from an entire watershed and sprayed it with herbicides for several years to suppress plant growth. As a control, adjacent watersheds were not logged. Over the past 50 years, ecologists have tracked how the ecosystem has responded to this disturbance. Without plants to take up water and nutrients, the movement of elements in the ecosystem changed dramatically. For example, the amount of water leaving the watershed in the stream increased several-fold. In addition, because the nitrates available in the soil were no longer being used by plants, there was a large increase in the amount of nitrates that leached out of the soil and into the streams. In forests that were not logged, there was a net gain of soil nitrogen over time at an annual rate of 1 to 3 kg per ha, due to precipitation and nitrogen fixation. In the logged forest, however, there was a net loss of nitrogen at an annual rate of 54 kg per ha. Because nitrates do not bind well to soil, they leached out of the soil and into the stream after logging.

Many other nutrients were also affected by the logging. For example, the researchers tracked the movement of calcium ions that come from precipitation and weathering of the bedrock. They found the vast majority of calcium ions came from the decomposition of detritus, whereas less than 10 percent of the calcium derived from weathering. When the region experienced years of acid precipitation, the inputs of calcium from precipitation and weathering could not keep pace with the calcium that leached from the soil. As a result, the forest experienced a net loss of calcium.

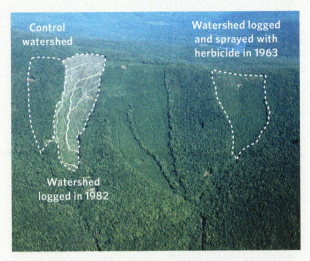

Today, the Hubbard Brook Experimental Forest continues to provide insights into nutrient cycling. For example, researchers recently reported that over a 5-decade period, the amount of nitrates leaving forested watersheds through the streams each year declined by more than 90 percent. Changes in the composition of tree species had only a small impact on the decline in nitrates leaving the watershed. This small impact was due to maple trees being naturally replaced by beech trees due to ecological succession; the leaves of beech trees decompose and release nitrates more slowly than the leaves of maple trees.

Climate change was responsible for about 40 percent of the reduction in nitrates in the stream; warmer temperatures in the late fall and early spring in recent years give plants a longer time to take up the nitrates, so fewer nitrates leach out of the soil and into the stream. The remaining 50 to 60 percent of the nitrate reduction over the 5 decades was due to a recovery from historic disturbances that removed more than 20 percent of the trees, including logging that took place in 1906 and 1917, a hurricane in 1938, and an ice storm in 1998. Just as we saw with the clear-cutting of a watershed, tree removal causes a short-term flush of nitrates out of the soils and into the stream. As the forests begin to come back, however, the soil recovers and holds on to more of the available nitrates. As the soil accumulates nitrates, the amount of nitrates leached into the stream declines.

Another surprising result came in 2016 when researchers reported on the effects of adding calcium to part of the forest in an attempt to reverse the long-term harmful effects of acid rain. The hypothesis was that the added calcium would improve tree growth and that the trees would, in turn, take up more nitrates from the soil. Surprisingly, they found that calcium caused a 30-fold greater release of nitrates from the soil and into the streams. The revised hypothesis is that the addition of calcium created a more favorable environment for the microbes, allowing the microbes to increase their breakdown of organic matter, including an increase in the release of nitrates.

The long-term experiment at Hubbard Brook shows us how natural events and human activities can dramatically alter the movement of elements within and between ecosystems in ways that can have major consequences for terrestrial and aquatic ecosystems.

Logging a watershed. To determine the role of plants in the cycling of nutrients, researchers at the Hubbard Brook Experimental Forest logged an entire watershed and then applied herbicides for several years to prevent plants from growing. Other watersheds were logged and allowed to regrow, while still others were left unmanipulated as controls. Photo by US Forest Service, Northern Research Station.

SOURCES:

Bernal, S., et al. 2012. Complex response of the forest nitrogen cycle to climate change. *PNAS* 109: 3406–3411.

Likens, G. E. 2004. Some perspectives on long-term biogeochemical research from the Hubbard Brook Ecosystem Study. *Ecology* 85: 2355–2362.

Rosi-Marshall, E. J., et al. 2016. Acid rain mitigation shifts a forested watershed from a net sink to a net source of nitrogen. *PNAS* 113: 7580–7583.

SUMMARY OF LEARNING OBJECTIVES

21.1 The hydrologic cycle moves many elements through ecosystems. Water evaporates from water bodies, soil, and plants and moves as water vapor into the atmosphere. This water vapor condenses into clouds and eventually falls back to Earth as precipitation. This precipitation can be taken up by plants, run across the surface of the ground, or infiltrate the groundwater. This water moves to streams and lakes and ultimately makes its way back to the oceans. Humans can alter this cycle by reducing infiltration due to the construction of impermeable surfaces, by logging trees and thereby increasing runoff, and by causing global warming, which increases the rate of evaporation.

Key Term: Hydrologic cycle

21.2 The carbon cycle is closely tied to the movement of energy. Carbon exists in the atmosphere in the form of CO_2. This CO_2 can be used by terrestrial producers and, after dissolving into the water, by aquatic producers. The producers, consumers, scavengers, detritivores, and decomposers in terrestrial and aquatic ecosystems can produce CO_2 when they respire. Carbon can also leave the water by sedimentation and be buried both in the water and on land. Buried carbon can be extracted in the form of fossil fuels. The combustion of fossil fuels and the combustion of organic matter during fires release CO_2 into the atmosphere. Humans can alter this cycle primarily by affecting the extraction and combustion of carbon.

21.3 Nitrogen cycles through ecosystems in many different forms. Nitrogen gas in the atmosphere can be converted into ammonia and nitrates by lightning, nitrogen-fixing bacteria, and the manufacture of fertilizers. Producers can take up these forms of nitrogen and assimilate them. This nitrogen is then transferred through terrestrial and aquatic food webs. During decomposition, nitrogen in organisms and their wastes can be converted to ammonia by the process of mineralization. Ammonia can be converted into nitrites and nitrates through the process of nitrification, and nitrates can be converted to nitrous oxide and nitrogen gas by the process of denitrification. Humans alter this cycle primarily by manufacturing and applying large amounts of fertilizer and combusting fossil fuels, which produces nitric oxide in the air that later mixes with precipitation and falls to the ground as nitrates. These activities alter the fertility of terrestrial and aquatic environments.

Key Terms: Nitrogen fixation, Nitrification, Mineralization, Denitrification

21.4 The phosphorus cycle moves between land and water. Most phosphorus is released by weathering of rocks. This phosphorus is taken up by terrestrial and aquatic producers, which pass it to consumers, scavengers, detritivores, and decomposers. The phosphorus from excretions and decomposed organisms dissolves in the water of soil or in the water of streams, rivers, lakes, and oceans. In the ocean, phosphorus precipitates into sediments that are slowly converted into rocks. Humans affect the phosphorus cycle primarily by mining rocks for fertilizer. This phosphorus-rich fertilizer can alter the fertility of terrestrial and aquatic habitats and lead to algal blooms and then eutrophication.

Key Terms: Eutrophication, Cultural eutrophication

21.5 In terrestrial ecosystems, most nutrients regenerate in the soil. The nutrients in terrestrial ecosystems are primarily regenerated in soils. Some nutrients such as phosphorus are regenerated by weathering of rocks, and all nutrients are regenerated by the decomposition of dead organic matter. Because decomposition rates are faster under warm temperatures and high precipitation, tropical ecosystems have high decomposition rates and low amounts of dead organic matter. Boreal and other cold ecosystems have low rates of decomposition and large amounts of dead organic matter.

Key Term: Watershed

21.6 In aquatic ecosystems, most nutrients regenerate in the sediments. In many streams and some wetlands, allochthonous inputs of leaves from the surrounding terrestrial environment are the major source of nutrients. The rate of leaf decomposition is determined, primarily, by the water temperature and the lignin content of the leaves. In rivers, lakes, and streams, much of the organic matter settles out of the water and onto the bottom sediments, where it decomposes. When lakes and oceans stratify, the movement of nutrients that have been released by decomposition is hindered from moving to the more productive surface waters. Although nitrogen and phosphorus are the most common limiting nutrients in aquatic ecosystems, some regions of the ocean are limited by the availability of other nutrients, including silicon and iron.

CRITICAL THINKING QUESTIONS

1. How does energy from the Sun drive the movement of water from the oceans to the continents and back to the oceans again?

2. How might the ocean reduce the effects of fossil fuel combustion on CO_2 concentrations in the atmosphere?

3. Why is methane gas commonly produced in swamps?

4. Given that the bottom of the ocean is anaerobic, what process in the nitrogen cycle is likely to be occurring in this location?

5. How might nitrogen-fixing bacteria living in symbiosis with a plant affect the types of environments in which the plant could live?

6. Why is the weathering of bedrock in a New Hampshire forest responsible for such a small fraction of the nutrients available to plants?

7. Why do tropical and temperate soils have different rates of nutrient regeneration?

8. Why do agricultural soils in boreal Canada retain their nutrients for many more years than agricultural soils in tropical South America?

9. How might global warming cause the release of CO_2 from boreal forest soils?

10. What is the likely chain of events in which the dumping of raw sewage into the Mississippi River leads to fish kills in the dead zone in the Gulf of Mexico?

GRAPHING THE DATA The Decomposition of Organic Matter

As we have seen in this chapter, dead organic matter commonly decomposes in a pattern that follows a negative exponential curve. To compare the decomposition rates of leaves from cherry trees and maple trees in a stream, scientists placed leaf bags into a stream and retrieved them over time. They conducted three replicates of this experiment, and obtained the data shown in the table.

For each species of leaf litter, calculate the mean amount of litter remaining at each time point and then graph the mean amount of litter remaining over time using a scatterplot.

	LEAF LITTER MASS (g)				
	0 DAYS	10 DAYS	30 DAYS	60 DAYS	100 DAYS
MAPLE					
REPLICATE 1	100	64	32	23	12
REPLICATE 2	100	60	30	20	10
REPLICATE 3	100	56	27	18	8
CHERRY					
REPLICATE 1	100	80	60	40	30
REPLICATE 2	100	78	63	44	33
REPLICATE 3	100	83	58	38	25

22 Landscape Ecology and Global Biodiversity

Can We Have Too Much Biodiversity?

The word biodiversity prompts several thoughts. We think of the diversity of evolved species that play unique roles in nature and collectively contribute to the proper functioning of ecosystems. We also may think about the many species around the world that are declining in abundance and some that are even going extinct. What is less appreciated is that while many places on Earth are losing species, these same places are simultaneously experiencing large increases in biodiversity. The reason for this apparent contradiction is that the number of extinctions is often far outweighed by the number of introduced, non-native species.

Given that islands have discrete boundaries, they are a good place to examine extinctions of native species and introductions of non-native species that have become naturalized over time (i.e., they have sustained populations in nature). As we have discussed in previous chapters, species on islands can be highly susceptible to extinctions due to relatively small population sizes, overharvesting by humans, and a lack of long-term coexistence with introduced competitors, predators, and pathogens.

In one study, the change in island biodiversity over time was quantified by examining the number of birds and vascular plants living on islands that are either native and still present, native but extinct, or non-native. The results showed that across multiple islands around the world, the number of bird species that have gone extinct since humans arrived is similar to the number of birds that have arrived and become naturalized, resulting in a new composition of species, but no net change in biodiversity. In contrast, the number of native plants that have gone extinct since humans arrived is far outweighed by the number of non-native plants that have arrived and become naturalized. While larger islands contained more native species and more naturalized species, the average island has about twice as many plant species today as it once had. For example, New Zealand has about 2,000 species of native plants, but today it has another 2,000 species of non-native, naturalized plants. It had no native land mammals except for bats, but today has several species of introduced land mammals.

The observations from islands also pertain to many other regions of the world. For example, California has experienced about 28 extinctions of native plant species, but more than 1,000 non-native plant species have become naturalized. Collectively, these results set up a paradox between the benefits of having high biodiversity in an ecosystem versus the desire to preserve the native species that live in each ecosystem. Put another way, the activities of humans, including the movements of plant and animal species around the world, may be causing the global decline in species at the global scale due to extinctions, but it can often cause no change or an increase in species richness at the local scale in many parts of the world—such as islands or the state of California—due to the large number of introduced species to these places. As a result, we need to consider not only the number of species and how they affect proper ecosystem function, but also the particular community composition in terms of native versus non-native species. Moreover, we need to understand how local and global biodiversity will change in future decades, including whether the naturalized species might eventually drive native species to extinction.

The patterns of biodiversity across the globe highlight the need to understand the processes that produce the historical causes of species richness and the fact that we sometimes need to think about ecology at large spatial scales to understand the distribution and composition of species on Earth. In this chapter, we will explore ecology at large spatial scales to examine the distribution of species both across landscapes and around the world.

> "While many places on Earth are losing species, these same places are simultaneously experiencing large increases in biodiversity."

SOURCES:
Sax, D. F., et al. 2002. Species invasions exceed extinctions on islands worldwide: A comparative study of plants and birds. *The American Naturalist* 160: 766–783.
Vellend, M. 2017. The biodiversity conservation paradox. *American Scientist* 105: 94–101.

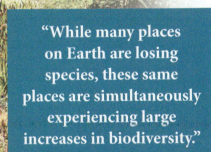

Species diversity of flowering plants in California. While California has experienced 28 extinctions of native plants, they have also experienced introductions of more than 1,000 non-native plants from around the world. Photo by George Oze/Alamy.

LEARNING OBJECTIVES

After reading this chapter, you should be able to:

22.1 Discuss how landscape ecology examines ecological patterns and processes at large spatial scales.

22.2 Explain why the number of species increases with geographic area.

22.3 Describe the equilibrium theory of island biogeography.

22.4 Highlight the causes for biodiversity being highest near the equator and lowest toward the poles.

22.5 Explain how the distribution of species around the world is affected by Earth's history.

Throughout this book, we have considered the role of physical conditions and of species interactions at particular locations. We have seen how these factors can affect the ecology of individuals, populations, communities, and ecosystems. When we explored these topics, we often found it helpful to focus on relatively homogeneous areas of land or water. However, as we moved across landscapes and across continents, we observed that terrestrial and aquatic ecosystems vary from place to place. In this chapter, we will consider much larger areas, from landscapes that include a variety of habitats to entire continents that contain a range of climates. In taking this large-scale approach, our goal is to understand why we find different numbers of species and often very different types of species in different places around the globe and why we sometimes find very similar types of species on widely separated continents. Once we understand the patterns of biodiversity, we can understand the processes that affect diversity and develop plans to conserve it, which is the topic of the next chapter.

22.1 Landscape ecology examines ecological patterns and processes at large spatial scales

If you were to look across an expansive landscape from the window of a plane, you would undoubtedly see a diversity of terrestrial and aquatic habitats comprising a wide range of sizes and shapes (see Figure 4.2). The field of **landscape ecology** considers the spatial arrangement of habitats at different scales and examines how they influence individuals, populations, communities, and ecosystems. In this section, we will discuss the sources of habitat heterogeneity and how this heterogeneity generates local and regional biodiversity.

CAUSES OF HABITAT HETEROGENEITY

The current heterogeneity of habitats reflects recent and historical events caused by both natural forces and human activities. Historical processes that have long-lasting influences on the current ecology of an area are known as **legacy effects.** An interesting legacy effect of glaciers that you can observe today is the presence of eskers, which are the remnants of long, winding streams of water that once flowed inside or under the glaciers. Over time, these glacial streams deposited soil and rock in their paths. After glaciers melt away, these old stream beds appear as long and winding hills (**Figure 22.1**). These hills harbor different microhabitats that favor unique plant and animal communities.

Landscape ecology The field of study that considers the spatial arrangement of habitats at different scales and examines how they influence individuals, populations, communities, and ecosystems.

Legacy effect A long-lasting influence of historical processes on the current ecology of an area.

Figure 22.1 A natural legacy effect of the landscape. This long, winding esker near Whitefish Lake, Northwest Territories, Canada, was formed by a stream that once ran through a glacier thousands of meters thick. Particles of soil settled into the bed of the stream, which today forms the long ridge of land that is the esker. Photo by George D. Lepp/Getty Images.

Natural forces continue to cause habitat heterogeneity in modern times. At both local and regional scales, catastrophes such as tornadoes, hurricanes, floods, mudslides, and fires can alter vegetation structure, which causes changes in populations and communities that depend on it. Although catastrophic events have always occurred naturally, human activity has influenced them. For example, natural fires in Yellowstone National Park were largely suppressed through much of the twentieth century. During the summer of 1988, however, hundreds of fires were ignited both by human activities and by natural causes such as lightning strikes, and these fires were exacerbated by windy conditions and a summer-long drought. Most fires burned relatively small areas of less than 40 ha, but a few of the fires burned much larger areas. In total, nearly 500,000 ha burned, leaving a mosaic of burned and unburned patches across the greater Yellowstone landscape (**Figure 22.2**).

As we saw in Chapter 18, some animals—for example, beavers and alligators—are ecosystem engineers and can alter the habitats in a landscape. Humans are the most extensive ecosystem engineers; they build homes, offices, and factories, construct dams and irrigation channels, channelize waterways for improved navigation, and clear forests for lumber, paper, and agriculture. Logging is a particularly clear example of a human activity that produces a mosaic of habitat types across the landscape. In the western United States, it is common practice to log medium-sized swaths of forest scattered throughout the landscape. This practice helps minimize soil erosion and other damaging effects of large-scale clear-cutting. It also produces a mosaic of forest patches of different ages that persists for many years.

Human habitations can also cause habitat heterogeneity. For instance, during the first century CE, the Romans built small villages and farms in France. These farms were abandoned by the fourth century and the land reverted to forest (**Figure 22.3a**). Researchers studied the soil conditions and plant species located between 0 and 500 m from the former Roman settlements. They discovered that sites closer to the settlements had higher soil pH, more available phosphorus, and greater plant species richness, including many weedy species. These legacy effects, which are shown in Figure 22.3b, were

(a)

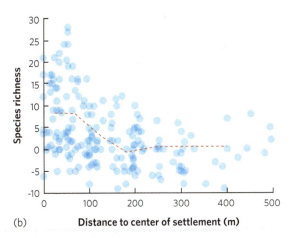

(b)

Figure 22.3 Legacy effects of humans on the landscape. (a) Ruins of Roman farming villages are present throughout northern France. These Gallo-Roman ruins are in the forest of Tronçais at the site of "Petit-Jardins" in Isle-et-Bardais, Allier. As the stone building materials weathered, they likely contributed calcium and phosphorus to the soil and altered the abiotic conditions. In addition, the Romans introduced numerous plant species. **(b)** Collectively, these changes created a legacy of more plant species living close to the old settlements with fewer species living 500 m away. The number of plant species represents the difference from the mean number of plant species found far from the settlements. The dashed line represents the changing mean species richness as one moves away from the center of the settlement. Data from E. Dambrine et al., Present forest biodiversity patterns in France related to former Roman agriculture, *Ecology* 88 (2007): 1430–1439. Photo by Laure Laüt-Taccoen.

Figure 22.2 Habitat heterogeneity after a fire. In Yellowstone National Park, hundreds of fires burned patches of the park and left behind a heterogeneous mix of burned and unburned areas. Photo courtesy Jim Peaco, Yellowstone National Park, National Park Service.

attributed to two causes: the slow breakdown of the ancient building materials contributed calcium and phosphorus to the soil; and the Romans introduced numerous plant species to the area. In short, human habitation from 1,600 years ago continues to have strong legacy effects on the modern forest.

RELATIONSHIPS BETWEEN HABITAT HETEROGENEITY AND SPECIES DIVERSITY

When we quantify the number of species in a given area, we include a greater variety of habitats at the landscape level than at the local scale and therefore generally observe a larger number of species. For example, ecologists in Vermont recently surveyed bird species along 27 streams that feed into Lake Champlain. For each stream, they measured the physical characteristics of the stream itself, such as depth and width, and the types of habitats in the riparian areas along the stream banks. While each stream supported an average of only 17 bird species, the entire landscape of streams contained 101 bird species. The number increased significantly because different bird groups preferred different habitat characteristics. For example, waterbirds had their highest species richness and abundance in shallow streams, whereas fish-eating birds were more abundant in larger streams with little agriculture along the stream bank because such streams contain more fish. In contrast, the richness and abundance of insect-eating birds were highest in areas containing a variety of habitat types, including shallow streams, meadows, and broadleaf forests. This mixture of meadow, forest, and stream habitats allows a greater variety of insect-eating shore birds and forest birds to exist and feed on the insects that emerge from the stream. Because the heterogeneity of habitat types across the landscape of streams supports a higher richness of bird species than any single habitat, conserving a variety of habitats over a large area is critical to the conservation of high bird species diversity.

LOCAL AND REGIONAL SPECIES DIVERSITY

The study of birds in the streams of Vermont highlights the fact that species diversity can be measured at different spatial scales. For example, if we consider the number of species in a relatively small area of homogeneous habitat, such as a stream, we are looking at **local diversity** or **alpha diversity.** If we consider the number of species in all the habitats that comprise a large geographic area, we would be looking at **regional diversity** or **gamma diversity.** In the Vermont stream study, the regional diversity would be the entire list of 101 birds that the researchers identified across all 27 streams.

If each species occurred in all habitats within a region, then the species diversity at the local and regional scales would be identical. However, as we saw in the case of the stream-dwelling birds of Vermont, different species prefer different habitats. Therefore, the number of species at a local scale is less than the number of species at the regional scale. Moreover, the list of species in each local habitat is different from each other. Ecologists refer to the number of species that differ in occurrence between two habitats as **beta diversity.** For example, imagine two streams in Vermont: Stream A contains five species not found in Stream B and Stream B contains three species not found in Stream A. Because the two streams differ by a total of eight species, the beta diversity is eight. The greater the difference in species between two habitats, the greater the beta diversity.

The collection of species that occurs within a region is called the **regional species pool.** The actual species that live in each local site depend on the species that exist in the regional pool and how well biotic and abiotic conditions at the local scale match the niche requirements of species in the regional pool. Therefore, species in the regional species pool are sorted among localities according to their adaptations and interactions, a process called **species sorting.**

An example of species sorting can be seen in an experiment in which researchers set up artificial wetlands and manipulated a variety of conditions, including the fertility of the soil and the amount of flooding the soil experienced. The researchers then sowed the seeds of 20 wetland plant species into each wetland to see which plants would germinate and persist over the next 5 years. Of the original 20 species, one species failed to germinate in any of the wetlands and five others were unable to persist. Of the remaining 14 species, each wetland contained only three to five species by the end of the experiment. Moreover, there were particular combinations of plant species that survived under each flooding and fertility condition. These results confirmed that differences in local conditions cause the sorting of species from the regional species pool.

CONCEPT CHECK

1. What are two sources of habitat heterogeneity?
2. Why is there a positive relationship between habitat heterogeneity and species richness?
3. Why do habitat preferences of species typically cause alpha diversity to be lower than gamma diversity?

Local diversity The number of species in a relatively small area of homogeneous habitat, such as a stream. *Also known as* **Alpha diversity**.

Regional diversity The number of species in all the habitats that comprise a large geographic area. *Also known as* **Gamma diversity**.

Beta diversity The number of species that differ in occurrence between two habitats.

Regional species pool The collection of species that occurs within a region.

Species sorting The process of sorting species in the regional pool among localities according to their adaptations and interactions.

22.2 The number of species increases with increased area

When we consider landscape ecology, we see that larger aquatic or terrestrial areas contain a greater number of species. In this section, we discuss examples of this phenomenon from a variety of systems and then explore the mathematical relationship between area size and species richness. We will then consider the underlying processes that cause the positive relationship between area size and species richness.

SPECIES–AREA RELATIONSHIPS

In a classic study of species diversity, Robert MacArthur and E. O. Wilson examined the patterns of species richness in taxonomic groups living on islands of different sizes. When they plotted the species richness of different islands against the size of the islands, they found that larger islands had a greater species richness. For example, as you can see in **Figure 22.4a**, the number of amphibian and reptile species living on islands in the West Indies increases with island size.

The observation that larger areas tend to contain more species has led to the concept of the *species–area curve*. The **species–area curve** is a graphical relationship in which increases in area (*A*) are associated with increases in the number of species (*S*). This curve can be described by the following equation:

$$S = cA^z$$

where *c* and *z* are constants fitted to the data. To make it easier to work with this equation in the form of a graph, we can take the logarithm of both sides:

$$\log S = \log c + z \log A$$

This is the equation for a straight line with a *y* intercept of log *c* and a slope of *z*. For example, Figure 22.4b plots the same amphibian and reptile data using axes on a log scale. A similar linear relationship can be seen in another collection of data by MacArthur and Wilson in which they examined the number of bird species living on the Sunda Islands in Malaysia, the Philippines, and New Guinea. As you can see in **Figure 22.5**, when the data are plotted on a log scale, the number of bird species linearly increases with island size.

Across many different groups of organisms, the relationship between log *A* and log *S* typically has a slope within the range of *z* = 0.20 to 0.35 across a wide range of areas, from 1 m² to a country of modest size. This relatively narrow range of slope values across different studies and different taxa suggests that the relationships between species richness and island area reflect similar processes.

Species–area curve A graphical relationship in which increases in area (*A*) are associated with increases in the number of species (*S*).

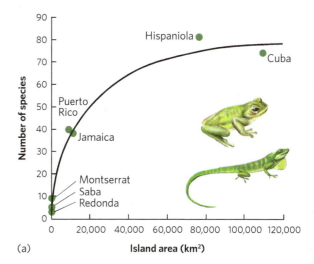

(a)

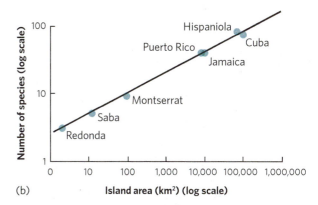

(b)

Figure 22.4 A species–area curve for amphibians and reptiles. On the islands of the West Indies, larger islands contain more species of amphibians and reptiles, such as the Jamaican green tree frog and the Jamaican anole. **(a)** When plotted as untransformed data, the relationship is a plateauing curve. **(b)** When plotted on a log scale, the relationship is a straight line. Data from R. H. MacArthur and E. O. Wilson, *The Theory of Island Biogeography* (Princeton University Press, 1967).

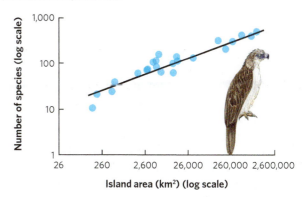

Figure 22.5 A species–area curve for birds. Among the Sunda Islands in Malaysia, Philippines, and New Guinea, islands with a larger area contain more species of birds, such as the Philippines Eagle, when both variables are plotted on a log scale. Data from R. H. MacArthur and E. O. Wilson, An equilibrium theory of insular zoogeography, *Evolution* 17 (1963): 373–387.

Oceanic islands provide good examples of the species–area relationship, but the pattern can also be observed in a wide range of ecosystems. For example, researchers in Ontario, Canada, surveyed the species living in 30 wetlands of different sizes. As shown in **Figure 22.6**, they found that as wetland area increased, there was a corresponding increase in the number of species of plants, amphibians, reptiles, birds, and mammals. As you can see, the positive correlation between area and species richness is common in nature.

HABITAT FRAGMENTATION

Habitat islands can form when natural processes such as fires and hurricanes cause habitats to become fragmented. Similarly, human activities have caused widespread fragmentation of large habitats throughout the world. For example, in the Central American nation of Costa Rica, the forests in 1940 covered much of the country and existed as a large, continuous habitat. As human populations and their associated activities increased, the forests were cleared on a continuing basis, as shown in **Figure 22.7**. By 2005, much of the forested area had been cleared and the forests that remained existed in many small fragments, which were essentially islands of forested habitat within a matrix of cleared lands.

Fragmentation of a large contiguous habitat creates several effects: the total amount of habitat decreases, the number of habitat patches increases, the average patch size decreases, the amount of edge habitat increases, and patch isolation increases. Conversely, in the habitat matrix between the fragments, such as cleared fields between forest fragments, the total area increases and the matrix becomes more continuous. Of particular interest is how isolated

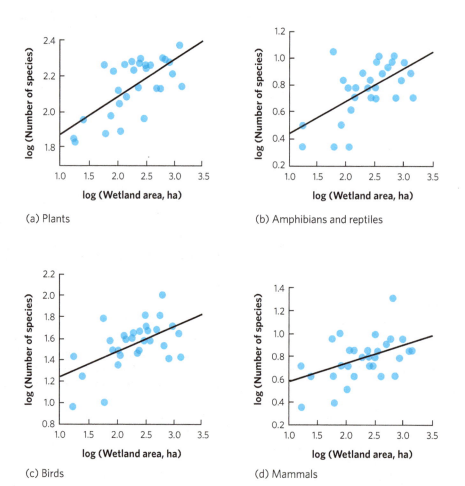

(a) Plants

(b) Amphibians and reptiles

(c) Birds

(d) Mammals

Figure 22.6 Species richness increases with habitat area. In a survey of 30 wetlands in Ontario, Canada, an increase in wetland area was found to be associated with an increase in species richness in **(a)** plants, **(b)** amphibians and reptiles, **(c)** birds, and **(d)** mammals. Data from C. S. Findlay and J. Houlahan, Anthropogenic correlates of species richness in southeastern Ontario wetlands, *Conservation Biology* 11 (1997): 1000–1009.

Figure 22.7 **The fragmentation of a contiguous forests into islands of forested habitat.** In Costa Rica, the forest—shown in dark green—was once largely continuous. Over time, it was cleared by human activities, and today exists only in a number of smaller fragments. After United Nations Environment Program, Food and Agricultural Program of the United Nations, and the United Nations Forum on Forests Secretariat, *Vital Forest Graphics* (2009).

habitat patches with different sizes and shapes influence biodiversity, and how habitat corridors and the quality of the matrix between habitat fragments affect the rate of species turnover.

Effects of Fragment Size

As we saw in the case of oceanic islands, the reduction in habitat size that comes with fragmentation commonly causes a decline in species diversity. This happens because each fragment supports smaller populations of species than existed in the original larger habitat and, as discussed in Chapter 13, smaller populations experience higher rates of extinction (see Figure 13.13). For example, in eastern Venezuela a large river was dammed to create a 4,300-km² lake known as Lago Guri. The region had been composed of a mixture of grazed fields and tropical forests, but after the dam caused the region to flood, hundreds of high points in the landscape became islands in the newly formed lake, as shown in **Figure 22.8a**. The smallest

islands did not contain enough prey to sustain the large vertebrate predators, so predators on the smallest islands went extinct. These extinctions caused a chain of events that affected the species that remained on the islands. The extinction of the large predators allowed the herbivores on the islands, such as leaf-cutter ants, iguana lizards, and howler monkeys (*Alouatta seniculus*), to increase. The increase in herbivores then consumed more of the island plants. As you can see in Figure 22.8b, greater abundance of herbivores on smaller islands caused higher mortality in existing sapling trees and reduced recruitment of new trees into the sapling stage.

Effects of Fragment Edges

As you may recall from Chapter 18, when habitats are fragmented, they produce ecotones, which are regions with distinct environmental conditions and unique species compositions over a relatively short distance.

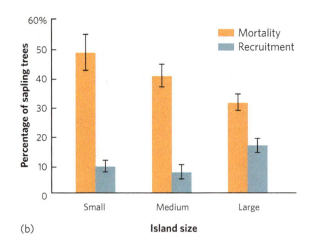

(a)

(b)

Figure 22.8 Fragmentation of terrestrial habitats from the flooding of Lago Guri. In eastern Venezuela, a dam was constructed to form Lago Guri in 1978 and the area behind the dam was flooded. **(a)** Flooding caused the hills of the region to become islands in the lake. **(b)** On the smallest islands, large predators were no longer present, so the populations of herbivores increased, with a major effect on the vegetation. Sapling trees on small islands experienced the highest mortality and the lowest recruitment. Error bars are standard errors. Data from J. Terborgh et al., Vegetation dynamics of predator-free land-bridge islands, *Journal of Ecology* 94 (2006): 253–263. Photo by Peter Langer/ DanitaDelimont.com.

Increases in fragmentation cause an increase in the amount of edge habitat compared to the amount of edge present in the original unfragmented habitat. For example, consider a single square habitat with an area of 1 ha compared to the same total area divided into 16 smaller square habitats. As illustrated in **Figure 22.9**, the single large habitat would have 400 m of edge, whereas the 16 smaller habitats would have 1,600 m of edge. In regard to fragment shape, round habitats have the lowest ratio of edge to area, while long, skinny ovals or rectangles have much higher edge-to-area ratios.

An increase in edge habitat changes both the abiotic conditions and the species composition of a habitat. In a fragmented forest, for instance, the edge of a newly created fragment experiences greater sunlight, warmer temperatures in the summer, and higher rates of evaporation. These changes can make the edge of the forest less suitable to many forest species and more suitable to others.

Given that fragmentation increases the amount of habitat edge, it also increases the abundance of those species that prefer edge habitat, and this can affect other species living in the fragment. For example, in the southern United States, the population of bronzed cowbirds (*Molothrus aeneus*) has increased with habitat fragmentation. This bird is considered a nest parasite because it reproduces by placing its eggs into the nests of other bird species, which serve as hosts and take care of the

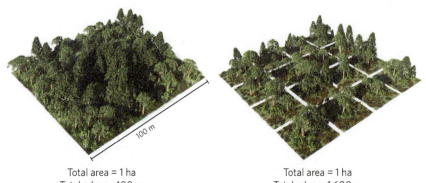

Total area = 1 ha
Total edge = 400 m

Total area = 1 ha
Total edge = 1,600 m

Figure 22.9 The effect of habitat fragmentation on edge habitat. If we have two habitats of the same total area, the habitat that is fragmented has much more edge.

Estimating the Number of Species in an Area

Eco TV

 macmillanlearning.com/ricklefsvideo

As we have seen, ecologists often need to estimate how many species exist in an area. Since counting every individual is rarely possible, we must take a sample of an area to estimate how many species are present. One way to estimate the number of species in a sample is to graph the number of species that we observe as we increase our sample size. One would expect that the more we sample, the closer we will get to knowing the actual number of species in an area. Eventually, the curve will reach a plateau so that any additional sampling will not discover additional species. A graph of the number of species observed in relation to the number of individuals sampled is known as a **species accumulation curve.**

A second way to estimate the number of species in an area is based on the number of rare species that are detected. When we sample many individuals in a community, we typically end up with a high number of individuals for each of the common species but we may detect only one or two individuals of rare species.

We can estimate the number of species actually living in an area (S) by sampling the community to determine how many species are observed (S_{obs}), how many species are represented by two individuals (f_2), and how many are represented by one individual (f_1). In essence, species represented by two individuals are likely to be more common in the community than species represented by single individuals. We can therefore estimate the number of actual species that live in a community by starting with how many species we have observed and multiplying this by a ratio that incorporates the number of species represented by one versus two individuals:

$$S = S_{obs} + \frac{f_1(f_1 - 1)}{2(f_2 + 1)}$$

As you can see in this equation, the estimated number of species in a community increases with an increase in the number of species represented by a single individual and decreases with an increase in the number of species represented by two individuals. For example, consider the adjacent table that lists the number of individuals observed for 12 species sampled from three pond communities.

Using the data for Community A, we can estimate the actual number of species. In this community, we have observed 12 species. Two individuals were observed for each of two species, whereas one individual was observed for each of three species. Using our equation above, we get

$$S = S_{obs} + \frac{f_1(f_1 - 1)}{2(f_2 + 1)}$$

$$S = 12 + \frac{3(3 - 1)}{2(2 + 1)}$$

$$S = 12 + \frac{6}{6}$$

$$S = 13$$

Species	Community A	Community B	Community C
WOOD FROGS	32	65	42
GREEN FROGS	45	24	45
NEWTS	12	14	5
SPOTTED SALAMANDERS	15	17	12
RAMSHORN SNAILS	2	23	67
POND SNAILS	25	14	35
AMPHIPODS	2	14	2
ISOPODS	1	1	2
FINGERNAIL CLAMS	9	1	0
DIVING BEETLES	1	1	0
WATER BOATMEN	5	1	0
DRAGONFLIES	1	1	0

Based on these results, we conclude that although we observed 12 species, we estimate that the community actually contains 13 species.

YOUR TURN Using the equation, estimate the number of species that are present in Community B and Community C. Based on your results, how does the number of species represented by single individuals affect the estimated number of species in the community?

Species accumulation curve A graph of the number of species observed in relation to the number of individuals sampled.

Figure 22.10 Edge habitat and nest parasitism. Bronzed cowbirds deposit their eggs in the nests of forest songbirds along the edges of habitat fragments, such as this nest of a Bewick's wren in south Texas. The much larger cowbird chick receives a disproportionate amount of food from host parents, and this causes a reduction in the number of host offspring that survive. Photo by Rolf Nussbaumer/DanitaDelimont.com.

cowbird chicks (**Figure 22.10**). This allows the cowbird to produce offspring without providing time-consuming parental care, but reduces the number of offspring the host bird can raise. While cowbirds spend much of their time living in fields, they enter the forest edge to find nests of forest birds in which they deposit their eggs. Researchers have found that as forest fragmentation has created more edge habitat, more nests are parasitized by the cowbirds. Subsequently, the reproduction and population sizes of the host species of songbirds have declined.

Fragmentation can also have consequences for human health. You may recall from Chapter 15 that human exposure to Lyme disease depends on a complex food web that starts with mice and chipmunks serving as hosts for newly hatched ticks that carry the pathogenic bacteria. In forest fragments in the northeastern United States, the abundance of many vertebrate animals has declined, but the abundance of the white-footed mouse has increased, probably because most of the mouse's competitors and predators cannot live in the smallest forest fragments. Surveys of mouse populations in forest fragments of different sizes show that the smallest forest fragments have the highest mouse densities, and thus the highest tick densities. It was also found that mice living in the smallest fragments have the highest proportion of ticks infected with the Lyme disease bacterium. In short, the fragmentation of forests due to human activities has created a landscape that makes us more likely to be exposed to Lyme disease.

Corridors, Connectivity, and Conservation

In Chapters 11 and 12, we touched on the importance of considering geographic scales larger than the local area. In our discussion of metapopulations, we emphasized that

many populations are divided into habitat fragments and that the regional populations persist because each patch is connected by the occasional dispersal of individuals between patches. We saw an example of this in Chapter 11 when we discussed the recovery of the collared lizard. This lizard depended on patches of open habitat, known as glades, as well as corridors between the glades through which they could disperse. Corridors facilitate movement, which can save declining populations of many species from local extinction. They also increase gene flow and genetic diversity within populations, which counteract the negative effects of genetic bottlenecks and genetic drift. Corridors can simply be pieces of preserved habitat or they can be constructed, such as corridors built to let animals cross a highway (**Figure 22.11**).

Although corridors can rescue declining populations by adding new colonists that bring genetic variation, they can also have unintended downsides. For example, corridors built to help conservation of a particular species can facilitate the movement of predators (including poachers), competitors, and pathogens that are harmful to conservation efforts. Therefore, resource managers must carefully consider the costs and benefits of developing corridors among habitat before spending time and money to implement this strategy.

The importance of corridors is greatest for those organisms that require a continuous connection to move between habitat fragments. However, organisms such as birds and flying insects can pass over stretches of inhospitable habitat matrix and therefore may not need a continuously connected corridor. Instead, these species can move between large patches of favorable habitat if they have access to small intervening patches where they can stop to rest or forage. These small intervening patches

Figure 22.11 Building corridors. Road construction through forests is one way that humans create fragmented large habitats. To help wildlife move between forest fragments, some roads, such as this one in Alberta, Canada, include overhead corridors with surfaces of soil and natural vegetation. Photo by JOEL SATORE/National Geographic Creative.

that dispersing organisms can use to move between large favorable habitats are known as **stepping stones.**

The role of habitat corridors and stepping stones has spurred major efforts at preservation. India, for example, is home to nearly 60 percent of all Asian elephants (*Elephas maximus*), which live in several national parks and protected areas. These areas are the fragmented remains of a much larger and contiguous habitat. The World Land Trust and the Wildlife Trust of India are working together to protect important corridors between the protected habitat fragments to ensure the long-term persistence of the elephants (**Figure 22.12**). Although elephants are charismatic animals that can draw attention to conservation needs, these corridors are likely to assist in the conservation of other charismatic species including tigers (*Panthera tigris*) and Himalayan black bears (*Ursus thibetanus*), as well as many less-charismatic species.

The Effects of Matrix Conditions on Dispersal

As you may recall from our discussion of metapopulations in Chapter 11, the quality of the matrix that lies between favorable habitat fragments is a key factor that helps determine whether organisms can move between these favorable fragments. The matrix may contain favorable conditions for the movement of organisms or it may be inhospitable. For example, researchers in Colorado mapped the locations of meadows that exist in a matrix of willow thickets and coniferous forests and then studied the movements of more than 6,000 individual butterflies from six taxonomic groups. Butterflies feed in the meadows, but when moving from one meadow to another, they must pass through the matrix. After capturing the butterflies in meadows and writing a unique number on their wings, the researchers released the butterflies. They then recaptured the butterflies to determine whether the butterflies exhibited

Figure 22.12 Wildlife corridors in India. The original habitat of Asian elephants is now fragmented into several protected areas. Conservation groups are currently trying to preserve strips of land between these protected areas to serve as corridors through which elephants and other species can move. Photo by Jagdeep Rajput/ DanitaDelimont.com.

a preference for flying through the willow-thicket matrix or through the coniferous-forest matrix. In two of the six taxonomic groups, there was no preference; one group rarely left its meadow, while the other group was composed of excellent fliers that easily navigated both matrix habitats. In contrast, individuals from the other four taxonomic groups were 3 to 12 times more likely to move between meadows through the willow-thicket matrix than through coniferous-forest matrix. This confirms that some species have distinct preferences for dispersing through particular matrices, and these preferences are related to their ability to disperse. From this example, we can see that the matrix may affect the movements of organisms among fragments.

CONCEPT CHECK

1. What does a species–area curve represent?
2. Why does habitat fragmentation lead to an increased abundance of species that prefer ecotones?
3. What is the relationship between increased habitat fragmentation and habitat edge?

22.3 The equilibrium theory of island biogeography incorporates habitat area and isolation

When Robert MacArthur and E. O. Wilson examined the relationship between island size and species richness, they noticed that although the size of the island was strongly associated with the number of species, there was a large amount of variation around the species–area curve that suggested other processes were also affecting the number of species living on the island. They considered the locations of the islands from which they collected data and observed that the distance from a source of colonizing species also affects the number of species living on an island. Islands closer to a source of colonizing species—for example, a mainland—appeared to receive more colonizing species. In this section, we will examine the combined effects of island area and isolation on species richness and then explore a graphical model that helps us understand these patterns. Finally, we will explore how this information can be used to design nature reserves.

THE EVIDENCE

To test their hypothesis that species richness is determined by both island area and isolation, MacArthur and Wilson collected data on different types of organisms from groups of islands throughout the world. For example, they examined the number of bird species that existed on 25 islands in the South Pacific. The nearest source of new species for these islands was the large island of New Guinea, and these islands

Stepping stones Small intervening habitat patches that dispersing organisms can use to move between large favorable habitats.

could be categorized as being near, intermediate, or far from New Guinea. When the researchers plotted the number of bird species for each island as a function of island area, shown in **Figure 22.13**, they found that both island area and the distance from New Guinea affected the number of species living on the island. For islands of a given distance from the mainland, large islands contain more species than small islands. For islands of a given area, near islands contain more bird species than far islands.

Although oceanic islands provide a good test of how colonization and extinction affect the number of species, we can also consider other types of distinct habitats that exist on continents. For example, James Brown and his colleagues examined the number of mammal species living on mountaintops in the southwestern United States. These mountaintops include alpine tundra and conifer forest habitats that are surrounded by a matrix of low-elevation habitats, including woodlands, grasslands, and desert scrub. Twenty-six species of mammals in the region prefer to live in the mountaintop habitats. Researchers asked whether the number of mammal species on each mountaintop was affected by the area of the mountaintop or the isolation of the mountaintop from two sites containing sources of colonizing species: the southern Rocky Mountains and the Mogollon Rim, a mountain range that cuts across northern Arizona. As you can see in **Figure 22.14**, the researchers discovered more species of mammals on mountaintops with the largest areas and fewer species of mammals on mountaintops that were farthest away from the two sources of colonizers.

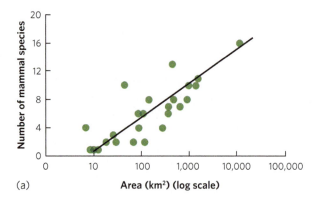

(a)

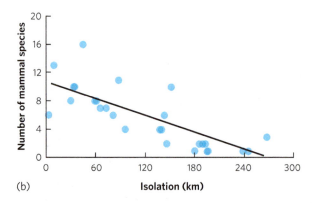

(b)

Figure 22.14 Effects of area and isolation on mountaintop mammals. For mammals that live on mountaintop habitats in the southwestern United States, researchers found **(a)** more mammal species living on mountaintops of greater area and **(b)** fewer mammal species living on mountaintops that are more distant from a source of colonists. Data from M. V. Lomolino et al., Island biogeography of montane forest mammals in the American Southwest, *Ecology* 70 (1989): 180–194.

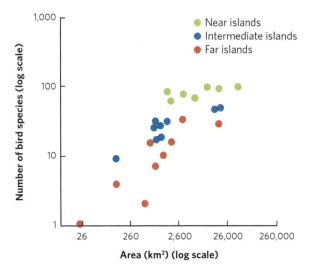

Figure 22.13 Effects of island area and isolation on bird species richness. In the South Pacific, researchers examined the number of birds living on islands that were various distances from New Guinea. On islands of a similar distance from New Guinea, large islands contained more bird species than small islands. On islands of a similar area, islands near to New Guinea contained more bird species than islands far from New Guinea. Data from R. H. MacArthur and E. O. Wilson, An equilibrium theory of insular zoogeography, *Evolution* 17 (1963): 373–387.

The observations on oceanic islands and mountaintop islands suggest that island area and isolation are important factors in determining species richness. However, a manipulative experiment was needed to demonstrate that these processes do, in fact, cause the observed pattern. In a classic experiment, E. O. Wilson and his graduate student Daniel Simberloff worked on a set of tiny islands in the Florida Keys. These islands typically contain a single mangrove tree and no animals other than insects, spiders, and other arthropods. Prior to the experiment, the team documented that islands located closer to sources of colonizing species contained more species than islands located far away from sources of colonizing species. Tents were built over selected islands, as shown in **Figure 22.15a**, and then the islands were fumigated with an insecticide that killed nearly every arthropod. After removing the tents, the researchers returned to the islands every few weeks for a year to track how many species recolonized the islands from large islands nearby. As you can see in Figure 22.15b, each island was rapidly recolonized by arthropods. Moreover, the final number of species was close to the original number of species prior to the fumigation and

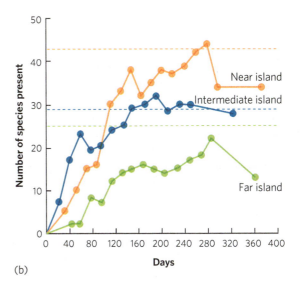

(a)

(b)

Figure 22.15 An experimental test of island biogeography theory. (a) Researchers in the Florida Keys constructed a scaffold frame around islands and covered the scaffold with tarps to act as a tent when fumigating the islands. Fumigating the islands removed most of the arthropods. **(b)** Over the course of a year, researchers returned to determine how many species of arthropods had recolonized. By the end of the experiment, the islands had nearly recovered their original number of species, as indicated by the dashed lines, with the near islands containing a higher number of species than the far islands. Data from D. S. Simberloff and E. O. Wilson, Experimental zoogeography of islands: The colonization of empty islands, *Ecology* 50 (1969): 278–296. Photo by Daniel Simberloff.

the more isolated islands once again contained fewer species of arthropods. The actual list of species was different, but the richness was similar. This experiment was important because it demonstrated that isolation was a key factor in determining how many species could live on an island.

THE THEORY

Based on repeated observations that both island size and island isolation affect the number of species living on an island, Robert MacArthur and E. O. Wilson developed the **equilibrium theory of island biogeography,** which states that the number of species on an island reflects a balance between the colonization of new species and the extinction of existing species.

To understand the equilibrium theory of island biogeography, we need first to understand the factors that affect the colonization of species onto an island and the extinction of species from an island. As we discussed in Chapter 11, species differ in their ability to disperse depending on their overall size and mode of locomotion. So now imagine that we have an island that is completely uninhabited by any species. With no species on the island, a large number of species on the nearby mainland could potentially colonize it. As a growing number of species colonize the island, the pool of species that have not yet colonized the island shrinks and those that remain on the mainland are probably not very good at dispersing. Therefore, the rate of new species colonizing an island declines as a function of how many species have already colonized the island. If the island contained every species that is found on the nearby mainland, the rate of new colonizing species would fall to zero. In **Figure 22.16,**

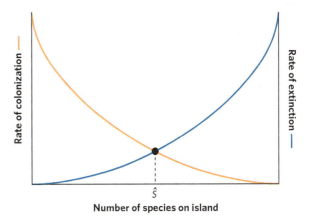

Figure 22.16 Achieving an equilibrium number of species on an island. As the number of species living on an island increases, the rate of colonization by new species from the regional pool declines. At the same time, the rate of extinctions of species living on the island increases. The equilibrium number of species, $\hat{S}$, occurs where the two curves cross and the opposing processes are balanced.

the relationship between species richness and rate of colonization by new species is shown as an orange line.

Next we need to consider the factors that affect the rate of extinction on an island. Since the extinction rate is expressed as the number of species that will go extinct over a period of time, the extinction rate must be affected by how

Equilibrium theory of island biogeography A theory stating that the number of species on an island reflects a balance between the colonization of new species and the extinction of existing species.

many species are present. In the simplest case, when there are no species, there can be no extinctions. As the island begins to be colonized, it holds a few species that could potentially go extinct. As more species live on the island, more species are subject to possible extinction, and so the extinction rate increases. The extinction rate should also be affected by harmful species interactions on the island. For example, competition, predation, and parasitism are all more likely to increase as the total number of species increases. The blue line in Figure 22.16 shows that the extinction rate ranges from zero when no species are present on the island to a maximum rate of extinction when the island contains every species that it can possibly support.

Given that the island continues to experience colonization of new species and extinction of existing species, these two opposing forces will eventually reach an equilibrium point where the rate of colonization is equal to the rate of extinction. On our graphical model, equilibrium occurs where the two curves cross. Following the dashed line down from the equilibrium point, we see that the number of species on the island at equilibrium is designated as $\hat{S}$ on the x axis. It is important to remember that the model predicts the number of species present at equilibrium, but not a particular composition of species at equilibrium. At equilibrium, there is a continuous turnover of species on the island; while new species colonize the island, others go extinct. We saw an example of this in the islands in the Florida Keys that were fumigated by Simberloff and Wilson. Over time, the species richness of the islands came to equilibrium, though the composition of species changed as the islands continued to experience extinctions and colonizations of invertebrate species.

Using the graphical model in Figure 22.16 as a foundation, we can make predictions about the combined effects of island size and isolation on the number of species on an island when it is at equilibrium. Let's first consider the effects of island size. We would expect smaller islands to sustain smaller populations of each species. As we have discussed in several previous chapters, smaller populations typically have higher extinction rates. Therefore, smaller islands should experience higher rates of extinction, as shown by the curve labeled "Small" in **Figure 22.17**. When an island is at equilibrium, extinction rates balance colonization rates. Therefore, the number of species on a small island at equilibrium should be lower than the number of species on a large island at equilibrium. If we follow the two equilibrium points on the figure down to the x axis, we can see that the number of species at equilibrium is lower for small islands than for large islands.

Next we can consider the effects of island isolation. An island that is near a continent should experience higher rates of colonization by new species than an island that is far from a continent. Once again using Figure 22.16 as our starting point, we can make predictions about how island isolation should affect the number of species at equilibrium. In **Figure 22.18**, we can see that the far island has a lower colonization curve than the near island. This creates two equilibrium points between colonization and extinction rates. If we follow the dashed lines down from each

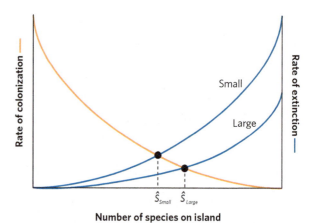

Figure 22.17 Effects of island size on the equilibrium number of species. Because smaller islands support smaller populations that are more prone to extinction, smaller islands have steeper extinction curves. As a result, smaller islands contain fewer species at equilibrium ($\hat{S}_{Small}$) than larger islands ($\hat{S}_{Large}$).

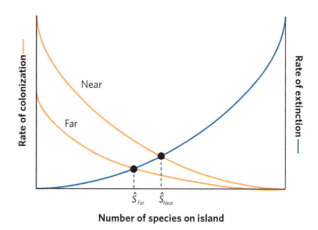

Figure 22.18 Effects of island isolation on the equilibrium number of species. Islands that are far from a source of new colonizing species experience lower rates of colonization than islands that are close. As a result, farther islands contain fewer species at equilibrium ($\hat{S}_{Far}$) than near islands ($\hat{S}_{Near}$).

equilibrium point, we see that the number of species at equilibrium is lower for far islands than for near islands.

When we combine the effects of island size and island isolation from a continent, we arrive at several predictions. Looking at **Figure 22.19**, we see that the combinations of large versus small islands and near versus far islands create four different possible equilibrium points. Once again, we can follow the dashed line down from each equilibrium point to determine how many species are predicted to be on a particular type of island when colonization and extinction rates are at equilibrium. Small islands that are far from a major source of new species should contain the fewest species. In contrast, large islands that are near a major source of new species should contain the most species. In short, the combination of island size and isolation determines the number of species that the island can hold.

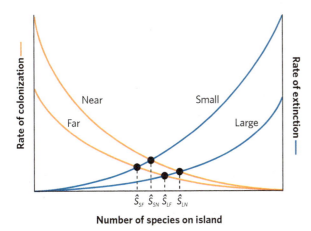

Figure 22.19 The combined effects of island size and isolation on species richness. When we simultaneously consider island size and isolation, we find that the small islands that are far from a continent should have the lowest number of species at equilibrium ($\hat{S}_{SF}$), whereas large islands that are close to a continent should have the highest number of species ($\hat{S}_{LN}$).

APPLYING THE THEORY TO THE DESIGN OF NATURE RESERVES

Understanding the effects of island size, shape, and distance from sources of colonizing species has helped scientists to design nature reserves to protect biodiversity, as depicted in **Figure 22.20.** For example, we know that large areas typically contain more species than small areas because they contain a greater diversity of habitats. Large areas can support larger populations of each species, which lowers the rate of species extinction. Therefore, setting aside large nature reserves will better protect biodiversity than will small nature reserves.

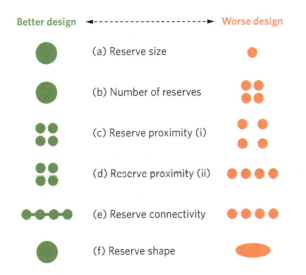

Figure 22.20 Nature reserve design. The number, shape, size, and proximity of reserves affect the probable success of preserving a region's biodiversity. As noted in this figure, some designs are likely to be better than others for preserving biodiversity. Adapted from J. M. Diamond, The island dilemma: Lessons of modern biogeographic studies for the design of nature reserves, *Biological Conservation* 7 (1975): 129–146, and J. C. Williams et al., Spatial attributes and reserve design models, *Environmental Modeling and Assessment* 10 (2005): 163–181.

Similarly, a single large reserve is the better option to preserve biodiversity than are several small reserves of the same total area. However, a single habitat poses several risks. When a species of conservation concern is located in a single large habitat, it is more likely that the species will be destroyed by natural disaster or disease.

When we are faced with creating multiple reserves, we need to consider the proximity of the reserves to each other. Reserves need to be close enough for organisms to disperse between them but far enough away to reduce the ability for predators and diseases to move between them. Setting this distance is a compromise between these positive and negative outcomes, and the best solution will differ among organisms with different dispersal abilities. Sometimes the ideal distance will be set for the species of greatest conservation interest. A similar challenge exists when reserves are joined by corridors that provide strips of hospitable habitat between adjacent reserves.

Because we know that increased edge can have detrimental effects on species that prefer to live in the interior of habitats, it is also important that we consider reserve shape. Round reserves have the lowest ratio of edge to area, while long, skinny ovals or rectangles have much higher edge to area ratios that will favor edge species. Although reserve size, number, shape, and proximity are all important considerations, often the ideal scenario is not possible; nature reserves typically represent a compromise between the areas that are desirable and the areas that are available.

CONCEPT CHECK

1. Why do mountaintop habitats function similarly to oceanic islands in regard to the theory of island biogeography?
2. Using the theory of island biogeography, why do we expect the number of species on an island to reach equilibrium?
3. What combination of island size and distance to the mainland would you expect to result in the greatest species richness?

22.4 Biodiversity is highest near the equator and declines toward the poles

We have examined the local and regional processes that can affect the number of species in a given location on Earth. However, patterns of biodiversity also exist at the global scale. One of the most striking patterns is that species richness of all taxa combined is highest near the tropics and declines toward the poles. For example, a hectare of forest typically has fewer than 5 species of trees in boreal regions, 10 to 30 species of trees in temperate regions, and up to 300 species of trees in tropical regions. These latitudinal trends in diversity are pervasive and extend even to the oceans. In this section, we will explore the patterns of diversity within and across latitudes. We will then discuss the two general hypotheses for these patterns of biodiversity and explore three important processes.

PATTERNS OF DIVERSITY

In the Northern Hemisphere, the number of species in most groups of animals and plants increases from north to south. For example, when a researcher in North America counted the number of mammal species living in blocks of habitat that were 241 km on a side, he found that there were fewer than 20 mammal species per block in northern Canada but more than 50 species per block in the southern United States. You can view these data in **Figure 22.21a**.

The number of mammals also increases as we move from east to west in North America. For instance, there are typically 50 to 75 species per sample block in the east, whereas there are 90 to 120 species per block in the west. Such a pattern is likely due to the greater amount of habitat heterogeneity in the extensive mountain ranges of western North America. This greater heterogeneity of environments in the west apparently provides suitable conditions for a greater number of species.

The pattern of species richness for birds in North America resembles the pattern for mammals, but the patterns for reptiles, trees, and amphibians are strikingly different, as you can see in Figures 22.21b–d. The richness of reptile species declines fairly uniformly as temperature decreases toward the north. However, trees and amphibians are more diverse in the moister eastern half of North America than in the drier, more mountainous western regions.

The general pattern of increased species richness as we move closer to the equator also holds true for the oceans. For example, in 2010 researchers compiled millions of records of marine organisms around the world, including whales, dolphins, fish, and corals. Using sample blocks of 880 km on a side, as shown in **Figure 22.22**, they found that the highest diversity occurred in the tropics including Central America and Southeast Asia, whereas the lowest diversity occurred near the poles where temperatures are colder. At a given latitude, they also found a higher diversity near the coasts and lower diversity in the open ocean.

PROCESSES THAT UNDERLIE PATTERNS OF DIVERSITY

Historically, ecologists have considered two general hypotheses for the decline in species richness as we move from the equator to the poles. According to one hypothesis, species are continually created over time and without limit. Because the world's current temperate and polar regions experienced repeated advances and retreats of glaciers during the Ice Age, species in these regions have been eliminated or driven toward refuges closer to the equator. In contrast, because the world's tropical regions did not experience glaciation, habitats in these areas have remained stable much longer and so have had more time to accumulate species.

The second hypothesis proposes that the number of species reflects an equilibrium between the processes that create new species and the processes that drive species extinct, similar to the state of equilibrium described by the theory of island biogeography. According to this hypothesis, the higher number of species in the tropics is the result of higher rates of speciation or lower rates of extinction, compared to the temperate and polar regions. Similarly, variation in the number of species

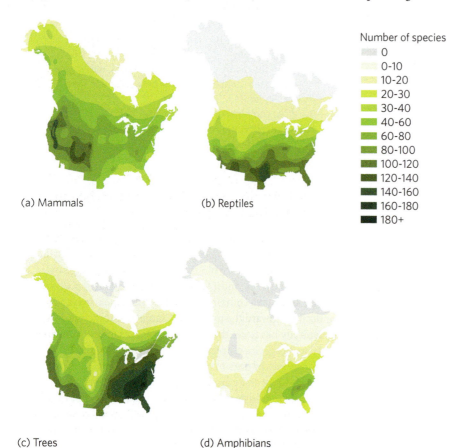

(a) Mammals　(b) Reptiles　(c) Trees　(d) Amphibians

Number of species: 0, 0-10, 10-20, 20-30, 30-40, 40-60, 60-80, 80-100, 100-120, 120-140, 140-160, 160-180, 180+

Figure 22.21 Patterns in North American diversity. When a researcher examined the number of species living in the United States and Canada, he found strong patterns in species richness. The contour lines on the map indicate the numbers of species found in blocks of land that were 241 km on a side. **(a)** Mammals increase in species richness from north to south and from east to west. **(b)** Reptiles increase in species richness from north to south. **(c, d)** For both trees and amphibians, the highest number of species occurs in the southeast, where there is a combination of high precipitation and warm temperatures. Data from D. J. Currie, Energy and large-scale spatial patterns of animal- and plant-species richness, *American Naturalist* 137 (1991): 27–49.

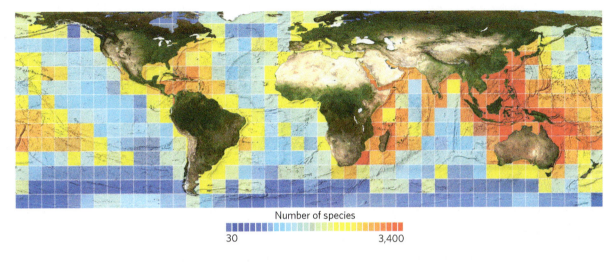

Number of species

30   3,400

Figure 22.22 Patterns of marine biodiversity. In the ocean, the greatest number of species exists in the tropics and declines toward the poles. At a given latitude, more species exist near the coasts and fewer species exist in the open ocean. The number of species is calculated using sample blocks of 880 km per side. After D. P. Tittensor et al., Global patterns and predictors of marine biodiversity across taxa, *Nature* 466 (2010): 1098–1103.

across a given latitude should also reflect an equilibrium between processes that create new species and destroy existing species. We will now consider three processes that play substantial roles in determining the number of species that occur in an area: ecological heterogeneity, solar energy and precipitation on land, and water temperatures in the oceans.

The Role of Ecological Heterogeneity

At any given latitude, we find more species in areas where there is greater ecological heterogeneity, such as heterogeneity in the soils and plant life. For example, grasslands contain vegetation that is less heterogeneous in growth form than shrublands or deciduous forests. As a result of these differences in habitat heterogeneity, surveys of breeding birds in North America find an average of 6 species in grasslands, 14 species in shrublands, and 24 species in floodplain deciduous forests. You may recall that we discussed a similar phenomenon in Chapter 18 when we looked at the positive relationship between the diversity of vegetation height and the diversity of birds (see Figure 18.13). This same pattern has also been observed in lizards of the southwestern United States, where lizard diversity is associated with vegetation diversity, and in the plants of South Africa, where higher plant diversity is associated with areas with more heterogenous soils.

Although more productive terrestrial habitats tend to have a greater number of species, habitat heterogeneity also plays a role in determining species richness when two habitats have similar levels of productivity. For instance, habitats with less variation in vegetation growth form, such as grasslands, have fewer animal species than habitats with similar productivity but more variation in vegetation. This principle also applies to plants. Marshes are highly productive but have a relatively uniform landscape and thus contain relatively few species of plants. While desert vegetation is less productive than marsh vegetation, the greater heterogeneity of the desert landscape provides room for more species of plants (**Figure 22.23**).

(a)

(b)

Figure 22.23 Habitat heterogeneity. Although marshes are more productive than deserts, the relative uniformity of the marsh environment has produced a low diversity of species, whereas the heterogeneous landscape of deserts has produced a higher diversity of species. Pictured here are **(a)** the White River Marsh Wildlife Area in Wisconsin and **(b)** the Sabino Canyon Recreation Area in Arizona. Photos by (a) Lee Wilcox, and (b) Ron Niebrugge/Alamy.

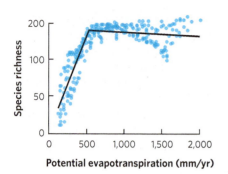

(a) Birds

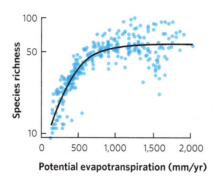

(b) Mammals

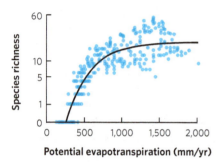

(c) Amphibians

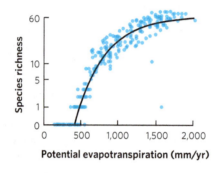

(d) Reptiles

Figure 22.24 The relationship between species diversity and potential evapotranspiration. For all four groups of vertebrates, the potential evapotranspiration of a sample is strongly correlated to the number of species found in the sample block. Data from D. J. Currie, Energy and large-scale spatial patterns of animal- and plant-species richness, *American Naturalist* 137 (1991): 27–49.

The Role of Solar Energy and Precipitation on Land

The number of species found at any location as one moves from the poles to the tropics is positively correlated to the amount of solar energy and precipitation at that location. Together, solar energy and precipitation can be measured as **potential evapotranspiration (PET). PET** is the amount of water that could be evaporated from the soil and transpired by plants, if water was abundant, given the average temperature and humidity of that location on Earth. Because PET integrates temperature and solar radiation, it provides an index to the overall energy input into the environment. As you can see in **Figure 22.24**, PET correlates with the species richness in North American vertebrates relatively well. In each group of vertebrates, the initial rise in PET is associated with a rise in species richness, which reflects increasing diversity from north to south within the continent. However, species richness eventually plateaus at very high levels of PET. These high PET levels occur as we move from east to west at middle latitudes where the increasing temperature fails to improve the capacity of the environment to support additional vertebrate species. In the arid western parts of the continent, for instance, increasing temperature eventually becomes a stressor.

The correlations between PET and species richness for terrestrial vertebrates have given rise to the idea that this correlation may actually represent a causal relationship between the two variables. This hypothesized causal relationship is known as the **energy–diversity hypothesis,** which states that sites with higher amounts of energy are able to support more species. Higher amounts of energy would also support higher abundances of individuals from each species, which should reduce the rate of extinction. Moreover, higher energy input might accelerate the rate of evolutionary change and thereby increase the rate of speciation. While these ideas are attractive, none of these mechanisms has yet been verified.

The Role of Water Temperature in the Oceans

As we have seen, biodiversity in marine environments is greater in the tropics than it is at higher latitudes. However, this pattern does not appear to be driven by greater productivity in the tropics. Though marine productivity is greatest in temperate latitudes (see Figure 20.6), the high productivity in those regions is seasonal; temperature differences and stratification of the water in temperate zones make nutrients readily available during the seasonal mixing of the water column but scarce during periods when the water is stratified. In contrast, tropical marine environments experience relatively stable temperatures that lead to relatively small nutrient fluctuations and low but steady productivity. Researchers have tested whether the patterns in species diversity across the oceans were better explained by marine productivity, mean water temperature, or variation in water temperature. They discovered that the only significant predictor of marine biodiversity across latitudes

Potential evapotranspiration (PET) The amount of water that could be evaporated from the soil and transpired by plants, given the average temperature and humidity.

Energy–diversity hypothesis A hypothesis that sites with higher amounts of energy are able to support more species.

was the mean temperature of the sea surface. Because a higher mean temperature is a measure of greater total energy, this pattern further supports the energy–diversity hypothesis.

The patterns of species richness across and within latitudes are potentially affected by all three processes of habitat heterogeneity, temperature and precipitation on land, and mean temperature in the oceans. In all three cases, the mechanisms involved suggest that the global distribution of species richness is the outcome of an equilibrium between the processes that create new species and the processes that cause the extinction of existing species.

CONCEPT CHECK

1. What is the geographic pattern of species richness for mammals in North America?
2. Why would increasing vegetation heterogeneity lead to increased species richness in birds?
3. Why might we expect potential evapotranspiration to be correlated to terrestrial species richness across latitudes?

22.5 The distribution of species around the world is also affected by Earth's history

When we consider the patterns of species diversity on Earth, we need to remember that Earth was formed 4.5 billion years ago and life arose during the first billion years. This means that the species on Earth today are the result of billions of years of evolution and that current diversity arose in response to past environmental conditions. In this section, we will examine how the movement of continents and historic changes in climate affected the distribution of species.

CONTINENTAL DRIFT

During Earth's history, the continents have repeatedly come together and drifted apart in a process called **continental drift.** Continental drift occurs because the continents are essentially giant islands of low-density rock that move by underlying convection currents of semi-molten material. About 250 Mya, all of Earth's landmasses were joined together as a single landmass, named **Pangaea,** illustrated in **Figure 22.25.** By 150 Mya, Pangaea had separated into a northern landmass, known as **Laurasia,** and a southern landmass, known as **Gondwana.** Laurasia subsequently split into North America, Europe, and Asia, while Gondwana split into South America, Africa, Antarctica, Australia, and India. Ultimately, India collided into Asia, which caused the land to rise and created the Himalayas. In the Northern Hemisphere, a widening Atlantic Ocean separated Europe from North America, but a land bridge had already formed on the other side of the world between North America and Asia. More recently, Europe and Africa joined about 17 Mya and a land bridge formed between North and South America at the Isthmus of Panama 3 to 6 Mya.

An important consequence of continental drift was the changing opportunities for dispersal among continents. Once separated, continents could independently evolve species in different regions of Earth. For instance, because Australia has experienced a long period of isolation from other continents, it has evolved many unique groups of species that include a wide variety of marsupial animals, such as koalas and kangaroos, and plants, such as eucalyptus trees. When continents later joined together, groups of organisms

---

Continental drift The movement of landmasses across the surface of Earth.

Pangaea The single landmass that existed on Earth about 250 Mya and subsequently split into Laurasia and Gondwana.

Laurasia The northern landmass that separated from Pangaea about 150 Mya and subsequently split into North America, Europe, and Asia.

Gondwana The southern landmass that separated from Pangaea about 150 Mya and subsequently split into South America, Africa, Antarctica, Australia, and India.

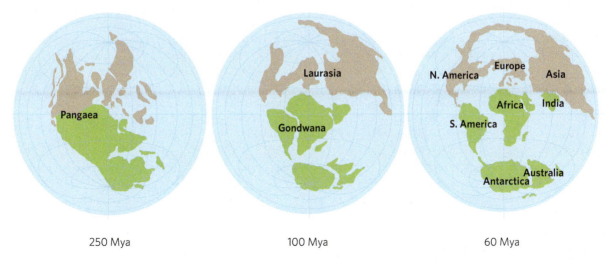

| 250 Mya | 100 Mya | 60 Mya |

Figure 22.25 Continental drift. About 250 Mya, all of Earth's landmasses were joined together into Pangaea. Pangaea later split into Gondwana and Laurasia, which, in turn, split into modern-day continents. Many of these landmasses subsequently joined together in new configurations.

that were unique to one landmass were able to move into new areas. You may recall from Chapter 3 that camels originated in North America about 30 Mya. When North America became connected to Asia at the Bering Land Bridge in Alaska, ancestral camels dispersed into Asia and then moved into Africa, where they diversified into the modern-day Bactrian camels and dromedary camels. At about the same time, other ancestral camels dispersed into South America, where they further diversified into llamas, guanacos, vicuñas, and alpacas. In short, the history of continents separating and coming together continues to affect the distributions of present-day animals and plants.

BIOGEOGRAPHIC REGIONS

Because continental drift has allowed individual continents independently to evolve groups of organisms for long periods, we can see patterns of species distributions on each continent. Alfred Wallace, a contemporary of Charles Darwin and co-discoverer of evolution by natural selection, first described these patterns and delineated six major biogeographic regions that are based on the distributions of animals. These regions, illustrated in **Figure 22.26**, are still recognized today. Botanists also recognize six major biogeographic regions, based on plant distributions, with boundaries that closely coincide with those regions based on animal distribution.

The Northern Hemisphere is divided into the **Nearctic region,** which roughly corresponds to North America, and the **Palearctic region,** which corresponds to Eurasia. Through most of the past 100 million years, the continents of these regions maintained connections across what is now

Greenland between North America and Eurasia and across the Bering Strait between Alaska and Russia. Consequently, these two regions share many groups of animals and plants. European forests seem familiar to tourists from North America, and vice versa; although few species are the same, both regions have representatives of many of the same genera and families.

The Southern Hemisphere is divided into four biogeographic regions. The **Neotropical region** corresponds to Central and South America and the **Afrotropical region,** also known as the **Ethiopian region,** corresponds to most of Africa. Farther to the east is the **Indomalayan region,** also known as the **Oriental region,** which includes India and Southeast Asia. The final biogeographic region is the **Australasian region,** which includes Australia, New Zealand, and New Guinea. In Wallace's delineation, the continent of Antarctica was not included.

Nearctic region The biogeographic region of the Northern Hemisphere that roughly corresponds to North America.

Palearctic region The biogeographic region of the Northern Hemisphere that corresponds to Eurasia.

Neotropical region The biogeographic region of the Southern Hemisphere that corresponds to South America.

Afrotropical region The biogeographic region of the Southern Hemisphere that corresponds to most of Africa. *Also known as* **Ethiopian region**.

Indomalayan region The biogeographic region of the Southern Hemisphere that corresponds to India and Southeast Asia. *Also known as* **Oriental region**.

Australasian region The biogeographic region of the Southern Hemisphere that corresponds to Australia, New Zealand, and New Guinea.

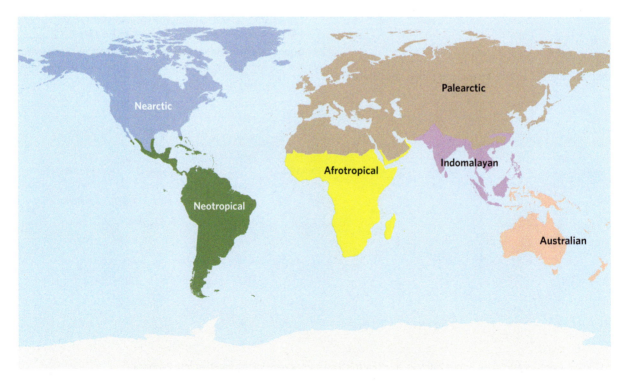

Figure 22.26 Biogeographic regions. Terrestrial regions of the world can be categorized in terms of distinctive groups of plants and animals that largely reflect the history of continental drift.

Each of these regions contains unique groups of species that reflect their long histories of isolation from the rest of the terrestrial world and the subsequent exchange of species after they were joined together. For example, when the Isthmus of Panama joined the Nearctic and Neotropical regions, there was a large, though uneven, exchange of mammals. Many North American lineages moved into South America and caused the extinction of most of the marsupial animals that had come to South America during an earlier connection with Australia. In contrast, only a few mammals moved into North America from South America; the most prominent are the opossums, which is the only marsupial species in the United States and Canada.

HISTORIC CLIMATE CHANGE

As we saw in Chapter 5, the position of the continents and the circulation of water around the continents influence the variation in climates around the world. Because the continents were drifting over the past 250 million years, it is not surprising that the climates of the world experienced dramatic changes. We know from fossil evidence that large portions of North America and Europe once had tropical climates. Tropical forests reached into Russia and Canada and warm temperate forests covered the Bering Land Bridge from Alaska to Asia. Moreover, the Antarctic land connection between South America and Australia supported temperate vegetation and abundant animal life. However, as Antarctica drifted over the South Pole and as North America and Eurasia gradually encircled the northern polar ocean, the Arctic Ocean became largely enclosed between North America and Eurasia. This created a circumpolar ocean current around Antarctica, which caused cooler temperatures at high latitudes. As a result, Earth's climates became more strongly differentiated from the equator to the poles. Tropical environments contracted into a narrow zone near the equator, and temperate and boreal climate zones expanded. These climate changes had profound effects on the geographic distributions of plants and animals.

About 2 Mya, the gradual cooling of the planet gave way to a series of dramatic oscillations in climate known as the Ice Age. Climate changes during this time had dramatic effects on habitats and organisms in most parts of the world. Alternating periods of cooling and warming caused the advance and retreat of ice sheets at high latitudes over much of the Northern Hemisphere and cycles of cool, dry climates and warm, wet climates in the tropics. Ice sheets came as far south as Ohio and Pennsylvania in North America and covered much of northern Europe, driving vegetation zones southward, possibly restricting tropical forests to isolated refuges where conditions remained moist.

A striking example of this disruption is the migration of forest trees in eastern North America and Europe. At the peak of the most recent glacial period, many tree species were restricted to southern refuges, but about 18,000 years ago after the ice began to retreat and the forests started to spread north again. Pollen grains, which are deposited in the lakes and bogs left by retreating glaciers, record the coming and going of plant species. These records show that the composition of plant associations changed as species migrated back north over different routes across the landscape. The migrations of some representative tree species from their southern refuges are mapped in **Figure 22.27**. As you can see in this figure, spruce trees shifted northward just behind the retreating glaciers. Oaks expanded out of their southern refuges to cover most of the eastern part of temperate North America, from southern Canada to the Gulf Coast. Hemlocks expanded into the Great Lakes Region and into central Canada.

The forests of Europe suffered even more extinctions from the spread of glaciers than those in North America because populations were blocked from shifting southward by the Alps

Thousands of years before present

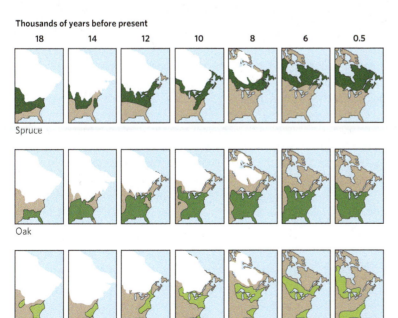

Spruce

Oak

Hemlock

Figure 22.27 Changing tree distribution following the retreat of glaciers. As the glaciers began retreating in North America 18,000 years ago, the trees began moving north to colonize the newly exposed habitats. Data from G. L. Jacobson et al., in W. F. Ruddiman and H. E. Wright, Jr. (eds.), *North America During Deglaciation* (Geological Society of America, 1987), pp. 277–288.

and the Mediterranean Sea. Several northern European tree species went extinct. Many species that survived were restricted to refuges in southern Europe, from which they expanded after the glaciers receded beginning about 18,000 years ago. Recent research suggests that many tree species in Europe have not yet expanded fully into their potential ranges, which suggests that the European flora has not returned to an equilibrium state. As we will see in the next chapter, the ability of species to respond to changing climates raises questions about how species will respond to global warming.

Throughout this chapter, we have seen that studying the distribution of species must be done at both the landscape scale and the global scale. In doing so, we see that biodiversity is maintained by large-scale factors including area, isolation, and historical events that continue to show their effects today. In "Ecology Today: Applying the Concepts," we will see that this knowledge has helped scientists design a massive nature reserve to help protect the biodiversity of Florida.

CONCEPT CHECK

1. How has continental drift affected the modern distribution of species on Earth?
2. How are biogeographic regions delineated?
3. How do scientists know that tree species have been moving north since the most recent ice age?

ECOLOGY TODAY   APPLYING THE CONCEPTS

Taking a Long Walk for Conservation

The Florida panther. Many species, such as this Florida panther—an endangered subspecies of mountain lion—are expected to benefit from the Florida Wildlife Corridor project that will link together large areas of protected habitats. Photo by ©Tom & Pat Leeson/AGE Fotostock.

As we have seen throughout this chapter, the distribution and persistence of biodiversity depend on a variety of factors. Knowing the factors responsible for maintaining diversity helps when designing nature reserves to protect biodiversity. Although we have a good idea about the features of an ideal nature reserve, in reality, the design of nature reserves and the corridors that connect them is largely determined by existing habitats, their configurations, and land ownership. An excellent example of this can be found in a conservation effort known as the Florida Wildlife Corridor, a visionary plan to create a continuous connection of protected habitat from the southern tip of Florida all the way up to Georgia. This would provide a large number of species, including endangered species such as the Florida panther, with a tremendous amount of land to inhabit for a long time to come. The ultimate goal is to connect a number of current federal nature reserves, including the Everglades National Park, the Big Cypress National Preserve, the Ocala National Forest, and the Okefenokee National Wildlife Refuge, with a variety of properties owned by the state of Florida, conservation organizations such as The Nature Conservancy, and private landowners including large cattle ranches.

With such a multitude of landowners, the Florida Wildlife Corridor is an excellent case study of the challenges presented by the design of nature reserves. It is as much a sociological problem as an ecological problem, so the promoters of the corridor had to think about a public relations campaign that was as big as the 1,600-km swath of land down the middle of the state. In 2012, four leaders of the project decided to travel the entire length of the proposed corridor in what they referred to as "1,000 miles in 100 days." Moving on foot and in kayaks, the conservationists traversed swamps, grasslands, and forests. They called attention to their effort by using daily photo and video updates on their websites. Along the way, they were joined by state and federal biologists, conservationists, politicians, and private ranch owners. At every stop, they emphasized that preserving this swath of land through the middle of Florida provided benefits not only to the organisms that lived in undeveloped lands, but also to the residents of Florida who rely on tourism as well as the water and agriculture that come out of this region. On Earth Day 2012, the 100-day expedition was completed at the Stephen C. Foster State Park in southern Georgia.

In 2015, the group did a second walk for conservation, from the Everglades to the western edge of the Florida Panhandle. On this trip, they observed a great deal of fragmented habitat, but it was clear that the possibility of protecting key habitats and corridors along this region of Florida still existed. Collectively, the Florida Wildlife Corridor is made up of 6.4 million ha, which is home to 42 federally endangered species. About 60 percent of the land is currently protected, and the group's goal is to protect an additional 120,000 ha of key habitats along the corridor by 2020.

The effort to protect the land in a north–south corridor through Florida continues today, with a growing momentum to find a way to protect the lands that lie between the state and federal reserves, while at the same time retaining the long-held traditions of agriculture and ranching on private lands. If such protection can be achieved, then the state will have a total reserve that is both large and interconnected, and this should result in the long-term persistence of species in the region.

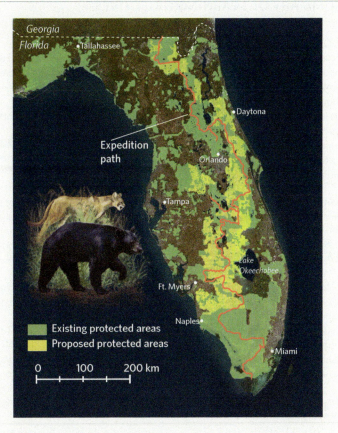

The Florida Wildlife Corridor. Conservationists hope to link public and private land through the state of Florida in a way that promotes the movement of wild organisms between habitats and facilitates their long-term persistence.

SOURCE:
The Florida Wildlife Corridor (http://www.floridawildlifecorridor.org/about).

SUMMARY OF LEARNING OBJECTIVES

22.1 Landscape ecology examines ecological patterns and processes at large spatial scales. Habitat heterogeneity exists across the landscape as a result of both natural processes and human activities, both in the past and in modern times. This heterogeneity allows more species to exist at the landscape scale than at local scales. Species diversity can change with spatial scale; we quantify alpha diversity at the local scale, gamma diversity at the regional scale, and beta diversity across habitats. The process of species sorting determines which species from the regional species pool are found in a local community.

Key Terms: Landscape ecology, Legacy effect, Local diversity, Regional diversity, Beta diversity, Regional species pool, Species sorting

22.2 The number of species increases with increased area. The number of species found in areas of different sizes can be described by the species–area curve. As the area sampled grows, we see an increase in species richness because we begin to sample new habitats that contain different species. Habitat fragmentation divides a large habitat into several small habitats and thereby increases the amount of habitat edge. These effects cause the decline in the abundance of many species, but favor those species that prefer habitat edges.

Key Terms: Species–area curve, Species accumulation curve, Stepping stones

22.3 The equilibrium theory of island biogeography incorporates both habitat area and isolation. Species diversity is highest on large islands that are close to a source of colonizing species because these islands have a low rate of species extinction and a high rate of colonization by new species. In contrast, species diversity is lowest on small islands that are far from a source of colonizing species because these islands have a high rate of species extinction and a low rate of colonization by new species.

Key Term: Equilibrium theory of island biogeography

22.4 Biodiversity is highest near the equator and declines toward the poles. Species diversity can also vary within a given latitude belt as a result of habitat heterogeneity, temperature, and precipitation. In terrestrial ecosystems, the best predictor of species diversity is potential evapotranspiration. The energy–diversity hypothesis states that sites with higher amounts of energy are able to support more species. In marine ecosystems, the best predictor of species diversity is the temperature at the surface of the sea.

Key Terms: Potential evapotranspiration (PET), Energy–diversity hypothesis

22.5 The distribution of species around the world is also affected by Earth's history. Continental drift is a major factor in determining the distribution of species. As continents have come together and separated, they have experienced unique evolutionary lineages that today are recognized as distinct biogeographic regions. The drift of the continents has also affected historic climate changes, including widespread changes in the distribution of tropical, temperate, and polar regions around the world.

Key Terms: Continental drift, Pangaea, Laurasia, Gondwana, Nearctic region, Palearctic region, Neotropical region, Afrotropical region, Indomalayan region, Australasian region

CRITICAL THINKING QUESTIONS

1. How can the fragmentation of a landscape have both positive and negative effects on biodiversity?

2. Why would it be important to consider the quality of the matrix that exists between habitat fragments when considering the movement of amphibian species between ponds?

3. How do local diversity and regional diversity affect the slope and intercept of the species–area relationship?

4. Compare and contrast alpha, gamma, and beta diversity.

5. Why might we expect that regions with high species diversity will also exhibit high niche diversity?

6. Why should small islands that are distant from a mainland have fewer plant species than large islands that are close to a mainland?

7. When North America and South America came together by continental drift, what may have caused the North American mammals that moved into South America to drive many South American mammals extinct?

8. In what ways might the age and area of a region affect its species richness?

9. How does knowledge of historic climatic patterns affect our interpretation of present-day patterns of species diversity?

10. What is the likely relationship between the length of time the modern continents have been connected and the similarity among the families and genera on each continent?

GRAPHING THE DATA Species Accumulation Curves

Ecologists often need to determine the number of species in an area they are sampling. As we have discussed in this chapter, they can use an equation to estimate the total number of species or they can graph a species accumulation curve that plots the sample size along the x axis and the number of species observed along the y axis. Using the data in the table, graph species accumulation curves for forest communities A and B.

Based on your graph, which community has been sampled sufficiently to provide a good estimate of the species richness that exists and which community requires more sampling? Explain your answer.

Number of Individuals Sampled	Number of Species Observed in Community A	Number of Species Observed in Community B
20	12	5
40	20	10
60	27	15
80	33	19
100	38	23
120	42	27
140	45	30
160	47	33
180	48	36
200	48	38

23 Conservation of Global Biodiversity

Protecting Hotspots of Biodiversity

 The biodiversity of the world faces a wide range of threats from a growing human population that has caused species to go extinct at a rapid rate. To reverse this downward spiral, conservationists seek ways to protect aquatic and terrestrial ecosystems so that threats from human activities can be reduced or eliminated. A common approach to protecting species is to protect their habitats. But there are limits to how much habitat can be protected: habitat protection often means that the habitat must be purchased and not all habitats are available for purchase. Also, limited funds are available for purchasing habitats, and political and economic factors often play a role in determining whether or not particular habitats can be set aside. With all these limitations, how should we prioritize the protection of the world's biodiversity?

In 1988, Norman Myers noted that small isolated areas such as tropical islands contain a large proportion of endemic species, which are species that have a relatively restricted distribution and are not found in other parts of the world. Therefore, a large proportion of the world's terrestrial species is located in a relatively small proportion of the world's land. Myers argued that we should concentrate our conservation efforts on these species-rich areas, which he defined as biodiversity hotspots, because saving these areas should save the most species.

Myers identified 10 locations on Earth as biodiversity hotspots. Shortly afterward, the group Conservation International adopted Myers's approach and decided that hotspots should be defined as areas containing at least 1,500 endemic plant species and experiencing at least 70 percent vegetation loss due to human activities. These criteria would identify areas of high species richness facing substantial threats. The group assumed that regions with high plant diversity also contained high animal diversity. Conservation International identified 34 biodiversity hotspots around the world, including the Caribbean Islands, Central America, the coast of California, the island of Madagascar, and several sites in the islands of Southeast Asia. Collectively, these locations represent 2.3 percent of the Earth's land surface, but contain 50 percent of the world's plants (more than 150,000 species) and 42 percent of the world's vertebrate animals (nearly 12,000 species).

A similar effort is under way for aquatic hotspots, with a particular focus on oceans. In the oceans, scientists have argued for hotspots in the open ocean, deep thermal vents (see Chapter 1), and coral reefs. However, many of the same debates occur in how to define marine hotspots. This can be particularly difficult in regions of the ocean where hot spots are seasonal, such as the seasonal upwelling of nutrients that causes periods of high productivity and greater biodiversity.

> **"A large proportion of the world's terrestrial species is located in a relatively small proportion of the world's land."**

While identifying hotspots based on high numbers of endemic species is certainly reasonable, some scientists have suggested other approaches. One such approach is to focus on areas that have high species richness without a focus on endemic species. For example, the Amazon rainforest has a very high number of species, but most are not endemic to small geographic areas. Another approach is to prioritize species-rich locations facing the highest current or projected threats of species extinctions, such as locations that have, or are projected to have, a rapidly growing human population. Each approach supports a different list of protection priorities. Focusing conservation priorities on places with high diversity, however, automatically excludes low-diversity locations with species that people care about, such as the bison, wolves, and grizzly bears that live in western North America. It also places an emphasis on species richness, rather than on the important functions that many ecosystems provide. As an example, although wetlands typically have low plant diversity, they are incredibly important for flood control and water filtration.

Conserving biodiversity. Efforts are being made to protect the world's biodiversity, including this tiger longwing butterfly (*Heliconius ismenius*) from Costa Rica. Photo by directphoto.bz/Alamy.

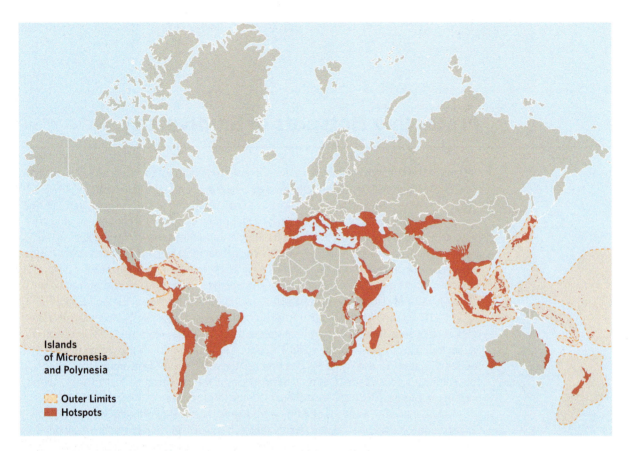

Biodiversity hotspots. Thirty-four biodiversity hotspots have been identified for terrestrial sites around the world. These sites contain at least 1,500 endemic plant species and have experienced at least a 70 percent decline in their vegetation. While the criteria used are based on plants, these areas also contain a high diversity of animals. Outer limits on the map indicate hotspot regions that include oceanic islands. http://specieslist.com/images/external/ci-hotspots.jpg.

It is clear that we can take a variety of approaches in how we prioritize the conservation of biodiversity. All these approaches seek to help us invest our limited resources to save the most biodiversity that we possibly can. In this chapter, we will focus on how biodiversity serves us, the causes of the decline in biodiversity, and the efforts that are being made to save it.

SOURCES:
Bacchetta, G., et al. 2012. A new method to set conservation priorities in biodiversity hotspots. *Plant Biosystems* 146: 638–648.
Marchese, C. 2015. Biodiversity hotspots: A shortcut for a more complicated concept. *Global Ecology and Conservation* 3: 297–309.
Myers, N., et al. 2000. Biodiversity hotspots for conservation priorities. *Nature* 403: 853–858.

LEARNING OBJECTIVES

23.1 Identify the value of biodiversity based on social, economic, and ecological considerations.

23.2 Explain why the current rate of extinction is unprecedented.

23.3 Describe the ways in which human activities are causing the loss of biodiversity.

23.4 Identify conservation efforts that can slow or reverse declines in biodiversity.

Throughout this book, we have examined factors that affect the distribution of species around the world. We have seen that these distributions are a result of the abiotic conditions a species can tolerate, the positive and negative interactions that occur among species, the ability to disperse to suitable habitats, and geologic processes that include the movement of the world's continents. We have also examined how human activities affect particular species and how conservation efforts try to minimize these impacts. In this final chapter, we take a broad view of the decline in biodiversity around the globe. We begin by considering the many different ways in which people value biodiversity. We will then compare current rates of biodiversity decline with historic rates and examine the ways in which human activities contribute to these declines. Finally, we will discuss efforts under way to slow or even reverse the decline of biodiversity.

23.1 The value of biodiversity arises from social, economic, and ecological considerations

A case for conserving the world's biodiversity can reflect a range of different values. For example, the **instrumental value of biodiversity** focuses on economic values that species can provide, such as lumber for building or crops for eating. In contrast, the **intrinsic value of biodiversity** recognizes that species have inherent value that is not tied to any economic benefit. Of course, species and ecosystems can have both instrumental and intrinsic values.

INSTRUMENTAL VALUES

The total economic benefit of biodiversity is difficult to estimate because many of the world's species remain undiscovered and the values of each species and ecosystem can be difficult to estimate. For example, the total economic benefit of biodiversity in the United States is estimated at $319 billion per year. For perspective, this is about 10 percent of U.S. annual gross domestic product. At the global level, estimates of the total benefit of biodiversity, including all of the ecosystem services provided, is $125 trillion. We can group the instrumental values of biodiversity into four categories of services: *provisioning, regulating, cultural,* and *supporting.*

Provisioning Services

Provisioning services are benefits of biodiversity that provide products humans use, including lumber, fur, meat, crops, water, and fiber. In many cases, plants and animals from the wild have been cultivated or domesticated and then selectively bred to enhance their valuable qualities. Provisions also include pharmaceutical chemicals that come from plants and animals; nearly 70 percent of the top 150 pharmaceutical drugs originate in chemicals that are produced in nature. A prominent example of the economic benefits of such a drug is the chemical Taxol, which is used to fight cancer. Today, Taxol is synthesized in the laboratory, but it originally came from the Pacific yew tree (*Taxus brevifolia*) (**Figure 23.1**). This single chemical currently generates more than $1.6 billion in annual sales around the world. Over the past 25 years, more than 800 natural chemicals have been identified in the search to provide treatments for everything from cancer to contraception, and there is no indication that the pace of these discoveries is slowing down.

Regulating Services

Regulating services are benefits of biodiversity that include climate regulation, flood control, and water purification. For example, wetlands absorb large amounts of water and so prevent flooding from water runoff during rainy periods. The plants living in the wetlands also remove contaminants from the water and make it more suitable for drinking. The CO_2 that is taken out of the air by producers on land and in the ocean is another regulating service. Of the 8 gigatons of carbon that are put into the air each year by human activities,

Figure 23.1 Provisioning services. The Pacific yew is one of many species that has served as a critical source of pharmaceutical chemicals that improve human health. Photo by inga spence/Alamy.

about 4 gigatons are taken out of the air by producers, which reduces the effect humans have on global temperatures due to global warming.

Cultural Services

Cultural services are benefits of biodiversity that provide aesthetic, spiritual, or recreational value. For example, cultural services include the benefits that people obtain when they go hiking, camping, boating, or birdwatching. People pay to visit beautiful natural areas, such as the Florida

---

Instrumental value of biodiversity The economic value a species can provide.

Intrinsic value of biodiversity A focus on the inherent value of a species, not tied to any economic benefit.

Provisioning services Benefits of biodiversity that humans use, including lumber, fur, meat, crops, water, and fiber.

Regulating services Benefits of biodiversity that include climate regulation, flood control, and water purification.

Cultural services Benefits of biodiversity that provide aesthetic, spiritual, or recreational value.

Everglades in the United States or Banff National Park in Canada. Sometimes areas are preserved because income from tourists can exceed what would be received from clearing a forest or from using the land for housing and industry. Many tropical countries have capitalized on this attraction by establishing parks and support services for tourists. In Palo Verde National Park in Costa Rica, for example, monkeys and beautiful tropical birds draw tourists to areas where the species are protected (**Figure 23.2**). Diversity itself is often the attraction in tropical rainforests and coral reefs because these ecosystems contain hundreds of different species of trees, birds, corals, or fish.

Supporting Services

Supporting services are benefits of biodiversity that allow ecosystems to exist, such as primary production, soil formation, and nutrient cycling. As we have seen in Chapters 20 and 21, these processes are essential for the existence of species and ecosystems. There would be no ecosystems without producers that capture the energy of the Sun and then transfer it to all other trophic levels. Similarly, the formation

Figure 23.2 Attracting tourists to Palo Verde National Park. Striking examples of biodiversity, such as this resplendent quetzal (*Pharomachrus mocinno*), attract tourists from around the world.
Photo by Brandon Lindblad/Shutterstock.

of soil and the cycling of nutrients both play key roles in the persistence of existing ecosystems.

INTRINSIC VALUES

In contrast to instrumental values, the intrinsic values of biodiversity do not provide any economic benefits to humans. Instead, people who place intrinsic value in biodiversity feel religious, moral, or ethical obligations to preserve the world's species. For example, a major motivation for efforts to bring back the bald eagle from the brink of extinction in the 1970s was that it was the national symbol of the United States and we had a moral obligation to prevent its extinction. However, it becomes very difficult to prioritize conservation efforts only by arguing that all species are intrinsically valuable. In reality, we can consider both instrumental and intrinsic values when deciding how to focus our conservation efforts.

CONCEPT CHECK

1. What are three provisioning services of biodiversity?
2. Why are regulating services considered to be an instrumental value of biodiversity?
3. Why is it difficult to assign an economic value to intrinsic values of biodiversity?

23.2 The current rate of extinction is unprecedented

The current number of species on Earth is difficult to estimate. We do know that 1.3 million species have received Latin names and about 15,000 new species are described each year. While estimates for the total number of species range from 3 to 100 million, depending on the assumptions used, most scientists agree that there are about 10 million species. Some species are declining in abundance and facing extinction as humans continue to alter terrestrial and aquatic ecosystems. However, as we have seen throughout this book, some extinctions are natural. Therefore, we need to understand the historic versus modern rates of extinctions. In this section, we will explore the past and present rates of extinction and then examine how specific groups of organisms are faring. As part of this discussion, we will consider both declines in species diversity and declines in genetic diversity.

BACKGROUND EXTINCTION RATES

Over the past 500 million years, the world has experienced five **mass extinction events,** which are defined as events in which at least 75 percent of the existing species go extinct within a 2-million-year period. During these events, large numbers of species, genera, and families around the world went extinct. For simplicity, **Figure 23.3a** illustrates the number of families that have gone extinct.

Supporting services Benefits of biodiversity that allow ecosystems to exist, such as primary production, soil formation, and nutrient cycling.

Mass extinction events Events in which at least 75 percent of the existing species go extinct within a 2-million-year period.

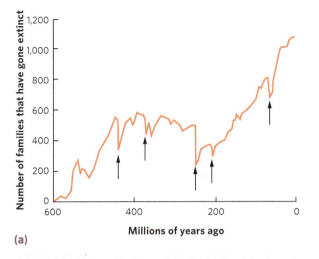

(a)

(b)

Figure 23.3 Mass extinctions. (a) Over the past 500 million years, there have been five mass extinctions. During these periods, the world experienced a significant decline in the number of families, which also means there were declines in the number of genera and species. Continued speciation during subsequent years has helped to offset these extinctions. **(b)** The fifth extinction is hypothesized to have been caused by volcanic eruptions, cooler climates, and a massive asteroid that struck the Yucatan Peninsula and put massive amounts of dust into the air and blocked the Sun's rays. Data from J. J. Sepkoski, Jr., Ten years in the library: New data confirm paleontological data, Paleobiology 19 (1993): 43–51. Photo by MARK GARLICK/SCIENCE PHOTO LIBRARY/Getty Images.

During the first mass extinction, about 443 Mya, most species lived in the oceans. An ice age caused sea levels to drop and the ocean chemistry to change, which resulted in 86 percent of species going extinct. The second mass extinction happened 359 Mya when much of the ocean lacked oxygen—for reasons that are unclear—and 75 percent of all species went extinct. During the third mass extinction—248 Mya—an astounding 96 percent of all species then present on Earth went extinct. Although researchers have constructed multiple hypotheses to explain this third mass extinction, we are still uncertain about the cause. The fourth mass extinction, which occurred 200 Mya, caused 80 percent of the world's species to go extinct. Hypotheses for the causes

of this fourth extinction include increased volcanic activity, asteroid collisions with Earth, and climate change.

The fifth mass extinction happened 65 Mya and is best known as the one that led to the extinction of dinosaurs. This event is attributed to several factors. First, volcanic eruptions and changes in climate caused long periods of cold weather. This was followed by a massive asteroid that struck the Yucatan Peninsula in Mexico. The asteroid is estimated to have been 10 km wide and struck with a force more than 1 billion times that of the atom bomb dropped on Hiroshima during World War II (Figure 23.3b). The explosion created a massive, 180-km-wide crater.

Scientists hypothesize that the explosion put so much dust into the atmosphere that it blocked the Sun's rays, making Earth much less hospitable to dinosaurs along with many other groups, such as flowering plants. During this time, 76 percent of the species on Earth went extinct.

As you can see from this history of natural mass extinction events, only a small percentage of all species that ever lived on Earth are present today. In fact, over the past 3.5 billion years, it is estimated that 4 billion species have existed on Earth and 99 percent of these species are now extinct. However, as you can see in Figure 23.3, after each mass extinction event, new species evolved and, overall, the number of species has increased with time.

A POSSIBLE SIXTH MASS EXTINCTION

We have seen that a mass extinction is defined as the extinction of 75 percent of species within a 2-million-year period. It is a widely held view in the scientific community that the increase in the human population during the past 10,000 years may have initiated a sixth mass extinction event. To evaluate this hypothesis, we need to quantify the rates of extinction during the first five mass extinctions and then compare them with the current rate of extinction. Researchers have addressed this question by looking at groups of organisms, such as mammals, for which there are very good fossil data on extinctions. When the extinction rate for mammals during the most recent 500 years is compared to the rate of mammal extinctions over 500-year intervals in the past, we see that the current extinction rate exceeds the historic extinction rate. In fact, the United Nations Convention on Biological Diversity estimates that the extinction rate during the last 50 years has been as much as 1,000 times higher than the historic rate. In the next section, we will explore the reasons for this change in rate. Should this rate continue for hundreds or thousands of years, it could qualify as a mass extinction event.

GLOBAL DECLINES IN SPECIES DIVERSITY

When we think about the decline in the world's biodiversity, we often focus on the last few centuries, from the Industrial Revolution to today—a time in which we have drastically altered our world. However, human impacts on biodiversity can be seen much farther back in time. For example, based on a rich fossil history of mammals, researchers have defined different geographic regions of North America and determined the number of fossil species in each region. With these data they created species–area curves for different time periods in North America, which we discussed in Chapter 22. Compared to the period before the arrival of humans—150,000 to 11,500 years ago—the species–area curve for the period after human arrival—11,500 to 500 years ago—was significantly lower, as you can see in **Figure 23.4**. Simply put, the arrival of humans coincided with a 15 to 42 percent decline in mammal diversity, depending on the geographic region examined. The losses include 56 species and 27 genera of large mammals, including a giant ground sloth, the saber-toothed tiger, and several species of horses, camels, elephants, and lions.

Explanations for these extinctions include rapid climate change following the retreat of the glaciers, hunting pressure from the human population, and epidemic diseases carried from Asia by domesticated animals. Many scientists hypothesize that most of the large mammals were driven extinct by humans who hunted them. We know that the diversity of mammal species had been substantially reduced during the period of human occupation of North America prior to the Industrial Revolution, although we can't be sure of the reason. Any current impacts are adding to previous extinctions.

Whether the world's current extinction rate of species will approach the magnitude of a mass extinction depends on how many of the species currently on Earth go extinct during the next few centuries. To assess how different groups of organisms are currently faring, the International Union

for Conservation of Nature (IUCN) has defined categories that describe whether a species is abundant, threatened, or extinct. *Extinct* describes a situation in which a species was known to be in the wild in the year 1500 but no individuals remain alive today. *Extinct in the wild* is a category used when the only individuals remaining are in captivity, such as animals living in zoos. *Threatened* species are those whose populations face a high risk of extinction in the future. This category includes species that are considered "endangered." *Near-threatened* species are those that will likely become threatened in the future. In contrast, *least-concern* species are those that have abundant populations and are not likely to become threatened in the future. In some cases, the status of a species has not been determined or there are simply insufficient data to make a reliable determination.

Assessing the status of species from a large taxonomic group is not easy. Many groups contain thousands of species, and often a large percentage of them have not been studied well enough to know if a species is abundant or declining. Making such a determination requires substantial time and money for each species. Currently, our best data to assess the decline in biodiversity are for conifers, birds, mammals, amphibians, fish, and reptiles. Assessments for these taxonomic groups were produced by the IUCN in 2017, and you can see a summary of these data in **Figure 23.5**.

Conifers

Conifers include pines, spruces, firs, cedars, and redwoods, and the survival prospects of 95 percent of these species have been assessed. This high level of assessment is possible, in part, because the group has a relatively low number of species and large trees and shrubs are more easily assessed for population declines. Of the 606 species of conifers, none has gone extinct. Of those species that have sufficient data for assessment, 50 percent are categorized as of least concern, 16 percent are near-threatened, and 34 percent are threatened.

Birds

Some of the best data on species assessments come from birds because they are relatively easy to monitor and they have been studied for a long time. The 2017 assessment found sufficient data for assessment of 99 percent of the more than 11,000 species of birds on Earth. Since the year 1500, 156 of these (1.4 percent) have gone extinct. Of the remaining species for which there are reliable data, 77 percent have sufficiently abundant populations to be categorized as least concern, 9 percent are near-threatened, and 13 percent are threatened with extinction.

Reptiles

Reptiles include snakes, lizards, and turtles. Of the more than 5,000 species of reptiles, 86 percent have sufficient data for assessment. Twenty-eight species (0.4 percent) have gone extinct during the last 500 years. Of those remaining with sufficient data for assessment, 68 percent are of least concern, 8 percent are near-threatened, and 24 percent are threatened.

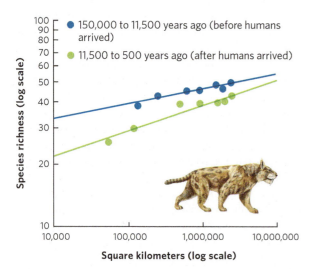

Figure 23.4 Declining North American mammals. When researchers created species–area curves of North American mammals from different geographic regions before and after human arrival, they discovered that from 11,500 to 500 years ago (i.e., after humans arrived), the number of mammals declined anywhere from 15 to 42 percent, depending on the geographic region examined.

Data from M. A. Carrasco et al., Quantifying the extent of North American mammal extinction relative to the pre-anthropogenic baseline, *PLoS One* 4 (2009): e8331.

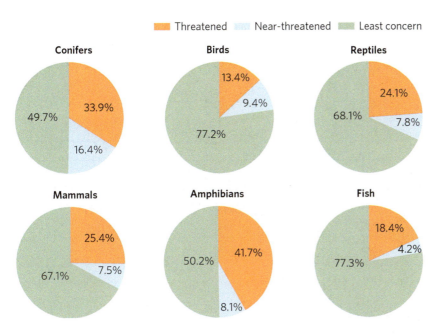

Threatened   Near-threatened   Least concern

Figure 23.5 The global status of conifers, birds, reptiles, mammals, amphibians, and fish. Species threatened by the risk of future extinction include 34 percent of conifers, 13 percent of birds, 25 percent of mammals, 42 percent of amphibians, 18 percent of fish, and 24 percent of reptiles. Data from International Union for Conservation of Nature (2017), https://goo.gl/ZJFSbI, https://goo.gl/SVYrfB.

Mammals

Of the 5,560 species of mammals that have lived on Earth since the year 1500, 86 percent have sufficient data to assess their status. During the past 500 years, 83 mammal species (1.5 percent) have gone extinct. Of those remaining for which there are reliable data, researchers found that 67 percent of the species are categorized as of least concern, 8 percent are near-threatened, and 25 percent are threatened.

Amphibians

Amphibians have been particularly hard hit in recent decades from several causes that include habitat loss and the spread of the deadly chytrid fungal disease that we discussed in Chapter 15. Of the more than 6,500 species of amphibians assessed, only 76 percent contained sufficient data to assess their conservation status. During the past 500 years, 33 species (0.5 percent) have gone extinct. Of the remaining species with sufficient data, 50 percent are of least concern, 8 percent are near-threatened, and a staggering 42 percent of amphibian species are threatened.

Because amphibians are not as visible as birds and mammals, scientists are still discovering new species at a rapid rate. For example, more than 3,000 new species of amphibians have been discovered in the past 25 years, which represents nearly half of the world's described amphibian species. This translates to a discovery of a new amphibian species every 2.5 days. One challenge with so many new discoveries is that little is known about the population status of these species. Researchers expect hundreds of more species to be discovered in the future, and therefore our estimates of each category will continue to be updated.

Fish

Like amphibians, fish have also experienced a tremendous number of declines. The IUCN considered more than 16,000 species of fish in its 2017 assessment and found that more than 3,000 species were found to be data deficient. Since the year 1500, 64 fish species are now extinct. Of those remaining with reliable data, 77 percent of the species are categorized as of least concern, 4 percent are near-threatened, and 18 percent are threatened.

The data from conifers, birds, reptiles, mammals, amphibians, and fish suggest that if human impacts on these species continue, we should expect a large number of species extinctions in future centuries. Although our best data come from these six groups, it is predicted that the patterns of decline observed in these groups are representative of many other groups for which data on the status of species are relatively poor. This prediction is supported by preliminary efforts to assess the status of other major groups. Although less than 10 percent of all species of flowering plants and insects have been assessed, about half of the currently assessed species are categorized as threatened.

As we have discussed in previous chapters, these declines in species richness are a concern not only because of the risk of losing species, but also because of the effect these declines have on communities and ecosystems. As we discussed in Chapter 17, decreases in the number of mycorhizal fungi cause decreases in the biomass of plants (see Figure 17.22). In Chapter 18, we also observed that a decline in species richness can cause communities to be less stable over time by affecting community resistance or resilience (see Figure 18.22). Declines in species richness can also cause a decline in the functioning of ecosystems. For example, a review of all research examining patterns between experimental manipulations of plant species richness and the aboveground biomass of plants—just one measure of an ecosystem's function—has found that commonly a positive relationship exists, although there are exceptions. You can see examples of these relationships in **Figure 23.6**.

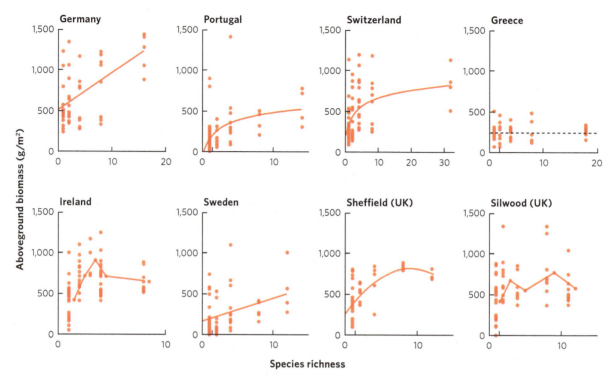

Figure 23.6 Effects of species richness on ecosystem function. Researchers have observed positive relationships between species richness and ecosystem function in seven of eight locations in Europe, although the shape of the relationship differs among locations. Data from D. U. Hooper et al., Effects of biodiversity of ecosystem functioning: A consensus of current knowledge, *Ecological Monographs* 75 (2005): 3–35.

GLOBAL DECLINES IN GENETIC DIVERSITY

In addition to the decline in species diversity around the world, we have also seen a decline in the genetic diversity of many species. As we discussed in Chapter 7, causes of decline in genetic diversity include declining population sizes, inbreeding depression, and the bottleneck effect. Smaller populations do not possess the same amount of genetic diversity as large populations. These declines in genetic diversity reduce the probability that a population contains genotypes able to survive changing environmental conditions, including changes in climate and emerging infectious diseases.

The decline in the genetic diversity of livestock and crops has a direct and immediate effect on humans. The primary livestock that we consume or use for labor and transportation include just seven species of mammals (cattle, pigs, sheep, goats, buffalo, horses, and donkeys) and four species of birds (chickens, turkeys, ducks, and geese). Humans have bred these species for a wide variety of different traits, including size, strength, quality of meat, and ability to persist under challenging environmental conditions such as drought and diseases (**Figure 23.7**).

During the past century, many livestock varieties have not been maintained because larger, modern livestock operations favor relatively few breeds that are the most productive in terms of meat or milk. In Europe, for instance, about half of the livestock breeds present in 1900 are now extinct. Of those breeds remaining, 43 percent are at a serious risk of extinction. In North America, 80 percent of the livestock

breeds that have been evaluated are either declining in abundance or facing extinction. Around the world, of the 7,000 breeds of 35 domesticated species of birds and mammals, more than 10 percent are already extinct and another 21 percent are at risk of extinction. Such a rapid decline in genetic diversity means there is considerably less diversity to draw from should we need to breed domesticated animals that can live in new locations or in changing environments, or that can withstand new diseases. Simply put, reduced genetic variation reduces our future options.

Declines in genetic diversity are also occurring in plant species that are important to humans. Humans historically consumed more than 7,000 species of plants, but today consume only about 150 species. Moreover, just 12 species of plants make up the vast majority of people's diets; among these are wheat, rice, and corn. In some cases—for example, corn—modern varieties look very different from their ancestors (**Figure 23.8**). In earlier times, humans bred varieties that grew well under particular local environmental conditions. However, as agricultural practices changed, irrigation and fertilizer made it possible to reduce the harshness of the growing environment, and small farms gave way to much larger operations that favored only the top-producing varieties. As a result, many of the older, local varieties of crops are no longer available. For example, U.S. farmers grew about 8,000 varieties of apples in 1900, but today, 95 percent of these varieties are extinct. Similarly, 80 percent of the corn varieties that existed in Mexico in 1930 and 90 percent of the wheat varieties that existed in China in 1949 are now gone.

Figure 23.7 Genetic diversity of chickens. Over the past few centuries, people have bred a wide variety of livestock breeds, such as chickens, to suit their particular needs or local environmental conditions. Today, most of these varieties are extinct because large livestock operations have focused on just a few breeds to maximize the production of livestock.

Figure 23.8 Genetic diversity of crops. (a) Corn originated from a wild plant in Mexico, known as teosinte, shown on the left. Modern cultivated corn is shown on the right. An ear of the F_1 hybrid is shown in the center. **(b)** From the teosinte ancestor, humans have bred a wide range of genetic varieties to perform well under different conditions, including these varieties grown in Oaxaca State, Mexico. Today, the tremendous variety of genetic crops is at risk of being lost. Photos by (a) John Doebley and (b) Philippe Psaila/Science Source.

Losing the genetic diversity of these crops reduces our options when we need to respond to challenges such as new pathogens that attack a crop. For example, in the 1970s, a fungus attacked cornfields in the southern United States and killed half the corn crop because the plants all came from a single variety. Fortunately, another variety of corn possessed a fungus-resistant gene, and breeders were able to produce a new variety that is resistant to the fungus.

To protect the genetic diversity of plant varieties, many countries have been archiving seed varieties of different crop species in thousands of storage facilities around the world. Concern that such facilities could be destroyed by natural disasters or war led to the construction of the Svalbard Global Seed Vault. Located on an island in the Arctic region north of the Norwegian mainland, the facility is a 125-m tunnel built into a mountain with rooms for seeds on each

side of the tunnel, as illustrated in **Figure 23.9**. This facility protects seeds from virtually all catastrophes. The vault has a total capacity of 1.5 million samples; as of 2013, the facility contained more than 700,000 samples of crop seeds from nearly every country. The Svalbard Global Seed Vault and many other seed storage facilities around the world preserve the ability to call upon the genetic diversity of plant species far into the future.

CONCEPT CHECK

1. What data can we use to determine if we are in the midst of a sixth extinction?
2. Which groups of animals are the most threatened?
3. Why have we lost the genetic diversity of many crop plants?

Each heat-sealed aluminum foil packet contains approximately 500 seeds.

Each storage box holds up to 500 seed packets.

Each vault holds hundreds of boxes.

Vaults

The concrete-lined cave is held at –18°C.

Control room

North Pole

Greenland   Svalbard

Iceland   Norway

Atlantic Ocean

Entrance

Figure 23.9 The Svalbard Global Seed Vault. This facility was built on an island north of the Norwegian mainland to preserve the genetic diversity of crop plants so that humans will be able to use this genetic variation long into the future.

23.3 Human activities are causing the loss of biodiversity

The current rapid decline in biodiversity is caused by the rapid increase in human populations and our many activities. Virtually all areas within temperate latitudes that are suitable for agriculture have been plowed or fenced, 35 percent of the land area is used for crops or permanent pastures, and countless additional hectares are grazed by livestock. Tropical forests are being logged at a rate of 10 million ha each year. Semiarid subtropical regions, particularly in sub-Saharan Africa, have been turned into deserts by overgrazing and harvesting firewood. Rivers and lakes are badly contaminated in many parts of the world. Gases from chemical industries and the burning of fossil fuels pollute our atmosphere. In this section, we take a global view of human impacts, including habitat loss, overharvesting, introduced species, pollution, and global climate change. While each of these factors is important, keep in mind that many of these factors occur simultaneously.

HABITAT LOSS

The destruction and degradation of habitat have been the largest cause of declining biodiversity. In the United States, for example, most old-growth forests were cut down in the eighteenth century and only a fraction of the original forest remains today. Of course, many of these forests have regrown. Logging of these new forests has typically continued using sustainable practices, although these younger forests do not provide habitat to all the same species as the original old-growth forests did. Today, many areas of the tropics are experiencing a similar pattern of deforestation. For instance, humans have cleared large forests on the island of Sumatra in Southeast Asia, such that only a small fraction of the original forest remains, as you can see in **Figure 23.10**. This deforestation has critically endangered many endemic birds and mammals, such as the Sumatran tiger (*Panthera tigris sumatrae*), the Sumatran ground cuckoo (*Carpococcyx viridis*), and the Sumatran orangutan (*Pongo abelii*; **Figure 23.11**). Because such endemic species live nowhere else in the world, intense conservation efforts have focused on saving them from extinction by protecting what little habitat remains.

To gain insight into the global scale of habitat change, researchers have assessed how forests have been changing in more modern times. From 1980 to 2000, continued loss of forests has occurred in many regions, including the Amazon, Russia, and Southeast Asia. However, there has been an increase in forests in the United States, Europe,

GLOBAL CHANGE

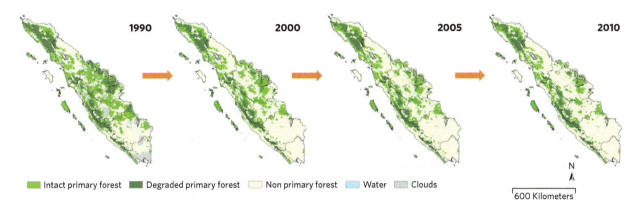

Intact primary forest   Degraded primary forest   Non primary forest   Water   Clouds

N

600 Kilometers

Figure 23.10 Deforestation of a tropical forest. The island of Sumatra was once widely covered in forests. Even during the past few decades, much of the forest has been cleared. Today, only a small fraction of the island contains intact primary forest (i.e., forests that have not been logged) and degraded primary forest (i.e., forests that have experienced some amount of logging). Because the forest cover is based on images from satellites, some land cover cannot be determined due to the presence of clouds. Maps by B. A. Margono, et al. 2012. Mapping and monitoring deforestation and forest degradation in Sumatra (Indonesia) using Landsat time series data sets from 1990 to 2010. *Environmental Research Letters* 7: 034010. Photo courtesy of Belinda Arunarwati Margono et al.

Figure 23.11 Sumatran orangutan. Habitat destruction of the island of Sumatra has caused the decline of many endemic species, including the Sumatran orangutan. Photo by Scubazoo/Alamy.

and Northeast Asia. You can view a map of these changes in **Figure 23.12**. The species composition that currently exists in regions that have experienced an increase in forest cover, however, is often quite different from what existed originally, particularly in cases where a single species of tree is planted due to its high commercial value.

As we have discussed in previous chapters, habitat loss also leads to smaller habitat sizes and increased habitat fragmentation. In Chapter 22, we saw that a reduction in habitat size can lead to reduced population sizes and make it more likely a population will go extinct. This process is thought to be a significant reason that many national parks have lost species of mammals over the past 50 years, despite being protected from most harmful human activities. In addition, fragmented habitats have a high proportion of edge habitat that can alter the abiotic conditions

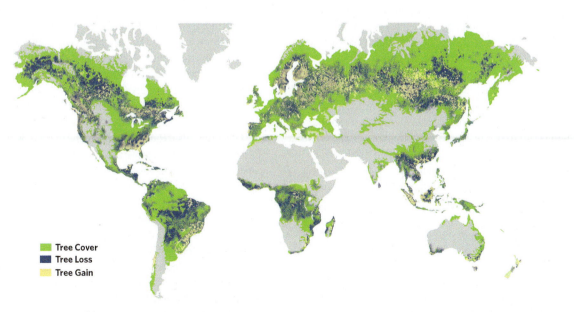

Tree Cover
Tree Loss
Tree Gain

Figure 23.12 Changes in forest cover. While some regions of the world experienced a decline in forest cover from 2001 to 2015, other regions experienced an increase. Data from Global Forest Watch, http://www.globalforestwatch.org/map. Source: Hansen/UMD/Google/USGS/NASA, accessed through Global Forest Watch

of the interior habitat and favor edge species. We saw an example of this in Chapter 22, when we discussed the nest parasite known as the bronzed cowbird, an edge specialist that parasitizes the nests of other songbirds and leads to their decline.

Forests are not the only habitats that have been changing as a result of human activities. For example, according to the National Park Service, the original tallgrass prairie once covered 69 million ha across the middle of North America. Less than 4 percent remains today. Because the remaining areas are small fragments, many local populations of prairie plants and animals have gone extinct. A similar story exists for wetland habitats. Scientists estimate that in the 1600s, wetlands covered more than 89 million ha in the lower 48 states, but drainage for agriculture and other uses has reduced the area of wetlands by more than half. In some places such as California, 90 percent of the original wetlands have been lost. As we have seen, these habitat losses have a large negative effect on the world's biodiversity.

OVERHARVESTING

 Human advances in techniques for logging trees, plowing grasslands, and capturing animals more efficiently have allowed us to harvest species at rapid rates and drive some species to extinction. For instance, during the past 3 centuries, commercial hunters in North America have hunted to extinction the Steller's sea cow (*Hydrodamalis gigas*), great auk (*Pinguinus impennis*), passenger pigeon (*Ectopistes migratorius*), and Labrador duck (*Camptorhynchus labradorius*). Each of these once-abundant species was valued for food or feathers, and they were easily killed.

Extinction caused by overhunting and overfishing is not a recent phenomenon. Wherever humans have colonized new regions, some elements of the fauna have suffered. For example, researchers examined the skeletal remains of animals in archaeological sites around the Mediterranean region to see how human diets changed over thousands of years. At a site in current-day Italy, they found that early human populations initially ate large quantities of tortoises and shellfish, which were easy to catch. As you can see in **Figure 23.13**, as supplies of those foods were depleted over time, people switched to hunting hares, partridges, and other small mammals and birds.

Similar scenarios occurred when humans colonized other parts of the world. When Australia was colonized 50,000 years ago, several species of large mammals, flightless birds, and a species of tortoise soon disappeared. On Madagascar, a large island off the southeastern coast of Africa, the arrival of humans about 1,500 years ago caused the demise of 14 species of lemurs and between 6 to 12 species of elephant birds— giant flightless birds found only on that island. At about the same time, a small population of fewer than 1,000 Polynesian colonists in New Zealand hunted 11 species of moas—another group of large flightless birds—so that these birds went extinct in less than a century. In each of these cases, humans encountered island species that were unaccustomed to

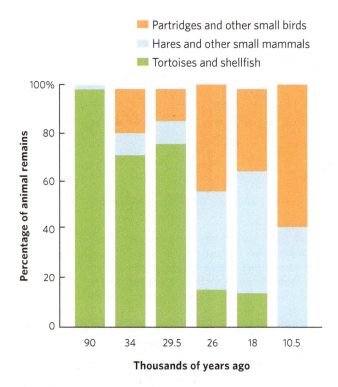

Figure 23.13 Historic overharvest. By examining the bones of consumed animals found in archaeological excavations, researchers have determined that early humans in Italy initially consumed animals that were easy to catch, such as tortoises and shellfish. Once these animals became rare due to overharvesting, people turned to eating hares, partridges, and other species of mammals and birds. Data from M. C. Stiner et al., Paleolithic population growth pulses evidenced by small animal exploitation, *Science* 283 (1999): 190–194.

humans and any other predation pressure. A failure to recognize danger and the lack of defensive strategies made these species particularly vulnerable to human hunting.

Overharvesting of species continues in modern times. In some cases, the harvesting is part of an illegal trade in plants and animals, an endeavor that is valued at $5 billion to $20 billion annually. For example, many animal skins are sold for furs and some cultures believe that certain animal body parts have medicinal value. Some species of rare trees, such as big-leaf mahogany (*Swietenia macrophylla*), are sold for their lumber, while species of rare flowers, such as endangered species of orchids, are sold for their beauty.

Governments frequently regulate the harvest of plants and animals in the wild to ensure that the harvested species will persist for future generations to enjoy. However, harvesting regulations must balance not only the good of the harvested population but also the human employment that the harvest supports. As a result, some regulations set harvest levels that do not stop populations from declining. In marine environments, for example, modern fishing techniques have made it much easier to harvest enormous numbers of fish and shellfish. These techniques include fishing lines that are several kilometers long with thousands of baited hooks, nets that can surround schools of fish up to 2 km in diameter and 250 m deep, and huge trawlers that can scrape large areas of the ocean bottom. The ability to fish more efficiently and to cover larger areas has caused a decline of many fish populations around the world.

When a commercially important fish species no longer has a population that can be fished, it is called a **collapsed fishery**.

For example, consider the case of the Atlantic cod, a species of fish caught by commercial trawlers. On the Grand Banks fishery off the coast of Newfoundland in Canada, the amount of cod caught from 1850 to 1960 slowly increased from 100,000 to 300,000 metric tons, as shown in **Figure 23.14**. During the 1970s and 1980s, new technologies—including advanced sonar, GPS, and larger trawlers—allowed a rapid increase in the number of cod caught, with a peak catch of 800,000 metric tons. However, the population crashed to very low levels in the early 1990s, leading the Canadian government to close the fishery in 1992. Despite low cod numbers, the Canadian government was under pressure from cod fishermen to allow continued fishing. The government agreed, but soon the number of fish fell so much that cod fishing had to be completely stopped and 35,000 Canadian fishermen lost their livelihood. In 2012, 20 years after the ban was imposed, scientists reported that there were finally signs that the cod population was beginning to rebound.

The decline in cod also occurred in New England. Because it was not as severe as the decline off the Newfoundland coast, fishing continued, but the quota of cod that could be caught was reduced. By 2010, the U.S. government greatly reduced the amount of cod that could be caught by commercial anglers, in the hope that the population would rebound. An assessment in 2011 found that the cod population had been very slow to respond, and so limits were reduced even further for 2013 through 2016. U.S. cod fishermen lobbied for higher cod quotas so that they could continue to make a living. However, the government biologists argued that if the cod limits were not lowered substantially, there would soon be no cod left to catch. This debate mirrored the Canadian experience of two decades earlier. Although reducing the

harvest of overharvested species has real economic impacts on those employed in the industry, failure to restrict the harvest hastens the decline of the species to levels at which there are no individuals left to harvest. Unfortunately, cod fishing in New England has continued to decline through 2016, with record low numbers of cod caught by commercial anglers.

There has been a steady increase in the percent of collapsed fisheries, and estimates are that approximately 14 percent of fisheries are now collapsed, as illustrated in **Figure 23.15**. Some regions, such as the eastern Bering Sea off the shores of Alaska, have very few collapsed fisheries. In contrast, collapsed fisheries occur in more than 25 percent of species assessed off the coast of the northeastern United States and more than 60 percent of species assessed off the coast of eastern Canada.

INTRODUCED SPECIES

Another cause of declines in biodiversity is the increasing number of species that are introduced from one region to another. Some of these introductions are intentional, such as when tropical plants are sold for houseplants in temperate parts of the world. Often, however, species are introduced by accident, as we saw in Chapter 15 with the many pathogens that have moved between continents and have caused emerging infectious diseases. Although only about 5 percent of introduced species become established in a new region, those that do can have a variety of effects. Some introduced species provide important benefits, such as the common honeybee that was introduced to North America from Europe in the 1600s. Other introduced species can have substantial negative effects on native species, like the brown tree snake introduced to the island of Guam, which we discussed in Chapter 14 (see Figure 14.2). In that case, the snake caused the decline or extinction of nine species of birds, three species of bats, and several species of lizards. In general, introduced species that compete with native species rarely cause extinction of native species, whereas introduced species that act as predators or pathogens on native species can cause large population declines and extinctions of native species.

Some of the most complete data on introduced species exist for the Nordic countries of Sweden, Finland, Norway, Denmark, and Iceland. As you can see in **Figure 23.16**, the number of introduced species in this area has rapidly increased since 1900. Across terrestrial, freshwater, and marine ecosystems, there are currently more than 1,600 introduced species in the Nordic region. Similarly, during the past 200 years in North America, thousands of species have been introduced, many of which have spread rapidly and are considered invasive species. According to the Center for Invasive Species and Ecosystem Health, North America currently has a large number of invasive species that include nearly 200 pathogens, 300 vertebrates, 500 insects, and 1,600 plants.

One of the highest-profile introduced species in the United States is the silver carp (*Hypophthalmichthys*

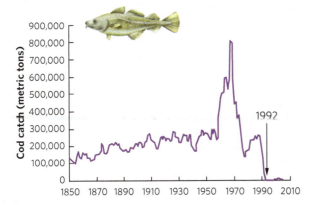

Figure 23.14 A collapse of the Atlantic cod fishery. From 1850 to 1960, there was a slow increase in the catch of Atlantic cod by commercial fishermen off the coast of Newfoundland in eastern Canada. New technologies in the 1970s and 1980s allowed much larger catches of cod, but this led to an overharvest of the fish and a collapse of the population in 1992 that persists today. After Millennium Ecosystem Assessment, *Ecosystems and human well-being: Synthesis* (Island Press, 2005).

Collapsed fishery When a fishery no longer has a population that can be fished.

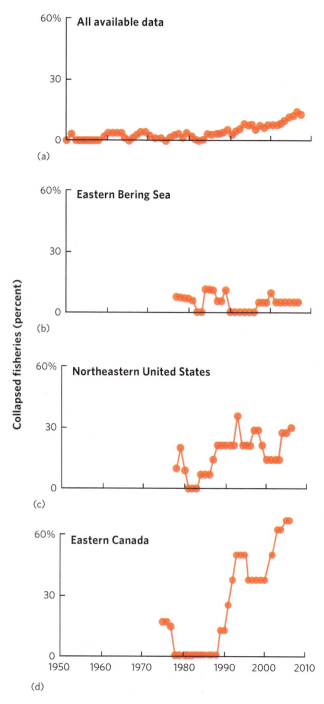

(a)

(b)

(c)

Collapsed fisheries (percent)

All available data

Eastern Bering Sea

Northeastern United States

Eastern Canada

(d)

1950 1960 1970 1980 1990 2000 2010

Figure 23.15 Collapsed fisheries. Over the past 60 years, there has been a steady increase in the percent of fish and other seafood that are categorized as collapsed. **(a)** Around the world, 14 percent of assessed species are considered collapsed. In different regions of North America, these percentages differ a great deal, including **(b)** a low percentage of collapsed fisheries in the eastern Bering Sea, **(c)** a modest percentage in the water off the coast of the northeastern United States, and **(d)** a high percentage off the coast of eastern Canada. Data from B. Worm et al., Rebuilding global fisheries, *Science* 325 (2009): 578–585.

molitrix), a species of fish that has been introduced around the world because it consumes excess algae in ponds used by water treatment plants and aquaculture operations. Brought to the United States in the 1970s, the carp escaped captivity in the 1980s when floodwaters washed them out of their

ponds and into the Mississippi River. The species rapidly spread through the Mississippi River and its tributaries, including the Illinois River. The primary concern was that the Illinois River connects the Mississippi River to Lake Michigan and, thus, the carp had the potential to invade the entire Great Lakes ecosystem. In 2010, the carp's DNA was detected in Lake Michigan, which suggests that the carp may be spreading throughout the Great Lakes. The silver carp is such a voracious consumer of algae that scientists worry it will compete with native consumers of algae, which serve as a key link in the food chain for many species of commercially important fish. The carp also has the unusual behavior of jumping out of the water when a boat passes by. Since the silver carp can reach a mass of 18 kg and jump up to 3 m out of the water, it poses a serious safety risk to boaters (**Figure 23.17**). It will take several years before we can assess the full impact of the silver carp on North American waters.

One introduced species with largely unappreciated negative effects is the domestic house cat. In 2013, researchers examined predation by free-ranging domestic cats in the United States and found that the cats killed 1.4 to 3.7 billion birds and 7 to 21 billion mammals each year. Collectively, these data suggest that introduced species have the potential to cause widespread effects on native species and ecosystems. As the movement of people, cargo, and species becomes more common among the regions of the world, the unique species compositions originally found in different regions are slowly become more similar, a process known as **biotic homogenization.**

POLLUTION

We have talked about many types of pollution throughout this book. For example, in Chapter 2, we discussed acid precipitation, and in Chapter 21, we looked at the effects of adding excess nutrients to bodies of water. Both examples demonstrate the harmful effects of pollutants on biodiversity.

Pesticides are a common type of pollutant. They include insecticides that kill insects and other invertebrate animals, herbicides that kill plants, and fungicides that kill fungi. These chemicals are designed to target a particular type of pest; ideally, they will not harm nonpests in the ecosystem. However, some pesticides do kill nonpests, either directly, by being toxic, or indirectly, by altering food webs, as we saw in the wetland study discussed at the end of Chapter 18. In that study, a very small amount of an insecticide, while not directly toxic to tadpoles, was toxic to zooplankton. The death of zooplankton set off a chain of events in the food web that prevented the tadpoles from obtaining enough food to metamorphose before the pond dried. By indirect effect, the insecticide caused the death of approximately half of the tadpoles.

A classic study on pesticides examined the role of the insecticide DDT on birds of prey. During the 1950s and 1960s in the United States, populations of many

Biotic homogenization The process by which unique species compositions originally found in different regions slowly become more similar due to the movement of people, cargo, and species.

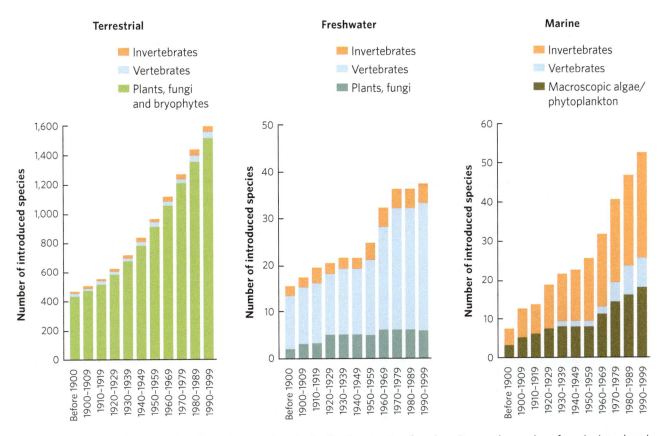

Figure 23.16 Increases in introduced species over time. In the Nordic countries of northern Europe, the number of species introduced to terrestrial, freshwater, and marine ecosystems has increased over time. Data from Secretariat of the Convention on Biological Diversity, *Global Biodiversity Outlook* 2 (2006), https://www.cbd.int/gbo2/.

Figure 23.17 Silver carp. Silver carp jump into the air and pose a danger to boaters like this biologist on the Illinois River near Starved Rock State Park. Photo by Chris Olds/U.S. Fish & Wildlife Service.

predatory birds, particularly the peregrine falcon (*Falco peregrinus*), bald eagle, osprey (*Pandion haliaetus*), and brown pelican (*Pelecanus occidentalis*), declined drastically. Several of these species disappeared from large areas and the peregrine falcon disappeared from the entire eastern United States. The causes of these population declines were traced to pollution of aquatic habitats by DDT, an insecticide

that was widely used to control crop pests and mosquito vectors of malaria after World War II. This insecticide was favored both because it was effective at killing insect pests and because it persisted in the environment, allowing it to continue killing pest insects for a long time.

When this pesticide enters a water body, it binds to particles, including algae, to become about 10 times more concentrated than it is in the water. When the algae are consumed by zooplankton, DDT accumulates in fat tissues until the zooplankton have a concentration that is about 800 times higher than in the water, as shown in **Figure 23.18**. The process of increasing the concentration of a contaminant as it moves up the food chain is known as **biomagnification**. When small fish eat the zooplankton and large fish eat the small fish, DDT is further concentrated about 30-fold. Finally, when a fish-eating bird such as an osprey consumes a large fish, DDT becomes concentrated another 10-fold. In short, at the top of the food chain, the insecticide is 276,000 times more concentrated in the body of the fish-eating bird than it was in the water. Such a high concentration in predatory birds interferes with their physiology in a way that causes the eggs they lay to have very thin shells. When the parents sit on these thin-shelled eggs, the egg shells break, and the embryos die.

Biomagnification The process by which the concentration of a contaminant increases as it moves up the food chain.

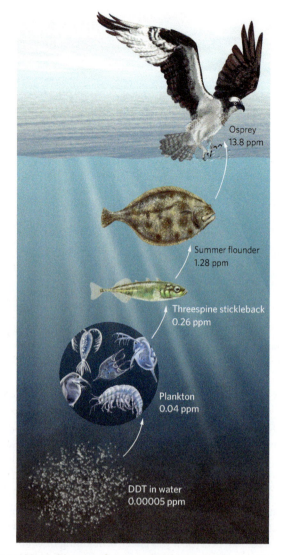

Osprey
13.8 ppm

Summer flounder
1.28 ppm

Threespine stickleback
0.26 ppm

Plankton
0.04 ppm

DDT in water
0.00005 ppm

Figure 23.18 Biomagnification of DDT. When DDT was widely used to control insects, it could be found in very low concentrations in the water. However, its concentration increases in particles with which it binds in the water, such as algae. At each higher trophic level, the insecticide becomes more concentrated. Data from G. M. Wood-well, C. F. Wurster, Jr., and P. A. Isaacson, DDT residues in an East Coast estuary: A case of biological concentration of a persistent insecticide, *Science* 156 (1967): 821–824.

This eggshell thinning caused the populations of predatory birds to plummet during the 1960s.

Understanding the connection between DDT and declining bird populations led the U.S. government to ban the use of DDT and related pesticides in 1972, although it is still used in other parts of the world. Fortunately, alternative insecticides have since been developed that do not persist and are not stored in the fat of animals and, as a result, do not biomagnify through a food web. With the help of many biologists who hand-reared hundreds of predatory birds, populations of species such as the peregrine falcon and bald eagle have rebounded. DDT is still used in many tropical countries, although now in a limited manner, such as in houses to control the mosquitoes that carry malaria.

GLOBAL CLIMATE CHANGE

Changing global climates play both future and current roles in the biodiversity of species. In Chapter 5, we discussed the greenhouse effect and how gases such as water vapor, CO_2, methane, and nitrous oxide allow our planet to be warmed naturally by absorbing infrared radiation emitted by Earth and then reradiating a portion back toward Earth. During the past century, human activities have increased the concentration of these greenhouse gases in the atmosphere (see "Ecology Today: Applying the Concepts" in Chapter 4), as well as the gases used as refrigerants known as chlorofluorocarbons (CFCs). One result of the increase in greenhouse gases has been an increase in the average temperature on Earth. From 1880, when the earliest measurements were taken, to 2013, the temperature on Earth has increased an average of 0.8°C, as illustrated in **Figure 23.19**. While this is an average, some parts of the world such as northern Canada and Alaska have experienced temperature increases as high as 4°C.

Global warming has caused a number of effects that can be readily observed. For example, we might expect warmer temperatures to cause more of the world's ice to melt. As we noted at the end of Chapter 5, the massive polar ice cap in the Arctic has decreased in mass by 45 percent during the past 30 years. Similarly, the decline in ice from 2002 to 2016 has been more than 1,500 metric gigatons for Antarctica and more than 3,500 gigatons for Greenland. You can view these data in parts a, b, and c of **Figure 23.20**. All of this melting ice, combined with an overall warming of the oceans that causes water to expand, should cause a rise in sea levels. Researchers have examined water gauges for ocean tides since 1870 and, as expected, have observed a steady increase in sea level; today, the sea level is 0.2 m higher than it was in 1870. During the past 20 years, the sea level has been rising more than 3 mm per year, as shown in Figure 23.20d. At this rate, the sea will rise 0.3 m every

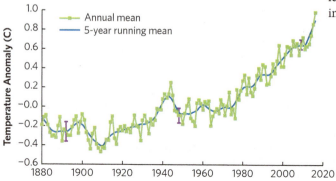

Figure 23.19 Global warming over time. Based on thousands of measurements around the world, scientists have observed a 0.8°C increase in the average temperature of our planet. Because of year-to-year variation in temperatures, the pattern of increasing temperatures is clearer when we examine the 5-year-running mean temperatures. Temperature anomaly is a comparison of each year's temperature compared to the mean temperature observed from 1951 to 1980. Data from http://data.giss.nasa.gov/gistemp/graphs_v3/.

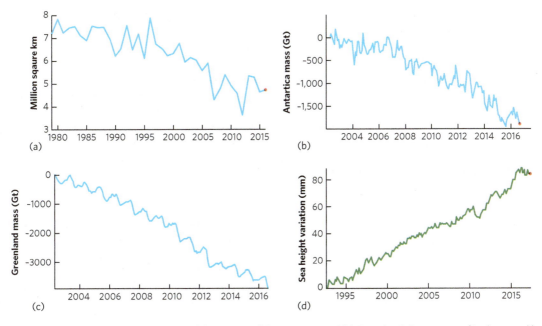

Figure 23.20 Global warming and melting ice. In **(a)** the Arctic, **(b)** Antarctica, and **(c)** Greenland, the amount of ice has steadily declined from 2002 to 2010. Data are plotted relative to the mean ice mass throughout the entire time period. **(d)** Measurements of sea level from 1993 to 2013 demonstrate a continual increase in sea level compared to the sea level in 1993. Data from http://climate.nasa.gov/key_indicators#globalTemp.

100 years, which is enough to dramatically affect habitats on islands and along coastlines.

As we have seen throughout this book, these changes in global temperatures are already affecting species. In Chapter 8, we discussed how many species of plants now flower earlier in the spring than they did in past decades and how some species of birds and amphibians now breed earlier in the year than they used to. In Chapter 11, we saw that warming ocean temperatures have caused a major shift in the fish species that live in the North Sea. Several of the species that historically lived in the North Sea have moved farther north to cooler waters, while a number of species that had lived in more southern waters moved up into the North Sea; this changed the species composition of the community.

Global climate change has not yet led to the widespread extinction of species. However, even using computer models, it is difficult to predict how temperature and precipitation will change over future decades. One of the critical factors is how much we continue to increase the amount of CO_2 in the atmosphere. As you can see in **Figure 23.21**, it is predicted that

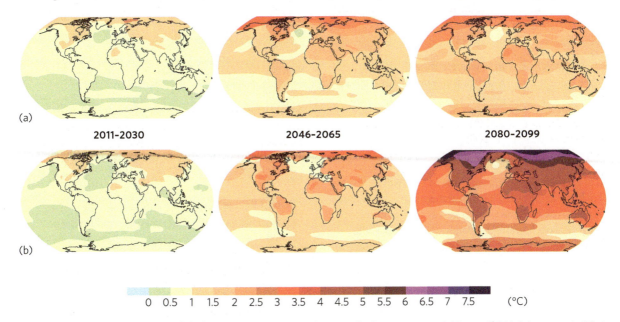

Figure 23.21 Predicted changes in global temperatures. Depending on whether we assume **(a)** low or **(b)** high increases in CO_2 in future years, the temperature is predicted to change. The changes in temperature are relative to the mean temperature observed from 1961 to 1990. Data from G. A. Meehl et al., Global climate projections, in S. Solomon et al. (Eds.), *Climate change 2007: The physical science basis. Contribution of Working Group I to the Fourth Assessment Report of the Intergovernmental Panel on Climate Change* (Cambridge University Press, 2007).

ANALYZING ECOLOGY

Contaminant Half-Lives

Eco TV
macmillanlearning.com/ricklefsvideo

As we have mentioned, an important factor in determining whether a contaminant such as a pesticide or radioactive compound affects the environment is the length of time the chemical persists in the environment. A helpful way to assess this persistence is to measure the time required for the chemical to break down to half of its original concentration, which is called its half-life. To calculate the **half-life**, we start by recognizing that most chemicals have a breakdown rate that follows a negative exponential curve, as shown in the accompanying figure for a hypothetical chemical. If we were to start with 8 mg of the chemical and then ask how many days must pass until the chemical breaks down to only 4 mg, we see that it takes 10 days. To break down by half again, from 4 to 2 mg, it takes another 10 days.

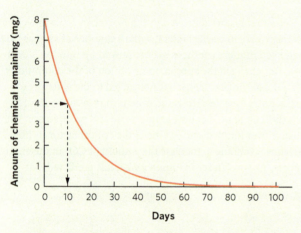

It is relatively easy to determine the half-life when we have a line graph. However, it is more difficult when

we only have two data points: the amount of a chemical we start with and the amount of chemical remaining at some later point in time. In such cases, we can make use of an equation that calculates half-lives:

$$t_{1/2} = t \ln (2) \div \ln (N_0 \div N_t)$$

where $t_{1/2}$ is the half-life of the chemical, N_0 is the initial amount of the chemical, and N_t is the amount of chemical after some amount of time (t) has elapsed. For example, using data from the figure, we begin with 8 mg of a chemical and after 30 days, we have 1 mg remaining. Using these data, we can calculate the half-life of the chemical:

$$t_{1/2} = 30 \ln (2) \div \ln (8 \div 1)$$

$$t_{1/2} = 20.8 \div 2.08$$

$$t_{1/2} = 10$$

Note that this is the same answer we obtained by estimating the half-life directly from the curve.

YOUR TURN Modern pesticides are often designed to break down rapidly so their effects on nontarget species in the ecosystem are minimized. Imagine that you spray a wetland with an insecticide to kill mosquitoes that carry the West Nile virus, and this causes a concentration in the water of 50 parts per billion (ppb). Twenty-four hours later, you sample the water and you find that the water now contains 10 ppb of the chemical. Using these data, calculate the chemical's half-life.

Half-life The time required for a chemical to break down to half of its original concentration.

---

a small increase in CO_2 will cause the northern latitudes to experience an additional 4°C rise in average temperatures by the end of this century. A large increase in CO_2 will raise temperatures by an additional 7°C. Researchers predict that these changes will cause extreme weather events such as hurricanes and droughts to occur more frequently. In addition, some regions of the world will receive more annual precipitation than is currently typical, while other regions will receive less. Although specific predictions vary with different models, the

distributions of organisms in nature will probably shift as the Earth's climates change over the next century.

CONCEPT CHECK

1. How has forested habitat been changing in North America?
2. What evidence indicates that overharvesting animals is not a recent problem?
3. Why do some pollutants biomagnify through a food web?

23.4 Conservation efforts can slow or reverse declines in biodiversity

We have seen that human activities have caused a decline in the world's biodiversity. We now look at what can be done to slow or even reverse these declines. Over the long term, we need to stabilize the size of the human population because human activities have caused most species declines during the past several centuries. Over the short term, we need to reduce human-caused sources of mortality and low reproduction in species so that these species can persist long into the future. During the past several decades, scientists have been developing effective strategies to preserve biodiversity, although these approaches are often neither easy nor inexpensive. In this section, we will focus on habitat protection, habitat management, reduced harvest, and species reintroductions.

HABITAT PROTECTION

Because the largest contributor to the loss of biodiversity is the loss of habitat, one of the major factors in conserving biodiversity has been habitat protection. The goal in protecting a habitat is commonly the preservation of a large enough area to support a **minimum viable population (MVP),** which is the smallest population size of a species that can persist in the face of environmental variation. The population must also be distributed widely enough to prevent local catastrophes, such as hurricanes and fires, from threatening

the entire species. At the same time, some degree of population subdivision may help prevent the spread of disease from one subpopulation to another.

The task of protecting a suitable habitat becomes more complicated when a population's habitat requirements change with the seasons or when it undertakes large-scale seasonal migrations. In the Serengeti ecosystem of East Africa, for example, patterns of rainfall distribution and plant growth vary seasonally. Huge populations of wildebeests, zebras, and gazelles migrate seasonally in search of suitable grazing areas. Because migratory populations use the entire area of the Serengeti ecosystem over the course of a year, the preservation of only one section would not meet their needs. For similar reasons, the massive herds of bison, also known as buffalo, can never be fully restored to North American prairies because their seasonal migration routes are now blocked by miles of fencing and agricultural fields. Bison survive in a few small reserves in the American West—most notably in the Greater Yellowstone Ecosystem—but most of the lands they once occupied can never be recovered.

This Greater Yellowstone Ecosystem, an 80,000-km^2 area that is centered on Yellowstone National Park in Wyoming, is one of the best-known examples of preserving a large collection of habitat and includes federal, state, and private lands from three states, as illustrated in **Figure 23.22**.

Minimum viable population (MVP) The smallest population size of a species that can persist in the face of environmental variation.

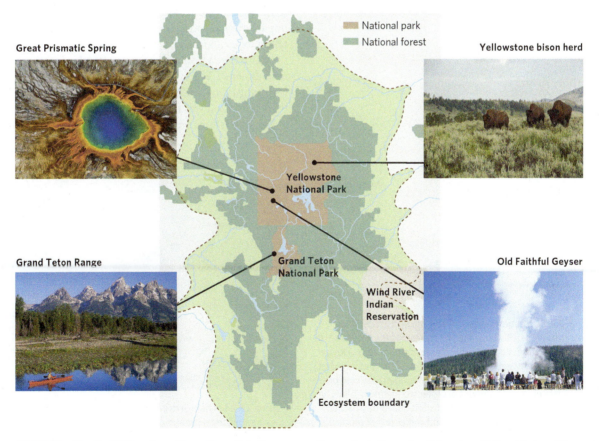

Figure 23.22 The Greater Yellowstone Ecosystem. To better preserve the species in the region, Yellowstone National Park and the many different types of land surrounding it are collectively managed as one large ecosystem. Photos by (a) Luis Castaneda/AGE footstock (Grand Prismatic Spring), (b) James Kay / DanitaDelimont.com (Grand Teton Range), (c) SuperStock/AGE Fotostock (Yellowstone bison herd), and (d) James Steinberg/Science Source (Old Faithful Geyser).

The management plan for the Greater Yellowstone Ecosystem seeks to maintain the region in a natural, self-sustaining condition. Natural forest fires should be allowed to burn, as they did in 1988 over half of Yellowstone National Park, and top predators, such as the grizzly bear and gray wolf, should be restored. As we will see in "Ecology Today: Applying the Concepts," these top predators are natural controls over populations of large grazers, which has far-reaching effects on the ecosystem.

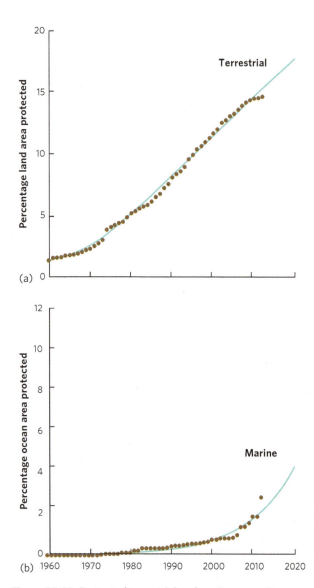

Figure 23.23 Protected terrestrial and marine areas. Dots represent measured data, while the line represents the projected trajectory of future protection. **(a)** Terrestrial ecosystems have a longer history of being protected, and this level of protection has increased over the past 40 years. **(b)** Marine ecosystems began receiving protection much later, but the amount of protected marine habitats has also grown over time. A small amount of additional terrestrial and marine habitat has been protected, but the initial year of protection is unknown so it is not shown. Data from Secretariat of the Convention on Biological Diversity, *Global Biodiversity Outlook 4* (2014), https://www.cbd.int/gbo/gbo4/publication/gbo4-en-hr.pdf, https://www.cbd.int/gbo2/.

The need to preserve habitat is recognized around the world, and many countries are setting aside such areas. The amount of terrestrial habitat that is being protected has continually increased over the past 40 years, as shown in **Figure 23.23a**. In fact, 57 percent of the world's countries have protected at least one-tenth of their land. Though the need to protect marine habitats has been recognized much more recently, as you can see in Figure 23.23b, it has also increased over time.

Even when lands are set aside, many countries cannot afford to protect them from squatters and poachers, and often governments allow mining and logging within protected lands. However, involving local people in the design and management of protected areas has been particularly effective since the benefits of conservation, which often include the income generated by ecotourism, become tangible and economically compelling to local people. The most successful efforts typically include large geographic areas, low human densities, a positive public attitude toward the preservation effort, and effective law enforcement to protect species from being poached.

REDUCED HARVESTING

When overharvesting is identified as a cause of decline in a species, reducing the harvest is an obvious approach to protection. However, this is complicated when it involves people's livelihoods. For example, we saw that the collapse of the Atlantic cod fishery had a major negative economic impact on the fishing industry. Although some species can take a long time to recover their population size, reducing the harvest of a declining species often leads to a return to abundance. For instance, the northern elephant seal (*Mirounga angustirostris*) was hunted so much that by the late 1800s it was thought to be extinct (**Figure 23.24**). Small populations were subsequently discovered in the

Figure 23.24 The northern elephant seal. Once hunted nearly to extinction, legal protection by Mexico and the United States has allowed the population to increase to more than 150,000 today. These seals are located on the coast of central California. Photo by Wild Nature Photos/Animals Animals/Earth Scenes.

1890s and the total population was estimated to be about 100 individuals. In the early 1900s, the seal was given protection status by the United States and Mexico. Once protected, the elephant seal population started to rebound rapidly, and today there are more than 150,000 individuals, distributed throughout much of the former range of this species in California and Mexico. Similar successes have occurred for the American crocodile (*Crocodylus acutus*), the whooping crane (*Grus americana*), and the bald eagle.

SPECIES REINTRODUCTIONS

Sometimes a species comes so close to extinction that it requires human intervention to bring it back from the brink. We saw an example of this when we discussed the recovery program for the black-footed ferret at the end of Chapter 13. Such efforts can require a great deal of effort and money, and the recovery may take decades.

A classic example of a species reintroduction is the case of the California condor, a large vulture that feeds on dead animals (**Figure 23.25**). During the latter half of the twentieth century, condor numbers dwindled due to a variety of causes, including illegal shooting and the illegal collection of eggs from condor nests. Moreover, some of the animals the condor scavenged and consumed contained lead bullet fragments or, in the case of coyotes and rodents, had been poisoned. When the condor population in southern California sank to fewer than 30 individuals in the late 1970s, biologists made the difficult decision to bring the entire population into captivity. From 1982 to 1987, the 22 remaining wild birds were captured and brought to special breeding facilities located at zoos in Los Angeles and San Diego, where they were protected from the threats that had taken such a toll in the wild.

Recovering the condor population is extremely challenging. Individuals take 6 to 8 years to reach maturity and in the wild typically lay only one egg per year. However, researchers discovered that if they removed the single egg, a female would lay another. In fact, a female condor could be tricked into laying up to three eggs per year. Biologists incubated the surplus eggs and reared the birds using puppets to feed the chicks to minimize human contact. As the chicks grew up, they were released back into the wild. At the same time, efforts were made to reduce the threats to the condors. These substantial efforts have paid off; in 2016, there were 446 condors alive both in captivity and in the wild following reintroductions into California, Arizona, and Mexico. In addition to increasing the condor population, this high-profile case heightened local residents' awareness of conservation issues and resulted in the preservation of large tracts of habitat in mountainous regions of southern California. People have also come to understand that restoring the condor population can be compatible with other land uses. Recreation does not have to be banned from condor habitats as long as human access to nesting sites is restricted. Legal

Figure 23.25 The California condor. This large vulture was on the brink of extinction in the 1980s when only 22 birds remained in the wild. A captive breeding program, followed by reintroductions into the wild, has increased that number to 400 birds today living both in captivity and in the wild. Photo by Diana Kosaric/age fotostock.

hunting doesn't harm condors as long as steel rather than lead bullets are used. Finally, ranching does not threaten condor populations so long as coyote and rodent control programs are condor-safe.

Species reintroductions can cost millions of dollars and are usually directed toward species that appeal to the public, such as black-footed ferrets, condors, and wolves. Some people may question the wisdom of spending millions of dollars to save one species. However, efforts made to save a single species, such as setting aside habitat and reducing the prevalence of poisons, often have positive effects on a large number of species. We will see an excellent example of this in "Ecology Today: Applying the Concepts," which details the ecosystem effects of reintroducing the gray wolf to Yellowstone National Park.

Throughout this book, we have examined the factors that determine the distribution and abundance of species throughout the world. We have seen that there is an amazing diversity of species on Earth that impresses us with its beauty and plays essential roles in the ecosystems upon which we depend. The decline in the world's biodiversity is at a critical stage because human activities threaten populations of many species. However, by appreciating the importance of the world's species and understanding their ecology, we can take the steps necessary to slow population declines and species extinctions, and to find ways to coexist.

CONCEPT CHECK

1. How has the amount of protected habitat around the world changed over recent decades?
2. How has the ban on harvesting marine mammals affected populations of northern elephant seals?
3. How can investment in a large, charismatic species also favor the conservation of other species?

ECOLOGY TODAY   APPLYING THE CONCEPTS

Returning Wolves to Yellowstone

The wolves and elk of Yellowstone. Following the reintroduction of wolves, the elk population has declined and some regions of the ecosystem have changed dramatically. Photo by Larry Thorngren.

As we have seen in this chapter, restoring biodiversity can be a difficult challenge, and nowhere has this been more apparent than in the attempt to return the gray wolf to the Greater Yellowstone Ecosystem. The challenges in this case include ecological, economic, social, and political factors. After all, not everyone is excited about bringing back a top predator with a reputation for being a bloodthirsty killer.

The gray wolf once roamed throughout most of North America, but fear of wolves and concerns about their preying on livestock led to government programs that eliminated the wolf from nearly all of the United States and southwestern Canada. These programs caused the wolf to be eradicated from all of the lower 48 states except for northern Minnesota by 1925. In the 1940s, Aldo Leopold, a prominent professor of wildlife management, first proposed returning the wolves to Yellowstone National Park. By the 1960s, public attitudes toward wolves started to shift as many people began to see the intrinsic value of restoring ecosystems to a more natural state. In 1973, the Endangered Species Act was passed, and it classified gray wolves as an endangered species, which required the development of a plan that would increase their abundance. Because Yellowstone was a large protected area, it was an obvious place for a possible reintroduction.

The plan to reintroduce wolves had supporters and detractors. Public opinion polls found that the general public and the tourists who visited the park favored the idea. However, local hunters worried that the wolves would reduce the populations of large game such as elk. Moreover, the local livestock ranchers worried about their herds because wolves sometimes kill cattle and sheep. For nearly 2 decades, the reintroduction of wolves was debated, studied, and pushed in different directions by politicians representing various constituents. Throughout the early 1990s, the U.S. Fish and Wildlife Service held 130 town meetings in the area and received thousands of written comments on a proposed reintroduction plan. They also estimated the impacts of a wolf reintroduction to Yellowstone. Their best estimate was that an eventual population of 100 wolves would kill about 20 cattle, 70 sheep, and 1,200 wild ungulates such as elk, deer, and antelope each year from a population of 95,000 wild ungulates. The economic benefit from increased tourism to the area because of the wolves was estimated at $23 million, which represented a substantial instrumental value. After a series of court hearings and repeated judgments in favor of the reintroduction plan, 31 wolves were captured in Canada and released into Yellowstone in 1995 and 1996.

Nearly 2 decades later, the effects of the reintroduction have been felt throughout the ecosystem. When the wolves were eradicated in the early 1900s, the elk population rapidly increased and their feeding nearly eliminated aspen and cottonwood trees along rivers. The return of the wolves, and their rapid increase to a population of more than 220 individuals by 2001, reduced the elk population by half. As a result, aspen and cottonwood trees are now thriving in some areas of the park. Some researchers have argued that the effects on aspen growth are due to a combination of density- and trait-mediated effects (see Chapter 18). It has been suggested that the trait-mediated effects result from wolves scaring the elk away from streams and rivers and into higher elevations, where the elks are safer from predation. However, others argue that there is insufficient evidence to suggest that such fear-induced habitat shifts have had a widespread effect on the growth of aspen and cottonwood seedlings.

Because aspens and cottonwoods are a favorite food of beavers, the increase in these trees has attracted more beavers into the region, which has led to an increase in beaver dams that form large ponds. In addition, the wolf reintroduction caused a severe reduction in the population of coyotes since they are prey for wolves. This likely had a cascading effect on the many prey species that coyotes consume. The abundant carcasses of elk and other prey of the wolves also benefited populations of scavengers such as ravens and golden eagles, which are once again common in Yellowstone. In short, the reintroduction of the wolf caused the ecosystem to move back to more closely resemble its earlier condition.

In 2008, the Yellowstone wolf population was deemed to have recovered well enough to be removed from the federal list of endangered species. The wolves are now hunted in limited numbers and the wolf population has been maintained at about 100 individuals from 2013 to 2015. It will be interesting to see what continued effects the wolves will have on the ecosystem.

The return of the wolf to the Greater Yellowstone Ecosystem demonstrates that restoring biodiversity is not a simple task. Restoration plans need to consider not only the ecology of the reintroduction but also related social, political, and economic factors. The plans must be evaluated by all interested parties and the process can often take a long time. With patience, however, species and ecosystems can eventually be restored.

(a)

(b)

(c)

Effects of wolves on the Yellowstone ecosystem. With the reduction of elk populations due to wolf introductions, aspen and cottonwood trees along some river banks have recovered, as seen in the foreground of the three photos. These photos were taken at Soda Butte Creek in **(a)** 1997, **(b)** 2001, and **(c)** 2010. Photos by (a) National Park Service; (b) and (c) William J. Ripple.

SOURCES:

Fritts, S. H., et al. 1997. Planning and implementing a reintroduction of wolves to Yellowstone National Park and Central Idaho. *Restoration Ecology* 5: 7–27.

Ripple, W. J., and R. L. Betscha. 2011. Trophic cascades in Yellowstone: The first 15 years after wolf reintroduction. *Biological Conservation* 145: 205–213.

Smith, D. W., et al. 2003. Yellowstone after wolves. *BioScience* 53: 330–340.

Kaufmann, M. J., et al. 2010. Are wolves saving Yellowstone's aspen? A landscape-level test of a behaviorally mediated trophic cascade. *Ecology* 91: 2742–2755.

Beschta, R. L., and W. J. Ripple. 2013. Are wolves saving Yellowstone's aspen? A landscape-level test of a behaviorally mediated trophic cascade: Comment. *Ecology* 94: 1420–1425.

Kaufmann, M. J., et al. 2013. Are wolves saving Yellowstone's aspen? A landscape-level test of a behaviorally mediated trophic cascade: Reply. *Ecology* 94: 1425–1431.

Beschta, R. L., and W. J. Ripple. 2015. Divergent patterns of riparian cottonwood recovery after the return of wolves in Yellowstone, USA. *Ecohydrology* 8: 58–66.

SUMMARY OF LEARNING OBJECTIVES

23.1 The value of biodiversity arises from social, economic, and ecological considerations. Instrumental values represent the material benefits that species and ecosystems provide to humans, including food, medicines, water filtration, and pollination. Intrinsic values recognize that species and ecosystems are valuable regardless of any benefit to humans.

Key Terms: Instrumental value of biodiversity, Intrinsic value of biodiversity, Provisioning services, Regulating services, Cultural services, Supporting services

23.2 Although extinction is a natural process, its current rate is unprecedented. Historically, there have been five mass extinctions, each followed by continued speciation. The current rate of species extinction is higher than background rates, which suggests that we may be in the early stages of a sixth mass extinction. Many species are also experiencing a rapid decline in genetic diversity, although efforts are being made to preserve the genetic diversity of livestock and crops.

Key Term: Mass extinction events

23.3 Human activities are causing the loss of biodiversity. One of the most important factors contributing to species declines is the loss of habitat. Other impacts include overharvesting, introducing exotic species, and polluting the environment with contaminants. Global warming and global climate change present a rising threat. Some changes have already occurred and species may not be able to accommodate more substantial climate shifts in the future.

Key Terms: Collapsed fishery, Biotic homogenization, Biomagnifications, Half-life

23.4 Conservation efforts can slow or reverse declines in biodiversity. Major efforts are being made to respond to the decline in biodiversity. Increasing amounts of terrestrial and marine habitat are being protected, harvest regulations are being adjusted to slow population declines, and species approaching extinction are being reintroduced where high-quality habitat exists and the threats can be reduced.

Key Term: Minimum viable population (MVP)

CRITICAL THINKING QUESTIONS

1. Why might different people or groups favor different criteria when prioritizing biodiversity hotspots?

2. Compare and contrast instrumental values versus intrinsic values of species and ecosystems.

3. How can economic benefits of biodiversity be used as an argument to protect biodiversity?

4. What steps can be taken to slow the sixth mass extinction?

5. Why should we be concerned with preserving both species diversity and genetic diversity?

6. Why have humans historically overharvested many species of animals?

7. Why might introduced competitors result in less extinction of native species than introduced predators?

8. Why do we need to consider the process of biomagnification when assessing the risk of a pesticide to wildlife?

9. Why is it difficult to predict which species will be driven extinct by global warming?

10. What are the ecological, economic, and social challenges that can arise when considering a species reintroduction?

GRAPHING THE DATA Stacked Bar Graphs

When we want to compare the distribution of categories within different groups, such as the conservation status of species of various taxa, we can either examine percentages or absolute numbers. As part of our discussion of the conservation status of conifers, birds, mammals, and amphibians, Figure 23.5 used pie charts to show differences in the percentages of species that were categorized as of least concern, near-threatened, and threatened. However, we can obtain a somewhat different perspective if we graph the absolute number in each category using a stacked bar graph, similar to Figure 23.13. A stacked bar graph allows us to display data in a way that stacks one set of data on top of another set of data within a category. This permits us to observe each category in terms of its separate and combined components.

Using the following data for the number of species of conifers, birds, mammals, and amphibians in each status category, create a stacked bar graph.

Conservation Status	Conifers	Birds	Mammals	Amphibians
LEAST CONCERN	333	7,677	3,124	2,392
NEAR-THREATENED	75	880	325	389
THREATENED	177	1,313	1,139	1,933

Compare your graph to the pie charts presented in Figure 23.5. What different insights do you gain by plotting the absolute numbers of species using a stacked bar graph that you could not obtain from a pie chart of proportions of species in each category?

Reading Graphs

Ecologists often find it useful to display the data they collect in graphs. Unlike a table that contains rows and columns of data, graphs help us to visualize patterns and trends in the data. These patterns help us draw conclusions from our observations. Because graphs are such a fundamental part of ecology and most other sciences, you should become familiar with the major types of graphs that ecologists use. In this section, we review different types of graphs. We also discuss how to create each type of graph and how to interpret the data that are presented in the graphs.

Scientists use graphs to present data and ideas

A graph is a tool that allows scientists to visualize data or ideas. Organizing information in the form of a graph can help us understand relationships more clearly. Throughout your study of ecology, you will encounter many different types of graphs. In this section, we will look at the most common types of graphs that ecologists use.

SCATTER PLOT GRAPHS

Although many of the graphs in this book may look different from each other, they all follow the same basic principles. Let's begin with an example in which researchers measured the mean temperature during the month of January in 56 cities throughout the United States. These cities span latitudes of 25° N to 50° N. The researchers wanted to determine if there was a relationship between the latitude of a city and the mean temperature in January.

We can examine this relationship by creating a *scatter plot graph,* as shown in **Figure A.1**. In the simplest form of a scatter plot graph, researchers look at two variables; they put the values of one variable on the *x* axis and the values of the

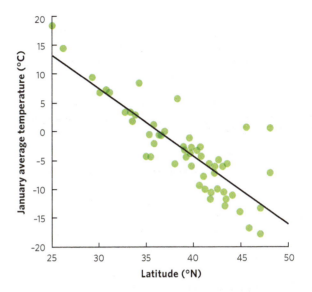

Figure A.2 (Text Figure from "Analyzing Ecology," Chapter 5) This scatter plot graph includes a line of best fit that helps us more easily see the relationship between latitude and mean January temperature for 56 cities in the United States.

other variable on the *y* axis. In our example, the latitude of the city is on the *x* axis and the mean January temperature of the city is on the *y* axis. By convention, the units of measurement tend to get larger as we move from left to right on the *x* axis and from bottom to top on the *y* axis. Using these two axes, the researchers could plot the latitude and temperature of each city on the scatter plot graph. As you can see in the figure, there is a pattern in the graph of mean January temperature declining as one moves north into the higher latitudes of the United States.

When two variables are graphed in this way, we can draw a line through the middle of the data points to describe the general trend of these points, as we have done in **Figure A.2**. Because such a line is drawn in a way that fits the general trend of the data, we call this the *line of best fit*. This line allows us to visualize a general trend. In our graph, the addition of a line of best fit makes it easier to identify a trend in the data; as we move from low to high latitudes, we observe lower January temperatures. This is known as a negative relationship between the two variables because as one variable increases, the other variable decreases.

When graphing data using a scatter plot graph, the line of best fit may be either straight or curved. For example, when researchers examined the conductivity of the water in ponds near roads, which is an indicator of the amount of road salt present, they found that as they moved away from the road, there was an initial rapid decrease in conductivity that eventually leveled off, as illustrated in **Figure A.3**. In this case, the line of best fit is a curved line.

Sometimes scientists collect data that follow a curved line but they wish to view the data in terms of a straight line in order to compare the slopes and intercepts of lines produced by different sets of data. The slopes and intercepts can be

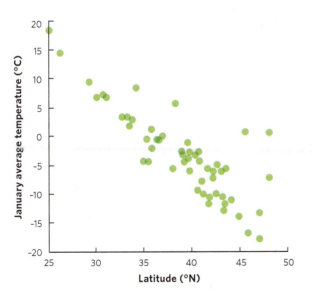

Figure A.1 (Text Figure from "Analyzing Ecology," Chapter 5) This scatter plot graph shows the relationship between latitude and mean January temperature for 56 cities in the United States.

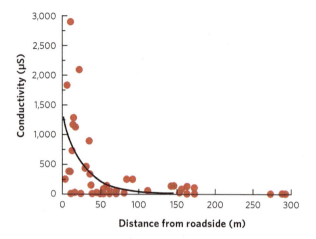

Figure A.3 (Text Figure 2.12) This scatter plot graph shows the relationship between the distance from a pond to the nearest road and the amount of salt in the pond's water, which is measured in terms of conductivity. The line of best fit is curved.

calculated using the equation for a line: $y = mx + b$, where y is the dependent variable, x is the independent variable, m is the slope, and b is the intercept. In many cases, data that follow a curved line can be transformed to follow a straight line by using axes that contain the values arranged on a logarithmic scale. For example, when scientists examined the number of amphibian and reptile species living on islands of different sizes in the West Indies, the resulting scatter plot graph illustrated that the number of species initially increased rapidly as island area increased and then eventually leveled off, as shown in **Figure A.4a**. When the exact same data are plotted on a scatter plot graph using axes with a log scale, the line of best fit changes from a curved line to a straight line, as you can see in Figure A.4b.

LINE GRAPHS

A *line graph* is used to display data that occur as a sequence of measurements over time or space. For example, scientists have estimated the number of humans living on Earth from 8,000 years ago through to the present time. Using all the available data points, a line graph can be used to connect each data point over time, as shown in **Figure A.5**. In contrast to a line of best fit that fits a straight or curved line through the middle of all data points, a line graph connects one data point to another, so it can be straight or curved, or it can move up and down as it follows the movement of the data points.

   Figure A.6 is an example of a line graph displaying data measurements over space. In this example, scientists tested soil samples for the concentration of chromium as they walked from one type of soil toward another type of soil. As they approached the serpentine soil, which is characterized by high concentrations of heavy metals, the concentration of chromium sharply increased.

   When a graph includes data points with a very large range of values, the size of the graph can become cumbersome. To keep the graph from becoming too large, we can use a break in the axis, as shown in **Figure A.7**. In this figure, we are examining the decline in the population of prairie chickens in Illinois from the 1860s to the 1990s. Note the double hatch marks between 25,000 and 12 million on the y axis and on the line that connects the data points. This indicates a break in the

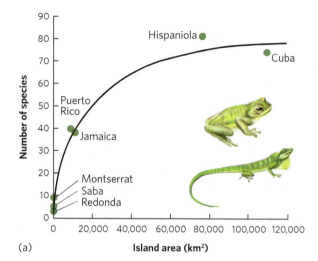

(a)

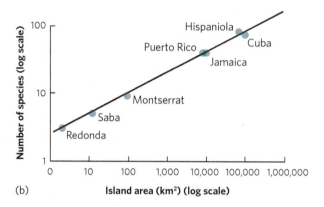

(b)

Figure A.4 (Text Figure 22.4) **(a)** This scatter plot shows the number of amphibian and reptile species on islands of different sizes in the West Indies. Note that the line of best fit is a curved line. **(b)** When the same data are plotted on a log scale, the line of best fit is a straight line.

scale. As you can see, the y axis initially increases from 0 to 20,000 birds, but after the double hatch mark, the scale jumps up to 12 million (note that the axis is in units of thousands of

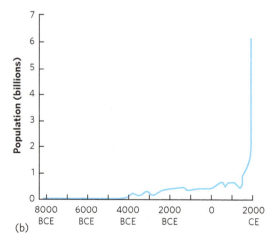

(b)

Figure A.5 (Text Figure 12.2b) This line graph shows population growth over time by connecting a series of many data points.

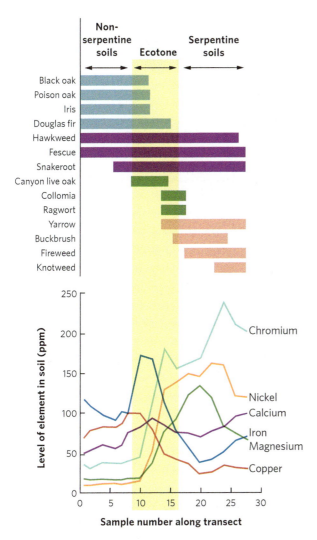

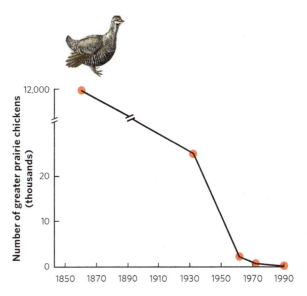

Figure A.7 (Text Figure 7.7a) This line graph illustrates the decline in the population of prairie chickens over time in Illinois. Because there is such a large difference in population size between 1860 and 1930, the researchers inserted a break in the *y* axis, and a corresponding break in the line connecting the data points, to allow a compressed figure that spans a wide range of *y*-axis values.

Figure A.6 (Text Figure 18.5) This line graph shows data measurement over space. It reflects changes in the concentration of several metals in the soil as researchers moved from nonserpentine soils to serpentine soils.

birds). The double hatch marks indicate that we are condensing the middle part of the *y* axis. Inserting an axis break is especially helpful when we wish to focus on how the data change after the double hatch mark. In Figure A.7, we have no data points between 25,000 and 12 million, so we can compress the *y* axis in this range to better view the change in population size over time.

Line graphs can also illustrate how several different variables change over time. When two variables contain different units or a different range of values, we can use two *y* axes. For example, **Figure A.8** presents data on changes in the

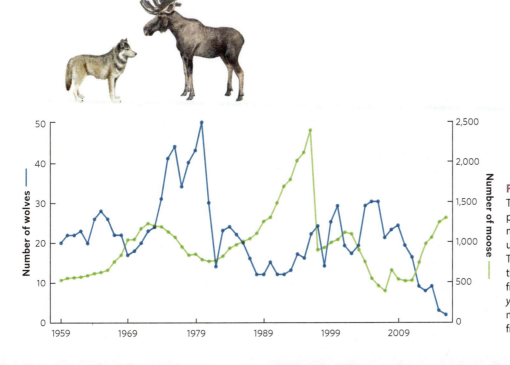

Figure A.8 (Text Figure 13.6) To illustrate changes in the population sizes of wolves and moose over time, researchers used a line graph with two *y* axes. The *y* axis on the left represents the number of wolves and ranges from 0 to 50. In contrast, the *y* axis on the right represents the number of moose and ranges from 0 to 2,500.

population sizes of two different animals on Isle Royale in the years 1955 to 2011. The left y axis represents the population changes in the wolf population, whereas the right y axis represents the population changes in the moose population during the same time span.

BAR GRAPHS

A *bar graph* plots numerical values that come from different categories. For example, researchers conducted an experiment on a species of bird known as the European magpie, which typically lays seven eggs. The researchers manipulated the number of eggs in a nest either by adding one or two eggs or by taking away one or two eggs. They then waited to see how many of the eggs hatched into birds that lived long enough to leave the nest as fledglings. The results of this experiment are illustrated in **Figure A.9**. As we can see in the figure, the researchers used the number of eggs as categories along the x axis and then plotted the number of offspring that survived to fledgling, which is represented by the y axis.

Often a bar graph is used to display mean values based on multiple observations. In such cases, scientists often like to show the amount of variation in the data that comprise each of the mean values. We can do this by adding *error bars* to a bar graph. As an example, researchers who were interested in the effects of inbreeding measured the number of eggs that snails laid when they had mates available, compared to when they did not have mates available. Because they measured the eggs laid by several different snails in each category, they created a graph, shown in **Figure A.10**, that illustrates the mean number of eggs laid. However, they also wanted to convey the amount of variation in eggs laid by these different individuals, so they also included error bars to each mean; in this case, the error bars represent standard errors of the means.

Figure A.11 shows an example of a bar graph that presents two sets of data for each category. In this example, researchers

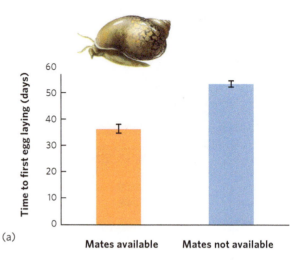

(a)

Figure A.10 (Text Figure 4.7) This bar graph shows the number of eggs produced by snails with mates available or not available. In this study, the researchers added error bars to convey the variation in egg laying among individuals. Error bars indicate standard errors of the means.

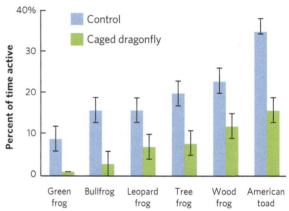

Figure A.11 (Text Figure 14.14) This bar graph shows activity of six tadpole species, each subdivided into two categories: tadpoles raised without predators or tadpoles raised in the presence of a caged predator. Error bars are standard errors.

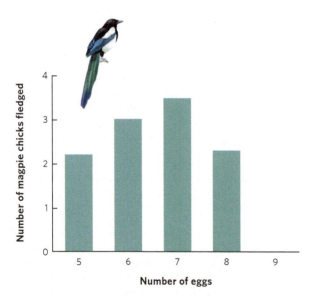

Figure A.9 (Text Figure 8.4) This bar graph shows the relationship between the number of eggs in nests of European magpies, plotted on the x axis, and the number of offspring that leave the nest as fledgling birds, plotted on the y axis.

measured the activity of six species of tadpoles. They made these measurements both in the presence and in the absence of predators. To include all this information, they drew two categories for each of the six species along the x axis: control or caged predator.

A bar graph is a very flexible tool and can be altered in several ways to accommodate data sets of different sizes or even several data sets that a researcher wishes to compare. When scientists measured the net primary productivity of different ecosystems, as shown in **Figure A.12**, they put the categories—various ecosystems—on the y axis and the plotted values—net primary productivity—on the x axis.

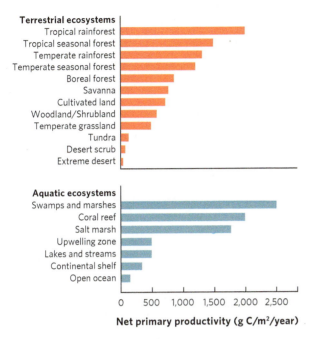

Figure A.12 (Text Figure 20.8) This bar graph shows net primary productivity of ecosystems around the world. In this case, the graph lists the categories on the *y* axis and plots the values on the *x* axis.

This orientation makes it easier to accommodate the relatively large amount of text needed to name each ecosystem.

PIE CHARTS

A *pie chart* is a graph represented by a circle with slices of various sizes representing categories within the whole pie. The entire pie represents 100 percent of the data and each slice is sized according to the percentage of the pie that it represents. For example, **Figure A.13** shows the percentage of various plant and animal groups from around the world that have been categorized as threatened, near-threatened, or of least concern from a conservation point of view. For each group of animals, each slice of the pie represents the percentage of species that fall within each conservation category.

TWO SPECIAL TYPES OF GRAPHS USED BY ECOLOGISTS

While scatter plot graphs, line graphs, bar graphs, and pie charts are used by many different types of scientists, ecologists also use two types of graphs that are not common in most other fields of science: *climate diagrams* and *age structure diagrams*. Although these two types of graphs are discussed within the text, we provide them here for review.

Climate Diagrams

Climate diagrams are used to illustrate the annual patterns of temperature and precipitation that help to determine the productivity of biomes on Earth. **Figure A.14** shows two hypothetical biomes. By graphing the average monthly temperature and precipitation of a biome, we can see how

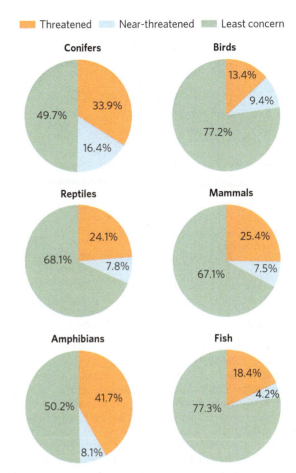

Figure A.13 (Text Figure 23.5) Pie charts illustrate the percentages that different groups compose. In this group of pie charts, we see the percentage of various plant and animal groups that fall into each of three conservation categories.

conditions in a biome vary during a typical year. We can also observe the specific time period when the temperature is warm enough for plants to grow. In the biome illustrated in Figure A.14a, the growing season—indicated by the shaded region on the *x* axis—is mid-March through mid-October. In Figure A.14b, the growing season is mid-April through mid-September.

In addition to identifying the growing season, climate diagrams show the relationships among precipitation, temperature, and plant growth. In Figure A.14a, the precipitation line is above the temperature line in every month. This means that water supply exceeds demand, so plant growth is more constrained by temperature than by precipitation throughout the entire year. In Figure A.14b, the precipitation line intersects the temperature line. At this point, the amount of precipitation available to plants equals the amount of water lost by plants through evapotranspiration. When the precipitation line falls below the temperature line from May through September, water demand exceeds supply and plant growth will be constrained more by precipitation than by temperature.

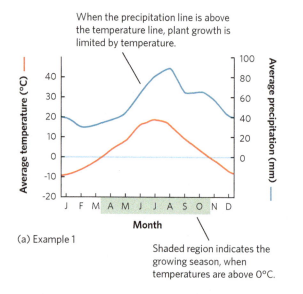

When the precipitation line is above the temperature line, plant growth is limited by temperature.

(a) Example 1

Shaded region indicates the growing season, when temperatures are above 0°C.

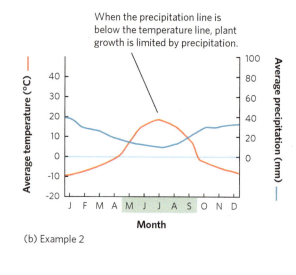

When the precipitation line is below the temperature line, plant growth is limited by precipitation.

(b) Example 2

Figure A.14 (Text Figure 6.4) This figure shows two hypothetical climate diagrams. Climate diagrams illustrate patterns of temperature and precipitation throughout the year, the length of the growing season, and whether plant growth is limited by temperature or precipitation.

Age Structure Diagrams

An age structure diagram is a visual representation of age distribution for both males and females in a country. **Figure A.15** presents two examples. Each horizontal bar of the diagram represents a 5-year-age group and the length of a given bar represents the number of males or females in that age group.

While every nation has a unique age structure, we can group countries very broadly into three categories. Figure A.15a shows a country with many more young people than older people. The age structure diagram of a country with this population will be in the shape of a pyramid, with its widest part at the bottom, moving toward the smallest at the top. Age structure diagrams with this shape are typical of countries in the developing world, such as Venezuela and India.

A country with less of a difference between the number of individuals in the younger and older age groups has an age structure diagram that looks more like a column. With fewer individuals in the younger age groups, we can deduce that the country has little or no population growth. Figure A.15b shows the age structure of the U.S. population, which is similar to the age structure of the populations of Canada, Australia, Sweden, and many other developed countries.

As you can see, we can use many different types of graphs to display data collected in ecological studies. This makes it easier to view patterns in our data and to reach the correct interpretation. With this knowledge of graph making, you should be prepared to interpret the graphs throughout the book and to do the graphing exercises at the end of each chapter.

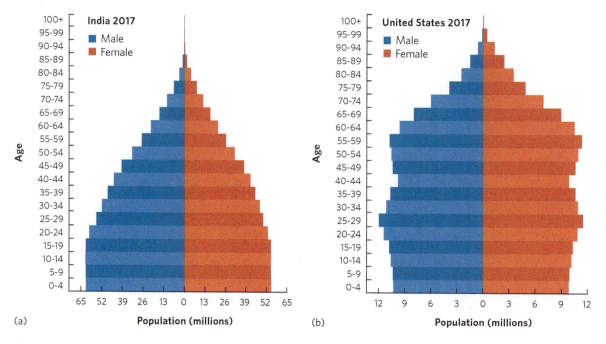

Figure A.15 (Text Figure 12.18) These diagrams portray the age structure of men and women in (a) India and (b) the United States.

Statistical Tables

STUDENT'S *T*-TEST DISTRIBUTION TABLE

You can identify the critical *t* value by selecting a column that contains your desired probability level and then selecting the row that has the appropriate degrees of freedom.

Degrees of freedom	Probability levels (α)		
	0.1	0.05	0.01
1	3.078	6.314	31.821
2	1.886	2.920	6.965
3	1.638	2.353	4.541
4	1.533	2.132	3.747
5	1.476	2.015	3.365
6	1.440	1.943	3.143
7	1.415	1.895	2.998
8	1.397	1.860	2.896
9	1.383	1.833	2.821
10	1.372	1.812	2.764
11	1.363	1.796	2.718
12	1.356	1.782	2.681
13	1.350	1.771	2.650
14	1.345	1.761	2.624
15	1.341	1.753	2.602
16	1.337	1.746	2.583
17	1.333	1.740	2.567
18	1.330	1.734	2.552
19	1.328	1.729	2.539
20	1.325	1.725	2.528
21	1.323	1.721	2.518
22	1.321	1.717	2.508
23	1.319	1.714	2.500
24	1.318	1.711	2.492
25	1.316	1.708	2.485
26	1.315	1.706	2.479
27	1.314	1.703	2.473
28	1.313	1.701	2.467
29	1.311	1.699	2.462
30	1.310	1.697	2.457
60	1.296	1.671	2.390
120	1.289	1.658	2.358
∞	1.282	1.645	2.326

CHI-SQUARE (X^2) DISTRIBUTION TABLE

You can identify the critical X^2 value by selecting a column that contains your desired probability level and then selecting the row that has the appropriate degrees of freedom.

Degrees of freedom	Probability levels (α) .100	.050	.010
1	2.706	3.841	6.635
2	4.605	5.991	9.210
3	6.251	7.815	11.345
4	7.779	9.488	13.277
5	9.236	11.071	15.086
6	10.645	12.592	16.812
7	12.017	14.067	18.475
8	13.362	15.507	20.090
9	14.684	16.919	21.666
10	15.987	18.307	23.209
11	17.275	19.675	24.725
12	18.549	21.026	26.217
13	19.812	22.362	27.688
14	21.064	23.685	29.141
15	22.307	24.996	30.578
16	23.542	26.296	32.000
17	24.769	27.587	33.409
18	25.989	28.869	34.805
19	27.204	30.144	36.191
20	28.412	31.410	37.566
21	29.615	32.671	38.932
22	30.813	33.924	40.289
23	32.007	35.172	41.638
24	33.196	36.415	42.980
25	34.382	37.652	44.314
26	35.563	38.885	45.642
27	36.741	40.113	46.963
28	37.916	41.337	48.278
29	39.087	42.557	49.588
30	40.256	43.773	50.892

Z-DISTRIBUTION TABLE

To determine if a Z value is significant, begin by selecting the row that represents the first two digits of your calculated Z test statistic and then select the column that represents the third digit of your calculated Z test statistic. This gives you a probability value. By subtracting this probability value from 1, you can determine the a value.

Z	0.00	0.01	0.02	0.03	0.04	0.05	0.06	0.07	0.08	0.09
0.0	0.500	0.504	0.508	0.512	0.516	0.520	0.524	0.528	0.532	0.536
0.1	0.540	0.544	0.548	0.552	0.556	0.560	0.564	0.568	0.571	0.575
0.2	0.579	0.583	0.587	0.591	0.595	0.599	0.603	0.606	0.610	0.614
0.3	0.618	0.622	0.626	0.629	0.633	0.637	0.641	0.644	0.648	0.652
0.4	0.655	0.659	0.663	0.666	0.670	0.674	0.677	0.681	0.684	0.688
0.5	0.692	0.695	0.699	0.702	0.705	0.709	0.712	0.716	0.719	0.722
0.6	0.726	0.729	0.732	0.736	0.739	0.742	0.745	0.749	0.752	0.755
0.7	0.758	0.761	0.764	0.767	0.770	0.773	0.776	0.779	0.782	0.785
0.8	0.788	0.791	0.794	0.797	0.800	0.802	0.805	0.808	0.811	0.813
0.9	0.816	0.819	0.821	0.824	0.826	0.829	0.832	0.834	0.837	0.839
1.0	0.841	0.844	0.846	0.849	0.851	0.853	0.855	0.858	0.860	0.862
1.1	0.864	0.867	0.869	0.871	0.873	0.875	0.877	0.879	0.881	0.883
1.2	0.885	0.887	0.889	0.891	0.893	0.894	0.896	0.898	0.900	0.902
1.3	0.903	0.905	0.907	0.908	0.910	0.912	0.913	0.915	0.916	0.918
1.4	0.919	0.921	0.922	0.924	0.925	0.927	0.928	0.929	0.931	0.932
1.5	0.933	0.935	0.936	0.937	0.938	0.939	0.941	0.942	0.943	0.944
1.6	0.945	0.946	0.947	0.948	0.950	0.951	0.952	0.953	0.954	0.955
1.7	0.955	0.956	0.957	0.958	0.959	0.960	0.961	0.962	0.963	0.963
1.8	0.964	0.965	0.966	0.966	0.967	0.968	0.969	0.969	0.970	0.971
1.9	0.971	0.972	0.973	0.973	0.974	0.974	0.975	0.976	0.976	0.977
2.0	0.977	0.978	0.978	0.979	0.979	0.980	0.980	0.981	0.981	0.982
2.1	0.982	0.983	0.983	0.983	0.984	0.984	0.985	0.985	0.985	0.986
2.2	0.986	0.986	0.987	0.987	0.988	0.988	0.988	0.988	0.989	0.989
2.3	0.989	0.990	0.990	0.990	0.990	0.991	0.991	0.991	0.991	0.992
2.4	0.992	0.992	0.992	0.993	0.993	0.993	0.993	0.993	0.993	0.994
2.5	0.994	0.994	0.994	0.994	0.995	0.995	0.995	0.995	0.995	0.995
2.6	0.995	0.996	0.996	0.996	0.996	0.996	0.996	0.996	0.996	0.996
2.7	0.997	0.997	0.997	0.997	0.997	0.997	0.997	0.997	0.997	0.997
2.8	0.997	0.998	0.998	0.998	0.998	0.998	0.998	0.998	0.998	0.998
2.9	0.998	0.998	0.998	0.998	0.998	0.998	0.999	0.999	0.999	0.999
3.0	0.999	0.999	0.999	0.999	0.999	0.999	0.999	0.999	0.999	0.999
3.1	0.999	0.999	0.999	0.999	0.999	0.999	0.999	0.999	0.999	0.999
3.2	0.999	0.999	0.999	0.999	0.999	0.999	0.999	1.000	1.000	1.000
3.3	1.000	1.000	1.000	1.000	1.000	1.000	1.000	1.000	1.000	1.000

Answers to "Analyzing Ecology" and "Graphing the Data"

Chapter 1

Analyzing Ecology

Insect abundance on uncaged trees:

    Mean = 3.0

    Variance = 1.0

Chapter 2

Analyzing Ecology

Standard deviation of the mean = 7.0

Standard error of the mean = 3.1

Graphing the Data

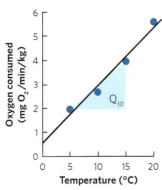

Based on the data, the Q_{10} between 5°C and 15°C is 2.0.
Based on the data, the Q_{10} between 10°C and 20°C is 2.07.

Chapter 3

Analyzing Ecology

Each of the three soil treatments would be considered a categorical variable because they fall into distinct categories. In contrast, if we had considered three treatments such as 100 percent sand, 70 percent sand, and 40 percent sand, then we would have a continuous variable.

Graphing the Data

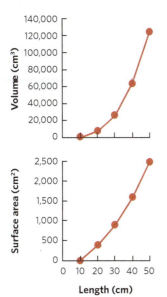

Chapter 4

Analyzing Ecology

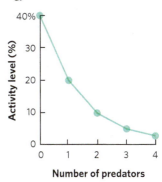

a. It is a negative correlation.

b. It is curvilinear.

Graphing the Data

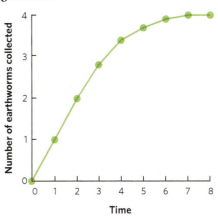

The data represent a correlation. The data also represent causation because increased time allows the birds to collect a greater number of earthworms.

Chapter 5

Analyzing Ecology

Given the following equation:

    Temperature = –1.2 × Latitude + 43

At a latitude of 10 degrees, the mean January temperature is

    Temperature = –1.2 × 10 + 43 = 31°C

At a latitude of 20 degrees, the mean January temperature is

    Temperature = –1.2 × 20 + 43 = 19°C

At a latitude of 30 degrees, the mean January temperature is

    Temperature = –1.2 × 30 + 43 = 7°C

Graphing the Data

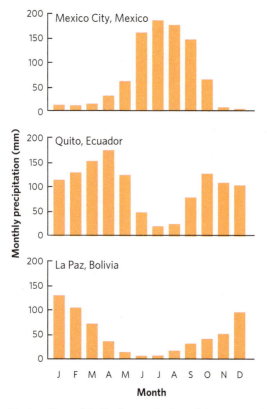

a. Mexico City and La Paz have a single peak in precipitation, whereas Quito has two peaks in precipitation.

b. Peak precipitation occurs when the ITCZ passes overhead. Because of the different latitudes of the three cities, the ITCZ passes over Mexico City and La Paz once each year, but it passes over Quito twice each year.

Chapter 6

Analyzing Ecology

Mean = 15.6

Median = 17

Mode = 17

The values differ because the mean measures the central tendency of the data by calculating the average value, while the median ranks the data and uses the middle data point; the mode uses the commonly occurring value.

Graphing the Data

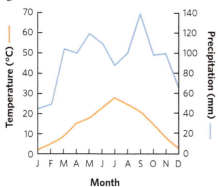

Chapter 7

Analyzing Ecology

To determine the response to selection (R), we multiply the strength of selection on a trait (S) and the heritability of the trait (h^2).

S	h^2	R
0.5	0.7	0.35
1	0.7	0.7
1.5	0.7	1.05
2	0.9	1.8
2	0.6	1.2
2	0.3	0.6
2	0	0

Based on the above results, the responses to selection (R) are greater whenever the strength of selection increases or the heritability of the trait increases.

Graphing the Data

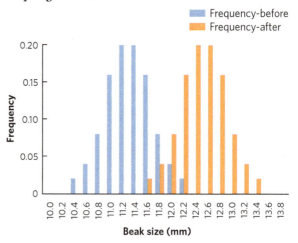

We can determine the mean beak size before or after selection by multiplying each beak depth by its frequency and then taking the sum of these products.

Mean beak size before selection = 11.3 mm

Mean beak size after selection = 12.5 mm

Because the mean beak depth has increased after selection, this set of data is an example of directional selection.

Chapter 8

Analyzing Ecology

For Population A:

$$R^2 = 1 - \left(\frac{(21-20)^2 + (19-20)^2 + (13-12)^2 + (11-12)^2 + (5-4)^2 + (3-4)^2}{(21-12)^2 + (19-12)^2 + (13-12)^2 + (11-12)^2 + (5-12)^2 + (3-12)^2} \right) =$$

$$1 - \left(\frac{6}{262} \right) = 0.98$$

For Population B:

$$R^2 = 1 - \left(\frac{(22-20)^2 + (18-20)^2 + (14-12)^2 + (10-12)^2 + (6-4)^2 + (2-4)^2}{(22-12)^2 + (18-12)^2 + (14-12)^2 + (10-12)^2 + (6-12)^2 + (2-12)^2} \right) =$$

$$1 - \left(\frac{24}{280} \right) = 0.91$$

For Population C:

$$R^2 = 1 - \left(\frac{(24-20)^2+(16-20)^2+(16-12)^2+(8-12)^2+(7-4)^2+(1-4)^2}{(24-12)^2+(16-12)^2+(16-12)^2+(8-12)^2+(7-12)^2+(1-12)^2} \right) =$$

$$1 - \left(\frac{82}{338} \right) = 0.76$$

Because of the higher coefficient of determination, Population A provides the strongest confidence that there is a negative relationship between seed size and seed number.

Graphing the Data

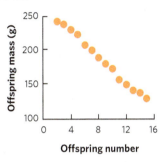

Lizards that produce a larger number of eggs do so by producing smaller eggs.

Chapter 9

Analyzing Ecology

If there are four males and five females:

Average male fitness = 10 gene copies ÷ 4 males
= 2.5 gene copies/male
Average female fitness = 10 gene copies ÷ 5 females
= 2.0 gene copies/female

If there are six males and five females:

Average male fitness = 10 gene copies ÷ 6 males
= 1.7 gene copies/male
Average female fitness = 10 gene copies ÷ 5 females
= 2.0 gene copies/female

Over the long term, if either sex becomes rare, it will have higher fitness, which will subsequently cause it to become more common. Over the long term, this process will favor an equal sex ratio.

Graphing the Data

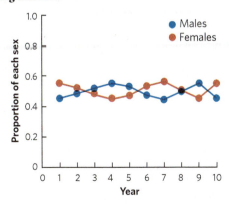

Whenever one of the sexes becomes rare, it subsequently becomes common. Over the long term, we tend toward an equal sex ratio.

Chapter 10

Analyzing Ecology

If the fitness that primary helpers give to their parents declines to 1.0, their indirect fitness declines to 0.32 and their inclusive fitness declines to 0.73. Under this scenario, the secondary helper strategy would be the most favored by natural selection.

MALE ROLE	YEAR 1 B_1	r_1	INDIRECT FITNESS	YEAR 2 B_2	r_2	P_{sm}	DIRECT FITNESS	INCLUSIVE FITNESS
PRIMARY HELPER	1.0 × 0.32		= 0.32	2.5 × 0.5 × 0.32			= 0.41	0.32 + 0.41 = 0.73
SECONDARY HELPER	1.3 × 0.00		= 0.00	2.5 × 0.5 × 0.67			= 0.84	0.00 + 0.84 = 0.84
DELAYER	0.0 × 0.00		= 0.00	2.5 × 0.5 × 0.23			= 0.29	0.00 + 0.29 = 0.29

Graphing the Data

	SCHOOL SIZE				
TRIAL	3	5	10	15	20
1	0.9	0.7	0.4	0.4	0.1
2	0.8	0.8	0.5	0.5	0.1
3	0.7	0.9	0.6	0.3	0.2
4	1.1	0.6	0.8	0.2	0.3
5	1.0	1.0	0.7	0.6	0.3
MEAN	0.90	0.80	0.60	0.40	0.20
STANDARD DEVIATION	0.16	0.16	0.16	0.16	0.10

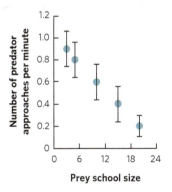

As the size of the prey school increases, there is a decline in the number of times a predator approaches.

Chapter 11

Analyzing Ecology

Given the equation:

$$N = M \times C \div R$$

$$N = 20 \times 48 \div 24$$

$$N = 40$$

Because these 40 crayfish were found in a 300 m² stretch of a stream, the density of the crayfish can be calculated as

40 crayfish ÷ 300 m² = 0.13 crayfish/m²

Graphing the Data

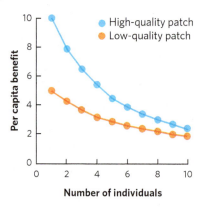

Once four individuals have arrived at the high-quality patch, the next individual to arrive would receive a larger per capita benefit by moving to the low-quality patch.

If there are 12 individuals, all individuals would gain the highest per capita benefit if 8 individuals remained in the high-quality patch and 4 individuals moved to the low-quality patch.

Chapter 12

Analyzing Ecology

AGE (x)	SURVIVAL RATE (n_x)	SURVIVOR-SHIP (l_x)	FECUNDITY (b_x)	$(l_x) \times (b_x)$	$(x) \times (l_x) \times (b_x)$
0	530	1.000	0.05	0.05	0.00
1	134	0.253	1.28	0.32	0.32
2	56	0.116	2.28	0.26	0.53
3	39	0.089	2.28	0.20	0.61
4	23	0.058	2.28	0.13	0.53
5	12	0.039	2.28	0.09	0.44
6	5	0.025	2.28	0.06	0.34
7	2	0.022	2.28	0.05	0.35

Net reproductive rate $(R_0) = \Sigma\, l_x b_x = 1.17$

Generation time $(T) = \dfrac{\Sigma x l_x b_x}{\Sigma l_x b_x} = \dfrac{3.13}{1.17} = 2.7$ years

Intrinsic rate of increase $(\lambda_a) = R_0^{\frac{1}{T}} = 1.17^{\frac{1}{2.7}} = 1.06$

Graphing the Data

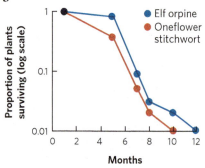

Chapter 13

Analyzing Ecology

The population sizes from year 1 through 15 are as follows:

YEAR	CHANGE IN POPULATION SIZE	TOTAL POPULATION SIZE
1		10.0
2	10.0	20.0
3	19.8	39.8
4	35.0	74.8
5	49.5	124.4
6	34.4	158.8
7	−42.6	116.2
8	−75.2	41.0
9	−7.3	33.7
10	21.9	55.6
11	40.5	96.1
12	47.0	143.1
13	6.2	149.2
14	−70.7	78.5
15	−42.5	36.0

Given that the product $rT = 1.1 \times 1 = 1.1$, we would expect the population will experience damped oscillations, which you can confirm from the graphed data.

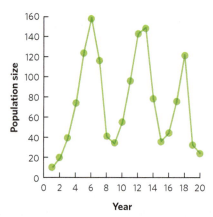

Graphing the Data

PROPORTION OF OCCUPIED PATCHES p	PROBABILITY OF EXTINCTION e	RATE OF EXTINCTION $e \times p$	PROBABILITY OF COLONIZATION c	RATE OF COLONIZATION $(c \times p) \times 1-p$
0.1	0.25	0.025	0.5	0.045
0.2	0.25	0.050	0.5	0.080
0.3	0.25	0.075	0.5	0.105
0.4	0.25	0.100	0.5	0.120
0.5	0.25	0.125	0.5	0.125
0.6	0.25	0.150	0.5	0.120
0.7	0.25	0.175	0.5	0.105
0.8	0.25	0.200	0.5	0.080
0.9	0.25	0.225	0.5	0.045
1	0.25	0.250	0.5	0

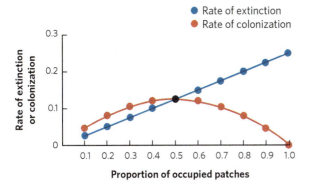

For these data, the rates of extinction and colonization come to equilibrium where the two lines meet when the proportion of occupied patches is 0.5. This can also be calculated using the equilibrium equation:

$$\hat{p} = 1 - \frac{e}{c} = 1 - \frac{0.25}{0.50} = 0.5$$

Chapter 14

Analyzing Ecology

When we repeatedly sample the two distributions, we can conclude that they are significantly different if the two means do not overlap 95 percent of the time. This would happen when the two distributions are approximately 2 standard deviations apart from each other.

Graphing the Data

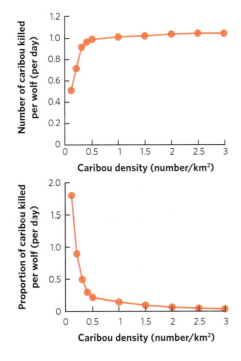

Because the number of caribou killed per wolf gradually slows as caribou density increases, the wolves most closely follow a type II functional response.

Chapter 15

Analyzing Ecology

To calculate the t-value:

$$t = \frac{\overline{X}_1 - \overline{X}_2}{\sqrt{\frac{s_1^2}{n_1} + \frac{s_2^2}{n_2}}} = \frac{10 - 5}{\sqrt{\frac{4}{8} + \frac{4}{8}}} = \frac{5}{\sqrt{1}} = 5$$

To calculate the degrees of freedom:

$$\text{degrees of freedom} = 8 + 8 - 2 = 14$$

Based on the statistical table for a t-test (see the Statistical Tables appendix), the critical value of t for 14 degrees of freedom and an alpha value of 0.05 is 2.1. Since our t-value exceeds the critical value from the table, we conclude that the two means are significantly different.

Graphing the Data

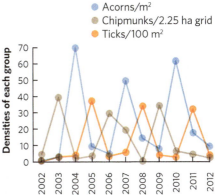

Based on the graph, one can see that the peak density of acorns is followed 1 year later by the peak density of chipmunks and 2 years later by the peak density of ticks.

Chapter 16

Analyzing Ecology

Use the following equation for a chi-square test:

$$\chi^2 = \sum_{i=1}^{n} \frac{(O_i - E_i)^2}{(E_i)}$$

$$\chi^2 = \frac{(12 - 10)^2}{10} + \frac{(8 - 10)^2}{10}$$

$$\chi^2 = 0.8$$

Degrees of freedom =
(number of observed categories – 1) = (2 – 1) = 1

Using the chi-square table in the Statistical Tables appendix, we can compare our calculated value (0.8) to the critical chi-square value when we have 1 degree of freedom and an alpha value of 0.05. The critical chi-square value is 3.841, which exceeds our chi-square value. Thus, we conclude that our observed distribution of mice in the bare zone and in the grass is not significantly different from an equal distribution.

Graphing the Data

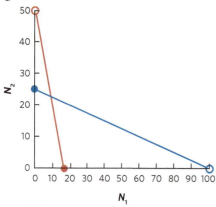

Based on this graph, the predicted outcome of this competition is that either of the two species can win, but it depends on the initial number of each species.

Chapter 17

Analyzing Ecology

$$U = R_1 - (n_1 (n_1 + 1) \div 2)$$
$$U = 40.5 - (8 (8 + 1) \div 2)$$
$$U = 40.5 - (72 \div 2)$$
$$U = 4.5$$

Using this value for R_1, we can now calculate z:

$$z = \frac{4.5 - 32}{9.52} = 2.89$$

When we look up the critical value of z in the Statistical Tables appendix, we use the absolute value of z. As a result, both R_1 and R_2 provide us with values of z that have the same absolute value.

Graphing the Data

NUMBER OF FUNGAL SPECIES	MEAN PHOSPHORUS REMAINING IN THE SOIL (MG P/KG SOIL)	STANDARD ERROR
0	16.0	0.6
2	11.0	0.6
4	9.0	0.6
8	5.0	0.6
14	4.0	0.6

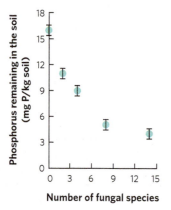

Chapter 18

Analyzing Ecology

Simpson's index

Community A:

$$\frac{1}{(0.24)^2 + (0.16)^2 + (0.08)^2 + (0.34)^2 + (0.18)^2} = \frac{1}{.24} = 4.21$$

Community B:

$$\frac{1}{(0.20)^2 + (0.20)^2 + (0.20)^2 + (0.20)^2 + (0.20)^2} = \frac{1}{.20} = 5.00$$

Community C:

$$\frac{1}{(0.25)^2 + (0.25)^2 + (0.25)^2 + (0.25)^2} = \frac{1}{.25} = 4.00$$

Shannon's index

Community A:

$$-[(0.24)(\ln 0.24) + (0.16)(\ln 0.16) + (0.8)(\ln 0.08) + (0.34)(\ln 0.34) + (0.18)(\ln 0.18)] = 1.51$$

Community B:

$$-[(0.20)(\ln 0.20) + (0.20)(\ln 0.20) + (0.20)(\ln 0.20) + (0.20)(\ln 0.20) + (0.20)(\ln 0.20)] = 1.61$$

Community C:

$$-[(0.25)(\ln 0.25) + (0.25)(\ln 0.25) + (0.25)(\ln 0.25) + (0.25)(\ln 0.25)] = 1.39$$

For both indexes, greater evenness with the same richness provides a higher index value. When evenness is identical, the community with the higher richness has a higher index value.

Graphing the Data

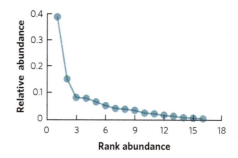

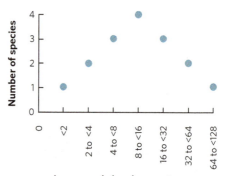

Chapter 19

Analyzing Ecology

Comparing Communities A and B:

$$J = \frac{3}{1 + 4 + 3} = 0.33$$

Comparing Communities A and C:

$$J = \frac{0}{4 + 6 + 0} = 0.00$$

Comparing Communities B and C:

$$J = \frac{4}{3 + 2 + 4} = 0.44$$

Based on these calculations, Communities A and B are the most similar to each other.

Graphing the Data

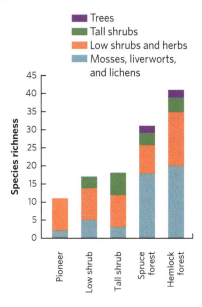

Over time, the composition of the seral stages changes, but total species richness continually increases.

Chapter 20

Analyzing Ecology

Lake ecosystem:

Consumption efficiency = 65%
Assimilation efficiency = 70%
Net production efficiency = 42%
Ecological efficiency = 19%

Stream ecosystem:

Consumption efficiency = 65%
Assimilation efficiency = 38%
Net production efficiency = 40%
Ecological efficiency = 10%

The two aquatic ecosystems have higher ecological efficiencies than the terrestrial ecosystem primarily due to higher consumption efficiencies.

In this example, the lower ecological efficiency in the stream ecosystem compared to the lake ecosystem is caused by a lower assimilation efficiency in the stream ecosystem.

Graphing the Data

TERRESTRIAL ECOSYSTEMS	NPP (g/m²/yr)	AREA (10⁶ km²)	TOTAL PRODUCTION (10¹² kg/yr)
Tropical rainforest	2,000	17.0	34.00
Tropical seasonal forest	1,500	7.5	11.25
Temperate rainforest	1,300	5.0	0.01
Temperate seasonal forest	1,200	7.0	8.40
Boreal forest	800	12.0	9.60
Savanna	700	15.0	10.50
Cultivated land	650	14.0	9.10
Woodland/shrubland	600	8.0	4.80
Temperate grassland	500	9.0	4.50
Tundra	140	8.0	1.12
Desert shrub	70	18.0	1.26
AQUATIC ECOSYSTEMS			
Swamp and marsh	2,500	2.0	5.00
Coral reef	2,000	0.6	1.20
Salt marsh	1,800	1.4	2.52
Upwelling zones	500	0.4	0.20
Lake and stream	500	2.5	1.25
Continental shelf	360	26.6	9.58
Open ocean	125	332.0	41.50

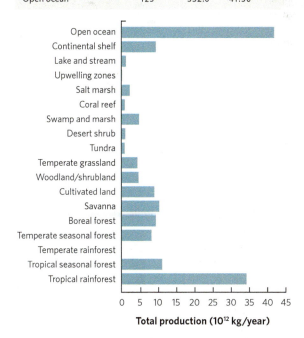

Some areas, such as the open ocean, do not have a high productivity per unit area, but the total area is so large that it contributes a relatively large amount to total production. In contrast, some highly productive ecosystems, such as swamps and marshes, do not cover a large area so they contribute a relatively small amount of total production.

Chapter 21

Analyzing Ecology

We can calculate these answers using the following equation: $m_t = m_o\,e^{-kt}$.

For $k = 0.05$:

After 10 days: $m_t = m_o\,e^{-kt} = 100\,e^{-(0.05)(10)} = 61$ grams

After 50 days: $m_t = m_o\,e^{-kt} = 100\,e^{-(0.05)(50)} = 8.2$ grams

After 100 days: $m_t = m_o\,e^{-kt} = 100\,e^{-(0.05)(100)} = 0.7$ grams

For $k = 0.10$:

After 10 days: $m_t = m_o\,e^{-kt} = 100\,e^{-(0.10)(10)} = 37$ grams

After 50 days: $m_t = m_o\,e^{-kt} = 100\,e^{-(0.10)(50)} = 0.7$ grams

After 100 days: $m_t = m_o\,e^{-kt} = 100\,e^{-(0.10)(100)} = 0.0$ grams

Graphing the Data

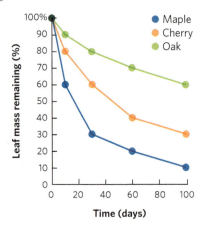

Based on this graph, maple leaves have the highest value of k.

Chapter 22

Analyzing Ecology

For this problem, we will use the following equation:

$$S = S_{obs} + \frac{f_i(f_i - 1)}{2(f_2 + 1)}$$

For Community B:

$$S = 12 + \frac{5 \times (5 - 1)}{2 \times (0 + 1)}$$

$$S = 12 + \frac{20}{2}$$

$$S = 22$$

For Community C:

$$S = 8 + \frac{0 \times (0 - 1)}{2 \times (2 + 1)}$$

$$S = 8 + \frac{0}{6}$$

$$S = 8$$

The greater the number of species represented by a single individual, the greater the estimated number of species in the community.

Graphing the Data

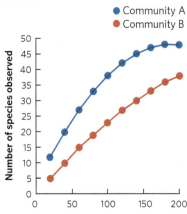

Community A has been sampled sufficiently enough to provide a good estimate of species richness, as evidenced by the fact that the species accumulation curve reaches a plateau. In contrast, the curve for Community B suggests that more sampling needs to be conducted before a plateau is achieved.

Chapter 23

Analyzing Ecology

We can make use of the equation that calculates half-lives:

$$t_{1/2} = t \ln(2) \div \ln(N_0 \div N_t)$$

$$t_{1/2} = (1) \ln(2) \div \ln(50\ \text{ppb} \div 10\ \text{ppb})$$

$$t_{1/2} = 0.69 \div 1.61$$

$$t_{1/2} = 0.43\ \text{days}$$

Graphing the Data

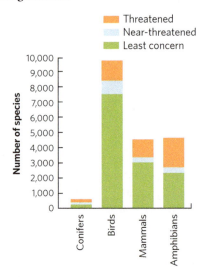

When we look at the absolute numbers in each category, we are able not only to have a sense of the proportion of species that fall within each category, but also to know the absolute number in each category. For example, the percentage of threatened mammals is about twice that of birds, but the total number of threatened birds and mammals is about the same. In addition, the percentage of threatened mammals and conifers is similar, but the actual number of threatened mammals is more than six times higher.

Glossary

Abundance The total number of individuals in a population that exist within a defined area.

Acclimation An environmentally induced change in an individual's physiology.

Acid deposition Acids deposited as rain and snow or as gases and particles that attach to the surfaces of plants, soil, and water. *Also known as* **Acid rain**.

Acidity The concentration of hydrogen ions in a solution.

Active transport The movement of molecules or ions through a membrane against a concentration gradient.

Adaptation A characteristic of an organism that makes it well suited to its environment.

Adiabatic cooling The cooling effect of reduced pressure on air as it rises higher in the atmosphere and expands.

Adiabatic heating The heating effect of increased pressure on air as it sinks toward the surface of Earth and decreases in volume.

Aestivation The shutting down of metabolic processes during the summer in response to hot or dry conditions.

Afrotropical region The biogeographic region of the Southern Hemisphere that corresponds to most of Africa. *Also known as* **Ethiopian region**.

Age structure In a population, the proportion of individuals that occurs in different age classes.

Albedo The fraction of solar energy reflected by an object.

Algal bloom A rapid increase in the growth of algae in aquatic habitats, typically due to an influx of nutrients.

Alleles Different forms of a particular gene.

Allelopathy A type of interference that occurs when organisms use chemicals to harm their competitors.

Allochthonous Inputs of organic matter, such as leaves, that come from outside of an ecosystem.

Allopatric speciation The evolution of new species through the process of geographic isolation. *Also known as* the **Shannon–Wiener index**.

Alternative stable state When a community is disturbed so much that the species composition and relative abundance of populations in the community change, and the new community structure is resistant to further change.

Altruism A social interaction that increases the fitness of the recipient and decreases the fitness of the donor.

Anaerobic Without oxygen. *Also known as* **Anoxic**.

Annual An organism that has a life span of one year.

Aphotic zone The area of the neritic and oceanic zones where the water is so deep that sunlight cannot penetrate.

Apparent competition When two species have a negative effect on each other through an enemy—including a predator, parasite, or herbivore.

Arbuscular mycorrhizal fungi A type of endomycorrhizal fungi that infects a tremendous number of plants, including apple trees, peach trees, coffee trees, and grasses.

Area- and volume-based surveys Surveys that define the boundaries of an area or volume and then count all the individuals in the space.

Artificial selection Selection in which humans decide which individuals will breed and the breeding is done with a preconceived goal for the traits of the population.

Asexual reproduction A reproduction mechanism in which progeny inherit DNA from a single parent.

Assimilated energy The portion of energy that a consumer digests and absorbs.

Assimilation efficiency The percentage of consumed energy that is assimilated.

Atmosphere The 600-km thick layer of air that surrounds the planet.

Atmospheric convection currents The circulations of air between the surface of Earth and the atmosphere.

Australasian region The biogeographic region of the Southern Hemisphere that corresponds to Australia, New Zealand, and New Guinea.

Autochthonous Inputs of organic matter that are produced by algae and aquatic plants inside an ecosystem.

Basic metapopulation model A model that describes a scenario in which there are patches of suitable habitat embedded within a matrix of unsuitable habitat.

Batesian mimicry When palatable species evolve warning coloration that resembles unpalatable species.

Benthic zone The area consisting of the sediments at the bottoms of lakes, ponds, and oceans.

Beta diversity The number of species that differ in occurrence between two habitats.

Bicarbonate ion (HCO_3^-) An anion formed by the dissociation of carbonic acid.

Binary fission Reproduction through duplication of genes followed by division of the cell into two identical cells.

Biomagnification The process by which the concentration of a contaminant increases as it moves up the food chain.

Biomass residence time The length of time that biomass remains in a given trophic level.

Biome A geographic region that contains communities composed of organisms with similar adaptations.

Biosphere approach An approach to ecology concerned with the largest scale in the hierarchy of ecological systems, including movements of air and water—and the energy and chemical elements they contain—over Earth's surface.

Biosphere All the ecosystems on Earth.

Biotic homogenization The process by which unique species compositions originally found in different regions slowly become more similar due to the movement of people, cargo, and species.

Blood shunting An adaptation that allows specific blood vessels to shut off so less of an animal's warm blood flows to the cold extremities.

Boreal forest A biome densely populated by evergreen needle-leaved trees, with a short growing season and severe winters. *Also known as* **Taiga**.

Bottleneck effect A reduction of genetic diversity in a population due to a large reduction in population size.

Bottom-up control When the abundances of trophic groups in nature are determined by the amount of energy available from the producers in a community.

Boundary layer A region of unstirred air or water that surrounds the surface of an object.

C_3 photosynthesis The most common photosynthetic pathway, in which CO_2 is initially assimilated into a three-carbon compound, glyceraldehyde 3-phosphate (G3P).

C_4 photosynthesis A photosynthetic pathway in which CO_2 is initially assimilated into a four-carbon compound, oxaloacetic acid (OAA).

Carbonate ion (CO_3^{2-}) An anion formed by the dissociation of carbonic acid.

Carrying capacity (K) The maximum population size that can be supported by the environment.

Caste Individuals within a social group sharing a specialized form of behavior.

Categorical variable A variable that falls into a distinct category or grouping. *Also known as* **Nominal variable.**

Cation exchange capacity The ability of a soil to retain cations.

Census A count of every individual in a population.

Central place foraging Foraging behavior in which acquired food is brought to a central place, such as a nest with young.

Chi-square test A statistical test that determines whether the number of observed events in different categories differs from an expected number of events, which is based on a particular hypothesis.

Chloroplasts Specialized cell organelles found in photosynthetic organisms.

Chromosomes Compact structures consisting of long strands of DNA that are wound around proteins.

Chronosequence A sequence of communities that exist over time at a given location.

Climate diagram A graph that plots the average monthly temperature and precipitation of a specific location on Earth.

Climate The typical atmospheric conditions that occur throughout the year, measured over many years.

Climax community The final seral stage in the process of succession.

Clones Individuals that descend asexually from the same parent and bear the same genotype.

Clustered dispersion A pattern of population dispersion in which individuals are aggregated in discrete groups.

Codominant When two alleles both contribute to the phenotype.

Coefficient of determination (R^2) An index that tells us how well data fit to a line.

Coefficient of relatedness The numerical probability of an individual and its relatives carrying copies of the same genes from a recent common ancestor.

Coevolution When two or more species affect each other's evolution.

Cohesion The mutual attraction among water molecules.

Cohesion–tension theory The mechanism of water movement from roots to leaves due to water cohesion and water tension.

Cohort life table A life table that follows a group of individuals born at the same time from birth to the death of the last individual.

Collapsed fishery When a fishery no longer has a population that can be fished.

Commensalism An interaction in which two species live in close association and one species receives a benefit, while the other experiences neither a benefit nor a cost.

Community All populations of species living together in a particular area.

Community approach An approach to ecology that emphasizes the diversity and relative abundances of different kinds of organisms living together in the same place.

Community resilience The time it takes after a disturbance for a community to return to its original state.

Community resistance The amount that a community changes when acted upon by some disturbance, such as the addition or removal of a species.

Community stability The ability of a community to maintain a particular structure.

Competition coefficients Variables that convert between the number of individuals of one species and the number of individuals of the other species.

Competition An interaction resulting in negative effects between two species that depend on the same limiting resource to survive, grow, and reproduce.

Competitive exclusion principle The principle that two species cannot coexist indefinitely when they are both limited by the same resource.

Concurrent circulation Movement of two fluids in the same direction on either side of a barrier through which heat or dissolved substances are exchanged.

Conduction The transfer of the kinetic energy of heat between substances that are in contact with one another.

Consumer An organism that obtains its energy from other organisms. *Also known as* **Heterotroph.**

Consumption efficiency The percentage of energy or biomass in a trophic level that is consumed by the next higher trophic level.

Continental drift The movement of landmasses across the surface of Earth.

Continuous variable A variable that can take on any numeric value, including values that are not whole numbers.

Control A manipulation that includes all aspects of an experiment except the factor of interest.

Convection The transfer of heat by the movement of liquids and gases.

Convergent evolution A phenomenon in which two species descended from unrelated ancestors look similar because they have evolved under similar selective forces.

Cooperation When the donor and the recipient of a social behavior both experience increased fitness from an interaction.

Coral bleaching Loss of color in corals as a result of the corals expelling their symbiotic algae.

Coral reef A marine biome found in warm, shallow waters that remain 20°C year-round.

Coriolis effect The deflection of an object's path due to the rotation of Earth.

Correlation A statistical description of how one variable changes in relation to another variable.

Cosmopolitan Species with very large geographic ranges that can span several continents.

Cost of meiosis The 50 percent reduction in the number of a parent's genes passed on to the next generation via sexual reproduction versus asexual reproduction.

Countercurrent circulation Movement of two fluids in opposite directions on either side of a barrier through which heat or dissolved substances are exchanged.

Crassulacean acid metabolism (CAM) A photosynthetic pathway in which the initial assimilation of carbon into a four-carbon compound occurs at night.

Crypsis Camouflage that either allows an individual to match its environment or breaks up the outline of an individual to blend in better with the background environment.

Cultural eutrophication An increase in the productivity of aquatic ecosystems caused by human activities.

Cultural services Benefits of biodiversity that provide aesthetic, spiritual, or recreational value.

Damped oscillations A pattern of population growth in which the population size initially oscillates, but the magnitude of the oscillations declines over time.

Decomposer Organisms that break down dead organic material into simpler elements and compounds that can be recycled through the ecosystem.

Delayed density dependence When density dependence occurs based on a population density at some time in the past.

Demographic stochasticity Variation in birth rates and death rates due to random differences among individuals.

Demography The study of populations.

Denitrification The process of converting nitrates into nitrogen gas.

Density dependent Factors that affect population size in relation to the population's density.

Density independent Factors that limit population size regardless of the population's density.

Density In a population, the number of individuals in a quantified area or volume.

Density-mediated indirect effect An indirect effect caused by changes in the density of an intermediate species.

Deoxyribonucleic acid (DNA) A molecule composed of two strands of nucleotides that are wound together into a shape known as a double helix.

Dependent variable A factor that is being changed.

Determinate growth A growth pattern in which an individual does not grow any more once it initiates reproduction.

Deterministic model A model that is designed to predict a result without accounting for random variation in population growth rate.

Detritivore An organism that feeds on dead organic matter and waste products that are collectively known as detritus.

Diapause A type of dormancy in insects that is associated with a period of unfavorable environmental conditions.

Die-off A substantial decline in density that typically goes well below the carrying capacity.

Dilution effect The reduced, or diluted, probability of predation to a single animal when it is in a group.

Dioecious Plants that contain either only male flowers or only female flowers on a single individual.

Direct effect An interaction between two species that does not involve other species.

Direct fitness The fitness that an individual gains by passing on copies of its genes to its offspring.

Direct selection Selection that favors direct fitness.

Directional selection When individuals with an extreme phenotype experience higher fitness than the average phenotype of the population.

Dispersal limitation The absence of a population from suitable habitat because of barriers to dispersal.

Dispersal The movement of individuals from one area to another.

Dispersion The spacing of individuals with respect to one another within the geographic range of a population.

Disruptive selection When individuals with either extreme phenotype experience higher fitness than individuals with an intermediate phenotype.

Dominance hierarchy A social ranking among individuals in a group, typically determined through fighting or other contests of strength or skill.

Dominant allele An allele that masks the expression of the other allele of a given gene.

Donor The individual who directs a behavior toward another individual as part of a social interaction.

Dormancy A condition in which organisms dramatically reduce their metabolic processes.

Doubling time The time required for a population to double in size.

Dry climate A climate characterized by low precipitation and a wide range of temperatures, commonly found at approximately 30° N and 30° S latitudes.

Dynamic steady state When the gains and losses of ecological systems are in balance.

Ecological efficiency The percentage of net production from one trophic level, compared to the next lower trophic level. *Also known as* **Food chain efficiency.**

Ecological envelope The range of ecological conditions that are predicted to be suitable for a species.

Ecological niche modeling The process of determining the suitable habitat conditions for a species.

Ecological stoichiometry The study of the balance of nutrients in ecological interactions, such as between an herbivore and a plant.

Ecological systems Biological entities that have their own internal processes and interact with their external surroundings.

Ecology The scientific study of the abundance and distribution of organisms in relation to other organisms and environmental conditions.

Ecosystem approach An approach to ecology that emphasizes the storage and transfer of energy and matter, including the various chemical elements essential to life.

Ecosystem One or more communities of living organisms interacting with their nonliving physical and chemical environments.

Ecotone A boundary created by sharp changes in environmental conditions over a relatively short distance, accompanied by a major change in the composition of species.

Ectomycorrhizal fungi Fungi characterized by hyphae that surround the roots of plants and enter between root cells but rarely enter the cells.

Ectoparasite A parasite that lives on the outside of an organism.

Ectotherm An organism with a body temperature that is largely determined by its external environment.

Egested energy The portion of consumed energy that is excreted or regurgitated.

El Niño–Southern Oscillation (ENSO) The periodic changes in winds and ocean currents in the South Pacific, causing weather changes throughout much of the world.

Electromagnetic radiation Energy from the Sun, packaged in small particle-like units called photons.

Emerging infectious disease A disease that is newly discovered or has been rare and then suddenly increases in occurrence.

Endemic species Species that live in a single, often isolated, location.

Endomycorrhizal fungi Fungi characterized by hyphal threads that extend far out into the soil and penetrate root cells between the cell wall and the cell membrane.

Endoparasite A parasite that lives inside an organism.

Endophytic fungi Fungi that live inside a plant's tissues.

Endotherm An organism that can generate sufficient metabolic heat to raise its body temperature higher than the external environment.

Energy residence time The length of time that energy remains in a given trophic level.

Energy–diversity hypothesis A hypothesis that sites with higher amounts of energy are able to support more species.

Environmental sex determination A process in which sex is determined largely by the environment.

Environmental stochasticity Variation in birth rates and death rates due to random changes in environmental conditions.

Epilimnion The surface layer of the water in a lake or pond.

Epistasis When the expression of one gene is controlled by another gene.

Equilibrium isocline The population size of one species that causes the population of another species to be stable. *Also known as* **Zero growth isocline.**

Equilibrium theory of island biogeography A theory stating that the number of species on an island reflects a balance between the colonization of new species and the extinction of existing species.

Estuary An area along the coast where the mouths of freshwater rivers mix with the salt water from oceans.

Eusocial A type of animal society in which individuals live in large groups with overlapping generations, cooperation in nest building and brood care, and reproductive dominance by one or a few individuals.

Eutrophication An increase in the productivity of aquatic ecosystems.

Evaporation The transformation of water from a liquid to a gaseous state with the input of heat energy.

Evenly spaced dispersion A pattern of dispersion of a population in which each individual maintains a uniform distance between itself and its neighbors.

Evolution Change in the genetic composition of a population over time.

Experimental unit The object to which we apply an experimental manipulation.

Exploitative competition Competition in which individuals consume and drive down the abundance of a resource to the point that other individuals cannot persist.

Exponential growth model A model of population growth in which the population increases continuously at an exponential rate.

Extra-pair copulations When an individual that has a social bond with a mate also breeds with other individuals.

Facilitation A mechanism of succession in which the presence of one species increases the probability that a second species can become established.

Facultative mutualists Two species that provide fitness benefits to each other, but whose interaction is not critical to the persistence of either species.

Fall turnover The vertical mixing of lake water that occurs in fall, assisted by winds that drive the surface currents.

Fecundity The number of offspring produced by an organism per reproductive episode.

Field capacity The maximum amount of water held by soil particles against the force of gravity.

Fire-maintained climax community A successional stage that persists as the final seral stage due to periodic fires.

Fitness The survival and reproduction of an individual.

Food chain A linear representation of how different species in a community feed on each other.

Food web A complex and realistic representation of how species feed on each other in a community.

Founder effect When a small number of individuals leave a large population to colonize a new area and bring with them only a small amount of genetic variation.

Frequency-dependent selection When the rarer phenotype in a population is favored by natural selection.

Freshwater wetland An aquatic biome that contains standing fresh water, or soils saturated with fresh water for at least part of the year, and which is shallow enough to have emergent vegetation throughout all depths.

Functional response The relationship between the density of prey and an individual predator's rate of food consumption.

Fundamental niche The range of abiotic conditions under which species can persist.

Gene pool The collection of alleles from all individuals in a population.

Generalist A species that interacts with many other species.

Generation time (T) The average time between the birth of an individual and the birth of its offspring.

Genetic drift A process that occurs when genetic variation is lost because of random variation in mating, mortality, fecundity, and inheritance.

Genotype The set of genes an organism carries.

Geographic range A measure of the total area covered by a population.

Geometric growth model A model of population growth that compares population sizes at regular time intervals.

Global climate change A phenomenon that refers to changes in Earth's climates, including global warming, changes in the global distribution of precipitation and temperature, changes in the intensity of storms, and altered ocean circulation.

Glycerol A chemical that prevents the hydrogen bonds of water from coming together to form ice unless the temperatures are well below freezing.

Glycoproteins A group of compounds that can be used to lower the freezing temperature of water.

Gondwana The southern landmass that separated from Pangaea about 150 Mya and subsequently split into South America, Africa, Antarctica, Australia, and India.

Good genes hypothesis The hypothesis that an individual chooses a mate that possesses a superior genotype.

Good health hypothesis The hypothesis that an individual chooses the healthiest mates.

Grazer-maintained climax community When a successional stage persists as the final seral stage due to intense grazing.

Greenhouse effect The process of solar radiation striking Earth, being converted to infrared radiation, and being absorbed and re-emitted by atmospheric gases.

Greenhouse gases Compounds in the atmosphere that absorb the infrared heat energy emitted by Earth and then emit some of the energy back toward Earth.

Gross primary productivity (GPP) The rate at which energy is captured and assimilated by producers in a given area.

Growing season The months in a location that are warm enough to allow plant growth.

Growth rate In a population, the number of new individuals that are produced in a given amount of time minus the number of individuals that die.

Guild Within a given trophic level, a group of species that feeds on similar items.

Gyre A large-scale water circulation pattern between continents.

Habitat corridor A strip of favorable habitat located between two large patches of habitat that facilitates dispersal.

Habitat fragmentation The process of breaking up large habitats into a number of smaller habitats.

Habitat The place, or physical setting, in which an organism lives.

Hadley cells The two circulation cells of air between the equator and 30° N and 30° S latitudes.

Half-life The time required for a chemical to break down to half of its original concentration.

Handling time The amount of time that a predator takes to consume a captured prey.

Haplodiploid A sex-determination system in which one sex is haploid and the other sex is diploid.

Herbivore An organism that consumes producers such as plants and algae.

Heritability The proportion of the total phenotypic variation that is caused by genetic variation.

Hermaphrodite An individual that produces both male and female gametes.

Heterozygous When an individual has two different alleles of a particular gene.

Hibernation A type of dormancy that occurs in mammals in which individuals reduce the energetic costs of being active by lowering their heart rate and decreasing their body temperature.

Homeostasis An organism's ability to maintain constant internal conditions in the face of a varying external environment.

Homeotherm An organism that maintains constant temperature conditions within its cells.

Homozygous When an individual has two identical alleles of a particular gene.

Horizon A distinct layer of soil.

Horizontal transmission When a parasite moves between individuals other than parents and their offspring.

Hydrologic cycle The movement of water through ecosystems and atmosphere.

Hyperosmotic When an organism has a higher solute concentration in its tissues than the surrounding water.

Hypolimnion The deeper layer of water in a lake or pond.

Hyposmotic When an organism has a lower solute concentration in its tissues than the surrounding water.

Hypothesis An idea that potentially explains a repeated observation.

Ideal free distribution When individuals distribute themselves among different habitats in a way that allows them to have the same per capita benefit.

Inbreeding depression The decrease in fitness caused by matings between close relatives due to offspring inheriting deleterious alleles from both the egg and the sperm.

Inclusive fitness The sum of direct fitness and indirect fitness.

Independent communities Communities in which species do not depend on each other to exist.

Independent variable A factor that causes other variables to change.

Indeterminate growth A growth pattern in which an individual continues to grow after it initiates reproduction.

Indirect effect An interaction between two species that involves one or more intermediate species.

Indirect fitness The fitness that an individual gains by helping relatives pass on copies of their genes.

Indirect selection Selection that favors indirect fitness. *Also known as* **Kin selection.**

Individual approach An approach to ecology that emphasizes the way in which an individual's morphology, physiology, and behavior enable it to survive in its environment.

Individual A living being; the most fundamental unit of ecology.

Indomalayan region The biogeographic region of the Southern Hemisphere that corresponds to India and Southeast Asia. *Also known as* **Oriental region.**

Industrial melanism A phenomenon in which industrial activities cause habitats to become darker due to pollution and, as a result, individuals possessing darker phenotypes are favored by selection.

Infection resistance The ability of a host to prevent an infection from occurring.

Infection tolerance The ability of a host to minimize the harm once an infection has occurred.

Inflection point The point on a sigmoidal growth curve at which the population achieves its highest growth rate.

Inhibition A mechanism of succession in which one species decreases the probability that a second species will become established.

Instrumental value of biodiversity The economic value a species can provide.

Interdependent communities Communities in which species depend on each other to exist.

Interference competition When competitors do not immediately consume resources but defend them.

Intermediate disturbance hypothesis The hypothesis that more species are present in a community that occasionally experiences disturbances than in a community that experiences frequent or rare disturbances.

Interspecific competition Competition among individuals of different species.

Intertidal zone A biome consisting of the narrow band of coastline between the levels of high tide and low tide.

Intertropical convergence zone (ITCZ) The area where the two Hadley cells converge and cause large amounts of precipitation.

Intraspecific competition Competition among individuals of the same species.

Intrinsic growth rate (r) The highest possible per capita growth rate for a population.

Intrinsic value of biodiversity A focus on the inherent value of a species, not tied to any economic benefit.

Introduced species A species that is introduced to a region of the world where it has not historically existed. *Also known as* **Exotic species** *or* **Non-native species.**

Invasive species An introduced species that spreads rapidly and has negative effects on other species, human recreation, or human economies.

Ions Atoms or groups of atoms that carry an electric charge.

Isozymes Different forms of an enzyme that catalyze a given reaction.

Iteroparity When organisms reproduce multiple times during their life.

Joint equilibrium point The point at which the equilibrium isoclines for predator and prey populations cross.

Joint population trajectory The simultaneous trajectory of predator and prey populations.

J-shaped curve The shape of exponential growth when graphed.

Keystone species A species that substantially affects the structure of communities despite the fact that individuals of the species might not be particularly numerous.

Lake An aquatic biome that is larger than a pond and is characterized by nonflowing fresh water with some area of water that is too deep for plants to rise above the water's surface.

Landscape ecology The field of study that considers the spatial arrangement of habitats at different scales and examines how they influence individuals, populations, communities, and ecosystems.

Landscape metapopulation model A population model that considers both differences in the quality of the suitable patches and the quality of the surrounding matrix.

Latent heat release When water vapor is converted back to liquid, water releases energy in the form of heat.

Laterization The breakdown of clay particles, which results in the leaching of silicon from the soil, leaving oxides of iron and aluminum to predominate throughout the soil profile.

Laurasia The northern landmass that separated from Pangaea about 150 Mya and subsequently split into North America, Europe, and Asia.

Law of conservation of energy Energy cannot be created or destroyed; it can only change form. *Also known as* **The first law of thermodynamics.**

Law of conservation of matter Matter cannot be created or destroyed; it can only change form.

Leaching A process in which groundwater removes some substances by dissolving them and moving them down through the soil to lower layers.

Legacy effect A long-lasting influence of historical processes on the current ecology of an area.

Lek The location of an animal aggregation to put on a display to attract the opposite sex.

Liebig's law of the minimum Law stating that a population increases until the supply of the most limiting resource prevents it from increasing further.

Life history The schedule of an organism's growth, development, reproduction, and survival.

Life tables Tables that contain class-specific survival and fecundity data.

Lifetime dispersal distance The average distance an individual moves from where it was hatched or born to where it reproduces.

Limnetic zone The open water beyond the littoral zone, where the dominant photosynthetic organisms are floating algae. *Also known as* **Pelagic zone.**

Line-transect surveys Surveys that count the number of individuals observed as one moves along a line.

Littoral zone The shallow area around the edge of a lake or pond containing rooted vegetation.

Local diversity The number of species in a relatively small area of homogeneous habitat, such as a stream. *Also known as* **Alpha diversity**.

Local mate competition When competition for mates occurs in a very limited area and only a few males are required to fertilize all the females.

Logistic growth model A growth model that describes slowing growth of populations at high densities.

Log-normal distribution A normal, or bell-shaped, distribution that uses a logarithmic scale on the *x*-axis.

Longevity The life span of an organism. *Also known as* **Life expectancy**.

Lotic Characterized by flowing fresh water.

Lotka–Volterra model A model of predator–prey interactions that incorporates oscillations in the abundances of predator and prey populations and shows predator numbers lagging behind those of their prey.

Macroevolution Evolution at higher levels of organization, including species, genera, families, orders, and phyla.

Mangrove swamp A biome that occurs along tropical and subtropical coasts and contains salt-tolerant trees with roots submerged in water.

Manipulation The factor that we want to vary in an experiment. *Also known as* **Treatment**.

Manipulative experiment A process by which a hypothesis is tested by altering a factor that is hypothesized to be an underlying cause of the phenomenon.

Mark-recapture survey A method of population estimation in which researchers capture and mark a subset of a population from an area, return it to the area, and then capture a second sample of the population after some time has passed.

Mass extinction events Events in which at least 75 percent of the existing species go extinct within a 2-million-year period.

Mate guarding A behavior in which one partner prevents the other partner from participating in extra-pair copulations.

Mathematical model A representation of a system with a set of equations that correspond to hypothesized relationships among the system's components.

Mating system The number of mates each individual has and the permanence of the relationship with those mates.

Matric potential The potential energy generated by the attractive forces between water molecules and soil particles. *Also known as* **Matrix potential**.

Mesopredators Relatively small carnivores that consume herbivores.

Microcosm A simplified ecological system that attempts to replicate the essential features of an ecological system in a laboratory or field setting.

Microevolution Evolution at the level of populations.

Microhabitat A specific location within a habitat that typically differs in environmental conditions from other parts of the habitat.

Migration The seasonal movement of animals from one region to another.

Mineralization The process of breaking down organic compounds into inorganic compounds.

Minimum viable population (MVP) The smallest population size of a species that can persist in the face of environmental variation.

Mixotroph An organism that obtains its energy from more than one source.

Moist continental mid-latitude climate A climate that exists at the interior of continents and is typically characterized by warm summers, cold winters, and moderate amounts of precipitation.

Moist subtropical mid-latitude climate A climate characterized by warm, dry summers and cold, wet winters.

Monoecious Plants that have separate male and female flowers on the same individual.

Monogamy A mating system in which a social bond between one male and one female persists through the period that is required for them to rear their offspring.

Müllerian mimicry When several unpalatable species evolve a similar pattern of warning coloration.

Mutation A random change in the sequence of nucleotides in regions of DNA that either comprise a gene or control the expression of a gene.

Mutualism An interaction between two species in which each species receives benefits from the other.

Mycorrhizal fungi Fungi that surround plant roots and help plants obtain water and minerals.

Natural experiment An approach to hypothesis testing that relies on natural variation in the environment.

Natural selection Change in the frequency of genes in a population through differential survival and reproduction of individuals that possess certain phenotypes.

Nearctic region The biogeographic region of the Northern Hemisphere that roughly corresponds to North America.

Negative density dependence When the rate of population growth decreases as population density increases.

Negative feedbacks The action of internal response mechanisms that restores a system to a desired state, or set point, when the system deviates from that state.

Neotropical region The biogeographic region of the Southern Hemisphere that corresponds to South America.

Neritic zone The ocean zone beyond the range of the lowest tidal level, and which extends to depths of about 200 m.

Net primary productivity (NPP) The rate at which energy is assimilated by producers and converted into producer biomass in a given area.

Net production efficiency The percentage of assimilated energy that is used for growth and reproduction.

Net reproductive rate The total number of female offspring that we expect an average female to produce over the course of her life.

Net secondary productivity The rate of consumer biomass accumulation in a given area.

Niche The range of abiotic and biotic conditions that an organism can tolerate.

Nitrification The final process in the nitrogen cycle, which converts ammonium (NH_4^+) or ammonia (NH_3) to nitrite (NO_2^-) and then to nitrate (NO_3^-).

Nitrogen fixation The process of converting atmospheric nitrogen into forms producers can use.

Nonrenewable resources Resources that are not regenerated.

Numerical response A change in the number of predators through population growth or population movement due to immigration or emigration.

Obligate mutualists Two species that provide fitness benefits to each other and require each other to persist.

Observations Information, including measurements, that is collected from organisms or the environment. *Also known as* **Data**.

Oceanic zone The ocean zone beyond the neritic zone.

Omnivore A species that feeds at several trophic levels.

Optimal foraging theory A model describing foraging behavior that provides the best balance between the costs and benefits of different foraging strategies.

Optimum The narrow range of environmental conditions to which an organism is best suited.

Osmoregulation The mechanisms that organisms use to maintain proper solute balance.

Osmosis The movement of water across a semipermeable membrane.

Osmotic potential The force with which an aqueous solution attracts water by osmosis.

Overshoot When a population grows beyond its carrying capacity.

Palearctic region The biogeographic region of the Northern Hemisphere that corresponds to Eurasia.

Pangaea The single landmass that existed on Earth about 250 Mya and subsequently split into Laurasia and Gondwana.

Parasite load The number of parasites of a given species that an individual host can harbor.

Parasite An organism that lives in or on another organism, but rarely kills it.

Parasitoid An organism that lives within and consumes the tissues of a living host, eventually killing it.

Parent material The layer of bedrock that underlies soil and plays a major role in determining the type of soil that will form above it.

Parental investment The amount of time and energy given to an offspring by its parents.

Parity The number of reproductive episodes an organism experiences.

Parthenogenesis A form of asexual reproduction in which an embryo is produced without fertilization.

Passive transport The movement of ions and small molecules through a membrane along a concentration gradient, from a location with many solutes to a location with few solutes.

Pathogen A parasite that causes disease in its host.

Perennial An organism that has a life span of more than one year.

Perfect flowers Flowers that contain both male and female parts.

Permafrost A phenomenon whereby layers of soil are permanently frozen

pH A measure of acidity or alkalinity; defined as $pH = -\log [H^+]$.

Phenotype An attribute of an organism, such as its behavior, morphology, or physiology.

Phenotypic plasticity The ability of a single genotype to produce multiple phenotypes.

Phenotypic trade-off A situation in which a given phenotype experiences higher fitness in one environment, whereas other phenotypes experience higher fitness in other environments.

Photic zone The area of the neritic and oceanic zones that contains sufficient light for photosynthesis by algae.

Photoperiod The amount of light that occurs each day.

Photorespiration The oxidation of carbohydrates to CO_2 and H_2O by rubisco, which reverses the light reactions of photosynthesis.

Photosynthetically active region Wavelengths of light that are suitable for photosynthesis.

Phylogenetic trees Hypothesized patterns of relatedness among different groups such as populations, species, or genera.

Pioneer species The earliest species to arrive at a site.

Pleiotropy When a single gene affects multiple traits.

Podsolization A process occurring in acidic soils typical of cool, moist regions, where clay particles break down in the E horizon, and their soluble ions are transported down to the lower B horizon.

Poikilotherm An organism that does not have constant body temperatures.

Polar cells The atmospheric convection currents that move air between 60° and 90° latitudes in the Northern and Southern Hemispheres.

Polar climate A climate that experiences very cold temperatures and relatively little precipitation.

Polyandry A mating system in which a female mates with more than one male.

Polygamy A mating system in which a single individual of one sex forms long-term social bonds with more than one individual of the opposite sex.

Polygenic When a single trait is affected by several genes.

Polygyny A mating system in which a male mates with more than one female.

Polyploid A species that contains three or more sets of chromosomes.

Pond An aquatic biome that is smaller than a lake and is characterized by nonflowing fresh water with some area of water that is too deep for plants to rise above the water's surface.

Population approach An approach to ecology that emphasizes variation over time and space in the number of individuals, the density of individuals, and the composition of individuals.

Population cycles Regular oscillation of population size over a long period of time.

Population The individuals of the same species living in a particular area.

Positive density dependence When the rate of population growth increases as population density increases. *Also known as* **Inverse density dependence** *or the* **Allee effect**.

Potential evapotranspiration (PET) The amount of water that could be evaporated from the soil and transpired by plants, given the average temperature and humidity.

Predator An organism that kills and partially or entirely consumes another individual.

Prediction A logical consequence of a hypothesis.

Primary consumer A species that eats producers.

Primary productivity The rate at which solar or chemical energy is captured and converted into chemical bonds by photosynthesis or chemosynthesis.

Primary sexual characteristics Traits related to fertilization.

Primary succession The development of communities in habitats that are initially devoid of plants and organic soil, such as sand dunes, lava flows, and bare rock.

Principle of allocation The observation that when resources are devoted to one body structure, physiological function, or behavior, they cannot be allotted to another.

Priority effect When the arrival of one species at a site affects the subsequent colonization of other species.

Producer An organism that uses photosynthesis to convert solar energy into organic compounds or uses chemosynthesis to convert chemical energy into organic compounds. *Also known as* **Autotroph**.

Profundal zone The area in a lake that is too deep to receive sunlight.

Promiscuity A mating system in which males mate with multiple females and females mate with multiple males and do not create a lasting social bond.

Provisioning services Benefits of biodiversity that humans use, including lumber, fur, meat, crops, water, and fiber.

Proximate hypothesis A hypothesis that addresses the immediate changes in an organism's hormones, physiology, nervous system, or muscular system.

Pyramid of biomass A trophic pyramid that represents the standing crop of organisms present in different trophic groups.

Pyramid of energy A trophic pyramid that displays the total energy existing at each trophic level.

Q_{10} The ratio of the rate of a physiological process at one temperature to its rate at a temperature 10°C cooler.

Queen The dominant, egg-laying female in eusocial insect societies.

Radiation The emission of electromagnetic energy by a surface.

Rain shadow A region with dry conditions found on the leeward side of a mountain range as a result of humid winds from the ocean, causing precipitation on the windward side.

Random assortment The process of making haploid gametes in which the combination of alleles that are placed into a given gamete could be any combination of those possessed by the diploid parent.

Random dispersion A pattern of dispersion of a population in which the position of each individual is independent of the position of other individuals in the population.

Randomization An aspect of experiment design in which every experimental unit has an equal chance of being assigned to a particular manipulation.

Rank-abundance curve A curve that plots the relative abundance of each species in a community in rank order from the most abundant species to the least abundant species.

Realized niche The range of abiotic and biotic conditions under which a species persists.

Recessive An allele whose expression is masked by the presence of another allele.

Recipient The individual who receives the behavior of a donor in a social interaction.

Recombination The reshuffling of genes that can occur as DNA is copied during meiosis and chromosomes exchange genetic material.

Red Queen hypothesis The hypothesis that sexual selection allows hosts to evolve at a rate that can counter the rapid evolution of parasites.

Regional diversity The number of species in all the habitats that comprise a large geographic area. *Also known as* **Gamma diversity**.

Regional species pool The collection of species that occurs within a region.

Regression A statistical tool that determines whether there is a relationship between two variables and that also describes the nature of that relationship.

Regulating services Benefits of biodiversity that include climate regulation, flood control, and water purification.

Relative abundance The proportion of individuals in a community represented by each species.

Remote sensing A technique that measures conditions on Earth from a distant location, typically using satellites or airplanes that take photographs of large areas of the globe.

Renewable resources Resources that are constantly regenerated.

Replication Being able to produce a similar outcome multiple times.

Rescue effect The phenomenon of dispersers supplementing a declining subpopulation that is headed toward extinction.

Reservoir species Species that can carry a parasite but do not succumb to the disease that the parasite causes in other species.

Resource Anything an organism consumes or uses that causes an increase in the growth rate of a population when it becomes more available.

Respired energy The portion of assimilated energy a consumer uses for respiration.

Riparian zone A band of terrestrial vegetation alongside rivers and streams that is influenced by seasonal flooding and elevated water tables.

Risk-sensitive foraging Foraging behavior that is influenced by the presence of predators.

River A wide channel of slow-flowing fresh water.

Root pressure When osmotic potential in the roots of a plant draws in water from the soil and forces it into the xylem elements.

RuBP carboxylase-oxidase An enzyme involved in photosynthesis that catalyzes the reaction of RuBP and CO_2 to form two molecules of glyceraldehyde 3-phosphate (G3P). *Also known as* **Rubisco**.

Runaway sexual selection When selection for preference of a sexual trait and selection for that trait continue to reinforce each other.

Salinization The process of repeated irrigation, which causes increased soil salinity.

Salt marsh A saltwater biome that contains nonwoody emergent vegetation.

Sample standard deviation A statistic that provides a standardized way of measuring how widely data are spread from the mean.

Sample variance A measurement that indicates the spread of data around the mean of a population when only a sample of the population has been measured.

Saturation point The limit of the amount of water vapor the air can contain.

Saturation The upper limit of solubility in water.

Scavenger An organism that consumes dead animals.

Sclerophyllous Vegetation that has small, durable leaves.

Search image A learned mental image that helps the predator locate and capture food.

Secondary consumer A species that eats primary consumers.

Secondary sexual characteristics Traits related to differences between the sexes in terms of body size, ornaments, color, and courtship.

Secondary succession The development of communities in habitats that have been disturbed and include no plants but still contain an organic soil.

Selection The process by which certain phenotypes are favored to survive and reproduce over other phenotypes.

Selfishness When the donor of a social behavior experiences increased fitness and the recipient experiences decreased fitness.

Self-thinning curve A graphical relationship that shows how decreases in population density over time lead to increases in the mass of each individual in the population.

Semelparity When organisms reproduce only once during their life.

Semipermeable membrane A membrane that allows only particular molecules to pass through.

Senescence A gradual decrease in fecundity and an increase in the probability of mortality.

Sequential hermaphrodites Individuals that possess male or female reproductive function and then switch to possess the other function.

Seral stage Each stage of community change during the process of succession.

Sexual dimorphism The difference in the phenotype between males and females of the same species.

Sexual reproduction A reproduction mechanism in which progeny inherit DNA from two parents.

Sexual selection Natural selection for sex-specific traits related to reproduction.

Shannon's index (H') A measurement of species diversity, given by the following formula:

$$H' = -\sum_{i=1}^{s}(p_i)(\ln p_i)$$

Also known as **Shannon-Wiener index**.

Simpson's index A measurement of species diversity, given by the following formula:

$$\frac{1}{\sum_{i=1}^{s}(p_i)^2}$$

Simultaneous hermaphrodites Individuals that possess male and female reproductive functions at the same time.

Sink subpopulations In low-quality habitats, subpopulations that rely on outside dispersers to maintain the subpopulation within a metapopulation.

Social behaviors Interactions with members of one's own species, including mates, offspring, other relatives, and unrelated individuals.

Soil The layer of chemically and biologically altered material that overlies bedrock or other unaltered material at Earth's surface.

Solar equator The latitude receiving the most direct rays of the Sun.

Solute A dissolved substance.

Source subpopulations In high-quality habitats, subpopulations that serve as a source of dispersers within a metapopulation.

Source–sink metapopulation model A population model that builds on the basic metapopulation model and accounts for the fact that not all patches of suitable habitat are of equal quality.

Spatial structure The pattern of density and spacing of individuals in a population.

Specialist A species that interacts with one other species or a few closely related species.

Speciation The evolution of new species.

Species accumulation curve A graph of the number of species observed in relation to the number of individuals sampled.

Species evenness A comparison of the relative abundance of each species in a community.

Species richness The number of species in a community.

Species sorting The process of sorting species in the regional pool among localities according to their adaptations and interactions.

Species Historically defined as a group of organisms that naturally interbreed with each other and produce fertile offspring. Current research demonstrates that no single definition can be applied to all organisms.

Species–area curve A graphical relationship in which increases in area (A) are associated with increases in the number of species (S).

Spitefulness When a social interaction reduces the fitness of both donor and recipient.

Spring turnover The vertical mixing of lake water that occurs in early spring, assisted by winds that drive the surface currents.

S-shaped curve The shape of the curve when a population is graphed over time using the logistic growth model.

Stabilizing selection When individuals with intermediate phenotypes have higher survival and reproductive success than those with extreme phenotypes.

Stable age distribution When the age structure of a population does not change over time.

Stable limit cycle A pattern of population growth in which the population size continues to exhibit large oscillations over time.

Standard error of the mean A measurement of variation in data that takes into account the number of replicates that were used to measure the standard deviation.

Standing crop The biomass of producers present in a given area of an ecosystem at a particular moment in time.

Static life table A life table that quantifies the survival and fecundity of all individuals in a population during a single time interval.

Stepping stones Small intervening habitat patches that dispersing organisms can use to move between large favorable habitats.

Stochastic model A model that incorporates random variation in population growth rate.

Stomata Small openings on the surface of leaves, which serve as the points of entry for CO_2 and exit points for water vapor.

Stratification The condition of a lake or pond when the warmer, less dense surface water floats on the cooler, denser water below.

Stream A narrow channel of fast-flowing fresh water. *Also known as* **Creek**.

Strength of selection The difference between the mean of the phenotypic distribution before selection and the mean after selection, measured in units of standard deviations.

Subpopulations When a larger population is broken up into smaller groups that live in isolated patches.

Subtropical desert A biome characterized by hot temperatures, scarce rainfall, long growing seasons, and sparse vegetation.

Succession The process by which the species composition of a community changes over time.

Supercooling A process in which glycoproteins in the blood impede ice formation by coating any ice crystals that begin to form.

Supporting services Benefits of biodiversity that allow ecosystems to exist, such as primary production, soil formation, and nutrient cycling.

Survey Counting a subset of the population.

Susceptible-Infected-Resistant (S-I-R) model The simplest model of infectious disease transmission that incorporates immunity.

Symbiotic relationship When two different types of organisms live in a close physical relationship.

Sympatric speciation The evolution of new species without geographic isolation.

Temperate grassland/cold desert A biome characterized by hot, dry summers and cold, harsh winters and dominated by grasses, nonwoody flowering plants, and drought-adapted shrubs.

Temperate rainforest A biome known for mild temperatures and abundant precipitation, dominated by evergreen forests.

Temperate seasonal forest A biome with moderate temperature and precipitation conditions, dominated by deciduous trees.

Temporal environmental variation The description of how environmental conditions change over time.

Territory Any area defended by one or more individuals against the intrusion of others.

Tertiary consumer A species that eats secondary consumers.

The handicap principle The principle that the greater the handicap an individual carries, the greater its ability to offset that handicap.

Thermal inertia The resistance to a change in temperature due to a large body volume.

Thermal optimum The range of temperatures within which organisms perform best.

Thermal pollution Discharging water that is too hot to sustain aquatic species.

Thermocline A middle depth of water in a lake or pond that experiences a rapid change in temperature over a relatively short distance in depth.

Thermohaline circulation A global pattern of surface- and deep-water currents that flow as a result of variations in temperature and salinity that change the density of water.

Thermophilic Heat-loving.

Thermoregulation The ability of an organism to control the temperature of its body.

Tolerance A mechanism of succession in which the probability that a species can become established depends on its dispersal ability and its ability to persist under the physical conditions of the environment.

Top predators Predators that typically consume both herbivores and mesopredators.

Top-down control When the abundance of trophic groups is determined by the existence of predators at the top of the food web.

Torpor A brief period of dormancy that occurs in birds and mammals in which individuals reduce their activity and their body temperature.

Trait-mediated indirect effect An indirect effect caused by changes in the traits of an intermediate species.

Transient climax community A climax community that is not persistent.

Transpiration The process by which leaves can generate water potential as water evaporates from the surfaces of leaf cells into the air spaces within the leaves.

Trophic cascade Indirect effects in a community that are initiated by a predator.

Trophic level A level in a food chain or food web of an ecosystem.

Trophic pyramid A chart composed of stacked rectangles representing the amount of energy or biomass in each trophic group.

Tropical climate A climate characterized by warm temperatures and high precipitation, occurring in regions near the equator.

Tropical rainforest A warm and rainy biome, characterized by multiple layers of lush vegetation.

Tropical seasonal forest A biome with warm temperatures and pronounced wet and dry seasons, dominated by deciduous trees that shed their leaves during the dry season.

t-test A statistical test that determines if the distributions of data from two groups are significantly different.

Tundra The coldest biome, characterized by a treeless expanse above permanently frozen soil.

Type I functional response A functional response in which a predator's rate of prey consumption increases in a linear fashion with an increase in prey density until satiation occurs.

Type II functional response A functional response in which a predator's rate of prey consumption begins to slow down as prey density increases and then plateaus when satiation occurs.

Type III functional response A functional response in which a predator exhibits low prey consumption under low prey densities, rapid consumption under moderate prey densities, and slowing prey consumption under high prey densities.

Ultimate hypothesis A hypothesis that addresses why an organism has evolved to respond in a certain way to its environment in terms of the fitness costs and benefits of the response.

Upwelling An upward movement of ocean water.

Variance of the mean A measurement that indicates the spread of data around the mean of a population when every member of the population has been measured.

Vector An organism that a parasite uses to disperse from one host to another.

Vegetative reproduction A form of asexual reproduction in which an individual is produced from the nonsexual tissues of a parent.

Vertical transmission When a parasite is transmitted from a parent to its offspring.

Viscosity The thickness of a fluid that causes objects to encounter resistance as they move through it.

Visible light Wavelengths in between infrared and ultraviolet radiation that are visible to the human eye.

Warning coloration A strategy in which distastefulness evolves in association with very conspicuous colors and patterns. Also known as aposematism.

Water potential A measure of water's potential energy.

Watershed An area of land that drains into a single stream or river.

Weather The variation in temperature and precipitation over periods of hours or days.

Weathering The physical and chemical alteration of rock material near Earth's surface.

Wilting point The water potential at which most plants can no longer retrieve water from the soil, which is about −1.5 MPa.

Woodland/shrubland A biome characterized by hot, dry summers and mild, wet winters, a combination that favors the growth of drought-tolerant grasses and shrubs.

Zero population growth isocline Population sizes at which a population experiences zero growth.

Index

Note: Page numbers followed by f and t indicate figures and tables, respectively. **Boldfaced** page numbers indicate key terms.